An Introduction to
q-analysis

An Introduction to
q-analysis

Warren P. Johnson

American Mathematical Society
Providence, Rhode Island

2010 *Mathematics Subject Classification.* Primary 05A30, 05A17, 11P84, 11P81, 33D15.

For additional information and updates on this book, visit
www.ams.org/bookpages/mbk-134

Library of Congress Cataloging-in-Publication Data
Names: Johnson, Warren Pierstorff, 1960- author.
Title: Introduction to q-analysis / Warren P. Johnson.
Description: Providence : American Mathematical Society, 2020. | Includes bibliographical references and index. | Summary:
Identifiers: LCCN 2020021142 | ISBN 9781470456238 (paperback) | ISBN 9781470462109 (ebook)
Subjects: LCSH: Combinatorial analysis. | Graph theory. | Number theory. | AMS: Combinatorics – Enumerative combinatorics – q-calculus and related topics. | Combinatorics – Enumerative combinatorics – Partitions of integers. | Number theory – Additive number theory; partitions – Partition identities; identities of Rogers-Ramanujan type. | Number theory – Additive number theory; partitions – Elementary theory of partitions. | Special functions.
Classification: LCC QA164 .J54 2020 — DDC 511/.6–dc23
LC record available at https://lccn.loc.gov/2020021142

Copying and reprinting. Individual readers of this publication, and nonprofit libraries acting for them, are permitted to make fair use of the material, such as to copy select pages for use in teaching or research. Permission is granted to quote brief passages from this publication in reviews, provided the customary acknowledgment of the source is given.

Republication, systematic copying, or multiple reproduction of any material in this publication is permitted only under license from the American Mathematical Society. Requests for permission to reuse portions of AMS publication content are handled by the Copyright Clearance Center. For more information, please visit www.ams.org/publications/pubpermissions.

Send requests for translation rights and licensed reprints to reprint-permission@ams.org.

© 2020 by the American Mathematical Society. All rights reserved.
The American Mathematical Society retains all rights
except those granted to the United States Government.
Printed in the United States of America.

∞ The paper used in this book is acid-free and falls within the guidelines
established to ensure permanence and durability.
Visit the AMS home page at https://www.ams.org/
10 9 8 7 6 5 4 3 2 1   25 24 23 22 21 20

Contents

An Introduction to q-analysis	xi
Chapter 1. Inversions	**1**
1.1. Stern's problem	1
Exercises	5
1.2. The q-factorial	7
Exercises	11
1.3. q-binomial coefficients	14
Exercises	19
1.4. Some identities for q-binomial coefficients	20
Exercises	23
1.5. Another property of q-binomial coefficients	25
Exercises	27
1.6. q-multinomial coefficients	29
Exercises	31
1.7. The Z-identity	33
Exercises	36
1.8. Bibliographical Notes	37
Chapter 2. q-binomial Theorems	**39**
2.1. A noncommutative q-binomial Theorem	39
Exercises	43
2.2. Potter's proof	45
Exercises	47
2.3. Rothe's q-binomial theorem	49
Exercises	53
2.4. The q-derivative	57
Exercises	59
2.5. Two q-binomial theorems of Gauss	61
Exercises	66
2.6. Jacobi's q-binomial theorem	71
Exercises	72
2.7. MacMahon's q-binomial theorem	74
Exercises	77
2.8. A partial fraction decomposition	79
Exercises	82
2.9. A curious q-identity of Euler, and some extensions	82
Exercises	86
2.10. The Chen–Chu–Gu identity	88
Exercises	91

2.11. Bibliographical Notes	91
Chapter 3. Partitions I: Elementary Theory	**93**
3.1. Partitions with distinct parts	93
Exercises	95
3.2. Partitions with repeated parts	98
Exercises	103
3.3. Ferrers diagrams	106
Exercises	113
3.4. q-binomial coefficients and partitions	116
Exercises	119
3.5. An identity of Euler, and its "finite" form	120
Exercises	126
3.6. Another identity of Euler, and its finite form	128
Exercises	130
3.7. The Cauchy/Crelle q-binomial series	132
Exercises	137
3.8. q-exponential functions	141
Exercises	145
3.9. Bibliographical Notes	148
Chapter 4. Partitions II: Geometric Theory	**149**
4.1. Euler's pentagonal number theorem	149
Exercises	153
4.2. Durfee squares	157
Exercises	162
4.3. Euler's pentagonal number theorem: Franklin's proof	164
Exercises	167
4.4. Divisor sums	167
Exercises	173
4.5. Sylvester's fishhook bijection	180
Exercises	187
4.6. Bibliographical Notes	188
Chapter 5. More q-identities: Jacobi, Gauss, and Heine	**191**
5.1. Jacobi's triple product	191
Exercises	195
5.2. Other proofs and related results	201
Exercises	205
5.3. The quintuple product identity	214
Exercises	218
5.4. Lebesgue's identity	221
Exercises	223
5.5. Basic hypergeometric series	227
Exercises	230
5.6. More $_2\phi_1$ identities	233
Exercises	236
5.7. The q-Pfaff–Saalschütz identity	239
Exercises	241

5.8.	Bibliographical Notes	243

Chapter 6. Ramanujan's $_1\psi_1$ Summation Formula — 247
  6.1. Ramanujan's formula — 247
  Exercises — 249
  6.2. Four proofs — 250
  Exercises — 253
  6.3. From the q-Pfaff–Saalschütz sum to Ramanujan's $_1\psi_1$ summation — 256
  Exercises — 259
  6.4. Another identity of Cauchy, and its finite form — 259
  Exercises — 260
  6.5. Cauchy's "mistaken identity" — 263
  Exercises — 265
  6.6. Ramanujan's formula again — 266
  Exercises — 268
  6.7. Bibliographical Notes — 268

Chapter 7. Sums of Squares — 271
  7.1. Cauchy's formula — 271
  Exercises — 272
  7.2. Sums of two squares — 276
  Exercises — 278
  7.3. Sums of four squares — 281
  Exercises — 286
  7.4. Bibliographical Notes — 288

Chapter 8. Ramanujan's Congruences — 289
  8.1. Ramanujan's congruences — 289
  Exercises — 291
  8.2. Ramanujan's "most beautiful" identity — 292
  Exercises — 298
  8.3. Ramanujan's congruences again — 300
  8.4. Bibliographical Notes — 303

Chapter 9. Some Combinatorial Results — 305
  9.1. Revisiting the q-factorial — 305
  Exercises — 309
  9.2. Revisiting the q-binomial coefficients — 311
  Exercises — 314
  9.3. Foata's bijection for q-multinomial coefficients — 316
  Exercises — 319
  9.4. MacMahon's proof — 319
  Exercises — 321
  9.5. q-derangement numbers — 323
  Exercises — 329
  9.6. q-Eulerian numbers and polynomials — 331
  Exercises — 338
  9.7. q-trigonometric functions — 338
  Exercises — 342
  9.8. Combinatorics of q-tangents and secants — 343

9.9. Bibliographical Notes	349
Chapter 10. The Rogers–Ramanujan Identities I: Schur	**351**
10.1. Schur's extension of Franklin's argument	351
Exercises	356
10.2. The Bressoud–Chapman proof	357
Exercises	361
10.3. The AKP and GIS identities	363
10.4. Schur's second partition theorem	365
Exercises	370
10.5. Bibliographical Notes	375
Chapter 11. The Rogers–Ramanujan Identities II: Rogers	**377**
11.1. Ramanujan's proof	377
Exercises	381
11.2. The Rogers–Ramanujan identities and partitions	383
Exercises	388
11.3. Rogers's second proof	388
Exercises	391
11.4. More identities of Rogers	394
Exercises	399
11.5. Rogers's identities and partitions	399
11.6. The Göllnitz–Gordon identities	403
Exercises	407
11.7. The Göllnitz–Gordon identities and partitions	412
Exercises	414
11.8. Bibliographical Notes	416
Chapter 12. The Rogers–Selberg Function	**417**
12.1. The Rogers–Selberg function	417
Exercises	419
12.2. Some applications	420
Exercises	423
12.3. The Selberg coefficients	423
Exercises	427
12.4. The case $k=3$	427
12.5. Explicit formulas for the Q functions	429
Exercises	430
12.6. Explicit formulas for $S_{3,i}(x)$	430
Exercises	431
12.7. The payoff for $k=3$	432
Exercises	434
12.8. Gordon's theorem	434
12.9. Bibliographical Notes	436
Chapter 13. Bailey's $_6\psi_6$ Sum	**437**
13.1. Bailey's formula	437
Exercises	439
13.2. Another proof of Ramanujan's "most beautiful" identity	442
13.3. Sums of eight squares and of eight triangular numbers	444

Exercises	447
13.4. Bailey's $_6\psi_6$ summation formula	449
Exercises	450
13.5. Askey's proof: Phase 1	454
Exercises	457
13.6. Askey's proof: Phase 2	457
Exercises	460
13.7. Askey's proof: Phase 3	460
Exercises	461
13.8. An integral	465
Exercises	470
13.9. Bailey's lemma	471
13.10. Watson's transformation	475
Exercises	479
13.11. Bibliographical Notes	481
Appendix A. A Brief Guide to Notation	483
Appendix B. Infinite Products	487
Exercises	491
Appendix C. Tannery's Theorem	495
Bibliography	501
Index of Names	513
Index of Topics	517

An Introduction to q-analysis

The phrase "q-analysis" was used in the first referee's report I ever got. While the subject of this book has a flavor all its own, and has been studied for almost 300 years, there is no term in common use that describes it really well. The closest standard name is "q-series", which is not bad—finite and infinite series occur almost everywhere, as does the letter q—but it is a little too restrictive. I think it needs another appellation, q-analysis is the best one I can think of, and I thank that anonymous referee for it (and an excellent report). Peter Paule used it in [**181**].

I have tried very hard to write a book that can be read by undergraduates. The prerequisites are minimal. One cannot have "the fear of all sums" that plagues many calculus students, but very little specific knowledge of calculus 2 will be required. In particular, you do not need an extensive knowledge of convergence tests, since for q-series the ratio test is nearly always appropriate. (The root test is marginally better in a few cases, and once in a while the nth term test is helpful.) Moreover, we will be much less concerned with when or whether an infinite series converges than with what it converges to.

We will also be seeing zillions of finite and infinite products of a certain kind (this is one reason why I don't want to just say "q-series"), but no prior knowledge of these is assumed. What little we need is developed in Appendix B, and even this can be skipped if one is willing to believe that the infinite products converge. Previous experience with mathematical induction would surely be helpful, but few if any subjects are as well suited to teach induction as q-analysis. The instructor should try to ensure that the students are comfortable with induction (or at least getting more comfortable) early in the term. It is used less often after the first two chapters, but it never goes away.

In the first section of the book it would help to know (or to be told) what $\binom{n}{2}$ means. No further knowledge of binomial coefficients is really necessary since we will make an extensive study of q-binomial coefficients and theorems in the first two chapters that will fill in any gaps. In a few places we will use complex numbers at the level of $e^{i\theta} = \cos\theta + i\sin\theta$. There are some allusions to deeper parts of complex analysis in Chapter 13.

Where to start studying q-analysis is not completely clear. In this book one could start with any of the first three chapters. Chronologically, the story begins with the pathbreaking work on partitions in Chapter 16 of the greatest mathematics book ever written, Leonhard Euler's *Introductio in Analysin Infinitorum*, which is the subject of my Chapter 3, and I have started before with what is now the first part of Chapter 2. The motivation for the current order is that I would rather work with finite sums (Chapter 2) before infinite ones (Chapter 3 and beyond), and I now prefer to discuss the combinatorial properties of q-binomial coefficients (Chapter 1) before q-binomial theorems (Chapter 2).

When I began this project many years ago, I intended that the q-derivative would play a larger role, but it now appears only in the optional sections 2.4 and 3.8, and in the last few sections of Chapter 9. The lovely little book [**154**] takes this point of view, which relieves me of some responsibility, and my own thinking has changed somewhat. While the formal analogies with ordinary calculus are undeniably beautiful, strictly speaking one can't go much beyond Euler that way, and I would rather develop Euler's theorems hand-in-hand with their combinatorial meaning.

There is, I believe, enough material here for two semesters. In a single semester, the instructor has a lot of flexibility. Aside from Chapter 3, the only really essential sections are 1.2, 1.3, two of 2.1–2.3, and 5.1. One can choose to emphasize the history of the subject, or its combinatorial aspects, or the applications to number theory; or one can just pick out the results that one finds the most beautiful. No other subject has as many beautiful formulas (in my possibly biased opinion), so one will get a lot of that no matter what one does. Here are some specific comments on each chapter:

I think one should try to resist the temptation to go too rapidly over the first several sections of Chapter 1, at least with undergraduates. I include Terquem's proof partly for historical reasons and partly to emphasize the cleverness of Rodrigues by comparison, but partly also as a device to slow myself down. Even though the ideas are initially very simple, nothing less than a parallel universe of mathematics is being constructed here, and students need time to become citizens of it. (The notation alone takes some getting used to. I have included a brief summary for reference in Appendix A.) Conjugate permutations and Rothe diagrams can be skipped, although they foreshadow the more important idea of conjugate partitions in Chapter 3. Section 1.6 is not vital, but it shows that sections 1.2 and 1.3 are hinting at something more general. A course emphasizing combinatorics should do sections 1.5 and 1.7.

The centerpiece of Chapter 2 is Rothe's q-binomial theorem in section 2.3. It is equivalent to the Potter–Schützenberger theorem of sections 2.1 and 2.2—perhaps less beautiful, but more useful. As much as anything else, what one wants to get out of the first two chapters is the sense that these two theorems and what I call the Fundamental Property of q-binomial coefficients (section 1.3) are really all saying the same thing.

One can do any two of sections 2.1–2.3. If you like both 2.1 and 2.2, then you can do Rothe's q-binomial theorem by Gruson's method (see the exercises in section 2.2) rather than as in section 2.3. (Instructors should in general be alive to the possibility of doing something in the exercises instead of or in addition to the text.) Gauss's identities in section 2.5 are largely of historical interest, but the second has an important connection with Sylvester's fishhook bijection in section 4.5, the first plays a minor role in section 9.7, and both are used in section 11.4. They are proved again in section 3.5, so one can at least postpone them. Although Jacobi's q-binomial theorem has historically been underrated, in my opinion, it could just be stated here, as there is a natural proof in section 3.6. Euler's theorem in section 2.9 also deserves to be better known, but the rest of the section can be skipped.

Sometimes one proves a q-identity by first establishing a finite form of it and then taking a limit. The limiting process often technically requires a little-known result from analysis called **Tannery's theorem**, which I have included in Appendix C, though the application is usually not made explicit. The parts of Chapter 2 not already mentioned present finite forms of identities yet to come. MacMahon's q-binomial theorem in section 2.7 is of some independent interest, but it is mainly a finite form of the Jacobi triple product, the most important identity appearing after Chapter 3, for which several other proofs are also given. MacMahon's identity might better be presented as in problem 7 of section 2.7 instead of as in the text. The partial fractions identity in section 2.8 has been included mainly to allow an instructor to do the number-theoretic applications in Chapter 7 without having to do Ramanujan's $_1\psi_1$ summation in Chapter 6. The Chen–Chu–Gu identity in section 2.10 is a finite form of the quintuple product identity in section 5.3, but that section is also optional, and it contains a second proof.

Chapter 3 introduces partitions, a subject at the intersection of combinatorics and number theory that pervades the rest of the book. The subset of q-identities with natural partition-theoretic interpretations is so large that partitions must be considered an essential part of q-analysis, not just an application of it. Analytically, nearly everything in Chapter 3 is a corollary of the Cauchy/Crelle series (my name; experts will know it as the infinite series version of the q-binomial theorem), which is not difficult to prove directly. However, it is not just the truth of the theorems that I want to establish, but their significance and their inevitability. (This I think is a word that mathematicians should use more often. I have seen it in Hardy's moving obituary of Ramanujan [**131**] and in Rota's equally beautiful essays [**204**] and [**205**]. It is also on the back cover of [**154**]. For Rota, it comes from Immanuel Kant.) I believe it is more illuminating to work from the bottom up here than from the top down, so I recommend that most of the material in the first seven sections of Chapter 3 be done in the order in which it appears. (In a graduate course, one might try to save some time here.) Franklin's "excesses" argument in section 3.4 can be skipped, as can the combinatorial proof of Cauchy/Crelle in section 3.7. The material on ee partitions in section 3.3 is included mostly as background for the Göllnitz–Gordon identities in Chapter 11.

Any course on q-analysis should include Euler's pentagonal number theorem, but one has several options. The historically minded reader might do section 4.1, which is essentially Euler's argument, but others may just note the recurrence in (4.1.4). The combinatorially minded reader should do sections 4.2 and 4.3, which present Franklin's gorgeous partition-theoretic proof; this is one of the greatest achievements of Sylvester's group at Johns Hopkins in the early 1880s and is highly recommended. The other attractive option is to wait for Jacobi's triple product identity in section 5.1, as the pentagonal number theorem is an easy corollary. Euler's theorem on divisor sums in section 4.4 explains why he worked so hard to prove the pentagonal number theorem, so it should be done by readers interested either in number theory or in history. The latter can skip the rest of the section. Another triumph of the Johns Hopkins school is Sylvester's fishhook bijection in section 4.5.

Section 5.1 proves Jacobi's triple product identity. None of the other sections in Chapter 5 is vital, although they are all interesting. The most important are sections 5.5 and 5.7, but section 5.3 is used in Chapter 8 and section 5.4 has a connection with section 4.5.

The next three chapters are shorter than the first five. Chapter 6 is devoted to Ramanujan's $_1\psi_1$ summation formula, which I think should be in any course. As evidence of this I give several different proofs. Section 6.2 has four, two in the text that rely on the q-Gauss sum from section 5.5, and two more in the problems that don't. Section 6.3 has a proof of the finite to infinite type, due to Michael Schlosser, that needs the q-Pfaff–Saalschütz identity from section 5.7. The last three sections outline another proof, due in part to Cauchy, that does not require Chapter 5 and develops Jacobi's triple product as a byproduct.

Chapter 7 contains applications of the $_1\psi_1$ to Jacobi's theorems on sums of two and four squares. As mentioned above, one can do this material without Chapter 6 if one does section 2.8, which leads to a simple proof of the relevant special case. But I think that most readers will like one or more of the proofs of the $_1\psi_1$ at least as well as the argument of section 2.8.

Chapter 8 is also on number theory, specifically congruence properties of partitions. The key theorem in the chapter was stated by Ramanujan, and has often been called his "most beautiful" identity. The approach to it given here requires the quintuple product identity from section 5.3. I might skip this chapter if I planned to cover section 13.2.

Chapter 9 returns to combinatorics. The first five sections are a natural continuation of Chapter 1, although some of the material requires the first few sections of Chapter 3. A highlight of this chapter is Foata's bijective proof in section 9.3 of MacMahon's theorem that the inversion number and the major index are equidistributed, and this requires only Chapter 1—it is not even necessary to do sections 9.1 and 9.2 first, although these simpler arguments may provide motivation. Section 9.4 gives MacMahon's original proof, and section 9.5 a pretty related result. The last three sections in this chapter use the q-derivative, with section 9.8 giving combinatorial properties of q-trigonometric functions.

Chapters 10–12 are on the Rogers–Ramanujan identities and related topics. These can be done anytime after the Jacobi triple product, in a variety of ways. Chapter 10 begins with Schur's combinatorial proof, an extension of Franklin's argument from Chapter 4. Perhaps the simplest proof, due to Robin Chapman, is in section 10.2. It is another finite to infinite type argument and can be viewed as a simplification of Schur's second proof. Ramanujan's proof is in section 11.1, and a version of one of Rogers's proofs is in section 11.3. Another of Rogers's proofs is nearly the same as Selberg's proof, which is in Chapter 12. At a minimum, I suggest doing one of the proofs and section 11.2, which interprets the Rogers–Ramanujan identities in terms of partitions. These three chapters also contain similar q-identities with some further material on partitions, for example the Göllnitz–Gordon identities, which are to the number 8 what the Rogers–Ramanujan identities are to the number 5, in sections 11.6 and 11.7.

Chapter 13 is on Bailey's "very well poised $_6\psi_6$ sum", probably the deepest result in the book. The first few sections focus on a special case, with a more elementary proof, that is still strong enough to give Ramanujan's "most beautiful" identity from Chapter 8 and Jacobi's eight square theorem as corollaries. I give Askey's proof of the $_6\psi_6$ formula, his similar evaluation of an integral, and finally Watson's transformation, another key fact about very well poised q-hypergeometric series. Following Andrews's fundamental survey paper from the mid 1970s, some of the exercises obtain the quintuple product and the two, four, and eight square theorems as corollaries of the $_6\psi_6$.

I should say something about the exercises. There are a lot of them, especially in the first five chapters. They are vital to learning the subject, and I have worked very hard on them. Many are routine, though there are more of these earlier in the book than later; some are even trivial. For example, anything that looks like "check equation so-and-so" is generally not difficult. There are also many longer exercises, including some extremely long ones which if fleshed out could be (and in several cases once were) entire sections. I have often broken the harder problems into several parts, and no doubt some readers will feel that I have overdone this. A student who feels spoon-fed might try guessing what the next step of the problem ought to be. This could lead you to a better proof. If it does, please write to me and tell me about it.

The manuscript has benefitted greatly from a detailed report by an anonymous reviewer. Despite that person's best efforts and mine, there are undoubtedly some remaining errors, obscurities, and other infelicities. Please write to me if you find any.

The three authors of [24] have each played an important role in my evolution as a mathematician. My debt to George Andrews will be obvious to the most casual reader of this book. He has profoundly affected my view of many of the subjects presented here, both personally and through his writings, and he has done me many kindnesses.

Ranjan Roy, who passed away while I was making the final edits to the book, was a trusted mentor, friend, colleague, and role model for many years. More than anyone else, he showed me the kind of career that I could have. It is also thanks to him that I first got to teach some of this material to a fondly remembered group of seven students at Beloit College in Fall 1996.

But I must dedicate the book to my late thesis advisor, Richard Askey. Dick rescued my mathematical career in my second year of graduate school, just by being himself, and he was very patient with me afterward throughout a slow process of development. No one else has had as much influence on my adult life.

CHAPTER 1

Inversions

1.1. Stern's problem

We begin our story in 1838. There were very few mathematical journals then; the best was August Leopold Crelle's *Journal für die Reine und Angewandte Mathematik*, which is still one of the leading journals today. Volume 18 of Crelle's *Journal* has a bit of filler on p. 100: three problems posed by Moritz Abraham Stern, who was a well-known mathematician at the time though he is little remembered today. (He was the first Jew to become a professor of mathematics at a Prussian university—at Göttingen, home of Gauss, who had been his thesis adviser—without converting.) Stern's third problem was: what is the total number of inversions in all of the permutations of $\{1, 2, \ldots, n\}$?

We can't hope to answer this question without first understanding it. A permutation of $\{1, 2, \ldots, n\}$ is, for our purposes, simply a list of these numbers in some order; *e.g.*, 3746152 is a permutation of $\{1, 2, 3, 4, 5, 6, 7\}$. An **inversion** in a permutation is a pair of (not necessarily consecutive) numbers that occur in decreasing order. In the permutation 15342, for example, 5, 3, and 4 all precede 2; 5 also precedes 3 and 4; and all other pairs of numbers appear in increasing order, so there are 5 inversions, namely 53, 54, 52, 32, and 42. Let's make a table of the permutations of $\{1, 2, 3\}$:

permutation	inversions	# inversions
123		0
132	32	1
213	21	1
231	21, 31	2
312	31, 32	2
321	32, 31, 21	3

We see that there are $0+1+1+2+2+3 = 9$ inversions in all the permutations of $\{1, 2, 3\}$. Let's introduce some notation: we let I_n equal the number Stern asked for, *i.e.*, I_n will denote the total number of inversions in all the permutations of $\{1, 2, \ldots, n\}$, so that $I_3 = 9$. Since there are only two permutations of $\{1, 2\}$, namely 12 and 21, the former with no inversion and the latter with one, we have $I_2 = 1$, and it is also clear that $I_1 = 0$. Stern worked out one more case, $I_4 = 72$. Let's check this. To get all the permutations of $\{1, 2, 3, 4\}$, we just have to insert 4 into the permutations of $\{1, 2, 3\}$ listed above, in every possible way. If we put 4 at the beginning, this creates three new inversions in each permutation:

permutation	inversions	permutation	inversions
4123	3 + 0	4231	3 + 2
4132	3 + 1	4312	3 + 2
4213	3 + 1	4321	3 + 3

There are $6 \times 3 = 18$ new inversions in these permutations, in addition to the 9 we already had, hence 27 inversions in all. Or we could put 4 into the second position:

permutation	inversions	permutation	inversions
1423	2 + 0	2431	2 + 2
1432	2 + 1	3412	2 + 2
2413	2 + 1	3421	2 + 3

Now there are $6 \times 2 = 12$ new inversions, and 9 old ones, so there are 21 inversions in these permutations. If 4 is put in the third position, then we have:

permutation	inversions	permutation	inversions
1243	1 + 0	2341	1 + 2
1342	1 + 1	3142	1 + 2
2143	1 + 1	3241	1 + 3

Now there are 6 new inversions and 9 old ones; and finally if 4 is put at the end of all the permutations in our original table, then there are no new inversions and the same 9 old ones:

permutation	inversions	permutation	inversions
1234	0	2314	2
1324	1	3124	2
2134	1	3214	3

Adding all this up we see that $I_4 = 27 + 21 + 15 + 9 = 72$. Let's write this arithmetic more suggestively:

$$I_4 = (18 + 9) + (12 + 9) + (6 + 9) + (0 + 9) = (18 + 12 + 6 + 0) + 4 \times 9$$
$$= 6(3 + 2 + 1 + 0) + 4 I_3 = 3!\,(3 + 2 + 1 + 0) + 4 I_3.$$

This is the case $n = 3$ of

(1.1.1) $\qquad I_{n+1} = n!\,\{n + (n-1) + (n-2) + \cdots + 2 + 1 + 0\} + (n+1)\,I_n.$

We can use the same argument as above to prove (1.1.1) in general. There are $n!$ permutations of $\{1, 2, \ldots, n\}$, and to get all the permutations of $\{1, 2, \ldots, n+1\}$ we have to insert $n+1$ into them in all possible ways. If we insert it in the kth position in all the permutations of $\{1, 2, \ldots, n\}$, this causes $n - k + 1$ new inversions in each permutation. Therefore there are $n!\,(n - k + 1)$ new inversions, and I_n old inversions, in all of the permutations of $\{1, 2, \ldots, n+1\}$ in which $n+1$ is in the kth position. Since k may be any of $1, 2, \ldots, n, n+1$, when we add up all the possibilities we get (1.1.1).

At an early stage of his career, the great French mathematician Joseph Liouville decided that there should be a French mathematical journal comparable to Crelle's, so he started the *Journal de Mathématiques Pures et Appliquées*, and in 1839 several solutions to Stern's problem appeared there. The first was by Olry Terquem, who later edited a journal himself, the *Nouvelles Annales de Mathématiques*; unlike Liouville's and Crelle's journals, it did not survive to the present day, but some important 19th century work appeared there. In volume 3 of Liouville's *Journal*, Terquem derives (1.1.1) (much less long-windedly than we did), and then uses it to work out a formula for I_n. Our first solution will be similar in spirit, though we will not follow Terquem exactly. As many readers probably know,

$$1 + 2 + 3 + \cdots + n = \frac{n(n+1)}{2} = \binom{n+1}{2}$$

(see the Exercises for proofs), so we may rewrite (1.1.1) as

$$I_{n+1} = (n+1)\,I_n + n!\binom{n+1}{2}$$

or as

(1.1.2) $$I_n = n\,I_{n-1} + n!\,\frac{n-1}{2}.$$

We may solve (1.1.2) for I_n by iteration—a technique we will also use many times. If we replace n by $n-1$ in (1.1.2), then it becomes

$$I_{n-1} = (n-1)\,I_{n-2} + (n-1)!\,\frac{n-2}{2},$$

and, substituting this into (1.1.2), we have

$$I_n = n\left\{(n-1)\,I_{n-2} + (n-1)!\,\frac{n-2}{2}\right\} + n!\,\frac{n-1}{2}$$

(1.1.3) $$= n(n-1)\,I_{n-2} + n!\left(\frac{n-2}{2} + \frac{n-1}{2}\right).$$

If we replace n by $n-2$ in (1.1.2), then it becomes

$$I_{n-2} = (n-2)\,I_{n-3} + (n-2)!\,\frac{n-3}{2},$$

and, substituting this into (1.1.3), we get

$$I_n = n(n-1)\left\{(n-2)\,I_{n-3} + (n-2)!\,\frac{n-3}{2}\right\} + n!\left(\frac{n-2}{2} + \frac{n-1}{2}\right)$$

$$= n(n-1)(n-2)\,I_{n-3} + n!\left(\frac{n-1}{2} + \frac{n-2}{2} + \frac{n-3}{2}\right).$$

By this time it is reasonable to guess that

(1.1.4) $$I_n = n(n-1)\cdots(n-k+1)\,I_{n-k} + \frac{n!}{2}\{(n-1) + (n-2) + \cdots + (n-k)\}.$$

This is the form we have found for $k = 2$ and $k = 3$, and for $k = 1$ it is just (1.1.2). (When $k = 0$ it says I_n equals itself. Problem 3 gives a better way to write it.) We prove (1.1.4) by induction on k, by exactly the same sort of calculation as above. If we replace n by $n-k$ in (1.1.2), then it becomes

$$I_{n-k} = (n-k)\,I_{n-k-1} + (n-k)!\,\frac{n-k-1}{2},$$

and if we substitute this in (1.1.4), then we get

$$I_n = n(n-1)\cdots(n-k+1)\left\{(n-k)I_{n-k-1} + (n-k)!\,\frac{n-k-1}{2}\right\}$$
$$+ \frac{n!}{2}\{(n-1)+(n-2)+\cdots+(n-k)\}$$
$$= n(n-1)\cdots(n-k+1)(n-k)I_{n-k-1}$$
$$+ \frac{n!}{2}\{(n-1)+(n-2)+\cdots+(n-k)+(n-k-1)\},$$

which is (1.1.4) with $k+1$ in place of k, so (1.1.4) is true by induction. Taking $k = n-1$ there we get

$$I_n = n(n-1)\cdots 2\,I_1 + \frac{n!}{2}\{(n-1)+(n-2)+\cdots+1\}.$$

But $I_1 = 0$, so the first term vanishes, and the second term simplifies to

$$(1.1.5) \qquad I_n = \frac{n!}{2}\binom{n}{2}.$$

One could hardly hope for a simpler formula, but one might wish for a simpler solution. A much easier proof was given by Olinde Rodrigues in volume 4 of Liouville's *Journal*. If $n < 2$, then there are no inversions, so (1.1.5) certainly holds in that case. If $n \geq 2$, let the **mate** of a permutation be the same permutation read backwards; thus, for example, the mate of 3416752 is 2576143. Then take each of the $n!$ permutations of $\{1,2,\ldots,n\}$, and consider it along with its mate. Call each pair of permutations a **couple**, so that the permutations of $\{1,2,3\}$ separate into the couples 123 and 321, 132 and 231, 213 and 312. The point is that any pair of numbers is inverted either in a given permutation or in its mate, but not both. For example: in 3416752, 6 is inverted with 5 and 2, but not with $1, 3, 4, 7$; whereas in 2576143, 6 is inverted with $1, 3, 4, 7$ but not with 2 and 5. The permutations of $\{1,2,\ldots,n\}$ separate into $\frac{n!}{2}$ couples. Each of the $\binom{n}{2}$ pairs of the numbers $\{1,2,\ldots,n\}$ is an inversion exactly once in each couple, and so (1.1.5) is true.

We close this section with a digression that foreshadows an important concept in Chapter 3. In 1800 Rothe defined the **conjugate** of a permutation π of $\{1,2,\ldots,n\}$ to be the permutation π' that has b in the ath position whenever a is in the bth position in π. Thus, for example, the conjugate of $\pi = 5316742$ is $\pi' = 3726145$, because 5 is first in π while 1 is fifth in π', 3 is second in π while 2 is third in π', and so forth. Then we have

THEOREM 1 (Rothe's theorem on conjugate permutations). *Conjugate permutations have the same number of inversions.*

In the above example π has $4+2+0+2+2+1+0 = 11$ inversions, and π' has $2+5+1+3+0+0+0 = 11$ inversions. In general, suppose a is in the bth position and c is in the dth position in π. Then b is in the ath position and d in the cth position in π'. We may as well assume that $b < d$, which leaves us with two possibilities. If $a < c$, then a and c are not inverted in π, and also b and d are not inverted in π' since b comes before d. If $a > c$, then a and c are inverted in π, and the smaller b comes after the larger d in π' so b and d are inverted there. So each inversion in π has a corresponding inversion in π', and vice versa. This proves Rothe's theorem.

Rothe's proof is more interesting, if possibly less convincing. He introduces something we will call the **Rothe diagram** of a permutation. We use the example above to illustrate it. Since π and π' have length 7, start with a labeled 7×7 array and represent the permutation 5316742 with dots in the appropriate rows:

```
    1  2  3  4  5  6  7
1               •
2         •
3   •
4                  •
5                     •
6         •
7      •
```

Note that the columns represent the conjugate permutation 3726145. Then represent the inversions with ∘s. 5 is inverted with all of 1–4, so we put four ∘s to the left of the bullet in the first row representing 5. 3 is inverted with 1 and 2, so we put two ∘s to the left of the bullet in the second row. 1 is not inverted with anything after it, so we do not use any ∘s in the third row. In the fourth row, 6 is inverted in 5316742 only with 4 and 2, so we only put ∘s in these columns. Similarly 7 is inverted only with 4 and 2, so only these columns get ∘s in the fifth row. In the sixth row 4 is inverted with 2, so we put a ∘ in the second column, and there are no ∘s in the last row:

```
    1  2  3  4  5  6  7
1   ∘  ∘  ∘  ∘  •
2   ∘  ∘  •
3   •
4      ∘     ∘     •
5      ∘     ∘        •
6      ∘     •
7      •
```

This is the Rothe diagram of the permutation 5316742. Rothe then observes that the ∘s in each column represent the inversions in the conjugate permutation 3726145: 3 is inverted with 2 and 1, 7 is inverted with everything but 3, 2 is inverted with 1, and 6 is inverted with 1,4,5. This always happens, by the same argument we used to prove Rothe's theorem.

Exercises

1. The most standard evaluation of $1 + 2 + \cdots + n$ is to write
$$\begin{aligned} S_n &= 1 + 2 + \cdots + n, \\ S_n &= n + (n-1) + \cdots + 1 \end{aligned}$$
and then add the columns. Explain why this gives
$$S_n = 1 + 2 + \cdots + n = \frac{n(n+1)}{2} = \binom{n+1}{2}.$$
This is often called the *Gauss trick* or some similar name. In his later years, Gauss liked to tell a story of having used it with $n = 100$ in elementary school.

2. Prove the result of problem 1 by induction on n.

3. A more careful statement of (1.1.4) is that, for any integer k with $0 \le k \le n$, we have

(1.1.6) $$I_n = \frac{n!}{(n-k)!} I_{n-k} + \frac{n!}{2}\left\{nk - \binom{k+1}{2}\right\}.$$

Prove that (1.1.6) is equivalent to (1.1.4).

4. Prove (1.1.6) by the same argument we used to prove (1.1.4).

5. Prove that (1.1.6) reduces to (1.1.5) if $k = n-1$ or if $k = n$. What if $k = n - 2$?

6. Terquem's solution of (1.1.1) is a *forward* iteration. Starting from $I_{n+1} = (n+1)I_n + (n+1)!\frac{n}{2}$, he proved that

$$I_{n+p} = \frac{(n+p)!}{n!} I_n + \frac{(n+p)!}{2}\left\{np + \binom{p}{2}\right\} \quad \text{for any nonnegative integer } p.$$

Then he set $n = 1$ to get $I_{p+1} = \frac{(p+1)!}{2}\binom{p+1}{2}$, and then he renamed $p+1$ as n to get (1.1.5). Fill in the details of this argument.

7. If we define $A_n = \frac{I_n}{n!}$, show that (1.1.2) becomes $A_n - A_{n-1} = \frac{n-1}{2}$, where $A_1 = 0$. What does

$$(A_n - A_{n-1}) + (A_{n-1} - A_{n-2}) + (A_{n-2} - A_{n-3}) + \cdots + (A_2 - A_1)$$

equal? (You should be able to give two good answers to this question.) Show how this gives another derivation of (1.1.5).

8. What is the *average* number of inversions that a permutation of $\{1, 2, \ldots, n\}$ has? Is the answer obvious—in other words, could you have guessed it if you didn't already know the answer to Stern's problem? If so, can you make this observation into a solution of Stern's problem?

9. Another nice solution of Stern's problem has been given by Emeric Deutsch (private communication). For each inversion in each permutation of $\{1, 2, \ldots, n\}$ we get to choose three things:
   (a) which two numbers are inverted,
   (b) which two positions those numbers occupy (the larger one coming first),
   (c) the positions of the other $n - 2$ numbers.
   Show how this gives (1.1.5).

10. A common technique for solving recurrence relations is to introduce a generating function. Here is an outline of such a solution of (1.1.1): if $f(x) = \sum_{n=0}^{\infty} I_n \frac{x^n}{n!}$, then (since $I_0 = 0$)

$$f(x) = \sum_{n=0}^{\infty} I_{n+1} \frac{x^{n+1}}{(n+1)!}$$

$$= \sum_{n=0}^{\infty} \left\{(n+1)I_n + (n+1)!\frac{n}{2}\right\} \frac{x^{n+1}}{(n+1)!}$$

$$= x\sum_{n=0}^{\infty} I_n \frac{x^n}{n!} + \sum_{n=0}^{\infty} \frac{n}{2} x^{n+1}$$

$$= x f(x) + \frac{x^2}{2}\sum_{n=0}^{\infty} nx^{n-1},$$

so
$$(1-x)\,f(x) = \frac{x^2}{2}\frac{1}{(1-x)^2} \quad \text{and hence} \quad f(x) = \frac{x^2}{2(1-x)^3}.$$
But
$$\frac{x^2}{2(1-x)^3} = \frac{x^2}{2}\sum_{k=0}^{\infty}\binom{k+2}{2}x^k = \frac{1}{2}\sum_{k=0}^{\infty}\binom{k+2}{2}x^{k+2} = \frac{1}{2}\sum_{n=2}^{\infty}\binom{n}{2}x^n,$$
so (1.1.5) holds. Fill in the details of this argument.

11. Draw the Rothe diagram of the permutation $\pi = 596381472$, find the conjugate permutation π', and check that they have the same number of inversions.

12. Some permutations are their own conjugates, like 1 and 321 and 4561237. If U_n denotes the number of self-conjugate permutations of $\{1, 2, \ldots, n\}$, show that $U_1 = 1$, $U_2 = 2$, $U_3 = 4$, and $U_4 = 10$.

13. Show that $U_5 = 26$, and that in general $U_{n+1} = U_n + n\,U_{n-1}$ for $n \geq 1$, where we define $U_0 = 1$. (**Hint:** Consider the different positions that $n+1$ can occupy in a self-conjugate permutation of $\{1, 2, \ldots, n, n+1\}$.)

14. As in problem 10, we can convert the recurrence relation of problem 13 into a generating function for the numbers U_n. Show that if $U_0 = 1$ and $U_{n+1} = U_n + n\,U_{n-1}$ for $n \geq 1$, and if $g(x) = \sum_{n=0}^{\infty} U_n \frac{x^n}{n!}$, then $g(0) = 1$ and $g'(x) = (1+x)\,g(x)$. Explain why this implies that

$$\sum_{n=0}^{\infty} U_n \frac{x^n}{n!} = e^{\frac{x^2}{2}+x}.$$

It is possible to "solve" this to get an explicit formula for U_n, but we leave this until section 1.6.

1.2. The q-factorial

While Rodrigues's proof shows that a seemingly difficult problem sometimes becomes easy when looked at the right way (a phenomenon that we can never have too many examples of), the main point of his paper was something else. Let's repeat our first table with an added column, and some obvious abbreviations:

perm	inv	q^{inv}	perm	inv	q^{inv}
123	0	q^0	231	2	q^2
132	1	q^1	312	2	q^2
213	1	q^1	321	3	q^3

Here q is a variable, and we just made the inversions column an exponent of q in the succeeding column. When we added up the inversions column we got 9; when we add up the new column we get

$$1 + q + q + q^2 + q^2 + q^3 = (1 + q + q^2) + (q + q^2 + q^3) = (1+q)(1+q+q^2).$$

Now think about what happens when we insert 4 into these permutations, as we did in section 1.1. Let's re-do some of those tables. If 4 is at the beginning, then

we have

perm	q^{inv}	perm	q^{inv}
4123	q^{3+0}	4231	q^{3+2}
4132	q^{3+1}	4312	q^{3+2}
4213	q^{3+1}	4321	q^{3+3}

where again q is a variable, and we just made the inversions column in the previous version of this table an exponent of q. Evidently there is a common factor of q^3, representing the three inversions caused by putting 4 at the beginning, and the total contribution from these permutations is $q^3(1+q)(1+q+q^2)$. If 4 is in the second position, then we have:

perm	q^{inv}	perm	q^{inv}
1423	$2+0$	2431	$2+2$
1432	$2+1$	3412	$2+2$
2413	$2+1$	3421	$2+3$

Here there is a common factor of q^2, since there are two inversions involving 4 in each of these permutations, which in all contribute $q^2(1+q)(1+q+q^2)$. If 4 is in the third position, then it will cause one inversion, and otherwise we will have the same inversions as before, so those permutations yield $q(1+q)(1+q+q^2)$. Finally, if 4 is at the end, then it causes no new inversions, so these permutations just contribute the same thing as the permutations of $\{1,2,3\}$, namely $(1+q)(1+q+q^2)$. In all, the permutations of $\{1,2,3,4\}$ therefore contribute $(1+q)(1+q+q^2)(1+q+q^2+q^3)$.

Something quite beautiful seems to be happening here, and we need to make a careful statement of just what that something is. First let's introduce some notation, so that we may describe it succinctly. We keep getting polynomials of the form $1 + q + q^2 + \cdots + q^{n-1}$ for various positive integers n, so let's define

$$[n]_q := 1 + q + q^2 + \cdots + q^{n-1} \quad \text{if } n \text{ is a positive integer.}$$

These polynomials are finite geometric series, which means that we can use a well-known trick to rewrite them. Note that

$$q[n]_q = q + q^2 + \cdots + q^{n-1} + q^n, \quad \text{and therefore} \quad [n]_q - q[n]_q = 1 - q^n.$$

If $q \neq 1$, then we can solve this for $[n]_q$: $[n]_q = \frac{1-q^n}{1-q}$ if n is a positive integer and $q \neq 1$. If $q = 1$, then $[n]_q$ is just n. This allows us to make a more general definition: if n is *any* real number (not necessarily a positive integer), then

$$[n]_q = \begin{cases} \frac{1-q^n}{1-q} & \text{if } q \neq 1 \\ n & \text{if } q = 1. \end{cases}$$

$[n]_q$ is called the *q*-**analogue** of the number n. n will almost always be a positive integer in what follows, but the more general definition allows us to conclude that $[0]_q = 0$; note also that $[1]_q = 1$.

Next, we define the q-analogue of $n!$ in the obvious way:

$$n!_q := [1]_q [2]_q \cdots [n]_q.$$

It is convenient to define $0!_q = 1$, just as $0!$ is defined to be 1. (As a general rule, one always defines an empty *sum* to be 0, and an empty *product* to be 1; that way you don't change something by adding an empty sum to it, or by multiplying it by an empty product. Since $[n]_q$ is essentially a sum and $n!_q$ essentially a product, in some sense this explains why $[0]_q$ ought to be 0 and $0!_q$ ought to be 1.)

Let us also pause here to bring in some notation that will be used extensively later on. If we write out the definition of $n!_q$ without using the $[k]_q$ notation it looks like

$$n!_q = \frac{1-q}{1-q} \frac{1-q^2}{1-q} \cdots \frac{1-q^n}{1-q}.$$

Products like the numerator of this fraction occur very often in this subject, so let's define

$$(q;q)_n := \begin{cases} (1-q)(1-q^2)\cdots(1-q^n) & \text{if } n \text{ is a positive integer,} \\ 1 & \text{if } n = 0, \end{cases}$$

so that

(1.2.1) $$(q;q)_n = (1-q)^n \, n!_q.$$

In later chapters we will generalize this notation by defining

$$(a;q)_n := \begin{cases} (1-a)(1-aq)(1-aq^2)\cdots(1-aq^{n-1}) & \text{if } n \text{ is a positive integer,} \\ 1 & \text{if } n = 0. \end{cases}$$

Now we are in a position to describe the theorem that we were seeing above. When we looked at all the permutations of $\{1,2,3\}$, counted all the inversions in each, made all those inversion numbers exponents of a variable q, and added all the terms together, we wound up with a polynomial that factored as $(1+q)(1+q+q^2)$. Since $[1]_q = 1$, we now see that this is the same thing as $[1]_q[2]_q[3]_q$, which is $3!_q$. Moreover, when we did the same thing with all the permutations of $\{1,2,3,4\}$ we wound up with $1(1+q)(1+q+q^2)(1+q+q^2+q^3)$, which is $4!_q$. The general result is

THEOREM 2 (Rodrigues's theorem). *If $\Pi(n)$ is the set of all permutations of $\{1,2,\ldots,n\}$, then*

$$n!_q = \sum_{\pi \in \Pi(n)} q^{\operatorname{inv} \pi}.$$

One can put Rodrigues's theorem in a slightly different form: f $r_k(n)$ denotes the number of permutations of $\{1,2,\ldots,n\}$ with exactly k inversions, then

$$n!_q = \sum_{k=0}^{\binom{n}{2}} r_k(n) \, q^k.$$

This simply combines like terms in the previous version of the theorem. The upper limit is $\binom{n}{2}$ because that is the largest number of inversions that a permutation of $\{1,2,\ldots,n\}$ can have.

We may prove Rodrigues's theorem by induction on n. We have done the cases $n = 3$ and $n = 4$ already, and we leave it to the reader to check it for smaller n. Assuming Rodrigues's theorem holds for n, consider the sum $\sum_{\pi \in \Pi(n+1)} q^{\operatorname{inv} \pi}$, where

$\Pi(n+1)$ is the set of all permutations of $\{1, 2, \ldots, n, n+1\}$. We treat this the same way as before—split it up according to where $n+1$ is in each permutation:

$$\sum_{\pi \in \Pi(n+1)} q^{\text{inv } \pi} = \sum_{\substack{\pi \in \Pi(n+1) \\ n+1 \text{ is first}}} q^{\text{inv } \pi} + \sum_{\substack{\pi \in \Pi(n+1) \\ n+1 \text{ is second}}} q^{\text{inv } \pi} + \cdots + \sum_{\substack{\pi \in \Pi(n+1) \\ n+1 \text{ is last}}} q^{\text{inv } \pi}$$

If $n+1$ is in the first position, then it is inverted with all of $\{1, 2, \ldots, n\}$, so there are n inversions involving $n+1$, and some other inversions not involving $n+1$. (We were calling these the "new inversions" and the "old inversions" respectively above.) For the inversions not involving $n+1$, we can completely ignore $n+1$, and permutations of $\{1, 2, \ldots, n, n+1\}$ in which $n+1$ is ignored are in effect just permutations of $\{1, 2, \ldots, n\}$. Therefore

$$\sum_{\substack{\pi \in \Pi(n+1) \\ n+1 \text{ is first}}} q^{\text{inv } \pi} = \sum_{\substack{\pi \in \Pi(n+1) \\ n+1 \text{ is first}}} q^{\text{inv } \pi \text{ involving } n+1} q^{\text{inv } \pi \text{ not involving } n+1}$$

$$= q^n \sum_{\substack{\pi \in \Pi(n+1) \\ n+1 \text{ is first}}} q^{\text{inv } \pi \text{ not involving } n+1}$$

$$= q^n \sum_{\pi \in \Pi(n)} q^{\text{inv } \pi} = q^n \, n!_q,$$

where the last step uses the induction hypothesis. Now

$$\sum_{\substack{\pi \in \Pi(n+1) \\ n+1 \text{ is second}}} q^{\text{inv } \pi}$$

is exactly the same, except that if $n+1$ is second in a permutation of $\{1, 2, \ldots, n, n+1\}$, then it is inverted with whatever $n-1$ numbers come after it. So

$$\sum_{\substack{\pi \in \Pi(n+1) \\ n+1 \text{ is second}}} q^{\text{inv } \pi} = \sum_{\substack{\pi \in \Pi(n+1) \\ n+1 \text{ is second}}} q^{\text{inv } \pi \text{ involving } n+1} q^{\text{inv } \pi \text{ not involving } n+1}$$

$$= q^{n-1} \sum_{\substack{\pi \in \Pi(n+1) \\ n+1 \text{ is second}}} q^{\text{inv } \pi \text{ not involving } n+1}$$

$$= q^{n-1} \sum_{\pi \in \Pi(n)} q^{\text{inv } \pi} = q^{n-1} \, n!_q$$

and in general we have

$$\sum_{\pi \in \Pi(n+1)} q^{\text{inv } \pi} = \sum_{\substack{\pi \in \Pi(n+1) \\ n+1 \text{ is first}}} q^{\text{inv } \pi} + \sum_{\substack{\pi \in \Pi(n+1) \\ n+1 \text{ is second}}} q^{\text{inv } \pi} + \cdots + \sum_{\substack{\pi \in \Pi(n+1) \\ n+1 \text{ is last}}} q^{\text{inv } \pi}$$

$$= q^n \, n!_q + q^{n-1} \, n!_q + \cdots + 1 \cdot n!_q$$

$$= \left(q^n + q^{n-1} + \cdots + 1\right) n!_q$$

$$= [n+1]_q \, n!_q = (n+1)!_q.$$

This proves Rodrigues's theorem by induction.

Exercises

1. Check that Rodrigues's theorem holds in the cases $n = 0, 1, 2$.
2. Explain why the largest number of inversions that a permutation of $\{1, 2, \ldots, n\}$ can have is $\binom{n}{2}$. Which permutation (or permutations) has (or have) this many inversions?
3. Using L'Hopital's rule or otherwise, evaluate $\lim\limits_{q \to 1} \frac{1-q^n}{1-q}$. Does the result make sense when compared with the definition of $[n]_q$?
4. Show that $[n]_q = [k]_q + q^k[n-k]_q$. This simple property is surprisingly useful.
5. Show that $[-n]_q = -q^{-n}[n]_q$.
6. Show that $[n]_q + q[k]_q[n-k-1]_q = [k+1]_q[n-k]_q$.
7. Show that $q^{a-1}[b+1]_q - [a+1]_q = q^{a+1}[b-1]_q - [a-1]_q$.
8. If n and k are positive integers, show that
$$\frac{[nk]_q}{[k]_q} = 1 + q^k + \cdots + q^{(n-1)k} = [n]_{q^k}.$$
9. Show that
$$\frac{1}{[k]_q} = \left[\frac{1}{k}\right]_{q^k}.$$
10. Show that
$$\sum_{k=1}^{n} q^{(k-1)^2}[2k-1]_q = [n^2]_q.$$
We will need the $q = 1$ case (the sum of the first n odd numbers is n^2) several times later.
11. Show that
$$\sum_{k=1}^{n} \frac{q^k}{[k]_q[k+1]_q} = \frac{q[n]_q}{[n+1]_q}.$$
What happens to this identity as $n \to \infty$? (The answer varies according to the size of $|q|$. There is one value of q for which this sum doesn't make sense. Which one?)
12. If m is a positive integer, show that
$$\sum_{k=1}^{n} \frac{q^k}{[k]_q[k+1]_q \cdots [k+m]_q} = \frac{1}{[m]_q}\left(\frac{1}{m!_q} - \frac{n!_q}{(n+m)!_q}\right).$$
What happens to this identity as $n \to \infty$? (The answer varies according to the size of $|q|$. There is one value of q for which this sum doesn't make sense. Which one?)
13. Show that
$$\sum_{k=1}^{n} \frac{q^{4k-1}}{[2k-1]_q[2k]_q[2k+2]_q[2k+3]_q} = \frac{q}{[3]_q[4]_q}\left(1 + \frac{1}{[3]_q} - \frac{1}{[2n+1]_q} - \frac{1}{[2n+3]_q}\right)$$
$$- \frac{1}{[2]_q[3]_q}\left(\frac{1}{[2]_q} - \frac{1}{[2n+2]_q}\right).$$

What happens to this identity as $n \to \infty$? (The answer varies according to the size of $|q|$. There is one value of q for which this sum doesn't make sense. Which one?)

14. Another interpretation of the q-factorial is in terms of **crossing diagrams**. There are six crossing diagrams from three nodes to three nodes, namely:

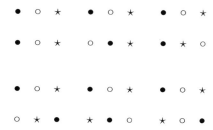

Connecting each node on the top line to the node of the same type on the bottom line we find $0, 1, 1, 2, 2, 3$ crossings respectively. If $c(\delta)$ denotes the number of crossings in the diagram δ, and if $\Delta(3)$ denotes the set of six diagrams above, then
$$\sum_{\delta \in \Delta(3)} q^{c(\delta)} = q^0 + q^1 + q^1 + q^2 + q^2 + q^3 = 3!_q.$$
Prove that, if $\Delta(n)$ is the set of all crossing diagrams from n nodes to n nodes, then $\sum_{\delta \in \Delta(n)} q^{c(\delta)} = n!_q$. There are two different ways that you might go about this: either directly, by induction on n; or by setting up a 1-1 correspondence between crossing diagrams and permutations and using Rodrigues's theorem.

15. (This problem is due to Christopher Hammond.) Suppose $F(a, b)$ satisfies
$$F(a, b) = F\left(\frac{a}{q + a(1-q)}, \frac{b}{q + b(1-q)}\right).$$
(i) Show that
$$F(a, b) = F\left(\frac{a}{q^n + a(1-q^n)}, \frac{b}{q^n + b(1-q^n)}\right)$$
for any nonnegative integer n.

(ii) Show that
$$A = \frac{a}{q + a(1-q)} \iff a = \frac{A}{q^{-1} + A(1-q^{-1})}.$$

(iii) Explain why (ii) implies that the result of (i) holds for *every* integer n.

(iv) How does the iteration in (i) behave as $n \to \infty$? Consider the cases $q > 1$, $q = 1$, $-1 < q < 1$, $q = -1$, and $q < -1$.

(v) How does the iteration behave as $n \to -\infty$?

16. Rodrigues used his theorem to give another solution to Stern's problem. Here is an outline of this solution:

(i)
$$\log\left(\sum_{\pi \in \Pi(n)} q^{\text{inv } \pi}\right) = \log(n!_q) = \sum_{k=1}^{n} \log\left(1 + q + q^2 + \cdots + q^{k-1}\right).$$

(ii) Taking the derivative of the left and right sides of (i) with respect to q, we get

$$\frac{\sum_{\pi \in \Pi(n)} (\text{inv } \pi) \, q^{(\text{inv } \pi)-1}}{\sum_{\pi \in \Pi(n)} q^{\text{inv } \pi}} = \sum_{k=1}^{n} \frac{1 + 2q + 3q^2 + \cdots + (k-1)q^{k-2}}{1 + q + q^2 + \cdots + q^{k-1}}.$$

(iii) Setting $q = 1$ in (ii) we get

$$\frac{\sum_{\pi \in \Pi(n)} \text{inv } \pi}{n!} = \sum_{k=1}^{n} \frac{1 + 2 + \cdots + (k-1)}{k} = \sum_{k=1}^{n} \frac{k-1}{2} = \frac{1}{2}\binom{n}{2}$$

from which the answer to Stern's problem follows. Fill in the details.

17. Recall the idea of self-conjugate permutations from problems 12–14 in the previous section, and define $U_n(q) = \sum_\pi q^{\text{inv } \pi}$, where the sum is over all self-conjugate permutations of $\{1, 2, \ldots, n\}$, and as usual inv π denotes the number of inversions in the permutation π. Also define $U_0(q) = 1$. Calculate the first several values of $U_n(q)$, and show that

$$U_{n+1}(q) = U_n(q) + \left(q + q^3 + q^5 + \cdots + q^{2n-1}\right) U_{n-1}(q) \quad \text{for } n \geq 1.$$

(Same hint as in problem 13 in the previous section.)

18. A **reciprocal** polynomial is one like $1 + 2q + 3q^2 + 2q^3 + q^4$ (which is $[3]_q^2$) where the coefficients 1-2-3-2-1 are the same when read backwards as when read forwards. More formally,

$$c_k x^k + c_{k+1} x^{k+1} + c_{k+2} x^{k+2} + \cdots + c_{n-2} x^{n-2} + c_{n-1} x^{n-1} + c_n x^n$$

is a reciprocal polynomial if $c_k = c_n$, $c_{k+1} = c_{n-1}$, $c_{k+2} = c_{n-2}$, and so on, i.e., $c_{k+j} = c_{n-j}$ for every j. It might be better to call them palindromic polynomials, but there is a good reason for the name reciprocal that we will see in a moment. Since any factors of x do not affect the reciprocal property, we may as well assume that the lowest power of x is zero. Note that if $p_n(x) = a_n x^n + a_{n-1} x^{n-1} + \cdots + a_1 x + a_0$, then

$$x^n p_n(\tfrac{1}{x}) = a_n + a_{n-1} x + \cdots + a_1 x^{n-1} + a_0 x^n,$$

so if $p_n(x)$ is a reciprocal polynomial of degree n, then $x^n p_n(\tfrac{1}{x}) = p_n(x)$.

(i) Explain why the q-numbers are reciprocal polynomials.

(ii) Show that the product of a reciprocal polynomial of degree m and a reciprocal polynomial of degree n is a reciprocal polynomial of degree $m + n$.

(iii) Use (i) and (ii) to show that the q-factorials are reciprocal polynomials.

(iv) Consider all the permutations of $\{1, 2, \ldots, n\}$ and think of inversions. What happens if we switch 1 and n, 2 and $n-1$, and generally j and $n+1-j$ for every j? Try to use this idea and Rodrigues's theorem to show that the q-factorials are reciprocal polynomials. The result of problem 2 may help.

(v) Let $p(x) = a_n x^n + \cdots + a_1 x + a_0$ be a polynomial of degree n. If there is a k such that $a_0 \leq a_1 \leq \cdots \leq a_k$ and $a_k \geq a_{k+1} \geq \cdots \geq a_n$, then $p(x)$ is said to be **unimodal**. For example, $3!_q = 1 + 2q + 2q^2 + q^3$ is unimodal with $k = 1$ or $k = 2$, and $4!_q = 1 + 3q + 5q^2 + 6q^3 + 5q^4 + 3q^5 + q^6$ is unimodal with $k = 3$.

It is generally not easy to prove that a polynomial (or a sequence of numbers) is unimodal. One problem is that an example like

$$(1+3q+q^2+q^3+q^4)(1+2q+q^2+q^3+q^4) = 1+5q+8q^2+7q^3+8q^4+7q^5+3q^6+2q^7+q^8$$

shows that the product of two unimodal polynomials need not be unimodal. However, the product of two reciprocal unimodal polynomials is a reciprocal unimodal polynomial. Prove this, and hence show that the q-factorial is unimodal.

1.3. q-binomial coefficients

In this section we'll play the same sort of game as before, but this time with "words" made from a two-letter "alphabet". For the moment, let's take the "letters" to be 0 and 1, where as usual $1 > 0$. In general, we'll consider sequences of k 0's and $n - k$ 1's, and as before we'll keep track of inversions; this time, an inversion will be any pair of a 1 and a 0 where the 1 comes before the 0.

The smallest interesting example is when $k = 2$ and $n = 4$, so we are looking at all the sequences of two 0's and two 1's. There are $\binom{4}{2} = 6$ such sequences, which are listed in the following table. In the second column of the table we count the number of inversions in each sequence, and make that number an exponent of q:

sequence	q^{inv}	sequence	q^{inv}
0011	q^0	1001	q^2
0101	q^1	1010	q^3
0110	q^2	1100	q^4

When we add the inversions column we get $1 + q + 2q^2 + q^3 + q^4$, which must be some sort of q-analogue of $\binom{4}{2}$. You might be a little surprised that it is not the same polynomial as $3!_q$, but if so your surprise will be temporary. Our experience with q-factorials suggests that we should try to factor it, which is not too hard:

$$1 + q + 2q^2 + q^3 + q^4 = \left(1 + q + q^2\right) + \left(q^2 + q^3 + q^4\right) = \left(1 + q + q^2\right)\left(1 + q^2\right).$$

We can write $1 + q + q^2 = [3]_q$, and problem 8 in the previous section suggests something we can do with the other factor:

$$1 + q + 2q^2 + q^3 + q^4 = [3]_q \left(1 + q^2\right) \frac{1+q}{1+q}$$
$$= [3]_q \frac{1 + q + q^2 + q^3}{1+q}$$
$$= \frac{[4]_q [3]_q}{[2]_q}.$$

Since $[1]_q = 1$, we may rewrite this further:

$$1 + q + 2q^2 + q^3 + q^4 = \frac{[4]_q [3]_q}{[2]_q [1]_q}$$
$$= \frac{[4]_q [3]_q}{[2]_q [1]_q} \frac{[2]_q [1]_q}{[2]_q [1]_q}$$
$$= \frac{4!_q}{2!_q \, 2!_q}.$$

This suggests what happens in general, but let's do one more example before we try to write down the general case. This time let's take $k = 4$ and $n = 6$, so we are looking at all the sequences of four 0's and two 1's. There are $\binom{6}{2} = 15$ such sequences, and again we'll count the number of inversions for each sequence and make that number an exponent of q:

sequence	q^{inv}	sequence	q^{inv}	sequence	q^{inv}
000011	q^0	001010	q^3	100010	q^5
000101	q^1	001100	q^4	011000	q^6
001001	q^2	010010	q^4	100100	q^6
000110	q^2	100001	q^4	101000	q^7
010001	q^3	010100	q^5	110000	q^8

When we add all this up we must get some sort of q-analogue of $\binom{6}{2}$, so we set

$$\binom{6}{2}_q := 1 + q + 2q^2 + 2q^3 + 3q^4 + 2q^5 + 2q^6 + q^7 + q^8,$$

which we should try to factor. Since $\binom{6}{2} = \frac{6!}{4!\,2!} = \frac{6 \cdot 5}{2}$, it is reasonable to hope that $[5]_q$ will be a factor. If we write

$$\binom{6}{2}_q = (1 + q + q^2 + q^3 + q^4)$$
$$+ (q^2 + q^3 + q^4 + q^5 + q^6)$$
$$+ (q^4 + q^5 + q^6 + q^7 + q^8)$$
$$= (1 + q + q^2 + q^3 + q^4)(1 + q^2 + q^4),$$

then we can see that it is. The other factor is

$$1 + q^2 + q^4 = (1 + q^2 + q^4)\frac{1+q}{1+q} = \frac{1 + q + q^2 + q^3 + q^4 + q^5}{1+q} = \frac{[6]_q}{[2]_q},$$

so we have

$$\binom{6}{2}_q = 1 + q + 2q^2 + 2q^3 + 3q^4 + 2q^5 + 2q^6 + q^7 + q^8$$
$$= \frac{[6]_q [5]_q}{[2]_q} = \frac{[6]_q [5]_q}{[2]_q [1]_q} \frac{[4]_q [3]_q [2]_q [1]_q}{[4]_q [3]_q [2]_q [1]_q}$$
$$= \frac{6!_q}{4!_q\, 2!_q}.$$

We are in the same position now as in the previous section: we have evidence that something beautiful is happening, and we need to try to describe it as best we can, and then prove our description is correct. We define q-binomial coefficients in the obvious way:

$$\binom{n}{k}_q = \begin{cases} \frac{n!_q}{k!_q\,(n-k)!_q}, & n, k \text{ integers, } 0 \leq k \leq n, \\ 0, & \text{otherwise.} \end{cases}$$

Equally well, we could define

$$\binom{n}{k}_q = \begin{cases} \frac{(q;q)_n}{(q;q)_k\,(q;q)_{n-k}}, & n, k \text{ integers, } 0 \leq k \leq n, \\ 0, & \text{otherwise,} \end{cases}$$

since the factors of $1-q$ by which $n!_q$ and $(q;q)_n$ differ all cancel. Note that the q-binomial coefficients are symmetric in k and $n-k$, as the ordinary binomial coefficients are; i.e., $\binom{n}{k}_q = \binom{n}{n-k}_q$. They are sometimes called Gaussian binomial coefficients instead, or Gaussian polynomials, since Gauss was the first to consider them, in a paper of 1808. Now we can state the theorem that the above examples suggest:

THEOREM 3 (Fundamental Property of q-binomial coefficients). *Let $S(k, n-k)$ denote the set of all sequences of k 0's and $n-k$ 1's. If σ is such a sequence, let $\operatorname{inv}\sigma$ denote the number of inversions in σ. Then*

$$\binom{n}{k}_q = \sum_{\sigma \in S(k,n-k)} q^{\operatorname{inv}\sigma}.$$

As with Rodrigues's theorem, this can be stated in a slightly different form, by combining like terms: if $c_j(k, n-k)$ denotes the number of sequences of k 0's and $n-k$ 1's with exactly j inversions, then

$$\binom{n}{k}_q = \sum_{j=0}^{k(n-k)} c_j(k, n-k)\, q^j.$$

Note that in particular the Fundamental Property implies that the q-binomial coefficients are *polynomials* in q. This is not at all obvious from the definition, which has a lot of denominator factors. The same sort of thing happens with the ordinary binomial coefficients

$$\binom{n}{k} = \begin{cases} \frac{n!}{k!\,(n-k)!}, & n, k \text{ integers}, \ 0 \leq k \leq n, \\ 0, & \text{otherwise}, \end{cases}$$

which look like they might be fractions but are actually integers.

We need a few preliminaries before we try to prove this. Suppose we take all these sequences of k 0's and $n-k$ 1's and perform the following steps:
(1) change every 0 to a 1 and every 1 to a 0,
(2) read the sequences backwards (from right to left) instead of forwards (from left to right),

The first step changes all the inversions to non-inversions, and vice versa. The second step changes all the non-inversions back to inversions, and vice versa. These two steps then produce sequences of $n-k$ 0's and k 1's with the same numbers of inversions as before, which shows that the sum side of the Fundamental Property is also symmetric in k and $n-k$. This doesn't prove that the sum equals $\binom{n}{k}_q$, but it is certainly encouraging—if the sum were not symmetric in k and $n-k$, then we would have *disproved* it.

Since the Pascal recurrence

$$\binom{n+1}{k} = \binom{n}{k-1} + \binom{n}{k}$$

for the binomial coefficients is so important, it is natural to ask whether something similar is true for the q-binomial coefficients. First let's work problem 4 from the last section. If $q \neq 1$, we have

$$[n]_q = \frac{1-q^n}{1-q} = \frac{1-q^k+q^k-q^n}{1-q} = \frac{1-q^k}{1-q} + q^k \frac{1-q^{n-k}}{1-q} = [k]_q + q^k[n-k]_q,$$

and this also holds if $q = 1$ (why?). Therefore

$$\binom{n+1}{k}_q = \frac{(n+1)!_q}{k!_q\,(n+1-k)!_q}$$

$$= \frac{n!_q}{k!_q\,(n+1-k)!_q}\left\{[k]_q + q^k\,[n+1-k]_q\right\}$$

$$= \frac{n!_q}{(k-1)!_q\,(n+1-k)!_q} + q^k\,\frac{n!_q}{k!_q\,(n-k)!_q}$$

(1.3.1)
$$= \binom{n}{k-1}_q + q^k \binom{n}{k}_q.$$

If we replace k by $n+1-k$ in (1.3.1), there results

$$\binom{n+1}{n+1-k}_q = \binom{n}{n+1-k-1}_q + q^{n+1-k} \binom{n}{n+1-k}_q.$$

Using the symmetry of the q-binomial coefficients this simplifies to

(1.3.2)
$$\binom{n+1}{k}_q = \binom{n}{k}_q + q^{n-k+1} \binom{n}{k-1}_q,$$

so that the q-binomial coefficients have not one but *two* q-Pascal recurrences. This has an interesting consequence that we will see in Chapter 2.

We are now ready to prove the Fundamental Property. The idea is to show that the sum there also satisfies (1.3.1), or (1.3.2). We do this first, and then talk about why this suffices for a proof. Consider $\sum_{\sigma \in S(k,n+1-k)} q^{\operatorname{inv}\sigma}$, the sum over all sequences of k 0's and $n+1-k$ 1's. Split the sum into two pieces, the sequences which begin with 0 and the sequences which begin with 1:

(1.3.3)
$$\sum_{\sigma \in S(k,n+1-k)} q^{\operatorname{inv}\sigma} = \sum_{\substack{\sigma \in S(k,n+1-k) \\ 0 \text{ comes first}}} q^{\operatorname{inv}\sigma} + \sum_{\substack{\sigma \in S(k,n+1-k) \\ 1 \text{ comes first}}} q^{\operatorname{inv}\sigma}.$$

If 0 comes first, it is not inverted with anything that comes after it, and what comes after it is a sequence of $k-1$ 0's and $n+1-k$ 1's, so the first sum on the right side of (1.3.3) is the same as $\sum_{\sigma \in S(k-1,n+1-k)} q^{\operatorname{inv}\sigma}$. If 1 comes first, the rest of the sequence consists of k 0's and $n-k$ 1's, and the initial 1 is inverted with all of the k 0's. Therefore the second sum on the right side of (1.3.3) is the same as $q^k \sum_{\sigma \in S(k,n-k)} q^{\operatorname{inv}\sigma}$, because this is what we get if we cut off the initial 1 and then put back in all the inversions we lose by so doing. Thus we can rewrite (1.3.3) as

(1.3.4)
$$\sum_{\sigma \in S(k,n+1-k)} q^{\operatorname{inv}\sigma} = \sum_{\sigma \in S(k-1,n+1-k)} q^{\operatorname{inv}\sigma} + q^k \sum_{\sigma \in S(k,n-k)} q^{\operatorname{inv}\sigma}.$$

This allows us to prove the Fundamental Property by induction on n, for suppose we knew that it was true for all sequences of 0's and 1's of length n. Then the right side of (1.3.4) equals

$$\binom{n}{k-1}_q + q^k \binom{n}{k}_q,$$

so according to (1.3.1) we have

$$\sum_{\sigma \in S(k,n+1-k)} q^{\text{inv } \sigma} = \sum_{\sigma \in S(k-1,n+1-k)} q^{\text{inv } \sigma} + q^k \sum_{\sigma \in S(k,n-k)} q^{\text{inv } \sigma}$$

$$= \binom{n}{k-1}_q + q^k \binom{n}{k}_q$$

$$= \binom{n+1}{k}_q$$

and the Fundamental Property is also true for $n+1$. Therefore, if the Fundamental Property is true for $\binom{0}{0}_q$, then it is true for all of the q-binomial coefficients. But there are no inversions in sequences of 0 0's and 0 1's, so the sum equals $q^0 = 1$, which equals $\binom{0}{0}_q$. This proves the Fundamental Property.

There is an interesting alternative characterization of the Fundamental Property that is due to Pólya. A **lattice path** in $\mathbb{R}^2$ starts at the origin and moves into the first quadrant using two kinds of steps: North (N), from (i,j) to $(i,j+1)$; or East (E), from (i,j) to $(i+1,j)$.

THEOREM 4 (Pólya's Property). *Consider all the lattice paths from the origin to the point $(k, n-k)$ in n steps, and let $A_{nk}(r)$ denote the number of such paths for which the area under the path, above the x-axis, and bounded on the right by the line $x = k$ is equal to r. Then*

$$\binom{n}{k}_q = \sum_{r=0}^{k(n-k)} A_{nk}(r) \, q^r.$$

This can be proved directly, in much the same way that we proved the Fundamental Property, but it is easier to derive it from the Fundamental Property. The point is that lattice paths are in 1-1 correspondence with sequences of 0's and 1's, with an East step corresponding to a 0 and a North step corresponding to a 1. The correspondence for $\binom{4}{2}_q$ is

0011

0110

1010

0101

1001

1100

Any pair of an N step followed (immediately or not) by an E step creates one unit ○ of area, so units of area correspond exactly to inversions, and this proves the theorem.

Exercises

1. Show that
$$\binom{m+2}{2}_q = q^2 \binom{m}{2}_q + \binom{2m+1}{1}_q.$$

2. Prove the Fundamental Property by induction using (1.3.2) instead of (1.3.1). (Split up the sequences of k 0's and $n+1-k$ 1's according to whether 0 comes last or 1 comes last.)

3. Explain why we can rewrite Pólya's Property in the following form. Let $\Lambda(k, n-k)$ denote the set of all lattice paths from the origin to $(k, n-k)$ in n steps. If λ is such a path, let $a(\lambda)$ denote the area between the path, the x-axis, and the line $x = k$. Then
$$\binom{n}{k}_q = \sum_{\lambda \in \Lambda(k, n-k)} q^{a(\lambda)}.$$

4. Give a direct proof of Pólya's Property by induction on n. You may prefer to prove it in the form given in the previous problem.

5. Use the Fundamental Property and the idea of switching 0's and 1's to show that the q-binomial coefficients are reciprocal polynomials. (See problem 13 from the previous section for the definition.) The q-binomial coefficients are also unimodal, but this is quite difficult to prove.

6. (a) Show that $k + \binom{n-k+1}{2} = \binom{n-k}{2} + n$. This is a useful lemma in a problem that comes much later.

    (b) Prove the following q-analogues of the result in (a):
$$[k]_q + q^k \binom{n-k+1}{2}_q = q^{k+2} \binom{n-k}{2}_q + [n]_q,$$
$$q^n [k]_q + q^{2k+1} \binom{n-k+1}{2}_q = q^{2k+1} \binom{n-k}{2}_q + q^n [n]_q,$$
$$q^n [k]_q + q^k \binom{n-k+1}{2}_q = q^{k+2} \binom{n-k}{2}_q + q^k [n]_q,$$
$$q^{2n} [k]_q + q^{2k+1} \binom{n-k+1}{2}_q = q^{2k+1} \binom{n-k}{2}_q + q^{n+k} [n]_q.$$

7. Let $Z_2(k)$ be the set of all sequences of 2 0's and k 1's which either begin with 0 or end with 0, or both. (For example, $Z_2(2)$ comprises the five sequences 0011,0101,0110,1010,1100.) Prove that $\sum_{\sigma \in Z_2(k)} q^{\text{inv}\,\sigma} = [2k+1]_q$. For more problems like this see the next set of exercises.

8. How are problems 1 and 7 related?

9. Show that
$$[n]_q [n]_{q^2} = q^{n-1} [n]_q + (1 + q^{n+1}) \binom{n}{2}_q.$$

10. Show that
$$\sum_{k=1}^{n} q^{3(n-k)} [2k-1]_q [2k-1]_{q^2} = \binom{2n+1}{3}_q.$$

11. Show that
$$\sum_{k=1}^{n} q^{3(n-k)}[2k]_q[2k]_{q^2} = \binom{2n+2}{3}_q.$$

12. Using the result of problem 7, or otherwise, prove that
$$\binom{n+2}{2}_q = [2n+1]_q + q^2[2n-3]_q + q^4[2n-7]_q + \cdots$$
$$= \sum_{j=0}^{\lfloor \frac{2n+1}{4} \rfloor} q^{2j}[2n+1-4j]_q.$$

(Here $\lfloor x \rfloor$ denotes the **floor** of x, which is the greatest integer $\leq x$. Its only purpose here is to stop the sum just before $2n+1-4j$ becomes negative.)

13. For integers k and n with $0 \leq k \leq n$, define the **Carlitz coefficients** $c_q(n,k)$ by the recurrence
$$c_q(n+1,k) = c_q(n,k-1) + (q^k - 1)c_q(n,k),$$
where $c_q(0,0) = 1$ and $c_q(n,k) = 0$ if k is negative or if $k > n$. (Note that this implies $c_q(n,k) = 0$ if n is negative.)

(i) Show that $c_q(n,0) = 0$ if $n > 0$.

(ii) Show that $c_q(n,n) = 1$ if $n \geq 0$.

(iii) The main reason why the Carlitz coefficients are interesting is that
$$\binom{n}{j}_q = \sum_{k=j}^{n} \binom{n}{k} c_q(k,j).$$
Prove this by induction on n.

(iv) Show that all the Carlitz coefficients are zero when $q = 1$ except for $c_q(n,n) = 1$.

(v) Show that $c_q(n, n-1) = [n]_q - n$ for $n \geq 0$.

(vi) Show that $c_q(n, n-2) = \binom{n}{2}_q + \binom{n}{2} - n[n-1]_q$ for $n \geq 0$.

(vii) Show that $c_q(n,1) = (q-1)^{n-1}$ for $n \geq 1$.

(viii) Show that, for $n \geq 1$,
$$c_q(n,2) = (q-1)^{n-2}\{1 + (q+1) + \cdots + (q+1)^{n-2}\}$$
$$= (q-1)^{n-2}\frac{(q+1)^{n-1} - 1}{q}.$$

(ix) Show that $c_q(n,k)$ is $(q-1)^{n-k}$ times a polynomial in q.

1.4. Some identities for q-binomial coefficients

In the last section we proved

(1.4.1)
$$\binom{n+1}{k}_q = \binom{n}{k}_q + q^{n-k+1}\binom{n}{k-1}_q.$$

1.4. SOME IDENTITIES FOR q-BINOMIAL COEFFICIENTS

It follows on replacing n by $n-1$ that

$$\binom{n}{k}_q = \binom{n-1}{k}_q + q^{n-k}\binom{n-1}{k-1}_q,$$

and substituting this into (1.4.1) we get

$$\binom{n+1}{k}_q = q^{n-k+1}\binom{n}{k-1}_q + q^{n-k}\binom{n-1}{k-1}_q + \binom{n-1}{k}_q.$$

Following Gauss, we can keep doing this. Since

$$\binom{n-1}{k}_q = \binom{n-2}{k}_q + q^{n-k-1}\binom{n-2}{k-1}_q,$$

substituting this in the above we get

$$\binom{n+1}{k}_q = q^{n-k+1}\binom{n}{k-1}_q + q^{n-k}\binom{n-1}{k-1}_q + q^{n-k-1}\binom{n-2}{k-1}_q + \binom{n-2}{k}_q.$$

If we do this $n-m+1$ times for a generic nonnegative integer $m \leq n$, we get

(1.4.2)
$$\binom{n+1}{k}_q = q^{n-k+1}\binom{n}{k-1}_q + q^{n-k}\binom{n-1}{k-1}_q + q^{n-k-1}\binom{n-2}{k-1}_q + \cdots$$
$$+ q^{m-k+1}\binom{m}{k-1}_q + \binom{m}{k}_q.$$

(We leave the proof of this to the reader.) The smallest positive integer for which (1.4.2) is interesting is $m = k-1$ (if m is smaller, the result is the same as it is for $m = k-1$, since the additional terms are all zero). Taking $m = k-1$ we get Gauss's summation theorem for the q-binomial coefficients:

$$\binom{n+1}{k}_q = q^{n-k+1}\binom{n}{k-1}_q + q^{n-k}\binom{n-1}{k-1}_q + q^{n-k-1}\binom{n-2}{k-1}_q + \cdots$$
$$+ q\binom{k}{k-1}_q + \binom{k-1}{k-1}_q.$$

Replacing k by $k+1$, this takes the form

$$\binom{n+1}{k+1}_q = q^{n-k}\binom{n}{k}_q + q^{n-k-1}\binom{n-1}{k}_q + q^{n-k-2}\binom{n-2}{k}_q + \cdots$$
$$+ q\binom{k+1}{k}_q + \binom{k}{k}_q,$$

or

(1.4.3)
$$\binom{n+1}{k+1}_q = \sum_{m=k}^{n} q^{m-k}\binom{m}{k}_q.$$

There is also a simple combinatorial proof of (1.4.3). We know that the left side generates sequences of $k+1$ 0's and $n-k$ 1's, keeping track of inversions. The right side is classifying these sequences by the position of the last 0. If the last 0 is at the end of the sequence, then this 0 is inverted with the $n-k$ 1's in the sequence, which also has k other 0's; so in this case we get $q^{n-k}\binom{n}{k}_q$ when we keep track of inversions. If the last two numbers in the sequence are 01, then this piece of it has $n-k-1$ inversions with the $n-k-1$ 1's in the rest of the sequence, which also has

k more 0's; so these sequences give us $q^{n-k-1}\binom{n-1}{k}_q$. The generic term in the sum looks like $q^{m-k}\binom{m}{k}_q$ for some integer m between k and n. This term corresponds to the sequences where the last 0 is followed by $n-m$ 1's. The trailing 1's cause no inversions, so this part of the sequence has only the inversions between the last 0 and the $m-k$ 1's in the first part of the sequence, which also has k 0's. The last term $\binom{k}{k}_q$ corresponds to the sequence where all the 1's are at the end.

We can play the same game with

$$(1.4.4) \qquad \binom{n+1}{k}_q = \binom{n}{k-1}_q + q^k \binom{n}{k}_q.$$

Replacing n by $n-1$ we have

$$\binom{n}{k}_q = \binom{n-1}{k-1}_q + q^k \binom{n-1}{k}_q,$$

and substituting this in (1.4.4) gives

$$\binom{n+1}{k}_q = \binom{n}{k-1}_q + q^k \left\{ \binom{n-1}{k-1}_q + q^k \binom{n-1}{k}_q \right\}$$
$$= \binom{n}{k-1}_q + q^k \binom{n-1}{k-1}_q + q^{2k} \binom{n-1}{k}_q.$$

As before, we can keep doing this repeatedly. If we do it m times we get

$$(1.4.5) \qquad \binom{n+1}{k}_q = \binom{n}{k-1}_q + q^k \binom{n-1}{k-1}_q + q^{2k} \binom{n-2}{k-1}_q + \cdots$$
$$+ q^{(m-1)k} \binom{n-m+1}{k-1}_q + q^{mk} \binom{n-m+1}{k}_q.$$

The largest value of m for which this is interesting is $m = n-k+1$, which gives

$$(1.4.6) \qquad \binom{n+1}{k}_q = \sum_{m=0}^{n-k+1} \binom{n-m}{k-1}_q q^{mk},$$

or, unfolded,

$$\binom{n+1}{k}_q = \binom{n}{k-1}_q + q^k \binom{n-1}{k-1}_q + q^{2k} \binom{n-2}{k-1}_q + \cdots + q^{(n-k+1)k} \binom{k-1}{k-1}_q.$$

We can also prove this by counting inversions in sequences. The left side generates sequences of k 0's and $n+1-k$ 1's, keeping track of inversions. The right side is classifying these sequences by where the first 0 occurs. If it is at the beginning of the sequence, this gives $\binom{n}{k-1}_q$ as before. If the sequence starts 10, then the first 1 is inverted with all of the k 0's; and in the rest of the sequence there are $k-1$ 0's and $n-k$ 1's. These sequences contribute the term $q^k \binom{n-1}{k-1}_q$, and so on. The generic term in the sum is $\binom{n-m}{k-1}_q q^{mk}$, which corresponds to the sequences that start with m 1's and then a 0. Each of the m 1's is inverted with the k 0's in the sequence, the rest of which consists of $k-1$ 0's and $n-m-k+1$ 1's. The last term in the sum, $q^{(n-k+1)k} \binom{k-1}{k-1}_q$, corresponds to the sequence where all the 1's are at the beginning.

We can also use this type of counting argument to prove

(1.4.7) $$\binom{m+n}{k}_q = \sum_j \binom{m}{k-j}_q \binom{n}{j}_q q^{j(m-k+j)}.$$

Consider sequences of k 0's and $m+n-k$ 1's, and count inversions. If we just do this as usual, we wind up with the left side of (1.4.7). The right side comes from breaking the sequences somewhere in the middle. Suppose we look at the first m numbers in each sequence separately, and the last n numbers separately also. The first piece comprises some number $k-j$ of 0's and $m-k+j$ 1's. The first q-binomial coefficient on the right side of (1.4.7) counts all the inversions among these m numbers. Similarly, the second piece comprises j 0's and $n-j$ 1's, and the term $\binom{n}{j}_q$ in (1.4.7) counts all the inversions among these n numbers.

Did we miss any inversions? Of course we did: any 1 among the first m numbers is inverted with any 0 among the last n numbers, and those are exactly the inversions we haven't counted yet. How many of these are there? Exactly as many as there are pairs of a 1 among the first m numbers (of which there are $m-k+j$) and a 0 among the last n numbers (of which there are j). So we need a factor of $q^{j(m-k+j)}$ to take care of these inversions, and now we have got all of them. Therefore the right side of (1.4.7) counts the same thing as the left side.

Because of the symmetry of the q-binomial coefficients, there are several different forms of (1.4.7), so it is better to remember how the argument goes than to memorize the formula. Another form occurs in problem 11. In Chapter 2 we will want to know that

(1.4.8) $$\sum_{j=0}^{b} \binom{a}{j+k}_q \binom{b}{j}_q q^{j(j+k)} = \binom{a+b}{a-k}_q,$$

and this is also an alternate form of (1.4.7). We know that $\binom{a}{j+k}_q$ generates sequences of $a-j-k$ 0's and $j+k$ 1's, keeping track of inversions, and that $\binom{b}{j}_q$ generates sequences of j 0's and $b-j$ 1's, keeping track of inversions. If the sequences generated by $\binom{b}{j}_q$ come after the ones generated by $\binom{a}{j+k}_q$, then there are $j(j+k)$ more inversions, between the 1's in the first sequences and the 0's in the second. When these types of sequences are combined in this order and inversions are counted as exponents of q, we get all the inversions in sequences of $a-k$ 0's and $b+k$ 1's, so this proves (1.4.8).

Exercises

1. Show that a special case of (1.4.3) is

$$\sum_{m=1}^{n} q^{m-1} [m]_q = \binom{n+1}{2}_q$$

which is a q-extension of the fundamental fact

(1.4.9) $$1 + 2 + \cdots + n = \binom{n+1}{2}.$$

2. Show that another q-extension of (1.4.9) is
$$\sum_{m=0}^{n-1} [n-m]_q q^{2m} = \binom{n+1}{2}_q.$$

3. Use (1.4.4) to prove (1.4.5) by induction on m.

4. Use (1.4.1) to prove (1.4.2) by induction on m. Note that this is a *downward* induction; you should assume it's true for m and prove it for $m-1$.

5. (a) Show that
$$\binom{m+1}{2}_q + q\binom{m}{2}_q = [m]_q^2.$$

   (b) Using (a) or otherwise, show that
$$\sum_{k=1}^{n} q^{k-1}[k]_q^2 = \binom{n+2}{3}_q + q^2\binom{n+1}{3}_q.$$

6. (a) Show that
$$\binom{m+2}{3}_q + 2[2]_q q \binom{m+1}{3}_q + q^3 \binom{m}{3}_q = [m]_q^3.$$

   (b) Using (a) or otherwise, show that
$$\sum_{k=1}^{n} q^{k-1}[k]_q^3 = \binom{n+3}{4}_q + 2[2]_q q^2 \binom{n+2}{4}_q + q^5 \binom{n+1}{4}_q.$$

7. If m is a nonnegative integer and n is a positive integer, show that
$$\sum_{k=0}^{m} \binom{n}{k}_q (-1)^k q^{\binom{k}{2}} = (-1)^m q^{\binom{m+1}{2}} \binom{n-1}{m}_q.$$
What happens if $n = 0$?

8. Let $Z_3(k)$ be the set of all sequences of 3 0's and k 1's which either begin with 0 or end with 0, or both. (For example, $Z_3(2)$ contains the nine sequences 00011, 00101, 01001, 01010, 00110, 01100, 10010, 10100, 11000.) Prove that
$$\sum_{\sigma \in Z_3(k)} q^{\mathrm{inv}\,\sigma} = [k+1]_q [k+1]_{q^2}.$$

9. For an integer $j \geq 1$, let $Z_j(k)$ be the set of all sequences of j 0's and k 1's which either begin with 0 or end with 0, or both. Prove that
$$\sum_{\sigma \in Z_j(k)} q^{\mathrm{inv}\,\sigma} = \frac{(j+k-2)!_q [2k+j-1]_q}{(j-1)!_q k!_q} = \binom{j+k-1}{k}_q \frac{[2k+j-1]_q}{[k+j-1]_q}.$$
Check the cases $j = 1, 2, 3$.

10. Use problem 9 to prove the following generalization of problem 12 in the previous section:
$$\binom{n+k}{k}_q = \sum_{j \geq 0} q^{kj} \binom{n+k-2j-1}{k-1}_q \frac{[2n+k-4j-1]_q}{[n+k-2j-1]_q}.$$

11. Let $P_n(q)$ be the $n \times n$ matrix whose ij^{th} entry is $\binom{i+j-2}{i-1}_q$ for $1 \leq i,j \leq n$. ($P_n(q)$ is a q-**Pascal** matrix.)

   (i) Explain why $P_n(q)$ is a symmetric matrix.

   (ii) Prove that $P_n(q) = L_n(q) D_n(q) L_n^T(q)$, where $L_n(q)$ is the $n \times n$ matrix whose ij^{th} entry is $\binom{i-1}{j-1}_q$, for $1 \leq i,j \leq n$, and $D_n(q)$ is the diagonal $n \times n$ matrix whose ii^{th} entry is $q^{(i-1)^2}$ for $1 \leq i \leq n$. **Hint:** Explain why the ij^{th} entry of $L_n(q) D_n(q) L_n^T(q)$ is

$$\sum_{k=1}^{n} \binom{i-1}{k-1}_q q^{(k-1)^2} \binom{j-1}{k-1}_q = \sum_{l=0}^{n-1} \binom{i-1}{l}_q \binom{j-1}{l}_q q^{l^2},$$

and then explain why this is $\binom{i+j-2}{i-1}_q$.

   (iii) Explain why it follows that the determinant of $P_n(q)$ is

$$q^{0^2+1^2+\cdots+(n-1)^2} = q^{\frac{n(n-1)(2n-1)}{6}}.$$

12. Recall that $(x;q)_n = (1-x)(1-xq)\cdots(1-xq^{n-1})$ with $(x;q)_0 = 1$, and let $A_n(x_1,\ldots,x_n;q)$ denote the determinant of the $n \times n$ matrix whose ij entry is $(x_i;q)_{j-1}$. Show that

$$A_n(x_1,\ldots,x_n;q) = q^{\binom{n}{3}} \prod_{1 \leq i < j \leq n} (x_i - x_j).$$

Hint: Use row and column operations to show that

$$A_n(x_1,\ldots,x_n;q) = q^{\binom{n-1}{2}}(x_1 - x_2)\cdots(x_1 - x_n) A_{n-1}(x_2,\ldots,x_n;q).$$

13. This problem outlines a different proof of the result of the previous problem. Justify each of the following statements:

   (i) Since $A_n(x_1,\ldots,x_n;q)$ becomes zero if any two of the variables x_k are the same, every expression $x_i - x_j$ with $1 \leq i < j \leq n$ has to be a factor.

   (ii) There can't be any more factors involving any of the variables x_k.

   (iii) Therefore, we only have to explain the factor $q^{\binom{n}{3}}$. We can find it by looking at the coefficient of $x_1^{n-1} x_2^{n-2} \cdots x_{n-1}^1 x_n^0$, which comes only from the "reverse diagonal" term in the expansion of the determinant. (How do the minus signs work themselves out?)

1.5. Another property of q-binomial coefficients

Suppose we divide (or **partition**) the set $\{1,2,3,4,5\}$ into a first subset of size two and a second subset of size three in all possible ways. (There are $\binom{5}{2} = 10$ ways, which we will list below.) Define a **between-set inversion** (which we may just call an inversion if the context is clear) to be any pair of numbers a,b where a is in the first set, b is in the second set, and $a > b$. For example, if $\{2,3\}$ is the first set, this makes $\{1,4,5\}$ the second set, and 2 and 3 are both inverted with 1 but not with 4 or 5, so there are 2 between-set inversions. We can also associate the sequence 21122 to these two sets, representing that 2 and 3 are in the first set (so we put 1 in the second and third positions) and 1, 4, and 5 are in the second set (so we put 2 in the first, fourth, and fifth positions). If we do this for all possible partitions, we get:

first set	second set	# inversions	sequence
$\{1,2\}$	$\{3,4,5\}$	0	11222
$\{1,3\}$	$\{2,4,5\}$	1	12122
$\{1,4\}$	$\{2,3,5\}$	2	12212
$\{1,5\}$	$\{2,3,4\}$	3	12221
$\{2,3\}$	$\{1,4,5\}$	2	21122
$\{2,4\}$	$\{1,3,5\}$	3	21212
$\{2,5\}$	$\{1,3,4\}$	4	21221
$\{3,4\}$	$\{1,2,5\}$	4	22112
$\{3,5\}$	$\{1,2,4\}$	5	22121
$\{4,5\}$	$\{1,2,3\}$	6	22211

As before, we now take all the numbers in the inversions column, make them exponents of q, and add. This gives

$$1 + q + 2q^2 + 2q^3 + 2q^4 + q^5 + q^6 = 1 + q + q^2 + q^3 + q^4 + q^2\left(1 + q + q^2 + q^3 + q^4\right)$$

$$= [5]_q(1+q^2) = [5]_q \frac{1+q+q^2+q^3}{1+q} = \frac{[5]_q[4]_q}{[2]_q} = \binom{5}{2}_q.$$

This isn't very surprising, especially when we look at the sequence column. Any pair a,b with a in the first set, b in the second set, and $a > b$ corresponds to a 2 in the bth position and an a in the 1st position in the corresponding sequence, which is an inversion in that sequence since $b < a$. This gives immediately the following theorem.

THEOREM 5. *Let $T(a,b)$ denote the set of all partitions of $\{1,2,\ldots,n\}$ into a first subset of size k and a second subset of size $n-k$, and define a between-set inversion as above. If $\operatorname{inv} t$ denotes the number of between-set inversions in the partition t, then*

(1.5.1) $$\binom{n}{k}_q = \sum_{t \in T(k,n-k)} q^{\operatorname{inv} t}.$$

This follows from the Fundamental Property of q-binomial coefficients by the correspondence outlined above. We could just leave it at that, but it is interesting to try to redevelop the theory from this new point of view. Suppose we take all the partitions of $\{1,2,\ldots,n\}$ into a first subset of size k and a second subset of size $n-k$, and perform the following steps:

(1) switch the first and second set,
(2) change each number in both sets from k to $n+1-k$.

The first step changes $T(k,n-k)$ into $T(n-k,k)$, and it changes all the inversions to non-inversions and vice versa. The second step changes all the non-inversions back to inversions, and vice versa. This shows that the sum side of (1.5.1) is symmetric in k and $n-k$, so if we took (1.5.1) as the *definition* of the q-binomial coefficients, it would prove that

$$\binom{n}{k}_q = \binom{n}{n-k}_q.$$

Suppose Santa Claus has n ordinary reindeer, and one other reindeer, Rudolph, with a red nose. Santa maintains two lists: one with the reindeer ranked by utility in foggy conditions, on which Rudolph is first; and one where they are ranked by utility in nonfoggy conditions, on which Rudolph is last. On Christmas Eve, Santa partitions the reindeer into a first set of k that will deliver presents, and a second set of $n - k + 1$ that will stay behind to guard the North Pole against intruders. Considering all possible partitions and counting between-set inversions, we will get $\binom{n+1}{k}_q$ using either list. (If $n = 4$ and $k = 2$, then two reindeer deliver presents and three remain behind, and the example with partitions of $\{1, 2, 3, 4, 5\}$ represents the foggy list if Rudolph is 1, and the nonfoggy list if Rudolph is 5.)

If we use the foggy list and Rudolph is in the first set, then he is not inverted with any of the reindeer in the second set. Therefore all the inversions in this case are between the second set and the $k - 1$ ordinary reindeer in the first set, and $\binom{n}{k-1}_q$ counts those. If Rudolph is in the second set, then he is inverted with all of the k reindeer in the first set, and the other inversions are between them and the $n - k$ ordinary reindeer in the second set, so $q^k \binom{n}{k}_q$ counts these and we have

$$\binom{n+1}{k}_q = \binom{n}{k-1}_q + q^k \binom{n}{k}_q,$$

which is (1.3.1). If we use the nonfoggy list we get (1.3.2) instead. We leave this as an exercise.

Finally, let's try to prove an identity like (1.4.7) using this approach. Consider partitions of $\{1, 2, \ldots, m+n\}$ into a first subset of size k and a second subset of size $m + n - k$. Let's call the numbers $\{1, 2, \ldots, m\}$ the *small* numbers, and $\{m+1, m+2, \ldots, m+n\}$ the *large* numbers. Suppose that j of the small numbers are in the first subset. Then the other $m - j$ small numbers are in the second subset, $k - j$ large numbers are in the first subset, and the other $n - k + j$ large numbers are in the second subset. Now $\binom{m}{j}_q$ takes care of the between-set inversions among the small numbers, $\binom{n}{k-j}_q$ takes care of the between-set inversions among the large numbers, and each of the $k - j$ large numbers in the first subset is inverted with each of the $m - j$ small numbers in the second subset. Summing over all the partitions and all values of j we get

(1.5.2) $$\binom{m+n}{k}_q = \sum_j \binom{m}{j}_q \binom{n}{k-j}_q q^{(k-j)(m-j)},$$

which is another equivalent form of (1.4.7).

Exercises

1. Prove (1.3.2) by considering the nonfoggy list.
2. Show that (1.5.2) is equivalent to (1.4.7).
3. Prove (1.4.3) by using between-set inversions.
4. Prove (1.4.6) by using between-set inversions.

5. By considering between-set inversions, prove that
$$\sum_{k=1}^{n} q^{2k-3}[2k-1]_q \left([2k-2]_q + q^{2k-1}\right) = \binom{2n+1}{3}_q.$$

   Hint: Consider partitions of $\{1, 2, \ldots, 2n+1\}$ into two subsets where the first set has three members, the largest of which is either $2k$ or $2k+1$. It may help to rewrite
$$q^{2k-3}[2k-1]_q \left([2k-2]_q + q^{2k-1}\right) = q^{2k-3}[2]_q \binom{2k-1}{2}_q + q^{4k-4}[2k-1]_q.$$

6. Show that
$$\binom{2n+1}{3}_q - \binom{2n-1}{3}_q = q^{2n-3}[2n-1]_q \left([2n-2]_q + q^{2n-1}\right).$$

   Using this or otherwise, prove the result of the previous problem by induction.

7. By considering between-set inversions, prove that
$$\sum_{k=1}^{n} q^{2k-2}[2k]_q \left([2k-1]_q + q^{2k}\right) = \binom{2n+2}{3}_q.$$

   Hint: Consider partitions of $\{1, 2, \ldots, 2n+2\}$ into two subsets where the first set has three members, the largest of which is either $2k+1$ or $2k+2$. It may help to rewrite
$$q^{2k-2}[2k]_q \left([2k-1]_q + q^{2k}\right) = q^{2k-2}[2]_q \binom{2k}{2}_q + q^{4k-2}[2k]_q.$$

8. Show that
$$\binom{2n+2}{3}_q - \binom{2n}{3}_q = q^{2n-2}[2n]_q \left([2n-1]_q + q^{2n}\right).$$

   Using this or otherwise, prove the result of the previous problem by induction.

9. Recall from problem 9 in section 1.3 that
$$[n]_q[n]_{q^2} = q^{n-1}[n]_q + (1 + q^{n+1})\binom{n}{2}_q.$$

   Use this to show that the left side is the generating function for between-set inversions of $\{1, 2, \ldots, n+2\}$ into two subsets where the first set has three elements and contains at least one of 1 and $n+2$.

10. Use the previous problem to give a combinatorial proof that
$$\sum_{k=1}^{n} q^{3(n-k)}[2k-1]_q[2k-1]_{q^2} = \binom{2n+1}{3}_q.$$

    The $k=1$ term represents the partition of $\{1, 2, \ldots, 2n+1\}$ into two subsets with first set $\{n, n+1, n+2\}$.

11. Use problem 9 to give a combinatorial proof that
$$\sum_{k=1}^{n} q^{3(n-k)}[2k]_q[2k]_{q^2} = \binom{2n+2}{3}_q.$$

The $k = 1$ term represents the partitions of $\{1, 2, \ldots, 2n + 2\}$ into two subsets where the first set comprises three of $\{n, n+1, n+2, n+3\}$.

1.6. q-multinomial coefficients

There is, in a sense, a remarkable coincidence in the material of sections 1.2 and 1.3, which we passed over without comment: the q-factorial, which came from counting inversions in permutations of $\{1, 2, \ldots, n\}$, implies the same definition of $\binom{n}{k}_q$ that counting inversions in sequences of k 0's and $n - k$ 1's does.

Since Rodrigues's theorem and the Fundamental Property of q-binomial coefficients are consistent with each other in this sense, there could be a result which contains both of them as particular cases. It is easy to see what a q-multinomial coefficient should be. If $k_1, \ldots, k_m$ are nonnegative integers that add up to n, then

$$(1.6.1) \qquad \binom{n}{k_1, \ldots, k_m}_q := \frac{n!_q}{k_1!_q \ldots k_m!_q} = \frac{(q;q)_n}{(q;q)_{k_1} \ldots (q;q)_{k_m}}.$$

Again, we will assume that the q-multinomial coefficient equals zero if $k_1, \ldots, k_m$ are not all nonnegative integers, or if they do not add up to n. Is (1.6.1) an interesting object? Suppose we look at all the sequences of length n made up of k_1 1's, k_2 2's, $\ldots$, k_m m's. Let's call the set of all such sequences $S_n(k_1, \ldots, k_m)$. As usual, an inversion will be any pair of numbers in the sequence where the larger number precedes the smaller. If m is two, we get the q-binomial coefficient (with a cosmetic difference: we were using sequences of 0's and 1's, and now we will be using 1's and 2's instead), and Rodrigues's result is the case where every k_i equals 1. In general, if we count inversions in this type of sequence, MacMahon proved that we get the q-multinomial coefficient.

THEOREM 6 (Fundamental Property of q-multinomial coefficients). *With the above notation,*

$$(1.6.2) \qquad \binom{n}{k_1, \ldots, k_m}_q = \sum_{\omega \in S_n(k_1, \ldots, k_m)} q^{\operatorname{inv} \omega}.$$

This is surprisingly easy to prove, by induction on m. We already know it when $m = 2$ (what happens when $m = 1$?), so we will assume it is true for m, and show that this means it must be true for $m + 1$ also. The argument is much like the one for the q-factorial: suppose we have k_{m+1} copies of $m + 1$, and n numbers altogether. For the moment, think of the $m + 1$'s as 1's and everything else as 0's. An $m + 1$ is inverted with anything that comes after it, except another $m + 1$. This is exactly the counting problem that the q-binomial coefficients solved. So if we just consider the inversions that the $m+1$'s are involved in, we get a factor $\binom{n}{k_{m+1}}_q$. This leaves $n - k_{m+1}$ elements that are not $m + 1$'s. Assume there are k_i i's for each i, $1 \leq i \leq m$. Then by induction the inversions among these give us a factor $\binom{n - k_{m+1}}{k_1, \ldots, k_m}_q$. The product of these accounts for all the possible inversions and is

$$\binom{n}{k_{m+1}}_q \binom{n - k_{m+1}}{k_1, \ldots, k_m}_q = \frac{n!_q \, (n - k_{m+1})!_q}{(n - k_{m+1})!_q \, k_{m+1}!_q \, k_1!_q \ldots k_m!_q}$$

$$= \binom{n}{k_1, \ldots, k_{m+1}}_q$$

which is just what we wanted. Thus (1.6.2) is true by induction.

The extension of the between-set inversion idea is straightforward. For example, suppose we partition $\{1, 2, 3, 4, 5, 6, 7, 8\}$ into a first subset $\{1, 6, 8\}$, a second subset $\{3, 4\}$, and a third subset $\{2, 5, 7\}$. There are 11 between-set inversions: 6 with each of 2, 3, 4, 5; 8 with all of these and 7; and 3 and 4 with 2. We can code this partition by writing down which subset each of 1–8 is in. This gives 13223131, where the 1 in the sixth position is inverted with the 2's and 3's in the second through fifth positions, the 1 in the eighth position is inverted with each of these and the 3 in the seventh position, and the 2's in the third and fourth positions are inverted with the 3 in the second position.

THEOREM 7 (Set partition property of q-multinomial coefficients). *Let $n = k_1 + \cdots + k_m$, and let $\Pi(k_1, \ldots, k_m)$ denote the collection of partitions of $\{1, \ldots, n\}$ into a first subset of size k_1, a second subset of size k_2, and so on, and a last subset of size k_m. If $\operatorname{inv} \pi$ denotes the number of between-set inversions in the partition π, then*

$$\binom{n}{k_1, \ldots, k_m}_q = \sum_{\pi \in \Pi(k_1, \ldots, k_m)} q^{\operatorname{inv} \pi}.$$

This follows immediately from the Fundamental Property and the above correspondence. A between-set inversion occurs whenever we have two numbers $a < b$ with b in an earlier set than a. In the corresponding code, a gets a higher number than b and that number occurs earlier in the sequence, so the inversions are preserved by the correspondence.

What about the Pascal recurrence for the q-multinomial coefficients? If $k_1 + \cdots + k_m = n + 1$ and each k_i is a nonnegative integer, then we have

(1.6.3)
$$\binom{n+1}{k_1, \ldots, k_m}_q = \binom{n}{k_1 - 1, k_2, \ldots, k_m}_q + q^{k_1} \binom{n}{k_1, k_2 - 1, k_3, \ldots, k_m}_q$$
$$+ q^{k_1 + k_2} \binom{n}{k_1, k_2, k_3 - 1, \ldots, k_m}_q + \cdots$$
$$+ q^{k_1 + \cdots + k_{m-1}} \binom{n}{k_1, \ldots, k_{m-1}, k_m - 1}_q.$$

To see this, all we have to do is think about sequences of length $n + 1$ of the above type, and consider the various possibilities for the first number in the sequence. If it is 1, it causes no inversions, and this corresponds to the first term on the right side of (1.6.3). If it is 2, then it is inverted with the k_1 1's in the sequence but nothing else, and this corresponds to the second term in (1.6.3), and so forth. If the first number in the sequence is m, then there are inversions with everything but the other $k_m - 1$ m's, and this corresponds to the last term in the sequence.

MacMahon's proof of (1.6.2) was essentially the above argument read backwards. He showed algebraically that the q-multinomial coefficients defined by (1.6.1) satisfy (1.6.3), and then used the above combinatorial reasoning to infer (1.6.2). This is similar to our first proof of the Fundamental Property of q-binomial coefficients.

We make another remark about (1.6.3) that will be used later. The left side of (1.6.3) is symmetric in $k_1, \ldots, k_m$, so the right side must be too, even if it doesn't look symmetric in the parameters. What this means is that there are many equivalent forms of (1.6.3); as stated, (1.6.3) considers the parameters in the order $k_1, \ldots, k_m$, but we can take them in any order we want to.

Exercises

1. Once we had q-factorials, the definition
$$\binom{n}{k}_q = \begin{cases} \frac{n!_q}{k!_q\,(n-k)!_q}, & n, k \text{ integers}, 0 \le k \le n, \\ 0, & \text{otherwise}, \end{cases}$$
was obvious, although it was not so clear that these objects would have the beautiful combinatorics asserted by the Fundamental Property of q-binomial coefficients. An alternative approach, which would especially make sense if we had not studied q-factorials first, would be to use the Fundamental Property as the *definition* of the q-binomial coefficients. To have a fully adequate theory, one would then have to be able to define q-factorials in terms of q-binomial coefficients. How would you do this? How would you define the ordinary factorial in terms of ordinary binomial coefficients?

2. What does (1.6.3) look like if we take the parameters in the order $k_m, \ldots, k_1$?

3. Let $I(k_1, k_2, \ldots, k_m)$ be the total number of inversions in all the sequences of k_1 1's, k_2 2's, ..., k_m m's. Generalize Rodrigues's solution to Stern's problem to show that
$$I(k_1, k_2, \ldots, k_m) = \frac{1}{2}\binom{k_1 + k_2 + \cdots + k_m}{k_1, k_2, \ldots, k_m} \sum_{1 \le i < j \le m} k_i\, k_j.$$
An alternate form of the answer is
$$I(k_1, \ldots, k_m) = \frac{1}{2}\binom{k_1 + k_2 + \cdots + k_m}{k_1, k_2, \ldots, k_m}$$
$$\times \left\{ \binom{k_1 + k_2 + \cdots + k_m}{2} - \binom{k_1}{2} - \cdots - \binom{k_m}{2} \right\}.$$
Explain why these are the same.

4. (a) Show that
$$\binom{n}{k}_q = \sum_j q^{j^2} \binom{k}{j}_q \binom{n-k}{j}_q.$$
The sum goes from $j = 0$ to the smaller of k and $n - k$.

   (b) Show that the result of (a) can be rewritten as

(1.6.4)
$$\binom{n}{k}_q^2 = \sum_j q^{j^2} \binom{n}{j, j, k-j, n-k-j}_q.$$

5. Prove (1.6.4) by using between-set inversions. (When $q = 1$, (1.6.4) has a simple combinatorial interpretation: you choose k things from a collection of n distinct things, while a friend independently chooses k things from the same collection; $k - j$ is the number of things that both of you chose. This idea can also be used for a general q, but the details are not quite straightforward.)

6. For nonnegative integers r and s define
$$f_j(r,s) = \sum_i (-1)^i q^{\binom{i+j}{2}} \binom{r+s-i}{i, r-i, s-i}_q,$$
where the sum goes from $i = 0$ to the smaller of r and s. Since there are no admissible values of i if r or s is negative, we have $f_j(r,s) = 0$ in that case.

   (i) Show that $f_j(r,0) = q^{\binom{j}{2}} = f_j(0,s)$.

   (ii) Use (1.6.3) to show

(1.6.5) $\qquad f_j(r+1, s) = f_j(r, s) + q^{r+1} f_j(r+1, s-1) - q^{r+j} f_j(r, s-1)$

   for $s \geq 0$. What is a good order for the parameters $i, r-i, s-i$?

   (iii) In later chapters we will want to know $f_j(r,s)$ for $j = 1$ and for $j = 0$. Using (ii) or otherwise, show that $f_1(r,1) = 1$ and $f_0(r,1) = q^r$ for all nonnegative integers r.

   (iv) Prove that $f_1(r,s) = 1$ and $f_0(r,s) = q^{rs}$ for all nonnegative integers r and s.

7. As in the previous problem, define
$$f_j(r,s) = \sum_i (-1)^i q^{\binom{i+j}{2}} \binom{r+s-i}{i, r-i, s-i}_q$$
for nonnegative integers r and s.

   (i) Use (1.6.3) to show

(1.6.6) $\qquad f_j(r+1, s) = f_j(r+1, s-1) + q^s f_j(r, s) - q^{s+j-1} f_j(r, s-1)$

   for $s \geq 0$.

   (ii) Use (1.6.6) to show that $f_1(r,s) = 1$ and $f_0(r,s) = q^{rs}$ for all nonnegative integers r and s.

8. With reference to the previous two problems, use either (1.6.5) or (1.6.6) to show that $f_2(r,s) = q[s+1]_q - q^{r+2}[s]_q$.

9. In problems 12–14 in section 1.1 we studied self-conjugate permutations, obtaining a recurrence relation and a generating function for the number U_n of self-conjugate permutations of $\{1, 2, \ldots, n\}$. It is possible to count U_n directly with a multinomial coefficient. A permutation of $\{1, 2, \ldots, n\}$ is self-conjugate if and only if it is $12 \ldots n$ with some number (possibly zero) of pairs of elements switched. Explain why this implies that

$$U_n = 1 + \binom{n}{2} + \binom{n}{2,2,n-4}\frac{1}{2} + \binom{n}{2,2,2,n-6}\frac{1}{3!} + \cdots$$

$$= \sum_{j=0}^{\lfloor n/2 \rfloor} \binom{n}{\underbrace{2, 2, \ldots, 2}_{j \text{ 2's}}, n-2j} \frac{1}{j!}.$$

The factorial arises because we don't care what order we choose the pairs of numbers in. Show that this expression simplifies to

(1.6.7) $\qquad U_n = \sum_{j=0}^{\lfloor n/2 \rfloor} \dfrac{n!}{j!\,(n-2j)!\,2^j}.$

10. In problem 14 of section 1.1 we showed that
$$\sum_{n=0}^{\infty} U_n \frac{x^n}{n!} = e^{\frac{x^2}{2}+x}.$$
By using the exponential series and the binomial theorem, show that this implies (1.6.7).

1.7. The Z-identity

The object of this section is to present a beautiful counting proof, by inversions, of a rather complicated q-binomial coefficient identity. In the $q = 1$ case this proof is due to Foata. Zeilberger figured out how to extend it to a q-analogue. Zeng pointed out that Zeilberger's argument would apply to a slightly more general identity, which we will call

THEOREM 8 (The Z-identity). *For any nonnegative integers a, b, c, d, e we have*
$$\binom{a+b+d+e}{a+d}_q \binom{a+c+d+e}{c+d}_q \binom{b+c+d+e}{b+d}_q$$
$$= \sum_m q^{(d-m)(e-m)} \binom{a+b+c+d+e+m}{a+m, b+m, c+m, d-m, e-m}_q,$$
where the sum is over all the values of m that make sense, namely the ones between the largest of $-a, -b, -c$ and the smaller of d and e.

We can think of the left side as follows: the first q-binomial coefficient generates sequences w_1 of $a+d$ 1's and $b+e$ 2's, keeping track of inversions; the second generates sequences w_2 of $a+e$ 1's and $c+d$ 3's, by inversions; and the third generates sequences w_3 of $b+d$ 2's and $c+e$ 3's by inversions.

Let's write down an example at this point. Suppose $a = 5$, $b = 4$, $c = 3$, $d = 2$, and $e = 1$. Then w_1 would have $5+2$ 1's and $4+1$ 2's; w_2 has $5+1$ 1's and $3+2$ 3's; and w_3 has $4+2$ 2's and $3+1$ 3's. Three sample sequences are:

$$111212212211$$
$$31331111331$$
$$2323322232.$$

Note that there are 14 inversions on the first line (in w_1), 18 on the second line (in w_2) and 14 on the third line (in w_3), so there are 46 inversions in the whole array. Thus the left side of the Z-identity equals
$$\sum_{\text{arrays}} q^{\text{inv } w_1 + \text{inv } w_2 + \text{inv } w_3}.$$

Next we need a different way of looking at these arrays. We could try reading them by columns, but there is a problem—the rows usually won't all have the same length, as in our example. Foata found a very clever way around this obstacle. If all three rows have a different number in them, then we take the column as it is. Thus in our example the first thing we get is $\binom{1}{3}{2}$. We cross out those entries in the array and proceed to the next column, where we do not see three different numbers, but rather two 1's and a 3. If we see two of the same number in a column, then we take those two numbers with a blank entry in the other row; thus $\binom{1}{1}$ in this case. We cross out the two numbers we took, but not the one we didn't, in the array and we keep going. Here's what you will get if you finish this example:

$$
\begin{array}{cccccccccccccc}
1 & 1 & & 1 & 2 & 1 & 2 & 2 & 1 & 2 & 2 & & 1 & 1 \\
3 & 1 & & 3 & 3 & 1 & 1 & 1 & & 1 & & & 3 & 3 & 1 \\
2 & & & & 3 & 2 & 3 & & & 3 & 2 & & 2 & 2 & 3 & 2
\end{array}
$$

There are only five kinds of columns that you can possibly get; this example has at least two of all five kinds. (We'll call these the five letters of the **Foata alphabet**.) The next thing we want to do is get some idea of how many of each kind of column we might get in general. Suppose there are

$$z_1 \text{ of } \begin{pmatrix} 1 \\ 1 \\ \end{pmatrix}, \quad z_2 \text{ of } \begin{pmatrix} 1 \\ 3 \\ 2 \end{pmatrix}, \quad z_3 \text{ of } \begin{pmatrix} 2 \\ \\ 2 \end{pmatrix}, \quad z_4 \text{ of } \begin{pmatrix} 2 \\ 1 \\ 3 \end{pmatrix}, \quad z_5 \text{ of } \begin{pmatrix} 3 \\ 3 \end{pmatrix}.$$

Then counting the number of 1's in the first row, we must have $z_1 + z_2 = a + d$, and counting the number of 2's in the first row we must have $z_3 + z_4 = b + e$. Looking at the second and third rows we get four other equations. Let's list all six equations:

(1.7.1)
$$\begin{aligned}
z_1 + z_2 &= a + d, & z_3 + z_4 &= b + e, \\
z_1 + z_4 &= a + e, & z_2 + z_5 &= c + d, \\
z_2 + z_3 &= b + d, & z_4 + z_5 &= c + e.
\end{aligned}$$

Since (1.7.1) is a system of six equations in the five unknowns z_1 through z_5, it is not obvious that there *is* a solution in general. But in fact (1.7.1) always has at least one solution, and it may have many. One way to see this is to add together the two equations containing a, the two containing b, and the two containing c. This gives us three more equations, namely

$$\begin{aligned}
2z_1 + z_2 + z_4 &= 2a + d + e, \\
2z_3 + z_2 + z_4 &= 2b + d + e, \\
2z_5 + z_2 + z_4 &= 2c + d + e,
\end{aligned}$$

from which we can deduce that

$$\frac{d + e - z_2 - z_4}{2} = z_1 - a = z_3 - b = z_5 - c.$$

Then set $m = z_1 - a = z_3 - b = z_5 - c$. Evidently we have $z_1 = a + m$, $z_3 = b + m$, and $z_5 = c + m$, and plugging back into our original system (1.7.1) we get that all the equations are satisfied if $z_2 = d - m$ and $z_4 = e - m$. So if m is any integer, positive or not, such that all of the z_i's are nonnegative, then we get a solution of (1.7.1). (Note that $m = 0$ will always work.) In the above example $z_1 = 4$, $z_2 = 3$, $z_3 = 3$, $z_4 = 2$, $z_5 = 2$, and $m = -1$. The argument so far is enough to prove that

$$\binom{a+b+d+e}{a+d}\binom{a+c+d+e}{c+d}\binom{b+c+d+e}{b+d}$$
$$= \sum_m \binom{a+b+c+d+e+m}{a+m, b+m, c+m, d-m, e-m}$$

and this is essentially what Foata proved with it.

1.7. THE Z-IDENTITY

What about inversions in the Foata alphabet? This was Zeilberger's contribution. To be able to talk about inversions at all, we have to choose an ordering of the "letters" in the alphabet. A good choice, for reasons that we will see soon, is

$$\begin{pmatrix}1\\1\\ \end{pmatrix} < \begin{pmatrix}1\\3\\2\end{pmatrix} < \begin{pmatrix}2\\ \\2\end{pmatrix} < \begin{pmatrix}2\\1\\3\end{pmatrix} < \begin{pmatrix}\\3\\3\end{pmatrix}.$$

With this ordering, there are 40 inversions in the transformed array:

$$\begin{array}{cccccccccccc} 1 & 1 & & 1 & 2 & 1 & 2 & 2 & 1 & 2 & 2 & & 1 & 1 \\ 3 & 1 & 3 & 3 & 1 & 1 & 1 & & 1 & & & 3 & 3 & 1 \\ 2 & & 3 & 2 & 3 & & 3 & 2 & & 2 & 2 & 3 & 2 \end{array}$$

The first $\begin{pmatrix}1\\3\\2\end{pmatrix}$ column is inverted with the four $\begin{pmatrix}1\\1\end{pmatrix}$ columns, the first $\begin{pmatrix}3\\3\end{pmatrix}$ column is inverted with every column that comes after it except the other $\begin{pmatrix}3\\3\end{pmatrix}$ column, and so forth. If you count all these inversions up, 40 is the number you wind up with. But the original array had 46 inversions, so we are off by 6 inversions. Why is this, and can we predict how many inversions we will be off by in general, so that we might fix the formula?

The $\begin{pmatrix}1\\1\end{pmatrix}$ column was a good choice to be the smallest one, since its entries are as small as possible, and by the same token the $\begin{pmatrix}3\\3\end{pmatrix}$ column was a good choice to be the biggest. Why put the $\begin{pmatrix}2\\ \\2\end{pmatrix}$ column in between $\begin{pmatrix}1\\3\\2\end{pmatrix}$ and $\begin{pmatrix}2\\1\\3\end{pmatrix}$? There is a good reason: it makes sense to have

$$\begin{pmatrix}1\\3\\2\end{pmatrix} < \begin{pmatrix}2\\ \\2\end{pmatrix}$$

since no entry of the former is larger than any entry in the same row of the latter. And it makes sense to have

$$\begin{pmatrix}2\\ \\2\end{pmatrix} < \begin{pmatrix}2\\1\\3\end{pmatrix}$$

since again no entry of the former is larger than any entry in the same row of the latter. Every pair of columns respects order in this way except one,

$$\begin{pmatrix}1\\3\\2\end{pmatrix} \quad \text{and} \quad \begin{pmatrix}2\\1\\3\end{pmatrix}$$

We said the former was less than the latter, which looks good in two out of the three rows but not in the middle row. Therefore we miss one inversion in the second array any time that we have a $\begin{pmatrix}2\\1\\3\end{pmatrix}$ column coming after a $\begin{pmatrix}1\\3\\2\end{pmatrix}$ column, because that gets counted as one inversion in the first array and no inversion in the second.

What if a $\begin{pmatrix}2\\1\\3\end{pmatrix}$ column comes *before* a $\begin{pmatrix}1\\3\\2\end{pmatrix}$ column? In the second array we do count that as an inversion, since we defined $\begin{pmatrix}1\\3\\2\end{pmatrix}$ to be less than $\begin{pmatrix}2\\1\\3\end{pmatrix}$. But in the first array it counts as two inversions, one in the first row and one in the third. Therefore we also miss one inversion any time a $\begin{pmatrix}2\\1\\3\end{pmatrix}$ column comes before a $\begin{pmatrix}1\\3\\2\end{pmatrix}$

column. In sum, any time there is a pair of a $\begin{pmatrix}1\\3\\2\end{pmatrix}$ column and a $\begin{pmatrix}2\\1\\3\end{pmatrix}$ column, no matter which order they come in, we get one less inversion when we count them by columns than when we count them by rows. Therefore, the second count will always be too small by the number of $\begin{pmatrix}1\\3\\2\end{pmatrix}$ columns times the number of $\begin{pmatrix}2\\1\\3\end{pmatrix}$ columns. In our example, there were three $\begin{pmatrix}1\\3\\2\end{pmatrix}$ columns and two $\begin{pmatrix}2\\1\\3\end{pmatrix}$ columns, so we should be off by $3 \times 2 = 6$ inversions, and indeed we were since we got 46 the first way and 40 the second way. In general we will be off by $z_2 \times z_4 = (d-m)(e-m)$, and that is why the factor $q^{(d-m)(e-m)}$ appears inside the sum along with the q-multinomial coefficient—the q-multinomial coefficient takes care of the inversions among the columns, and the power of q picks up the missing ones. This proves the Z-identity. As Zeng also pointed out, a different proof can be given using the q–Pfaff-Saalschütz identity, which we will see in Chapter 5.

Exercises

1. If $a = 3$, $b = 4$, $c = 2$, $d = 4$, $e = 3$, then a possible triple of sequences is:

    11212221211221
    311133131331
    2322222332233.

    Count the number of inversions in each sequence. What "word" in the Foata alphabet corresponds to these sequences? How many inversions does it have? Is this the "right" number of inversions? Explain.

2. If $a = 1$, $b = 2$, $c = 3$, $d = 4$, $e = 5$, then a possible triple of sequences is:

    221222112112
    1311333313113
    32232233332323.

    Count the number of inversions in each sequence. What "word" in the Foata alphabet corresponds to these sequences? How many inversions does it have? Is this the "right" number of inversions? Explain.

3. Make up your own examples like problems 2 and 3 and solve them.

4. Show that, as a special case of the Z-identity, we have

    $$\binom{n}{k}_q^3 = \sum_j q^{(k-j)(n-k-j)} \binom{n+j}{j,j,j,k-j,n-k-j}_q.$$

5. Solve the system of equations (1.7.1) by whatever method you learned in linear algebra (probably elimination and back substitution), and show that your solution is equivalent to the one in the text.

1.8. Bibliographical Notes

It is worth stopping for a moment to say something more about August Leopold Crelle. Nothing is more important for the health of mathematics than the support and encouragement of promising young mathematicians. Crelle was fortunate to meet Abel and Jacobi, two young men with a great deal to say, at a time when he was trying to get a journal off the ground, but they were fortunate to meet him too. Even much later, when he was old and his journal well-established, Crelle helped get Eisenstein's career off to a flying start by publishing his first 25(!) papers in 1844. Jacobi would have become great in any case, but Abel and Eisenstein both died very young, and we have Crelle to thank for the fact that most of their work appeared in their lifetimes, rather than posthumously. No doubt it was also much easier for Abel, Jacobi, Eisenstein and many others to work in the knowledge that their ideas would be heard. Although a mediocre mathematician himself, few people did more for mathematics in the second quarter of the 19th century than Crelle. To have his name attached to a great journal from 1826 through the foreseeable future is nothing less than he deserves.

Stern's problem appears in [**226**]. Terquem's solution is in [**235**], and Rodrigues's two solutions are in [**196**]. The "couples" argument is also given on p. 94 of Netto's book [**174**], where there are references to Stern and Terquem, but not Rodrigues. Subsequent pages of Netto's book study the coefficients of the q-factorial, in effect, but without ever introducing it. If Netto ever read [**196**], then he must have forgotten it by the time he wrote his book. Otherwise his fourth chapter would have been much better.

The inversion concept goes back to Cramer's pioneering work on determinants in 1750 in the appendix of his book [**84**]; the number of inversions in a given permutation equals the number of row exchanges needed to make the corresponding permutation matrix into the identity. However, Cramer called them *dérangements*. Laplace called them *variations* instead when he wrote about determinants in 1772 [**161**], and Stern called them *Variationen* when he posed his problem, but by that time the term "inversions" was common in France, and was used by both Terquem and Rodrigues in their solutions. It was introduced by Gergonne in 1813 in an expository paper [**119**] on determinants based on Laplace's work, and popularized in Garnier's textbook [**111**] of the following year. See the first volume of Muir's great history of determinants [**173**] for more details. The connection between determinants and inversions led Muir to rediscover the q-factorial in 1899 [**172**]. Rothe's theorem on conjugate permutations dates back to 1800 [**207**], and is also discussed in [**173**]. Gergonne published [**119**] in his *Annales de Mathématiques Pures et Appliquées*, which was a precursor of Liouville's *Journal*. For the Liouville-Gergonne relationship see [**165**].

q-binomial coefficients appear for the first time in Gauss's paper [**115**]. Curiously, he never wrote down the fundamental recurrence (1.3.1); perhaps he thought it insufficiently different from (1.3.2) to be worth noting. The result of problem 8 in section 1.2 is in [**116**]. The history of the Fundamental Property of q-binomial coefficients is very strange. I can find no evidence that it was known, in that form, before MacMahon published the Fundamental Property of q-multinomial coefficients in [**167**], where he called the number of inversions the "superior index".

The Fundamental Property of q-multinomial coefficients also appears in the introduction (and only there; presumably it was added at the last minute) to the second volume of MacMahon's book [**168**], without proof but with a reference to [**169**]. There MacMahon gives the same proof as in [**167**], now using the word "inversion". Neither [**169**] nor [**168**] has a reference to [**167**]. All this is made still more curious by the fact that MacMahon knew earlier a related but harder result about the major index—see Chapter 9.

The Fundamental Property of q-binomial coefficients has been attributed to Netto's book, but I have not found it there, and neither did MacMahon. However, an equivalent result on partitions was proved by Cayley in 1855. We will see it in Chapter 3.

The Fundamental Property of q-multinomial coefficients was rediscovered by Carlitz [**64**], and Pólya [**186**], [**187**] seems to have known it at about the same time. As far as I know, [**64**] is the first paper with a reference to [**196**]. Carlitz says there that Charles A. Church pointed out Rodrigues's paper to him. Pólya's Property appears in [**185**] and [**188**].

The Z-identity, in a slightly different form, is equation (4) in [**253**]. It is slightly more general than Zeilberger's identity [**249**], which is a q-analogue of Foata's identity from [**106**]. For the q-Pfaff–Saalschütz identity see Chapter 5.

For the unimodality of the q-binomial coefficients see two other papers of Zeilberger. In [**250**] he gives a prize-winning account of a combinatorial proof of Kathy O'Hara. Another approach is in [**251**], and there are several related papers in the same volume. I have chosen to denote the q-binomial coefficients by $\binom{n}{k}_q$ rather than $\begin{bmatrix}n\\k\end{bmatrix}$ or $\begin{bmatrix}n\\k\end{bmatrix}_q$ for reasons discussed in Knuth's beautiful paper [**159**], in which he proposes a notation for the Stirling numbers of the second kind (or Stirling subset numbers) that I have followed ever since. The reasons for using $\binom{n}{k}_q$ have to do with the Stirling numbers of the first kind.

Pallavi Jayawant read a preliminary version of this chapter and pointed out several errors. I have referred to M. A. Stern in print at least once as "Maximilian Stern", and I cannot now remember why. It is apparently true that some of his friends called him "Max", but "Moritz Abraham Stern" is the more proper name. I thank Brian Hayes for the correction.

CHAPTER 2

q-binomial Theorems

2.1. A noncommutative q-binomial Theorem

There are quite a few results that could be described as q-binomial theorems, of which the oldest date back to the early 1800s. The most natural one from the point of view of inversions is however much more recent, and we begin with it.

Suppose we expand $(x+y)^n$ *without* assuming that xy is the same thing as yx. If you have seen matrices or groups, then you know that there are things that do not commute with each other. You see this in your everyday life also—it can make a big difference what order you do things in. For example, two things you might do with a bucket of water are

- leave it outside on a cold night,
- empty it on somebody.

The order is significant. Indeed, part of the reason why linear algebra is such an important subject is that it can deal with phenomena that are noncommutative, such as rotations (about different axes), reflections, projections, permutations, and so on. Far from being a defect, the noncommutativity of matrix multiplication is a boon.

With this sermon out of the way, consider $(x+y)^n$, which means

$$\overbrace{(x+y)(x+y)\cdots(x+y)}^{n \text{ factors}},$$

Because of the noncommutativity of x and y, we have to be careful about multiplying this out. We will get

$$xxx\cdots x + yxx\cdots x + xyxx\cdots x + yyxx\cdots x + \cdots + xyy\cdots y + yy\cdots y$$

which consists of one copy of every possible arrangement of n factors which could be either x or y. Let's denote the sum of all the different products of m x's and n y's by $V_{m,n}$; so that, for example,

$$V_{2,2} = xxyy + xyxy + xyyx + yxxy + yxyx + yyxx.$$

We'll define $V_{0,0} = 1$ and $V_{m,n} = 0$ if m or n is negative. Let's look specifically at $(x+y)^4$:

$$\begin{aligned}(x+y)^4 &= xxxx + \{xxxy + xxyx + xyxx + yxxx\} \\ &\quad + \{xxyy + xyxy + xyyx + yxxy + yxyx + yyxx\} \\ &\quad + \{xyyy + yxyy + yyxy + yyyx\} + yyyy \\ &= V_{4,0} + V_{3,1} + V_{2,2} + V_{1,3} + V_{0,4}.\end{aligned}$$

In exactly the same way we have

THEOREM 9 (The totally noncommutative binomial theorem). *If we make no assumption whatever about the commutativity of x and y, then*

$$(x+y)^n = \sum_{k=0}^{n} V_{k,n-k}$$

with $V_{i,j}$ defined as above.

When x and y commute, then all of the terms of $V_{k,n-k}$ are equal to $x^k y^{n-k}$, and there are $\binom{n}{k}$ such terms (why?), and we get back the ordinary binomial theorem. Let's try a different assumption: suppose $yx = qxy$, where q is a variable that commutes with x and y. What happens to the totally noncommutative binomial theorem then?

To see this, let's look at the $(x+y)^4$ example above. The first and last terms give us no trouble; we can write them as x^4 and y^4 respectively. Let's try to write the second term, $xxxy + xxyx + xyxx + yxxx$, in the form (something)x^3y. $xxxy$ is already in that form, but in the other terms we have to use $yx = qxy$ to move the y to the right of all the x's. If we use $yx = qxy$ once on the term $yxxx$, we will get $(yx)xx = (qxy)xx = qxyxx$. Therefore $yxxx + xyxx = (1+q)xyxx$, and if we use $yx = qxy$ on this, we get $(1+q)x(yx)x = (1+q)x(qxy)x = (q+q^2)xxyx$. Therefore $xxyx + xyxx + yxxx = (1+q+q^2)xxyx$, and using $yx = qxy$ one last time gives us $(1+q+q^2)xx(yx) = (1+q+q^2)xx(qxy) = (q+q^2+q^3)xxxy$. Therefore

$$xxxy + xxyx + xyxx + yxxx = (1+q+q^2+q^3)xxxy = [4]_q x^3 y.$$

In the same way,

$$xyyy + yxyy + yyxy + yyyx = (1+q+q^2+q^3)xyyy = [4]_q xy^3.$$

Do you see what's happening yet? Let's finish the $(x+y)^4$ example by working out $xxyy + xyxy + xyyx + yxxy + yxyx + yyxx$. In each term, we have to move all the y's to the right of all the x's, and each time we move a y past an x we pick up a q. Therefore each term equals x^2y^2 times q to a power, where the exponent of q is the number of times we had to move a y past an x in that term. Think back to the beginning of section 1.3, where we said

> In this section we'll play the same sort of game as before, but this time with "words" in a two-letter "alphabet". For the moment, let's take the "letters" to be 0 and 1, where as usual $1 > 0$. In general, we'll consider sequences of k 0's and $n-k$ 1's, and as before we'll keep track of inversions; this time, an inversion will be any pair of a 1 and a 0 where the 1 comes before the 0.

Suppose we take the "letters" to be x and y instead, where we think of xy as the natural order and of yx as inverted. This is exactly the problem we are considering now; the powers of q for each sequence are

sequence	q^{inv}	sequence	q^{inv}
$xxyy$	q^0	$yxxy$	q^2
$xyxy$	q^1	$yxyx$	q^3
$xyyx$	q^2	$yyxx$	q^4

and we see that, if $yx = qxy$, then

$$xxyy+xyxy+xyyx+yxxy+yxyx+yyxx = (1+q+2q^2+q^3+q^4)xxyy = \binom{4}{2}_q x^2y^2.$$

In general we have

LEMMA 1. *If $V_{i,j}$ is as above and $yx = qxy$, then*

$$V_{k,n-k} = \binom{n}{k}_q x^k y^{n-k}.$$

This is just a restatement of the Fundamental Property of q-binomial coefficients. For it is clear that we must get $x^k y^{n-k}$ times some function of q, and as we argued above, each term of $V_{k,n-k}$ contributes q to a power which is the number of inversions (y's in front of x's) in that term. Hence the Fundamental Property tells us that this function of q must be $\binom{n}{k}_q$. Combining the lemma with the totally noncommutative binomial theorem we get

THEOREM 10 (The Potter–Schützenberger q-binomial theorem). *If $yx = qxy$ and all other quantities are commuting, then*

$$(2.1.1) \qquad (x+y)^n = \sum_{k=0}^{n} \binom{n}{k}_q x^k y^{n-k}.$$

Although we already have a fully satisfactory proof of (2.1.1), let's prove it by induction on n as well. For this we recall the q-Pascal recurrences

$$(2.1.2) \qquad \binom{n+1}{k}_q = \binom{n}{k-1}_q + q^k \binom{n}{k}_q$$

$$(2.1.3) \qquad \qquad\quad = q^{n-k+1} \binom{n}{k-1}_q + \binom{n}{k}_q$$

from section 1.3. We have already checked the theorem for $n = 4$, and we leave it to the reader to check it for smaller n. Assuming it is true for n, we have

$$(2.1.4) \qquad (x+y)^{n+1} = (x+y)(x+y)^n$$

$$(2.1.5) \qquad \qquad\quad = (x+y) \sum_{k=0}^{n} \binom{n}{k}_q x^k y^{n-k}$$

$$(2.1.6) \qquad \qquad\quad = \sum_{k=0}^{n} \binom{n}{k}_q x^{k+1} y^{n-k} + \sum_{k=0}^{n} \binom{n}{k}_q y x^k y^{n-k}.$$

In the second sum we use the fact that, if $yx = qxy$, then $yx^k = q^k x^k y$ for any nonnegative integer k. We also replace k by $j-1$ in the first sum; if $0 \leq k \leq n$, then $1 \leq j \leq n+1$. Now

$$(x+y)^{n+1} = \sum_{j=1}^{n+1} \binom{n}{j-1}_q x^j y^{n+1-j} + \sum_{k=0}^{n} \binom{n}{k}_q q^k x^k y^{n+1-k}.$$

If we replace k by j in the second sum, and add a 0 term to each sum, then this becomes

$$(x+y)^{n+1} = \sum_{j=0}^{n+1} \binom{n}{j-1}_q x^j y^{n+1-j} + \sum_{j=0}^{n+1} \binom{n}{j}_q q^j x^j y^{n+1-j}$$

$$= \sum_{j=0}^{n+1} \binom{n+1}{j}_q x^j y^{n+1-j},$$

where we used (2.1.2) in the last step. This shows that if (2.1.1) is true for n, then it is also true for $n+1$. Since it is trivially true for $n=0$, it is true for all nonnegative integers n.

We can also get a multinomial version of the Potter–Schützenberger theorem. Let $x_1, \ldots, x_m$ be m noncommuting variables such that

$$x_j x_i = \begin{cases} q x_i x_j, & \text{if } j > i, \\ x_i x_j, & \text{if } j = i, \\ q^{-1} x_i x_j, & \text{if } j < i. \end{cases}$$

Then

(2.1.7) $$(x_1 + \cdots + x_m)^n = \sum_{k_1 + \cdots + k_m = n} \binom{n}{k_1, \ldots, k_m}_q x_1^{k_1} \ldots x_m^{k_m}.$$

If we multiply the left side of (2.1.7) out and use the commutation relations to put each term in the form $x_1^{k_1} \ldots x_m^{k_m}$, we will get a factor of q anytime there is an inversion among the subscripts of the x_i's. Thus (2.1.7) follows from the Fundamental Property of q-multinomial coefficients.

This theorem has traditionally been attributed to Schützenberger, who proved it in 1953, but it had also appeared in a note by Potter three years earlier, as was pointed out by Olga Holtz. One could also argue that it was not so much a new theorem as a new and striking way to state the Fundamental Property of q-binomial coefficients. Beautiful as it is, it does have one serious drawback—if $q \neq 1$, one cannot replace x or y by numbers (except 0, which is not interesting), because this would violate $yx = qxy$. It is natural to wish for a "commutative" q-binomial theorem that would be free from this objection, and we will return to this soon. We conclude this section by exploring the consequences of the obvious identity $(x+y)^{m+n} = (x+y)^m (x+y)^n$ when $yx = qxy$. Expanding on both sides, we get

$$\sum_{k=0}^{m+n} \binom{m+n}{k}_q x^k y^{m+n-k} = \sum_{i=0}^{m} \binom{m}{i}_q x^i y^{m-i} \sum_{j=0}^{n} \binom{n}{j}_q x^j y^{n-j}$$

$$= \sum_{i=0}^{m} \sum_{j=0}^{n} \binom{m}{i}_q \binom{n}{j}_q x^i y^{m-i} x^j y^{n-j}.$$

Now we use $yx = qxy$ repeatedly to move the factor x^j through the factor y^{m-i}, so that we get all the x's together and all the y's together. Each x has to go through each y, so we must use $yx = qxy$ a total of $(m-i)j$ times to accomplish this, thereby picking up a factor of $q^{(m-i)j}$. So now we have

$$\sum_{k=0}^{m+n}\binom{m+n}{k}_q x^k y^{m+n-k} = \sum_{i=0}^{m}\sum_{j=0}^{n}\binom{m}{i}_q \binom{n}{j}_q x^{i+j} y^{m+n-(i+j)} q^{(m-i)j}$$

$$= \sum_{k=0}^{m+n} x^k y^{m+n-k} \sum_{i+j=k}\binom{m}{i}_q \binom{n}{j}_q q^{(m-i)j}.$$

By equating coefficients of $x^k y^{m+n-k}$, it follows that

$$\binom{m+n}{k}_q = \sum_j \binom{m}{k-j}_q \binom{n}{j}_q q^{j(m-k+j)}.$$

We proved the same identity by counting inversions in Chapter 1.

Exercises

1. Prove that if $yx = qxy$, then $yx^k = q^k x^k y$ for any nonnegative integer k.
2. Explain why $V_{m,n} = xV_{m-1,n} + yV_{m,n-1}$. Check the definition—does this still hold if m or n are 0?
3. Explain why $V_{m,n} = V_{m-1,n}x + V_{m,n-1}y$. Does this still hold if m or n are 0?
4. Use either problem 2 or problem 3 to prove the totally noncommutative binomial theorem by induction on n.
5. What happens to the results of problems 2 and 3 if you set $x = y = 1$?
6. Rewrite the result of problem 2 as $V_{k,n+1-k} = xV_{k-1,n+1-k} + yV_{k,n-k}$, and use it along with (2.1.2) to prove the Fundamental Lemma by induction on n.
7. Rewrite the result of problem 3 as $V_{k,n+1-k} = V_{k-1,n+1-k}x + V_{k,n-k}y$, and use it along with (2.1.3) to prove the Fundamental Lemma by induction on n.
8. Prove the Potter–Schützenberger q-binomial theorem by induction using (2.1.3) instead of (2.1.2). (Start with $(x+y)^{n+1} = (x+y)^n (x+y)$.)
9. Consider functions of a variable t. Two things we might conceivably want to do to a typical $f(t)$ are
   (a) replace t by qt,
   (b) take the derivative with respect to t,
   Let η be an operator that does the first, and D an operator which does the second; in other words, $Df(t) = f'(t)$ and $\eta f(t) = f(qt)$. Explain why $\eta D f(t) := \eta(Df(t)) = f'(qt)$ but $D\eta f(t) := D(\eta f(t)) = q f'(qt)$. Conclude that $D\eta = q\eta D$.
10. Let x and q be commuting variables, and define the matrices

$$A = \begin{pmatrix} 1 & x \\ -\frac{1}{x} & -1 \end{pmatrix}, \quad B = \begin{pmatrix} 1+q & (1-q)x \\ \frac{1-q}{x} & 1+q \end{pmatrix}, \quad C = \begin{pmatrix} q+1 & (q-1)x \\ \frac{q-1}{x} & q+1 \end{pmatrix}.$$

Show that $BA = qAB = 2qA = AC = qCA$.

11. (This exercise is due to Greg Henderson.) Suppose D is a diagonal 2×2 matrix, M is an arbitrary 2×2 matrix, and $DM = qMD$. Show that the only nontrivial

possibilities for D and M (in other words, the only cases in which $DM = qMD$ with a generic value of q and without both sides being the zero matrix) are

$$D = \begin{pmatrix} qb & 0 \\ 0 & b \end{pmatrix} \quad \text{and} \quad M = \begin{pmatrix} 0 & x \\ 0 & 0 \end{pmatrix}$$

or

$$D = \begin{pmatrix} a & 0 \\ 0 & qa \end{pmatrix} \quad \text{and} \quad M = \begin{pmatrix} 0 & 0 \\ y & 0 \end{pmatrix}.$$

12. Let

$$X = \begin{pmatrix} 0 & 0 & 0 & 0 \\ 1 & 0 & 0 & 0 \\ 0 & 1 & 0 & 0 \\ 0 & 0 & 1 & 0 \end{pmatrix} \quad \text{and} \quad Y = \begin{pmatrix} 1 & 0 & 0 & 0 \\ 0 & q & 0 & 0 \\ 0 & 0 & q^2 & 0 \\ 0 & 0 & 0 & q^3 \end{pmatrix}.$$

Show that $YX = qXY$, and generalize to $n \times n$ matrices.

13. What happens if you replace X in the previous problem by its transpose?

14. Prove the following generalization of the Potter–Schützenberger theorem: if $yx = qxy$ and $yw = qwy$ and all other pairs of variables commute, then

$$(x+y)^n = \sum_{k=0}^{n} \binom{n}{k}_q (x+w)(x+wq) \cdots (x+wq^{k-1})(y-w)^{n-k},$$

where $(x+w)(x+wq) \cdots (x+wq^{k-1}) = 1$ if $k = 0$. What happens when $q \to 1$?

15. Try to think of a particularly striking real-world example of two things that produce dramatically different results depending on which order they are done in. Here's a nice one that I heard from Georgia Benkart:
    - stick your head out a window,
    - open the window.

    Cooking may be a good source of examples.

16. Gian-Carlo Rota, Bruce Sagan and Paul R. Stein found a very interesting generalization of the notion of a derivative to a noncommutative ring; in other words, to an expression like $V_{m,n}$, where we have a commutative addition and a possibly noncommutative multiplication. For example, suppose we want the **cyclic derivative** of $axxbaxcb$ with respect to x, where none of the letters x, a, b, c necessarily commutes with any of the others. The cyclic derivative operator D_x consists of two other operators C and T_x.

    Step 1: First apply the **cycling** operator C to get

    $$C\,axxbaxcb = axxbaxcb + baxxbaxc + cbaxxbax + xcbaxxba$$
    $$+ axcbaxxb + baxcbaxx + xbaxcbax + xxbaxcba,$$

    where we tear off one letter at a time from the back and put it on the front until we get back to where we started. This step is independent of which letter the derivative is being taken with respect to.

    Step 2: Apply the **truncation** operator T_x, which annihilates any term that doesn't begin with x, and removes the initial x from any term that does.

In this example, this gives

$$D_x\, axxbaxcb = T_x\, C\, axxbaxcb$$
$$= 0 + 0 + 0 + cbaxxba + 0 + 0 + baxcbax + xbaxcba$$
(2.1.8) $$= cbaxxba + baxcbax + xbaxcba.$$

We also extend the definition of D_x by linearity, so that $D_x\,(u+v) = D_x\,u + D_x\,v$ and, for example, $D_x\,5u = 5\,D_x\,u$. Note that this means a number is treated differently from a letter, and numbers are assumed to commute with letters.

(i) If all the letters commute with each other, what does (2.1.8) say?

(ii) Show that $C\,V_{m,n} = (m+n)V_{m,n}$.

(iii) Show that $D_x\,V_{m,n} = (m+n)V_{m-1,n}$.

(iv) Use (iii) to show that, even with no assumption about the commutativity of x and y, we have $D_x(x+y)^n = n(x+y)^{n-1}$. The result of problem 2 might help.

2.2. Potter's proof

For the algebraically inclined reader, the Potter–Schützenberger q-binomial theorem (and its q-multinomial generalization) is the central result in inversion theory; the relation $yx = qxy$ is precisely what is needed to describe algebraically the combinatorial idea of an inversion.

Although we have preferred to take a combinatorial approach, Potter derived (2.1.1) purely algebraically, without having to guess the answer and prove it by induction, and in a rather simple way. As we will see in the exercises, this argument in essence dates back to Gruson in 1814. It has often been rediscovered.

The problem, again, is to expand $(x+y)^n$ under the assumptions that $yx = qxy$ and that q commutes with x and y. Let's set

(2.2.1) $$(x+y)^n = \sum_{k=0}^{n} C(n,k,q)\, x^k\, y^{n-k}$$

and try to force out the coefficients $C(n,k,q)$. The first step is to rederive the fundamental recurrences (2.1.2) and (2.1.3). According to (2.2.1), we have

(2.2.2) $$(x+y)^{n+1} = \sum_{k=0}^{n+1} C(n+1,k,q)\, x^k\, y^{n+1-k}$$

and also

$$(x+y)^{n+1} = (x+y)(x+y)^n$$
$$= (x+y)\sum_{k=0}^{n} C(n,k,q)\, x^k\, y^{n-k}$$
$$= \sum_{k=0}^{n} C(n,k,q)\, x^{k+1}\, y^{n-k} + \sum_{k=0}^{n} C(n,k,q)\, y\, x^k\, y^{n-k}.$$

Change $k+1$ to k in the first sum, and in the second use $yx = qxy$ k times to move the lone y past x^k to where its $n-k$ brothers and sisters are. This gives us $yx^k = q^k x^k y$, and we have

$$(x+y)^{n+1} = \sum_{k=1}^{n+1} C(n, k-1, q)\, x^k\, y^{n+1-k} + \sum_{k=0}^{n} C(n, k, q)\, q^k\, x^k\, y^{n+1-k}.$$

Comparing this with (2.2.2) we get

(2.2.3) $$C(n+1, k, q) = C(n, k-1, q) + q^k\, C(n, k, q),$$

which should look familiar. We also have

$$(x+y)^{n+1} = (x+y)^n (x+y)$$
$$= \left(\sum_{k=0}^{n} C(n, k, q)\, x^k\, y^{n-k}\right)(x+y)$$
$$= \sum_{k=0}^{n} C(n, k, q)\, x^k\, y^{n-k}\, x + \sum_{k=0}^{n} C(n, k, q)\, x^k\, y^{n+1-k}.$$

In the first sum, use $yx = qxy$ $n-k$ times, so that $y^{n-k} x = q^{n-k} x y^{n-k}$. Then change $k+1$ to k in the first sum. There results

$$(x+y)^{n+1} = \sum_{k=1}^{n+1} C(n, k-1, q) q^{n-k+1}\, x^k\, y^{n+1-k} + \sum_{k=0}^{n} C(n, k, q)\, x^k\, y^{n+1-k}.$$

Comparing this with (2.2.2) we get

(2.2.4) $$C(n+1, k, q) = C(n, k, q) + q^{n-k+1} C(n, k-1, q),$$

again as hoped.

Combining (2.2.3) and (2.2.4) we have

$$C(n, k-1, q) + q^k\, C(n, k, q) = q^{n-k+1}\, C(n, k-1, q) + C(n, k, q),$$

or

$$C(n, k, q)(1 - q^k) = C(n, k-1, q)(1 - q^{n-k+1}),$$

or

(2.2.5) $$C(n, k, q) = C(n, k-1, q)\, \frac{1 - q^{n-k+1}}{1 - q^k}.$$

If we replace k by $k-1$ in (2.2.5) we get

$$C(n, k-1, q) = C(n, k-2, q)\, \frac{1 - q^{n-k+2}}{1 - q^{k-1}}$$

and substituting this into the right side of (2.2.5) gives

$$C(n, k, q) = C(n, k-2, q)\, \frac{1 - q^{n-k+1}}{1 - q^k}\, \frac{1 - q^{n-k+2}}{1 - q^{k-1}}.$$

Continuing in this way, after k iterations we will get

(2.2.6) $$C(n, k, q) = C(n, 0, q)\, \frac{(1 - q^{n-k+1})(1 - q^{n-k+2}) \ldots (1 - q^n)}{(1 - q^k)(1 - q^{k-1}) \ldots (1 - q)}.$$

By the definition (2.2.1), $\binom{n}{0}_q$ is the coefficient of y^n in the expansion of $(x+y)^n$, which is just y^n no matter what we assume about the commutation of x and y. This forces $C(n, 0, q) = 1$ and proves that

$$C(n, k, q) = \frac{(q;q)_n}{(q;q)_k (q;q)_{n-k}} = \binom{n}{k}_q = \frac{n!_q}{k!_q \, (n-k)!_q}.$$

Thus we have derived (again)

THEOREM 11 (The Potter–Schützenberger q-binomial theorem). *If $yx = qxy$ and all other quantities are commuting, then*

$$(x+y)^n = \sum_{k=0}^{n} \binom{n}{k}_q x^k y^{n-k},$$

where $\binom{n}{k}_q = \frac{(q;q)_n}{(q;q)_k \, (q;q)_{n-k}}$ if k and n are integers with $0 \le k \le n$ and is 0 otherwise.

One of the interesting features of this argument (a point stressed by Dick Askey) is that one *cannot* use a simpler version of it to derive the ordinary binomial theorem algebraically in the same way. It fails at the step where we equated the two different q-Pascal recurrences: for $q = 1$, since there is only one Pascal recurrence, the algebra collapses to $0 = 0$, which is true but useless.

Exercises

1. In 1814, Johann Philipp Gruson used a similar argument to prove Rothe's q-binomial theorem, the subject of the next section. He wanted to find the coefficients A_k (which should really be called $A_{n,k}(q)$, but we will follow his notation) in the expansion

$$Y := (x+z)(x+zq) \cdots (x+zq^{n-1}) = \sum_{k=0}^{n} A_k x^{n-k} z^k.$$

(i) Explain why $A_0 = 1$.

Gruson sets $P = (x+zq)(x+zq^2) \cdots (x+zq^n)$ and notes that

$$P = \sum_{k=0}^{n} A_k x^{n-k} (zq)^k \quad \text{and} \quad P = \frac{x+zq^n}{x+z} Y.$$

Then he sets

$$\frac{Y}{x+z} = \sum_{k=0}^{n-1} B_k x^{n-1-k} z^k.$$

(ii) Show that $B_0 = A_0 = 1$, that $B_{n-1} = A_n$, and that $A_k = B_k + B_{k-1}$ for $1 \le k \le n-1$.

(iii) Explain why $P = (x + zq^n) \sum_{k=0}^{n-1} B_k x^{n-1-k} z^k$.

(iv) Show that (iii) again implies $B_0 = A_0 = 1$ and $B_{n-1} = A_n$, and that it also implies $A_k q^k = B_k + B_{k-1} q^n$ for $1 \le k \le n-1$.

(v) Use (ii) and (iv) to show that
$$B_k = \frac{q^n - q^k}{q^k - 1} B_{k-1} \quad \text{and} \quad A_k = \frac{q^n - 1}{q^k - 1} B_{k-1}.$$

(vi) Show that (v) implies
$$B_k = \frac{(q^n - q)(q^n - q^2) \cdots (q^n - q^k)}{(q - 1)(q^2 - 1) \cdots (q^k - 1)}.$$

(vii) Show that (v) and (vi) imply
$$A_k = \frac{(q^n - 1)(q^n - q) \cdots (q^n - q^{k-1})}{(q - 1)(q^2 - 1) \cdots (q^k - 1)}.$$

(viii) As far as Gruson was concerned, this completes the proof, but show that he's proved
$$(x + z)(x + zq)(x + zq^2) \cdots (x + zq^{n-1}) = \sum_{k=0}^n \binom{n}{k}_q q^{\binom{k}{2}} x^{n-k} z^k.$$

This is Rothe's q-binomial theorem.

2. Show that
$$\frac{(q^n - 1)(q^n - q) \cdots (q^n - q^{k-1})}{(q - 1)(q^2 - 1) \cdots (q^k - 1)} + q^n \frac{(q^n - 1)(q^n - q) \cdots (q^n - q^{k-2})}{(q - 1)(q^2 - 1) \cdots (q^{k-1} - 1)}$$
$$= \frac{(q^{n+1} - 1)(q^{n+1} - q) \cdots (q^{n+1} - q^{k-1})}{(q - 1)(q^2 - 1) \cdots (q^k - 1)}.$$

3. Gruson wrote the result of problem 1 as
$$(x + z)(x + zq)(x + zq^2) \cdots (x + zq^{n-1}) = x^n + \frac{q^n - 1}{q - 1} x^{n-1} z$$
$$+ \frac{(q^n - 1)(q^n - q)}{(q - 1)(q^2 - 1)} x^{n-2} z^2 + \frac{(q^n - 1)(q^n - q)(q^n - q^2)}{(q - 1)(q^2 - 1)(q^3 - 1)} x^{n-3} z^3 + \cdots$$
$$+ \frac{(q^n - 1)(q^n - q) \cdots (q^n - q^{k-1})}{(q - 1)(q^2 - 1) \cdots (q^k - 1)} x^{n-k} z^k + \cdots + q^{\frac{n(n-1)}{2}} z^n.$$

He also proved this by induction, using the result of problem 2. Reconstruct his argument.

4. The second person to use a Gruson/Potter type argument, a few years after Gruson, was Ferdinand Schweins. In this problem we outline Schweins's argument with a simplified notation. Consider the product
$$S_n(y) := (1 + y)(1 + yq) \cdots (1 + yq^{n-1}),$$
which means 1 if $n = 0$. We want to expand $S_n(y)$ in powers of y. Define the coefficient of y^k by $s(n, k)$ (which will, of course, be a function of q). The only k's that we can possibly get are the ones between 0 and n, so we have

(2.2.7) $$S_n(y) = (1 + y)(1 + yq) \cdots (1 + yq^{n-1}) = \sum_{k=0}^n s(n, k) \, y^k.$$

Clearly $s(n,0) = 1$, and we can assume $s(n,k) = 0$ if $k > n$ or $k < 0$. Schweins observes (or would observe) that

(2.2.8) $$S_n(y) = (1+y)S_{n-1}(yq)$$
(2.2.9) $$= S_{n-1}(y)(1+yq^{n-1}).$$

(i) (2.2.8) implies that

$$\sum_{k=0}^{n} s(n,k) y^k = (1+y) \sum_{k=0}^{n-1} s(n-1,k) y^k q^k.$$

Show that this leads to

$$s(n,k) = s(n-1,k)q^k + s(n-1,k-1)q^{k-1}.$$

(ii) (2.2.9) implies that

$$\sum_{k=0}^{n} s(n,k) y^k = (1+yq^{n-1}) \sum_{k=0}^{n-1} s(n-1,k) y^k.$$

Show that this leads to

$$s(n,k) = s(n-1,k) + s(n-1,k-1)q^{n-1}.$$

(iii) Show that comparing the results of (i) and (ii) and changing $n-1$ to n gives

$$s(n,k) = q^{k-1} \frac{1-q^{n-k+1}}{1-q^k} s(n,k-1).$$

(iv) Show that iterating the result of (iii) down to $k = 0$ gives $s(n,k) = \binom{n}{k}_q q^{\binom{k}{2}}$. This proves that

(2.2.10) $$(1+y)(1+yq)\cdots(1+yq^{n-1}) = \sum_{k=0}^{n} \binom{n}{k}_q q^{\binom{k}{2}} y^k.$$

(v) To make (2.2.10) superficially more general we can replace y by $\frac{x}{a}$ and multiply through by a^n. Show that this gives

$$(a+x)(a+xq)\cdots(a+xq^{n-1}) = \sum_{k=0}^{n} \binom{n}{k}_q q^{\binom{k}{2}} a^{n-k} x^k.$$

This is Rothe's q-binomial theorem.

2.3. Rothe's q-binomial theorem

George Andrews showed that the Potter–Schützenberger theorem can be used to derive a much older commutative q-binomial theorem. (Another way of doing this was given earlier by Johann Cigler; see problem 12.) Andrews's idea is to replace x by xy and y by ay, where a commutes with x, y and q. We can do this as long as these replacements are consistent with the condition $yx = qxy$, so we have to check that $(ay)(xy) = q(xy)(ay)$, and this does hold, since the left side is $a(yx)y = a(qxy)y$, which is the same as the right side since a commutes with everything. The Potter–Schützenberger theorem now reads

$$(2.3.1) \qquad (xy + ay)^n = \sum_{k=0}^{n} \binom{n}{k}_q (xy)^k (ay)^{n-k},$$

where $yx = qxy$. The next step is to factor all the y's out to the right on both sides. Let's look at the sum side first. $(ay)^{n-k} = a^{n-k} y^{n-k}$ since a and y commute. What about $(xy)^k$? By definition, this means $xyxyxy \ldots xy$, with k factors of x and k factors of y, and we want to factor all the y's to the right. There are $k-1$ x's to the right of the first y, $k-2$ x's to the right of the second y, and so on, so we have to use $yx = qxy$ a total of $(k-1) + (k-2) + (k-3) + \cdots + 1 + 0 = \binom{k}{2}$ times to get all the y's to the right of all the x's. Therefore, the right side of (2.3.1) equals

$$(2.3.2) \qquad \sum_{k=0}^{n} \binom{n}{k}_q q^{\binom{k}{2}} x^k y^k a^{n-k} y^{n-k} = \left(\sum_{k=0}^{n} \binom{n}{k}_q q^{\binom{k}{2}} x^k a^{n-k} \right) y^n.$$

Note that the y^n can come out of the sum, since it does not depend on k. On the other hand, the left side of (2.3.1) is

$$\overbrace{((x+a)y)((x+a)y)\ldots((x+a)y)}^{n \text{ factors}}.$$

Again, we will use $yx = qxy$ to try to move all the y's to the right. To see how this is going to work, start with the leftmost y, and imagine trying to move it to where the next y is. Just looking at these factors, we have

$$(x+a)y(x+a)y = (x+a)(yx + ya)y = (x+a)(qxy + ay)y = (x+a)(qx+a)y^2.$$

With n factors, we wind up with

$$(2.3.3) \qquad \overbrace{(x+a)y(x+a)y\ldots(x+a)y}^{n \text{ factors}} = (x+a)(xq+a)(xq^2+a)\ldots(xq^{n-1}+a)y^n.$$

The expression in (2.3.2) equals the expression in (2.3.3), since they both came from (2.3.1). Now just cancel the factors of y^n, and we have proved

THEOREM 12 (Rothe's q-binomial theorem). *If a, x and q are arbitrary commuting variables, then*

$$(x+a)(xq+a)(xq^2+a)\ldots(xq^{n-1}+a) = \sum_{k=0}^{n} \binom{n}{k}_q q^{\binom{k}{2}} x^k a^{n-k},$$

where the product on the left is defined to be 1 if $n = 0$.

This dates back to 1811, and is the oldest "true" q-analogue of the binomial theorem. (Gauss's identities in section 2.5 are three years older, but they are only q-versions of special cases of the binomial theorem. Gauss did eventually find Rothe's theorem as well, but he never published it.)

2.3. ROTHE'S q-BINOMIAL THEOREM

For another proof of Rothe's theorem, set $r_n(x, a)$ equal to the sum there. Then, applying (2.1.2) to $r_{n+1}(x, a)$, we get

$$r_{n+1}(x, a) = \sum_{k=0}^{n+1} \binom{n+1}{k}_q q^{\binom{k}{2}} x^k a^{n+1-k}$$

$$= \sum_{k=0}^{n+1} \left\{ \binom{n}{k-1}_q + q^k \binom{n}{k}_q \right\} q^{\binom{k}{2}} x^k a^{n+1-k}$$

$$= \sum_{k=1}^{n+1} \binom{n}{k-1}_q q^{\binom{k}{2}} x^k a^{n+1-k} + \sum_{k=0}^{n} \binom{n}{k}_q q^{\binom{k}{2}} (xq)^k a^{n+1-k}$$

$$= \sum_{j=0}^{n} \binom{n}{j}_q q^{\binom{j+1}{2}} x^{j+1} a^{n-j} + \sum_{j=0}^{n} \binom{n}{j}_q q^{\binom{j}{2}} (xq)^j a^{n+1-j},$$

where we replaced k by $j+1$ in one of the last two sums, and simply renamed k as j in the other. Using the fact that $\binom{j+1}{2} = \binom{j}{2} + j$ for any nonnegative integer j, we have

$$r_{n+1}(x, a) = \sum_{j=0}^{n} \binom{n}{j}_q q^{\binom{j}{2}} q^j x^{j+1} a^{n-j} + \sum_{j=0}^{n} \binom{n}{j}_q q^{\binom{j}{2}} (xq)^j a^{n+1-j}$$

$$= \sum_{j=0}^{n} \binom{n}{j}_q q^{\binom{j}{2}} (xq)^j a^{n-j} (x + a)$$

$$= (a + x) \, r_n(xq, a).$$

Iterating this, we get

$$r_{n+1}(x, a) = (a + x) \, r_n(xq, a)$$
$$= (a + x) \left\{ (a + xq) \, r_{n-1}(xq^2, a) \right\}$$
$$= (a + x)(a + xq) \left\{ (a + xq^2) \, r_{n-2}(xq^3, a) \right\}$$
$$= \text{and so forth}$$
$$= (a + x)(a + xq)(a + xq^2) \cdots (a + xq^n) \, r_0(xq^{n+1}, a).$$

But $r_0(u, v) = 1$ for any choice of u and v, so this proves Rothe's theorem with $n + 1$ in place of n.

If we set $a = 1$ and $x = -1$ in Rothe's theorem, then the product side has a factor of $1 - 1$ unless $n = 0$, so we have

(2.3.4) $$\sum_{j=0}^{n} \binom{n}{j}_q q^{\binom{j}{2}} (-1)^j = \begin{cases} 1 & \text{if } n = 0, \\ 0 & \text{if } n > 0. \end{cases}$$

This enables us to invert the matrix $L_n(q)$ from one of the problems in Chapter 1, whose ij^{th} entry was $\binom{i-1}{j-1}_q$. We claim that the ij^{th} entry of $L_n^{-1}(q)$ is

(2.3.5) $$(-1)^{i-j} \binom{i-1}{j-1}_q q^{\binom{i-j}{2}}.$$

For example, (2.3.5) says that the inverse of

$$L_4(q) = \begin{pmatrix} 1 & 0 & 0 & 0 \\ 1 & 1 & 0 & 0 \\ 1 & 1+q & 1 & 0 \\ 1 & 1+q+q^2 & 1+q+q^2 & 1 \end{pmatrix}$$

is

$$L_4^{-1}(q) = \begin{pmatrix} 1 & 0 & 0 & 0 \\ -1 & 1 & 0 & 0 \\ q & -1-q & 1 & 0 \\ -q^3 & q+q^2+q^3 & -1-q-q^2 & 1 \end{pmatrix}$$

and one can easily check this. If we multiply $L_n(q)$ by the matrix whose ij^{th} entry is as claimed in (2.3.5), we get a matrix whose ij^{th} entry is

$$\sum_{k=1}^{n} \binom{i-1}{k-1}_q (-1)^{k-j} \binom{k-1}{j-1}_q q^{\binom{k-j}{2}},$$

so we have to show that this is 1 if $i = j$ and 0 otherwise. There are no nonzero terms in the sum if $j > i$, by definition of the q-binomial coefficient, so we certainly get zero in that case. If $j = i$, we get one nonzero term, when $k = i = j$, and this term equals 1. So the only case that requires an argument is when $j < i$, when the sum runs over all k between j and i. Here we can write

$$\binom{i-1}{k-1}_q \binom{k-1}{j-1}_q = \binom{i-1}{j-1}_q \binom{i-j}{k-j}_q$$

and if we reindex the sum by letting $l = k - j$, then we have

$$\binom{i-1}{j-1}_q \sum_{l=0}^{i-j} \binom{i-j}{l}_q (-1)^l q^{\binom{l}{2}}$$

which equals 0 by (2.3.4) unless $i = j$, when it equals 1. This proves (2.3.5). We could now write down the inverse of the q-Pascal matrix $P_n(q)$, since $P_n(q) = L_n(q)D_n(q)L_n(q)^T$ and we know how to invert all three factors, but the result is not particularly nice.

The following curious theorem has a similar proof and is sometimes useful.

THEOREM 13. If A_n and B_n are two sequences, then

(2.3.6) $\quad A_n = \sum_{k=0}^{n} \binom{n}{k}_q B_k \quad$ if and only if $\quad B_n = \sum_{k=0}^{n} \binom{n}{k}_q (-1)^{n-k} q^{\binom{n-k}{2}} A_k.$

We prove one direction and leave the other as an exercise. Change n to k and k to j in the first equation and substitute it into the second to get the double sum

$$\sum_{k=0}^{n} \sum_{j=0}^{k} \binom{n}{k}_q \binom{k}{j}_q (-1)^{n-k} q^{\binom{n-k}{2}} B_j.$$

Changing the order of summation and rewriting the q-binomial coefficients as above this becomes

$$\sum_{j=0}^{n} \binom{n}{j}_q B_j \sum_{k=j}^{n} \binom{n-j}{n-k}_q (-1)^{n-k} q^{\binom{n-k}{2}}.$$

Again reindexing the inner sum by letting $l = n - k$, we get

$$\sum_{j=0}^{n} \binom{n}{j}_q B_j \sum_{l=0}^{n-j} \binom{n-j}{l}_q (-1)^l q^{\binom{l}{2}}.$$

By (2.3.4) the inner sum is zero unless $j = n$, and this expression collapses to B_n, as desired. Therefore the formula for A_n in (2.3.6) implies the formula for B_n.

Exercises

Note: The notation $(x; q)_n = (1-x)(1-xq)(1-xq^2)\cdots(1-xq^{n-1})$ mentioned in Chapter 1, with $(x; q)_0 = 1$, is used many times in the problems below.

1. Show that

(2.3.7) $$(x; q)_n = \sum_{k=0}^{n} \binom{n}{k}_q q^{\binom{k}{2}} (-1)^k x^k.$$

You can either prove this directly, imitating the proof of Rothe's q-binomial theorem; or derive it as a special case of Rothe's theorem. The latter is less work but the former may be more instructive.

2. Show that

$$(-q; q)_n = \sum_{k=0}^{n} \binom{n}{k}_q q^{\binom{k+1}{2}}.$$

(Either directly, or as a special case of Rothe's q-binomial theorem, or as a special case of (2.3.7).)

3. Prove Rothe's q-binomial theorem using (2.1.3) instead of (2.1.2).

4. Prove that

$$\sum_{k=0}^{n} \frac{q^k}{(q; q)_k} = \frac{1}{(q; q)_n}.$$

We will have a simple combinatorial explanation of this identity in Chapter 3.

5. Prove that the formula for B_n in (2.3.6) implies the one for A_n.

6. Show that (2.3.6) may be rewritten in the more symmetric form

$$A_n = \sum_{k=0}^{n} \binom{n}{k}_q (-1)^k C_k \iff C_n = \sum_{k=0}^{n} \binom{n}{k}_q (-1)^k q^{\binom{n-k}{2}} A_k.$$

In the $q = 1$ case, this form of Theorem 13 has the advantage that we only need to prove one direction. (Why?)

7. Prove that

$$\sum_{k=0}^{n} \frac{(-1)^k q^{\binom{k}{2}}}{(q; q)_k} = \frac{(-1)^n q^{\binom{n+1}{2}}}{(q; q)_n}.$$

8. Show that if $0 \leq m \leq n$, then

$$\sum_{k=0}^{m} q^k \left(q^{k+1}; q\right)_{n-k} = \left(q^{m+1}; q\right)_{n-m}.$$

What happens when $m = n$?

9. Show that
$$\sum_{k=0}^{n} q^k \left(q^{k+1}; q\right)_{n-k} = 1$$
by writing $q^k = 1 - (1 - q^k)$.

10. Set $\quad S_n(x) = \sum_{k=0}^{n} \binom{n}{k}_q \frac{q^{k^2} x^k}{(xq; q)_k}$.

(i) Show that $S_0(x) = 1$.

(ii) Show that $\quad S_n(x) = \dfrac{S_{n-1}(xq)}{1 - xq} \quad$ for $n \geq 1$.

(iii) Explain why (i) and (ii) imply that $\quad S_n(x) = \dfrac{1}{(xq; q)_n}$.

(iv) What happens to this identity when $q \to 1$?

11. Define an operator η by $\eta f(x) = f(qx)$, as in problem 9 in section 2.1. Consider the product $(x + x^2\eta)^n x$ for a nonnegative integer n. When $n = 0$ this is just x. When $n = 1$ it is
$$(x + x^2\eta) x = x^2 + x^2 \cdot \eta x = x^2 + x^2 \cdot xq = x^2(1 + xq) = x^2 + x^3 q,$$
and when $n = 2$ it is
$$(x + x^2\eta)^2 x = (x + x^2\eta)(x^2 + x^3 q)$$
$$= x^3 + x^4 q + x^2 \cdot \eta x^2 + x^2 \cdot \eta x^3 q = x^3 + x^4 q + x^2(xq)^2 + x^2(xq)^3 q$$
$$= x^3 + x^4 q(1 + q) + x^5 q^4 = x^3 \left(1 + [2]_q xq + x^2 q^4\right).$$

Use induction on n to prove **Andrews's q-binomial theorem**
$$(x + x^2\eta)^n x = x^{n+1} \sum_{k=0}^{n} \binom{n}{k}_q q^{k^2} x^k.$$
What does this reduce to if $q = 1$?

12. This problem outlines Cigler's derivation of Rothe's q-binomial theorem from the Potter–Schützenberger q-binomial theorem. It is in essence the same as Andrews's method, but the details are superficially different. Cigler starts by writing the Potter–Schützenberger theorem as

(2.3.8) $$(A + B)^n = \sum_{k=0}^{n} \binom{n}{k}_q A^k B^{n-k},$$

where $BA = qAB$.

(i) He takes $A = x\eta$ and $B = a\eta$, where a is independent of x and η is the shift operator on functions of x; i.e., $\eta f(x) = f(qx)$, as in problem 11. By applying both sides to a generic function $f(x)$, show that these choices of A and B give $BA = qAB$.

(ii) Show by induction on n that
$$(x\eta + a\eta)^n 1 = ((x + a)\eta)^n 1 = (x + a)(xq + a) \cdots (xq^{n-1} + a).$$

This even holds for $n = 0$ if we define the right side to be 1 in that case, as usual.

(iii) Show that
$$(x\eta)^k (a\eta)^{n-k} 1 = x^k q^{\binom{k}{2}} a^{n-k}.$$

(iv) Using (ii) and (iii), show that applying (2.3.8) to the function 1 with A and B as in (i) gives Rothe's q-binomial theorem.

13. Define
$$S_n(a,b) = \sum_{k=0}^{n} \binom{n}{k}_q (-a)^k q^{\binom{k}{2}} (b;q)_k \left(abq^k;q\right)_{n-k}.$$

(i) Show that $S_0(a,b) = 1$ and $S_1(a,b) = 1 - a$.

(ii) Show that
$$S_2(a,b) = (1-a)(1-aq).$$

(iii) Show that
$$S_{n+1}(a,b) = (1-a)S_n(aq,b).$$

(iv) Use (iii) to prove that
$$(2.3.9) \qquad S_n(a,b) = \sum_{k=0}^{n} \binom{n}{k}_q (-a)^k q^{\binom{k}{2}} (b;q)_k \left(abq^k;q\right)_{n-k} = (a;q)_n.$$

14. For an alternative proof of (2.3.9), start with
$$S_n(a,bq) = \sum_{k=0}^{n} \binom{n}{k}_q (-a)^k q^{\binom{k}{2}} (bq;q)_k \left(abq^{k+1};q\right)_{n-k}.$$

(i) Note that the last factor of $\left(abq^{k+1};q\right)_{n-k}$ is $1 - abq^n = 1 - abq^k + abq^k \left(1 - q^{n-k}\right)$, and the last factor of $(bq;q)_k$ is $1 - bq^k = 1 - b + b\left(1 - q^k\right)$. By using these two facts, show that
$$S_n(a,bq) = S_n(a,b) + \sum_{k=1}^{n} \binom{n}{k}_q (-a)^k q^{\binom{k}{2}} b(1-q^k)(bq;q)_{k-1} \left(abq^k;q\right)_{n-k}$$
$$+ \sum_{k=0}^{n-1} \binom{n}{k}_q (-a)^k q^{\binom{k}{2}} abq^k (1-q^{n-k})(bq;q)_k \left(abq^{k+1};q\right)_{n-k-1}.$$

Then show that the last two sums cancel each other.

(ii) From (i) we have $S_n(a,bq) = S_n(a,b)$ for arbitrary b and q. Assuming $|q| < 1$, explain why using this repeatedly gives $S_n(a,b) = S_n(a,0)$, and then explain how we know that $S_n(a,0) = (a;q)_n$.

15. Define
$$T_n(b,c) = \sum_{k=0}^{n} \binom{n}{k}_q (-1)^k q^{\binom{k+1}{2}} (b;q)_k \left(cq^k;q\right)_{n-k} q^{-nk}.$$

(i) Show that $T_0(b,c) = 1$ and $T_1(b,c) = b - c$.

(ii) Show that
$$T_2(b,c) = (b-c)(b-cq).$$

(iii) Show that
$$T_{n+1}(b,c) = (b-c)T_n(b,cq).$$

(iv) Use (iii) to prove that

$$T_n(b,c) = \sum_{k=0}^{n} \binom{n}{k}_q (-1)^k q^{\binom{k+1}{2}} (b;q)_k \left(cq^k;q\right)_{n-k} q^{-nk}$$

(2.3.10)
$$= (b-c)(b-cq)\cdots(b-cq^{n-1}).$$

16. Let n be a nonnegative integer. Prove that

(2.3.11)
$$\frac{(q^{-n};q)_k}{(q;q)_k} = (-1)^k q^{\binom{k}{2}} \binom{n}{k}_q q^{-nk}.$$

Hence show that (2.3.9) can be rewritten as

(2.3.12)
$$\sum_{k=0}^{n} \frac{(q^{-n};q)_k(b;q)_k}{(ab;q)_k(q;q)_k} (aq^n)^k = \frac{(a;q)_n}{(ab;q)_n}$$

and (2.3.10) can be rewritten as

(2.3.13)
$$\sum_{k=0}^{n} \frac{(q^{-n};q)_k(b;q)_k}{(c;q)_k(q;q)_k} q^k = \frac{(b-c)(b-cq)\cdots(b-cq^{n-1})}{(c;q)_n}.$$

17. (2.3.12) and (2.3.13) are really the same identity. To see this, first show that

(2.3.14)
$$(x;q^{-1})_m = (-1)^m q^{-\binom{m}{2}} \left(\frac{1}{x};q\right)_m x^m.$$

Then change q to q^{-1} in (2.3.12), use (2.3.14) six times, and finally replace a and b by $\frac{1}{a}$ and $\frac{1}{b}$ respectively. You should get (2.3.13) with c replaced by ab. Both (2.3.12) and (2.3.13) are often called the q-Chu–Vandermonde sum.

18. (a) Let m be a nonnegative integer. Show that setting $b = q^{-m}$ and $c = q^{-m-n}$ in (2.3.13) gives

$$\sum_{k} \frac{(q^{-m};q)_k(q^{-n};q)_k}{(q^{-m-n};q)_k(q;q)_k} q^k = \frac{1}{\binom{m+n}{m}_q},$$

where the upper limit on the sum is the smaller of m and n.

(b) Using (2.3.11) or otherwise, show that the left side of this can be rewritten as

$$\frac{1}{\binom{m+n}{m}_q} \sum_{k} (-1)^k q^{\binom{k+1}{2}} \binom{m+n-k}{k,m-k,n-k}_q.$$

This gives an alternative proof of the result of problem 6 in section 1.6.

19. This problem outlines a proof of **Agarwal's q-binomial theorem**

(2.3.15)
$$\sum_{j=0}^{m} \binom{n}{j}_q (-1)^j q^{\binom{j}{2}} (1-aq^{2j}) \frac{(a;q)_j}{(aq^{n+1};q)_j} = \binom{n-1}{m}_q (-1)^m q^{\binom{m+1}{2}} \frac{(a;q)_{m+1}}{(aq^{n+1};q)_m},$$

if $m \geq 0$ and $n \geq 1$, where the sum equals 1 if $n = 0$. This is a surprising generalization of problem 7 in section 1.4.

(i) Verify (2.3.15) for $m = 0$ and $m = 1$.

(ii) Using (2.1.2) and (2.1.3), or otherwise, show that

$$\binom{n}{m}_q (1-aq^{2m}) - \binom{n-1}{m-1}_q (1-aq^{n+m}) = \binom{n-1}{m}_q q^m (1-aq^m).$$

(iii) Using (ii) or otherwise, prove (2.3.15) by induction on m. **Hint:** (ii) is particularly well suited for going from $m-1$ to m in (2.3.15).

20. (a) Prove the following generalization of the Potter–Schützenberger theorem: if $yx = qxy$ and all other pairs of variables commute, then

$$(x+y+b)\left((x+y)q+b\right)\left((x+y)q^2+b\right)\cdots\left((x+y)q^{n-1}+b\right)$$
$$= \sum_{k=0}^{n} \binom{n}{k}_q q^{\binom{k}{2}} x^k (yq^k+b)(yq^{k+1}+b)\cdots(yq^{n-1}+b),$$

where an empty product (e.g., $(yq^k+b)(yq^{k+1}+b)\cdots(yq^{n-1}+b)$ when $k=n$) equals 1 as usual.

(b) Show that this reduces to the Potter–Schützenberger theorem if $b = 0$.

(c) What happens to it when $y = 0$?

2.4. The q-derivative

Recall that the **derivative** of a function $f(x)$ is the function $f'(x)$ defined by

$$f'(x) = \lim_{h \to 0} \frac{f(x+h) - f(x)}{h}.$$

In an old-fashioned subject called the **calculus of finite differences** one considers this object without the limit—that is, an operator Δ_h is defined by

$$\Delta_h f(x) = \frac{f(x+h) - f(x)}{h}.$$

Δ_h is sometimes called the **forward difference** with stepsize h. Note that

(2.4.1) $$\Delta_{-h} f(x) = \frac{f(x-h) - f(x)}{-h} = \frac{f(x) - f(x-h)}{h}.$$

Δ_{-h} is sometimes called the **backward difference** with stepsize h. Note that the derivative could just as well be defined by

$$f'(x) = \lim_{h \to 0} \Delta_{-h} f(x) = \lim_{h \to 0} \frac{f(x) - f(x-h)}{h}.$$

The calculus of finite differences is still an interesting subject today—for instance, one can use it to prove Rothe's generalized binomial theorem—but now we just want to rewrite (2.4.1) so that it fits naturally into q-analysis. For this we just have to set $x - h = qx$, so that $h = x(1-q)$. Making this change on the right side of (2.4.1) we get the q-**derivative** $\mathbf{D}_q$ of a function $f(x)$:

(2.4.2) $$\mathbf{D}_q f(x) = \frac{f(x) - f(qx)}{x(1-q)},$$

where $q \neq 1$. If $q \to 1$, then $h \to 0$ and the q-derivative becomes the ordinary derivative. A few simple observations are enough to suggest that (2.4.2) is a good definition. It is linear, like the ordinary derivative; that is, if a and b are independent of x, then

$$\mathbf{D}_q \{a f(x) + b g(x)\} = a \mathbf{D}_q f(x) + b \mathbf{D}_q g(x).$$

It has a power rule as good as we could hope for:

(2.4.3) $$\mathbf{D}_q x^n = \frac{x^n - x^n q^n}{x(1-q)} = [n]_q x^{n-1}.$$

Note also that if a is constant, then
$$\mathbf{D}_q (ax)^n = a^n \, \mathbf{D}_q \, x^n = a^n \, [n]_q x^{n-1} = a[n]_q(ax)^{n-1}.$$
This suggests that the q-derivative has a property akin to the simplest case of the chain rule, namely
$$\frac{d}{dx} f(ax) = af'(ax)$$
if a is constant. To express it, we need an alternate notation for the q-derivative, so we define

(2.4.4) $$f^*(x) = \mathbf{D}_q f(x).$$

Then
$$\mathbf{D}_q f(ax) = \frac{f(ax) - f(aqx)}{x(1-q)} = a \frac{f(ax) - f(aqx)}{(ax)(1-q)},$$
which implies that

(2.4.5) $$\mathbf{D}_q f(ax) = a \, f^*(ax).$$

This is used in a few of the problems for this section, with a equal to a power of q. What keeps the q-derivative from being really important is that it does not have a better chain rule than this.

The q-derivative of a constant C is
$$\mathbf{D}_q C = \frac{C - C}{x(1-q)} = 0.$$
Moreover, if $f^*(x) = 0$, then $f(x) = f(xq)$ for an arbitrary x and q, which can only be true if $f(x)$ is constant. More explicitly, assuming $f(x)$ is nice near $x = 0$ and $|q| < 1$, $f(x) = f(xq)$ implies
$$f(x) = f(xq) = f(xq^2) = f(xq^3) = \cdots = f(0),$$
so whatever value $f(x)$ has at $x = 0$, it has the same value for any other x. (The sudden appearance of the assumption $|q| < 1$ may seem off-putting, but we will see it many more times, starting in Chapter 3. We will nearly always assume $|q| < 1$ whenever some infinite process is involved. Here we iterated the equation $f(x) = f(xq)$ infinitely many times.)

Assuming n is a nonnegative integer, iterating (2.4.3) gives

(2.4.6) $$\mathbf{D}_q^k x^n = [n]_q [n-1]_q \cdots [n-k+1]_q x^{n-k} = \frac{n!_q}{(n-k)!_q} x^{n-k} \quad \text{if } k \leq n,$$

and we can get a formal q-analogue of Taylor's theorem from this. If $k < n$, then the right side of (2.4.6) is zero if $x = 0$. If $k > n$, then the right side of (2.4.6) is zero for any x since eventually we will have taken the q-derivative of a constant. So the only value of k for which the right side of (2.4.6) is not zero at $x = 0$ is $k = n$, in which case (2.4.6) reduces to $\mathbf{D}_q^n x^n = n!_q$. In summary,

(2.4.7) $$\mathbf{D}_q^k x^n \Big|_{x=0} = k!_q \, \delta_{nk}.$$

(Here δ_{nk} is the so-called **Kronecker delta**, which equals 1 if $n = k$ and is zero otherwise.) Now suppose we want to expand a function $f(x)$ in powers of x, say

(2.4.8) $$f(x) = \sum_{n=0}^{\infty} c_n x^n.$$

How do we find the coefficient c_k? In calculus, we would find c_k (in principle) by taking k derivatives of $f(x)$ and then setting $x = 0$. In this context, what could be more natural than to take k q-derivatives of (2.4.8)? This kills off all the terms with $n < k$, and by (2.4.6) it results in

$$\mathbf{D}_q^k f(x) = \sum_{n=k}^{\infty} c_n \frac{n!_q}{(n-k)!_q} x^{n-k}.$$

Setting $x = 0$ here and recalling (2.4.7), only the $n = k$ term survives and we have

$$\mathbf{D}_q^k f(x)\Big|_{x=0} = k!_q\, c_k.$$

It follows that

$$c_k = \frac{\mathbf{D}_q^k f(x)\big|_{x=0}}{k!_q}$$

for each k, and so our formal q-Taylor theorem is

(2.4.9) $$f(x) = \sum_{n=0}^{\infty} \left(\mathbf{D}_q^n f(x)\big|_{x=0}\right) \frac{x^n}{n!_q}.$$

We use the word "formal" because we have not discussed convergence, which for a general $f(x)$ would be a delicate question. We can gain some confidence in (2.4.9) by observing that it reduces to the ordinary Taylor's theorem when $q \to 1$.

As long as there are no convergence issues, (2.4.9) is incontestably true. We can have complete faith in the following special case: if $P_n(x)$ is a polynomial of degree n in x, then

(2.4.10) $$P_n(x) = \sum_{k=0}^{n} \left(\mathbf{D}_q^k P_n(x)\big|_{x=0}\right) \frac{x^k}{k!_q}.$$

In the exercises we will use (2.4.10) to give two more proofs of Rothe's q-binomial theorem.

Exercises

(**Reminder:** $f^*(x)$ denotes the q-derivative of $f(x)$.)

1. Show that the q-derivative has two simple product rules, namely

(2.4.11) $$\mathbf{D}_q f(x)g(x) = f(x)\, g^*(x) + f^*(x)\, g(qx)$$
(2.4.12) $$= f^*(x)\, g(x) + f(qx)\, g^*(x).$$

(Note that you can just prove one of these, and then use the symmetry in f and g to get the other.)

2. Explain why we must also have the following forms of the q-product rule:

$$\mathbf{D}_q f(x)\, g(x) = \frac{1}{2} \{f(x)\, g^*(x) + f^*(x)\, g(x) + f(qx)\, g^*(x) + g(qx)\, f^*(x)\}$$
$$= t\, [f(x)\, g^*(x) + g(qx)\, f^*(x)] + (1-t)\, [g(x)\, f^*(x) + f(qx)\, g^*(x)],$$

where t is any real number in the latter. (In the former $t = \tfrac{1}{2}$.)

3. Prove the q-reciprocal rule

(2.4.13) $$\mathbf{D}_q \frac{1}{g(x)} = \frac{-g^*(x)}{g(x)g(qx)}.$$

4. Either directly or by using problems 1 and 3, prove the q-quotient rules

(2.4.14) $$\mathbf{D}_q \frac{f(x)}{g(x)} = \frac{g(x)f^*(x) - f(x)g^*(x)}{g(x)g(xq)}$$

and

(2.4.15) $$\mathbf{D}_q \frac{f(x)}{g(x)} = \frac{g(xq)f^*(x) - f(xq)g^*(x)}{g(x)g(xq)}.$$

5. Show that
$$\mathbf{D}_q \phi(x)\phi(xq)\phi(xq^2)\cdots\phi(xq^{n-1}) = [n]_q \phi(xq)\phi(xq^2)\cdots\phi(xq^{n-1})\mathbf{D}_{q^n}\phi(x),$$
where $\mathbf{D}_{q^n}\phi(x)$ denotes the q-derivative of $\phi(x)$ with q^n in place of q. What does this reduce to if $q \to 1$?

6. Let $P_n(x) = (a+x)(a+xq)\cdots(a+xq^{n-1})$, where $P_0(x) = 1$.
    (i) Show that $\mathbf{D}_q P_n(x) = [n]_q P_{n-1}(xq)$.
    (ii) Either directly or with the aid of (2.4.5), show that
    $$\mathbf{D}_q^k P_n(x) = q^{\binom{k}{2}} \frac{n!_q}{(n-k)!_q} P_{n-k}(xq^k)$$
    for $0 \le k \le n$.
    (iii) Use the result of (ii) and (2.4.10) to prove Rothe's q-binomial theorem
    $$(a+x)(a+xq)\cdots(a+xq^{n-1}) = \sum_{k=0}^{n} \binom{n}{k}_q q^{\binom{k}{2}} a^{n-k} x^k.$$

7. For a slight variation on problem 6, let $P_n(x) = (x+a)(x+aq)\cdots(x+aq^{n-1})$, where $P_0(x) = 1$.
    (i) Show that $\mathbf{D}_q P_n(x) = [n]_q P_{n-1}(x)$, and explain why this implies that
    $$\mathbf{D}_q^k P_n(x) = \frac{n!_q}{(n-k)!_q} P_{n-k}(x)$$
    for $0 \le k \le n$.
    (ii) Use (i) and (2.4.10) to prove Rothe's q-binomial theorem in the form
    $$(x+a)(x+aq)\cdots(x+aq^{n-1}) = \sum_{k=0}^{n} \binom{n}{k}_q q^{\binom{n-k}{2}} a^{n-k} x^k.$$

8. One can write down a formula for the nth q-derivative of a function. Show that
$$\mathbf{D}_q^n f(x) = \frac{1}{x^n q^{\binom{n}{2}}(1-q)^n} \sum_{k=0}^{n} \binom{n}{k}_q q^{\binom{n-k}{2}} (-1)^k f(xq^k).$$

Note that this does not reduce to anything interesting when $q \to 1$.

9. (a) Prove the three-function q-product rule
$$\mathbf{D}_q f(x)g(x)h(x) = f(x)g(x)h^*(x) + f(x)g^*(x)h(qx) + f^*(x)g(xq)h(xq).$$
(b) Prove the k-function q-product rule
$$\mathbf{D}_q f_1(x)\cdots f_{k-1}(x)f_k(x) = f_1(x)\cdots f_{k-1}(x)f_k^*(x) + f_1(x)\cdots f_{k-1}^*(x)f_k(qx)$$
$$+ \cdots + f_1(x)f_2^*(x)\cdots f_k(qx) + f_1^*(x)f_2(qx)\cdots f_k(qx).$$

10. Prove the q-Leibniz rule
$$\mathbf{D}_q^n f(x)g(x) = \sum_{k=0}^{n} \binom{n}{k}_q f^{(k)}(x)g^{(n-k)}(q^k x),$$
where $f^{(k)}(x)$ denotes the kth q-derivative of $f(x)$.

11. The q-Leibniz rule in problem 10 has one drawback: it appears not to be symmetric in f and g, even though it must be. What formula does problem 8 give for $\mathbf{D}_q^n f(x)g(x)$?

12. A common generalization of problems 9 and 10 is
$$\mathbf{D}_q^n f_1(x)\cdots f_{k-1}(x)f_k(x)$$
$$= \sum \binom{n}{b_1,\ldots,b_k}_q f_1^{(b_1)}(x)f_2^{(b_2)}(q^{b_1}x)\cdots f_k^{(b_k)}(q^{b_1+\cdots+b_{k-1}}x),$$
where $f^{(k)}(x)$ again denotes the kth q-derivative of $f(x)$, and the sum is over all ordered k-tuples $(b_1, b_2, \ldots, b_k)$ of nonnegative integers that add up to n.

2.5. Two q-binomial theorems of Gauss

The first theorems that look like a q-version of the binomial theorem occur already in the paper in which Gauss introduced the q-binomial coefficients. He was looking for something like (2.3.4), and he did not realize until later that the factor $q^{\binom{j}{2}}$ has to be there to get such a nice result. Instead he considered $\sum_{j=0}^{m} \binom{m}{j}_q (-1)^j$, which is obviously zero if m is odd, say $m = 2n + 1$, because
$$(-1)^j \binom{2n+1}{j}_q = -(-1)^{2n+1-j} \binom{2n+1}{2n+1-j}_q$$
so the terms cancel in pairs. In the even case, let's define
$$g_n(q) = \sum_{j=0}^{2n} \binom{2n}{j}_q (-1)^j.$$
Then $g_0(q) = 1$ and $g_1(q) = 1 - [2]_q + 1 = 2 - [2]_q = 1 - q$, and we also have
$$g_2(q) = \sum_{j=0}^{4} \binom{4}{j}_q (-1)^j = 1 - [4]_q + \binom{4}{2}_q - [4]_q + 1$$
$$= 2(1 - [4]_q) + 1 + q + 2q^2 + q^3 + q^4$$
$$= 1 + q + 2q^2 + q^3 + q^4 - 2(q + q^2 + q^3)$$
$$= 1 - q - q^3 + q^4 = (1-q)(1-q^3).$$

Although these sums reduce to zero for a positive n when $q = 1$, they are not zero in general, and in fact it appears as though $g_n(q) = (1-q)(1-q^3)\cdots(1-q^{2n-1})$. If you try to prove this by induction, you run into another interesting fact. Consider the sum

$$G_n(q) = \sum_{j=0}^{2n} \binom{2n}{j}_q (-q)^j,$$

which should also reduce to zero for a positive n and $q = 1$. Again we have $G_0(q) = 1$, and $G_1(q) = 1 - q[2]_q + q^2 = 1 + q^2 - q(1+q) = 1 - q$. Coincidence? Let's try

$$G_2(q) = \sum_{j=0}^{4} \binom{4}{j}_q (-q)^j = 1 - q[4]_q + \binom{4}{2}_q q^2 - q^3[4]_q + q^4$$
$$= 1 + q^4 - [4]_q q(1+q^2) + q^2\left(1 + q + 2q^2 + q^3 + q^4\right)$$
$$= 1 + q^4 - q(1+q^2)([4]_q - q[3]_q)$$
$$= 1 - q - q^3 + q^4 = (1-q)(1-q^3).$$

This suggests that $g_n(q) = G_n(q)$, which is not too hard to prove. Note that

$$\sum_{j=0}^{2n} \binom{2n}{j}_q (-1)^j \left(1 - q^j\right) = \sum_{j=0}^{2n} \frac{(2n)!_q}{j!_q\,(2n-j)!_q} (-1)^j [j]_q (1-q).$$

When $j = 0$ the corresponding term is 0, because of the $1 - q^j$ on the left or the $[j]_q$ on the right. Therefore the sum on the right can start at $j = 1$, and

$$\sum_{j=0}^{2n} \binom{2n}{j}_q (-1)^j \left(1 - q^j\right) = \sum_{j=1}^{2n} \frac{(2n)!_q}{j!_q\,(2n-j)!_q} (-1)^j [j]_q (1-q)$$
$$= \sum_{j=1}^{2n} \frac{(2n-1)!_q}{(j-1)!_q\,(2n-j)!_q} (-1)^j [2n]_q (1-q)$$
$$= (1 - q^{2n}) \sum_{j=1}^{2n} \binom{2n-1}{j-1}_q (-1)^j$$
$$= 0 \quad \text{by symmetry of the } q\text{-binomial coefficients.}$$

Hence

$$\sum_{j=0}^{2n} \binom{2n}{j}_q (-1)^j = \sum_{j=0}^{2n} \binom{2n}{j}_q (-1)^j q^j,$$

so $g_n(q) = G_n(q)$, or in other words

$$g_n(q) = \sum_{j=0}^{2n} \binom{2n}{j}_q (-1)^j = \sum_{j=0}^{2n} \binom{2n}{j}_q (-q)^j.$$

2.5. TWO q-BINOMIAL THEOREMS OF GAUSS

We now apply the recurrences (2.1.2) and (2.1.3) to $g_{n+1}(q)$. We have

$$g_{n+1}(q) = \sum_{j=0}^{2n+2} \binom{2n+2}{j}_q (-1)^j$$

$$= \sum_{j=0}^{2n+2} \left\{ \binom{2n+1}{j-1}_q + q^j \binom{2n+1}{j}_q \right\} (-1)^j$$

$$= \sum_{j=0}^{2n+1} \binom{2n+1}{j}_q (-q)^j,$$

where we used (2.1.2) first, threw away one of the sums because it equals zero by symmetry, and also discarded the $j = 2n+2$ term in the last sum because it equals zero. Next, apply (2.1.3) to this:

$$g_{n+1}(q) = \sum_{j=0}^{2n+1} \binom{2n+1}{j}_q (-q)^j$$

$$= \sum_{j=0}^{2n+1} \left\{ \binom{2n}{j}_q + q^{2n-j+1} \binom{2n}{j-1}_q \right\} (-q)^j$$

$$= \sum_{j=0}^{2n} \binom{2n}{j}_q (-q)^j + \sum_{j=1}^{2n+1} \binom{2n}{j-1}_q (-1)^j q^{2n+1}.$$

Again in each of the last two sums we discarded a zero term. Note that the first of these sums is $g_n(q)$. In the second, replace j by $i+1$. Since $1 \le j \le 2n+1$ we have $1 \le i+1 \le 2n+1$, or in other words $0 \le i \le 2n$. Then

$$g_{n+1}(q) = g_n(q) + q^{2n+1} \sum_{i=0}^{2n} \binom{2n}{i}_q (-1)^{i+1}$$

$$= g_n(q) - q^{2n+1} \sum_{i=0}^{2n} \binom{2n}{i}_q (-1)^i$$

$$= g_n(q) - q^{2n+1} g_n(q) = \left(1 - q^{2n+1}\right) g_n(q).$$

Replacing n by $n-1$ this becomes $g_n(q) = \left(1 - q^{2n-1}\right) g_{n-1}(q)$. Now we just iterate this:

$$g_{n-1}(q) = \left(1 - q^{2(n-1)-1}\right) g_{n-2}(q)$$
$$= \left(1 - q^{2n-3}\right) g_{n-2}(q),$$

so

$$g_n(q) = \left(1 - q^{2n-1}\right) \left(1 - q^{2n-3}\right) g_{n-2}(q),$$

and so forth; eventually we reach

$$g_n(q) = \left(1 - q^{2n-1}\right) \left(1 - q^{2n-3}\right) \cdots (1-q) g_0(q).$$

But $g_0(q) = 1$, so

$$g_n(q) = \left(1 - q^{2n-1}\right) \left(1 - q^{2n-3}\right) \cdots (1-q).$$

We can write this result compactly by extending some of our previous notation. Recall that for a nonnegative integer n,

$$(2.5.1) \qquad (a;q)_n := \begin{cases} (1-a)(1-aq)(1-aq^2)\cdots(1-aq^{n-1}) & \text{if } n \geq 1, \\ 1 & \text{if } n = 0. \end{cases}$$

$(a;q)_n$ is called a q-**shifted factorial**. Note that

$$(a;q^2)_n = (1-a)(1-aq^2)(1-aq^4)\cdots(1-aq^{2n-2}) \quad \text{if } n \geq 1,$$

so in particular

$$(q;q^2)_n = (1-q)(1-q^3)(1-q^5)\cdots(1-q^{2n-1}) \quad \text{if } n \geq 1,$$

and therefore

$$(2.5.2) \qquad (q;q^2)_n = \sum_{j=0}^{2n} \binom{2n}{j}_q (-1)^j$$

$$(2.5.3) \qquad = \sum_{j=0}^{2n} \binom{2n}{j}_q (-q)^j$$

for any nonnegative integer n, and also

$$(2.5.4) \qquad \sum_{j=0}^{2n-1} \binom{2n+1}{j}_q (-1)^j = 0,$$

$$(2.5.5) \qquad \sum_{j=0}^{2n-1} \binom{2n+1}{j}_q (-q)^j = (q;q^2)_n$$

for any positive integer n. The parts of this that Gauss did are (2.5.2) and (2.5.4).

Gauss's second theorem is a beautiful (if somewhat peculiar) generalization of the binomial coefficient sum $\sum_{k=0}^{n} \binom{n}{k} = 2^n$. We state it in a slightly different form than Gauss did, but our proof will be essentially the same as his. We'll call the sum we wish to evaluate

$$G_n(q) = \sum_{k=0}^{n} \binom{n}{k}_{q^2} q^k,$$

where $\binom{n}{k}_{q^2}$ means the q-binomial coefficient with q replaced by q^2; in other words,

$$\binom{n}{k}_{q^2} = \frac{(q^2;q^2)_n}{(q^2;q^2)_k \, (q^2;q^2)_{n-k}} = \frac{(1-q^2)(1-q^4)\cdots(1-q^{2n})}{(1-q^2)\cdots(1-q^{2k})(1-q^2)\cdots(1-q^{2n-2k})}.$$

Replacing q by q^2 and n by $n-1$, the recurrence (2.1.3) becomes

$$(2.5.6) \qquad \binom{n}{k}_{q^2} = \binom{n-1}{k}_{q^2} + q^{2n-2k}\binom{n-1}{k-1}_{q^2}.$$

2.5. TWO q-BINOMIAL THEOREMS OF GAUSS

Applying this to $G_n(q)$ we have

$$G_n(q) = \sum_{k=0}^{n} \left\{ \binom{n-1}{k}_{q^2} + q^{2n-2k} \binom{n-1}{k-1}_{q^2} \right\} q^k$$

$$= \sum_{k=0}^{n-1} \binom{n-1}{k}_{q^2} q^k + \sum_{k=1}^{n} \binom{n-1}{k-1}_{q^2} q^{2n-k},$$

where as usual we discarded a term that equals zero from each of the last two sums. The first of these is just $G_{n-1}(q)$, and in the second we replace k by $n - j$. Since $1 \leq k \leq n$ we have $1 \leq n - j \leq n$, which translates into $0 \leq j \leq n - 1$; note also that

$$\binom{n-1}{k-1}_{q^2} = \binom{n-1}{n-j-1}_{q^2} = \binom{n-1}{j}_{q^2}.$$

Then we have

$$G_n(q) = G_{n-1}(q) + \sum_{j=0}^{n-1} \binom{n-1}{j}_{q^2} q^{2n-(n-j)}$$

$$= G_{n-1}(q) + q^n \sum_{j=0}^{n-1} \binom{n-1}{j}_{q^2} q^j$$

(2.5.7) $$= G_{n-1}(q) + q^n G_{n-1}(q) = (1 + q^n) G_{n-1}(q).$$

Once again we can solve this for $G_n(q)$ by iteration. Replacing n by $n - 1$ it says that

$$G_{n-1}(q) = \left(1 + q^{n-1}\right) G_{n-2}(q),$$

and substituting this into (2.5.7) gives

$$G_n(q) = (1 + q^n)\left(1 + q^{n-1}\right) G_{n-2}(q).$$

Continuing in this way we eventually reach

$$G_n(q) = (1 + q^n)\left(1 + q^{n-1}\right)\left(1 + q^{n-2}\right) \cdots (1 + q) G_0(q),$$

and $G_0(q) = 1$. This proves Gauss's identity

(2.5.8) $$\sum_{k=0}^{n} \binom{n}{k}_{q^2} q^k = (1 + q^n)\left(1 + q^{n-1}\right)\left(1 + q^{n-2}\right) \cdots (1 + q) = (-q; q)_n.$$

Note that from problem 2 in the previous section we also have

(2.5.9) $$\sum_{k=0}^{n} \binom{n}{k}_q q^{\binom{k+1}{2}} = (-q; q)_n,$$

a very different expansion of the same product.

Exercises

1. Sometimes it is convenient to have Rothe's q-binomial theorem in the form

(2.5.10) $$(-zq;q^2)_n = \sum_{k=0}^{n} \binom{n}{k}_{q^2} q^{k^2} z^k.$$

   Show that this is equivalent to (2.3.7).

2. Gauss's proof of (2.5.2) and (2.5.4), while similar to ours in spirit, is slightly different in detail, and in fact slightly easier since he was not trying to prove (2.5.3) and (2.5.5) at the same time. Here is an outline of it:

$$h_m(q) = \sum_{j=0}^{m} \binom{m}{j}_q (-1)^j$$

$$= \sum_{j=0}^{m-1} \binom{m-1}{j}_q (-1)^j + \sum_{j=1}^{m} \binom{m-1}{j-1}_q (-1)^j q^{m-j}$$

$$= \sum_{k=0}^{m-1} \binom{m-1}{k}_q (-1)^k + \sum_{k=0}^{m-1} \binom{m-1}{k}_q (-1)^{k+1} q^{m-k-1}$$

$$= \sum_{k=0}^{m-2} \binom{m-1}{k}_q (-1)^k \left(1 - q^{m-k-1}\right).$$

   (Why is the upper limit on the last sum $m-2$ instead of $m-1$?) Also,

$$\binom{m-1}{k}_q \left(1 - q^{m-k-1}\right) = \binom{m-2}{k}_q \left(1 - q^{m-1}\right)$$

   and therefore $h_m(q) = \left(1 - q^{m-1}\right) h_{m-2}(q)$. Since $h_0(q) = 1$ and $h_1(q) = 0$, (2.5.2) and (2.5.4) follow. Fill in the details of this argument.

3. Define q-Fibonacci numbers $F_n(q)$ by $F_0(q) = 1 = F_1(q)$ and

$$F_{n+1}(q) = F_n(q) + q^n F_{n-1}(q) \quad \text{if } n \geq 1.$$

   (i) Use (2.1.3) to show that $F_n(q) = \sum_k \binom{n-k}{k}_q q^{k^2}$. The sum here is over $0 \leq k \leq \lfloor \frac{n}{2} \rfloor$, with the same notation as in problem 12 from section 1.3.

   (ii) Use (2.1.3) to show that

$$F_{n+k}(q) = \sum_{j=0}^{k} \binom{k}{j}_q q^{nj} F_{n-j}(q) \quad \text{if } n \geq k \geq 0.$$

4. For integers k and n with $0 \leq k \leq n$, define

$$S_{n,k}(q) = \sum_{j=0}^{k} \binom{n}{j}_{q^2} q^{\binom{j}{2}} (-1)^{n-j} (-q;q)_j (1 - q^{2n-j}).$$

   By induction on k (or otherwise), show that

$$S_{n,k}(q) = (-1)^{n-k} q^{\binom{k+1}{2}} \frac{(q^{2n-2k};q^2)_{k+1}}{(q;q)_k}.$$

Explain why this means that

$$S_{n,n}(q) = \sum_{j=0}^{n} \binom{n}{j}_{q^2} q^{\binom{j}{2}}(-1)^{n-j}(-q;q)_j(1-q^{2n-j}) = 0.$$

5. For a nonnegative integer n, define

$$R_n(q) = \sum_{k=0}^{n} \binom{n}{k}_{q^2} q^{\binom{k}{2}}(-1)^{n-k}(-q;q)_k.$$

Use (2.5.6) to show that $R_n(q) = q^{2n-1} R_{n-1}(q) - S_{n-1,n-1}(q)$ for $n \geq 1$. Use this and problem 4 to show that $R_n(q) = q^{n^2}$.

6. Problem 5 was to show that

$$\sum_{k=0}^{n} \binom{n}{k}_{q^2} q^{\binom{k}{2}}(-1)^{n-k}(-q;q)_k = q^{n^2}.$$

What happens to this identity when $q \to 1$?

7. Prove that

(2.5.11) $$\sum_j \binom{n}{j}_q \binom{n}{2k-j}_q (-1)^{k-j} q^{(k-j)^2} = \binom{n}{k}_{q^2}.$$

Hint: Use $(x^2; q^2)_n = (x;q)_n(-x;q)_n$ and (2.3.7). The sum goes over all j for which the q-binomial coefficients are not zero. (Which ones are these?) Perhaps a nicer form of (2.5.11) is

(2.5.12) $$\sum_j \binom{n}{k+j}_q \binom{n}{k-j}_q (-1)^j q^{j^2} = \binom{n}{k}_{q^2},$$

where the sum is again over all j for which the q-binomial coefficients are nonzero. Note that unlike (2.5.11), this may include some negative values of j.

8. Check either (2.5.11) or (2.5.12) in the case $n = 4$, $k = 2$.

9. For a nonnegative integer n, define

$$A_n(q) = \sum_k q^{2k^2+k} \binom{2n}{n+k}_{q^2},$$

$$B_n(q) = \sum_k q^{2k^2+k} \binom{2n+1}{n+k+1}_{q^2},$$

where the sums go over the full natural range of the q-binomial coefficients; the first from $k = -n$ to $k = n$ and the second from $k = -n-1$ to $k = n$.

(i) Explain why we also have

$$A_n(q) = \sum_k q^{2k^2-k} \binom{2n}{n+k}_{q^2},$$

$$B_n(q) = \sum_k q^{2k^2-k} \binom{2n+1}{n+k}_{q^2}.$$

(ii) Show that $A_0(q) = 1$ and $B_0(q) = 1 + q$.
(iii) Show that $B_n(q) = (1 + q^{2n+1}) A_n(q)$ for $n \geq 0$.
(iv) Show that $A_n(q) = (1 + q^{2n}) B_{n-1}(q)$ for $n \geq 1$.

(v) Conclude from (ii)–(iv) that $A_n(q) = (-q;q)_{2n}$ and $B_n(q) = (-q;q)_{2n+1}$ for $n \geq 0$.

10. For a nonnegative integer n, define
$$E_n(q) = \sum_k (-1)^k q^{\frac{1}{2}(3k^2+k)} \binom{2n}{n+k}_{q^3},$$
$$F_n(q) = \sum_k (-1)^k q^{\frac{1}{2}(3k^2+k)} \binom{2n+1}{n+k+1}_{q^3},$$
$$G_n(q) = \sum_k (-1)^k q^{\frac{1}{2}(3k^2+k)} \binom{2n+1}{n+k}_{q^3},$$
where the sums again go over the full natural range of the q-binomial coefficients.

(i) Show that $E_0(q) = 1$, $F_0(q) = 1 - q$, and $G_0(q) = 1 - q^2$.

(ii) Show that $F_n(q) = (1 - q^{3n+1}) E_n(q)$ and $G_n(q) = (1 - q^{3n+2}) E_n(q)$ for $n \geq 0$.

(iii) Show that $E_n(q) = (1 - q^{3n-1}) F_{n-1}(q) = (1 - q^{3n-2}) G_{n-1}(q)$ for $n \geq 1$.

(iv) Show that $E_n(q) = (q;q^3)_n (q^2;q^3)_n$ for $n \geq 0$.

(v) Show that $F_n(q) = (q;q^3)_{n+1} (q^2;q^3)_n$ and $G_n(q) = (q;q^3)_n (q^2;q^3)_{n+1}$ for $n \geq 0$.

11. For a nonnegative integer n, define
$$S_n(q) = \sum_k (-1)^k q^{k^2} \binom{2n+1}{n+k}_q,$$
$$T_n(q) = \sum_k (-1)^k q^{k^2+k} \binom{2n+1}{n+k+1}_q,$$
$$U_n(q) = \sum_k (-1)^k q^{k^2} \binom{2n}{n+k}_q,$$
$$V_n(q) = \sum_k (-1)^k q^{k^2+k} \binom{2n}{n+k}_q,$$
where the sums again go over the full natural range of the q-binomial coefficients.

(i) Show that $S_0(q) = 1 - q$, $T_0(q) = 0$, and $U_0(q) = 1 = V_0(q)$.

(ii) By changing k to $-j-1$ or otherwise, show that $T_n(q) = 0$ for all nonnegative integers n.

(iii) Show that $U_{n+1}(q) = S_n(q) + q^{n+1} T_n(q) = S_n(q)$ for $n \geq 0$.

(iv) Show that $T_n(q) = V_n(q) - q^n U_n(q)$ for $n \geq 0$, and hence $V_n(q) = q^n U_n(q)$ for $n \geq 0$.

(v) Show that $S_n(q) = U_n(q) - q^{n+1} V_n(q)$ for $n \geq 0$.

(vi) Conclude from (i)–(v) that $S_n(q) = (q;q^2)_{n+1}$, $U_n(q) = (q;q^2)_n$, and $V_n(q) = q^n(q;q^2)_n$ for $n \geq 0$.

12. For a nonnegative integer n, define

$$W_n(q) = \sum_k (-1)^k q^{\frac{k(3k-1)}{2}} \binom{2n+1}{n+k}_q,$$

$$X_n(q) = \sum_k (-1)^k q^{\frac{3k(k-1)}{2}} \binom{2n+1}{n+k}_q,$$

$$Y_n(q) = \sum_k (-1)^k q^{\frac{k(3k-1)}{2}} \binom{2n}{n+k}_q,$$

$$Z_n(q) = \sum_k (-1)^k q^{\frac{3k(k-1)}{2}} \binom{2n}{n+k}_q,$$

where the sums again go over the full natural range of the q-binomial coefficients.

(i) Show that $W_0(q) = 1 - q$, $X_0(q) = 0$, and $Y_0(q) = 1 = Z_0(q)$.

(ii) By changing k to $-j - 1$ or otherwise, show that $X_n(q) = 0$ for all nonnegative integers n.

(iii) Show that

$$Y_n(q) = \sum_k (-1)^k q^{\frac{k(3k+1)}{2}} \binom{2n}{n+k}_q,$$

$$Z_n(q) = \sum_k (-1)^k q^{\frac{3k(k+1)}{2}} \binom{2n}{n+k}_q.$$

(iv) Show that $Y_{n+1}(q) = \left(1 + q^{n+1}\right) W_n(q)$ for $n \geq 0$.

(v) Show that $X_n(q) = q^n Y_n(q) - Z_n(q)$ for $n \geq 0$, and hence $Z_n(q) = q^n Y_n(q)$ for $n \geq 0$.

(vi) Show that $W_n(q) = Y_n(q) - q^{n+1} Z_n(q)$ for $n \geq 0$.

(vii) Conclude from (i)–(vi) that

$$W_n(q) = \frac{(q;q)_{2n+1}}{(q;q)_n}, \quad Y_n(q) = \frac{(q;q)_{2n}}{(q;q)_n}, \quad Z_n(q) = q^n \frac{(q;q)_{2n}}{(q;q)_n} \quad \text{for } n \geq 0.$$

This problem is used in section 11.3.

13. This problem outlines a proof of **Rowell's identity**

(2.5.13) $$\sum_{k=0}^n \binom{n}{k}_q (-a;q)_k q^{\binom{k+1}{2}} = \sum_{j=0}^n \binom{n}{j}_{q^2} (-q;q)_{n-j} q^{j^2} a^j.$$

(i) Explain why we can rewrite the left side of (2.5.13) as

$$\sum_{k=0}^n \binom{n}{k}_q q^{\binom{k+1}{2}} \sum_{j=0}^k \binom{k}{j}_q q^{\binom{j}{2}} a^j.$$

(ii) Explain why we can rewrite (i) as

$$\sum_{j=0}^n \binom{n}{j}_q a^j q^{\binom{j}{2}} \sum_{i=0}^{n-j} \binom{n-j}{i}_q q^{\binom{i+j+1}{2}} = \sum_{j=0}^n \binom{n}{j}_q q^{j^2} a^j \sum_{i=0}^{n-j} \binom{n-j}{i}_q q^{\binom{i+1}{2}} q^{ij}.$$

(iii) Explain why we can rewrite the right side of (ii) as
$$\sum_{j=0}^{n} \binom{n}{j}_q q^{j^2} a^j (-q^{j+1}; q)_{n-j} = \sum_{j=0}^{n} \binom{n}{j}_q q^{j^2} a^j \frac{(-q; q)_n}{(-q; q)_j}.$$

(iv) Show that the right side of (iii) can be rewritten as the right side of (2.5.13).

(v) Show that (2.5.13) reduces to (2.5.9) if $a = 0$.

14. This problem and the next outline a proof of the surprisingly tricky identity

(2.5.14) $$\sum_{k=0}^{2n} \frac{(-1)^k (q^{2n-k+1}; q)_k}{(a; q)_k} = \sum_{k=0}^{n} \frac{(-1)^k q^{k^2} (q^{2n-2k+2}; q^2)_k}{(aq; q^2)_k}.$$

Denote the left side by $L_n(a)$ and the right side by $R_n(a)$.

(i) Show that $L_0(a) = 1 = R_0(a)$ and $L_1(a) = R_1(a)$.

(ii) Show that

$$\frac{(1 - q^{2n+1})(1 - q^{2n+2})}{1 - aq} L_n(aq^2) = \sum_{j=1}^{2n+1} \frac{(-1)^{j+1} (q^{2n-j+2}; q)_{j+1}}{(aq; q)_j}.$$

(iii) Show that

$$\sum_{j=1}^{2n+1} \frac{(-1)^{j+1} (q^{2n-j+2}; q)_{j+1}}{(aq; q)_j} = a + (1 - a)L_{n+1}(a) - q^{2n+2}.$$

Hence $L_n(a)$ satisfies the recurrence

$$a + (1 - a)L_{n+1}(a) = q^{2n+2} + \frac{(1 - q^{2n+1})(1 - q^{2n+2})}{1 - aq} L_n(aq^2).$$

15. We now want to show that $R_n(a)$ satisfies the same recurrence, which is harder. Set

$$S_n(a) = \sum_{k=0}^{n} \frac{(-1)^k q^{k^2} (q^{2n-2k+2}; q^2)_{k+1}}{(aq; q^2)_k}.$$

Note that we could extend the sum to $k = n + 1$, because this just adds a zero term (why?).

(i) By writing

$$S_n(a) = \sum_{k=0}^{n} \frac{(-1)^k q^{k^2} (q^{2n-2k+2}; q^2)_{k+1} (1 - aq^{2k+1})}{(aq; q^2)_{k+1}}$$

and splitting the numerator, show that

$$S_n(a) = a\, R_{n+1}(a) - a + \frac{1 - q^{2n+2}}{1 - aq} R_n(aq^2).$$

(ii) By writing

$$S_n(a) = \sum_{k=0}^{n+1} \frac{(-1)^k q^{k^2} (q^{2n-2k+4}; q^2)_k (1 - q^{2n-2k+2})}{(aq; q^2)_k}$$

and splitting the numerator, show that
$$S_n(a) = R_{n+1}(a) - q^{2n+2} + q^{2n+1} \frac{1-q^{2n+2}}{1-aq} R_n(aq^2).$$

(iii) Show that (i) and (ii) imply
$$a + (1-a)R_{n+1}(a) = q^{2n+2} + \frac{(1-q^{2n+1})(1-q^{2n+2})}{1-aq} R_n(aq^2).$$

Explain why this together with the previous problem proves (2.5.14).

16. Show that both sides of (2.5.14) reduce to $(q;q^2)_n$ if $a = q$.

2.6. Jacobi's q-binomial theorem

A generalization of Rothe's q-binomial theorem was found by Jacobi.

THEOREM 14 (Jacobi's q-binomial theorem). *For all a, b, c and all nonnegative integers n,*

$$(b-a)(b-aq)\cdots(b-aq^{n-1})$$
$$= \sum_{k=0}^{n} \binom{n}{k}_q (b-c)(b-cq)\cdots(b-cq^{n-k-1})(c-a)(c-aq)\cdots(c-aq^{k-1}),$$

where $(b-a)(b-aq)\cdots(b-aq^{n-1}) = 1$ if $n = 0$.

Jacobi proved this when $b = 1$, but the general case follows by rescaling. We used this phrase in section 2.3, and we will use it again later, so we pause for a moment to explain it. Suppose we can prove Jacobi's theorem when $b = 1$, i.e., suppose we can prove that

$$(1-w)(1-wq)\cdots(1-wq^{n-1})$$
$$= \sum_{k=0}^{n} \binom{n}{k}_q (1-v)(1-vq)\cdots(1-vq^{n-k-1})(v-w)(v-wq)\cdots(v-wq^{k-1})$$

for all v and w, which is what Jacobi actually proved. Replace w by $\frac{a}{b}$ and v by $\frac{c}{b}$ to get

$$\left(1-\frac{a}{b}\right)\left(1-\frac{aq}{b}\right)\cdots\left(1-\frac{aq^{n-1}}{b}\right)$$
$$= \sum_{k=0}^{n} \binom{n}{k}_q \left(1-\frac{c}{b}\right)\left(1-\frac{cq}{b}\right)\cdots\left(1-\frac{cq^{n-k-1}}{b}\right) \frac{c-a}{b}\frac{c-aq}{b}\cdots\frac{c-aq^{k-1}}{b}.$$

If we now multiply through by b^n, then we get Jacobi's theorem for a general b.

We will prove Jacobi's theorem by induction on n. (Jacobi's proof was different, and we will see it in the next chapter.) It holds by definition if $n = 0$, when both sides are 1. If $n = 1$ it says $b - a = (b-c) + (c-a)$, which is a true statement. We leave it to the reader to verify it in the case $n = 2$, which already requires a bit of algebra. Assume it is true for n, and consider the sum

$$\sum_{k=0}^{n+1} \binom{n+1}{k}_q (b-c)(b-cq)\cdots(b-cq^{n-k})(c-a)(c-aq)\cdots(c-aq^{k-1}).$$

By (2.1.3) this equals

$$\sum_{k=0}^{n} \binom{n}{k}_q (b-c)(b-cq)\cdots(b-cq^{n-k})(c-a)(c-aq)\cdots(c-aq^{k-1})$$

$$+ \sum_{k=1}^{n+1} \binom{n}{k-1}_q q^{n-k+1}(b-c)(b-cq)\cdots(b-cq^{n-k})(c-a)(c-aq)\cdots(c-aq^{k-1})$$

$$= \sum_{j=0}^{n} \binom{n}{j}_q (b-c)\cdots(b-cq^{n-j-1})(b-cq^{n-j})(c-a)\cdots(c-aq^{j-1})$$

$$+ \sum_{j=0}^{n} \binom{n}{j}_q q^{n-j}(b-c)\cdots(b-cq^{n-j-1})(c-a)\cdots(c-aq^{j-1})(c-aq^j).$$

Combining these two sums we get

$$\sum_{j=0}^{n} \binom{n}{j}_q (b-c)\cdots(b-cq^{n-j-1})(c-a)\cdots(c-aq^{j-1}) \left\{ b - cq^{n-j} + q^{n-j}(c-aq^j) \right\},$$

and the term in braces simplifies to $b - aq^n$, which can be taken outside the sum since it does not depend on the summation index j. Then we have

$$\sum_{k=0}^{n+1} \binom{n+1}{k}_q (b-c)(b-cq)\cdots(b-cq^{n-k})(c-a)(c-aq)\cdots(c-aq^{k-1})$$

$$= (b-aq^n) \sum_{j=0}^{n} \binom{n}{j}_q (b-c)\cdots(b-cq^{n-j-1})(c-a)\cdots(c-aq^{j-1})$$

$$= (b-aq^n) \left\{ (b-a)(b-aq)\cdots(b-aq^{n-1}) \right\},$$

where we used the induction assumption to do the last step. This proves Jacobi's theorem.

Exercises

1. What happens to Jacobi's q-binomial theorem when $q = 1$?
2. Show that Jacobi's q-binomial theorem reduces to Rothe's q-binomial theorem when $c = 0$.
3. Prove Jacobi's q-binomial theorem using (2.1.2) instead of (2.1.3).
4. If $b \neq 0$, show that Jacobi's q-binomial theorem can be written as

$$\left(\frac{a}{b}; q\right)_n = \sum_{k=0}^{n} \binom{n}{k}_q \left(\frac{c}{b}; q\right)_{n-k} \left(\frac{c}{b}\right)^k \left(\frac{a}{c}; q\right)_k.$$

5. Show that the identity of problem 4 is equivalent to

(2.6.1) $$(uv; q)_n = \sum_{k=0}^{n} \binom{n}{k}_q (u; q)_{n-k} u^k (v; q)_k.$$

This is the form of Jacobi's q-binomial theorem that most often occurs in the literature.

6. Prove (2.6.1) directly by induction on n.

7. Is it true that
$$(uv;q)_n = \sum_{k=0}^{n} \binom{n}{k}_q (u;q)_{n-k}\, v^{n-k}\, (v;q)_k?$$
Explain.

8. By induction or as a special case of Jacobi's q-binomial theorem, or by using Theorem 13, prove that
$$x^n = \sum_{k=0}^{n} \binom{n}{k}_q (x-1)(x-q)\cdots(x-q^{k-1}).$$

9. Prove by induction on n that
$$(a+b)^n = \sum_{k=0}^{n} \binom{n}{k}_q a^k b\, (a(1-q)+b)\,(a(1-q^2)+b)\cdots(a(1-q^{n-k-1})+b).$$
What does this reduce to if $q \to 1$?

10. Prove the result of problem 9 by setting $v=0$ and $u = \dfrac{a}{a+b}$ in (2.6.1).

11. The result of problem 9 and the result of problem 20 in section 2.3 have a common generalization: if $yx = qxy$ and $ya = qay$ and all other pairs of variables commute, then
$$(x+y+a+b)\,((x+y)q+a+b)\,((x+y)q^2+a+b)\cdots((x+y)q^{n-1}+a+b)$$
$$= \sum_{k=0}^{n} \binom{n}{k}_q \left\{ \begin{array}{l} (x+a)(xq+a)\cdots(xq^{k-1}+a)(yq^k+b) \\ \times (yq^{k+1}+a(1-q)+b)\cdots(yq^{n-1}+a(1-q^{n-k-1})+b) \end{array} \right\}$$
with the usual conventions about empty products. Prove this by induction on n via the following outline (or otherwise):

(i) Show that $yq^m(xq^k+a) = (xq^k+a)yq^{m+1}$.

(ii) Show that
$$(yq^m+a(1-q^{m-k})+b)(xq^k+a) = (xq^k+a)(yq^{m+1}+a(1-q^{m-k})+b).$$

(iii) Think of
$$(x+y+a+b)\,((x+y)q+a+b)\cdots((x+y)q^{n-1}+a+b)\,((x+y)q^n+a+b)$$
as
$$\{(x+y+a+b)\,((x+y)q+a+b)\cdots((x+y)q^{n-1}+a+b)\}\,((x+y)q^n+a+b)$$
and write
$$(x+y)q^n + a + b = \{yq^n + a(1-q^{n-k}) + b\} + q^{n-k}(xq^k+a).$$
Split
$$\{(x+y+a+b)\,((x+y)q+a+b)\cdots((x+y)q^{n-1}+a+b)\}\,((x+y)q^n+a+b)$$
into two sums, and use (ii) to move $q^{n-k}(xq^k+a)$ into position in one of them. Use (2.1.3) to combine the sums together.

2.7. MacMahon's q-binomial theorem

If we take $a = 1$ in Rothe's q-binomial theorem, it becomes

$$(2.7.1) \qquad (-x;q)_n = \sum_{k=0}^n \binom{n}{k}_q q^{\binom{k}{2}} x^k.$$

This is equivalent not only to problem 1 in section 2.3, but also to Rothe's q-binomial theorem; see problem 6.

There is an interesting q-binomial theorem due to MacMahon, which looks like two instances of (2.7.1) multiplied together:

THEOREM 15 (MacMahon's q-binomial theorem).

$$(-qx;q^2)_a \left(-\frac{q}{x};q^2\right)_b = \sum_{k=-b}^{a} \binom{a+b}{a-k}_{q^2} q^{k^2} x^k.$$

When $a = b$ this was known to Cauchy and Gauss. It too is not really more general than (2.7.1), as problem 7 shows. We single it out for attention here because it is a finite form of the celebrated Jacobi triple product identity, which we will discuss in Chapter 5.

We will give two proofs of MacMahon's theorem. There are three more in problems 7–9, and you might like one of them better. Our first proof is direct but a bit complicated. Let $f_{a,b}(x)$ be defined by the left side of MacMahon's theorem, i.e.,

$$f_{a,b}(x) = (-qx;q^2)_a \left(-\frac{q}{x};q^2\right)_b$$
$$= (1+qx)(1+q^3x)\ldots(1+q^{2a-1}x)\left(1+\frac{q}{x}\right)\left(1+\frac{q^3}{x}\right)\ldots\left(1+\frac{q^{2b-1}}{x}\right).$$

We want to try to expand $f_{a,b}(x)$ in powers of x. What powers could we possibly get? The highest power we could get is x^a, which would come from the product of all the $q^{\text{something}}x$ terms in the first bunch of a factors, and all the 1's in the second bunch of b factors. Similarly, the lowest power of x we could get is x^{-b}. Therefore,

$$f_{a,b}(x) = \sum_{k=-b}^{a} c_k(a,b)\, x^k$$

for some coefficients $c_k(a,b)$, which we are trying to find. We will use a standard trick in this subject to do it. Notice that the powers of q advance in steps of 2. Therefore, $f_{a,b}(xq^2)$ should have most of the same factors as $f_{a,b}(x)$, so we look at $f_{a,b}(xq^2)$ and compare it to $f_{a,b}(x)$. Now

$$f_{a,b}(xq^2) = (-q^3x;q^2)_a \left(-\frac{1}{qx};q^2\right)_b$$
$$= (1+q^3x)(1+q^5x)\ldots(1+q^{2a+1}x)\left(1+\frac{1}{qx}\right)\left(1+\frac{q}{x}\right)\ldots\left(1+\frac{q^{2b-3}}{x}\right)$$
$$= \sum_{k=-b}^{a} c_k(a,b)\, x^k\, q^{2k}.$$

We see that $f_{a,b}(xq^2)$ does have a lot of factors in common with $f_{a,b}(x)$. What exactly are the differences? $f_{a,b}(xq^2)$ has two factors that $f_{a,b}(x)$ lacks, namely

$(1+q^{2a+1}x)$ and $\left(1+\frac{1}{qx}\right)$. Also, $f_{a,b}(xq^2)$ lacks two factors that $f_{a,b}(x)$ has, namely $(1+qx)$ and $\left(1+\frac{q^{2b-1}}{x}\right)$. Therefore,

$$\frac{f_{a,b}(xq^2)}{f_{a,b}(x)} = \frac{(1+q^{2a+1}x)\left(1+\frac{1}{qx}\right)}{(1+qx)\left(1+\frac{q^{2b-1}}{x}\right)} = \frac{1+q^{2a+1}x}{qx+q^{2b}},$$

where we multiplied the second fraction by qx on top and bottom. It follows that

$$f_{a,b}(xq^2)(qx+q^{2b}) = f_{a,b}(x)(1+q^{2a+1}x),$$

which is really what we were after. What we do now is substitute the proposed expansion of $f_{a,b}(x)$ in and get an equation for the coefficients $c_k(a,b)$. We have

$$(qx+q^{2b})\sum_{k=-b}^{a} c_k(a,b)\, x^k\, q^{2k} = (1+q^{2a+1}x)\sum_{k=-b}^{a} c_k(a,b)\, x^k.$$

Distributing and rearranging, this says

$$\sum_{k=-b}^{a} c_k(a,b)\, x^{k+1}\left(q^{2k+1}-q^{2a+1}\right) = \sum_{k=-b}^{a} c_k(a,b)\, x^k\left(1-q^{2k+2b}\right).$$

Notice that the term where $k=a$ on the left is zero, and so is the term where $k=-b$ on the right. These things have to be true, since there is no x^{a+1} term on the right, and no x^{-b} term on the left. This also means that we can change k to $k-1$ on the left with impunity, and then compare coefficients of x^k. The result is that

$$c_{k-1}(a,b)\, q^{2k-1}\left(1-q^{2a-2k+2}\right) = c_k(a,b)\left(1-q^{2k+2b}\right),$$

which we rewrite as

(2.7.2) $$c_k(a,b) = c_{k-1}(a,b)\, q^{2k-1}\, \frac{1-q^{2a-2k+2}}{1-q^{2k+2b}}.$$

Here k could be either positive or negative. Assume for now that it is positive. What we do next is similar to some things we have done before. We will iterate (2.7.2) to get down from $c_k(a,b)$ to $c_0(a,b)$. (It is natural to wonder whether this will actually help, but sometimes it's best to calculate first and ask questions later.) If we replace k by $k-1$ in (2.7.2), it becomes

$$c_{k-1}(a,b) = c_{k-2}(a,b)\, q^{2k-3}\, \frac{1-q^{2a-2k+4}}{1-q^{2k+2b-2}}.$$

Putting this back into (2.7.2) gives us

$$c_k(a,b) = c_{k-2}(a,b)\, q^{(2k-1)+(2k-3)}\, \frac{1-q^{2a-2k+2}}{1-q^{2k+2b}}\, \frac{1-q^{2a-2k+4}}{1-q^{2k+2b-2}}.$$

Repeating this trick k times gets us to

$$c_k(a,b) = c_0(a,b)\, q^{(2k-1)+(2k-3)+\cdots+3+1}\, \frac{(1-q^{2a-2k+2})(1-q^{2a-2k+4})\ldots(1-q^{2a})}{(1-q^{2k+2b})(1-q^{2k+2b-2})\ldots(1-q^{2b+2})}.$$

Can we do anything to make this look any better? The exponent of q simplifies to k^2 (see the first several problems). We can improve the appearance of the fraction by multiplying top and bottom by

$$(q^2;q^2)_{a-k} = (1-q^2)(1-q^4)\ldots(1-q^{2a-2k}).$$

Then the numerator is just equal to $(q^2;q^2)_a$. We do a similar thing with the denominator; here the right thing to multiply top and bottom by is $(q^2;q^2)_b$, and the result is that

$$(2.7.3) \qquad c_k(a,b) = c_0(a,b) \, q^{k^2} \, \frac{(q^2;q^2)_a \, (q^2;q^2)_b}{(q^2;q^2)_{a-k} \, (q^2;q^2)_{b+k}},$$

at least under our assumption that k is positive. Note however that (2.7.3) is certainly also true if $k = 0$, since all it says then is that $c_0(a,b) = c_0(a,b)$. Do we know what $c_0(a,b)$ is? No, we don't, at least not without some more work. Then what good is (2.7.3)?

To answer this, we ask ourselves, are there any $c_k(a,b)$ that we *do* know, or can easily find? It is not too hard to find $c_a(a,b)$. We were remarking on this before we started the calculation, when we asked what the highest power of x that we could possibly get was. The answer was x^a, which comes from the product

$$(qx)(q^3 x)(q^5 x) \ldots (q^{2a-1} x)(1)(1) \ldots (1) = q^{a^2} \, x^a.$$

Therefore $c_a(a,b) = q^{a^2}$. But according to (2.7.3),

$$c_a(a,b) = c_0(a,b) \, q^{a^2} \, \frac{(q^2;q^2)_a \, (q^2;q^2)_b}{(q^2;q^2)_{b+a}}$$

and so

$$c_0(a,b) = \frac{(q^2;q^2)_{b+a}}{(q^2;q^2)_a \, (q^2;q^2)_b} = \binom{a+b}{a}_{q^2}.$$

If we use this in (2.7.3), most conveniently in the form $c_0(a,b) \, (q^2;q^2)_a \, (q^2;q^2)_b = (q^2;q^2)_{a+b}$, it becomes

$$(2.7.4) \qquad c_k(a,b) = q^{k^2} \, \frac{(q^2;q^2)_{a+b}}{(q^2;q^2)_{a-k} \, (q^2;q^2)_{b+k}} = q^{k^2} \binom{a+b}{a-k}_{q^2},$$

at least for $k \geq 0$.

What about for $k < 0$? We don't want to go through another argument like this if we don't have to, and there is a cheap way of getting the $k < 0$ case from what we've done already. Go back to the definition of $f_{a,b}(x)$, and observe that if we change x to $\frac{1}{x}$, this interchanges the roles of a and b. So, if we switch a and b at the same time, we would get our original function back, or in other words

$$f_{b,a}\left(\frac{1}{x}\right) = f_{a,b}(x).$$

If we translate this into a statement about the c_k's, it reads

$$\sum_{k=-b}^{a} c_k(a,b) \, x^k = \sum_{k=-a}^{b} c_k(b,a) \, x^{-k} = \overset{\text{backwards to } -b}{\sum_{k=a}^{-b}} c_{-k}(b,a) \, x^k$$

which says that $c_k(a,b) = c_{-k}(b,a)$. Now suppose that $k \geq 0$. Then from this and (2.7.4),

$$c_{-k}(b,a) = q^{k^2} \binom{a+b}{a-k}_q$$

and if we change $-k$ to k here, since $(-k)^2 = k^2$, we have

$$c_k(b,a) = q^{k^2} \binom{a+b}{a+k}_q$$

if $k \leq 0$, or
$$c_k(a,b) = q^{k^2} \binom{a+b}{b+k}_q$$
if $k \leq 0$. But this is exactly the same expression as (2.7.4), and therefore $c_k(a,b)$ is given by (2.7.4) for all values of k. This proves MacMahon's q-binomial theorem.

Another proof comes from the observation that MacMahon's theorem appears to be two versions of (2.7.1) multiplied together. If we change q to q^2 there and then replace x by xq, we get

(2.7.5) $$(-xq;q^2)_n = \sum_{k=0}^{n} \binom{n}{k}_{q^2} q^{k^2} x^k.$$

This implies

(2.7.6)
$$(-xq;q^2)_a = \sum_{i=0}^{a} \binom{a}{i}_{q^2} q^{i^2} x^i,$$
$$\left(-\frac{q}{x};q^2\right)_b = \sum_{j=0}^{b} \binom{b}{j}_{q^2} q^{j^2} x^{-j}.$$

(We leave the details of (2.7.5) and (2.7.6) to the reader.) Multiplying the two identities in (2.7.6) together gives

(2.7.7) $$(-qx;q^2)_a \left(-\frac{q}{x};q^2\right)_b = \sum_{i=0}^{a}\sum_{j=0}^{b} \binom{a}{i}_{q^2}\binom{b}{j}_{q^2} q^{i^2+j^2} x^{i-j}.$$

Now replace i by $j+k$, so that $i-j=k$, and k will be the new index on the outer sum. The smallest possible value of k is $0-b = -b$, and the largest is $a-0 = a$ (why?). Making these replacements in (2.7.7) we get

(2.7.8) $$(-qx;q^2)_a \left(-\frac{q}{x};q^2\right)_b = \sum_{k=-b}^{a} \left(\sum_{j=0}^{b} \binom{a}{j+k}_{q^2}\binom{b}{j}_{q^2} q^{2j(j+k)}\right) q^{k^2} x^k.$$

The inner sum is (4.7) from Chapter 1 with q^2 in place of q, so
$$\sum_{j=0}^{b} \binom{a}{j+k}_{q^2}\binom{b}{j}_{q^2} q^{2j(j+k)} = \binom{a+b}{a-k}_{q^2},$$
and substituting this in (2.7.8) proves MacMahon's q-binomial theorem.

Exercises

1. One way to see that $1+3+5+\cdots+(2n-1) = n^2$ is to use the reversing trick from problem 1, section 1.1. Explain.
2. Another method is to use $1+2+\cdots+n = \frac{n(n+1)}{2}$. Explain why this implies that
$$1+2+3+\cdots+2n = n(2n+1) \quad \text{and} \quad 2+4+6+\cdots+2n = n(n+1),$$
and how $1+3+5+\cdots+(2n-1) = n^2$ follows from these.

3. Explain how $1+3+5+\cdots+(2n-1) = n^2$ follows from $2+4+6+\cdots+2n = n(n+1)$ alone.
4. Prove (2.7.5) and (2.7.6).
5. Give a direct proof of (2.7.1) by iteration and/or induction.
6. Rothe's q-binomial theorem follows easily from (2.7.1) by rescaling: assuming (2.7.1) is true, replace x by $\frac{x}{a}$ and then multiply through by a^n. Check the details.
7. Here is an outline of a proof of MacMahon's q-binomial theorem by rewriting (2.7.1):

    (i) Show that replacing n by $a+b$ and q by q^2 in (2.7.1) gives
    $$(1+x)(1+xq^2)(1+xq^4)\ldots(1+xq^{2a+2b-2}) = \sum_{j=0}^{a+b} \binom{a+b}{j}_{q^2} q^{j(j-1)} x^j.$$

    (ii) Replace x by $\frac{x}{q^{2b-1}}$ in the identity in (i).

    (iii) Multiply both sides of the identity in (ii) by $q^{b^2} x^{-b}$. On the left side, take $q^{b^2} x^{-b}$ in the form
    $$\frac{q^{2b-1}}{x} \frac{q^{2b-3}}{x} \frac{q^{2b-5}}{x} \cdots \frac{q}{x}$$
    and put each of these factors in an appropriate place.

    (iv) Reindex the sum in (iii) by replacing $j - b$ by k. The result should be MacMahon's q-binomial theorem. Fill in the details.

8. Prove MacMahon's q-binomial theorem by induction on b.
9. Prove MacMahon's q-binomial theorem by induction on a.
10. Using MacMahon's q-binomial theorem or otherwise, show that
    $$\sum_{k=-b}^{a} \binom{a+b}{a-k}_q (-1)^k q^{\frac{k(k+1)}{2}} = \begin{cases} 0 & \text{if } b > 0, \\ (q;q)_a & \text{if } b = 0. \end{cases}$$

11. This problem outlines a derivation of an interesting consequence of MacMahon's q-binomial theorem due to Hirschhorn.

    (i) Take $a = n$ and $b = n+1$ in MacMahon's theorem to get
    $$(-qx; q^2)_n \left(-\frac{q}{x}; q^2\right)_{n+1} = \sum_{k=-n-1}^{-1} \binom{2n+1}{n-k}_{q^2} q^{k^2} x^k + \sum_{k=0}^{n} \binom{2n+1}{n-k}_{q^2} q^{k^2} x^k.$$

    (ii) Reindex the first sum on the right side by letting $k = -j - 1$, and the second sum by letting $k = j$. Show that this gives
    $$(-qx; q^2)_n \left(-\frac{q}{x}; q^2\right)_{n+1} = \sum_{j=0}^{n} \binom{2n+1}{n-j}_{q^2} q^{j^2} x^j \left[1 + \left(\frac{q}{x}\right)^{2j+1}\right].$$

    (iii) Show that dividing both sides of (ii) by $1 + \frac{q}{x}$ and then letting $x \to -q$ gives
    $$(q^2; q^2)_n^2 = \sum_{j=0}^{n} (-1)^j (2j+1) q^{j^2+j} \binom{2n+1}{n-j}_{q^2},$$

or

(2.7.9) $$(q;q)_n^2 = \sum_{j=0}^{n}(-1)^j(2j+1)q^{\binom{j+1}{2}}\binom{2n+1}{n-j}_q$$

after replacing q^2 by q.

12. Show that (2.7.9) can be rewritten as

(2.7.10) $$2(q;q)_n^2 = \sum_{k=-n-1}^{n}(-1)^k(2k+1)q^{\frac{k(k+1)}{2}}\binom{2n+1}{n-k}_q.$$

This form goes over the full natural range of the q-binomial coefficient, so the limits can be left off the sum.

13. Another proof of (2.7.9) proves the equivalent (2.7.10) by induction together with the companion identity

(2.7.11) $$\sum_{k=-n}^{n}(-1)^k(2k+1)q^{\frac{k(k+1)}{2}}\binom{2n}{n-k}_q = \begin{cases} 2(q;q)_n(q;q)_{n-1} & \text{if } n \geq 1, \\ 1 & \text{if } n = 0. \end{cases}$$

(i) Verify (2.7.10) and (2.7.11) for $n = 0$ and $n = 1$.

(ii) Set

$$H_n(q) = \sum_k (-1)^k(2k+1)q^{\frac{k(k+1)}{2}}\binom{2n+1}{n-k}_q,$$

$$J_n(q) = \sum_k (-1)^k(2k+1)q^{\frac{k(k+1)}{2}}\binom{2n}{n-k}_q$$

for $n \geq 0$. Show that $H_n(q) = (1-q^n)J_n(q)$ for $n \geq 1$. Problem 10 should help.

(iii) Show that $J_n(q) = (1-q^n)H_{n-1}(q)$ for $n \geq 1$. Again Problem 10 should help.

(iv) Use (i)–(iii) to prove (2.7.10) and (2.7.11).

2.8. A partial fraction decomposition

In this section we derive a result superficially similar to MacMahon's q-binomial theorem that we will have a use for in Chapter 7. Consider the product

$$p_n(x) = \frac{(ax;q)_n \left(\frac{q}{ax};q\right)_n}{(x;q)_{n+1}\left(\frac{q}{x};q\right)_n}$$

for a nonnegative integer n. Note that if we multiply top and bottom by x^n, then we have a polynomial of degree $2n$ divided by a polynomial of degree $2n+1$. Therefore we should be able to expand $p_n(x)$ in partial fractions, just as we would do in integral calculus. Because $(x;q)_{n+1}$ has roots of multiplicity 1 at $x = q^{-k}$ for $0 \leq k \leq n$ and $\left(\frac{q}{x};q\right)_n$ has zeros of multiplicity 1 at $x = q^k$ for $1 \leq k \leq n$, we have

(2.8.1) $$p_n(x) = \sum_{k=0}^{n}\frac{A_k}{x-q^{-k}} + \sum_{k=1}^{n}\frac{B_k}{x-q^k}$$

for some coefficients A_k and B_k, which presumably depend also on n, a, and q, but not on x. If we multiply (2.8.1) by $x - q^{-k}$ and let $x \to q^{-k}$, then all of the terms tend to zero except one and we have

$$A_k = \lim_{x \to q^{-k}} (x - q^{-k}) p_n(x).$$

It is straightforward to compute most of this limit: for three of the four sets of factors we can just plug in and get

$$\frac{(aq^{-k};q)_n \left(\frac{q^{k+1}}{a};q\right)_n}{(q^{k+1};q)_n}.$$

It is convenient to rewrite the fourth set of factors as

$$(x;q)_{n+1} = (1-x)\cdots(1-xq^{k-1})(1-xq^k)(1-xq^{k+1})\cdots(1-xq^n),$$

and we can plug into all but one of these. Then

$$A_k = \frac{(aq^{-k};q)_n \left(\frac{q^{k+1}}{a};q\right)_n}{(q^{k+1};q)_n} \times \frac{1}{(1-q^{-k})\cdots(1-q^{-1})(1-q)\cdots(1-q^{n-k})}$$

$$\times \lim_{x \to q^{-k}} \frac{x - q^{-k}}{1 - xq^k} \frac{q^k}{q^k}$$

$$= \frac{-1}{q^k} \frac{(aq^{-k};q)_n \left(\frac{q^{k+1}}{a};q\right)_n}{(q;q)_{n-k}(q^{k+1};q)_n} \times \frac{1}{(1-q^{-k})\cdots(1-q^{-1})}.$$

To simplify this we rewrite

$$(aq^{-k};q)_n = (1-aq^{-k})\cdots(1-aq^{-1})(1-a)(1-aq)\cdots(1-aq^{n-k-1})$$
$$= (1-aq^{-k})\cdots(1-aq^{-1})(a;q)_{n-k},$$

so that

$$A_k = \frac{-1}{q^k} \frac{(a;q)_{n-k} \left(\frac{q^{k+1}}{a};q\right)_n}{(q;q)_{n-k}(q^{k+1};q)_n} \times \frac{(1-aq^{-k})\cdots(1-aq^{-1})}{(1-q^{-k})\cdots(1-q^{-1})}.$$

Now

$$\frac{(1-aq^{-k})\cdots(1-aq^{-1})}{(1-q^{-k})\cdots(1-q^{-1})} = \frac{(1-aq^{-k})\cdots(1-aq^{-1})}{(1-q^{-k})\cdots(1-q^{-1})} \frac{q^{1+2+\cdots+k}}{q^{1+2+\cdots+k}}$$

$$= \frac{(q^k - a)\cdots(q-a)}{(q^k - 1)\cdots(q-1)}$$

$$= \frac{(a-q)\cdots(a-q^k)}{(1-q)\cdots(1-q^k)}$$

$$= \frac{a^k \left(\frac{q}{a};q\right)_k}{(q;q)_k},$$

so
(2.8.2)

$$A_k = -\left(\frac{a}{q}\right)^k \frac{(a;q)_{n-k} \left(\frac{q}{a};q\right)_k \left(\frac{q^{k+1}}{a};q\right)_n}{(q;q)_{n-k}(q;q)_k(q^{k+1};q)_n} = -\left(\frac{a}{q}\right)^k \frac{(a;q)_{n-k} \left(\frac{q}{a};q\right)_{n+k}}{(q;q)_{n-k}(q;q)_{n+k}}.$$

2.8. A PARTIAL FRACTION DECOMPOSITION

Note also that

(2.8.3) $$-\left(\frac{a}{q}\right)^k \frac{1}{x-q^{-k}} = \frac{-a^k}{xq^k - 1} = \frac{a^k}{1-xq^k}.$$

Similarly we can compute B_k as
$$B_k = \lim_{x \to q^k} (x - q^k) p_n(x).$$

Again we can plug right in for three of the four sets of factors to get
$$\frac{(aq^k; q)_n \left(\frac{q^{1-k}}{a}; q\right)_n}{(q^k; q)_{n+1}}.$$

For the fourth set it is convenient to rewrite
$$\left(\frac{q}{x}; q\right)_n = \left(1 - \frac{q}{x}\right) \cdots \left(1 - \frac{q^{k-1}}{x}\right)\left(1 - \frac{q^k}{x}\right)\left(1 - \frac{q^{k+1}}{x}\right) \cdots \left(1 - \frac{q^n}{x}\right),$$

and we can plug into all but one of these factors. Hence
$$B_k = \frac{(aq^k; q)_n \left(\frac{q^{1-k}}{a}; q\right)_n}{(q^k; q)_{n+1}} \times \frac{1}{(1-q^{1-k})\cdots(1-q^{-1})(1-q)\cdots(1-q^{n-k})}$$
$$\times \lim_{x \to q^k} \frac{x - q^k}{1 - \frac{q^k}{x}} \frac{x}{x}$$
$$= q^k \frac{(aq^k; q)_n \left(\frac{q^{1-k}}{a}; q\right)_n}{(q; q)_{n-k}(q^k; q)_{n+1}} \times \frac{1}{(1-q^{1-k})\cdots(1-q^{-1})}.$$

To simplify this we rewrite
$$\left(\frac{q^{1-k}}{a}; q\right)_n = \left(1 - \frac{q^{1-k}}{a}\right) \cdots \left(1 - \frac{q}{a}\right)\left(1 - \frac{1}{a}\right)\left(\frac{q}{a}; q\right)_{n-k},$$

so that
$$B_k = q^k \frac{(aq^k; q)_n \left(\frac{q}{a}; q\right)_{n-k}}{(q; q)_{n-k}(q^k; q)_{n+1}} \times \frac{\left(1 - \frac{q^{1-k}}{a}\right) \cdots \left(1 - \frac{q}{a}\right)\left(1 - \frac{1}{a}\right)}{(1-q^{1-k})\cdots(1-q^{-1})}.$$

Now
$$\frac{\left(1 - \frac{q^{1-k}}{a}\right) \cdots \left(1 - \frac{q}{a}\right)\left(1 - \frac{1}{a}\right)}{(1-q^{1-k})\cdots(1-q^{-1})} = \frac{\left(1 - \frac{q^{1-k}}{a}\right) \cdots \left(1 - \frac{q}{a}\right)\left(1 - \frac{1}{a}\right)}{(1-q^{1-k})\cdots(1-q^{-1})} \frac{a^k \, q^{1+2+\cdots+(k-1)}}{a^k \, q^{1+2+\cdots+(k-1)}}$$
$$= \frac{1}{a^k} \frac{(aq^{k-1} - 1)\cdots(aq - 1)(a - 1)}{(q^{k-1} - 1)\cdots(q - 1)}$$
$$= \frac{-1}{a^k} \frac{(a; q)_k}{(q; q)_{k-1}},$$

so
$$B_k = -\left(\frac{q}{a}\right)^k \frac{(a; q)_k (aq^k; q)_n \left(\frac{q}{a}; q\right)_{n-k}}{(q; q)_{n-k}(q; q)_{k-1}(q^k; q)_{n+1}} = -\left(\frac{q}{a}\right)^k \frac{(a; q)_{n+k} \left(\frac{q}{a}; q\right)_{n-k}}{(q; q)_{n-k}(q; q)_{n+k}}.$$

Comparing this with (2.8.2) we see that $B_k = A_{-k}$, and hence (2.8.1) becomes
$$p_n(x) = \sum_{k=-n}^{n} \frac{A_k}{x - q^{-k}}.$$

In view of (2.8.3), we have proved the following identity.

THEOREM 16. *For any nonnegative integer n, and all x except $q^{-n}, \ldots, q^n$, we have*
$$\frac{(ax;q)_n \left(\frac{q}{ax};q\right)_n}{(x;q)_{n+1}\left(\frac{q}{x};q\right)_n} = \sum_{k=-n}^{n} \frac{(a;q)_{n-k}\left(\frac{q}{a};q\right)_{n+k}}{(q;q)_{n-k}(q;q)_{n+k}} \frac{a^k}{1-xq^k}.$$

Exercises

1. Show similarly that
$$\frac{1}{(x;q)_{n+1}\left(\frac{q}{x};q\right)_n} = \frac{1}{(1-x)(q;q)_n^2} + \sum_{k=1}^{n} \frac{(-1)^k q^{\binom{k+1}{2}}}{(q;q)_{n+k}(q;q)_{n-k}} \left(\frac{1}{1-xq^k} + \frac{1}{q^k-x}\right).$$

2.9. A curious q-identity of Euler, and some extensions

In this section we discuss a strange q-identity of Euler that has not received much attention. Since it is most naturally written with a q-binomial coefficient it could be considered the oldest q-binomial theorem of all, but, although it fits more naturally in this chapter than anywhere else, there are good reasons not to call it a q-binomial theorem: (i) Euler did not think of it that way; and (ii) it does not really reduce to a special case of the binomial theorem when $q = 1$. For a positive integer n, let's look at

(2.9.1) $$E_n(q) := \sum_{k=1}^{n} \binom{n}{k}_q (q;q)_{k-1}.$$

Note that there is quite a bit of cancellation inside the sum, so we could instead write
$$E_n(q) = \sum_{k=1}^{n} \frac{(q^{n-k+1};q)_k}{1-q^k},$$
at least if $q \neq 1$, but we prefer the form (2.9.1). Let's write out a few instances of it. We have $E_1(q) = \binom{1}{1}_q (q;q)_0 = 1$, which doesn't provide much of a clue. More illuminating are
$$E_2(q) = \sum_{k=1}^{2} \binom{2}{k}_q (q;q)_{k-1} = [2]_q + (q;q)_1 = 1+q+1-q = 2$$
and
$$E_3(q) = \sum_{k=1}^{3} \binom{3}{k}_q (q;q)_{k-1} = [3]_q + [3]_q(1-q) + (1-q)(1-q^2)$$
$$= (1+q+q^2) + (1-q^3) + (1-q-q^2+q^3) = 3.$$

Thus, apparently, we have

2.9. A CURIOUS q-IDENTITY OF EULER, AND SOME EXTENSIONS

THEOREM 17 (Euler). *For any nonnegative integer n,*

(2.9.2) $$\sum_{k=1}^{n} \binom{n}{k}_q (q;q)_{k-1} = n.$$

(Both sides are zero if $n = 0$.)

Surprising as it may be that all the q's on the left side miraculously cancel, this is not too hard to prove by induction. Denoting the left side of (2.9.2) by $E_n(q)$, as in (2.9.1), we have to prove that $E_{n+1}(q) = n + 1$ from the assumption that $E_n(q) = n$. Using (2.1.2) we have

$$E_{n+1}(q) = \sum_{k=1}^{n+1} \binom{n+1}{k}_q (q;q)_{k-1} = \sum_{k=1}^{n+1} \left\{ \binom{n}{k-1}_q + q^k \binom{n}{k}_q \right\} (q;q)_{k-1}$$

$$= \sum_{k=1}^{n+1} \binom{n}{k-1}_q (q;q)_{k-1} + \sum_{k=1}^{n} \binom{n}{k}_q q^k (q;q)_{k-1}.$$

We reindex the first sum by renaming $k - 1$ as j. In the second sum we write $q^k = 1 - (1 - q^k)$, the point being that $(q;q)_{k-1}(1 - q^k) = (q;q)_k$. Then we have

$$E_{n+1}(q) = \sum_{j=0}^{n} \binom{n}{j}_q (q;q)_j + \sum_{k=1}^{n} \binom{n}{k}_q (q;q)_{k-1} - \sum_{k=1}^{n} \binom{n}{k}_q (q;q)_k.$$

The first and third sums almost cancel, but not quite—the third sum cancels every term of the first except the $j = 0$ term, which is $\binom{n}{0}_q (q;q)_0 = 1$. Therefore

$$E_{n+1}(q) = 1 + \sum_{k=1}^{n} \binom{n}{k}_q (q;q)_{k-1} = 1 + E_n(q),$$

and so $E_n(q) = n$ implies $E_{n+1}(q) = n + 1$. This proves Euler's theorem.

Euler's proof is much different, and we will outline it in Chapter 4. The remarkable thing about (2.9.2) is that the right side is independent of q, which implies that the left side must be too. Therefore, if we replace q by q^{-1} on the left side of (2.9.2) we must still have a true theorem. After some rather tedious algebra, which we leave as an exercise, this gives

(2.9.3) $$\sum_{k=1}^{n} \binom{n}{k}_q (-1)^{k-1} q^{\binom{k+1}{2} - nk} (q;q)_{k-1} = n.$$

This is in some ways closer to Euler's form of the identity, although he had $\frac{1}{a}$ in place of q. While (at least when expressed in our notation) it is not as pretty as (2.9.2), we will have an application for (2.9.3) in Chapter 4.

We conclude this section with a generalization of Euler's identity. (For another, see problem 9.) For a positive integer n, define

$$f_n(a,b) := \sum_{j=1}^{n} \binom{n}{j}_q (b-a)(b-aq) \cdots (b - aq^{n-j-1})(b^j - a^j)(q;q)_{j-1},$$

where, because of the factors $(q;q)_{j-1}$ and $b^j - a^j$, it is natural to take $f_0(a,b) = 0$. Also, as usual,

$$(b-a)(b-aq) \cdots (b - aq^{k-1}) = 1 \quad \text{if } k = 0.$$

We won't state the identity we are after right away—you will soon be able to guess what it is. Euler's identity is the case $b = 1$, $a = 0$.

For small values of n we have $f_1(a,b) = b - a$ and $f_2(a,b) = 2(b-a)(b-aq)$. We try to relate $f_{n+1}(a,b)$ to $f_n(a,b)$:

$$f_{n+1}(a,b) = \sum_{j=1}^{n+1} \binom{n+1}{j}_q (b-a)\cdots(b-aq^{n-j})(b^j - a^j)(q;q)_{j-1}$$

$$= \sum_{j=1}^{n+1} \left\{ \binom{n}{j-1}_q + q^j \binom{n}{j}_q \right\} (b-a)\cdots(b-aq^{n-j})(b^j - a^j)(q;q)_{j-1}$$

$$= \sum_{j=1}^{n+1} \binom{n}{j-1}_q (b-a)\cdots(b-aq^{n-j})(b^j - a^j)(q;q)_{j-1}$$

$$+ \sum_{j=1}^{n+1} \binom{n}{j}_q (b-a)\cdots(b-aq^{n-j-1})(b-aq^{n-j}) q^j (b^j - a^j)(q;q)_{j-1}.$$

Reindex the first sum above by replacing j by $k+1$, and in the second replace j by k:

$$f_{n+1}(a,b) = \sum_{k=0}^{n} \binom{n}{k}_q (b-a)\cdots(b-aq^{n-k-1})(b^{k+1} - a^{k+1})(q;q)_k$$

$$+ \sum_{k=1}^{n} \binom{n}{k}_q (b-a)\cdots(b-aq^{n-k-1})(bq^k - aq^n)(b^k - a^k)(q;q)_{k-1}$$

$$= S_1 + S_2,$$

where we have called the first piece of $f_{n+1}(a,b)$ above S_1 and the second piece S_2. We split up the factor $bq^k - aq^n$ in S_2:

$$S_2 = \sum_{k=1}^{n} \binom{n}{k}_q (b-a)\cdots(b-aq^{n-k-1}) bq^k (b^k - a^k)(q;q)_{k-1}$$

$$- \sum_{k=1}^{n} \binom{n}{k}_q (b-a)\cdots(b-aq^{n-k-1}) aq^n (b^k - a^k)(q;q)_{k-1}$$

$$= \sum_{k=1}^{n} \binom{n}{k}_q (b-a)\cdots(b-aq^{n-k-1}) bq^k (b^k - a^k)(q;q)_{k-1} - aq^n f_n(a,b).$$

In S_1 we rewrite the factor $b^{k+1} - a^{k+1}$ as

$$b^{k+1} - a^{k+1} = b(b^k - a^k) + a^k(b - a).$$

2.9. A CURIOUS q-IDENTITY OF EULER, AND SOME EXTENSIONS

Hence
$$S_1 = \sum_{k=0}^{n} \binom{n}{k}_q (b-a)\cdots(b-aq^{n-k-1})(b^k-a^k) b (q;q)_k$$
$$+ \sum_{k=0}^{n} \binom{n}{k}_q (b-a)\cdots(b-aq^{n-k-1}) a^k (b-a) (q;q)_k$$
$$= \sum_{k=1}^{n} \binom{n}{k}_q (b-a)\cdots(b-aq^{n-k-1})(b^k-a^k)(q;q)_{k-1} b(1-q^k)$$
$$+ (b-a)\sum_{k=0}^{n} \binom{n}{k}_q (b-a)\cdots(b-aq^{n-k-1}) a^k (q;q)_k.$$

Note that the $k=0$ term in the first sum was zero, so we threw it away. Now we put all these pieces back together:

$$f_{n+1}(a,b) = S_1 + S_2$$
$$= \sum_{k=1}^{n} \binom{n}{k}_q (b-a)\cdots(b-aq^{n-k-1})(b^k-a^k)(q;q)_{k-1} b(1-q^k)$$
$$+ \sum_{k=1}^{n} \binom{n}{k}_q (b-a)\cdots(b-aq^{n-k-1}) bq^k (b^k-a^k)(q;q)_{k-1} - aq^n f_n(a,b)$$
$$+ (b-a)\sum_{k=0}^{n} \binom{n}{k}_q (b-a)\cdots(b-aq^{n-k-1}) a^k (q;q)_k.$$

If we combine the first two sums above we get
$$\sum_{k=1}^{n} \binom{n}{k}_q (b-a)\cdots(b-aq^{n-k-1})(b^k-a^k)(q;q)_{k-1} \{b - bq^k + bq^k\}$$
$$= b\sum_{k=1}^{n} \binom{n}{k}_q (b-a)\cdots(b-aq^{n-k-1})(b^k-a^k)(q;q)_{k-1}$$
$$= b f_n(a,b).$$

Therefore the first three sums above, taken together, are equal to $(b-aq^n) f_n(a,b)$. We rewrite the fourth sum as
$$(b-a)\sum_{k=0}^{n} \binom{n}{k}_q (b-a)\cdots(b-aq^{n-k-1})(a-aq)(a-aq^2)\cdots(a-aq^k).$$

This equals $(b-a) \times (b-aq)(b-aq^2)\cdots(b-aq^n)$ by Jacobi's q-binomial theorem from section 2.6. Hence we finally have

(2.9.4) $\qquad f_{n+1}(a,b) = (b-a)(b-aq)\cdots(b-aq^n) + (b-aq^n) f_n(a,b).$

From this we can evaluate $f_n(a,b)$ by iteration. For a generic integer k between 1 and n, we have
(2.9.5)
$$f_n(a,b) = k(b-a)(b-aq)\cdots(b-aq^{n-1}) + (b-aq^{n-1})\cdots(b-aq^{n-k}) f_{n-k}(a,b).$$

(We leave this as an exercise.) If we take $k = n - 1$ here we get

$$\begin{aligned} f_n(a,b) &= (n-1)(b-a)(b-aq)\cdots(b-aq^{n-1}) + (b-aq^{n-1})\cdots(b-aq)f_1(a,b) \\ &= (n-1)(b-a)(b-aq)\cdots(b-aq^{n-1}) + (b-a)(b-aq)\cdots(b-aq^{n-1}) \\ &= n(b-a)(b-aq)\cdots(b-aq^{n-1}). \end{aligned}$$

In other words, we have proved that

$$(2.9.6) \quad \sum_{j=1}^{n} \binom{n}{j}_q (b-a)(b-aq)\cdots(b-aq^{n-j-1})(b^j - a^j)(q;q)_{j-1}$$

$$= n(b-a)(b-aq)\cdots(b-aq^{n-1})$$

for any nonnegative integer n. (Note that this is even true if $n = 0$.)

Exercises

1. What happens to Euler's theorem (2.9.2) when $q \to 1$? What happens to (2.9.6)?
2. Show directly that

$$\sum_{k=1}^{4} \binom{4}{k}_q (q;q)_{k-1} = 4.$$

3. Show that $n!_{q^{-1}} = q^{-\binom{n}{2}} n!_q$.
4. Using problem 3, or directly, show that $(q^{-1}; q^{-1})_n = (-1)^n q^{-\binom{n+1}{2}} (q;q)_n$.
5. Use problem 3 or problem 4 to show that

$$\binom{n}{k}_{q^{-1}} = q^{k(k-n)} \binom{n}{k}_q.$$

6. Use some combination of problems 3–5 to derive (2.9.3) from (2.9.2).
7. Use (2.1.3) and the result of problem 9 in section 2.3 to give an alternative proof of (2.9.2).
8. Use (2.1.2) and the result of problem 7 in section 2.3 to give an alternative proof of (2.9.3).
9. For a positive integer $j \leq n$, set

$$S_{n,j}(q) := \sum_{k=j}^{n} \binom{n}{k}_q (q^j; q)_{k-j}.$$

(i) Show that

$$S_{n,j}(q) - S_{n-1,j}(q) = \binom{n-1}{j-1}_q.$$

(ii) Explain why (i) implies that

$$(2.9.7) \quad \sum_{k=j}^{n} \binom{n}{k}_q (q^j; q)_{k-j} = \sum_{k=j}^{n} \binom{k-1}{j-1}_q.$$

What happens if $j = 1$?

EXERCISES

10. From Chapter 1 we know that

$$\binom{n}{j}_q = \sum_{k=j}^{n} \binom{k-1}{j-1}_q q^{k-j}.$$

This was a q-analogue of the diagonal property of Pascal's triangle. What happens if you subtract it from (2.9.7)?

11. Since the facts $1+2+3+\cdots+(n-1) = \binom{n}{2}$ and $1+2+3+\cdots+n = \binom{n+1}{2}$ are used approximately 27,843 times in this book, it is reasonable to wonder about $[1]_q + [2]_q + \cdots + [n]_q$. Show that

$$[1]_q + [2]_q + \cdots + [n]_q = \frac{n - q[n]_q}{1-q} = \frac{n - (n+1)q + q^{n+1}}{(1-q)^2}.$$

12. What formula does (2.9.7) give for the sum in the previous problem?

13. Here is another curious identity of Euler: for a positive integer k and a nonnegative integer n, show that

$$\sum_{j=1}^{k} q^{j-1} (q^j; q)_n = \frac{(q^k; q)_{n+1}}{1 - q^{n+1}}$$

by induction on k.

14. Euler's proof of the result of the previous problem did not use induction. Instead, Euler defined

$$Z = \sum_{j=1}^{k} q^{j-1} (q^j; q)_n$$

and looked at $(1 - q^{n+1}) Z$. Show that

$$q^{j-1} (q^j; q)_n (1 - q^{n+1}) = (q^j; q)_{n+1} - (q^{j-1}; q)_{n+1}$$

(make sure you check the case $j = 1$), and explain why Euler could conclude from this that

$$Z = \frac{(q^k; q)_{n+1}}{1 - q^{n+1}}.$$

15. Use (2.9.4) to prove (2.9.5) by induction on k.

16. What happens if we set $k = n$ in (2.9.5)? Would this be easier than setting $k = n - 1$? Why do you think we didn't do it this way?

17. (a) Show that

$$(x; q)_k = \sum_{j=1}^{k} (-1)^{j-1} q^{\binom{j}{2}} \binom{k}{j}_q (1 - x^j).$$

(b) Show that this can be rewritten as

$$\frac{(x; q)_k}{1 - q^k} = \sum_{j=1}^{k} (-1)^{j-1} q^{\binom{j}{2}} \binom{k-1}{j-1}_q \frac{1 - x^j}{1 - q^j}.$$

(c) Use (b) to show that

$$\sum_{k=1}^{n} \frac{q^k (x; q)_k}{1 - q^k} = \sum_{j=1}^{n} (-1)^{j-1} q^{\binom{j+1}{2}} \binom{n}{j}_q \frac{1 - x^j}{1 - q^j}.$$

18. What happens if you set $x = q$ in the result of part (c) of the previous problem?

2.10. The Chen–Chu–Gu identity

In this section we prove a q-binomial identity of Chen, Chu, and Gu that will be useful in Chapter 5. Consider the sum

(2.10.1) $$C_n(x) = \sum_{j=0}^{n} \binom{n}{j}_q \frac{q^{j^2} x^j (1 + xq^j)}{(x^2 q^j; q)_{n+1}}.$$

We have
$$C_0(x) = \frac{1+x}{1-x^2} = \frac{1}{1-x}$$
and
$$C_1(x) = \sum_{j=0}^{1} \binom{1}{j}_q \frac{q^{j^2} x^j (1 + xq^j)}{(x^2 q^j; q)_2}$$
$$= \frac{1+x}{(1-x^2)(1-x^2 q)} + \frac{xq(1+xq)}{(1-x^2 q)(1-x^2 q^2)}$$
$$= \frac{1}{1-x^2 q} \left[\frac{1}{1-x} + \frac{xq}{1-xq} \right]$$
$$= \frac{1}{1-x^2 q} \frac{1 - xq + xq(1-x)}{(1-x)(1-xq)}$$
$$= \frac{1}{1-x^2 q} \frac{1 - x^2 q}{(1-x)(1-xq)} = \frac{1}{(1-x)(1-xq)}.$$

A long calculation (exercise) shows further that
$$C_2(x) = \frac{1}{(1-x)(1-xq)(1-xq^2)}.$$

This suggests that

(2.10.2) $$C_n(x) = \sum_{j=0}^{n} \binom{n}{j}_q \frac{q^{j^2} x^j (1 + xq^j)}{(x^2 q^j; q)_{n+1}} = \frac{1}{(x; q)_{n+1}}.$$

This seems to be rather tricky to prove; the simplification is miraculous in both numerator and denominator. We start by observing from (2.10.1) that

$$C_{n-1}(xq) = \sum_{j=0}^{n-1} \binom{n-1}{j}_q \frac{q^{j(j+1)} x^j (1 + xq^{j+1})}{(x^2 q^{j+2}; q)_n}$$
$$= \sum_{j=1}^{n} \binom{n-1}{j-1}_q \frac{q^{j(j-1)} x^{j-1} (1 + xq^j)}{(x^2 q^{j+1}; q)_n}$$

(2.10.3) $$= \sum_{j=1}^{n} \binom{n-1}{j-1}_q \frac{q^{j(j-1)} x^{j-1}}{(x^2 q^{j+1}; q)_n} + \sum_{j=1}^{n} \binom{n-1}{j-1}_q \frac{q^{j^2} x^j}{(x^2 q^{j+1}; q)_n}.$$

Next, note again from (2.10.1) that
$$x\, C_n(x) = \sum_{j=0}^{n} \binom{n}{j}_q \frac{q^{j^2} x^j (x + x^2 q^j)}{(x^2 q^j; q)_{n+1}}$$

and rewrite $x + x^2 q^j = 1 + x - (1 - x^2 q^j)$. The point of doing this is that $1 - x^2 q^j$ will cancel with the first factor of the denominator. Then

$$x\, C_n(x) = \sum_{j=0}^n \binom{n}{j}_q \frac{q^{j^2} x^j (1+x)}{(x^2 q^j; q)_{n+1}} - \sum_{j=0}^n \binom{n}{j}_q \frac{q^{j^2} x^j (1 - x^2 q^j)}{(x^2 q^j; q)_{n+1}}$$

(2.10.4)
$$= \sum_{j=0}^n \binom{n}{j}_q \frac{q^{j^2} x^j (1+x)}{(x^2 q^j; q)_{n+1}} - \sum_{j=0}^n \binom{n}{j}_q \frac{q^{j^2} x^j}{(x^2 q^{j+1}; q)_n}.$$

The last sum in (2.10.4) now looks like it might combine nicely with (2.10.3). Let's rewrite the $1 + x$ in the other sum in (2.10.4) as $1 + xq^j + x(1 - q^j)$. Then

$$\sum_{j=0}^n \binom{n}{j}_q \frac{q^{j^2} x^j (1+x)}{(x^2 q^j; q)_{n+1}} = \sum_{j=0}^n \binom{n}{j}_q \frac{q^{j^2} x^j (1 + xq^j)}{(x^2 q^j; q)_{n+1}} + \sum_{j=0}^n \binom{n}{j}_q \frac{q^{j^2} x^{j+1}(1 - q^j)}{(x^2 q^j; q)_{n+1}}.$$

The point is that now the middle sum is $C_n(x)$. The last sum can start at $j = 1$, since the factor $1 - q^j$ makes the $j = 0$ term zero. We also have

$$\binom{n}{j}_q (1 - q^j) = \binom{n-1}{j-1}_q (1 - q^n),$$

so (2.10.4) becomes

(2.10.5) $\displaystyle x\, C_n(x) = C_n(x) + \sum_{j=1}^n \binom{n-1}{j-1}_q \frac{q^{j^2} x^{j+1}(1 - q^n)}{(x^2 q^j; q)_{n+1}} - \sum_{j=0}^n \binom{n}{j}_q \frac{q^{j^2} x^j}{(x^2 q^{j+1}; q)_n},$

and now both of the remaining sums look like they might combine nicely with (2.10.3). We rewrite the first one as

(2.10.6) $\displaystyle \sum_{j=1}^n \binom{n-1}{j-1}_q \frac{q^{j^2 - j} x^{j-1}\, x^2 q^j (1 - q^n)}{(x^2 q^j; q)_{n+1}}$

to make it look more like the first sum in (2.10.3). Next, rewrite $x^2 q^j (1 - q^n) = (1 - x^2 q^{j+n}) - (1 - x^2 q^j)$ here. Note that one of these two groups cancels with the last factor of $(x^2 q^j; q)_{n+1}$, and the other cancels with the first factor. Then (2.10.6) becomes

$$\sum_{j=1}^n \binom{n-1}{j-1}_q \frac{q^{j^2-j} x^{j-1}(1 - x^2 q^{j+n})}{(x^2 q^j; q)_{n+1}} - \sum_{j=1}^n \binom{n-1}{j-1}_q \frac{q^{j^2-j} x^{j-1}(1 - x^2 q^j)}{(x^2 q^j; q)_{n+1}}$$

$$= \sum_{j=1}^n \binom{n-1}{j-1}_q \frac{q^{j^2-j} x^{j-1}}{(x^2 q^j; q)_n} - \sum_{j=1}^n \binom{n-1}{j-1}_q \frac{q^{j^2-j} x^{j-1}}{(x^2 q^{j+1}; q)_n},$$

and the last sum is the negative of the first sum in (2.10.3). Adding (2.10.3) and (2.10.5) together, we therefore get

$$x\, C_n(x) + C_{n-1}(xq) = C_n(x) - \sum_{j=0}^n \binom{n}{j}_q \frac{q^{j^2} x^j}{(x^2 q^{j+1}; q)_n}$$

$$+ \sum_{j=1}^n \binom{n-1}{j-1}_q \frac{q^{j^2-j} x^{j-1}}{(x^2 q^j; q)_n} + \sum_{j=1}^n \binom{n-1}{j-1}_q \frac{q^{j^2} x^j}{(x^2 q^{j+1}; q)_n},$$

or, reindexing the next-to-last sum,

$$x\,C_n(x) + C_{n-1}(xq) = C_n(x) - \sum_{j=0}^{n}\binom{n}{j}_q \frac{q^{j^2}x^j}{(x^2q^{j+1};q)_n}$$
$$+ \sum_{j=0}^{n-1}\binom{n-1}{j}_q \frac{q^{j^2+j}x^j}{(x^2q^{j+1};q)_n} + \sum_{j=1}^{n}\binom{n-1}{j-1}_q \frac{q^{j^2}x^j}{(x^2q^{j+1};q)_n}.$$

Since

$$\binom{n}{j}_q = \binom{n-1}{j-1}_q + q^j\binom{n-1}{j}_q,$$

the three sums cancel, and we finally have $x\,C_n(x) + C_{n-1}(xq) = C_n(x)$, or

(2.10.7)
$$C_n(x) = \frac{C_{n-1}(xq)}{1-x}.$$

With this in hand, it is easy to prove (2.10.2) by iteration or induction.

In essence, (2.10.2) is already the Chen–Chu–Gu identity, but for future use we want to rewrite it to look similar to MacMahon's q-binomial theorem. First we replace n by $m+n$ and j by $m+k$ to get

$$\frac{1}{(x;q)_{n+m+1}} = \sum_{k=-m}^{n}\binom{m+n}{m+k}_q \frac{q^{(m+k)^2}x^{m+k}(1+xq^{m+k})}{(x^2q^{m+k};q)_{n+m+1}}.$$

Next set $x = -z/q^m$ here, which gives

$$\frac{1}{\left(-\frac{z}{q^m};q\right)_{n+m+1}} = \sum_{k=-m}^{n}\binom{m+n}{m+k}_q \frac{q^{m^2+2mk+k^2}(-z)^{m+k}q^{-m(m+k)}(1-zq^k)}{\left(\frac{z^2 q^{m+k}}{q^{2m}};q\right)_{n+m+1}}$$

(2.10.8)
$$= \sum_{k=-m}^{n}\binom{m+n}{m+k}_q \frac{q^{k(m+k)}(-1)^{m+k}z^{m+k}(1-zq^k)}{(z^2 q^{k-m};q)_{n+m+1}}.$$

On the left side we have

$$\left(-\frac{z}{q^m};q\right)_{n+m+1} = \left(1+\frac{z}{q^m}\right)\cdots\left(1+\frac{z}{q}\right)(1+z)(1+zq)\cdots(1+zq^n)$$

and the last $n+1$ factors are $(-z;q)_{n+1}$. It is convenient to rewrite the others as

$$\frac{z}{q^m}\left(\frac{q^m}{z}+1\right)\frac{z}{q^{m-1}}\left(\frac{q^{m-1}}{z}+1\right)\cdots\frac{z}{q}\left(\frac{q}{z}+1\right)$$
$$= \frac{z^m}{q^{1+2+\cdots+m}}\left(1+\frac{q}{z}\right)\left(1+\frac{q^2}{z}\right)\cdots\left(1+\frac{q^m}{z}\right) = \frac{z^m}{q^{\binom{m+1}{2}}}\left(-\frac{q}{z};q\right)_m,$$

so the left side of (2.10.8) is

(2.10.9)
$$\frac{q^{\binom{m+1}{2}}}{z^m\left(-\frac{q}{z};q\right)_m(-z;q)_{n+1}}.$$

We rewrite $(z^2 q^{k-m};q)_{n+m+1}$ in a similar way:

$$(z^2 q^{k-m};q)_{n+m+1} = (1-z^2q^{k-m})\cdots(1-z^2q^{-1})(1-z^2)(1-z^2q)\cdots(1-z^2q^{n+k})$$

and the last $n+k+1$ factors are $(z^2;q)_{n+k+1}$. It is convenient to rewrite the others as

$$\left(1-\frac{z^2}{q^{m-k}}\right)\cdots\left(1-\frac{z^2}{q}\right) = \frac{-z^2}{q^{m-k}}\left(1-\frac{q^{m-k}}{z^2}\right)\cdots\left(-\frac{z^2}{q}\right)\left(1-\frac{q}{z^2}\right)$$

$$= \frac{(-z^2)^{m-k}}{q^{1+2+\cdots+(m-k)}}\left(1-\frac{q}{z^2}\right)\cdots\left(1-\frac{q^{m-k}}{z^2}\right)$$

$$= \frac{(-1)^{m-k}z^{2m-2k}}{q^{\binom{m-k+1}{2}}}\left(\frac{q}{z^2};q\right)_{m-k}.$$

Using this and (2.10.9) in (2.10.8), the powers of -1 cancel and we have

$$\frac{q^{\binom{m+1}{2}}}{z^m\left(-\frac{q}{z};q\right)_m(-z;q)_{n+1}} = \sum_{k=-m}^{n}\binom{m+n}{m+k}_q\frac{q^{k(m+k)+\binom{m-k+1}{2}}z^{3k-m}(1-zq^k)}{\left(\frac{q}{z^2};q\right)_{m-k}(z^2;q)_{n+k+1}}$$

or

$$\frac{1}{\left(-\frac{q}{z};q\right)_m(-z;q)_{n+1}} = \sum_{k=-m}^{n}\binom{m+n}{m+k}_q\frac{q^{k(m+k)+\binom{m-k+1}{2}-\binom{m+1}{2}}z^{3k}(1-zq^k)}{\left(\frac{q}{z^2};q\right)_{m-k}(z^2;q)_{n+k+1}}.$$

Simplifying the exponent of q (exercise) we finally have

THEOREM 18 (The Chen–Chu–Gu identity). *For any nonnegative integers m and n, for all q, and for all $z \neq 0$, we have*

$$\frac{1}{\left(-\frac{q}{z};q\right)_m(-z;q)_{n+1}} = \sum_{k=-m}^{n}\binom{m+n}{m+k}_q\frac{q^{\frac{k(3k-1)}{2}}z^{3k}(1-zq^k)}{\left(\frac{q}{z^2};q\right)_{m-k}(z^2;q)_{n+k+1}}.$$

Exercises

1. Complete the proof of (2.10.2) by iteration or induction from (2.10.7).
2. Complete the proof of the Chen–Chu–Gu identity by showing that

$$k(m+k) + \binom{m-k+1}{2} - \binom{m+1}{2} = \frac{k(3k-1)}{2}.$$

3. Show directly that $C_2(x) = 1/(1-x)(1-xq)(1-xq^2)$.
4. What happens to (2.10.2) when $q=1$?

2.11. Bibliographical Notes

The Potter–Schützenberger q-binomial theorem is due to Potter [**189**] and Schützenberger [**216**], a few years apart. See also the article [**141**].

Gauss's two q-binomial theorems in section 2.5 come from [**115**]. The Rothe q-binomial theorem was stated, without proof and with a misprint, in the introduction to [**208**], as a sample of material that was cut from the book due to space limitations imposed by his publisher. He probably had a proof, and if he had written it out for publication then he might well have found the misprint. Gruson gave the first two published proofs in [**129**]. The introduction of Schweins's book [**218**] mentions the work of Gauss, Rothe, and Gruson.

This was the second time that Rothe found a significant generalization of the binomial theorem. In his 1793 thesis [**206**] he proved

THEOREM 19 (Rothe's generalized binomial theorem). *If*
$$R_n(x;h,w) = x\,(x+h+nw)\,(x+2h+nw)\cdots(x+(n-1)h+nw)$$
for $n \geq 1$ and $R_0(x;h,w) = 1$, then
$$R_n(x+y;h,w) = \sum_{k=0}^{n} \binom{n}{k} R_k(x;h,w)\, R_{n-k}(y;h,w).$$

More precisely, he proved the case $h = -1$ of this, but the general case then follows by rescaling. It reduces to the ordinary binomial theorem if h and w are both zero.

George Andrews communicated his remark to me shortly after I arrived at Penn State in 1993. I published it in [**152**], which also contains some of the generalized q-binomial theorems in the exercises in this chapter (and others that were too complicated to include here). Cigler's version of it is in [**78**]. Jacobi's q-binomial theorem is in [**150**], and MacMahon's q-binomial theorem is in his classic book [**168**]. The symmetric case with $a = b = n$ dates back to Gauss [**117**] and Cauchy [**68**]. Hirschhorn's identity (2.7.9) is in [**138**], with a different proof, and in section 1.8 of his beautiful recent book [**140**] as an exercise. That section also gave rise to some of the problems in section 2.5, as did Paule's lovely paper [**181**], with which there is some overlap.

Euler's theorem in section 2.9 comes from [**96**], and seems to be much less well known than his other work on q-analysis, the subject of the next chapter. The Chen–Chu–Gu identity is in [**75**], and I learned it from [**80**]. Its significance will become apparent in Chapter 5.

The cyclic derivative in the last problem in section 2.1 comes from [?RSS]. Although the connection to q-analysis is tenuous, the idea is so striking that I could not resist including it. Agarwal's q-binomial theorem in problem 19 in section 2.3 comes from [**2**], and is also mentioned in [**24**]. Andrews's q-binomial theorem from problem 11 in section 2.3 is in [**21**] and Rowell's identity from problem 13 in section 2.5 in [**209**]. Problems 14–16 in section 2.5 are adapted from [**26**]. Problem 1 in section 2.8 comes from [**25**].

CHAPTER 3

Partitions I: Elementary Theory

3.1. Partitions with distinct parts

We begin this chapter still farther back in time than we did in Chapter 1. Quite possibly the greatest mathematics book ever written is Leonhard Euler's *Introductio in analysin infinitorum*, which John Blanton, in his English translation of 1988, rendered as *Introduction to Analysis of the Infinite*. Writing in 1748, Euler showed as deep an understanding of the formal aspects of infinite series as anyone ever has. (The word "formal" in this context is short for "neglecting convergence issues", which would not be treated seriously until the 19th century.) Euler's book contains a wealth of material that is still of interest today. Here we are primarily concerned with his Chapter XVI, which essentially initiated the subject of partitions.

Consider the infinite product

$$(1+q)(1+q^2)(1+q^3)(1+q^4)(1+q^5)(1+q^6)(1+q^7)(1+q^8)\cdots$$

which, by an extension of the notation of Chapter 1, we could abbreviate as $(-q;q)_\infty$. (We will make a formal definition a little later; we also put off for the moment the question of whether such an infinite product really makes sense.) As Euler said (in Blanton's translation), "We ask about the form if the factors are actually multiplied." We will get a 1, from the product of all the 1's. We will get a q, from the q in the first factor times all the other 1's; and a q^2, from the q^2 in the second factor times all the other 1's. Interesting things start to happen at q^3, for we get two of those, one from the q^3 in the third factor times all the other 1's; and one from the q in the first factor times the q^2 in the second factor times all the other 1's. We get two copies of q^4, namely q^4 and $q^3 \cdot q$; and three copies of q^5, which come from q^5, $q^4 \cdot q$, and $q^3 \cdot q^2$. We do one more term before trying to describe things more generally: we get four copies of q^6, from q^6, $q^5 \cdot q$, $q^4 \cdot q^2$, and $q^3 \cdot q^2 \cdot q$; this corresponds to the fact that we can write 6 as a sum of distinct positive integers in four different ways: 6, 5 + 1, 4 + 2, and 3 + 2 + 1. So we see that the question "how many copies of q^7 (say) will we get?" is equivalent to asking "how many ways can we make 7 as a sum of positive integers, if we can only use each integer once in each sum?" This last restriction is essential (for now)—we do not want to allow $6 = 3 + 3$, because we do not get a term $q^3 \cdot q^3$ when we work out the infinite product, since it contains only one factor which has q^3. For 7 we have $7 = 7$, $7 = 6+1$, $7 = 5+2$, $7 = 4+3$, and $7 = 4+2+1$. To summarize the calculations so far, we have shown that

$$(-q;q)_\infty = 1 + q + q^2 + 2q^3 + 2q^4 + 3q^5 + 4q^6 + 5q^7 + \ldots.$$

More importantly, we have argued that the coefficient of q^n in $(-q;q)_\infty$ equals the number of ways that n can be written as a sum of distinct positive integers.

It is time for more careful definitions. Extending the notation for q-shifted factorials from Chapters 1 and 2, we define

$$(x;q)_\infty = \lim_{n\to\infty}(x;q)_n = \lim_{n\to\infty}(1-x)(1-xq)(1-xq^2)\cdots(1-xq^{n-1})$$

$$= \lim_{n\to\infty}\prod_{k=1}^{n}\left(1-xq^{k-1}\right) = \prod_{k=1}^{\infty}\left(1-xq^{k-1}\right) = \prod_{k=0}^{\infty}\left(1-xq^k\right),$$

where we also used a standard notation for products, analogous to the sigma notation for infinite series. In Appendix B it is proved that this product converges for every x as long as $|q|<1$. These products are ubiquitous in q-analysis, so from here on we will nearly always assume $|q|<1$.

A **partition** of a positive integer n is an unordered sum of positive integers that add up to n; thus, for example, $23+16+7+7+4+1+1+1$ is a partition of 60. Each summand is called a **part**, so that (for example) 16 is one of the parts of the above partition. When we say that the sum is *unordered*, we mean that $1+16+1+7+4+23+1+7$ does not count as a different partition of 60, since the parts are all the same as those of $23+16+7+7+4+1+1+1$. We will usually write the parts in nonincreasing order. Note that we do allow a part to occur more than once, as 7 and 1 do in this example. When no part occurs more than once we say that the parts are **distinct**. If $p_D(n)$ denotes the number of partitions of n using only distinct parts, we saw above that

$$(3.1.1) \quad (1+q)(1+q^2)(1+q^3)(1+q^4)(1+q^5)(1+q^6)\cdots = (-q;q)_\infty = \sum_{n=0}^{\infty} p_D(n)\, q^n.$$

The case $n=0$ is problematic; by convention we say that there is one empty partition of 0, with no parts, and since there are no repeated parts in an empty partition, $p_D(0)=1$. A verbalization of (3.1.1) is that "$(-q;q)_\infty$ ("minus q base q sub ∞") is the generating function for partitions into distinct parts." If P is a partition, it is sometimes convenient to denote the number that P partitions as $|P|$, so that $|23+16+7+7+4+1+1+1|=60$.

It is also interesting to consider a finite product of the same form, say

$$(1+q)(1+q^2)(1+q^3)\cdots(1+q^n) = (-q;q)_n,$$

and here we do not need $|q|<1$. If the factors are multiplied out, the coefficient of q^m in the result will be the number of partitions of m using only distinct parts from among $\{1,2,3,\ldots,n\}$, by exactly the same reasoning as before. We will return to this product later.

Other refinements are also possible. For example, consider

$$(-q;q^2)_\infty = (1+q)(1+q^3)(1+q^5)(1+q^7)\cdots = \prod_{n=0}^{\infty}\left(1+q^{2n+1}\right).$$

If we multiply the factors out we get

$$1+q+q^3+q^4+q^5+q^6+q^7+2q^8+2q^9+2q^{10}+2q^{11}+3q^{12}+3q^{13}+3q^{14}+4q^{15}+\ldots$$

where the coefficient of q^n counts the number of partitions of n into parts that are both odd and distinct; for example, 15 can be written as 15 itself or as $11+3+1$,

$9 + 5 + 1$, or $7 + 5 + 3$, which is why the coefficient of q^{15} above is 4. If we denote the number of partitions of n into distinct odd parts by $p_{OD}(n)$, then we have

$$(-q; q^2)_\infty = \sum_{n=0}^{\infty} p_{OD}(n)\, q^n.$$

In words, $(-q; q^2)_\infty$ is the generating function for partitions into distinct odd parts. We could also take a finite version of this product: $(-q; q^2)_n$ is the generating function for partitions into distinct odd parts which do not exceed $2n - 1$.

We conclude this section by proving a simple theorem which will be used in the next section; namely, that every positive integer m can be written uniquely as a sum of distinct powers of 2. First we prove another little lemma.

LEMMA 2. *If n is a positive integer, then*

$$[2^n]_q = (1+q)(1+q^2)(1+q^4)(1+q^8)\ldots\left(1+q^{2^{n-1}}\right).$$

If $q = 1$ this says

$$2^n = 2 \cdot 2 \cdot 2 \cdots 2,$$

with n factors of 2 on the right, which is certainly true, so we can assume $q \neq 1$. In this case we have

$$\frac{1-q^2}{1-q} \cdot \frac{1-q^4}{1-q^2} \cdot \frac{1-q^8}{1-q^4} \cdots \frac{1-q^{2^n}}{1-q^{2^{n-1}}} = \frac{1-q^{2^n}}{1-q} = [2^n]_q.$$

If we take $|q| < 1$ in Lemma 2 and let $n \to \infty$, we get

$$\frac{1}{1-q} = (1+q)(1+q^2)(1+q^4)(1+q^8)\ldots\left(1+q^{2^k}\right)\ldots.$$

But if $|q| < 1$, then also

$$\frac{1}{1-q} = 1 + q + q^2 + q^3 + q^4 + q^5 + q^6 + \ldots.$$

Combining these last two identities we have that, if $|q| < 1$,

$$(1+q)(1+q^2)(1+q^4)(1+q^8)\cdots = 1 + q + q^2 + q^3 + q^4 + q^5 + \ldots.$$

The left side of this is the generating function for partitions into distinct parts that are powers of 2. Since the coefficient of q^n on the right is 1 for any nonnegative integer n, it follows that there is one and only one partition of n as a sum of distinct powers of 2 for any positive integer n, which is what we wanted to prove. Of course, anyone who knows anything about computer science knows this—it's the point of binary arithmetic.

Exercises

1. Explain why $(-q^2; q^2)_\infty$ is the generating function for partitions with distinct even parts. In other words, if $p_{ED}(n)$ denotes the number of partitions of n into parts which are even and distinct, explain why

$$(-q^2; q^2)_\infty = (1+q^2)(1+q^4)(1+q^6)\cdots = \sum_{n=0}^{\infty} p_{ED}(n)\, q^n.$$

2. Explain why $p_{ED}(2n) = p_D(n)$.

3. How would you write 47 as a sum of distinct powers of 2? How about 156? Try enough examples that you understand how to do it in general.

4. Use the idea from problem 3 to prove by induction that any positive integer can be written as a sum of distinct powers of 2 in exactly one way. To show uniqueness you can use the fact that $1 + 2 + 2^2 + \cdots + 2^n < 2^{n+1}$. (Why is this true? How does it help?)

5. As Euler pointed out, the result in problem 4 (proved by Euler's method in the text) has an interesting interpretation: it says that if you have a balance and a set of weights that has all the powers of 2 (*i.e.*, a 1 gram weight, a 2 gram weight, a 4 gram weight, an 8 gram weight, and so on), then you can weigh any object that weighs an integer number of grams; or, to put it another way, you can weigh any object to within a fraction of a gram. Euler goes on to say that in fact (as was apparently well known at the time) you only need a set of weights that has all the powers of 3 (a 1 gram weight, a 3 gram weight, a 9 gram weight, a 27 gram weight, and so on), if you are willing to put weights on both sides of the balance. For example, you can weigh a 47 gram object by putting it on one side of the balance along with weights of 27 grams, 9 grams and 1 gram; and putting weights of 81 grams and 3 grams on the other side.

   Euler proved this by considering the infinite product

   $$\left(x^{-1} + 1 + x\right)\left(x^{-3} + 1 + x^3\right)\left(x^{-9} + 1 + x^9\right)\left(x^{-27} + 1 + x^{27}\right)\cdots.$$

   There are some difficulties with this product, which we will address in the problems in Appendix B. We consider instead a finite version of it,

   $$\left(x^{-1} + 1 + x\right)\left(x^{-3} + 1 + x^3\right)\cdots,\left(x^{-3^{n-1}} + 1 + x^{3^{n-1}}\right),$$

   for a general positive integer n. Show that for any $n \geq 1$,

   $$\left(x^{-1} + 1 + x\right)\left(x^{-3} + 1 + x^3\right)\cdots\left(x^{-3^{n-1}} + 1 + x^{3^{n-1}}\right) = \sum_{j=\frac{1-3^n}{2}}^{\frac{3^n-1}{2}} x^j.$$

   Explain why this implies that any object weighing $\frac{3^n-1}{2}$ grams or less can be weighed with weights of $1, 3, 9, 27, \ldots, 3^{n-1}$ grams, in exactly one way. Since n can be as large as we please, it follows that any object with a finite weight can be weighed with a sufficiently large set of weights that are distinct powers of 3, in exactly one way.

6. Prove the result of problem 5 by an argument similar to problem 4. In other words, prove by induction that any integer n can be written uniquely as

   $$n = \sum_{k=0}^{m} a_k \, 3^k,$$

   where each a_k is either 0 or ± 1 and m is the largest value of k for which a_k is not zero, unless $n = 0$ in which case every a_k is zero. For example, if $n = 1$, then $m = 0$ and $a_0 = 1$, and if $n = 2$, then $m = 1$, $a_1 = 1$, and $a_0 = -1$. If $n = 46$, then we can write $46 = 81 - 27 - 9 + 1$, so $m = 4$, $a_4 = 1$, $a_3 = a_2 = -1$, $a_1 = 0$, and $a_0 = 1$. (If we can prove this for a positive n, it follows easily for the corresponding negative n; why?)

7. Jacobi considered the infinite product

$$T = \frac{1-q}{1+q}\left(\frac{1-q^2}{1+q^2}\right)^{\frac{1}{2}}\left(\frac{1-q^4}{1+q^4}\right)^{\frac{1}{4}}\left(\frac{1-q^8}{1+q^8}\right)^{\frac{1}{8}}\left(\frac{1-q^{16}}{1+q^{16}}\right)^{\frac{1}{16}}\left(\frac{1-q^{32}}{1+q^{32}}\right)^{\frac{1}{32}}\cdots$$

under the usual assumption that $|q| < 1$. He gave two proofs that $T = (1-q)^2$. The first one is more or less as follows: define T_n to be the product of the first n factors of T; i.e.,

$$T_1 = \frac{1-q}{1+q}, \quad T_2 = \frac{1-q}{1+q}\left(\frac{1-q^2}{1+q^2}\right)^{\frac{1}{2}}, \quad T_3 = \frac{1-q}{1+q}\left(\frac{1-q^2}{1+q^2}\right)^{\frac{1}{2}}\left(\frac{1-q^4}{1+q^4}\right)^{\frac{1}{4}},$$

and so on. Prove that for any positive integer n we have

$$\left(1-q^{2^n}\right)^{\frac{1}{2^{n-1}}} T_n = (1-q)^2.$$

Why does this imply Jacobi's result?

8. Jacobi's second proof that $T = (1-q)^2$ is sneakier: he rewrote the definition of T as

$$T = (1-q)\frac{\left(1-q^2\right)^{\frac{1}{2}}}{1+q}\frac{\left(1-q^4\right)^{\frac{1}{4}}}{\left(1+q^2\right)^{\frac{1}{2}}}\frac{\left(1-q^8\right)^{\frac{1}{8}}}{\left(1+q^4\right)^{\frac{1}{4}}}\frac{\left(1-q^{16}\right)^{\frac{1}{16}}}{\left(1+q^8\right)^{\frac{1}{8}}}\frac{\left(1-q^{32}\right)^{\frac{1}{32}}}{\left(1+q^{16}\right)^{\frac{1}{16}}}\cdots.$$

Show that this implies $T = (1-q)\sqrt{T}$. Why does this give Jacobi's result?

9. Let q be a positive real number, and define a sequence $a_n(q)$ by $a_0(q) = q + q^{-1}$ and $a_n(q) = a_{n-1}^2(q) - 2$ for $n \geq 1$. Show that

$$a_n(q) = q^{2^n} + q^{-2^n} \quad \text{for } n \geq 0.$$

10. Let $a_n(q)$ be as in the previous problem with $q \neq 1$, and consider the product

$$P_n(q) = \prod_{k=0}^{n-1}\left(1 - \frac{1}{a_k(q)}\right) \quad \text{for } n \geq 1.$$

Show that

$$P_n(q) = \frac{(q-1)\left(q^2-1\right)\left(q^{3\cdot 2^n}-1\right)}{\left(q^3-1\right)\left(q^{2^n}-1\right)\left(q^{2^{n+1}}-1\right)}.$$

11. Let $a_n(q)$ be as above and consider the infinite product

$$P(q) = \prod_{k=0}^{\infty}\left(1 - \frac{1}{a_k(q)}\right).$$

What does $P(q)$ converge to when $q > 1$? What does it converge to when $0 < q < 1$? What happens when $q = 1$? Explain.

3.2. Partitions with repeated parts

So far we have seen that $(-q;q)_\infty$ is the generating function for partitions with distinct parts, and that $(-q;q)_n$ is the generating function for partitions with distinct parts less than or equal to n. Since partitions in general need not have distinct parts, we may wonder whether there is a function which generates all the partitions, without regard to whether the parts are distinct. We claim that $1/(q;q)_\infty$ is such a function. The infinite product again converges if and only if $|q| < 1$, and this restriction on q is vital for another reason also. We have

$$\frac{1}{(q;q)_\infty} = \frac{1}{1-q}\frac{1}{1-q^2}\frac{1}{1-q^3}\frac{1}{1-q^4}\frac{1}{1-q^5}\frac{1}{1-q^6}\cdots$$

and we expand all these factors using the geometric series

$$\frac{1}{1-r} = 1 + r + r^2 + r^3 + r^4 + \ldots \quad \text{if } |r| < 1.$$

Thus

$$\frac{1}{(q;q)_\infty} = (1+q+q^2+q^3+q^4+\ldots)(1+q^2+q^4+\ldots)(1+q^3+\ldots)(1+q^4+\ldots)\ldots$$

and again we ask what happens if we actually multiply this out. We get a 1, from the product of all the 1's, and a q, from the q in the first factor times all the other 1's. We get two copies of q^2, one from the first term and one from the second. There are three copies of q^3; one purely from the first term, one purely from the third, and one from the q^2 in the second term times the q in the first. These terms correspond respectively to the partitions $1+1+1$, 3, and $2+1$, which are all the partitions of 3. You can check that we get five copies of q^4 and seven copies of q^5, and so forth, and that there are five partitions of 4: 4, $3+1$, $2+2$, $2+1+1$, and $1+1+1+1$; and seven partitions of 5: 5, $4+1$, $3+2$, $3+1+1$, $2+2+1$, $2+1+1+1$, and $1+1+1+1+1$.

Why should this product be generating all the partitions? To see this, we rewrite it a bit differently. The right way to think of it for our present purposes is that the first factor generates all the 1's, and the second factor generates all the 2's, and so on:

$$\frac{1}{(q;q)_\infty} = (1+q^1+q^{1+1}+q^{1+1+1}+q^{1+1+1+1}+q^{1+1+1+1+1}+q^{1+1+1+1+1+1}+\ldots)$$
$$\times (1+q^2+q^{2+2}+q^{2+2+2}+\ldots)(1+q^3+q^{3+3}+\ldots)$$
$$\times (1+q^4+\ldots)(1+q^5+\ldots)(1+q^6+\ldots)\ldots.$$

Now look at the q^6 term in the product. We get a pure q^6, from the q^6 in the last factor times all the other 1's. We get $q^5 \cdot q^1$, from the fifth factor and the first. We get $q^4 \cdot q^2$ from the fourth factor and the second, and $q^4 \cdot q^{1+1}$ from the fourth factor and the first. The third, second, and first factors give us $q^3 \cdot q^2 \cdot q^1$. The second factor by itself contributes q^{2+2+2}, and the second and first factors together give $q^{2+2} \cdot q^{1+1}$ as well as $q^2 \cdot q^{1+1+1+1}$. Finally the first factor by itself contributes $q^{1+1+1+1+1+1}$. These terms correspond precisely to all the partitions of 6. In general, if $p(n)$ denotes the number of partitions of n, then

(3.2.1) $$\frac{1}{(q;q)_\infty} = \sum_{n=0}^\infty p(n)\, q^n.$$

(As usual, we count an empty partition of 0 once to get the first term.) We can also take a finite form of the generating function: $1/(q;q)_n$ leaves us only $\{1, 2, \ldots, n\}$ as possible parts, so it generates partitions whose parts are less than or equal to n; in other words, the coefficient of q^m in $1/(q;q)_n$ equals the number of partitions of m using (possibly repeated) parts that do not exceed n. For example, the coefficient of q^7 in $1/(q;q)_4$ must be 11, because there are 11 partitions of 7 using only parts that are less than or equal to 4, namely

$$\begin{array}{lll}
4+3 & 4+2+1 & 4+1+1+1 \\
3+3+1 & 3+2+2 & 3+2+1+1 \\
3+1+1+1+1 & 2+2+2+1 & 2+2+1+1+1 \\
2+1+1+1+1+1 & 1+1+1+1+1+1+1.
\end{array}$$

Note that although the product $(q;q)_n$ makes sense for any value of q, we still need our usual assumption $|q| < 1$ to expand $1/(q;q)_n$ as above.

Many refinements of this idea are possible. A famous example is

$$(3.2.2) \qquad \frac{1}{(q;q^2)_\infty} = \frac{1}{(1-q)(1-q^3)(1-q^5)(1-q^7)\cdots} = \sum_{n=0}^{\infty} p_O(n)\, q^n,$$

where $p_O(n)$ denotes the number of partitions of n using only odd parts. Let's work out the first several terms of this:

$$\frac{1}{(q;q^2)_\infty} = (1 + q + q^2 + q^3 + q^4 + q^5 + q^6 + q^7 + q^8 + q^9 + \ldots)$$
$$\times (1 + q^3 + q^6 + q^9 + \ldots)(1 + q^5 + \ldots)(1 + q^7 + \ldots)(1 + q^9 + \ldots)\cdots$$
$$= 1 + q + q^2 + 2q^3 + 2q^4 + 3q^5 + 4q^6 + 5q^7 + 6q^8 + 8q^9 + \ldots.$$

We'll check the q^9 term and leave the others to you: we can get a q^9 from the q^9 in the first term; or from the q^6 in the first term times the q^3 in the second or vice versa; or from the q^4 in the first term times the q^5 in the third; or from the q^2 in the first term times the q^7 in the fourth; or from the q in the first term times the q^3 in the second term times the q^5 in the third; or from the q^9 in the second or fifth term. These correspond respectively to the partitions $1+1+1+1+1+1+1+1+1$, $3+1+1+1+1+1+1$, $3+3+1+1+1$, $5+1+1+1+1$, $7+1+1$, $5+3+1$, $3+3+3$, and 9, which are all the ways of getting 9 using only odd parts. The first real surprise in partition theory is that this appears to be the same series as we got from $(-q;q)_\infty$ in section 3.1. That is, it appears that

$$(-q;q)_\infty = 1 + q + q^2 + 2q^3 + 2q^4 + 3q^5 + 4q^6 + 5q^7 + 6q^8 + \cdots = \frac{1}{(q;q^2)_\infty}.$$

If so, since

$$(-q;q)_\infty = \sum_{n=0}^{\infty} p_D(n)\, q^n \qquad \text{and} \qquad \frac{1}{(q;q^2)_\infty} = \sum_{n=0}^{\infty} p_O(n)\, q^n,$$

this would imply that $p_D(n) = p_O(n)$ for all n; in other words, it would imply

THEOREM 20 (Euler's "odd equals distinct" theorem). *If n is any nonnegative integer, there are exactly as many partitions of n using only odd parts as there are with distinct parts.*

We first give Euler's proof, and then a very interesting alternative proof due to Glaisher. One of the main ideas in Euler's argument has already appeared: since this theorem was suggested to us by the fact that it looks as though $(-q;q)_\infty$ might be equal to $1/(q;q^2)_\infty$, we should try to prove that they really are equal—if they are, then they must have the same expansion in powers of q, and Euler's theorem would follow. It is easier than you might expect to show that these two infinite products are the same:

$$(-q;q)_\infty = (1+q)(1+q^2)(1+q^3)(1+q^4)(1+q^5)(1+q^6)\cdots$$
$$= \frac{1-q^2}{1-q}\frac{1-q^4}{1-q^2}\frac{1-q^6}{1-q^3}\frac{1-q^8}{1-q^4}\frac{1-q^{10}}{1-q^5}\frac{1-q^{12}}{1-q^6}\cdots$$

Now cancel the numerator factors against the corresponding factors in the denominator. In what we've written you can see that $1-q^2$, $1-q^4$, and $1-q^6$ will all cancel; and, since we have an infinite product, every numerator factor will eventually cancel. We rephrase this in a form that some readers might find more palatable: the numerator above is

$$(1-q^2)(1-q^4)(1-q^6)(1-q^8)(1-q^{10})(1-q^{12})\cdots, \quad \text{which equals } (q^2;q^2)_\infty.$$

The denominator is

$$(q;q)_\infty = (1-q)(1-q^2)(1-q^3)(1-q^4)(1-q^5)(1-q^6)(1-q^7)(1-q^8)\ldots$$
$$= \{(1-q)(1-q^3)(1-q^5)(1-q^7)\ldots\}\{(1-q^2)(1-q^4)(1-q^6)(1-q^8)\ldots\}$$
$$= (q;q^2)_\infty (q^2;q^2)_\infty.$$

It follows that

$$(-q;q)_\infty = (1+q)(1+q^2)(1+q^3)(1+q^4)(1+q^5)(1+q^6)\cdots$$
$$= \frac{1-q^2}{1-q}\frac{1-q^4}{1-q^2}\frac{1-q^6}{1-q^3}\frac{1-q^8}{1-q^4}\frac{1-q^{10}}{1-q^5}\frac{1-q^{12}}{1-q^6}\cdots$$
$$= \frac{(q^2;q^2)_\infty}{(q;q^2)_\infty (q^2;q^2)_\infty}$$

(3.2.3)
$$= \frac{1}{(q;q^2)_\infty}.$$

This proves Euler's theorem. Glaisher's proof is quite different. His idea was to set up a 1-1 correspondence (or "bijection") between the two types of partitions. For example, there are ten partitions of 10 using distinct parts, and also (as we now know) ten partitions using only odd parts. Here they are:

Distinct	Odd
10	5+5
9+1	9+1
8+2	1+1+1+1+1+1+1+1+1+1
7+3	7+3
7+2+1	7+1+1+1
6+4	3+3+1+1+1+1
6+3+1	3+3+3+1
5+4+1	5+1+1+1+1+1
5+3+2	5+3+1+1
4+3+2+1	3+1+1+1+1+1+1+1

The reason for putting the "Odd" column in this order will appear presently. Suppose we have a partition of 10 with distinct parts. If all the parts are also odd, then we don't have to do anything—this happened with $9+1$ and with $7+3$—but most of the time there is at least one even part, and we have somehow to get rid of the evenness. We use the obvious fact that any even number can be written uniquely as a power of 2 times an odd number. For example, $48 = 16 \cdot 3$; the form is unique since to find it we just factor out the largest possible power of 2. So we write any even part in the "Distinct" column in this form, and then replace it by that many copies of the odd factor. That is, $10 = 2 \cdot 5$, so replace 10 by $5+5$, and $8 = 8 \cdot 1$, so 8 is replaced by a sum of eight 1's, and so forth. Doing this with all the even parts in the "Distinct" column we get the "Odd" column.

We can also describe how to get from the "Odd" column back to the "Distinct" column. If we start with a partition with only odd parts, it might happen (if we're lucky) that the parts are also distinct, as with $9+1$ and $7+3$. Most of the time, though, there will be repeated parts, and we have to figure out how to get rid of the repetition. We especially want a method of doing this which undoes what we were doing above. We illustrate this method with the example $3+1+1+1+1+1+1+1$: there is only one 3, so we leave that alone, but we must do something about the repeated 1's. There are seven 1's, which we think of as $7 \cdot 1$; in general, if we have m copies of some odd part p, we start by thinking of them as $m \cdot p$. Then we write m as a sum of distinct powers of 2, which we know from the previous section that we can do in exactly one way; when $m = 7$, we write that as $4+2+1$. Finally, then, $1+1+1+1+1+1+1$ gets replaced by $4+2+1$. For the example $3+3+3+1$, we leave the 1 alone and combine the 3's into $3 \cdot 3$, so that m and p are both 3. Take the 3 which is m and write it as a sum of distinct powers of 2, namely $2+1$; and then replace $3 \cdot 3$ by $(2+1) \cdot 3 = 6+3$. Thus the partition corresponding to $3+3+3+1$ is $6+3+1$.

Euler's "odd equals distinct" theorem is the oldest of a vast number of theorems of the form "there are exactly as many partitions of n of type A as there are of type B". Some of these theorems are rather easy to prove, while others are very hard, as we will see in the chapters to come. Here is another one which is not too difficult: let $T_1(n)$ denote the number of partitions of n where odd parts may be repeated, but the even parts (if any) must be distinct. Let $T_2(n)$ denote the number of partitions of n in which no part can be used more than three times. Then we claim that $T_1(n) = T_2(n)$. To prove this we write down the generating function for each type. Distinct even parts come from

$$(1+q^2)(1+q^4)(1+q^6)(1+q^8)\cdots = (-q^2;q^2)_\infty,$$

and possibly repeated odd parts (as we saw above) from $1/(q;q^2)_\infty$, so the generating function for $T_1(n)$ is

$$\frac{(-q^2;q^2)_\infty}{(q;q^2)_\infty} = \sum_{n=0}^{\infty} T_1(n)\, q^n.$$

The generating function for $T_2(n)$, where each part can be used up to three times, is

$$(1+q+q^2+q^3)(1+q^2+q^4+q^6)(1+q^3+q^6+q^9)(1+q^4+q^8+q^{12})\cdots = \sum_{n=0}^{\infty} T_2(n)\, q^n.$$

We'll try to show that these two generating functions are actually the same. We can rewrite the second one as

$$\sum_{n=0}^{\infty} T_2(n)\, q^n = \frac{1-q^4}{1-q}\frac{1-q^8}{1-q^2}\frac{1-q^{12}}{1-q^3}\frac{1-q^{16}}{1-q^4}\frac{1-q^{20}}{1-q^5}\cdots$$

$$= \frac{(q^4;q^4)_\infty}{(q;q)_\infty}.$$

As we observed above, $(q;q)_\infty = (q;q^2)_\infty (q^2;q^2)_\infty$, and replacing q by q^2 here we also have $(q^2;q^2)_\infty = (q^2;q^4)_\infty (q^4;q^4)_\infty$. Using each of these in turn, we can say that

$$\sum_{n=0}^{\infty} T_2(n)\, q^n = \frac{(q^4;q^4)_\infty}{(q;q)_\infty}$$

$$= \frac{(q^4;q^4)_\infty}{(q;q^2)_\infty (q^2;q^2)_\infty}$$

$$= \frac{(q^4;q^4)_\infty}{(q;q^2)_\infty (q^2;q^4)_\infty (q^4;q^4)_\infty}$$

$$= \frac{1}{(q;q^2)_\infty (q^2;q^4)_\infty}.$$

By replacing q by q^2 in Euler's "odd equals distinct" theorem we get $\frac{1}{(q^2;q^4)_\infty} = (-q^2;q^2)_\infty$, so that

$$\sum_{n=0}^{\infty} T_2(n)\, q^n = \frac{(q^4;q^4)_\infty}{(q;q)_\infty}$$

$$= \frac{1}{(q;q^2)_\infty (q^2;q^4)_\infty}$$

$$= \frac{(-q^2;q^2)_\infty}{(q;q^2)_\infty}$$

$$= \sum_{n=0}^{\infty} T_1(n)\, q^n.$$

It follows that $T_1(n) = T_2(n)$ for all nonnegative integers n.

In preparation for another theorem of this type, we conclude this section by writing down the generating function for partitions in which the largest repeated part is k for some $k \geq 1$. Any larger parts must be distinct, so they are generated by $\left(-q^{k+1};q\right)_\infty$. Any smaller part may be repeated, giving $1/(q;q)_{k-1}$. Finally, k must be repeated, and to ensure this we need a factor of $q^{k+k}/(1-q^k)$. Hence the generating function for these partitions is

$$\frac{1}{(q;q)_{k-1}} \frac{q^{2k}}{1-q^k} \left(-q^{k+1};q\right)_\infty = q^{2k} \frac{\left(-q^{k+1};q\right)_\infty}{(q;q)_k}$$

$$= q^{2k} \frac{\left(-q^{k+1};q\right)_\infty}{(q;q)_k} \frac{(-q;q)_k}{(-q;q)_k}$$

(3.2.4)
$$= q^{2k} \frac{(-q;q)_\infty}{(q^2;q^2)_k} = \frac{q^{2k}}{(q;q^2)_\infty (q^2;q^2)_k}.$$

Exercises

1. Let $T_3(n)$ denote the number of partitions of n using only parts that are not multiples of 4. What is the generating function for $T_3(n)$? Show that $T_3(n) = T_1(n) = T_2(n)$, with $T_1(n)$ and $T_2(n)$ as defined above.

2. Show that the generating function for $T_2(n)$ may be rewritten as
$$\sum_{n=0}^{\infty} T_2(n)\, q^n = (1+q)\left(1+q^2\right)\left(1+q^2\right)\left(1+q^4\right)\left(1+q^3\right)\left(1+q^6\right)\cdots$$
$$= (-q;q)_\infty \left(-q^2;q^2\right)_\infty.$$

3. Show that the generating function for $T_2(n)$ may be further rewritten as
$$\sum_{n=0}^{\infty} T_2(n)\, q^n = (1+q)\left(1+q^2\right)^2 \left(1+q^3\right)\left(1+q^4\right)^2 \left(1+q^5\right)\left(1+q^6\right)^2 \cdots$$
$$= \left(-q;q^2\right)_\infty \left(-q^2;q^2\right)_\infty^2.$$

4. Some combinatorial structures are most easily described using colors. A natural way to interpret the generating function in problem 3 is by using partitions where the odd parts must be red but the even parts may be either red or blue, and there are no repeated parts except possibly for a red part and a blue part of the same size. Let $T_4(n)$ denote the number of partitions of n of this type; thus, for example, $T_4(5) = 6$, because we may have

$$5_r \quad \text{or} \quad 4_r + 1_r \quad \text{or} \quad 4_b + 1_r$$
$$\text{or} \quad 3_r + 2_r \quad \text{or} \quad 3_r + 2_b \quad \text{or} \quad 2_r + 2_b + 1_r,$$

where p_r denotes a red part and p_b a blue one. Write down the 6 partitions counted by $T_2(5)$ and try to set up a natural 1-1 correspondence between them and the 6 partitions above.

5. Extend the 1-1 correspondence you found in problem 4 to a bijective proof that $T_2(n) = T_4(n)$ for all positive integers n.

6. Give a bijective proof that $T_4(n) = T_1(n)$ for all positive integers n. (**Hint:** Glaisher.) Together with problem 5, this constructs a bijective proof that $T_1(n) = T_2(n)$ for all positive integers n.

7. Let $M_1(n)$ be the number of partitions of n in which no part appears exactly once. Explain why
$$\sum_{n=0}^{\infty} M_1(n)\, q^n = \left(1 + \frac{q^2}{1-q}\right)\left(1 + \frac{q^4}{1-q^2}\right)\left(1 + \frac{q^6}{1-q^3}\right)\left(1 + \frac{q^8}{1-q^4}\right)\cdots.$$

8. Show that the generating function in problem 7 can be rewritten as
$$\sum_{n=0}^{\infty} M_1(n)\, q^n = \frac{(-q^3;q^3)_\infty}{(q^2;q^2)_\infty} = \frac{1}{(q^2;q^2)_\infty\, (q^3;q^6)_\infty}.$$

9. Let $M_2(n)$ be the number of partitions of n in which all parts are congruent to 0, 2, 3, or 4 mod 6; in other words, all the parts are either divisible by 6 or have remainder 2, 3, or 4 when divided by 6. Explain why problems 7 and 8 show that $M_1(n) = M_2(n)$.

10. There are several finite forms of (3.2.3). The most interesting one will reappear in the next section (it was also in Chapter 2), but here are two straightforward ones. Prove that

$$(-q;q)_{2n} = \frac{(q^{2n+2};q^2)_n}{(q;q^2)_n} \quad \text{and} \quad (-q;q)_{2n-1} = \frac{(q^{2n};q^2)_n}{(q;q^2)_n},$$

the former for $n \geq 0$ and the latter for $n \geq 1$.

11. If $|q| < 1$, show that

$$\sum_{j=0}^{\infty}\sum_{k=0}^{\infty} q^{k+\binom{k+j+1}{2}} = \frac{1}{1-q}.$$

Also see the next problem.

12. (a) The identity in problem 11 has a simple number-theoretic interpretation. Explain why it implies that any nonnegative integer n can be written in the form $n = k + \binom{k+j+1}{2}$ in exactly one way, where k and j are nonnegative integers.

(b) The representation of n given above is easy to find by a greedy algorithm. First choose $k + j$ as large as possible such that $n \geq \binom{k+j+1}{2}$, then set $k = n - \binom{k+j+1}{2}$, which also determines j. For example, if $n = 38$ we have $\binom{9}{2} = 36$ and $\binom{10}{2} = 45$, so $k + j = 8$. Then $k = 38 - 36 = 2$, so $j = 8 - 2 = 6$. Find the representation of 86 by this algorithm. Choose some small three-digit number and find its representation.

(c) The algorithm in (b) obviously finds a nonnegative k. Show that it also finds a nonnegative j. **Hint**: If j is negative, then $k + j + 1 \leq k$.

(d) The existence part of (a) follows from (c). Give a similar argument to show the uniqueness. (If $k+j$ is too large, then k is negative, and if $k+j$ is too small, then j is negative.)

13. Show that the left side of (3.2.4) can be rewritten as the right side.

14. For a given nonnegative integer k and a given positive integer d, call a **Smoot** partition with parameters k and d one in which k is the largest part that occurs at least d times. Show that the generating function for Smoot partitions with parameters k and d is

(3.2.5)
$$\frac{q^{dk}}{(q^d;q^d)_k} \frac{(q^d;q^d)_\infty}{(q;q)_\infty}.$$

The derivation of the generating function for $T_2(n)$ might help.

15. Show that (3.2.5) reduces to (3.2.4) if $d = 2$. What happens to (3.2.5) if $d = 1$?

16. (a) By considering even and odd n, show that

$$\sum_{n=0}^{\infty} \frac{(-1)^n q^{\binom{n+1}{2}}}{(-q;q)_n} = \sum_{m=0}^{\infty} \frac{q^{\binom{2m+1}{2}}}{(-q;q)_{2m+1}}.$$

(b) Show similarly that

$$\sum_{n=0}^{\infty} \frac{(-1)^n q^{\binom{n+1}{2}}}{(-q;q)_n} = 1 - \sum_{m=1}^{\infty} \frac{q^{\binom{2m}{2}}}{(-q;q)_{2m}}.$$

(c) Explain why (a) and (b) imply that

$$\sum_{k=1}^{\infty} \frac{q^{\binom{k}{2}}}{(-q;q)_k} = 1.$$

17. Show by induction that

$$\sum_{k=1}^{n} \frac{q^{\binom{k}{2}}}{(-q;q)_k} = 1 - \frac{q^{\binom{n+1}{2}}}{(-q;q)_n},$$

and explain why (assuming as usual that $|q| < 1$) this implies the result of part (c) of the previous problem.

18. Show by induction that

$$\sum_{k=1}^{n} \frac{(-1)^{k-1} q^{(k-1)^2}}{(q;q^2)_k} = 1 - \frac{(-1)^n q^{n^2}}{(q;q^2)_n}.$$

Assuming as usual that $|q| < 1$, what happens to this as $n \to \infty$?

19. Let s be a positive integer and $|q| < 1$. We want to evaluate the sum

$$O_s(q) = \sum_{k=1}^{\infty} \frac{q^{ks}}{\binom{s+k}{s+1}_q}.$$

(i) Show that we can rewrite

$$O_s(q) = (q;q)_{s+1} \sum_{k=1}^{\infty} \frac{q^{ks}}{(q^k;q)_{s+1}}.$$

(ii) By multiplying the right side of (i) by

$$\frac{1-q^s}{1-q^s} = \frac{1 - q^{k+s} - q^s(1-q^k)}{1-q^s},$$

show that

$$O_s(q) = \frac{(q;q)_{s+1}}{1-q^s} \sum_{k=1}^{\infty} \left[\frac{q^{ks}}{(q^k;q)_s} - \frac{q^{(k+1)s}}{(q^{k+1};q)_s} \right].$$

(iii) Explain why this gives

$$O_s(q) = \frac{(q;q)_{s+1}}{1-q^s} \frac{q^s}{(q;q)_s} = q^s \frac{1-q^{s+1}}{1-q^s}.$$

20. Show similarly that if s is a positive integer and $|q| < 1$, then

$$\sum_{k=1}^{\infty} \frac{q^k}{\binom{s+k}{s+1}_q} = \frac{1-q^{s+1}}{1-q^s} \left[1 - (q;q)_s \right].$$

21. For $|q| < 1$, consider the sum

$$S(x) = \sum_{k=1}^{\infty} \frac{x^k (q;q)_{k-1}}{(xq;q)_k}.$$

(i) Show that the series converges if $|x| < 1$.

(ii) By multiplying top and bottom by $1-x$ and using a trick like (ii) in problem 19, show that

$$S(x) = \frac{x}{1-x} \quad \text{for } |x| < 1.$$

(iii) Use (ii) to give an alternate solution of problem 19.

22. In one of his papers Jacobi writes that not only do we have

$$(1+q)(1+q^2)(1+q^3)(1+q^4)\cdots = \frac{1}{(1-q)(1-q^3)(1-q^5)(1-q^7)\cdots},$$

as we know from Euler, but also

$$(1+q)(1+q^2)(1+q^3)(1+q^4)\cdots$$

$$= \frac{(1-q^2)(1-q^4)(1-q^6)(1-q^8)\cdots}{(1-q)(1-q^2)(1-q^3)(1-q^4)\cdots}$$

$$= \frac{(1+q)(1+q^3)(1-q^4)(1+q^5)(1+q^7)(1-q^8)\cdots}{(1-q^2)(1-q^4)(1-q^6)(1-q^8)(1-q^{10})(1-q^{12})\cdots}$$

$$= \frac{(1-q)(1-q^2)(1-q^3)(1-q^4)(1-q^5)(1-q^6)\cdots}{(1-q)^2(1-q^2)(1-q^3)^2(1-q^4)(1-q^5)^2(1-q^6)\cdots}$$

$$= \frac{(1+q)(1-q^2)(1+q^3)(1-q^4)(1+q^5)(1-q^6)\cdots}{(1-q^2)^2(1-q^4)(1-q^6)^2(1-q^8)(1-q^{10})^2(1-q^{12})\cdots}$$

$$= \frac{(1-q^4)(1-q^8)(1-q^{12})(1-q^{16})(1-q^{20})(1-q^{24})\cdots}{(1-q)(1-q^3)(1-q^4)(1-q^5)(1-q^7)(1-q^8)\cdots}$$

$$= \frac{(1+q)(1+q^2)(1-q^3)(1+q^4)(1+q^5)(1-q^6)\cdots}{(1-q^3)^2(1-q^6)(1-q^9)^2(1-q^{12})(1-q^{15})^2(1-q^{18})\cdots}$$

$$= \frac{(1+q^3)(1+q^9)(1-q^{12})(1+q^{15})(1+q^{21})(1-q^{24})\cdots}{(1-q)(1-q^5)(1-q^6)(1-q^7)(1-q^{11})(1-q^{12})\cdots}.$$

Prove this, and write each fraction in the q-shifted factorial notation. (See problem 29 in section 5.1 for answers to the latter.) The pattern of the denominator of the last fraction may not be obvious; it is supposed to continue

$$(1-q^{13})(1-q^{17})(1-q^{18})(1-q^{19})(1-q^{23})(1-q^{24})(1-q^{25})\cdots.$$

3.3. Ferrers diagrams

There is a very simple way to draw a "picture" of a partition, which is usually called its **Ferrers diagram** (or Ferrers graph). The idea is to represent each part by a row of dots corresponding to the part size. For example, the usual Ferrers diagram of the partition $7+6+6+3+3+1$ is:

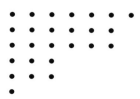

The partition obtained by reading the diagram by columns, rather than by rows, is called the **conjugate** of the original partition. Thus the conjugate of $7 + 6 + 6 + 3 + 3 + 1$ is $6 + 5 + 5 + 3 + 3 + 3 + 1$ and its Ferrers diagram is:

We saw a similar idea in the Rothe diagram of a permutation in Chapter 1, but Ferrers diagrams turn out to be more useful. As with permutations, it is possible to give a nongraphical definition of the conjugate: given a partition π whose largest part is k, set a_i equal to the number of parts of π which are greater than or equal to i; then $a_1 + a_2 + \cdots + a_k$ is the partition conjugate to π. But the graphical definition is much more easily apprehended. Since the largest part of a partition equals the number of parts of its conjugate, and vice versa, we have

THEOREM 21. *Let $p(n; i, j)$ denote the number of partitions of n into at most i parts, each of which is at most j. Then $p(n; i, j)$ is symmetric in i and j: $p(n; i, j) = p(n; j, i)$. Moreover, the number of partitions of n with at most i parts equals the number of partitions of n whose largest part is i.*

Both statements follow immediately by taking conjugates. An important corollary is

COROLLARY 1. *The generating function for partitions whose parts are at most n, namely $1/(q;q)_n$, is also the generating function for partitions with at most n parts.*

Occasionally, it is convenient to rephrase a result like this as: $1/(q;q)_n$ is the generating function for partitions with exactly n parts, some of which might be 0. We have not allowed 0 to be a part up till now, and when we do so, as here, it will be only temporary. If we have such a partition and we add 1 to each part, then we create a partition with n nonzero parts. It follows that $q^n/(q;q)_n$ is the generating function for partitions with exactly n parts. Combining these last two results, we get a pretty little summation theorem, which we saw in section 2.3, problem 4:

$$(3.3.1) \qquad \frac{1}{(q;q)_n} = \sum_{k=0}^{n} \frac{q^k}{(q;q)_k}.$$

For we know that the left side generates partitions with at most n parts. Such a partition has exactly k parts for some k between 0 and n, so the right side generates the same partitions as the left side. It is also quite easy to prove (3.3.1) by induction on n.

In a similar way, $1/(q^2;q^2)_n$ is the generating function for partitions with at most n parts, all of which are even, or with exactly n parts, some of which might be 0 but all of which are even. By adding 1 to each part we arrive at n nonzero parts, all of which are odd, and so we conclude that $q^n/(q^2;q^2)_n$ is the generating function for partitions with exactly n parts, all of which are odd. See problem 7 for an application of this.

Some partitions are their own conjugates, for example 9+8+6+6+6+5+2+2+1:

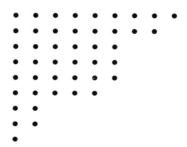

Such partitions are called **self-conjugate**. If we read their graphs by moving left across the first row and then down the first column, then left across the second row and down the second column, and so forth, we get partitions into distinct odd parts—for the above graph this way of reading gives $17 + 13 + 7 + 5 + 3$:

```
• • • • • • • • •
• * * * * * * *
• * ◊ ◊ ◊ ◊
• * ◊ ★ ★ ★
• * ◊ ★ ○ ○
• * ◊ ★ ○
• *
• *
•
```

The parts are odd because each row is the same length as the corresponding column, but the corner where they meet is only counted once. They are distinct because each row is at least one unit shorter than the row above when we read this way, and similarly for the columns. It follows that $(-q;q^2)_\infty$, which is the generating function for partitions into distinct odd parts, is also the generating function for self-conjugate partitions; and that $(-q;q^2)_n$, which is the generating function for partitions into distinct odd parts $\leq 2n-1$, is also the generating function for self-conjugate partitions whose maximum part size is n.

This idea gives a completely transparent proof that the sum of the first n odd numbers is n^2, i.e., $1 + 3 + 5 + \cdots + (2n-1) = n^2$, a fact that we used once in Chapter 2 and will use again later. We can depict $1 + 3 + 5 + 7 + 9 + 11 = 6^2$ as:

```
• • • • • •
• * * * * *
• * ◊ ◊ ◊ ◊
• * ◊ ★ ★ ★
• * ◊ ★ ○ ○
• * ◊ ★ ○ ⊙
```

There is a similar proof of

(3.3.2) $$1 + 2 + 3 + \cdots + n = \frac{n(n+1)}{2} = \binom{n+1}{2}.$$

We can depict $1+2+3+\cdots+n+[1+2+3+\cdots+(n-1)]$ as a square comprising two triangles; for example, $1+2+3+4+5+6+1+2+3+4+5 = 36 = 6^2$ looks like:

For this reason, numbers of the form (3.3.2) are called **triangular** numbers, and we will see them again in Chapters 7 and 13. It follows that

$$1 + 2 + 3 + \cdots + n + [1 + 2 + 3 + \cdots + (n-1)] = n^2.$$

Adding n to both sides and dividing by 2 we get (3.3.2). Alternatively, we could subtract n from both sides and divide by 2 to get (3.3.2) with $n-1$ in place of n.

The conjugate idea allows us to write down the generating function for partitions with exactly k even parts (and any number of odd parts). Suppose the even parts are $14 + 8 + 6$, with Ferrers diagram:

Then the conjugate is $3+3+3+3+3+3+2+2+1+1+1+1+1+1+1$, in which the largest part is 3, it occurs at least twice, and every smaller part occurs an even number of times. By combining the even parts in pairs this becomes $6+6+6+4+2+2+2$, in which the largest part is 6 and every smaller part is even. If we treat k even parts the same way, we get a partition with largest part $2k$ and all the smaller parts even, so the generating function for these k even parts is

$$\frac{q^{2k}}{1-q^{2k}} \frac{1}{(q^2;q^2)_{k-1}} = \frac{q^{2k}}{(q^2;q^2)_k}.$$

Since the odd parts are unrestricted, the generating function for partitions with exactly k even parts is

$$\frac{q^{2k}}{(q^2;q^2)_k} \frac{1}{(q;q^2)_\infty}.$$

Comparing this with (3.2.4), we have proved

THEOREM 22 (The Andrews–Deutsch theorem). *There are exactly as many partitions of a positive integer n with k even parts as there are partitions of n whose largest repeated part is k.*

More generally, call a **Yang** partition with parameters k and d one in which exactly k parts are divisible by d. The generating function for the parts *not* divisible by d is

$$\frac{1}{(1-q)\cdots(1-q^{d-1})(1-q^{d+1})\cdots(1-q^{2d-1})(1-q^{2d+1})\cdots} = \frac{(q^d;q^d)_\infty}{(q;q)_\infty}.$$

For the k parts that are divisible by d, take for example $16 + 12 + 12 + 12 + 8 + 8$ with $k = 6$ and $d = 4$. The Ferrers diagram is

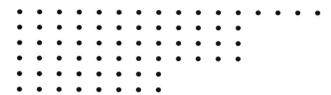

Then the conjugate is $6+6+6+6+6+6+6+6+4+4+4+4+1+1+1+1$, in which the largest part is 6, it occurs at least four times (it occurs eight times in this example because the original partition had no 4's), and every smaller part occurs a number of times divisible by 4. By combining the parts in groups of four this becomes $24 + 24 + 16 + 4$, in which the largest part is 24 and every smaller part is a multiple of four. It follows that the generating function for the k parts divisible by d is

$$\frac{q^{kd}}{1-q^{kd}} \frac{1}{(q^d;q^d)_{k-1}} = \frac{q^{kd}}{(q^d;q^d)_k},$$

and hence that the generating function for Yang partitions with parameters k and d is

$$\frac{q^{kd}}{(q^d;q^d)_k} \frac{(q^d;q^d)_\infty}{(q;q)_\infty}.$$

Comparing this with (3.2.5), we have proved

THEOREM 23 (The Smoot–Yang theorem). *There are exactly as many partitions of a positive integer n with exactly k parts divisible by d as there are partitions of n in which k is the largest part that occurs at least d times.*

For a variation on the conjugate idea, let an **ee partition** be a partition where the even parts may be repeated but the odd parts, if any, must be distinct. If the smallest part is odd, then we still have an ee partition (perhaps with fewer parts) if we subtract 1 from it and 2 from all the other parts. If the smallest part is even, then we still have an ee partition if we subtract 2 from all the parts. This suggests an algorithm for ee partitions, which we'll call the ee algorithm. We illustrate it with the example $18 + 11 + 8 + 8 + 8 + 7 + 3$. Call the two subtractions above the odd step and the even step, respectively. Since the smallest part is odd, perform the odd step, *i.e.*, write $18 + 11 + 8 + 8 + 8 + 7 + 3$ as the array:

$$\begin{array}{ccccccc} 2 & 2 & 2 & 2 & 2 & 2 & 1 \\ 16 & 9 & 6 & 6 & 6 & 5 & 2 \end{array}$$

The smallest part on the second line is even, so perform the even step on it:

$$\begin{array}{ccccccc} 2 & 2 & 2 & 2 & 2 & 2 & 1 \\ 2 & 2 & 2 & 2 & 2 & 2 & 2 \\ 14 & 7 & 4 & 4 & 4 & 3 & \end{array}$$

The smallest part on the third line is odd, so we perform the odd step again, after which the smallest part on the fourth line will be even, causing us to perform the even step:

$$
\begin{array}{cccccc}
2 & 2 & 2 & 2 & 2 & 1 \\
2 & 2 & 2 & 2 & 2 & 2 \\
2 & 2 & 2 & 2 & 1 \\
12 & 5 & 2 & 2 & 2 & 2
\end{array}
\longrightarrow
\begin{array}{ccccccc}
2 & 2 & 2 & 2 & 2 & 1 \\
2 & 2 & 2 & 2 & 2 & 2 \\
2 & 2 & 2 & 2 & 1 \\
2 & 2 & 2 & 2 & 2 \\
10 & 3
\end{array}
$$

Performing again the odd step and then the even step we have:

$$
\begin{array}{ccccccc}
2 & 2 & 2 & 2 & 2 & 1 \\
2 & 2 & 2 & 2 & 2 & 2 \\
2 & 2 & 2 & 2 & 1 \\
2 & 2 & 2 & 2 & 2 \\
2 & 1 \\
8 & 2
\end{array}
\longrightarrow
\begin{array}{ccccccc}
2 & 2 & 2 & 2 & 2 & 1 \\
2 & 2 & 2 & 2 & 2 & 2 \\
2 & 2 & 2 & 2 & 1 \\
2 & 2 & 2 & 2 & 2 \\
2 & 1 \\
2 & 2 \\
6
\end{array}
$$

Performing the even step three more times we finally have:

$$
\begin{array}{ccccccc}
2 & 2 & 2 & 2 & 2 & 1 \\
2 & 2 & 2 & 2 & 2 & 2 \\
2 & 2 & 2 & 2 & 1 \\
2 & 2 & 2 & 2 & 2 \\
2 & 1 \\
2 & 2 \\
2 \\
2 \\
2
\end{array}
$$

It's clear that there is at most one 1 in each row and that it comes at the end. Because we take 1 out of an odd number only once, when it is the smallest part left, there is also at most one 1 in each column, and it appears as high in the column as it can be while remaining on the end of its row. If we now move all the 1's down to the bottom of their columns, we have

$$
\begin{array}{cccccc}
2 & 2 & 2 & 2 & 2 & 2 \\
2 & 2 & 2 & 2 & 2 & 1 \\
2 & 2 & 2 & 2 & 2 \\
2 & 2 & 2 & 2 & 1 \\
2 & 2 \\
2 & 1 \\
2 \\
2 \\
2
\end{array}
$$

Reading this by rows we get the partition $14 + 13 + 12 + 11 + 4 + 3 + 2 + 2 + 2$. The ee algorithm must give distinct odd parts, because there is only one 1 in each row; this remains true even after moving the 1's down, because this can always be done by exchanging rows. It also implies a restriction on the part size: if we started with at most n parts (here $n \geq 7$), the even parts must be $\leq 2n$ and the odd parts must be $\leq 2n-1$. The ee algorithm is easily reversible: given a partition with even parts $\leq 2n$ and distinct odd parts $\leq 2n-1$, write it as above, broken apart into 2's with 1's on the end for the odd parts. All we really have to do after that is read the columns, but to reverse the algorithm we should move the 1's up the columns as high as they can go while remaining on the end of their rows. This gives exactly the sort of array that the ee algorithm produces for an ee partition, so reading the columns that's what we must get. It follows that the generating function for ee partitions with at most n parts is

$$(3.3.3) \qquad \frac{(-q;q^2)_n}{(q^2;q^2)_n}.$$

It is also interesting to give an analytic proof of (3.3.3). The generating function for ee partitions with exactly one part is clearly

$$q + q^2 + q^3 + q^4 + \cdots = \frac{q}{1-q},$$

because the ee restriction is no restriction at all in this case, so the generating function for ee partitions with at most one part is (adding in the empty partition)

$$1 + \frac{q}{1-q} = \frac{1}{1-q} = \frac{1+q}{1-q^2},$$

in agreement with (3.3.3).

Let's try to construct the generating function for ee partitions with exactly two parts and smallest part $2k-1$. If we subtract $2k-1$ from it and $2k$ from the other part (note that this is what the ee algorithm would do, in k steps) we have an ee partition with at most one part. Therefore

$$q^{2k-1+2k} \frac{1}{1-q} = \frac{q^{4k-1}}{1-q}$$

is the generating function for ee partitions with exactly two parts and smallest part $2k-1$. If instead the smallest part is $2k$, subtracting $2k$ from both parts (which is again what the ee algorithm would do in k steps) we have an ee partition with at most one part. Therefore

$$q^{2k+2k} \frac{1}{1-q} = \frac{q^{4k}}{1-q}$$

is the generating function for ee partitions with exactly two parts and smallest part $2k$, and hence the generating function for ee partitions with exactly two parts and smallest part either $2k-1$ or $2k$ is

$$q^{4k-1} \frac{1+q}{1-q}.$$

Summing this over all $k \geq 1$ we get

$$\frac{1+q}{1-q} (q^3 + q^7 + q^{11} + \ldots) = \frac{q^3(1+q)}{(1-q)(1-q^4)},$$

which must be the generating function for ee partitions with exactly two parts. Hence the generating function for ee partitions with at most two parts is

$$\frac{1}{1-q} + \frac{q^3(1+q)}{(1-q)(1-q^4)} = \frac{1-q^4+q^3+q^4}{(1-q)(1-q^4)} = \frac{1+q^3}{(1-q)(1-q^4)} = \frac{(1+q)(1+q^3)}{(1-q^2)(1-q^4)},$$

in agreement with (3.3.3).

Continuing in this way, suppose we have verified (3.3.3) up to n. It follows that the generating function for ee partitions with exactly n parts must be

(3.3.4) $$\frac{(-q;q^2)_n}{(q^2;q^2)_n} - \frac{(-q;q^2)_{n-1}}{(q^2;q^2)_{n-1}} = \frac{q^{2n-1}(1+q)(-q;q^2)_{n-1}}{(q^2;q^2)_n}.$$

If we have an ee partition with exactly $n+1$ parts and smallest part $2k-1$, then subtracting $2k-1$ from it and $2k$ from the other n parts we get an ee partition with at most n parts; as we also do if we subtract $2k$ from an ee partition with exactly $n+1$ parts and smallest part $2k$. Therefore, by induction and (3.3.3), the generating function for ee partitions with exactly $n+1$ parts and smallest part either $2k-1$ or $2k$ is

$$q^{2k-1}(1+q)q^{2nk}\frac{(-q;q^2)_n}{(q^2;q^2)_n} = q^{2k(n+1)-1}(1+q)\frac{(-q;q^2)_n}{(q^2;q^2)_n}.$$

Summing this for $k \geq 1$ we get

$$\frac{(1+q)(-q;q^2)_n}{(q^2;q^2)_n}\left(q^{2n+1} + q^{4n+3} + q^{6n+5} + \ldots\right) = \frac{(1+q)(-q;q^2)_n}{(q^2;q^2)_n}\frac{q^{2n+1}}{1-q^{2n+2}},$$

which is (3.3.4) with $n+1$ in place of n. Adding this to (3.3.3) we must get (3.3.3) with $n+1$ in place of n, by the same arithmetic as in (3.3.4). This proves (3.3.3).

Exercises

1. For a self-conjugate partition, what can you say about the Ferrers diagram?
2. What can you say about the partitions which are the conjugates of the ones counted by $M_1(n)$ in problem 7 of the previous section?
3. (This problem was suggested by Elena Warters.) Suppose n is a prime number, and consider the partitions counted by $M_1(n)$. Are there any whose conjugates are also counted by $M_1(n)$? Are there any that are self-conjugate?
4. Recall Gauss's identity (2.5.8)

$$(-q;q)_n = \sum_{k=0}^{n}\binom{n}{k}_{q^2} q^k.$$

Explain why letting $n \to \infty$ here gives (at least formally)

$$(-q;q)_\infty = \sum_{k=0}^{\infty}\frac{q^k}{(q^2;q^2)_k}.$$

5. In the identity in problem 4, what partitions does the left side generate? What partitions does the right side generate? What does this mean?
6. A generalization of (3.3.1) is

(3.3.5) $$\frac{1}{(x;q)_n} = 1 + \sum_{k=0}^{n-1}\frac{xq^k}{(x;q)_{k+1}}.$$

Prove this by induction on n.

7. Explain why
$$\frac{1}{(q;q)_\infty} = \sum_{k=0}^{\infty} \frac{q^k}{(q;q)_k}.$$

8. We know that $(-q;q)_n$ is the generating function for partitions with distinct parts $\leq n$. By considering the largest part in each partition, show that

(3.3.6) $$(-q;q)_n = 1 + \sum_{k=1}^{n} q^k (-q;q)_{k-1}.$$

9. By considering the smallest part in each partition instead, show that

(3.3.7) $$(-q;q)_n = 1 + \sum_{k=1}^{n} q^k (-q^{k+1};q)_{n-k}.$$

10. Show that (3.3.7) can be rewritten as
$$\sum_{k=1}^{n} \frac{q^k}{(-q;q)_k} = 1 - \frac{1}{(-q;q)_n},$$
and that this is a special case of (3.3.3).

11. By writing $xq^k = 1 + xq^k - 1$, prove the following generalization of (3.3.6):

(3.3.8) $$(-x;q)_n = 1 + x \sum_{k=0}^{n-1} q^k (-x;q)_k.$$

12. Prove (3.3.8) by induction on n. **Hint:** $(-x;q)_{n+1} = (1+x)(-xq;q)_n$.

13. Verify (3.3.4).

14. Explain why
$$\frac{(-q;q^2)_n}{(q^2;q^2)_n} = 1 + \sum_{k=1}^{n} \frac{q^{2k-1}(1+q)(-q;q^2)_{k-1}}{(q^2;q^2)_k}.$$

15. Explain why
$$\frac{(-q;q^2)_\infty}{(q^2;q^2)_\infty} = 1 + \sum_{k=1}^{\infty} \frac{q^{2k-1}(1+q)(-q;q^2)_{k-1}}{(q^2;q^2)_k}.$$

16. An **oo partition** is a partition where the odd parts may be repeated but the even parts, if any, must be distinct. Explain why the generating function for oo partitions with exactly n parts must be
$$q^n \frac{(-q;q^2)_n}{(q^2;q^2)_n}.$$

17. Explain why
$$\sum_{n=0}^{\infty} q^n \frac{(-q;q^2)_n}{(q^2;q^2)_n} = \frac{(-q^2;q^2)_\infty}{(q;q^2)_\infty}.$$

EXERCISES

18. Use induction on n to prove **Euler's lemma**: if $e_1, \ldots, e_n$ are arbitrary and $E_k = 1 + e_k$ for each k, then
$$E_1 E_2 \cdots E_n = 1 + e_1 + E_1 e_2 + E_1 E_2 e_3 + \cdots + E_1 E_2 \cdots E_{n-1} e_n$$
for any $n \geq 1$.

19. Suppose that $\{a_k\}$ is a sequence of complex numbers that is completely arbitrary except that $a_k \neq -1$ for all k. By writing $a_k = 1 + a_k - 1$, by induction, or otherwise, show that

(3.3.9)
$$\sum_{k=1}^{n} \frac{a_k}{(1+a_1)(1+a_2)\cdots(1+a_k)} = 1 - \frac{1}{(1+a_1)(1+a_2)\cdots(1+a_n)}.$$

20. Show that (3.3.9) and Euler's lemma are equivalent. (**Hint:** There is a simple relationship between e_k and a_k.)

21. Another form of this fact is **Nicole's identity**
$$\frac{1}{z} - \frac{z_1 \cdots z_n}{z(z+z_1)\cdots(z+z_n)} = \sum_{k=1}^{n} \frac{z_1 \cdots z_{k-1}}{(z+z_1)\cdots(z+z_k)},$$
where the only restriction is that $z, z+z_1, \ldots, z+z_n$ are all nonzero. Prove this, by induction or otherwise, and show that it is equivalent to (3.3.9).

22. Yet another form of the same fact is that for any nonzero numbers $b_1, \ldots, b_n$ we have
$$\frac{b_1 - 1}{b_1} + \frac{b_2 - 1}{b_1 b_2} + \frac{b_3 - 1}{b_1 b_2 b_3} + \cdots + \frac{b_n - 1}{b_1 \cdots b_n} = 1 - \frac{1}{b_1 \cdots b_n}.$$
Prove this and show it is equivalent to one of the other forms.

23. By taking $e_k = xq^{k-1}(b-a)/(1-bxq^{k-1})$ in Euler's lemma, or by taking $a_k = x(a-b)q^{k-1}/(1-axq^{k-1})$ in (3.3.7), or otherwise, show that
$$\frac{(ax;q)_n}{(bx;q)_n} = 1 + x(b-a) \sum_{k=1}^{n} \frac{(ax;q)_{k-1}}{(bx;q)_k} q^{k-1}.$$
Note that (3.3.5) and (3.3.8) are special cases of this.

24. By taking $e_k = xq^{n-k}(b-a)/(1-bxq^{n-k})$ in Euler's lemma, or otherwise, show that
$$\frac{(ax;q)_n}{(bx;q)_n} = 1 + x(b-a) \sum_{k=1}^{n} \frac{(axq^k;q)_{n-k}}{(bxq^{k-1};q)_{n-k+1}} q^{k-1}.$$
Note that (3.3.7) is a special case of this.

25. Take $E_k = \frac{[k+m]_q}{[k]_q}$ for a nonnegative integer m in Euler's lemma. What is e_k, and what does Euler's lemma say?

26. By taking
$$a_k = \frac{xq^{k-1}\left(a_1 + a_2 - b_1 - b_2 + (b_1 b_2 - a_1 a_2)xq^{k-1}\right)}{(1 - a_1 xq^{k-1})(1 - a_2 xq^{k-1})}$$

in (4.1.1), or otherwise, show that

$$1 + \sum_{k=1}^{n} \frac{\left(b_1 + b_2 - a_1 - a_2 + (a_1 a_2 - b_1 b_2) x q^{k-1}\right) x q^{k-1} (a_1 x; q)_{k-1} (a_2 x; q)_{k-1}}{(b_1 x; q)_k (b_2 x; q)_k}$$

$$= \frac{(a_1 x; q)_n (a_2 x; q)_n}{(b_1 x; q)_n (b_2 x; q)_n}.$$

27. Suppose $\{a_k\}$ is a positive sequence and $\sum_{k=1}^{\infty} a_k$ diverges. What can you say about

$$\sum_{k=1}^{\infty} \frac{a_k}{(1+a_1)(1+a_2)\cdots(1+a_k)}?$$

28. What happens if we take $a_i = aq^{i-1}$ in the previous problem where $q > 1$? What does it become if you replace q by q^{-1} to get back to the usual case of a small q?

3.4. q-binomial coefficients and partitions

We rephrase something that we proved in section 3.3: if $p(n; i, j)$ denotes the number of partitions of n into at most i parts, each at most j, then the generating function of the $p(n; i, j)$, namely

$$f_{i,j}(q) = \sum_{n=0}^{ij} p(n; i, j) q^n,$$

is a symmetric function of i and j. In fact, it is a function that we are very familiar with:

THEOREM 24 (Cayley's theorem). *The generating function for partitions into at most i parts, each at most j, is*

$$\sum_{n=0}^{ij} p(n; i, j) q^n = \binom{i+j}{i}_q.$$

There are many ways to prove this. Actually, we were quite close to proving it in Chapter 1, as it is more or less the same thing as Pólya's Property of the q-binomial coefficients. The 1-1 correspondence between sequences of 0's and 1's and lattice paths given there essentially constructs a Ferrers diagram of a partition, rotated by 180° from the standard orientation. For example, consider the sequence 1011000110010. Let's count how many inversions it has by counting how many 0's come after each 1: there are 7 0's after the first 1; 6 0's after the second and third 1's; 3 0's after the fourth and fifth 1's, and 1 0 after the last 1. Thus this sequence corresponds to the partition $7 + 6 + 6 + 3 + 3 + 1$. If we instead count inversions by counting how many 1's come before each 0, working from right to left we have 6 1's before the last 0; 5 1's before the second and third from last 0's; 3 1's before the fourth, fifth and sixth from last 0's, and 1 1 before the first 0—thus the sequence also corresponds to the partition $6 + 5 + 5 + 3 + 3 + 3 + 1$, which is the conjugate of $7 + 6 + 6 + 3 + 3 + 1$ as we saw above. If you take a piece of graph paper, draw an x and y axis, and start from the origin with a 1 marking a step in the positive

y-direction and a 0 marking a step in the positive x-direction, the boxes between the path and the x-axis will be (up to reorientation) a Ferrers diagram.

For another proof we can use either of the q-Pascal recurrences, which we restate here with a slight change in notation:

(3.4.1) $$\binom{i+j}{i}_q = \binom{i+j-1}{i-1}_q + q^i \binom{i+j-1}{i}_q,$$

(3.4.2) $$\binom{i+j}{i}_q = \binom{i+j-1}{i}_q + q^j \binom{i+j-1}{i-1}_q.$$

We are trying to argue that $\binom{i+j}{i}_q$ is the generating function for partitions with at most i parts each at most j. We get (3.4.1) by asking "are there really i parts?" If not, then there are at most $i-1$ parts, and we get the first term on the right side of (3.4.1). If there are i parts, then subtract 1 from each of them. Now we have at most i parts, each of which is at most $j-1$, and we get the last term in (3.4.1) (with the factor q^i making up for the i 1's which we subtracted).

We get (3.4.2) by asking "is there really a part of size j?" If not, then the maximum part size is really $j-1$, and we get the first term on the right side of (3.4.2). If so, then aside from this part we have at most $i-1$ parts, each of which is at most j, and we get the last term in (3.4.2) (with the factor q^j accounting for the one part of size j which we pulled out).

Note that Cayley's theorem implies that $\binom{n}{k}_q$ is the generating function for partitions into at most k parts, each at most $n-k$.

Cayley's theorem is often attributed to Sylvester, who published an epoch-making paper on partitions in the early 1880s. His proof was rather different, and we will see it in the next section. In fact, Cayley published the theorem in a paper of 1855; his argument was the same as Sylvester's. (More accurately, Sylvester's argument was the same as Cayley's. Cayley and Sylvester were close friends of long standing.) One of Sylvester's students, Fabian Franklin, devised still another proof, based upon the idea of the "excesses" of a partition. Let the kth excess of a partition be the largest part minus the $(k+1)^{\text{th}}$ largest part. For example, for the partition $7+6+6+3+3+1$, the first and second excesses are both $7-6=1$, the third and fourth excesses are $7-3=4$, and the fifth excess is $7-1=6$.

Recall that $1/(q;q)_j$ is the generating function for partitions with at most j parts. For convenience, let's again recast it as the generating function for partitions with exactly j parts, some of which might be 0. Suppose we have a partition with j parts in this sense and whose first excess is greater than i. There is a 1-1 correspondence between these partitions and ordinary partitions with at most j parts, because we could subtract $i+1$ from the largest part and it would still be at least as big as any other part. It follows that the generating function for partitions with j parts and whose first excess is greater than i is $q^{i+1}/(q;q)_j$. It further follows that the generating function for partitions with at most j parts whose first excess is not greater than i is

$$\frac{1}{(q;q)_j} - \frac{q^{i+1}}{(q;q)_j} = \frac{1-q^{i+1}}{(q;q)_j}.$$

Next, consider partitions with j parts (some possibly 0) whose first excess is less than or equal to i and whose second excess (largest part minus third largest) is greater than i. If we subtract $i+1$ from the largest part and 1 from the second largest, then what was the second largest part is now definitely the largest (because the first excess was less than i), and what was the largest part is now second largest (because it is still at least as big as the third largest), so we should switch the first and second parts to get the standard form with decreasing parts. If the first excess was originally e_1, then it is now $i - e_1$, so it is still between 0 and i inclusive. Therefore the generating function for these partitions is

$$\frac{q^{i+2}(1 - q^{i+1})}{(q;q)_j}.$$

It follows that the generating function for partitions with at most j parts and with first and second excesses both less than or equal to i is

$$\frac{1 - q^{i+1}}{(q;q)_j} - \frac{q^{i+2}(1 - q^{i+1})}{(q;q)_j} = \frac{(1 - q^{i+1})(1 - q^{i+2})}{(q;q)_j}.$$

Note that if the second excess is $\leq i$, then the first excess is automatically $\leq i$.

We keep doing this argument over and over. We do one more step, which will indicate how the general case goes. Now that we have the generating function for partitions whose second excess is less than or equal to i, we construct the generating function for such partitions whose third excess is greater than i. This time we subtract $i + 1$ from the first part, and 1 from each of the second and third parts. Since the first two excesses were no more than i, the second part is now the biggest, and the third part second biggest, and the first part third biggest. Switching second to first, third to second, and first to third we get a generic partition with first and second excesses still $\leq i$. It follows that

$$\frac{q^{i+3}(1 - q^{i+1})(1 - q^{i+2})}{(q;q)_j}$$

is the generating function for these partitions, and therefore that

$$\frac{(1 - q^{i+1})(1 - q^{i+2})}{(q;q)_j} - \frac{q^{i+3}(1 - q^{i+1})(1 - q^{i+2})}{(q;q)_j} = \frac{(1 - q^{i+1})(1 - q^{i+2})(1 - q^{i+3})}{(q;q)_j}$$

is the generating function for partitions with at most j parts and third excess $\leq i$. After k iterations (where $k \leq j - 1$) we find that

$$\frac{q^{i+k}(1 - q^{i+1})(1 - q^{i+2}) \ldots (1 - q^{i+k-1})}{(q;q)_j}$$

is the generating function for partitions with at most j parts, k^{th} excess $> i$, and $k - 1^{\text{th}}$ excess $\leq i$; and hence

$$\frac{(1 - q^{i+1})(1 - q^{i+2}) \ldots (1 - q^{i+k})}{(q;q)_j}$$

is the generating function for partitions with at most j parts and k^{th} excess $\leq i$. Since there are exactly j parts (some of which could be 0), it seems that we can use this argument $j - 1$ times before we run out of parts:

$$\frac{(1 - q^{i+1})(1 - q^{i+2}) \ldots (1 - q^{i+j-1})}{(q;q)_j}$$

is the generating function for partitions with at most j parts where the $j - 1^{\text{th}}$ excess is $\leq i$. But we can use it one more time, thinking of the largest part itself as a "last excess": if the largest part is $> i$ and all the excesses are $\leq i$, then subtract

$i+1$ from the largest part and 1 from all the other parts, and then rotate the parts as before: the transformed second part is the new largest part, the old third part is the new second part, and so forth, with the former largest part now the smallest part. All the excesses of the transformed partition are $\leq i$, so

$$\frac{q^{i+j}(1-q^{i+1})(1-q^{i+2})\cdots(1-q^{i+j-1})}{(q;q)_j}$$

is the generating function for partitions with at most j parts, largest part $> i$ and $j-1^{\text{th}}$ excess $\leq i$. Hence

$$\frac{(1-q^{i+1})(1-q^{i+2})\cdots(1-q^{i+j})}{(q;q)_j}$$

is the generating function for partitions with at most j parts where the largest part is $\leq i$ (and therefore all the excesses are too). This proves Cayley's theorem.

Exercises

1. Recall (1.4.3):

$$\binom{n+1}{k+1}_q = \binom{k}{k}_q + q\binom{k+1}{k}_q + q^2\binom{k+2}{k}_q + \cdots + q^{n-k}\binom{n}{k}_q.$$

   Prove this by counting partitions.

2. Explain why we can rewrite (1.4.5) as

$$\binom{n}{k}_q = q^{k(n-k)}\binom{k-1}{k-1}_q + q^{k(n-k-1)}\binom{k}{k-1}_q + q^{k(n-k-2)}\binom{k+1}{k-1}_q + \cdots$$
$$+ q^k\binom{n-2}{k-1}_q + \binom{n-1}{k-1}_q.$$

3. Prove the result of problem 2 by counting partitions.

4. Here is another nice combinatorial characterization of the q-binomial coefficients. Let

$$S(n,k) = \sum_{S_j} q^{s_j},$$

   where the sum is over all k-element subsets S_j of $\{1, 2, \ldots, n\}$, and s_j is the sum of the members of S_j. Then

$$S(n,k) = \binom{n}{k}_q q^{\binom{k+1}{2}}.$$

   Prove this by recurrence and induction, separating the k-element subsets of $\{1, 2, \ldots, n+1\}$ into those that contain $n+1$ and those that do not.

5. Prove the result of the previous problem by taking each k-element subset of $\{1, 2, \ldots, n\}$ and subtracting 1 from the smallest element, 2 from the next smallest, and so on, finally subtracting k from the largest.

6. Call a partition *tight* if the difference between the largest part and the smallest is either 0 or 1. For example, the tight partitions of 6 are 6, $3+3$, $2+2+2$, $2+2+1+1$, $2+1+1+1+1$, and $1+1+1+1+1+1$. How many tight partitions of n are there? (The answer is easy to guess. One way to prove it is to consider how tight partitions of n are related to tight partitions of $n+1$.)

7. Continuing the previous problem, prove that for each k with $1 \le k \le n$, there is exactly one tight partition of n with exactly k parts.

8. It is possible to use the material of this section to construct a q-analogue of Stern's problem, but it's kind of artificial. For example, let's look at the inversion 94 in the permutation 691384275. We say that the **number weight** w_1 of an inversion is the larger number minus 2, plus the smaller number minus 1, so the number weight of 94 is $(9-2)+(4-1) = 7+3 = 10$. We say that the **position weight** w_2 of an inversion is the position of the smaller number minus 2, plus the position of the larger number minus 1. Since the smaller number 4 is in the sixth position and the larger number 9 is in the second position, the position weight of 94 is $(6-2)+(2-1) = 4+1 = 5$. Finally, we say that the **permutation weight** w_3 of an inversion is the number of inversions in the rest of the permutation. If we delete 94 from 691384275 we are left with 6138275, which has 9 inversions. Then the total weight of the inversion 94 in the permutation 691384275 is $w_1 + w_2 + w_3 = 10 + 5 + 9 = 24$.

(i) Look at all the inversions in the permutations of $\{1,2,3\}$ and find the weight of each (w_3 will be zero for all of them; why?). Make all these weights exponents of q and add them all up. You should get $1 + 2q + 3q^2 + 2q^3 + q^4$. Show that this equals
$$[3]_q^2, \quad \text{which equals} \quad \frac{3!_q}{[2]_q}\binom{3}{2}_q.$$

(ii) If $I(n)$ denotes the set of all inversions in all the permutations of $\{1,2,\ldots,n\}$ and wt ι denotes the weight of the inversion ι, prove that
$$\sum_{\iota \in I(n)} q^{\text{wt }\iota} = \binom{n}{2}_q \binom{n}{2}_q (n-2)!_q = \frac{n!_q}{[2]_q}\binom{n}{2}_q.$$

When $q=1$ this becomes Deutsch's solution of Stern's problem from problem 9 of section 1.1.

3.5. An identity of Euler, and its "finite" form

Now that we know a few things about generating functions for partitions, we can use that knowledge to help us understand more complex generating functions. Assuming $|x|$ and $|q|$ are both less than 1, let's expand

$$\frac{1}{(x;q)_\infty} = \frac{1}{(1-x)(1-xq)(1-xq^2)(1-xq^3)(1-xq^4)\ldots}$$
$$= \left(1+x+x^2+x^3+\ldots\right)\left(1+xq+x^2q^2+x^3q^3+\ldots\right)$$
$$\times \left(1+xq^2+x^2q^4+x^3q^6+\ldots\right)\left(1+xq^3+x^2q^6+x^3q^9+\ldots\right)\cdots$$

As usual, we ask about the form this will have when we multiply all the factors together. We will get a 1, from the product of all the 1's. The coefficient of x will be $1+q+q^2+q^3+\cdots = 1/(1-q)$.

3.5. AN IDENTITY OF EULER, AND ITS "FINITE" FORM

How could we get an x^2? We get one from the x^2 in the first factor times all the other 1's. We get one from the x in the first factor times the xq in the second. We get one from the x^2q^2 in the second factor times all the other 1's, but this is not the only way we can get x^2q^2; we can also get it from the x in the first factor times the xq^2 in the third. We get two different x^2q^2's because there are two different partitions of 2, 2 itself and $1+1$. The former corresponds to $x \cdot xq^2$, and the latter to the x^2q^2 in the second factor.

Think of the product the way we thought about $1/(q;q)_\infty$: the first factor $1+x+x^2+x^3+\ldots$ generates 0's (or missing parts). The factor $1+xq+x^2q^2+x^3q^3+\ldots$ generates 1's, since every x has a q^1 along with it. The factor $1+xq^2+x^2q^4+x^3q^6+\ldots$ generates 2's, since every x comes with a q^2; and so on.

If we think of the product this way, we can see that the coefficient of x^2 is going to be the generating function for all partitions with at most two parts. (There might be less than two, if we take either the x or the x^2 from the first factor, which has no q's.) For we get a lot of terms $xq^a \cdot xq^b$, corresponding to the partition $a+b$; and also some terms x^2q^{2m}, which correspond to $m+m$. But we know what this generating function is, from section 3.3: the coefficient of x^2 is $1/(q;q)_2$.

Similarly, the coefficient of x^3 will be the generating function for partitions with at most three parts, which is $1/(q;q)_3$, and so on. Since the coefficient of x^k is $1/(q;q)_k$ for every nonnegative integer k, we get

THEOREM 25 (Euler). *If $|q| < 1$ and $|x| < 1$, then*

(3.5.1) $$\sum_{k=0}^{\infty} \frac{x^k}{(q;q)_k} = \frac{1}{(x;q)_\infty}.$$

We had to assume $|x| < 1$ above to be able to expand all the factors as geometric series. We also need $x \neq 1$ at least to avoid a zero in the denominator on the right. Obviously equivalent is

(3.5.2) $$\sum_{k=0}^{\infty} \frac{x^k q^k}{(q;q)_k} = \frac{1}{(xq;q)_\infty}.$$

One might prefer this form since now the coefficient of x^k is the generating function for partitions with exactly k parts, as in (3.3.1).

It is easy to give an analytic proof of (3.5.1). We start with the series

$$f(x) := \sum_{k=0}^{\infty} \frac{x^k}{(q;q)_k}$$

and use the same trick that Euler used. (Our argument is not quite Euler's, for he started with the product, and used the trick to find the series. We will outline his proof in one of the problems.) Euler's idea was to replace x by xq, and see how different the new function is from the old one. The difference is

$$f(x) - f(xq) = \sum_{k=0}^{\infty} \frac{x^k}{(q;q)_k} - \sum_{k=0}^{\infty} \frac{q^k x^k}{(q;q)_k}$$

$$= \sum_{k=1}^{\infty} \frac{x^k (1-q^k)}{(q;q)_k} \quad \text{(the } k=0 \text{ terms cancel)}$$

$$= \sum_{k=1}^{\infty} x \frac{x^{k-1}}{(q;q)_{k-1}}$$

$$= x \sum_{j=0}^{\infty} \frac{x^j}{(q;q)_j} \quad \text{(after renaming } k\text{)}$$

$$= x f(x).$$

It follows that $f(x) = f(xq)/(1-x)$. If we iterate this relation, we get

$$f(x) = \frac{f(xq)}{1-x}$$

$$= \frac{1}{1-x} \left\{ \frac{f(xq^2)}{1-xq} \right\}$$

$$= \frac{1}{(1-x)(1-xq)} \left\{ \frac{f(xq^3)}{1-xq^2} \right\}$$

$$= \ldots$$

$$= \frac{f(xq^n)}{(x;q)_n}$$

after n iterations. If we let $n \to \infty$ here, we will get an infinite product in the denominator, which will converge, as we have seen, if $|q| < 1$. If $|q| < 1$, then the numerator also converges, to $f(0)$, which, from the series definition of $f(x)$, equals 1. This proves Euler's theorem.

The "finite" form of this was proved by Cauchy. If we expand $1/(x;q)_{n+1}$ as above we get

$$\frac{1}{(x;q)_{n+1}} = \frac{1}{(1-x)(1-xq)(1-xq^2)\cdots(1-xq^n)}$$
$$= \left(1 + x + x^2 + \ldots\right)\left(1 + xq + x^2q^2 + \ldots\right)$$
$$\times \left(1 + xq^2 + x^2q^4 + \ldots\right)\cdots\left(1 + xq^n + x^2q^{2n} + \ldots\right).$$

When we multiply this out, in the first place we get a 1. The coefficient of x is $1 + q + q^2 + \cdots + q^n = [n+1]_q$, which is the generating function for partitions with at most one part which is at most n. The coefficient of x^2 will be the sum of all terms of the form $q^a q^b$, where $0 \le a \le b \le n$; or in other words the generating function for partitions with at most two parts each at most n, and so forth. The coefficient of x^k will be the generating function for partitions into at most k parts, each of which is at most n. But we know what this generating function is, from section 3.3. This proves

3.5. AN IDENTITY OF EULER, AND ITS "FINITE" FORM

THEOREM 26 (Cauchy). *If n is any nonnegative integer and $|q|$ and $|x|$ are both less than 1, then*

$$(3.5.3) \qquad \frac{1}{(x;q)_{n+1}} = \sum_{k=0}^{\infty} \binom{n+k}{k}_q x^k.$$

The $q = 1$ case of this will be useful in two places in Chapter 8. The proof given by Cayley, and later by Sylvester, of the fact that $\binom{n+k}{k}_q$ is the generating function for partitions into at most k parts each at most n is the above argument read backwards: since the coefficient of x^k in the product on the left must be the generating function for partitions of this type, and since Cauchy was able to prove (by a different method, which we will see later) that this coefficient is $\binom{n+k}{k}_q$, it must be the generating function we want.

There are several different analytic proofs of Cauchy's theorem. One is to start from the product side and force out the coefficients of its series expansion. To do this we set

$$\frac{1}{(x;q)_{n+1}} = \sum_{k=0}^{\infty} c_{n,k}(q)\, x^k$$

and try to work out what $c_{n,k}(q)$ has to be. We use Euler's trick of replacing x by xq:

$$\frac{1}{(xq;q)_{n+1}} = \sum_{k=0}^{\infty} c_{n,k}(q)\, x^k\, q^k.$$

Now

$$(1-x)\frac{1}{(x;q)_{n+1}} = \frac{1}{(1-xq)(1-xq^2)\cdots(1-xq^n)} = (1 - xq^{n+1})\frac{1}{(xq;q)_{n+1}},$$

so

$$(1-x)\sum_{k=0}^{\infty} c_{n,k}(q)\, x^k = (1 - xq^{n+1})\sum_{k=0}^{\infty} c_{n,k}(q)\, x^k\, q^k.$$

Distributing this we have

$$\sum_{k=0}^{\infty} c_{n,k}(q)\, x^k - \sum_{k=0}^{\infty} c_{n,k}(q)\, x^{k+1} = \sum_{k=0}^{\infty} c_{n,k}(q)\, x^k\, q^k - \sum_{k=0}^{\infty} c_{n,k}(q)\, x^{k+1}\, q^{n+k+1},$$

which we can rearrange to

$$\sum_{k=0}^{\infty} c_{n,k}(q)\, x^k \left(1 - q^k\right) = \sum_{k=0}^{\infty} c_{n,k}(q)\, x^{k+1}\left(1 - q^{n+k+1}\right).$$

We make a few little changes in this equation. Replace $k+1$ by j in the second sum, which will then run over $1 \leq j < \infty$. For consistency we also replace k by j in the first sum, and note that the $k = 0$ (or $j = 0$) term has the factor $1 - q^0$, so it equals zero and we can discard it. Then we have

$$\sum_{j=1}^{\infty} c_{n,j}(q)\, x^j \left(1 - q^j\right) = \sum_{j=1}^{\infty} c_{n,j-1}(q)\, x^j \left(1 - q^{n+j}\right).$$

Renaming j back as k and equating coefficients of x^k, we have

$$c_{n,k}(q)\left(1 - q^k\right) = c_{n,k-1}(q)\left(1 - q^{n+k}\right),$$

or

(3.5.4) $$c_{n,k}(q) = c_{n,k-1}(q) \frac{1-q^{n+k}}{1-q^k}.$$

Now we iterate this down to $k = 0$, as we have done before. Replacing k by $k-1$ in (3.5.4) we have

$$c_{n,k-1}(q) = c_{n,k-2}(q) \frac{1-q^{n+k-1}}{1-q^{k-1}},$$

and plugging this into (3.5.4) gives

$$c_{n,k}(q) = c_{n,k-2}(q) \frac{1-q^{n+k}}{1-q^k} \frac{1-q^{n+k-1}}{1-q^{k-1}}.$$

If we keep doing this, we will eventually reach

$$c_{n,k}(q) = c_{n,0}(q) \frac{(1-q^{n+k})(1-q^{n+k-1})\cdots(1-q^{n+1})}{(1-q^k)(1-q^{k-1})\cdots(1-q)}.$$

But $c_{n,0}(q)$ is the coefficient of x^0 in the expansion of $1/(x;q)_{n+1}$, which is clearly equal to 1, so we finally have

$$\begin{aligned}
c_{n,k}(q) &= \frac{(1-q^{n+k})(1-q^{n+k-1})\cdots(1-q^{n+1})}{(1-q^k)(1-q^{k-1})\cdots(1-q)} \\
&= \frac{(1-q^{n+k})(1-q^{n+k-1})\cdots(1-q^{n+1})}{(1-q^k)(1-q^{k-1})\cdots(1-q)} \frac{(1-q^n)(1-q^{n-1})\cdots(1-q)}{(1-q^n)(1-q^{n-1})\cdots(1-q)} \\
&= \frac{(q;q)_{n+k}}{(q;q)_n (q;q)_k} = \binom{n+k}{k}_q
\end{aligned}$$

and this proves Cauchy's theorem.

Euler's theorem was employed by Eduard Heine to give alternative proofs of the two q-binomial identities of Gauss from section 2.5. First observe that $(x;q)_\infty (-x;q)_\infty = (x^2;q^2)_\infty$. (We leave this as an exercise.) Next, by Euler's identity with x replaced by x^2 and q by q^2, we have

$$\frac{1}{(x^2;q^2)_\infty} = \sum_{m=0}^{\infty} \frac{x^{2m}}{(q^2;q^2)_m}.$$

On the other hand, we have

$$\begin{aligned}
\frac{1}{(x^2;q^2)_\infty} &= \frac{1}{(x;q)_\infty} \frac{1}{(-x;q)_\infty} \\
&= \sum_{k=0}^{\infty} \frac{x^k}{(q;q)_k} \sum_{j=0}^{\infty} \frac{(-1)^j x^j}{(q;q)_j} \\
&= \sum_{n=0}^{\infty} \sum_{j+k=n} \frac{x^n}{(q;q)_n} \binom{n}{j}_q (-1)^j \\
&= \sum_{n=0}^{\infty} \frac{x^n}{(q;q)_n} \sum_{j=0}^{n} \binom{n}{j}_q (-1)^j
\end{aligned}$$

and therefore

$$\sum_{m=0}^{\infty} \frac{x^{2m}}{(q^2;q^2)_m} = \sum_{n=0}^{\infty} \frac{x^n}{(q;q)_n} \sum_{j=0}^{n} \binom{n}{j}_q (-1)^j.$$

3.5. AN IDENTITY OF EULER, AND ITS "FINITE" FORM

This can only be true if the coefficient of x^r is the same on each side for every r. But there are no odd powers on the left side, and hence

$$\sum_{j=0}^{n} \binom{n}{j}_q (-1)^j = 0 \quad \text{if } n \text{ is odd.}$$

(As we observed before, this is clear without this argument by the symmetry of the q-binomial coefficients.) Since this implies that the right side has no odd powers either, we can replace n by $2m$ there to get

$$\sum_{m=0}^{\infty} \frac{x^{2m}}{(q^2;q^2)_m} = \sum_{m=0}^{\infty} \frac{x^{2m}}{(q;q)_{2m}} \sum_{j=0}^{2m} \binom{2m}{j}_q (-1)^j.$$

It follows that

$$\frac{1}{(q^2;q^2)_m} = \frac{1}{(q;q)_{2m}} \sum_{j=0}^{2m} \binom{2m}{j}_q (-1)^j,$$

or in other words

(3.5.5) $$\sum_{j=0}^{2m} \binom{2m}{j}_q (-1)^j = \frac{(q;q)_{2m}}{(q^2;q^2)_m} = (q;q^2)_m.$$

This proves Gauss's first identity (2.5.2) (we leave it to the reader to check the last step). For the second identity, start with $(x;q^2)_\infty (xq;q^2)_\infty = (x;q)_\infty$, which we again leave as an exercise. Then we have

$$\sum_{n=0}^{\infty} \frac{x^n}{(q;q)_n} = \frac{1}{(x;q)_\infty}$$

$$= \frac{1}{(x;q^2)_\infty} \frac{1}{(xq;q^2)_\infty}$$

$$= \sum_{j=0}^{\infty} \frac{x^j}{(q^2;q^2)_j} \sum_{k=0}^{\infty} \frac{x^k q^k}{(q^2;q^2)_k}$$

$$= \sum_{n=0}^{\infty} \sum_{j+k=n} \frac{x^n}{(q^2;q^2)_n} \binom{n}{k}_{q^2} q^k$$

$$= \sum_{n=0}^{\infty} \frac{x^n}{(q^2;q^2)_n} \sum_{k=0}^{n} \binom{n}{k}_{q^2} q^k.$$

Again the coefficients of x^n on both sides of this must be the same, so

$$\frac{1}{(q;q)_n} = \frac{1}{(q^2;q^2)_n} \sum_{k=0}^{n} \binom{n}{k}_{q^2} q^k,$$

or in other words

$$\sum_{k=0}^{n} \binom{n}{k}_{q^2} q^k = \frac{(q^2;q^2)_n}{(q;q)_n} = (-q;q)_n,$$

which is Gauss's identity (2.5.8).

Exercises

1. Verify that $(x;q)_\infty (-x;q)_\infty = (x^2;q^2)_\infty$.
2. Verify that $\frac{(q;q)_{2m}}{(q^2;q^2)_m} = (q;q^2)_m$.
3. Verify that $(x;q^2)_\infty (xq;q^2)_\infty = (x;q)_\infty$.
4. Is there an identity you could prove which would settle problems 2 and 3 both at once?
5. Complete the proof of Gauss's second identity by verifying that
$$\frac{(q^2;q^2)_n}{(q;q)_n} = (-q;q)_n.$$
6. Is there an identity you could prove which would settle problems 1 and 5 both at once?
7. Use Euler's identity to prove that
$$\frac{1}{(xq;q^2)_\infty} = \sum_{n=0}^\infty \frac{(xq)^n}{(q^2;q^2)_n}.$$
8. Prove the result of problem 7 by counting partitions.
9. Here is an outline of Euler's proof of his identity (3.5.1). He sought to determine the coefficients in the expansion
$$\frac{1}{(x;q)_\infty} = \sum_{n=0}^\infty A_n x^n.$$

(i)
$$\frac{1}{(xq;q)_\infty} = \sum_{n=0}^\infty A_n q^n x^n.$$

(ii) On the other hand,
$$\frac{1}{(xq;q)_\infty} = \frac{1-x}{(x;q)_\infty} = (1-x) \sum_{n=0}^\infty A_n x^n.$$

(iii) Therefore,
$$\sum_{n=0}^\infty A_n q^n x^n = (1-x) \sum_{n=0}^\infty A_n x^n = \sum_{n=0}^\infty A_n x^n - \sum_{n=0}^\infty A_n x^{n+1}.$$

(iv) Therefore,
$$\sum_{n=0}^\infty A_n x^n (1-q^n) = \sum_{n=0}^\infty A_n x^{n+1} = \sum_{n=1}^\infty A_{n-1} x^n.$$

Note that the $n=0$ term on the left side is zero (why?).

(v) Therefore $A_n = A_{n-1}/(1-q^n)$ if $n \geq 1$.
(vi) Therefore $A_n = A_0/(q;q)_n = 1/(q;q)_n$.
This proves (3.5.1). Fill in the details.

10. Prove Cauchy's theorem by induction on n using
$$\binom{n+k}{k}_q = \binom{n+k-1}{k-1}_q + q^k \binom{n+k-1}{k}_q.$$

11. Prove Cauchy's theorem by induction on n using
$$\binom{n+k}{k}_q = \binom{n+k-1}{k}_q + q^n \binom{n+k-1}{k-1}_q.$$

12. If $|q| < 1$ and $|z| < 1$, show that

(3.5.6) $$\sum_{n=0}^{\infty}(-a;q)_n z^n = \sum_{k=0}^{\infty} \frac{q^{\binom{k}{2}} a^k z^k}{(z;q)_{k+1}}.$$

Start by using Cauchy's theorem to expand $1/(z;q)_{k+1}$. The right side converges even without $|z| < 1$, but the left side does not.

13. For another proof of (3.5.6), set $f(z)$ equal to the right side.
   (i) Show that $(1-z)f(z) = 1 + az\,f(zq)$.
   (ii) If $f(z) = \sum_{n=0}^{\infty} c_n(a,q) z^n$, use (i) to show that $c_0(a,q) = 1$, and $c_n(a,q) = (1 + aq^{n-1})c_{n-1}(a,q)$ for $n \geq 1$. Then explain why this proves (3.5.6).

14. (a) Show that
$$\sum_{n=0}^{\infty}(x+yq)(x+yq^2)\cdots(x+yq^n)q^{n+1} = \sum_{k=0}^{\infty} \frac{y^k q^{\binom{k+2}{2}}}{(xq;q)_{k+1}}.$$

(b) Using part (a), or otherwise, show that
$$1 + \sum_{n=0}^{\infty}(x+q)(x+q^2)\cdots(x+q^n)q^{n+1} = \sum_{k=0}^{\infty} \frac{q^{\binom{k+1}{2}}}{(xq;q)_k}.$$

15. Show that an equivalent form of Cauchy's theorem is

(3.5.7) $$\frac{x^k}{(x;q)_{k+1}} = \sum_{m=k}^{\infty} \binom{m}{k}_q x^m.$$

16. We introduced q-Fibonacci numbers
$$F_n(q) = \sum_k \binom{n-k}{k}_q q^{k^2}$$

in problem 3 of section 2.5. Use Cauchy's theorem to prove that

(3.5.8) $$\sum_{k=0}^{\infty} \frac{q^{k^2} x^{2k}}{(x;q)_{k+1}} = \sum_{n=0}^{\infty} F_n(q) x^n,$$

assuming that both sides converge.

17. Show that the left side of (3.5.8) converges for any x (except those that make the denominator zero) if $|q| < 1$. If $q = 1$, show that the left side converges when $|x^2/(1-x)| < 1$.

18. If $q = 1$, show that (3.5.8) reduces to
$$\sum_{n=0}^{\infty} F_n x^n = \frac{1}{1 - x - x^2},$$
where the F_n's are the ordinary Fibonacci numbers.

19. Show that, if $f(x)$ denotes either side of (3.5.8), then $f(x)$ satisfies

(3.5.9) $\qquad (1 - x) f(x) = 1 + x^2 q \, f(xq).$

(Use the recurrence $F_{n+1}(q) = F_n(q) + q^n F_{n-1}(q)$ to do the right side of (3.5.9); this holds for all $n \geq 0$ if we define $F_{-1}(q) = 0$.) If you are willing to believe that there is only one function $f(x)$ satisfying (3.5.9) which is finite when $x = 0$, then this gives an alternate proof of (3.5.8). Show that if $f(x)$ satisfies (3.5.9) and $f(0)$ is not infinite, then $f(0) = 1$.

20. There *is* only one function $f(x)$ satisfying (3.5.9) which is finite when $x = 0$, because we can solve (3.5.9) for $f(x)$ by iteration. Use (3.5.9) to show that
$$f(x) = \frac{q^{n^2} x^{2n}}{(x; q)_n} f(xq^n) + \sum_{k=0}^{n-1} \frac{q^{k^2} x^{2k}}{(x; q)_{k+1}}$$
for any $n \geq 1$ (and even for $n = 0$ since an empty sum equals 0). Show that this reduces to the left side of (3.5.8) as $n \to \infty$ if $f(0)$ is finite.

3.6. Another identity of Euler, and its finite form

In this section we'll do just what we did in the last section, but now for partitions with distinct parts. What do we get if we multiply out the factors in $(-x; q)_\infty$, assuming $|q| < 1$ for convergence as usual? We have

$$\begin{aligned}(-x; q)_\infty &= (1+x)(1+xq)(1+xq^2)(1+xq^3)(1+xq^4)(1+xq^5)(1+xq^6) \cdots \\ &= 1 + x(1 + q + q^2 + q^3 + \ldots) + x^2(1 \cdot q + 1 \cdot q^2 + q \cdot q^2 + 1 \cdot q^3 + \ldots) \\ &\quad + x^3(1 \cdot q \cdot q^2 + 1 \cdot q \cdot q^3 + \ldots) + x^4(1 \cdot q \cdot q^2 \cdot q^3 + \ldots) + \ldots.\end{aligned}$$

The coefficient of x is evidently $1/(1-q)$. The coefficient of x^2 is the sum of all the possible products of two distinct nonnegative powers of q; in other words, it is the generating function for partitions with at most two parts, both different. To be a bit more precise, it is the generating function for partitions with exactly two distinct parts, where zero is allowed to be a part, but can only be used once in each partition just like all the other parts. Similarly the coefficient of x^k is the generating function for partitions with exactly k distinct parts, one of which might be zero.

We know what the generating function for partitions with at most k parts is, namely $1/(q; q)_k$. From this we deduced the generating function for partitions with exactly k parts, namely $q^k/(q; q)_k$, by adding 1 to each part, which replaced any missing parts by 1's. Suppose we instead add 0 to the smallest part, 1 to the next smallest, and so forth, finally adding $k-1$ to the largest part. We still might have a 0 part, but now we can only have one such part. Moreover, the differences between consecutive parts have all increased by one, so we have now a partition with k distinct parts, one of which could be zero. Since $0 + 1 + 2 + \cdots + (k-1) = \binom{k}{2}$,

3.6. ANOTHER IDENTITY OF EULER, AND ITS FINITE FORM

it follows that the generating function for partitions with exactly k distinct parts, one of which might be zero, is $q^{\binom{k}{2}}/(q;q)_k$. This proves

THEOREM 27 (Euler). *If $|q| < 1$, then*

$$(3.6.1) \qquad \sum_{k=0}^{\infty} \frac{q^{\binom{k}{2}} x^k}{(q;q)_k} = (-x;q)_\infty.$$

Again, one can also give an analytic proof of this. We leave it to the reader to check that (due to the quadratic power of q) the series converges for all x if $|q| < 1$. As before, we set $g(x)$ equal to the series and consider

$$g(x) - g(xq) = \sum_{k=1}^{\infty}(1-q^k)\frac{q^{\binom{k}{2}} x^k}{(q;q)_k}$$

$$= x\sum_{k=1}^{\infty} \frac{q^{\binom{k}{2}} x^{k-1}}{(q;q)_{k-1}} = x\sum_{j=0}^{\infty} \frac{q^{\binom{j+1}{2}} x^j}{(q;q)_j}$$

$$= x\sum_{j=0}^{\infty} \frac{q^{\binom{j}{2}} (xq)^j}{(q;q)_j} = x\, g(xq).$$

Therefore $g(x) = (1+x)\, g(xq)$, and, iterating,

$$g(x) = (1+x)\left\{(1+xq)\, g(xq^2)\right\}$$
$$= (1+x)(1+xq)\left\{(1+xq^2)\, g(xq^3)\right\}$$
$$= \text{and so on}$$
$$= (-x;q)_n\, g(xq^n) \quad \text{for all nonnegative integers } n$$
$$= (-x;q)_\infty\, g(0) \quad \text{in the limit as } n \to \infty$$
$$= (-x;q)_\infty \quad \text{since } g(0) = 1.$$

Euler's proof was similar to this in spirit, but he started with the product side and derived the series. We outline it in problem 3.

Recall that Euler's identity of the previous section had a "finite form", due to Cauchy, which was another infinite series. The finite form of (3.6.1) is truly finite—in fact, it is nothing but Rothe's q-binomial theorem. For

$$(-x;q)_n = (1+x)(1+xq)(1+xq^2)(1+xq^3)\cdots(1+xq^{n-1})$$
$$= 1 + x(1+q+q^2+q^3+\cdots+q^{n-1})$$
$$+ x^2(1\cdot q + 1\cdot q^2 + q\cdot q^2 + 1\cdot q^3 + \cdots + q^{n-2}\cdot q^{n-1})$$
$$+ x^3(1\cdot q\cdot q^2 + \cdots + q^{n-3}\cdot q^{n-2}\cdot q^{n-1}) + \cdots + x^n(1\cdot q\cdot q^2\cdots q^{n-1})$$

and now the coefficient of x^k is the generating function for partitions with exactly k distinct parts, one of which might be zero and all of which are at most $n-1$. If we subtract 0 from the smallest part, 1 from the next smallest, and so forth, finally subtracting $k-1$ from the largest, then we obtain a partition with at most k parts, which may not now be distinct since the gap between successive parts has gone down by one. The maximum part size is now $n-1-(k-1) = n-k$. In other words, by taking $q^{\binom{k}{2}}$ out of the generating function in the same way as before, we have arrived at the generating function for partitions into at most k parts, each of which is at most $n - k$, and we know that this generating function is just $\binom{n}{k}_q$. It follows that the coefficient of x^k in $(-x;q)_n$ is $\binom{n}{k}_q q^{\binom{k}{2}}$, or in other words

$$(-x;q)_n = \sum_{k=0}^{n} \binom{n}{k}_q q^{\binom{k}{2}} x^k, \qquad (3.6.2)$$

and we have proved Rothe's q-binomial theorem by counting partitions.

Exercises

1. Explain why the generating function for partitions with exactly n distinct parts, none of which is zero, is

$$\frac{q^{\binom{n+1}{2}}}{(q;q)_n}. \qquad (3.6.3)$$

2. Use problem 1 to give a proof of the identity

$$(-xq;q)_\infty = \sum_{n=0}^{\infty} \frac{q^{\binom{n+1}{2}} x^n}{(q;q)_n} \qquad (3.6.4)$$

by counting partitions. Technically, this is the identity that Euler proved. It is to (3.6.1) what (3.5.2) is to (3.5.1).

3. If

$$(-x;q)_\infty = \sum_{n=0}^{\infty} B_n x^n,$$

determine the coefficients B_n by using the equation

$$\sum_{n=0}^{\infty} B_n x^n = (-x;q)_\infty = (1+x)(-xq;q)_\infty = (1+x)\sum_{n=0}^{\infty} B_n q^n x^n,$$

similarly to the argument outlined in problem 9 in the previous section. This was Euler's proof of (3.6.1).

4. If

$$(-x;q)_n = \sum_{k=0}^{n} C_{n,k} x^k,$$

determine the coefficients $C_{n,k}$ by using the equation

$$\sum_{k=0}^{n+1} C_{n+1,k} x^k = (-x;q)_{n+1} = (1+x)(-xq;q)_n = (1+x)\sum_{k=0}^{n} C_{n,k} q^k x^k.$$

This is still another proof of Rothe's q-binomial theorem.

5. Use Euler's identity to prove

$$(-zq;q^2)_\infty = \sum_{n=0}^{\infty} \frac{q^{n^2} z^n}{(q^2;q^2)_n}. \qquad (3.6.5)$$

This was one of two identities Jacobi used to prove his triple product identity, which we will meet later.

6. Prove (3.6.5) by using the method of problem 3.

EXERCISES

7. Prove (3.6.5) as in the "analytic proof" of Euler's identity in this section.

8. What kind of partitions does the left side of (3.6.5) generate? Prove (3.6.5) by counting partitions.

9. (a) By using (3.6.2) and (3.5.1), or otherwise, show that

$$\sum_{n=0}^{\infty} \frac{(-aq^{n+1};q)_n}{(q;q)_n} q^n = \frac{1}{(q;q)_\infty} \sum_{k=0}^{\infty} a^k q^{3\binom{k+1}{2}}.$$

(b) Show that (a) can be rewritten as

$$\sum_{k=0}^{\infty} a^k q^{3\binom{k+1}{2}} = (-a;q)_\infty (q;q)_\infty \sum_{n=0}^{\infty} \frac{q^n}{(-a;q)_{n+1}(q;q)_n}.$$

10. (a) Show that changing q to q^2 in (3.6.1) and then setting $x = aq$ gives

$$(-aq;q^2)_\infty = \sum_{k=0}^{\infty} \frac{q^{k^2} a^k}{(q^2;q^2)_k}.$$

(b) Show that

$$\left(1 + \tfrac{q}{a}\right)(-aq;q^2)_\infty = \sum_{k=-1}^{\infty} \frac{q^{k^2} a^k}{(q^2;q^2)_{k+1}}.$$

(c) Prove by induction on n that

$$\left(-\tfrac{q}{a};q^2\right)_n (-aq;q^2)_\infty = \sum_{k=-n}^{\infty} \frac{q^{k^2} a^k}{(q^2;q^2)_{k+n}}.$$

(d) What happens to the result of (c) if we let $n \to \infty$? (The answer is that we get Jacobi's triple product identity, which we will see in section 5.1. This derivation of it is due to Michael Hirschhorn.)

11. Suppose $R(x)$ is a function satisfying $R(0) = 1$ and $R(x) = R(xq) + xq\,R(xq^2)$. Show that the power series expansion of $R(x)$ is

$$R(x) = \sum_{n=0}^{\infty} \frac{q^{n^2} x^n}{(q;q)_n}.$$

This will be used in Chapters 10–12.

12. Let

$$A(x) = \sum_{k=0}^{\infty} \frac{q^{k^2+k} x^k}{(q^2;q^2)_k} (-xq^{2k+1};q^2)_\infty \quad \text{and} \quad B(x) = \sum_{k=0}^{\infty} \frac{q^{k^2} x^k}{(q^2;q^2)_k} (-xq^{2k+2};q^2)_\infty.$$

(i) Show that $A(x) - xqA(xq^2) = B(xq)$ and $B(x) - xqB(xq^2) = A(xq)$. For the latter, you might find it easier to show that $B(x) - A(xq) = xqB(xq^2)$.

(ii) Let $C(x) = A(x) - B(x)$. Show that $C(x) + C(xq) = xqC(xq^2)$.

(iii) Show that $C(0) = 0$.

(iv) If

$$C(x) = \sum_{n=0}^{\infty} c_n x^n,$$

use (ii) to show that
$$c_n = c_{n-1} \frac{q^{2n-1}}{1+q^n} \quad \text{for } n \geq 1.$$
What is c_0?

(v) Show that
$$A(x) = B(x) = \sum_{n=0}^{\infty} \frac{q^{n^2} x^n}{(q;q)_n}.$$

3.7. The Cauchy/Crelle q-binomial series

We come now to the most important identity in this chapter. As we have seen, Euler was able to work out the expansions of $(-x;q)_\infty$ and $1/(x;q)_\infty$ in powers of x. We rephrase these two series here:

THEOREM 28 (Euler's partition identities). *If $|q| < 1$, then*

(3.7.1) $$(ax;q)_\infty = \sum_{j=0}^{\infty} \frac{q^{\binom{j}{2}} (-a)^j x^j}{(q;q)_j},$$

(3.7.2) $$\frac{1}{(bx;q)_\infty} = \sum_{k=0}^{\infty} \frac{b^k x^k}{(q;q)_k},$$

where (3.7.1) *holds for any a and x and* (3.7.2) *requires that $|bx| < 1$.*

Apparently it did not occur to Euler to try to expand $(ax;q)_\infty/(bx;q)_\infty$ in powers of x. If it had, he would have found that his method would still work. Instead, this expansion was found by four different mathematicians in the 1840s: Cauchy, Eisenstein, Jacobi, and Heine, in that order. Two of these, Cauchy and Jacobi, are among the greatest mathematicians who have ever lived, and Eisenstein's talents were of the same magnitude, though he accomplished less because he died at 29. Heine was also an excellent mathematician, if not quite on the same level, but his paper was much the most important of the four because it introduced q-hypergeometric series, which we will discuss in the last few sections of Chapter 5.

The identity in question is usually called the q-binomial theorem. We will have two equivalent versions of it; since there are so many other results which are also called the q-binomial theorem, we shall refer to one of them as the q-binomial series, and the other as the Cauchy/Crelle series. (Eisenstein, Jacobi, and Heine all published their versions of the series in Crelle's *Journal*, and it seems likely that each read the others' work. Whether any of them had read Cauchy's paper is unclear. One suspects that Jacobi had, because he and Cauchy were both writing a lot of papers on determinants in this period.) Our first proof will use the method Euler would have used, which is the method that Cauchy and Jacobi actually did use. We set
$$\frac{(ax;q)_\infty}{(bx;q)_\infty} = \sum_{n=0}^{\infty} D_n x^n$$
and attempt to work out the coefficients D_n. We have
$$\frac{(axq;q)_\infty}{(bxq;q)_\infty} = \sum_{n=0}^{\infty} D_n q^n x^n$$

3.7. THE CAUCHY/CRELLE q-BINOMIAL SERIES

and so
$$\sum_{n=0}^{\infty} D_n x^n = \frac{(ax;q)_\infty}{(bx;q)_\infty} = \frac{1-ax}{1-bx} \frac{(axq;q)_\infty}{(bxq;q)_\infty} = \frac{1-ax}{1-bx} \sum_{n=0}^{\infty} D_n q^n x^n.$$

From the extremes of this string of equalities we have
$$(1-bx)\sum_{n=0}^{\infty} D_n x^n = (1-ax)\sum_{n=0}^{\infty} D_n q^n x^n.$$

Distributing the linear factors we get
$$\sum_{n=0}^{\infty} D_n x^n - \sum_{n=0}^{\infty} D_n b x^{n+1} = \sum_{n=0}^{\infty} D_n q^n x^n - \sum_{n=0}^{\infty} D_n a q^n x^{n+1}$$

which we rearrange as
$$\sum_{n=0}^{\infty} D_n x^n (1-q^n) = \sum_{n=0}^{\infty} D_n x^{n+1} (b-aq^n) = \sum_{n=1}^{\infty} D_{n-1} x^n (b-aq^{n-1}).$$

Note that the $n=0$ term on the left side is zero. Equating coefficients of x^n we see that
$$D_n (1-q^n) = D_{n-1} (b-aq^{n-1}) \quad \text{if } n \geq 1.$$

We rewrite this as
$$D_n = D_{n-1} \frac{b - aq^{n-1}}{1 - q^n}$$

and iterate it:
$$D_n = D_0 \frac{(b-aq^{n-1})(b-aq^{n-2})\cdots(b-aq)(b-a)}{(1-q^n)(1-q^{n-1})\cdots(1-q^2)(1-q)}.$$

Since $D_0 = 1$ (one way to see this is to set $x=0$ in the series we started with), this proves

THEOREM 29 (The Cauchy/Crelle series). *If $|q| < 1$ and $|bx| < 1$, then*

$$(3.7.3) \qquad \frac{(ax;q)_\infty}{(bx;q)_\infty} = \sum_{n=0}^{\infty} \frac{(b-aq^{n-1})(b-aq^{n-2})\cdots(b-aq)(b-a)}{(q;q)_n} x^n,$$

where $(b-aq^{n-1})(b-aq^{n-2})\cdots(b-aq)(b-a) = 1$ *if* $n=0$.

(The assumption $|bx| < 1$ is needed for convergence of the series.) If $b=1$ this can be written more compactly:

THEOREM 30 (The q-binomial series). *If $|q|$ and $|x|$ are both less than 1, then*

$$(3.7.4) \qquad \frac{(ax;q)_\infty}{(x;q)_\infty} = \sum_{n=0}^{\infty} \frac{(a;q)_n}{(q;q)_n} x^n.$$

Although none of its four discoverers stated it this way, this is the form in which the q-binomial series most often occurs in the literature. It is equivalent to the Cauchy/Crelle q-binomial series by a rescaling argument: replace x by bx and a by $\frac{a}{b}$ in the q-binomial series and it becomes the Cauchy/Crelle series. It is almost the form Heine found: he had q^α in place of a. On the other hand, Eisenstein had q^α in place of b in the Cauchy/Crelle series with $a=1$.

One might also find the Cauchy/Crelle series by multiplying (3.7.1) and (3.7.2) together:

$$\frac{(ax;q)_\infty}{(bx;q)_\infty} = \left(\sum_{j=0}^\infty \frac{q^{\binom{j}{2}}(-a)^j x^j}{(q;q)_j}\right)\left(\sum_{k=0}^\infty \frac{b^k x^k}{(q;q)_k}\right)$$

$$= \sum_{n=0}^\infty \sum_{j+k=n} \frac{x^n}{(q;q)_n} \binom{n}{j}_q q^{\binom{j}{2}}(-a)^j b^k$$

$$= \sum_{n=0}^\infty \frac{x^n}{(q;q)_n} \sum_{j=0}^n \binom{n}{j}_q q^{\binom{j}{2}}(-a)^j b^{n-j}.$$

According to Rothe's q-binomial theorem,

$$\sum_{j=0}^n \binom{n}{j}_q q^{\binom{j}{2}}(-a)^j b^{n-j} = (b-a)(b-aq)\cdots(b-aq^{n-1}),$$

and thus we get the Cauchy/Crelle series again.

As Cauchy pointed out, his series contains four of our earlier identities as special cases. It obviously reduces to Euler's

(3.7.5) $$\sum_{k=0}^\infty \frac{x^k}{(q;q)_k} = \frac{1}{(x;q)_\infty}$$

if $a = 0$ and $b = 1$; and to Euler's

(3.7.6) $$\sum_{k=0}^\infty \frac{q^{\binom{k}{2}} x^k}{(q;q)_k} = (-x;q)_\infty$$

if $b = 0$ and $a = -1$. Cauchy's proof of

$$\frac{1}{(x;q)_{n+1}} = \sum_{k=0}^\infty \binom{n+k}{k}_q x^k$$

was to set $a = q^{n+1}$ and $b = 1$ in the Cauchy/Crelle series, and he also derived Rothe's q-binomial theorem by setting $a = -1$ and $b = -q^n$. Cauchy knew of Euler's work, but he did not know that Rothe had beaten him to his q-binomial theorem.

Jacobi used the Cauchy/Crelle series to prove the result we called Jacobi's q-binomial theorem in section 2.6. We have

$$\frac{(ax;q)_\infty}{(bx;q)_\infty} = \frac{(cx;q)_\infty}{(bx;q)_\infty}\frac{(ax;q)_\infty}{(cx;q)_\infty}$$

$$= \left(\sum_{j=0}^\infty \frac{(b-c)\cdots(b-cq^{j-1})}{(q;q)_j} x^j\right)\left(\sum_{k=0}^\infty \frac{(c-a)\cdots(c-aq^{k-1})}{(q;q)_k} x^k\right)$$

$$= \sum_{n=0}^\infty \sum_{j+k=n} \frac{x^n}{(q;q)_n}\binom{n}{k}_q (b-c)\cdots(b-cq^{j-1})(c-a)\cdots(c-aq^{k-1})$$

$$= \sum_{n=0}^\infty \frac{x^n}{(q;q)_n} \sum_{k=0}^n \binom{n}{k}_q (b-c)\cdots(b-cq^{n-k-1})(c-a)\cdots(c-aq^{k-1}).$$

On the other hand,
$$\frac{(ax;q)_\infty}{(bx;q)_\infty} = \sum_{n=0}^\infty \frac{(b-a)(b-aq)\cdots(b-aq^{n-1})}{(q;q)_n} x^n.$$
Equating coefficients of $\dfrac{x^n}{(q;q)_n}$ we get
$$(b-a)(b-aq)\cdots(b-aq^{n-1})$$
$$= \sum_{k=0}^n \binom{n}{k}_q (b-c)(b-cq)\cdots(b-cq^{n-k-1})(c-a)(c-aq)\cdots(c-aq^{k-1}),$$
which is Jacobi's q-binomial theorem, proved by his method. Alternatively, we could start with Jacobi's q-binomial theorem and read the argument backwards to get a rather silly proof that
$$\frac{(ax;q)_\infty}{(bx;q)_\infty} = \frac{(cx;q)_\infty}{(bx;q)_\infty} \frac{(ax;q)_\infty}{(cx;q)_\infty}.$$
We have not yet explained why one would call
$$\frac{(ax;q)_\infty}{(x;q)_\infty} = \sum_{n=0}^\infty \frac{(a;q)_n}{(q;q)_n} x^n$$
the q-binomial series. This is easiest to see from Heine's version:
$$\frac{(q^\alpha x;q)_\infty}{(x;q)_\infty} = \sum_{n=0}^\infty \frac{(q^\alpha;q)_n}{(q;q)_n} x^n.$$
If we let $q \to 1$ here, then
$$\frac{(q^\alpha;q)_n}{(q;q)_n} = \frac{1-q^\alpha}{1-q} \frac{1-q^{\alpha+1}}{1-q^2} \cdots \frac{1-q^{\alpha+n-1}}{1-q^n}$$
$$= \frac{1-q^\alpha}{1-q} \frac{1-q^{\alpha+1}}{1-q} \cdots \frac{1-q^{\alpha+n-1}}{1-q} \frac{1-q}{1-q} \frac{1-q}{1-q^2} \cdots \frac{1-q}{1-q^n}$$
$$\to \alpha(\alpha+1)\cdots(\alpha+n-1) \cdot 1 \cdot \frac{1}{2} \cdots \frac{1}{n}$$
$$= \frac{\alpha(\alpha+1)\cdots(\alpha+n-1)}{n!}.$$
Thus the series becomes
$$\sum_{n=0}^\infty \alpha(\alpha+1)\cdots(\alpha+n-1) \frac{x^n}{n!},$$
which is the Taylor series expansion of the general binomial $(1-x)^{-\alpha}$, converging to it if $|x| < 1$.

It may be unfair to ask why a certain mathematician *didn't* find a particular result, but sometimes it is hard to resist. Euler could easily have derived the Cauchy/Crelle series if he had only thought to look for it. Why didn't he? (It is very likely but not absolutely certain that he didn't. Hardy, in his book on Ramanujan, says that the formula "probably goes back to Euler", though he does not give a reference, and no one has ever been able to find one as far as I know. Euler wrote an extraordinary amount of mathematics, and it is not completely out of the question that the Cauchy/Crelle series may be somewhere in his collected papers, or possibly in his letters, which were numerous and have received less attention. I have looked at a lot of his work.)

There is, however, a good reason why he might not have found it: the special cases of it that he found have very natural interpretations in partition theory; whereas it is less clear what the Cauchy/Crelle series has to do with partitions. To conclude this section we will derive the Cauchy/Crelle series by partition counting.

To this end, it is convenient to replace a by $-r$, so that we are trying to interpret $(-rx;q)_\infty/(bx;q)_\infty$ as a partition series. We can say that the numerator generates partitions with distinct red parts, in that the coefficient of $r^k x^k$ when the numerator is expanded is the generating function for partitions with at most k distinct parts, which we will imagine are red. We can say that the denominator generates partitions with (not necessarily distinct) blue parts, in that the coefficient of $b^j x^j$ when the denominator is expanded is the generating function for partitions with at most j parts, which we will imagine are blue. But we know what these two generating functions are, namely

$$\frac{q^{\binom{k}{2}}}{(q;q)_k} \quad \text{and} \quad \frac{1}{(q;q)_j}, \quad \text{respectively.}$$

When the whole fraction is expanded in powers of x, every x comes with either an r or a b; hence if a term has factors x^n and r^k, then it also has a factor b^{n-k}. From above, the coefficient of $x^n b^k r^{n-k}$ is the generating function for partitions with at most n parts, which may be either blue or red. The red parts must be distinct and there are no more than k of them. The blue parts need not be distinct, but there are at most $n-k$ of them. Hence the coefficient of $x^n b^k r^{n-k}$ must be

$$\frac{q^{\binom{k}{2}}}{(q;q)_k} \frac{1}{(q;q)_{n-k}} = \frac{q^{\binom{k}{2}}}{(q;q)_n} \binom{n}{k}_q.$$

Since k may have any value between 0 and n, the coefficient of x^n is

$$\sum_{k=0}^{n} \binom{n}{k}_q \frac{q^{\binom{k}{2}}}{(q;q)_n} b^k r^{n-k}.$$

By Rothe's q-binomial theorem this coefficient equals

$$\frac{(b+r)(b+rq)\cdots(b+rq^{n-1})}{(q;q)_n},$$

and we have proved that

$$\frac{(-rx;q)_\infty}{(bx;q)_\infty} = \sum_{n=0}^{\infty} (b+r)(b+rq)\cdots(b+rq^{n-1}) \frac{x^n}{(q;q)_n},$$

which is just a restatement of the Cauchy/Crelle series. We needed to use Rothe's q-binomial theorem along the way, but since we proved that earlier by counting partitions, this proof is wholly partition-theoretic. Note that it is otherwise identical with our second proof of the Cauchy/Crelle series, where we multiplied two series together—we simply redid every step of that argument by counting partitions rather than by algebra.

Exercises

1. Check the details of Cauchy's proof of (3.7.5).
2. Check the details of Cauchy's proof of (3.7.6).
3. Here is an outline of another derivation of the Cauchy/Crelle series. We set
$$f(a,b) = \sum_{n=0}^{\infty} \frac{(b-a)(b-aq)\cdots(b-aq^{n-1})}{(q;q)_n} x^n,$$
where $|q| < 1$, $|bx| < 1$, and $(b-a)(b-aq)\cdots(b-aq^{n-1}) = 1$ if $n = 0$.

   (i) Explain why $f(bq, b) = 1/(1 - bx)$.

   (ii) Rewrite
$$f(a,b) = 1 + \sum_{n=1}^{\infty} \frac{(b-a)(b-aq)\cdots(b-aq^{n-1})}{(q;q)_n} x^n.$$

   (iii) Rewrite $b - a = (b - aq^n) - a(1 - q^n)$ in (ii), split the series into two pieces, reindex one of them, and recombine. The result should be $f(a,b) = (1 - ax) f(aq, b)$.

   (iv) Iterate the result of (iii) to get $f(a,b) = (ax; q)_\infty f(0, b)$.

   (v) We know $f(0, b)$ from one of Euler's identities, but we can also get it out of this argument. What expression does the result of (iv) give for $f(bq, b)$? Use this and (i) to find $f(0, b)$.

   (vi) Use (iv) and (v) to conclude that $f(a, b) = (ax; q)_\infty / (bx; q)_\infty$.

   (vii) Which identity of Euler would give us $f(0, b)$? Why was step (ii) necessary?

4. Here is an outline of yet another derivation of the Cauchy/Crelle series, similar to the first proof in the text but not quite the same. For $|bx| < 1$ and $|q| < 1$, let
$$\frac{(ax; q)_\infty}{(bx; q)_\infty} = \sum_{n=0}^{\infty} c_n(a, b) x^n.$$

   (i) By setting $x = 0$, or otherwise, explain why $c_0(a, b) = 1$.

   (ii) Show that
$$\frac{(ax;q)_\infty}{(bx;q)_\infty} - \frac{(axq;q)_\infty}{(bxq;q)_\infty} = x(b-a) \frac{(axq;q)_\infty}{(bx;q)_\infty}.$$

   (iii) Explain why this implies $(1 - q^n) c_n(a, b) = (b - a) c_{n-1}(aq, b)$ for $n \geq 1$.

   (iv) Explain why this implies
$$c_n(a,b) = \frac{(b-a)(b-aq)\cdots(b-aq^{n-1})}{(q;q)_n}$$
for $n \geq 0$.

5. Use the Cauchy/Crelle series to show that
$$\frac{(-q; q^2)_\infty}{(q^2; q^2)_\infty} = 1 + \sum_{n=1}^{\infty} \frac{q^{2n-1}(1+q)(-q; q^2)_{n-1}}{(q^2; q^2)_n}.$$

This was problem 15 in section 3.3.

6. Use the Cauchy/Crelle series to show that
$$\frac{(-q^2;q^2)_\infty}{(q;q^2)_\infty} = \sum_{n=0}^{\infty} q^n \frac{(-q;q^2)_n}{(q^2;q^2)_n}.$$
This was problem 17 in section 3.3.

7. Suppose that $|q| < 1$, that $D(x)$ is a function that is not infinite at $x = 0$, and that

(3.7.7) $$D(x) = \frac{1}{1-x} - \frac{x}{1-x} D(xq).$$

(i) Show that
$$D(x) = \frac{1}{1-x} - \frac{x}{(1-x)(1-xq)} + \frac{x^2 q}{(1-x)(1-xq)} D(xq^2).$$

(ii) Show that
$$D(x) = \frac{1}{1-x} - \frac{x}{(1-x)(1-xq)} + \frac{x^2 q}{(1-x)(1-xq)(1-xq^2)} - \frac{x^3 q^3}{(1-x)(1-xq)(1-xq^2)} D(xq^3).$$

(iii) For any positive integer n, show that

(3.7.8) $$D(x) = \sum_{k=0}^{n-1} \frac{(-x)^k q^{\binom{k}{2}}}{(x;q)_{k+1}} + \frac{(-x)^n q^{\binom{n}{2}}}{(x;q)_n} D(xq^n).$$

(iv) Explain why (3.7.8) implies that
$$D(x) = \sum_{k=0}^{\infty} \frac{(-x)^k q^{\binom{k}{2}}}{(x;q)_{k+1}}$$
is the unique solution of (3.7.7) when $|q| < 1$ and $D(0)$ is not infinite.

(v) Show that $D(x) = 1$ is also a solution of (3.7.7), and hence

(3.7.9) $$\sum_{k=0}^{\infty} \frac{(-x)^k q^{\binom{k}{2}}}{(x;q)_{k+1}} = 1.$$

8. For $n \geq 0$, define
$$D_n(x,q) = \sum_{k=0}^{n} \frac{(-x)^k q^{\binom{k}{2}}}{(x;q)_{k+1}}.$$

Prove by induction on n that
$$D_n(x,q) = 1 - \frac{(-x)^{n+1} q^{\binom{n+1}{2}}}{(x;q)_{n+1}},$$
and (assuming as usual that $|q| < 1$) explain why this implies (3.7.9).

9. Prove (3.7.9) by using (3.5.7) from problem 15 in section 3.5.

10. George Andrews has observed that

(a)
$$1 = 1 + \sum_{k=0}^{\infty} \frac{(-1)^k q^{\binom{k}{2}}}{(q;q)_k},$$

(b)
$$2 = 1 + \sum_{k=0}^{\infty} \frac{(-1)^k q^{k^2}}{(q;q^2)_{k+1}},$$

(c)
$$3 = 1 + \sum_{k=0}^{\infty} \frac{q^{\binom{k}{2}}}{(-q;q)_k},$$

and that (3.7.9) affords easy proofs of these identities. To prove (a), start by observing that an equivalent form of (3.7.9) is

$$\sum_{k=0}^{\infty} \frac{(-x)^k q^{\binom{k}{2}}}{(xq;q)_k} = 1 - x.$$

To prove (b), start by replacing q by q^2 in (3.7.9).

11. What other identity implies that

$$\sum_{k=0}^{\infty} \frac{(-1)^k q^{\binom{k}{2}}}{(q;q)_k} = 0?$$

12. Show that

$$\sum_{k=0}^{n} \frac{(-1)^k q^{\binom{k}{2}}}{(q;q)_k} = \frac{(-1)^n q^{\binom{n+1}{2}}}{(q;q)_n}.$$

This is a finite form of (a) in problem 10. Finite forms of (b) and (c) were in problems 17 and 18 in section 3.2.

13. Part (a) of problem 10 gives us a rare example of an interesting infinite series that converges to zero. Here is another one. Define

$$S_n(x;q) = 1 + \sum_{k=1}^{n} (1+q^k) \frac{(x-1)(x-q)\cdots(x-q^{k-1})}{(xq;q)_k},$$

where $S_0(x;q) = 1$.

(i) Prove that

$$S_n(x;q) = \frac{(x-q)(x-q^2)\cdots(x-q^n)}{(xq;q)_n} = \frac{(\frac{q}{x};q)_n}{(qx;q)_n} x^n$$

for any nonnegative integer n.

(ii) If $|q| < 1$ and $|x| < 1$, explain why

$$1 + \sum_{k=1}^{\infty} (1+q^k) \frac{(x-1)(x-q)\cdots(x-q^{k-1})}{(xq;q)_k} = 0.$$

14. *The worst proof yet of the Cauchy/Crelle series.* This problem outlines a derivation of the Cauchy/Crelle series from Taylor's theorem

(3.7.10)
$$f(x) = \sum_{n=0}^{\infty} f^{(n)}(0) \frac{x^n}{n!},$$

where $f^{(n)}(0)$ denotes the ordinary n^{th} derivative of $f(x)$ evaluated at $x = 0$. For convenience, we will denote

$$f_{a,b}(x) = \frac{(ax;q)_\infty}{(bx;q)_\infty}.$$

Note that (3.7.3) would be established if we can show that

(3.7.11) $$\left.\frac{d^n}{dx^n} f_{a,b}(x)\right|_{x=0} = f_{a,b}^{(n)}(0) = \frac{n!\,(b-a)(b-aq)\cdots(b-aq^{n-1})}{(q;q)_n},$$

where the right side means 1 if $n = 0$.

(i) Show that the logarithmic derivative of $f_{a,b}(x)$ with respect to x is

$$\frac{f'_{a,b}(x)}{f_{a,b}(x)} = \sum_{j=0}^{\infty} \frac{-aq^j}{1 - axq^j} - \sum_{j=0}^{\infty} \frac{-bq^j}{1 - bxq^j}.$$

(ii) If we denote

$$g_{a,b}(x) = \sum_{j=0}^{\infty} \frac{bq^j}{1 - bxq^j} - \sum_{j=0}^{\infty} \frac{aq^j}{1 - axq^j},$$

explain why we have

$$f'_{a,b}(x) = f_{a,b}(x) g_{a,b}(x),$$

and use this to show that $f'_{a,b}(0) = \frac{b-a}{1-q}$.

(iii) Show that

$$\frac{d^{k-1}}{dx^{k-1}} \frac{bq^j}{1 - bxq^j} = \frac{(k-1)!\,(bq^j)^k}{(1 - bxq^j)^k}$$

for $k \geq 1$, and use this to show that

$$g_{a,b}^{(k-1)}(0) = (k-1)!\,\frac{b^k - a^k}{1 - q^k}.$$

(iv) Use (ii) and (iii) to show that

$$f''_{a,b}(0) = \frac{2(b-a)(b-aq)}{(1-q)(1-q^2)}.$$

(v) By taking $q = 1$ in the q-Leibniz rule from problem 11 in section 2.5, show that

$$f_{a,b}^{(n)}(0) = \sum_{k=1}^{n} \binom{n-1}{k-1} f_{a,b}^{(n-k)}(0) g_{a,b}^{(k-1)}(0).$$

(vi) Use the result of (v) and the mysterious identity (2.9.6) from section 2.9 to prove (3.7.11) by induction on n.

3.8. q-exponential functions

In this section we use the q-derivative to give an alternative approach to the identities of the last few sections. In a modern calculus course, the next thing one learns to differentiate after the power functions is the exponential function. One finds that the previously mysterious function e^x is interesting because it is its own derivative. Moreover, the only functions with this property are constant multiples of e^x, so e^x is the only one that equals 1 when $x = 0$. So it is now natural, at least if we have read section 2.4, to ask for a function $f(x)$ that has $f(0) = 1$ and is its own q-derivative. This means that all the higher q-derivatives of $f(x)$ will be 1 when $x = 0$, so the right side of (2.4.9) becomes

$$(3.8.1) \qquad \sum_{n=0}^{\infty} \frac{x^n}{n!_q}, \quad \text{which we will denote by } e_q(x).$$

It is easy to check directly that $e_q(x)$ is its own q-derivative; we leave this as an exercise. Unlike the ordinary exponential function, however, the series in (3.8.1) does not converge for every x. Under our standard assumption that $|q| < 1$, it converges for $|x| < 1/|1-q|$; we leave this as an exercise also. So we assume from now on when we talk about $e_q(x)$ that $|x|$ is small enough for the series (3.8.1) to converge.

We can get another expression for $e_q(x)$ by solving the q-differential equation

$$(3.8.2) \qquad \mathbf{D}_q f(x) = f(x), \quad \text{which is equivalent to} \quad f(x) = \frac{f(x) - f(qx)}{x(1-q)}.$$

A little algebra (which we leave as an exercise) allows us to rewrite (3.8.2) as

$$f(x) = \frac{f(qx)}{1 - x(1-q)}.$$

Iterating this n times we get

$$f(x) = \frac{f(xq^n)}{(1 - x(1-q))(1 - xq(1-q)) \cdots (1 - xq^{n-1}(1-q))} = \frac{f(xq^n)}{(x(1-q); q)_n}.$$

Letting $n \to \infty$, using the assumptions $|q| < 1$ and $f(0) = 1$, we have

$$f(x) = \frac{1}{(x(1-q); q)_\infty}.$$

Since $f(x)$ is the q-exponential function $e_q(x)$, it follows that

$$(3.8.3) \qquad \frac{1}{(x(1-q); q)_\infty} = e_q(x) = \sum_{n=0}^{\infty} \frac{x^n}{n!_q}.$$

Does (3.8.3) look impressive? It really isn't—it's just Euler's theorem (3.5.1), namely

$$(3.8.4) \qquad \sum_{n=0}^{\infty} \frac{x^n}{(q;q)_n} = \frac{1}{(x;q)_\infty}$$

if $|q| < 1$ and $|x| < 1$, in an uglier notation. In Chapter 1 we observed that

$$n!_q (1-q)^n = (q;q)_n,$$

and if we use this to replace $(q;q)_n$ in (3.5.1) and then replace x by $x(1-q)$, we have (3.8.3). Thus the q-derivative is essentially just a rephrasing of Euler's old trick of relating $f(x)$ to $f(xq)$.

Let's try to solve the q-differential equation

(3.8.5) $\quad \mathbf{D}_q f(x) = f(qx), \quad$ which is equivalent to $\quad f(qx) = \dfrac{f(x) - f(qx)}{x(1-q)}.$

A little algebra (which we leave as an exercise) allows us to rewrite (3.8.5) as
$$f(x) = (1 + x(1-q)) f(qx).$$
Iterating this n times we get
$$f(x) = (1 + x(1-q))(1 + xq(1-q)) \cdots (1 + xq^{n-1}(1-q)) f(xq^n)$$
$$= (-x(1-q); q)_n f(xq^n).$$
Letting $n \to \infty$ and using the assumption $|q| < 1$ we get
$$f(x) = (-x(1-q); q)_\infty f(0).$$
As with ordinary first order differential equations, there is typically an arbitrary constant in the general solution of a first order q-differential equation. The simplest assumption we could make is $f(0) = 0$, but we see that in that case $f(x)$ would be identically zero. The next simplest assumption would be $f(0) = 1$, and this is reasonable because a solution of (3.8.5) should be another q-analogue of the exponential function. So let's denote the unique solution of (3.8.5) satisfying $f(0) = 1$ by $E_q(x)$. Then we just proved that

(3.8.6) $\quad\quad\quad\quad\quad\quad E_q(x) = (-x(1-q); q)_\infty.$

We would like to have an infinite series expansion of $E_q(x)$ so that we could make a statement like (3.8.3). There are several ways we could proceed. The cheap way would be to look back at section 3.6 to see how to expand (3.8.6), but it is more interesting to try to use the q-Taylor theorem (2.4.9). To do this, we need to be able to evaluate the nth q-derivative of $E_q(x)$ at $x = 0$, and there are two alternatives. We could try to compute q-derivatives of the infinite product in (3.8.6), or we could hope to find them from the functional equation (3.8.5). The former is not as hard as it sounds (see problem 6), but let's try the latter. We need to be able to find
$$\mathbf{D}_q^n f(x) \big|_{x=0}, \quad \text{where} \quad f^*(x) = \mathbf{D}_q f(x) = f(qx)$$
and $f(0) = 1$. Using (2.4.5), we have
$$\mathbf{D}_q^2 f(x) = \mathbf{D}_q f(qx) = q f^*(qx) = q f(q^2 x).$$
Using (2.4.5) again, we have
$$\mathbf{D}_q^3 f(x) = \mathbf{D}_q q f(q^2 x) = q^3 f^*(q^2 x) = q^3 f(q^3 x),$$
and it looks as though

(3.8.7) $\quad\quad\quad\quad\quad\quad \mathbf{D}_q^n f(x) = q^{\binom{n}{2}} f(q^n x).$

We know (3.8.7) already for $n = 0, 1, 2, 3$. Assuming it holds for n, we have
$$\mathbf{D}_q^{n+1} f(x) = \mathbf{D}_q q^{\binom{n}{2}} f(q^n x).$$
By (2.4.5) and (3.8.5) this becomes
$$\mathbf{D}_q^{n+1} f(x) = q^{\binom{n}{2}+n} f^*(q^n x) = q^{\binom{n+1}{2}} f(q^{n+1} x),$$

which is (3.8.7) with $n+1$ in place of n, so (3.8.7) holds for every nonnegative integer n. Setting $n=0$ there we have

$$\mathbf{D}_q^n f(x)\big|_{x=0} = q^{\binom{n}{2}} f(0) = q^{\binom{n}{2}}.$$

Therefore the q-Taylor series of $f(x) = E_q(x)$ is

$$\sum_{n=0}^{\infty} q^{\binom{n}{2}} \frac{x^n}{n!_q},$$

and, assuming convergence, we have

(3.8.8) $$\sum_{n=0}^{\infty} q^{\binom{n}{2}} \frac{x^n}{n!_q} = E_q(x) = (-x(1-q); q)_\infty.$$

This series converges for every x when $|q| < 1$, so it could be considered a better q-analogue of the exponential series than (3.8.3). It is Euler's theorem (3.6.1) with x replaced by $x(1-q)$.

There are two relations between the q-exponential functions $e_q(x)$ and $E_q(x)$. One is fairly obvious, while the other is more subtle. From the infinite product expansions it is clear that $E_q(x)$ and $e_q(x)$ are almost reciprocals of each other:

$$e_q(-x) = \frac{1}{(-x(1-q); q)_\infty} = \frac{1}{E_q(x)}.$$

For the second relation we start by observing that

$$q^{\binom{n}{2}} n!_{q^{-1}} = q^{1+2+\cdots+(n-1)}(1+q^{-1})\cdots(1+q^{-1}+q^{-2}+\cdots+q^{1-n})$$
$$= (q+1)(q^2+q+1)\ldots(q^{n-1}+q^{n-2}+\cdots+q+1)$$
(3.8.9) $$= n!_q.$$

If we change q to q^{-1} in either $e_q(x)$ or $E_q(x)$, the series still make sense if x is sufficiently small (see problem 4), and by (3.8.8) we have

$$E_{q^{-1}}(x) = \sum_{n=0}^{\infty} q^{-\binom{n}{2}} \frac{x^n}{n!_{q^{-1}}}$$
$$= \sum_{n=0}^{\infty} \frac{x^n}{q^{\binom{n}{2}} n!_{q^{-1}}}$$
$$= \sum_{n=0}^{\infty} \frac{x^n}{n!_q} = e_q(x).$$

Evidently we also have $e_{q^{-1}}(x) = E_q(x)$ (why?). One of the characteristic properties of an exponential function is that the reciprocal is found by changing x to $-x$, and we see that these q-exponential functions almost have the same property:

$$e_q(-x) = \frac{1}{e_{q^{-1}}(x)} \quad \text{and} \quad E_q(-x) = \frac{1}{E_{q^{-1}}(x)}.$$

Recall that

$$\mathbf{D}_q\, e_q(x) = e_q(x) \quad \text{and} \quad \mathbf{D}_q\, E_q(x) = E_q(xq).$$

What would a function look like if it satisfied the equation

(3.8.10) $$\mathbf{D}_q\, f(x) = u\, f(x) + v\, f(xq)$$

for some constants u and v? Let's also assume that $f(0) = 1$, as with the two q-exponential functions we already have. As before, we can solve (3.8.10) by iteration. A little algebra (exercise) shows that it is equivalent to

$$f(x) = f(xq) \frac{1 + vx(1-q)}{1 - ux(1-q)}.$$

By induction we then have

(3.8.11) $$f(x) = f(xq^n) \frac{(-vx(1-q); q)_n}{(ux(1-q); q)_n}$$

for any nonnegative integer n. Returning to our usual assumption that $|q| < 1$, and recalling that $f(0) = 1$, letting $n \to \infty$ in (3.8.11) gives

$$f(x) = \frac{(-vx(1-q); q)_\infty}{(ux(1-q); q)_\infty}$$

as the unique solution of (3.8.10) with $f(0) = 1$. Now we would like to get an identity like (3.8.8) or (3.8.3) for this function. It is convenient to rename it first: we have shown that

$$g(x; u, v) := \frac{(-vx(1-q); q)_\infty}{(ux(1-q); q)_\infty}$$

is the unique solution of

(3.8.12) $\quad \mathbf{D}_q\, g(x; u, v) = u\, g(x; u, v) + v\, g(xq; u, v) \quad \text{with} \quad g(0; u, v) = 1.$

On the other hand, if we compute the q-derivative of $g(x; u, v)$ directly we get

$$\mathbf{D}_q\, g(x; u, v) = \frac{\frac{(-vx(1-q); q)_\infty}{(ux(1-q); q)_\infty} - \frac{(-vxq(1-q); q)_\infty}{(uxq(1-q); q)_\infty}}{x(1-q)}$$

$$= \frac{(-vxq(1-q); q)_\infty}{(ux(1-q); q)_\infty} \frac{1 + vx(1-q) - (1 - ux(1-q))}{x(1-q)}.$$

The first of the two fractions on the previous line is $g(x; u, vq)$, and the second simplifies to $u + v$, so we have

$$\mathbf{D}_q\, g(x; u, v) = (u + v)\, g(x; u, vq).$$

Iterating this n times gives

$$\mathbf{D}_q^n\, g(x; u, v) = (u+v)(u+vq) \cdots (u+vq^{n-1})\, g(x; u, vq^n)$$

for any nonnegative integer n (where $(u+v)(u+vq) \cdots (u+vq^{n-1}) = 1$ if $n = 0$), and hence

$$\mathbf{D}_q^n\, g(x; u, v)\big|_{x=0} = (u+v)(u+vq) \cdots (u+vq^{n-1})\, g(0; u, vq^n)$$
(3.8.13) $$= (u+v)(u+vq) \cdots (u+vq^{n-1}).$$

The q-Taylor theorem (2.4.9) therefore implies that

$$g(x; u, v) = \sum_{n=0}^{\infty} (u+v)(u+vq) \cdots (u+vq^{n-1}) \frac{x^n}{n!_q},$$

and so our analogue of (3.8.8) and (3.8.3) is

(3.8.14) $\quad \dfrac{(-vx(1-q); q)_\infty}{(ux(1-q); q)_\infty} = g(x; u, v) = \displaystyle\sum_{n=0}^{\infty} (u+v)(u+vq) \cdots (u+vq^{n-1}) \dfrac{x^n}{n!_q}.$

This is just the Cauchy/Crelle series in a different notation. If we replace x by $x(1-q)$ in (3.7.3) and take $b = u$ and $a = -v$, the result is (3.8.14).

Exercises

1. Show that if
$$f(x) = \frac{f(x) - f(qx)}{x(1-q)},$$
then $f(x) = f(qx)/(1 - x(1-q))$, and that if
$$f(qx) = \frac{f(x) - f(qx)}{x(1-q)},$$
then $f(x) = (1 + x(1-q)) f(qx)$.

2. Use the ratio test to show that if $|q| < 1$, then
$$\sum_{n=0}^{\infty} \frac{x^n}{n!_q}$$
converges for $|x| < 1/|1-q|$.

3. Use the ratio test to show that if $|q| < 1$, then
$$\sum_{n=0}^{\infty} q^{\binom{n}{2}} \frac{x^n}{n!_q}$$
converges for all x.

4. Since the series in problems 2 and 3 reduce to e^x if $q = 1$, they converge for all x then (as is easily checked by the ratio test). We never assume q is large, but what if we did? Show that if $q > 1$, then the series in problem 2 converges for all x, while the series in problem 3 converges for $|x| < 1/|1 - \frac{1}{q}|$.

5. Show directly that
$$\mathbf{D}_q \sum_{n=0}^{\infty} \frac{x^n}{n!_q} = \sum_{n=0}^{\infty} \frac{x^n}{n!_q}.$$

6. (a) Show directly that
$$\mathbf{D}_q \frac{1}{(x(1-q); q)_{\infty}} = \frac{1}{(x(1-q); q)_{\infty}}.$$
   (b) Use (a) and (2.4.9) to prove (3.8.3).

7. (a) Show that if a is independent of x, then
$$\mathbf{D}_q (ax; q)_{\infty} = -\frac{a(axq; q)_{\infty}}{1-q}.$$
In particular,
$$\mathbf{D}_q (-x(1-q); q)_{\infty} = (-xq(1-q); q)_{\infty}.$$
   (b) Iterate the second result of part (a) to show that
$$\mathbf{D}_q^n (-x(1-q); q)_{\infty} = q^{\binom{n}{2}} (-xq^n(1-q); q)_{\infty}.$$
   (c) Use (b) and (2.4.5) to prove (3.8.8).

8. (a) Show that
$$\mathbf{D}_q \frac{1}{(x;q)_{n+1}} = \frac{[n+1]_q}{(x;q)_{n+2}}.$$
   (b) Show that for any nonnegative integer k we have
$$\mathbf{D}_q^k \frac{1}{(x;q)_{n+1}} = \frac{[n+1]_q[n+2]_q \cdots [n+k]_q}{(x;q)_{n+k+1}} = \frac{(n+k)!_q}{n!_q} \frac{1}{(x;q)_{n+k+1}}.$$
   (c) Use (b) and (2.4.9) to prove that
$$\frac{1}{(x;q)_{n+1}} = \sum_{k=0}^{\infty} \binom{n+k}{k}_q x^k.$$
   This is Cauchy's theorem (3.5.3).

9. Suppose $|q| < 1$, and let $f(x)$ be a function that is finite at $x = 0$. If

(3.8.15) $$F(x) = x(1-q) \sum_{n=0}^{\infty} q^n f(xq^n),$$

   show that $\mathbf{D}_q F(x) = f(x)$. Since $F(0) = 0$, F is called the q-**integral** of f from 0 to x.

10. If a is any nonnegative number and $|q| < 1$, show directly from (3.8.15) that the q-integral of x^a is $x^{a+1}/[a+1]_q$.

11. Show directly from (3.8.15) that the q-integral of
$$e_q(x) = \sum_{m=0}^{\infty} \frac{x^m}{m!_q}$$
    is $e_q(x) - 1$.

12. It is surprisingly hard to show directly from (3.8.15) that the q-integral of
$$e_q(x) = \frac{1}{(x(1-q);q)_\infty}$$
    is $e_q(x) - 1$, but it is a nice application of several of the identities in sections 3.5 and 3.6.

    (i) We have to evaluate
$$x(1-q) \sum_{n=0}^{\infty} q^n e_q(xq^n) = x(1-q) \sum_{n=0}^{\infty} \frac{q^n}{(xq^n(1-q);q)_\infty}.$$
    Show that this can be rewritten as
$$\frac{x(1-q)}{(x(1-q);q)_\infty} \sum_{n=0}^{\infty} q^n (x(1-q);q)_n.$$

    (ii) Show that (i) can be rewritten as
$$\frac{x(1-q)}{(x(1-q);q)_\infty} \sum_{n=0}^{\infty} q^n \sum_{k=0}^{n} \binom{n}{k}_q (-1)^k q^{\binom{k}{2}} x^k (1-q)^k.$$

    (iii) Show that (ii) can be rewritten as
$$\frac{x(1-q)}{(x(1-q);q)_\infty} \sum_{k=0}^{\infty} (-1)^k q^{\binom{k}{2}} x^k (1-q)^k \sum_{n=k}^{\infty} \binom{n}{k}_q q^n.$$

(iv) After setting $n = m + k$ in (iii), show that it becomes
$$\frac{1}{(x(1-q);q)_\infty} \sum_{k=0}^\infty (-1)^k q^{\binom{k+1}{2}} \frac{x^{k+1}(1-q)^{k+1}}{(q;q)_{k+1}}.$$

(v) Show that (iv) can be rewritten as
$$\frac{1}{(x(1-q);q)_\infty} \left[1 - \sum_{j=0}^\infty (-1)^j q^{\binom{j}{2}} \frac{x^j(1-q)^j}{(q;q)_j} \right]$$
and that this expression equals $e_q(x) - 1$.

13. Suppose that $yx = qxy$, and that t and q commute with x and y and each other. Show that $e_q(xt)\, e_q(yt) = e_q((x+y)t)$. This seems to be what Schützenberger wanted his (and Potter's) q-binomial theorem for.

14. Show directly from the series expansion
$$g(x; u, v) = \sum_{n=0}^\infty (u+v)(u+vq) \cdots (u+vq^{n-1}) \frac{x^n}{n!_q}$$
that $\mathbf{D}_q g(x; u, v) = u\, g(x; u, v) + v\, g(xq; u, v)$.

15. The higher q-derivatives of $g(x; u, v)$ can also be computed directly from the q-differential equation (3.8.12), with the aid of (2.4.5). For convenience we suppress the u and v from the notation: if
$$\mathbf{D}_q g(x) = u\, g(x) + v\, g(xq),$$
show that

(i) $\mathbf{D}_q^2 g(x) = u^2 g(x) + [2]_q uv\, g(xq) + qv^2\, g(xq^2)$.

(ii) $\mathbf{D}_q^3 g(x) = u^3 g(x) + [3]_q u^2 v\, g(xq) + [3]_q quv^2\, g(xq^2) + q^3 v^3\, g(xq^3)$.

(iii) $\mathbf{D}_q^n g(x) = \sum_{k=0}^n \binom{n}{k}_q q^{\binom{k}{2}} u^k v^{n-k} g(xq^k)$ for any nonnegative integer n.

(iv) Assuming that $g(0) = 1$, explain why (iii) implies (3.8.13).

16. (a) Show that
$$\mathbf{D}_q \frac{(ax;q)_n}{(bx;q)_n} = [n]_q(b-a) \frac{(axq;q)_{n-1}}{(bx;q)_{n+1}}.$$

(b) Calculate $\mathbf{D}_q^2 \frac{(ax;q)_n}{(bx;q)_n}$. Since this answer is not nearly as nice as the answer to (a), we should not expect $(ax;q)_n/(bx;q)_n$ to have a nice expansion in powers of x.

17. (a) Show that
$$\mathbf{D}_q \frac{x^n}{(bx;q)_n} = \frac{[n]_q x^{n-1}}{(bx;q)_{n+1}}.$$
What does this reduce to if $q \to 1$?

(b) Show more generally that
$$\mathbf{D}_q \frac{x^m}{(bx;q)_n} = \frac{x^{m-1}([m]_q + bxq^n[n-m]_q)}{(bx;q)_{n+1}}.$$
Therefore the result in (a) does not iterate nicely.

(c) Nevertheless, $x^m/(bx;q)_n$ has a nice expansion in powers of x. Starting from the result of problem 8(c), or otherwise, show that

$$\frac{x^m}{(bx;q)_n} = \sum_{k=m}^{\infty} \binom{n+k-m-1}{n-1}_q x^k b^{k-m}.$$

3.9. Bibliographical Notes

Though Leibniz had thought about them, nothing of significance was done on partitions before Euler's work. Chapter 16 of his great book [**94**] is still as good a place as any to start learning about partitions. Some of this chapter is also in the papers [**93**] and [**97**]. The standard reference for partitions since 1976 has been Andrews's book [**15**]. More recently he has written the more elementary book [**30**] with Kimmo Eriksson, which is an excellent alternative reference for many of the topics presented here. Most of what I know about partitions I learned from [**15**], or from Andrews's other writings, or from him directly. In particular, the original plan of this chapter and the next was loosely based on the introductory part of [**16**], and several of the exercises are lifted from [**15**]. Problem 7 in section 73. is adapted from a problem on the 1990 Putnam exam that he presumably contributed. Problems 7 and 8 in section 1 come from section 52 of Jacobi's *Fundamenta Nova* [**148**]. The last three problems in that section are adapted from one on the 2014 Putnam exam. Problem 11 in section 3.2 is adapted from a problem on the 1960 Putnam Exam, and problems 6 and 7 in section 3.4 from one on the 2003 Putnam exam. Problem 22 in section 3.2 comes from Jacobi's paper [**151**], and problem 19 in that section from [**176**]. Euler's lemma—problem 18 in section 3.3—is Propositio 1 in [**100**], except that it is stated for infinite products there. It becomes problem 5 on the 1952 Putnam exam if we replace each e_j by $-a_j$ and rearrange. Nicole's identity in problem 21 [**175**],[**159**] is equivalent and even older. The further equivalent forms in problems 22 and 19 come from [**156**] and [**158**] respectively.

Glaisher's bijection appears in [**123**]. Cayley's theorem appears in [**71**], and much later in Sylvester's paper [**232**], which is the other great work on partitions (besides [**94**]) before 1900. Franklin's argument at the end of section 3.4 is in section 20 of [**232**]. Ferrers diagrams first appeared in print in [**227**], a short paper published by Sylvester in 1853. (There is an interesting footnote in this paper concerning the astronomer John Couch Adams, which is amplified upon in [**155**].) Sylvester publicized Ferrers' idea on several other occasions [**228**], [**229**], [**231**], but it was probably not until [**232**] that it really became well-known. Norman Ferrers coedited the *Quarterly Journal of Pure and Applied Mathematics* with Sylvester at one time. The Andrews–Deutsch and Smoot–Yang theorems are taken from [**170**], which also has a bijective proof of the Smoot–Yang theorem.

Problem 10 in section 3.6 comes from [**139**] and section 1.3 of [**140**]. The references for the Cauchy/Crelle series are [**67**], [**90**], [**150**] and [**137**]. Ramanujan also knew it [**192**], and gave the same proof as Cauchy and Jacobi.

CHAPTER 4

Partitions II: Geometric Theory

4.1. Euler's pentagonal number theorem

So far we have seen all of Euler's major contributions to partition theory except one: his so-called pentagonal number theorem. Just as we started Chapter 3 with Euler's remarkably successful attempt to expand $(-q;q)_\infty$ in powers of q, we start this chapter with his rather more laborious attempt to expand $(q;q)_\infty$:

$$\begin{aligned}(q;q)_\infty &= (1-q)(1-q^2)(1-q^3)(1-q^4)(1-q^5)(1-q^6)(1-q^7)(1-q^8)\cdots\\&= 1 - q - q^2 - q^3 + q^3 - q^4 + q^4 - q^5 + q^5 + q^5 - q^6 + q^6 + q^6 - q^6\\&\quad - q^7 + q^7 + q^7 + q^7 - q^7 - q^8 + q^8 + q^8 + q^8 - q^8 - q^8 \cdots\\&= 1 - q - q^2 + q^5 + q^7 + \ldots.\end{aligned}$$

On the evidence of the first several terms, we can see that some powers of q appear with a $+$, some with a $-$, and some not at all. Euler wanted to know which powers were going to show up, and with what signs. It is also not at all obvious that we won't eventually get something like $5q^{37}$; *i.e.*, the coefficients of the first several powers are all 0 or ± 1, but that doesn't necessarily mean they are all like that—although it turns out that they are. Euler wrote out many more terms, and found that

$$(q;q)_\infty = 1 - q - q^2 + q^5 + q^7 - q^{12} - q^{15} + q^{22} + q^{26} - q^{35} - q^{40} + \ldots.$$

To see what the pattern of the exponents is he looked at the differences between them, which are

$$1, 1, 3, 2, 5, 3, 7, 4, 9, 5, 11, 6, 13, 7, 15, 8, \ldots.$$

This told Euler that the exponents must be quadratic (see problem 2), and he soon came up with the formula $n(3n \pm 1)/2$ for them. He knew that the sequence $n(3n-1)/2$ which, starting with $n=1$, goes $1, 5, 12, 22, 35, \ldots$, were the pentagonal numbers (see problem 1), and $n(3n+1)/2$ can be thought of as a pentagonal number with a negative n since

$$\frac{(-n)\{3(-n)+1\}}{2} = \frac{n(3n-1)}{2}.$$

So Euler knew pretty quickly what the theorem should be, but proving it was another matter. The argument he eventually found began with Euler's lemma from problem 18 of section 3.3. Taking $E_i = 1 - q^i$ and $e_i = -q^i$ there we get

$$(q;q)_n = 1 - q - (1-q)q^2 - (1-q)(1-q^2)q^3 - \ldots$$
$$- (1-q)(1-q^2)\cdots(1-q^{n-1})q^n$$

(4.1.1)
$$= 1 - q - \sum_{k=1}^{n-1} (q;q)_k \, q^{k+1}.$$

It is not difficult to prove this directly; see problem 3. If $|q| < 1$, then both sides converge as $n \to \infty$, and we have

$$(q;q)_\infty = 1 - q - \sum_{k=1}^{\infty} (q;q)_k \, q^{k+1}$$
$$= 1 - q - (1-q)q^2 - (1-q)(1-q^2)q^3$$
$$- (1-q)(1-q^2)(1-q^3)q^4 - (1-q)(1-q^2)(1-q^3)(1-q^4)q^5 - \ldots.$$

Next, distribute all the factors of $1 - q$ here:

$$(q;q)_\infty = 1 - q - \{q^2 - q^3\} - \{(1-q^2)q^3 - (1-q^2)q^4\}$$
$$- \{(1-q^2)(1-q^3)q^4 - (1-q^2)(1-q^3)q^5\} - \ldots$$
$$= 1 - q - q^2 + q^3\left(1 - (1-q^2)\right) + q^4(1-q^2)\left(1 - (1-q^3)\right)$$
$$+ q^5(1-q^2)(1-q^3)\left(1 - (1-q^4)\right) - \ldots$$
$$= 1 - q - q^2 + q^5 + (1-q^2)q^7 + (1-q^2)(1-q^3)q^9 + \ldots$$
$$= 1 - q - q^2 + q^5 + \sum_{k=1}^{\infty} (q^2;q)_k \, q^{2k+5}.$$

Writing this out broadly we have

$$(q;q)_\infty = 1-q-q^2+q^5+(1-q^2)q^7+(1-q^2)(1-q^3)q^9+(1-q^2)(1-q^3)(1-q^4)q^{11}+\ldots,$$

and now we distribute all the factors of $1 - q^2$:

$$(q;q)_\infty = 1 - q - q^2 + q^5 + \{q^7 - q^9\} + \{(1-q^3)q^9 - (1-q^3)q^{11}\}$$
$$+ \{(1-q^3)(1-q^4)q^{11} - (1-q^3)(1-q^4)q^{13}\} + \ldots$$
$$= 1 - q - q^2 + q^5 + q^7 + q^9(-1 + 1 - q^3)$$
$$+ q^{11}(1-q^3)(-1 + 1 - q^4) + q^{13}(1-q^3)(1-q^4)(-1 + 1 - q^5) + \ldots$$
$$= 1 - q - q^2 + q^5 + q^7 - q^{12} - (1-q^3)q^{15} + (1-q^3)(1-q^4)q^{18} - \ldots$$
$$= 1 - q - q^2 + q^5 + q^7 - q^{12} - \sum_{k=1}^{\infty} (q^3;q)_k \, q^{3k+12}.$$

In general we have

LEMMA 3. *For any positive integer n,*

(4.1.2)
$$(q;q)_\infty = \sum_{k=1-n}^{n} (-1)^k \, q^{\frac{k(3k-1)}{2}} + (-1)^n \, q^{\frac{n(3n-1)}{2}} \sum_{k=1}^{\infty} (q^n;q)_k \, q^{nk}.$$

4.1. EULER'S PENTAGONAL NUMBER THEOREM

We have seen the cases $n = 1, 2, 3$ of this already. We prove it by induction on n by the same method as before. Consider the sum

$$\sum_{k=1}^{\infty} (q^n; q)_k q^{nk} = (1-q^n)q^n + (1-q^n)(1-q^{n+1})q^{2n} + (1-q^n)(1-q^{n+1})(1-q^{n+2})q^{3n}$$
$$+ (1-q^n)(1-q^{n+1})(1-q^{n+2})(1-q^{n+3})q^{4n} + \ldots$$

Distribute each of the factors of $1 - q^n$:

$$(1-q^n)q^n + (1-q^n)(1-q^{n+1})q^{2n} + (1-q^n)(1-q^{n+1})(1-q^{n+2})q^{3n}$$
$$+ (1-q^n)(1-q^{n+1})(1-q^{n+2})(1-q^{n+3})q^{4n} + \ldots$$
$$= q^n + \{q^{2n}(1-q^{n+1}) - q^{3n}(1-q^{n+1})\}$$
$$+ \{q^{3n}(1-q^{n+1})(1-q^{n+2}) - q^{4n}(1-q^{n+1})(1-q^{n+2})\}$$
$$+ \left\{ \begin{matrix} q^{4n}(1-q^{n+1})(1-q^{n+2})(1-q^{n+3}) \\ -q^{5n}(1-q^{n+1})(1-q^{n+2})(1-q^{n+3}) \end{matrix} \right\} + \ldots$$
$$= q^n + (1-q^{n+1} - 1)q^{2n} + (1-q^{n+1})(1-q^{n+2} - 1)q^{3n}$$
$$+ (1-q^{n+1})(1-q^{n+2})(1-q^{n+3} - 1)q^{4n} + \ldots$$
$$= q^n - q^{3n+1} - (1-q^{n+1})q^{4n+2} - (1-q^{n+1})(1-q^{n+2})q^{5n+3} - \ldots$$
$$= q^n - q^{3n+1} - \sum_{k=1}^{\infty} (q^{n+1}; q)_k q^{(n+1)k + 3n+1}$$
$$= q^n - q^{3n+1} - q^{3n+1} \sum_{k=1}^{\infty} (q^{n+1}; q)_k q^{(n+1)k}.$$

Substituting this in (4.1.2) we get

$$(q; q)_\infty = \sum_{k=1-n}^{n} (-1)^k q^{\frac{k(3k-1)}{2}} + (-1)^n q^{n + \frac{n(3n-1)}{2}} + (-1)^{n+1} q^{3n+1 + \frac{n(3n-1)}{2}}$$
$$+ (-1)^{n+1} q^{3n+1 + \frac{n(3n-1)}{2}} \sum_{k=1}^{\infty} (q^{n+1}; q)_k q^{(n+1)k}.$$

Now

$$3n + 1 + \frac{n(3n-1)}{2} = \frac{3n^2 + 5n + 2}{2} = \frac{(n+1)(3n+2)}{2} = \frac{(n+1)(3(n+1) - 1)}{2}$$

and

$$n + \frac{n(3n-1)}{2} = \frac{n(3n+1)}{2} = \frac{(-n)(3(-n) - 1)}{2},$$

so the two terms above that aren't inside a sum are the $k = n + 1$ and $k = -n$ terms respectively of the first sum, and we have

$$(q; q)_\infty = \sum_{k=-n}^{n+1} (-1)^k q^{\frac{k(3k-1)}{2}} + (-1)^{n+1} q^{\frac{(n+1)(3(n+1)-1)}{2}} \sum_{k=1}^{\infty} (q^{n+1}; q)_k q^{(n+1)k},$$

which is (4.1.2) with $n+1$ in place of n. Hence (4.1.2) holds for $n+1$ if it holds for n, and since it holds for $n = 1, 2, 3$, it holds for all positive integers n.

Now we let $n \to \infty$ in (4.1.2). The series

$$\sum_{k=1}^{\infty} (q^n; q)_k \, q^{nk}$$

converges (we leave this as an exercise), so the second term in (4.1.2) tends to zero as $n \to \infty$ because of the factor $q^{\frac{n(3n-1)}{2}}$, since $|q| < 1$. Thus we get

THEOREM 31 (Euler's pentagonal number theorem). *If $|q| < 1$, then*

$$(q; q)_\infty = \sum_{k=-\infty}^{\infty} (-1)^k q^{\frac{k(3k-1)}{2}}$$

$$= 1 + \sum_{k=1}^{\infty} (-1)^k \left(q^{\frac{k(3k-1)}{2}} + q^{\frac{k(3k+1)}{2}} \right)$$

$$= 1 + \sum_{k=1}^{\infty} (-1)^k q^{\frac{k(3k-1)}{2}} \left(1 + q^k \right).$$

Euler worked so hard to prove this because he had two applications in mind. We'll see the second one in section 4.4. For the first one, recall that if $p(j)$ is the number of partitions of the integer j, then

(4.1.3) $$\frac{1}{(q; q)_\infty} = \sum_{j=0}^{\infty} p(j) \, q^j.$$

Multiplying this by $(q; q)_\infty$ and expanding the right side by the pentagonal number theorem, we get

$$1 = \left(\sum_{j=0}^{\infty} p(j) \, q^j \right) \left(1 - q - q^2 + q^5 + q^7 - q^{12} - q^{15} + q^{22} + q^{26} - - + + \ldots \right).$$

If $n \geq 1$, then the coefficient of q^n on the right side must be zero. This coefficient is evidently

$$p(n) - p(n-1) - p(n-2) + p(n-5) + p(n-7) - p(n-12) - p(n-15) \ldots,$$

so

(4.1.4) $p(n) = p(n-1) + p(n-2) - p(n-5) - p(n-7) + p(n-12) + p(n-15) - - + + \ldots,$

where the series alternates between pairs of positive terms and pairs of negative terms, and continues until we drop below $p(0)$. For example,

$$p(38) = p(37) + p(36) - p(33) - p(31) + p(26) + p(23) - p(16) - p(12) + p(3).$$

The series stops at this point because the next term would be $p(-2)$, which is not part of (4.1.3). The recurrence (4.1.4) gives a reasonably good way of constructing a table of values of $p(n)$; certainly much better than direct counting of all the partitions.

Exercises

1. To see why $1, 5, 12, 22, 35, \ldots$ are called the pentagonal numbers, start by drawing a dot. Then draw four more dots to make a pentagon. Then draw seven more dots to create a pentagon whose sides are twice as long, and which has the original pentagon in one corner. Then draw ten more dots to create a pentagon with sides three times as long as the original, and so on. The sequence $1, 5, 12, 22, \ldots$ is really $1, 1 + 4, 1 + 4 + 7, 1 + 4 + 7 + 10, \ldots$. Show that

$$1 + 4 + 7 + 10 + \cdots + (3n-2) = \frac{n(3n-1)}{2}.$$

2. The first forward difference of a function $f(n)$ is $f(n+1) - f(n)$ and the first backward difference is $f(n) - f(n-1)$, and the second difference is the difference of the first difference. Show that a quadratic function has constant second differences (either forward or backward, whichever you prefer). This is how Euler knew that the sequence $1, 5, 12, 22, 35, \ldots$ would be quadratic.

3. Give a direct proof of (4.1.1) by rewriting $q^{k+1} = 1 - (1 - q^{k+1})$.

4. Prove that the series

$$\sum_{k=1}^{\infty} (q^n; q)_k \, q^{nk}$$

converges for any positive integer n if $|q| < 1$. Does it converge if $n = 0$?

5. Show that the last two lines of the statement of Euler's pentagonal number theorem above are equivalent to the first line.

6. Compute $p(n)$ up through $p(5)$ by counting all the relevant partitions. (You can check your answers in Chapter 3; recall that $p(0) = 1$ by convention.) Then use (4.1.4) to compute $p(6)$ through $p(15)$.

7. Show that if $|q| < 1$, then

(4.1.5) $$q \left(q^{24}; q^{24} \right)_{\infty} = \sum_{n=-\infty}^{\infty} (-1)^n q^{(6n-1)^2}.$$

This was pointed out by Gauss, and will be used in section 8.2.

8. Show that if $|q| < 1$, then

$$(x; q)_{\infty} = 1 - x \sum_{k=0}^{\infty} q^k \, (x; q)_k.$$

You can start by applying Euler's lemma to $(x; q)_n$, or by imitating problem 3.

9. Show that for any nonnegative integer j we have

$$1 - \sum_{k=j}^{\infty} xq^k \, (xq^j; q)_{k-j} = (1 - xq^j) \left[1 - \sum_{k=j+1}^{\infty} xq^k \, (xq^{j+1}; q)_{k-(j+1)} \right]$$

and use this to give an alternative proof of the result of problem 8.

10. In another paper Euler considered the function

$$S(x; q) = \sum_{k=1}^{\infty} \frac{q^k}{q^k - 1} \left(\frac{x}{q}; q \right)_k.$$

Show that the series converges for all x if $|q| < 1$ and that

$$1 + S(xq; q) - S(x; q) = 1 - x \sum_{k=0}^{\infty} q^k (x; q)_k = (x; q)_\infty.$$

11. Suppose we take $x = q^{-n}$ in the result of problem 11, where n is a nonnegative integer. Explain why this gives

$$1 = \sum_{k=0}^{n} q^{k-n} (q^{-n}; q)_k.$$

12. Show that if we replace q by q^{-1} in the result of problem 11 we get

$$\frac{1}{(q;q)_n} = \sum_{j=0}^{n} \frac{q^j}{(q;q)_j},$$

which we have seen already in the previous two chapters.

13. Suppose we take $x = q^{-n}$ in the result of problem 10, where n is a nonnegative integer. Explain why this gives

$$1 + \sum_{k=1}^{n} \frac{q^k}{q^k - 1} (q^{-n}; q)_k = \sum_{k=1}^{n+1} \frac{q^k}{q^k - 1} \left(q^{-(n+1)}; q\right)_k.$$

14. Show that if we replace q by q^{-1} in the result of problem 13 we get

$$1 + \sum_{k=1}^{n} \binom{n}{k}_q (q;q)_{k-1} = \sum_{k=1}^{n+1} \binom{n+1}{k}_q (q;q)_{k-1},$$

and use this fact to show that

$$\sum_{k=1}^{n} \binom{n}{k}_q (q;q)_{k-1} = n.$$

We saw this result in section 2.9; this is essentially Euler's proof of it.

15. In the same paper that engendered problems 10–14, Euler considered the sum

$$f_m(q) = \sum_{k=1}^{\infty} (q^k; q)_m q^{k-1}$$

for $|q| < 1$ and a nonnegative integer m.

(i) Show that the series for $f_m(q)$ converges.

(ii) For nonnegative integers m and n define

$$f_{m,n}(q) = \sum_{k=1}^{n} (q^k; q)_m q^{k-1}.$$

Use induction on n to show that

$$f_{m,n}(q) = \frac{(q^n; q)_{m+1}}{1 - q^{m+1}}.$$

(iii) Show that $f_m(q) = 1/(1 - q^{m+1})$. What does this say if $m = 0$?

16. Euler's proof of (ii) in the previous problem did not use induction. Instead he wrote

$$(1-q^{m+1})f_{m,n}(q) = \sum_{k=1}^{n}(q^k;q)_m \left(q^{k-1} - q^{k+m}\right)$$

$$= \sum_{k=1}^{n}(q^k;q)_m \left((1-q^{k+m}) - (1-q^{k-1})\right).$$

Finish the calculation to show that $(1-q^{m+1})f_{m,n}(q) = (q^n;q)_{m+1}$.

17. Legendre gave an interesting variation of Euler's proof of the pentagonal number theorem. Here is an outline of it:

(i) Legendre started with

$$(q;q)_\infty = 1 + \sum_{k=1}^{\infty} \frac{(-1)^k q^{\frac{k(k+1)}{2}}}{(q;q)_k},$$

which follows from one of the identities in Chapter 3. Which one? Explain.

(ii) Show that, if $k \geq 1$, then

$$\frac{q^{\frac{k(k+1)}{2}}}{(q;q)_k} = \frac{q^{\frac{k(k+1)}{2}}}{(q;q)_{k-1}} + \frac{q^{\frac{k(k+3)}{2}}}{(q;q)_k}.$$

(iii) Using (i) and (ii), show that

$$(q;q)_\infty = 1 - q + \sum_{k=2}^{\infty} \frac{(-1)^k q^{\frac{k(k+1)}{2}}}{(q;q)_{k-1}} + \sum_{k=1}^{\infty} \frac{(-1)^k q^{\frac{k(k+3)}{2}}}{(q;q)_k}.$$

(iv) By reindexing the first sum in (iii) and combining it with the second, show that

$$(q;q)_\infty = 1 - q + \sum_{k=1}^{\infty}(-1)^k \frac{q^{\frac{k(k+3)}{2}}(1-q)}{(q;q)_k}.$$

(v) Show that the identity in (iv) can be rewritten as

$$(q;q)_\infty = 1 - q - q^2 + \sum_{k=1}^{\infty}(-1)^{k+1} \frac{q^{\frac{(k+1)(k+4)}{2}}}{(q^2;q)_k}.$$

This completes the first iteration of Legendre's argument.

(vi) For the second iteration of Legendre's argument, show that if $k \geq 1$, then

$$\frac{q^{\frac{(k+1)(k+4)}{2}}}{(q^2;q)_k} = \frac{q^{\frac{(k+1)(k+4)}{2}}}{(q^2;q)_{k-1}} + \frac{q^{\frac{(k+1)(k+6)}{2}}}{(q^2;q)_k},$$

and repeat the same steps as above. The final result should be

$$(q;q)_\infty = 1 - q - q^2 + q^5 + q^7 + \sum_{k=1}^{\infty}(-1)^k \frac{q^{\frac{(k+2)(k+7)}{2}}}{(q^3;q)_k}.$$

(vii) Legendre contented himself with one or two more iterations, but there is a general result here that we can prove, namely

(4.1.6) $$(q;q)_\infty = \sum_{k=-n}^{n}(-1)^k q^{\frac{k(3k-1)}{2}} + \sum_{k=1}^{\infty}(-1)^{k+n} \frac{q^{\frac{(k+n)(k+3n+1)}{2}}}{(q^{n+1};q)_k}$$

for all nonnegative integers n. Show that (i), (v), and (vi) are the cases $n = 0, 1, 2$ respectively of (4.1.6).

(viii) Show that
$$\frac{q^{\frac{(k+n)(k+3n+1)}{2}}}{(q^{n+1};q)_k} = \frac{q^{\frac{(k+n)(k+3n+1)}{2}}}{(q^{n+1};q)_{k-1}} + \frac{q^{\frac{(k+n)(k+3n+3)}{2}}}{(q^{n+1};q)_k},$$
and use this to prove (4.1.6) by induction on n by the same procedure as in parts (iii)–(v).

(ix) Show that (4.1.6) can be rewritten as
$$(4.1.7) \qquad (q;q)_\infty = \sum_{k=-n}^{n} (-1)^k q^{\frac{k(3k-1)}{2}} + (-1)^n q^{\frac{n(3n+1)}{2}} \sum_{k=1}^{\infty} (-1)^k \frac{q^{2nk} q^{\frac{k(k+1)}{2}}}{(q^{n+1};q)_k}.$$

(x) Show that the series
$$\sum_{k=1}^{\infty} (-1)^k \frac{q^{2nk} q^{\frac{k(k+1)}{2}}}{(q^{n+1};q)_k}$$
converges for any n, under our usual assumption that $|q| < 1$.

(xi) Explain why (x) implies that if we let $n \to \infty$ in (4.1.7), we get Euler's pentagonal number theorem.

This is a long and challenging problem, but Nick Bartlett got all the way through it as a freshman, so it can be done. Eric Moss has done the next problem.

18. Jacobi also gave an interesting variation of Euler's proof of the pentagonal number theorem. Here is an outline of it:

(i) Jacobi started by defining, for a nonnegative integer m,
$$f_m(z) = 1 - z + z(1-z)(1-zq) + z^2(1-z)(1-zq)(1-zq^2) + \ldots$$
$$+ z^{m-1}(1-z)(1-zq)\cdots(1-zq^{m-1}) = \sum_{k=1}^{m} (z;q)_k z^{k-1},$$
where $f_0(z) := 0$. Write out $f_1(z)$, $f_2(z)$, and $f_3(z)$, and show that
$$f_2(z) = 1 - z^2 q - z^2(1-zq) \quad \text{and} \quad f_3(z) + z^3 q^2 f_1(zq) = 1 - z^2 q - z^3(1-zq)(1-zq^2).$$

(ii) Show that $f_m(z) = f_{m-1}(z) + (z;q)_m z^{m-1}$ if $m \geq 1$.

(iii) Jacobi found a less obvious recurrence relation for $f_m(z)$ than the one in (ii). By induction on m or otherwise, show that
$$(4.1.8) \qquad f_m(z) + z^3 q^2 f_{m-2}(zq) = 1 - z^2 q - z^m (zq;q)_{m-1} \quad \text{for } m \geq 2.$$
You did the cases $m = 2, 3$ already in (i).

(iv) Use (4.1.8) and induction on n to show that
$$f_{2n}(z) = \sum_{k=0}^{n-1} (-1)^k q^{\frac{k(3k+1)}{2}} z^{3k} + \sum_{k=1}^{n} (-1)^k q^{\frac{k(3k-1)}{2}} z^{3k-1}$$
$$- \sum_{k=0}^{n-1} (-1)^k q^{2nk-\binom{k}{2}} z^{2n+k} (1-zq^{k+1})(1-zq^{k+2})\cdots(1-zq^{2n-1-k})$$

for $n \geq 1$.

(v) The formula for $f_{2n-1}(z)$ is similar but has an extra term. Show that

$$f_{2n-1}(z) = \sum_{k=0}^{n-1}(-1)^k q^{\frac{k(3k+1)}{2}} z^{3k} + \sum_{k=1}^{n-1}(-1)^k q^{\frac{k(3k-1)}{2}} z^{3k-1} + (-1)^n q^{3\binom{n}{2}} z^{3n-2}$$

$$- \sum_{k=0}^{n-2}(-1)^k q^{(2n-1)k - \binom{k}{2}} z^{2n-1+k}(1-zq^{k+1})(1-zq^{k+2})\cdots(1-zq^{2n-2-k})$$

for $n \geq 1$.

(vi) Show that the results of (iv) and (v) can be combined into

(4.1.9) $\quad f_m(z) = \sum_{k=0}^{\lfloor \frac{m-1}{2} \rfloor}(-1)^k q^{\frac{k(3k+1)}{2}} z^{3k} + \sum_{k=1}^{\lfloor \frac{m}{2} \rfloor}(-1)^k q^{\frac{k(3k-1)}{2}} z^{3k-1}$

$$+ \frac{1-(-1)^m}{2}(-1)^{\lceil \frac{m}{2} \rceil} q^{3\binom{\lceil \frac{m}{2} \rceil}{2}} z^{m+\lfloor \frac{m}{2} \rfloor}$$

$$- \sum_{k=0}^{\lfloor \frac{m-2}{2} \rfloor}(-1)^k q^{mk - \binom{k}{2}} z^{m+k}(1-zq^{k+1})(1-zq^{k+2})\cdots(1-zq^{m-1-k})$$

for $m \geq 1$, where $\lfloor x \rfloor$ denotes the **floor** of x, which is the greatest integer $\leq x$, and $\lceil x \rceil$ denotes the **ceiling** of x, which is the smallest integer $\geq x$.

(vii) Assuming as usual that $|q| < 1$, define

$$f(z) = \lim_{m \to \infty} f_m(z) = \sum_{k=1}^{\infty}(z;q)_k z^{k-1}.$$

Show that the series converges if $|z| < 1$. It also converges if $z = 1$. Why?

(viii) If $|z| < 1$, the last sum in (4.1.9) approaches zero as $m \to \infty$ because of the factors z^m and q^{mk}. Thus Jacobi could conclude that if $|z| < 1$, then

(4.1.10) $\quad f(z) = \sum_{k=1}^{\infty}(-1)^k q^{\frac{k(3k-1)}{2}} z^{3k-1} + \sum_{k=0}^{\infty}(-1)^k q^{\frac{k(3k+1)}{2}} z^{3k}.$

But if $z = 1$ we have to be a little more careful; (4.1.10) is false if $z = 1$. The terms of the last series in (4.1.9) still go to zero as $m \to \infty$ because of the m in the exponent of q, but the $k = 0$ term doesn't have this exponent, so it survives. Why does this imply Euler's pentagonal number theorem?

4.2. Durfee squares

One of the greatest British mathematicians in the 19th century was James Joseph Sylvester. He was an exception to the oft-stated "rule" that mathematicians tend to do their best work when they are young, as perhaps the best period of his career was the several years he spent at Johns Hopkins University around 1880, when he was past 60. One of his students there was W. P. Durfee, for whom the Durfee square is named. (It might be the earliest mathematical concept to be named after an American.)

The Durfee square of a partition is simply the largest square that fits in its Ferrers diagram. For example, the partition $8+5+4+3+3+1$ has a 3×3 Durfee square:

```
* * * • • • • •
* * * • •
* * * •
• • •
• • •
•
```

Is this a useful concept? In Chapter 5 we will discuss a famous theorem of Jacobi called the triple product identity, from several points of view. To do Jacobi's argument we will need to prove

THEOREM 32 (Jacobi's Durfee square identity). *If* $|q| < 1$ *and* $x \neq q^{-1}, q^{-2}, \ldots,$ *then*

$$(4.2.1) \qquad \frac{1}{(xq;q)_\infty} = \sum_{k=0}^{\infty} \frac{q^{k^2} x^k}{(q;q)_k (xq;q)_k}.$$

We know from (3.6.4) that the left side of (4.2.1) is the generating function for partitions; *i.e.*, the coefficient of x^n in (4.2.1) is the generating function for partitions with exactly n parts. Suppose a partition with n parts has a Durfee square of side k. Then it has two other components, one below the square and the other to the right. It is possible that one or both of these could be empty, *e.g.*, the partition $4+4+4+4$

```
* * * *
* * * *
* * * *
* * * *
```

has a Durfee square of side 4, and nothing else.

The lower component of a partition whose Durfee square is $k \times k$ is a partition into some number of parts, all of which are at most k. We look at the conjugate of the other (right) component, which is also a partition whose parts are at most k.

A Durfee square of side k contributes $x^k q^{k^2}$ to the generating function. The (conjugate of the) right component has the other pieces of the k parts in the square, and these pieces are all $\leq k$, so the right component contributes $1/(q;q)_k$ to the generating function. Finally the lower component contributes some more parts which are $\leq k$, and the generating function for it is $1/(xq;q)_k$. (It has an x too because the whole partition probably has more than k parts.) Summing over all k we get (4.2.1). We will see how Jacobi proved (4.2.1) in Chapter 5. For another proof see problem 7.

By the same sort of argument we can generalize this. (However, see problem 1.) Given a nonnegative integer r, we could look instead for the largest $k \times (k+r)$ rectangle that fits in the Ferrers diagram. This rectangle contributes $x^k q^{k(k+r)}$ to the generating function. The right component again has the remaining pieces of the largest k parts, so the conjugate is a partition with parts $\leq k$, so it again contributes $1/(q;q)_k$ to the generating function. The lower component has the remaining parts, all of which are $\leq k+r$, so the generating function for it is $1/(xq;q)_{k+r}$. Therefore

we have

(4.2.2) $$\frac{1}{(xq;q)_\infty} = \sum_{k=0}^{\infty} \frac{q^{k(k+r)} x^k}{(q;q)_k (xq;q)_{k+r}}.$$

We turn next to partitions with distinct parts. The partition $9+6+5+4+3+1$ has a Durfee square of side 4:

```
*  *  *  *  •  •  •  ◇  ◇
*  *  *  *  •  •
*  *  *  *  •
*  *  *  *
◇  ◇  ◇
◇
```

Note that in this case we can add a 3×3 isosceles triangle to the right of the Durfee square. We will call it the **Franklin triangle**, after Fabian Franklin, another of Sylvester's students. (This is not a standard name; we use it for reasons that will become clear in the next section.) Whenever we have a Durfee square of side k in a partition with distinct parts, we will always have a Franklin triangle whose side is either $k-1$ or k, according to whether the bottom of the Durfee square is a complete part, as in the above example; or only part of a part, as for example with $8+6+4+3+2+1$:

```
*  *  *  •  •  •  ◇  ◇
*  *  *  •  •  ◇
*  *  *  •
◇  ◇  ◇
◇  ◇
◇
```

Here the Durfee square and the Franklin triangle both have side 3.

Let's try the same argument as above on partitions with distinct parts. The coefficient of x^n in $(-xq;q)_\infty$ is the generating function for partitions with exactly n parts which are all distinct. Suppose such a partition has a Durfee square of side k. We consider two cases, as in the above examples. The first case is that the Franklin triangle has side $k-1$. Then the square and the triangle contribute $x^k q^{k^2 + \binom{k}{2}}$ to the generating function. To the right of the triangle are the remaining pieces of the first k parts, which are themselves, when read diagonally, a partition whose parts are at most $k-1$. These pieces therefore contribute $1/(q;q)_{k-1}$ to the generating function. Finally, below the square are some more parts whose size is at most $k-1$ (not k, because the bottom of the Durfee square is a complete part of size k). Since these parts are distinct, they contribute $(-xq;q)_{k-1}$ to the generating function. So the generating function for the first case is

(4.2.3) $$\sum_{k=1}^{\infty} x^k q^{k^2 + \binom{k}{2}} \frac{(-xq;q)_{k-1}}{(q;q)_{k-1}}.$$

In the second case the Durfee square and the Franklin triangle both have side k. This time they contribute $x^k q^{k^2 + \binom{k+1}{2}}$ to the generating function. To the right of the triangle are the remaining pieces of the first k parts, which are now a partition whose parts are at most k when we read them diagonally; and below the square are some distinct parts of size $\leq k$. These contribute $1/(q;q)_k$ and $(-xq;q)_k$ respectively to the generating function for the second case, which is

(4.2.4) $$\sum_{k=0}^{\infty} x^k\, q^{k^2+\binom{k+1}{2}}\, \frac{(-xq;q)_k}{(q;q)_k}.$$

Combining (4.2.3) and (4.2.4) we have (leaving some of the steps as exercises)

THEOREM 33 (Sylvester's identity). *For $|q| < 1$ and for all x,*

(4.2.5) $$(-xq;q)_\infty = 1 + \sum_{k=1}^{\infty} x^k\, q^{\frac{k(3k-1)}{2}}\, \frac{(-xq;q)_{k-1}}{(q;q)_k}\, (1+xq^{2k}).$$

This identity is interesting historically because, as remarked by George Andrews, it "holds the distinction of being the first q-series identity whose first proof was purely combinatorial." It is interesting for another reason, as noted by Sylvester: if we set $x = -1$ in (4.2.5), it reduces to

(4.2.6) $$(q;q)_\infty = 1 + \sum_{k=1}^{\infty} (-1)^k\, q^{\frac{k(3k-1)}{2}}\, (1+q^k),$$

which is equivalent to Euler's pentagonal number theorem.

We can rewrite (4.2.5) in a slightly different form, which is interesting for a reason we'll see in problem 12:

$$(-xq;q)_\infty = 1 + \sum_{k=1}^{\infty} x^k\, q^{\frac{k(3k-1)}{2}}\, \frac{(-xq;q)_{k-1}}{(q;q)_k}\, \{(1-q^k) + q^k(1+xq^k)\}$$

$$= 1 + \sum_{k=1}^{\infty} x^k\, q^{\frac{k(3k-1)}{2}}\, \frac{(-xq;q)_{k-1}}{(q;q)_{k-1}} + \sum_{k=1}^{\infty} x^k\, q^{\frac{k(3k+1)}{2}}\, \frac{(-xq;q)_k}{(q;q)_k}$$

$$= \sum_{k=0}^{\infty} x^k\, q^{\frac{k(3k+1)}{2}}\, \frac{(-xq;q)_k}{(q;q)_k} + \sum_{k=0}^{\infty} x^{k+1}\, q^{\frac{(k+1)(3k+2)}{2}}\, \frac{(-xq;q)_k}{(q;q)_k},$$

where on the last line we added the 1 to one of the sums, reindexed the other, and switched them. Combining the last two sums we get

(4.2.7) $$(-xq;q)_\infty = \sum_{k=0}^{\infty} x^k\, q^{\frac{k(3k+1)}{2}}\, \frac{(-xq;q)_k}{(q;q)_k}\, (1+xq^{2k+1}).$$

Presumably, the second q-series identity whose first proof was purely combinatorial was a finite form of (4.2.5) which comes right after it in Sylvester's paper. This time we want to find a similar expansion of $(-xq;q)_n$, which we know is the generating function for partitions with distinct parts each no larger than n. The proof is much the same as that of (4.2.5), but we will make another distinction: whether there is an **underpart** of the same size as the Durfee square right below it. So we'll consider four cases, though one of them is trivial and two of the others could be combined. Case 0 is the trivial one, the empty partition of zero, which contributes 1 to the generating function.

Case 1 is the underpart case: suppose there is a Durfee square of side j and a part of size j below it (the underpart). The Durfee square contributes $x^j q^{j^2}$ and the underpart xq^j to the generating function. Since the parts are distinct, the Franklin triangle must have side length j, so it contributes $q^{1+2+\cdots+j} = q^{\binom{j+1}{2}}$. There are,

4.2. DURFEE SQUARES

in general, more distinct parts of size at most $j-1$ below the underpart, and they contribute $(-xq;q)_{j-1}$. Finally, the first j parts, which were at most n, have had j subtracted from them (the Durfee square); also the largest part has had another j subtracted from it (the Franklin triangle), and so on. What remains of them after the square and triangle have been removed are at most j parts each at most $n-j-j=n-2j$. So these parts contribute $\binom{n-j}{j}_q$ to the generating function (by Cayley's theorem), and hence the generating function for case 1 is

$$(4.2.8) \qquad W_n(x) := \sum_{j\geq 1} \binom{n-j}{j}_q (-xq;q)_{j-1} q^{\frac{3j(j+1)}{2}} x^{j+1}.$$

In case 2 there is a Durfee square of side j, a Franklin triangle of side j, but this time no underpart of size j below the square. The analysis here is exactly the same as in case 1, with the only difference being that there is no factor of xq^j now since there is no underpart. Hence the generating function for case 2 is

$$V_n(x) := \sum_{j\geq 1} \binom{n-j}{j}_q (-xq;q)_{j-1} q^{\frac{j(3j+1)}{2}} x^j.$$

We could combine this with cases 1 and 0 to get

$$(4.2.9) \quad 1 + \sum_{j\geq 1} \binom{n-j}{j}_q (-xq;q)_j q^{\frac{j(3j+1)}{2}} x^j = \sum_{j\geq 0} \binom{n-j}{j}_q (-xq;q)_j q^{\frac{j(3j+1)}{2}} x^j.$$

This corresponds to the second case in the proof of (4.2.5).

In case 3 there is a Durfee square of side j and a Franklin triangle of side $j-1$, which means that there cannot be an underpart of size j (why?). The square still contributes $x^j q^{j^2}$ to the generating function, but the Franklin triangle now only contributes $q^{0+1+\cdots+(j-1)} = q^{\binom{j}{2}}$. The parts below the square again contribute $(-xq;q)_{j-1}$. Finally, the first j parts, which were at most n, have had j subtracted from them, and the largest part has had another $j-1$ subtracted from it, and so on. Since the Franklin triangle has a smaller side than the Durfee square and the parts are distinct, the jth part must be j exactly; there can't be anything to it but the bottom of the square. This means that what remains of the first j parts after the square and triangle have been removed are at most $j-1$ parts each at most $n-j-(j-1) = n-2j+1$. So these parts contribute $\binom{n-j}{j-1}_q$ to the generating function for this case, which is

$$U_n(x) := \sum_{j\geq 1} \binom{n-j}{j-1}_q (-xq;q)_{j-1} q^{\frac{j(3j-1)}{2}} x^j.$$

Combining this with (4.2.9), we have proved that
(4.2.10)
$$(-xq;q)_n = \sum_{j\geq 1} \binom{n-j}{j-1}_q (-xq;q)_{j-1} q^{\frac{j(3j-1)}{2}} x^j + \sum_{j\geq 0} \binom{n-j}{j}_q (-xq;q)_j q^{\frac{j(3j+1)}{2}} x^j.$$

However, Sylvester chose to write this differently. Rather than combining cases 0,1,2 to get (4.2.9), he combined cases 2 and 3 to get

$$(4.2.11) \qquad T_n(x) := U_n(x) + V_n(x) = \sum_{j\geq 1} \binom{n+1-j}{j}_q (-xq;q)_{j-1} q^{\frac{j(3j-1)}{2}} x^j,$$

and so his form of (4.2.10) is

(4.2.12)
$$(-xq;q)_n = 1 + \sum_{j\geq 1} \binom{n+1-j}{j}_q (-xq;q)_{j-1} q^{\frac{j(3j-1)}{2}} x^j$$
$$+ \sum_{j\geq 1} \binom{n-j}{j}_q (-xq;q)_{j-1} q^{\frac{3j(j+1)}{2}} x^{j+1}.$$

Exercises

1. Show that (4.2.2) follows from (4.2.1) by changing x to xq^r.
2. Show that
$$\sum_{k=0}^{\infty} \frac{q^{k^2}}{(q;q)_k^2} = \frac{1}{(q;q)_\infty} = \sum_{k=0}^{\infty} \frac{q^{k^2+k}}{(q;q)_k (q;q)_{k+1}}.$$
3. Show that
$$\frac{2}{(q;q)_\infty} = \sum_{k=0}^{\infty} \frac{q^{k^2-k}}{(q;q)_k^2} \quad \text{and} \quad \frac{1+q}{(q;q)_\infty} = \sum_{k=0}^{\infty} \frac{q^{k^2}}{(q;q)_k (q;q)_{k+1}}.$$
4. For $n \geq 0$, define
$$A_n = \sum_{k=0}^{n} \binom{n}{k}_q \frac{q^{k^2}}{(q;q)_k},$$
and for $n \geq 1$ define
$$B_n = \sum_{k=1}^{n} \binom{n}{k}_q \frac{q^{k^2-k}}{(q;q)_{k-1}} \quad \text{and} \quad C_n = \sum_{k=1}^{n} \binom{n-1}{k-1}_q \frac{q^{k^2-k}}{(q;q)_k}.$$

   (i) Calculate A_0, A_1, A_2, B_1, B_2, C_1, and C_2.
   After doing (i), it is reasonable to hope that
   (a) $A_n = \dfrac{1}{(q;q)_n}$ and (b) $B_n = \dfrac{1}{(q;q)_{n-1}}$ and (c) $C_n = \dfrac{1}{(q;q)_n}$,
   and the rest of the problem proves this.

   (ii) Show that $(1-q^n) C_n = B_n$.

   (iii) Show that $A_n = A_{n-1} + q^n C_n$ and $B_n = B_{n-1} + q^{n-1} A_{n-1}$.

   (iv) Knowing (ii) and (iii), there is a painless proof of (a)–(c) by induction. Assuming (a) and (c) hold for $n-1$, explain why (b) holds for $n-1$, and then why (a)–(c) must hold for n.

5. Prove (a) from problem 4 by a Durfee square argument.
6. Prove (b) and (c) from problem 4 as in problem 5, but using a $k \times (k-1)$ rectangle (with both orientations) instead of a square.
7. Prove Jacobi's Durfee square identity (4.2.1) by expanding
$$\frac{1}{(xq;q)_k}$$
via Cauchy's identity (3.5.3) (note that it can only be used for $k \geq 1$) and using (c) from problem 4.

8. Prove the identity

(4.2.13) $$\frac{1}{(xq;q)_n} = \sum_{k=0}^{n} \binom{n}{k}_q \frac{q^{k^2} x^k}{(xq;q)_k}$$

from problem 10 in section 2.3 by a Durfee square argument. This is a finite form of Jacobi's Durfee square identity (4.2.1).

9. Show that combining (4.2.3) and (4.2.4) gives the right side of (4.2.5).
10. Assuming that $|q| < 1$, show that the series in (4.2.5) converges for all x.
11. Show that (4.2.5) reduces to (4.2.6) when $x = -1$.
12. The quotation from Andrews begs the question: how hard is it to prove (4.2.5) analytically? Cayley published such a proof in Sylvester's *American Journal of Mathematics* a few years after Sylvester's identity appeared there. Here is an outline of it:

    (i) First set $F(x)$ equal to the right side of (4.2.5), and note that $F(x)$ also equals the right side of (4.2.7). What is the $k=0$ term of (4.2.7)?

    (ii) Rewrite the right side of (4.2.7) to show that

    $$F(x) = (1+xq)\left(1 + \sum_{k=1}^{\infty} x^k q^{\frac{k(3k+1)}{2}} \frac{(-xq^2;q)_{k-1}}{(q;q)_k}(1+xq^{2k+1})\right),$$

    and explain why this implies $F(x) = (1+xq)F(xq)$.

    (iii) Prove that $F(x) = (-xq;q)_\infty$ by iterating the result of (ii).

13. Imitate the proof of (4.2.7) to show that, if

    $$G(x) = 1 + \sum_{n=1}^{\infty} (-1)^n x^{2n} q^{\frac{n(5n-1)}{2}} \frac{(xq;q)_{n-1}}{(q;q)_n}(1-xq^{2n}),$$

    then also

    $$G(x) = \sum_{n=0}^{\infty} (-1)^n x^{2n} q^{\frac{n(5n+1)}{2}} \frac{(xq;q)_n}{(q;q)_n}(1-x^2 q^{4n+2}).$$

    We will use this fact in Chapter 11.

14. Redo the proof of (4.2.5) by using the four cases in the proof of (4.2.10) and (4.2.12). What form of (4.2.5) do you get if you combine the cases as in (4.2.12)?

15. Andrews worked out an analytic proof of (4.2.12) similar to Cayley's argument in problem 12. Define $S_n(x) = 1 + T_n(x) + W_n(x)$, where $W_n(x)$ was defined in (4.2.8) and $T_n(x)$ in (4.2.11). Note that $S_0(x) = 1 + 0 + 0 = 1$.

    (i) Show that

(4.2.14) $$T_n(x) + W_n(x) = \sum_{j \geq 1} \frac{(q;q)_{n-j}(-xq;q)_{j-1}}{(q;q)_{n-2j+1}(q;q)_j} q^{\frac{j(3j-1)}{2}} x^j \left\{1 - q^{n+1-j} + xq^{2j}(1-q^{n+1-2j})\right\}.$$

    (ii) Show that

(4.2.15) $$1 - q^{n+1-j} + xq^{2j}(1-q^{n+1-2j}) = 1 - q^j + q^j(1-q^{n+1-2j})(1+xq^j).$$

(You can start by adding and subtracting q^j.)

(iii) By using (4.2.15) in (4.2.14), show that

$$(4.2.16) \quad S_n(x) = 1 + \sum_{j \geq 1} \binom{n-j}{j-1}_q (-xq;q)_{j-1} q^{\frac{j(3j-1)}{2}} x^j$$

$$+ \sum_{j \geq 1} \binom{n-j}{j}_q (-xq;q)_j q^{\frac{j(3j-1)}{2}} (xq)^j.$$

(iv) Show that the last sum in (4.2.16) is $(1+xq)T_{n-1}(xq)$.

(v) Show that the rest of (4.2.16) is

$$1 + xq + \sum_{j \geq 1} \binom{n-j-1}{j}_q (-xq;q)_j q^{\frac{(3j+2)(j+1)}{2}} x^{j+1},$$

and that this equals $1 + xq + (1+xq)W_{n-1}(xq)$.

(vi) From (iv) and (v) it follows that $S_n(x) = (1+xq)S_{n-1}(xq)$, and hence (4.2.12) holds. Explain.

4.3. Euler's pentagonal number theorem: Franklin's proof

Recall that $(-q;q)_\infty$ is the generating function for partitions with distinct parts. $(q;q)_\infty$ is the same product, except that the signs are all different, and we get lots of cancellation:

$$(q;q)_\infty = (1-q)(1-q^2)(1-q^3)(1-q^4)(1-q^5)(1-q^6)(1-q^7)(1-q^8)\cdots$$
$$= 1 - q - q^2 + \{-q^3 + q^{2+1}\} + \{-q^4 + q^{3+1}\}$$
$$+ \{-q^5 + q^{4+1} + q^{3+2}\}$$
$$+ \{-q^6 + q^{5+1} + q^{4+2} - q^{3+2+1}\} + \cdots.$$

We still get a term for each partition into distinct parts, but some of the terms are positive and some negative. More precisely, when the number of factors is even the sign is positive, and when it is odd the sign is negative. This has two interesting consequences. We know from (3.6.3) that

$$\frac{q^{\binom{k+1}{2}}}{(q;q)_k}$$

is the generating function for partitions with exactly k distinct parts, so

$$(4.3.1) \quad (q;q)_\infty = \sum_{k=0}^{\infty} \frac{(-1)^k q^{\binom{k+1}{2}}}{(q;q)_k}.$$

It also tells us that the coefficient of q^n in $(q;q)_\infty$ equals the number of partitions of n into an even number of distinct parts minus the number of partitions of n into an odd number of distinct parts, which implies

THEOREM 34 (Legendre's pentagonal number theorem). *There are exactly as many partitions of n into an even number of distinct parts as into an odd number of distinct parts, unless n is a number of the form $\frac{k(3k-1)}{2}$ or $\frac{k(3k+1)}{2}$ (a pentagonal number), in which case there is one more or one less partition with an even number of distinct parts according to whether k is even or odd.*

4.3. EULER'S PENTAGONAL NUMBER THEOREM: FRANKLIN'S PROOF

Euler seems not to have noticed this, but it was pointed out by Legendre in his *Theorie des Nombres*. Could one prove this form of the pentagonal number theorem directly, without using generating functions? This is what Franklin accomplished.

Let's start with the example $8 + 7 + 6 + 4 + 3$, a partition of 28 with an odd number of distinct parts, and try to transform it to a partition of 28 with an even number of distinct parts in a simple way. We distinguish four regions: the Durfee square, the Franklin triangle, the "lower region" below the square, and the "upper region" to the right of the triangle. We think of the latter as diagonals parallel to the triangle.

```
  *  *  *  *  •  •  •  ◇
  *  *  *  *  •  •  ◇
  *  *  *  *  •  ◇
  *  *  *  *
  ★  ★  ★
```

If we made the diagonal of ◇'s a new row in the lower region, we would get a partition with two 3's, but if we take the row of ★'s in the lower region and make it a new diagonal in the upper region, we get:

```
  *  *  *  *  •  •  •  ◇  ★
  *  *  *  *  •  •  ◇  ★
  *  *  *  *  •  ◇  ★
  *  *  *  *
```

This is $9 + 8 + 7 + 4$, which is a partition of 28 with an even number of distinct parts. Now we cannot always apply this transformation. We certainly could not apply it to $9 + 8 + 7 + 4$, which does not have a part in the lower region. It also would not apply to $9 + 6 + 3 + 2$

because it would give

```
  *  *  *  •  •  ◇  ★  ○  ×  ⊙
  *  *  *  •  ◇  ★              ⊙
  *  *  *
```

which is not a legitimate Ferrers diagram. The general rule is

PROPOSITION 1 (Franklin's rule). *If the smallest part in the lower region is no larger than the last diagonal in the upper region, then make it the new last diagonal in the upper region. If the smallest part in the lower region is bigger than the last diagonal in the upper region, or if there is no smallest part in the lower region, then make the last diagonal in the upper region the new smallest part in the lower region.*

In the case of $9 + 8 + 7 + 4$, when there is no part in the lower region, this tells us to move the diagonal of ★'s down there, giving us back $8 + 7 + 6 + 4 + 3$. In the case of $9+6+3+2$, it says we should move the $\times$ down below the row of $\odot$'s

to get $8+6+3+2+1$. The rule obviously changes the number of parts by one, so it always pairs one with an even number of parts with one with an odd number of parts. It is also an involution, a fancy word which means that if you apply it twice, you get back where you started; as we saw with $8+7+6+4+3$. The parts remain distinct because the last diagonal in the upper region is never as large as a side of the Durfee square.

Let's look at all the partitions of 7 with distinct parts. We have:

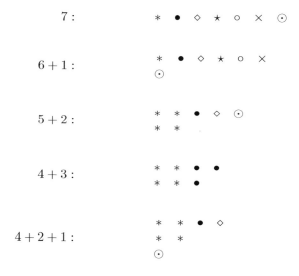

For 7 we can move the $\odot$ to the lower region to get $6+1$. For $6+1$ the smallest part in the lower region is the same size as the last diagonal, so we move the $\odot$ to the upper region as the new last diagonal, which gives 7. For $5+2$ we move the $\odot$ to the lower region to get $4+2+1$, and for $4+2+1$ the smallest part in the lower region is again the same size as the last diagonal, so it becomes the new last diagonal and we get back $5+2$.

But for $4+3$ there is nothing we can do, since both the lower and upper regions are empty! This is the only way that Franklin's rule can fail. In order for this to occur, a partition must consist of a Durfee square and a Franklin triangle, and nothing else, which can happen in two different ways: the square and the triangle could have the same side length, as with $4+3$; or the square could be one unit longer, as with $3+2$:

```
    *   *   ●
    *   *
```

Suppose the Durfee square has side k. If the Franklin triangle also has side k, then it represents $1 + 2 + 3 + \cdots + k = \frac{k(k+1)}{2}$, and the partition has the form
$$(k+1) + (k+2) + (k+3) + \cdots + (k+k) = k^2 + \frac{k(k+1)}{2} = \frac{k(3k+1)}{2}.$$
If the Franklin triangle has side $k - 1$, then it represents $1 + 2 + 3 + \cdots + (k-1) = \frac{k(k-1)}{2}$, and the partition has the form
$$(k+0) + (k+1) + (k+2) + \cdots + (k+(k-1)) = k^2 + \frac{k(k-1)}{2} = \frac{k(3k-1)}{2}.$$
These are the partitions that Franklin's bijection does not apply to. Since they have k parts, this means there is an extra partition of this form with an even number of parts if k is even, and an extra one of this form with an odd number of parts if k is odd. This is exactly the way that Legendre restated the pentagonal number theorem, so the proof is complete. As Igor Pak has pointed out, Franklin's argument also proves

THEOREM 35 (Fine's pentagonal number theorem). *There are exactly as many partitions of n into distinct parts where the largest part is even as into distinct parts where the largest part is odd, unless n is a number of the form $\frac{k(3k+1)}{2}$ or $\frac{k(3k-1)}{2}$, in which case there is one more or one less, respectively, with the largest part even.*

The "one more" with largest part even is
$$(k+1) + (k+2) + (k+3) + \cdots + (k+k),$$
and the extra one with largest part odd is
$$(k+0) + (k+1) + (k+2) + \cdots + (k+(k-1)).$$

Exercises

1. There are ten partitions of 10 with distinct parts. Draw the Ferrers diagram with Durfee square and Franklin triangle for each of them, and say which ones match up under Franklin's bijection.
2. There are fifteen partitions of 12 with distinct parts. Draw the Ferrers diagram with Durfee square and Franklin triangle for each of them. Which ones match up under Franklin's bijection, and which one is left over?
3. There are twenty-seven partitions of 15 with distinct parts. Draw the Ferrers diagram with Durfee square and Franklin triangle for each of them. Which ones match up under Franklin's bijection, and which one is left over?
4. Of which identity from Chapter 3 is (4.3.1) a special case?

4.4. Divisor sums

We start with a simple observation due to Lambert, a contemporary of Euler's. Consider the sum
$$\sum_{k=1}^{\infty} \frac{q^k}{1-q^k}$$

and expand the denominator into a geometric series:

$$\sum_{k=1}^{\infty} \frac{q^k}{1-q^k} = \sum_{k=1}^{\infty} \left(q^k + q^{2k} + q^{3k} + \ldots\right) = \sum_{k=1}^{\infty} \sum_{j=1}^{\infty} q^{jk}.$$

If we set this double sum equal to a single sum, *i.e.*,

$$\sum_{k=1}^{\infty} \frac{q^k}{1-q^k} = \sum_{k=1}^{\infty} \sum_{j=1}^{\infty} q^{jk} = \sum_{n=1}^{\infty} d(n) q^n,$$

what are the coefficients $d(n)$? Let's look at the case $n = 6$. We will get a q^6 term from all the ways that jk can equal 6 for positive integers j and k, and this can happen in four ways: $j = 1$ and $k = 6$; $j = 2$ and $k = 3$; and the same two combinations with j and k switched. In other words, we get one q^6 term for each of the four divisors of 6, and clearly this is true in general: the number of ways that $jk = n$ is the number of k's that divide n. We have proved

THEOREM 36 (Lambert's theorem). *If $|q| < 1$ and $d(n)$ is the number of divisors of n, then*

(4.4.1) $$\sum_{k=1}^{\infty} \frac{q^k}{1-q^k} = \sum_{n=1}^{\infty} d(n) q^n.$$

By exactly the same argument we have

(4.4.2) $$\sum_{k=1}^{\infty} \frac{f(k) q^k}{1-q^k} = \sum_{n=1}^{\infty} q^n \sum_{k|n} f(k),$$

where $k|n$ denotes that the sum is over all the divisors k of n, and $f(n)$ is arbitrary as long as it does not grow so rapidly as to make the series diverge. A series of this type is called a **Lambert series**, for obvious reasons. When $f(n)$ is identically 1, (4.4.2) reduces to (4.4.1). Another interesting case is $f(n) = n$, when (4.4.2) becomes

(4.4.3) $$\sum_{k=1}^{\infty} \frac{k q^k}{1-q^k} = \sum_{n=1}^{\infty} \sigma(n) q^n,$$

where $\sigma(n)$ denotes the sum of the divisors of n. Lambert knew this too, but Euler knew it before him. Our main object in this section is to prove a remarkable theorem of Euler about $\sigma(n)$, but let's find some elementary properties of this function first. If n is a prime number, p then its only divisors are 1 and p, so $\sigma(p) = 1 + p$. If $n = p^2$ for some prime number p, then the divisors are $1, p, p^2$, so $\sigma(p^2) = 1 + p + p^2$, and more generally

$$\sigma(p^b) = 1 + p + p^2 + \cdots + p^b = \frac{p^{b+1} - 1}{p - 1}$$

if p is prime and b is a nonnegative integer. (Note that this is even right for $b = 0$.) But we can do better than this. If n is any integer greater than 1, we can write $n = p_1^{b_1} p_2^{b_2} \cdots p_m^{b_m}$ for distinct primes $p_1, \ldots, p_m$ and positive integers $b_1, \ldots, b_m$, and we can see that the divisors of n are all the numbers $p_1^{a_1} p_2^{a_2} \cdots p_m^{a_m}$, where

$0 \leq a_i \leq b_i$ for each i, $1 \leq i \leq m$. It follows that

$$\sigma(p_1^{b_1} p_2^{b_2} \cdots p_m^{b_m}) = \sum_{a_1=0}^{b_1} \cdots \sum_{a_m=0}^{b_m} p_1^{a_1} p_2^{a_2} \cdots p_m^{a_m}$$

$$= \frac{p_1^{b_1+1} - 1}{p_1 - 1} \cdots \frac{p_m^{b_m+1} - 1}{p_m - 1}$$

(4.4.4)
$$= \prod_{i=1}^{m} \frac{p_i^{b_i+1} - 1}{p_i - 1}.$$

In number theory, a function $f(n)$ defined on the positive integers is called **multiplicative** if $f(mn) = f(m) f(n)$ whenever m and n are relatively prime, so that their greatest common divisor is 1. Since this means they have no prime factor in common, (4.4.4) shows that $\sigma(n)$ is multiplicative.

A major reason why Euler tried so hard to prove his pentagonal number theorem

(4.4.5)
$$(q;q)_\infty = 1 + \sum_{k=1}^{\infty} (-1)^k \left(q^{\frac{k(3k-1)}{2}} + q^{\frac{k(3k+1)}{2}} \right)$$

is that he wanted to apply it to prove something about $\sigma(n)$. Following Euler we calculate the logarithmic derivative of $(q;q)_\infty$ with respect to q:

$$\frac{d}{dq} \log (q;q)_\infty = \frac{d}{dq} \sum_{n=1}^{\infty} \log(1 - q^n) = \sum_{n=1}^{\infty} \frac{-nq^{n-1}}{1 - q^n}.$$

If we multiply this by $-q$ we will have the sum in (4.4.3). This suggests that something good might happen if we perform the same operations on the other side of (4.4.5). Taking the logarithmic derivative we get

$$\frac{\sum_{k=1}^{\infty} (-1)^k \left(\frac{k(3k-1)}{2} q^{\frac{k(3k-1)}{2}-1} + \frac{k(3k+1)}{2} q^{\frac{k(3k+1)}{2}-1} \right)}{1 + \sum_{k=1}^{\infty} (-1)^k \left(q^{\frac{k(3k-1)}{2}} + q^{\frac{k(3k+1)}{2}} \right)}.$$

Multiplying this by $-q$ and using (4.4.3) we finally have

$$\sum_{n=1}^{\infty} \sigma(n) q^n = \frac{\sum_{k=1}^{\infty} (-1)^{k-1} \left(\frac{k(3k-1)}{2} q^{\frac{k(3k-1)}{2}} + \frac{k(3k+1)}{2} q^{\frac{k(3k+1)}{2}} \right)}{1 + \sum_{k=1}^{\infty} (-1)^k \left(q^{\frac{k(3k-1)}{2}} + q^{\frac{k(3k+1)}{2}} \right)},$$

or

(4.4.6)
$$\left(\sum_{j=1}^{\infty} \sigma(j) q^j \right) \left(1 + \sum_{k=1}^{\infty} (-1)^k \left(q^{\frac{k(3k-1)}{2}} + q^{\frac{k(3k+1)}{2}} \right) \right)$$

$$= \sum_{m=1}^{\infty} (-1)^{m-1} \left(\frac{m(3m-1)}{2} q^{\frac{m(3m-1)}{2}} + \frac{m(3m+1)}{2} q^{\frac{m(3m+1)}{2}} \right).$$

To figure out what (4.4.6) is trying to tell us, we consider the coefficient of q^n on both sides. Let's first suppose that n is not a pentagonal number. In that case the coefficient of q^n on the right side is zero, so the coefficient on the left side must be too. The left side of (4.4.6) is

$$\left(1 - q - q^2 + q^5 + q^7 - q^{12} - q^{15} + q^{22} + q^{26} - q^{35} - q^{40} \ldots\right) \sum_{j=1}^{\infty} \sigma(j) q^j,$$

and we see that the coefficient of q^n is

$$\sigma(n) - \sigma(n-1) - \sigma(n-2) + \sigma(n-5) + \sigma(n-7) - \sigma(n-12) - \sigma(n-15) + \sigma(n-22) \ldots.$$

Since this equals zero, it follows that if n is not a pentagonal number, then

$$\sigma(n) = \sigma(n-1) + \sigma(n-2) - \sigma(n-5) - \sigma(n-7) + \sigma(n-12) + \sigma(n-15) - \sigma(n-22) \ldots,$$

where we keep subtracting pentagonal numbers p from n until we drop below zero, and the sign pattern is $+ + - - + + - - + + - - \ldots$. Note that $\sigma(0)$ does not occur in the series, for two reasons: by assumption n is not a pentagonal number, so $n - p$ is never zero; and even if it was, we would not include this term since the first sum on the left side of (4.4.6) starts at $j = 1$.

What if n is a pentagonal number? The only thing that changes is that the coefficient of q^n on the right side of (4.4.6) is now $\pm n$. The sign may seem troublesome until we observe that it must be the opposite of the sign of the q^n term in

$$1 - q - q^2 + q^5 + q^7 - q^{12} - q^{15} + q^{22} + q^{26} - q^{35} - q^{40} + q^{51} + q^{57} \ldots.$$

This means that if we move this term to the other side, it must continue the same sign pattern as before. We have proved

THEOREM 37 (Euler's divisor dum theorem). *If n is any positive integer and $\sigma(n)$ denotes the sum of the divisors of n, then*

$$\sigma(n) = \sigma(n-1) + \sigma(n-2) - \sigma(n-5) - \sigma(n-7) + \sigma(n-12) + \sigma(n-15) - \sigma(n-22) \ldots,$$

where we keep subtracting pentagonal numbers p from n until we reach or drop below zero, the sign pattern on the right side is $+ + - - + + - - + + - - \ldots$, and if we do reach $\sigma(0)$, which happens if and only if n is itself a pentagonal number, we substitute n for it with whichever sign continues the pattern.

Euler knew this amazing result for years before he could prove it, just as he knew the pentagonal number theorem. Let's do a couple of examples of it. We have

(4.4.7) $$\begin{aligned}\sigma(56) = {} & \sigma(55) + \sigma(54) - \sigma(51) - \sigma(49) + \sigma(44) + \sigma(41) \\ & - \sigma(34) - \sigma(30) + \sigma(21) + \sigma(16) - \sigma(5),\end{aligned}$$

the sequence ending here because the next pentagonal number is 57. To check this we use (4.4.4) or its special cases and the multiplicativity of $\sigma(n)$. Since

$$\sigma(p^b) = \frac{p^{b+1} - 1}{p - 1},$$

4.4. DIVISOR SUMS

we have $\sigma(5) = 6$, $\sigma(16) = 31$, $\sigma(41) = 42$, and $\sigma(49) = 57$. The multiplicative property gives

$$\sigma(56) = \sigma(7)\sigma(8) = 8 \times 15 = 120,$$
$$\sigma(55) = \sigma(5)\sigma(11) = 6 \times 12 = 72,$$
$$\sigma(54) = \sigma(2)\sigma(27) = 3 \times 40 = 120,$$
$$\sigma(51) = \sigma(3)\sigma(17) = 4 \times 18 = 72,$$
$$\sigma(44) = \sigma(4)\sigma(11) = 7 \times 12 = 84,$$
$$\sigma(34) = \sigma(2)\sigma(17) = 3 \times 18 = 54,$$
$$\sigma(30) = \sigma(5)\sigma(6) = 6 \times 12 = 72,$$
$$\sigma(21) = \sigma(3)\sigma(7) = 4 \times 8 = 32.$$

Then the right side of (4.4.7) is

$$72 + 120 - 72 - 57 + 84 + 42 - 54 - 72 + 32 + 31 - 6 = 120 = \sigma(56),$$

so Euler's theorem checks. (Note, however, that no one in their right mind would compute $\sigma(56)$ this way unless they were trying to check Euler's theorem.) Let's try $\sigma(57)$ also. Here we have

$$(4.4.8) \quad \begin{aligned} \sigma(57) = {} & \sigma(56) + \sigma(55) - \sigma(52) - \sigma(50) + \sigma(45) + \sigma(42) \\ & - \sigma(35) - \sigma(31) + \sigma(22) + \sigma(17) - \sigma(6) - \sigma(0), \end{aligned}$$

where $\sigma(0)$ is to be interpreted as 57 in this case. Using multiplicativity to compute the other values we get

$$120 + 72 - 98 - 93 + 78 + 96 - 48 - 32 + 36 + 18 - 12 - 57 = 80 = \sigma(57).$$

Euler did the examples 101 and 301.

We devote the remainder of this section to two variations of (4.4.1). One is

THEOREM 38 (Uchimura's theorem). *If $|q| < 1$ and $d(n)$ is the number of divisors of n, then*

$$(4.4.9) \quad \sum_{k=1}^{\infty} d(k) q^k = (q;q)_\infty \sum_{n=1}^{\infty} \frac{n\, q^n}{(q;q)_n}.$$

To see this we apply Euler's identity (2.9.2) from section 2.9 to the sum on the right:

$$\sum_{n=1}^{\infty} \frac{n\, q^n}{(q;q)_n} = \sum_{n=1}^{\infty} \frac{q^n}{(q;q)_n} \sum_{k=1}^{n} \binom{n}{k}_q (q;q)_{k-1}$$

$$= \sum_{n=1}^{\infty} \sum_{k=1}^{n} \frac{q^n}{(1-q^k)(q;q)_{n-k}}$$

$$= \sum_{n=1}^{\infty} \sum_{k=1}^{n} \frac{q^k}{1-q^k} \frac{q^{n-k}}{(q;q)_{n-k}}.$$

The double sum goes over all n and k with $1 \leq k \leq n < \infty$, so we can just as well write it as

$$\sum_{k=1}^{\infty}\sum_{n=k}^{\infty} \frac{q^k}{1-q^k} q^{n-k}(q;q)_{n-k} = \sum_{k=1}^{\infty} \frac{q^k}{1-q^k} \sum_{n=k}^{\infty} \frac{q^{n-k}}{(q;q)_{n-k}}$$

$$= \sum_{k=1}^{\infty} \frac{q^k}{1-q^k} \sum_{j=0}^{\infty} \frac{q^j}{(q;q)_j}.$$

Several identities from Chapter 3 tell us that

$$\sum_{j=0}^{\infty} \frac{q^j}{(q;q)_j} = \frac{1}{(q;q)_{\infty}},$$

so we have proved that

$$\sum_{n=1}^{\infty} \frac{n q^n}{(q;q)_n} = \sum_{k=1}^{\infty} \frac{q^k}{1-q^k} \sum_{j=0}^{\infty} \frac{q^j}{(q;q)_j}$$

$$= \sum_{k=1}^{\infty} \frac{q^k}{1-q^k} \frac{1}{(q;q)_{\infty}}$$

$$= \frac{1}{(q;q)_{\infty}} \sum_{k=1}^{\infty} \frac{q^k}{1-q^k}.$$

In view of Lambert's theorem (4.4.1), this is equivalent to Uchimura's theorem. A similar identity was stated without proof much earlier by Eisenstein.

THEOREM 39 (Eisenstein's theorem). *If $|q| < 1$ and $d(n)$ is the number of divisors of n, then*

(4.4.10) $$(q;q)_{\infty} \sum_{k=1}^{\infty} d(k) q^k = \sum_{n=1}^{\infty} (-1)^{n-1} \frac{n q^{\binom{n+1}{2}}}{(q;q)_n}.$$

The proof is very similar to that of (4.4.9), though perhaps slightly harder. We start with the sum on the right and apply the variant form (2.9.3) of Euler's theorem from section 2.9:

$$\sum_{n=1}^{\infty} (-1)^{n-1} \frac{n q^{\binom{n+1}{2}}}{(q;q)_n} = \sum_{n=1}^{\infty} \frac{(-1)^{n-1} q^{\binom{n+1}{2}}}{(q;q)_n} \sum_{k=1}^{n} \binom{n}{k}_q (-1)^{k-1} q^{\binom{k+1}{2}-nk} (q;q)_{k-1}$$

$$= \sum_{n=1}^{\infty} \sum_{k=1}^{n} \frac{(-1)^{n-k}}{(q;q)_{n-k}} \frac{q^{\binom{n+1}{2}+\binom{k+1}{2}-nk}}{1-q^k}.$$

Now

$$\binom{n+1}{2} + \binom{k+1}{2} - nk = \frac{n^2+n+k^2+k-2nk}{2} = \frac{(n-k)^2+n+k}{2}$$

$$= \frac{(n-k)^2+(n-k)+2k}{2} = k + \binom{n-k+1}{2}.$$

Using this above and changing the order of summation as before we have

$$\sum_{n=1}^{\infty}(-1)^{n-1}\frac{n\,q^{\binom{n+1}{2}}}{(q;q)_n} = \sum_{k=1}^{\infty}\sum_{n=k}^{\infty}\frac{q^k}{1-q^k}\frac{(-1)^{n-k}q^{\binom{n-k+1}{2}}}{(q;q)_{n-k}}$$

$$= \sum_{k=1}^{\infty}\frac{q^k}{1-q^k}\sum_{j=0}^{\infty}\frac{(-1)^j q^{\binom{j+1}{2}}}{(q;q)_j}.$$

Using (4.3.1) from the previous section we have

$$\sum_{n=1}^{\infty}(-1)^{n-1}\frac{n\,q^{\binom{n+1}{2}}}{(q;q)_n} = \sum_{k=1}^{\infty}\frac{q^k}{1-q^k}(q;q)_\infty = (q;q)_\infty\sum_{k=1}^{\infty}\frac{q^k}{1-q^k}.$$

Eisenstein's theorem (4.4.10) now follows from Lambert's theorem (4.4.1).

Exercises

1. This interesting problem is popular in mathematics education circles. A version of it was problem B-4 on the 1967 William Lowell Putnam Mathematical Competition. Quigley High School has 420 seniors, each of whom has a private locker. Miraculously, all 420 show up for school one day. Less miraculously, they find all 420 lockers closed, and they form a queue. The first student in line opens all the lockers. The second student closes the second, fourth, sixth, eighth, ..., and four hundred twentieth lockers. The third student closes the third locker, opens the sixth, closes the ninth, opens the twelfth, and so on. In general the nth student goes to all the lockers whose numbers are multiples of n and changes them—closing them if they are open, and opening them if they are closed. (Note that this leaves the second half of the class with rather little to do.) After all 420 students have gone through the lockers, which ones are open? Explain.

2. Fundamentally, the size of the senior class doesn't affect the answer to problem 1, but why is 420 a good choice?

3. You may be wondering what problem 1 has to do with q-analysis. There is another q-identity for the divisor sum due to Clausen, namely

$$\sum_{k=1}^{\infty} q^{k^2}\frac{1+q^k}{1-q^k} = \sum_{n=1}^{\infty} d(n)\,q^n.$$

Show this by expanding $(1+q^k)/(1-q^k)$ into a series and reasoning as in problem 1 (or otherwise). **Hint:** Which divisors does the $k=1$ term $q(1+q)/(1-q)$ take care of?

4. Assuming as usual that $|q|<1$, show that

$$\sum_{n=1}^{\infty}\frac{n\,q^n}{1-q^n}$$

converges.

5. Assuming as usual that $|q|<1$, show that

$$\sum_{n=1}^{\infty}\frac{n!\,q^n}{1-q^n}$$

does not converge. In other words, (4.4.2) does not apply if $f(n) = n!$.

6. Assuming as usual that $|q| < 1$, show that
$$-\log\left\{(1-q)(1-q^2)^{\frac{1}{2}}(1-q^3)^{\frac{1}{3}}\cdots\right\} = \sum_{n=1}^{\infty} \frac{q^n}{n} d(n).$$

7. Students often wonder why 1 is not considered a prime number. Euler pointed out one reason for it, having to do with $\sigma(n)$. What do you think it was?

8. Check the values of $\sigma(n)$ in the calculation of $\sigma(57)$.

9. Check Euler's divisor sum theorem for $n = 38$ and $n = 40$.

10. Check Euler's divisor sum theorem for his two examples, $n = 101$ (which is prime) and $n = 301 = 43 \times 7$.

11. Assuming $|q| < 1$, show that
$$\frac{q}{1-q} = \frac{q}{1-q^2} + \frac{q^2}{1-q^4} + \frac{q^4}{1-q^8} + \frac{q^8}{1-q^{16}} + \frac{q^{16}}{1-q^{32}} + \cdots$$
by expanding all the terms on both sides into geometric series.

12. Show the result of problem 11 by finding the sum of the first n terms on the right side and letting $n \to \infty$.

13. Show that the method of problem 12 still works if $|q| > 1$, but the sum is different.

14. (This problem was suggested by Emil Lalov.) Another interesting case of (4.4.2) is $f(k) = \varphi(k)$, the Euler phi function, which counts the number of positive integers $\leq k$ and relatively prime to k. Like $\sigma(k)$, $\varphi(k)$ is a multiplicative function, and clearly we have $\varphi(p) = p - 1$ for any prime number p. Assuming as usual that $|q| < 1$, show that
$$\sum_{k=1}^{\infty} \frac{\varphi(k)\, q^k}{1-q^k} = \sum_{n=1}^{\infty} n q^n = \frac{q}{(1-q)^2}.$$

Hint: To show the first equality, let $S_k(n)$ be the set of integers m with $1 \leq m \leq n$ whose greatest common divisor with n is k, and explain why there are exactly $\varphi(\frac{n}{k})$ numbers in $S_k(n)$.

15. This problem treats Uchimura's proof of (4.4.9).

(i) In the first place, Uchimura prefers to write his theorem as

(4.4.11) $$\sum_{n=1}^{\infty} nq^n \left(q^{n+1}; q\right)_{\infty} = \sum_{n=1}^{\infty} \frac{q^n}{1-q^n} = \sum_{n=1}^{\infty} d(n)q^n.$$

Explain why this is equivalent to (4.4.9).

(ii) Uchimura proves (4.4.11) by means of two auxiliary theorems. One is

(4.4.12) $$a\sum_{n=1}^{\infty} nq^n \left(aq^{n+1}; q\right)_{\infty} = \sum_{j=1}^{\infty} \frac{(-1)^{j-1} a^j q^{\binom{j+1}{2}}}{(q;q)_j (1-q^j)}.$$

To see this, first expand $\left(aq^{n+1}; q\right)_{\infty}$ by (3.6.1). After changing orders of summation and possibly renaming a summation index, show that this gives
$$\sum_{j=1}^{\infty} \frac{(-1)^{j-1} a^j q^{\binom{j}{2}}}{(q;q)_{j-1}} \sum_{n=1}^{\infty} n q^{nj}.$$

Use the second equality in problem 14 to do the inner sum and complete the proof of (4.4.12).

(iii) A second auxiliary theorem is

(4.4.13) $$\sum_{k=1}^{\infty} \frac{bq^k}{1-bq^k} = \sum_{j=1}^{\infty} \frac{(-1)^{j-1} b^j q^{\binom{j+1}{2}}}{(bq;q)_j (1-q^j)}.$$

To see this, denote the right side of (4.4.13) by $U(b,q)$ and explain why

$$U(b,q) - U(bq,q) = \sum_{j=1}^{\infty} \frac{(-1)^{j-1} b^j q^{\binom{j+1}{2}}}{(bq;q)_{j+1}}$$

$$= \sum_{j=1}^{\infty} \frac{(-1)^{j-1} b^j q^{\binom{j+1}{2}} \left[(1-bq^{j+1}) + bq^{j+1}\right]}{(bq;q)_{j+1}}$$

$$= \sum_{j=1}^{\infty} \frac{(-1)^{j-1} b^j q^{\binom{j+1}{2}}}{(bq;q)_j} - \sum_{j=1}^{\infty} \frac{(-1)^j b^{j+1} q^{\binom{j+2}{2}}}{(bq;q)_{j+1}}$$

$$= \frac{bq}{1-bq}.$$

Having made the right side telescope, we now do the same on the left side. Since $U(0,q) = 0$ (why?), and assuming as usual that $|q| < 1$, explain why we have

$$U(b,q) = [U(b,q) - U(bq,q)] + [U(bq,q) - U(bq^2,q)] + [U(bq^2,q) - U(bq^3,q)]$$

$$+ \cdots = \sum_{k=1}^{\infty} \frac{bq^k}{1-bq^k}.$$

(iv) Explain how (4.4.11) follows from (4.4.12) and (4.4.13).

(v) Uchimura uses a slightly different left side in (4.4.13). Assuming that $|b| \le 1$ and $|q| < 1$, show that

$$\sum_{n=1}^{\infty} \frac{b^n q^n}{1-q^n} = \sum_{k=1}^{\infty} \frac{bq^k}{1-bq^k}.$$

(Expand one side using a geometric series, change the order of summation, and resum.)

16. Show that the series

$$\sum_{n=1}^{\infty} \left(\frac{q^n}{1-q^n} - \frac{zq^n}{1-zq^n}\right) \left(\frac{1}{1-zq} + \frac{1}{1-zq^2} + \cdots + \frac{1}{1-zq^n}\right)$$

converges if $|q| < 1$, by using the ratio test or otherwise.

17. For $|q| < 1$, define $F(z)$ to be the series in problem 16, which was introduced by Bailey to treat the Lambert series in (viii) below.

(i) Show that

$$F(0) = \sum_{n=1}^{\infty} \frac{nq^n}{1-q^n}.$$

(ii) Show that $F(z) - F(zq) = -zq/(1-zq)^2$.

(iii) Explain why it follows from (ii) that

$$F(z) - F(zq^n) = \sum_{k=1}^{n} \frac{-zq^k}{(1-zq^k)^2}$$

for any positive integer n.

(iv) Explain why it follows from (i) and (iii) that

$$F(z) = \sum_{n=1}^{\infty} \frac{nq^n}{1-q^n} - \sum_{k=1}^{\infty} \frac{zq^k}{(1-zq^k)^2}.$$

(v) Recall from problem 14 that

$$\sum_{n=1}^{\infty} nx^n = \frac{x}{(1-x)^2}$$

for $|x| < 1$. Take $x = zq^k$ here and change the order of summation to show that

$$\sum_{k=1}^{\infty} \frac{zq^k}{(1-zq^k)^2} = \sum_{n=1}^{\infty} \frac{nz^n q^n}{1-q^n}.$$

(vi) Explain why it follows from (v) that for $|q| < 1$

$$\sum_{n=1}^{\infty} \left(\frac{q^n}{1-q^n} - \frac{zq^n}{1-zq^n} \right) \left(\frac{1}{1-zq} + \frac{1}{1-zq^2} + \cdots + \frac{1}{1-zq^n} \right)$$
$$= \sum_{n=1}^{\infty} \frac{nq^n(1-z^n)}{1-q^n}.$$

(vii) Show that the left side in (vi) can be rewritten as

$$\sum_{n=1}^{\infty} \frac{q^n(1-z)}{(1-q^n)(1-zq^n)} \left(\frac{1}{1-zq} + \frac{1}{1-zq^2} + \cdots + \frac{1}{1-zq^n} \right).$$

(viii) After changing the left side of (vi) to (vii), divide both sides by $1 - z$ and then let $z \to 1$. Show that for $|q| < 1$ this gives

$$\sum_{n=1}^{\infty} \frac{q^n}{(1-q^n)^2} \left(\frac{1}{1-q} + \frac{1}{1-q^2} + \cdots + \frac{1}{1-q^n} \right) = \sum_{n=1}^{\infty} \frac{n^2 q^n}{1-q^n}.$$

(ix) What does (4.4.2) say about the Lambert series on the right side of (viii)?

18. Bailey conceived another approach to part (viii) of the previous problem. For $|q| < 1$ and $|z| < 1$, define

$$B(z) = \sum_{n=1}^{\infty} \frac{(q;q)_{n-1} z^n}{(1-q^n)(z;q)_n}.$$

(i) Show that this series converges, given $|q| < 1$ and $|z| < 1$.

(ii) Show that

$$B(z) - B(zq) = \sum_{n=1}^{\infty} \frac{(q;q)_{n-1} z^n}{(z;q)_{n+1}}.$$

(iii) The series on the right side of (ii) can be made to telescope. By multiplying top and bottom by $1 - z$, and taking $1 - z = 1 - zq^n - z(1 - q^n)$ on top, show that

$$B(z) - B(zq) = \frac{z}{(1-z)^2}.$$

(iv) Explain why (iii) implies that

$$B(z) - B(zq^n) = \sum_{k=0}^{n-1} \frac{zq^k}{(1 - zq^k)^2}$$

for any positive integer n.

(v) Explain why (iv) implies that

$$B(z) = \sum_{k=0}^{\infty} \frac{zq^k}{(1 - zq^k)^2}.$$

Show also that this series only needs $|q| < 1$ for convergence.

(vi) As in part (vi) of the previous problem, show that

$$B(z) = \sum_{n=1}^{\infty} \frac{nz^n}{1 - q^n}.$$

(vii) So far the details have been easier than in the previous problem, and we have established that if $|q| < 1$ and $|z| < 1$, then

$$\sum_{n=1}^{\infty} \frac{(q;q)_{n-1} z^n}{(1 - q^n)(z;q)_n} = \sum_{n=1}^{\infty} \frac{nz^n}{1 - q^n}.$$

It is clear how to get the right side to look something like the right side of (ix) in the previous problem: take the derivative with respect to z and then multiply by z. The difficulty is that we have to do the same operations to the left side. Show that this gives

$$(4.4.14) \quad \sum_{n=1}^{\infty} \frac{(q;q)_{n-1} z^n}{(1 - q^n)(z;q)_n} \left(\frac{1}{1-z} + \frac{1}{1-zq} + \cdots + \frac{1}{1-zq^{n-1}} \right) = \sum_{n=1}^{\infty} \frac{n^2 z^n}{1 - q^n}.$$

Logarithmic differentiation might help.

(viii) Show that if $z = q$, then (4.4.14) reduces to

$$\sum_{n=1}^{\infty} \frac{q^n}{(1 - q^n)^2} \left(\frac{1}{1 - q} + \frac{1}{1 - q^2} + \cdots + \frac{1}{1 - q^n} \right) = \sum_{n=1}^{\infty} \frac{n^2 q^n}{1 - q^n}.$$

19. Bailey went on to consider the series

$$\sum_{n=1}^{\infty} \left(\frac{q^n}{(1-q^n)^2} - \frac{zq^n}{(1-zq^n)^2} \right) \left(\begin{array}{c} \dfrac{1}{1-q} + \dfrac{1}{1-q^2} + \cdots + \dfrac{1}{1-q^n} \\ + \dfrac{z}{1-z} + \dfrac{zq}{1-zq} + \cdots + \dfrac{zq^{n-1}}{1-zq^{n-1}} \end{array} \right).$$

(i) Show that the series converges if $|q| < 1$.

(ii) Denote the series by $H(z)$, where $|q| < 1$. Show that

$$H(z) - H(zq) = \sum_{n=1}^{\infty} \frac{q^n}{(1-q^n)^2} \left[\frac{z}{1-z} - \frac{zq^n}{1-zq^n} \right]$$
$$- \sum_{n=1}^{\infty} \frac{zq^n}{(1-zq^n)^2} \left[\frac{1}{1-q^n} + \frac{z}{1-z} \right]$$

and that these two sums cancel each other.

(iii) After (ii) we know that $H(z) = H(zq)$. Explain why it follows that $H(z) = H(zq^m)$ for any nonnegative integer m.

(iv) Explain why it follows that $H(z) = H(0)$. (Thus $H(z)$ does not actually depend on z.)

(v) Explain why it follows from one of the previous two problems that

$$H(z) = \sum_{n=1}^{\infty} \frac{n^2 q^n}{1 - q^n}.$$

20. Show that

$$\sum_{n=1}^{\infty} \frac{n^2 q^n}{1-q^n} = \sum_{n=1}^{\infty} \left(\frac{q^n}{(1-q^n)^2} - \frac{q^{n+1}}{(1-q^{n+1})^2} \right) \left(\frac{1+q}{1-q} + \frac{1+q^2}{1-q^2} + \cdots + \frac{1+q^n}{1-q^n} \right)$$
$$= \sum_{n=1}^{\infty} \left(\frac{q^n}{(1-q^n)^2} + \frac{q^{n+1}}{(1+q^{n+1})^2} \right) \left(\frac{1+q^2}{1-q^2} + \frac{1+q^4}{1-q^4} + \cdots + \frac{1+q^{2n}}{1-q^{2n}} \right).$$

Besides $z = 0$, these are the other two special cases that Bailey gives of the result of the previous problem.

21. In a subsequent paper Bailey uses the result of problem 19 in a much more involved way.

(i) Show that setting $z = -1$ there gives

$$\sum_{n=1}^{\infty} \frac{n^2 q^n}{1 - q^n}$$
$$= 2 \sum_{n=1}^{\infty} \frac{q^n(1+q^{2n})}{(1-q^{2n})^2} \left(-\frac{1}{2} + \frac{1}{1-q^n} + \frac{1+q^2}{1-q^2} + \frac{1+q^4}{1-q^4} + \cdots + \frac{1+q^{2n-2}}{1-q^{2n-2}} \right).$$

(ii) Explain why (i) is equivalent to

$$\sum_{k=1}^{\infty} \frac{4k^2 q^{2k}}{1-q^{2k}} + \sum_{k=0}^{\infty} \frac{(2k+1)^2 q^{2k+1}}{1-q^{2k+1}}$$
$$= 2 \sum_{k=1}^{\infty} \frac{q^{2k}(1+q^{4k})}{(1-q^{4k})^2} \left(-\frac{1}{2} + \frac{1}{1-q^{2k}} + \frac{1+q^2}{1-q^2} + \frac{1+q^4}{1-q^4} + \cdots + \frac{1+q^{4k-2}}{1-q^{4k-2}} \right)$$
$$+ 2 \sum_{k=0}^{\infty} \frac{q^{2k+1}(1+q^{4k+2})}{(1-q^{4k+2})^2} \left(-\frac{1}{2} + \frac{1}{1-q^{2k+1}} + \frac{1+q^2}{1-q^2} + \frac{1+q^4}{1-q^4} + \cdots + \frac{1+q^{4k}}{1-q^{4k}} \right).$$

(iii) The reason for rewriting (i) as (ii) is that Bailey now changes q to $-q$ in (ii) and subtracts the result from (ii). Show that this gives

$$\sum_{k=0}^{\infty} \frac{(2k+1)^2 q^{2k+1}}{1-q^{4k+2}} = \sum_{k=0}^{\infty} \frac{q^{2k+1}(1+q^{4k+2})}{(1-q^{4k+2})^2} \left(-1 + \frac{2}{1-q^{4k+2}} \right)$$

$$+ 2 \sum_{k=1}^{\infty} \frac{q^{2k+1}(1+q^{4k+2})}{(1-q^{4k+2})^2} \left(\frac{1+q^2}{1-q^2} + \frac{1+q^4}{1-q^4} + \cdots + \frac{1+q^{4k}}{1-q^{4k}} \right)$$

after some simplification.

(iv) The simple fact $\dfrac{1+q^{2j}}{1-q^{2j}} = \dfrac{2}{1-q^{2j}} - 1$ is used in each direction in the rest of Bailey's derivation. Using it once gives

$$\sum_{k=0}^{\infty} \frac{(2k+1)^2 q^{2k+1}}{1-q^{4k+2}} = \sum_{k=0}^{\infty} \frac{q^{2k+1}(1+q^{4k+2})^2}{(1-q^{4k+2})^3}$$

$$+ 2 \sum_{k=1}^{\infty} \frac{q^{2k+1}(1+q^{4k+2})}{(1-q^{4k+2})^2} \left(\frac{1+q^2}{1-q^2} + \frac{1+q^4}{1-q^4} + \cdots + \frac{1+q^{4k}}{1-q^{4k}} \right).$$

Show that

$$\sum_{k=0}^{\infty} \frac{q^{2k+1}(1+q^{4k+2})^2}{(1-q^{4k+2})^3} = \sum_{k=0}^{\infty} \frac{q^{2k+1}}{1-q^{4k+2}} + 4 \sum_{k=0}^{\infty} \frac{q^{6k+3}}{(1-q^{4k+2})^3}.$$

(v) Explain why (iv) implies that

$$\sum_{k=1}^{\infty} \frac{2k(k+1)q^{2k+1}}{1-q^{4k+2}} = 2 \sum_{j=0}^{\infty} \frac{q^{6j+3}}{(1-q^{4j+2})^3}$$

$$+ \sum_{k=1}^{\infty} \frac{q^{2k+1}(1+q^{4k+2})}{(1-q^{4k+2})^2} \left(\frac{1+q^2}{1-q^2} + \frac{1+q^4}{1-q^4} + \cdots + \frac{1+q^{4k}}{1-q^{4k}} \right),$$

where we changed the summation index to j in one of the sums for a reason that will appear in the next part.

(vi) There is a surprising simplification in (v). From Cauchy's identity (3.5.3) in section 3.5 we know that

$$\frac{1}{(1-z)^3} = \sum_{j=0}^{\infty} \binom{j+2}{2} z^j \quad \text{if } |z| < 1,$$

and hence

$$\left(\frac{x}{1-x^2} \right)^3 = \sum_{j=0}^{\infty} \binom{j+2}{2} x^{2j+3} = \sum_{k=1}^{\infty} \binom{k+1}{2} x^{2k+1}.$$

By taking $x = q^{2j+1}$ here and changing the order of summation, show that

$$2 \sum_{j=0}^{\infty} \frac{q^{6j+3}}{(1-q^{4j+2})^3} = \sum_{k=1}^{\infty} \frac{k(k+1)q^{2k+1}}{1-q^{4k+2}},$$

and hence (v) becomes

$$\sum_{k=1}^{\infty} \frac{k(k+1)q^{2k+1}}{1-q^{4k+2}} = \sum_{k=1}^{\infty} \frac{q^{2k+1}(1+q^{4k+2})}{(1-q^{4k+2})^2} \left(\frac{1+q^2}{1-q^2} + \frac{1+q^4}{1-q^4} + \cdots + \frac{1+q^{4k}}{1-q^{4k}} \right).$$

(vii) We could stop here, but Bailey uses the simple fact in (iv) in the other direction on this. Show that doing so and dividing by 2 again gives

$$\sum_{k=1}^{\infty} \frac{q^{2k+1}(1+q^{4k+2})}{(1-q^{4k+2})^2} \left(\frac{1}{1-q^2} + \frac{1}{1-q^4} + \cdots + \frac{1}{1-q^{4k}} \right)$$
$$= \sum_{k=1}^{\infty} \binom{k+1}{2} \frac{q^{2k+1}}{1-q^{4k+2}} + \sum_{k=1}^{\infty} \frac{kq^{2k+1}(1+q^{4k+2})}{(1-q^{4k+2})^2}.$$

4.5. Sylvester's fishhook bijection

Consider a partition with odd parts, say $21 + 19 + 15 + 15 + 15 + 9 + 7 + 7 + 3 + 1 + 1 + 1$, and write the Ferrers diagram as a series of right angles:

```
• • • • • • • • • • • •
• * * * * * * * * * *
• * ◇ ◇ ◇ ◇ ◇ ◇ ◇ ◇
• * ◇ ★ ★ ★ ★ ★ ★ ★
• * ◇ ★ ○ ○ ○ ○ ○ ○ ○
• * ◇ ★ ○ × × × × ×
• * ◇ ★ ○ × ⊙ ⊙ ⊙ ⊙
• * ◇ ★ ○ × ⊙ ◁ ◁ ◁ ◁
• * ◇ ★ ○ × ⊙ ◁ ⊕ ⊕
• * ◇ ★ ○ × ⊙ ◁ ⊕ ▷
• *     ★ ○         ◁         ⊖
        ○                         ⋈
```

This is not a standard Ferrers diagram because of the gaps in the last two rows and columns; that the parts are not all distinct causes this. But it is only an intermediate step toward a new partition. The oddness of the parts means it is still symmetric about its main diagonal. We reread the graph as a series of "fishhooks": the first fishhook goes all the way up the main diagonal, and then across the top row. The second fishhook goes up the diagonal right below the main one (which begins with ⊕), and then down the first column. The third fishhook goes up the diagonal right above the main one (which also begins with ⊕) to the second row, and then across. The fourth fishhook goes up the lower of the two diagonals which begin with ◁ to the second column, and then down. The fifth fishhook goes up the other diagonal which begins with ◁ to the third row, and then across; and so on

4.5. SYLVESTER'S FISHHOOK BIJECTION

until every node is contained in a fishhook:

```
•  •  •  •  •  •  •  •  •  •  •
⊙  •  *  *  *  *  *  *  *  *  *
⊙  ⊙  •  *  ◇  ◇  ◇  ◇  ◇  ◇
⊙  ×  ⊙  •  *  ◇  ★  ★  ★  ★
⊙  ×  ×  ⊙  •  *  ◇  ★  ○  ○  ○  ○
⊙  ×  ⊕  ×  ⊙  •  *  ◇  ★  ○
⊙  ×  ⊕  ⊕  ×  ⊙  •  *  ◇  ★
⊙  ×  ⊕  ◁  ⊕  ×  ⊙  •  *  ◇  ★
⊙  ×  ⊕  ◁  ◁  ⊕  ×  ⊙  •  *
⊙  ×  ⊕  ◁  ⊖  ◁  ⊕  ×  ⊙  •
⊙  ×     ◁  ⊖        ⊕           •
            ⊖
```

Now we write down the lengths of the fishhooks: $22 + 18 + 16 + 14 + 11 + 10 + 9 + 6 + 5 + 3$. This is a partition into distinct parts, although it is not the same partition into distinct parts that the Glaisher bijection would give us (see problem 1). It has an extra property that the Glaisher bijection lacks: $21 + 19 + 15 + 15 + 15 + 9 + 7 + 7 + 3 + 1 + 1 + 1$ has 7 different odd parts $(21, 19, 15, 9, 7, 3, 1)$, and correspondingly $22 + 18 + 16 + 14 + 11 + 10 + 9 + 6 + 5 + 3$ has 7 sequences $(22, 18, 16, 14, 11 + 10 + 9, 6 + 5, 3)$. In this section we want to understand this bijection.

To see why this procedure changes odd parts into distinct ones, we describe the fishhooks in general. Let's call n the **radius** of the odd part $2n+1$. Then the length of the first fishhook is the number of parts of the original partition (because each part has one node on the main diagonal) plus the radius of the largest part. The second fishhook is the number of parts ≥ 3 plus the radius of the largest part minus 1. The third fishhook is the number of parts ≥ 3 minus 1, plus the radius of the second largest part minus 1. To describe what happens in general, let's introduce some notation. Given a partition into odd parts, let P_i denote the number of parts whose size is at least i, let R_j denote the radius of the jth largest part, and let F_k denote the length of the kth fishhook. Then we have

$$F_1 = P_1 + R_1, \qquad F_2 = P_3 + (R_1 - 1),$$
$$F_3 = (P_3 - 1) + (R_2 - 1), \qquad F_4 = (P_5 - 1) + (R_2 - 2),$$
$$F_5 = (P_5 - 2) + (R_3 - 2), \qquad F_6 = (P_7 - 2) + (R_3 - 3),$$
$$F_7 = (P_7 - 3) + (R_4 - 3), \qquad F_8 = (P_9 - 3) + (R_4 - 4),$$

and so forth; in general

(4.5.1)
$$F_{2n-1} = (P_{2n-1} - (n-1)) + (R_n - (n-1)),$$
$$F_{2n} = (P_{2n+1} - (n-1)) + (R_n - n).$$

It is clear from these relations that F_k is always strictly bigger than F_{k+1}, so this does give us distinct parts. The relation $F_1 = P_1 + R_1$ implies

THEOREM 40 (Fine's refinement of Euler's partition theorem). *There are exactly as many partitions of n into distinct parts with largest part k as there are partitions of n into odd parts such that the largest part plus twice the number of parts equals $2k + 1$.*

To see this, note that F_1 is the largest part on the distinct side and that on the odd side P_1 is the number of parts and, since R_1 is the largest radius, $2R_1 + 1$ is the largest odd part. Since $F_1 = P_1 + R_1$ we have $2F_1 + 1 = 2P_1 + 2R_1 + 1$, which is precisely Fine's theorem.

Fine also pointed out another refinement of Euler's partition theorem that follows from $F_1 = P_1 + R_1$. Suppose n, the number we are partitioning, is even. Then a partition of n with all parts odd must have an even number of parts, so that P_1 is even. Hence F_1 and R_1 are either both even or both odd. If R_1 is even, then call it $2j$, so that the largest part on the odd side has the form $2(2j)+1 = 4j+1$ (in other words, it is $\equiv 1 \pmod 4$); while the largest number on the distinct side, F_1, is even. If F_1 is odd, then R_1 is odd, so $R_1 = 2i + 1$ for some i, and therefore the largest part on the odd side has the form $2(2i+1)+1 = 4i+3$ (it is $\equiv 3 \pmod 4$).

Conversely, suppose n is odd. Then a partition of n with all parts odd must have an odd number of parts, so that P_1 is odd. Hence one of F_1 and R_1 is even and the other odd. In this case, if the largest part on the distinct side F_1 is odd, then the largest part on the odd side has the form $4j + 1$, and if F_1 is even then the largest part on the odd side has the form $4i + 3$. We state this formally as

THEOREM 41 (Another Fine refinement of Euler's partition theorem). *If n is even, there are exactly as many partitions of n into odd parts where the largest part is $\equiv 1 \pmod 4$ as there are partitions of n into distinct parts where the largest part is even. Also, if n is even, there are exactly as many partitions of n into odd parts where the largest part is $\equiv 3 \pmod 4$ as there are partitions of n into distinct parts where the largest part is odd.*

Conversely, if n is odd there are exactly as many partitions of n into odd parts where the largest part is $\equiv 1 \pmod 4$ as there are partitions of n into distinct parts where the largest part is odd; and exactly as many partitions of n into odd parts where the largest part is $\equiv 3 \pmod 4$ as there are partitions of n into distinct parts where the largest part is even.

Further information can be gleaned from the relations (5.1). For example, we have $F_1 - F_2 - 1 = P_1 - P_3$, which tells us how many 1's the original partition had. We have $F_2 - F_3 - 1 = R_1 - R_2$, which tells us how much bigger the first part is than the second, and so forth; in general

$$(4.5.2) \quad F_{2n-1} - F_{2n} - 1 = P_{2n-1} - P_{2n+1} \quad \text{and} \quad F_{2n} - F_{2n+1} - 1 = R_n - R_{n+1}.$$

These relations show us that if any two consecutive parts in the original partition are different, then there is a corresponding difference of at least 2 between consecutive fishhooks, and this proves

THEOREM 42 (Sylvester's refinement of Euler's "odd equals distinct" theorem). *There are exactly as many partitions of n into odd parts with k different sizes as there are partitions of n into distinct parts which have k sequences of consecutive parts.*

We have used the word "bijection" to describe Sylvester's procedure. In order to justify that word, we have to be able to reconstruct the original partition from the new one. This is easy enough if we have the new parts in the form of fishhooks, but if all we know is the length of each hook, how do we know what shape they should have? We make some observations that will help us understand this. First, if the number of fishhooks is even, then the last one bends vertically, and the one

4.5. SYLVESTER'S FISHHOOK BIJECTION

before it horizontally, and the one before it vertically, and so forth; whereas if the number of fishhooks is odd, then the last one bends horizontally, and they alternate before that. Moreover, the last fishhook cannot actually bend at all: if there is an even number of hooks, then the last one must just be a column, with no diagonal piece; because if it did have a diagonal piece, then there would be a symmetric diagonal piece above the main diagonal which would be part of a shorter fishhook, which is a contradiction. If there is an odd number of hooks, then the last one can't have a flat piece—it must consist solely of a diagonal piece. If it did have a flat piece, it would be (part of) a row, and there would be a symmetric column on the other side of the main diagonal which would be part of a shorter fishhook, and this again contradicts the assumption that we are looking at the last fishhook.

We now know enough to be able to run Sylvester's bijection backwards. Suppose we want to know what partition $15 + 11 + 8 + 5 + 4$ came from. It has four sequences $(15, 11, 8, 5+4)$, so there must be four different odd part sizes; and it has an odd number of fishhooks, so the last one must be just a diagonal. This means that

$$4 = F_5 = (P_5 - 2) + (R_3 - 2) = 4 + 0,$$

so $P_5 = 6$ and $R_3 = 2$. Then there are six parts ≥ 5, and the radius of the third largest part is 2, so the third largest part is 5. Hence the fourth, fifth, and sixth largest parts are also 5. We keep working our way up through the fishhooks from here. We have

$$5 = F_4 = (P_5 - 1) + (R_2 - 2),$$

and since we know $P_5 = 6$, this tells us that

$$5 = F_4 = 5 + (R_2 - 2) = 5 + 0,$$

which gives us the shape of the fourth fishhook (it is also just a diagonal) and says $R_2 = 2$, so the second largest part is 5. Next,

$$8 = F_3 = (P_3 - 1) + (R_2 - 1),$$

and we know $R_2 = 2$, so

$$8 = F_3 = (P_3 - 1) + 1 = 7 + 1,$$

which gives us the shape of the third fishhook and tells us $P_3 = 8$. Then there are eight parts ≥ 3, and since $P_5 = 6$, we can conclude that exactly two of the parts are equal to 3. Next,

$$11 = F_2 = P_3 + (R_1 - 1) = 8 + (R_1 - 1) = 8 + 3,$$

which gives us the shape of the second fishhook and says $R_1 = 4$, so the largest part is 9. Finally,

$$15 = F_1 = P_1 + R_1 = P_1 + 4 = 11 + 4,$$

which gives us the shape of the first fishhook and tells us there are 11 parts in all. Since eight parts are ≥ 3, there are exactly three 1's.

We can now reconstruct the partition in either of two ways. We know the shapes of all the fishhooks, so we can just glue them together and read the diagram. But we also know what all the parts are from the above calculations: the largest part is 9, and then there are five 5's, two 3's, and three 1's. The diagram is:

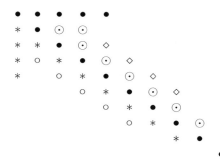

Let's do one more example, $21 + 19 + 16 + 13 + 12 + 8 + 6 + 3$. This has eight fishhooks, so the last one is just a column, which means that

$$3 = F_8 = (P_9 - 3) + (R_4 - 4) = 1 + 2,$$

so $P_9 = 4$ and $R_4 = 6$; there are four parts ≥ 9, and the fourth largest part is 13, so there are no 9's or 11's. Next,

$$6 = F_7 = (P_7 - 3) + (R_4 - 3) = (P_7 - 3) + 3 = 3 + 3,$$

so $P_7 = 6$, which implies that there are exactly two 7's since $P_9 = 4$.

$$8 = F_6 = (P_7 - 2) + (R_3 - 3) = 4 + (R_3 - 3) = 4 + 4,$$

so $R_3 = 7$ and the third largest part is 15.

$$12 = F_5 = (P_5 - 2) + (R_3 - 2) = (P_5 - 2) + 5 = 7 + 5,$$

so $P_5 = 9$, which implies that there are exactly three 5's since $P_7 = 6$.

$$13 = F_4 = (P_5 - 1) + (R_2 - 2) = 8 + (R_2 - 2) = 8 + 5,$$

so $R_2 = 7$ and the second largest part is also 15.

$$16 = F_3 = (P_3 - 1) + (R_2 - 1) = (P_3 - 1) + 6 = 10 + 6,$$

so $P_3 = 11$, which implies that there are exactly two 3's since $P_5 = 9$.

$$19 = F_2 = P_3 + (R_1 - 1) = 11 + (R_1 - 1) = 11 + 8,$$

so $R_1 = 9$ and the largest part is 19. Finally,

$$21 = F_1 = P_1 + R_1 = P_1 + 9 = 12 + 9,$$

so there are twelve parts, only one of which is a 1 since $P_3 = 11$. The partition is $19 + 15 + 15 + 13 + 7 + 7 + 5 + 5 + 5 + 3 + 3 + 1$, and the diagram is:

Recall Gauss's identity

$$(4.5.3) \qquad (-q;q)_n = \sum_{k=0}^{n} \binom{n}{k}_{q^2} q^k$$

from section 2.5. The left side generates partitions with distinct parts which do not exceed n. What about the right side? It follows from Cayley's theorem that $\binom{n}{k}_{q^2}$ generates partitions into at most k even parts, each at most $2n - 2k$. Think of the q^k in (4.5.3) as $q^{1+1+\cdots+1}$. Thus we have k 1's to play with, so let's add 1 to each of the at most k even parts (or exactly k even parts, some of which might be 0) above. This gives us exactly k parts, all odd and at most $2n - 2k + 1$. Or, in the terminology of the fishhook bijection, it gives us exactly k odd parts each with radius at most $n - k$, so that the maximum length of the corresponding fishhook is $k + (n - k) = n$. Therefore the right side of (4.5.3) is the generating function for the kinds of partitions with odd parts that we get by applying the inverse fishhook bijection to the partitions with distinct parts generated by the left side. In other words, the fishhook bijection gives us a combinatorial proof of Gauss's identity. It is a little surprising that it takes such a tricky argument to prove what had seemed like a fairly simple identity, but a simpler combinatorial proof is not known.

It is also interesting to try to find the q-identity that corresponds to Sylvester's refinement of Euler's "odd equals distinct" theorem. As George Andrews pointed out, it is not too hard to find the generating function on the odd side. We claim it is

$$\left(1 + bq + bq^2 + bq^3 + \ldots\right)\left(1 + bq^3 + bq^6 + bq^9 + \ldots\right)$$
$$\times \left(1 + bq^5 + bq^{10} + bq^{15} + \ldots\right)\left(1 + bq^7 + bq^{14} + bq^{21} + \ldots\right)\cdots,$$

in the sense that the coefficient of $b^k q^n$ in this product is the number of partitions of n using only odd parts and only k different ones. The first factor is the source of all the 1's, the second of all the 3's, the third of all the 5's, the fourth of all the 7's, and so on. Taking the 1 from a factor corresponds to not using any of the parts that go with that factor, and taking any other term from a factor gives us exactly one b, no matter how many copies of that part we use. We can rewrite this

expression as
$$\left(1+\frac{bq}{1-q}\right)\left(1+\frac{bq^3}{1-q^3}\right)\left(1+\frac{bq^5}{1-q^5}\right)\left(1+\frac{bq^7}{1-q^7}\right)\cdots,$$
which is
$$(4.5.4) \quad \prod_{m=0}^{\infty}\left(1+\frac{bq^{2m+1}}{1-q^{2m+1}}\right) = \prod_{m=0}^{\infty}\frac{1-q^{2m+1}+bq^{2m+1}}{1-q^{2m+1}} = \frac{((1-b)q;q^2)_{\infty}}{(q;q^2)_{\infty}}.$$

Andrews was able to identify the generating function of the distinct side of Sylvester's theorem as
$$(4.5.5) \quad \sum_{m=0}^{\infty}\frac{((1-b)q;q)_{m+1}}{(q;q)_m}q^{\binom{m+1}{2}}$$
and asked for a simple explanation, which was furnished by Ramamani and Venkatachaliengar. We can rewrite (4.5.5) as
$$\sum_{m=0}^{\infty}\frac{((1-b)q;q)_m}{(q;q)_m}q^{\binom{m+1}{2}}\left(1+(b-1)q^{m+1}\right)$$
$$=\sum_{m=0}^{\infty}\frac{((1-b)q;q)_m}{(q;q)_m}q^{\binom{m+1}{2}}+(b-1)\sum_{m=0}^{\infty}\frac{((1-b)q;q)_m}{(q;q)_m}q^{\binom{m+2}{2}}.$$

Reindexing the last sum and breaking off the $m=0$ term from the one before it, this becomes
$$1+\sum_{m=1}^{\infty}\frac{((1-b)q;q)_m}{(q;q)_m}q^{\binom{m+1}{2}}+(b-1)\sum_{m=1}^{\infty}\frac{((1-b)q;q)_{m-1}}{(q;q)_{m-1}}q^{\binom{m+1}{2}},$$
and combining these two sums we get
$$(4.5.6) \quad 1+b\sum_{m=1}^{\infty}\frac{((1-b)q;q)_{m-1}}{(q;q)_m}q^{\binom{m+1}{2}}.$$

Now this is
$$1+\sum_{m=1}^{\infty}bq^{\binom{m+1}{2}}\frac{1+(b-1)q}{1-q}\frac{1+(b-1)q^2}{1-q^2}\cdots\frac{1+(b-1)q^{m-1}}{1-q^{m-1}}\frac{1}{1-q^m}$$
$$=1+\sum_{m=1}^{\infty}bq^{1+2+\cdots+m}\left(1+\frac{bq}{1-q}\right)\left(1+\frac{bq^2}{1-q^2}\right)\cdots\left(1+\frac{bq^{m-1}}{1-q^{m-1}}\right)\frac{1}{1-q^m}.$$

Distributing the exponents of q and expanding each fraction we get
$$(4.5.7) \quad 1+b\sum_{m=1}^{\infty}\left[\begin{array}{c}(q+bq^2+bq^3+\ldots)\left(q^2+bq^4+bq^6+\ldots\right)\cdots\\ \times\left(q^{m-1}+bq^{2m-2}+bq^{3m-3}+\ldots\right)\left(q^m+q^{2m}+q^{3m}+\ldots\right)\end{array}\right]$$
as an equivalent form of (4.5.5) and (4.5.6). Again the first factor inside the sum in (4.5.7) is the source of all the 1's, the second factor of all the 2's, and so on. For a given m, the partitions generated by
$$b\left[\begin{array}{c}(q+bq^2+bq^3+\ldots)\left(q^2+bq^4+bq^6+\ldots\right)\cdots\\ \times\left(q^{m-1}+bq^{2m-2}+bq^{3m-3}+\ldots\right)\left(q^m+q^{2m}+q^{3m}+\ldots\right)\end{array}\right]$$
have at least one copy of each of $1,2,\ldots,m$, and if they have more than one j for $1\le j\le m-1$ (but not m), then a b comes with the overage. Therefore the

coefficient of $b^k q^n$ counts the number of partitions of n with p_1 1's, p_2 2's, and so on up to p_m m's, where each of $p_1, \ldots, p_m$ is at least 1, and exactly $k-1$ of $p_1, \ldots, p_{m-1}$ are larger than 1.

Let's write down an example at this point. If $m = 9$, a partition of 81 of this kind is $9+8+8+7+6+6+6+6+5+5+4+3+2+2+2+1+1$, with $k = 6$ since five of the parts from 1 through 8 are repeated. Writing the parts as columns, the Ferrers diagram is

and the conjugate is $17 + 15 + 12 + 11 + 10 + 8 + 4 + 3 + 1$, with distinct parts and $k = 6$ different sequences of parts. The parts of the conjugate must be distinct in any example of this kind, because the original partition does not skip any part sizes. The number of sequences must be k, because any p_j larger than 1, except for p_m, stops a sequence. To see this, start from the smallest parts of the original partition and work up. Because there are two 1's, the sequence beginning with 17 stops after one term, and so does the sequence beginning with 15 since there are three 2's. Since there is only one 3 and one 4, the sequence starting with 12 continues to 11 and 10, before stopping because there are two 5's. The sequence starting with 8 stops there because there are four 6's. The sequence starting with 4 continues to 3 because there is only one 7, but stops there since there are two 8's. If the largest part is 9, then having more copies of it just makes all the parts uniformly larger without affecting the sequences. Hence the conjugates of the partitions generated by (4.5.7), and therefore by (4.5.6) and (4.5.5), are precisely the ones counted by the distinct side of Sylvester's theorem. Combining it with (4.5.4) and (4.5.5), we must therefore have

$$\sum_{m=0}^{\infty} \frac{((1-b)q; q)_{m+1}}{(q; q)_m} q^{\binom{m+1}{2}} = \frac{((1-b)q; q^2)_\infty}{(q; q^2)_\infty}.$$

We can simplify this by replacing $(1-b)q$ by a, which leaves

(4.5.8) $$\sum_{m=0}^{\infty} \frac{(a; q)_{m+1}}{(q; q)_m} q^{\binom{m+1}{2}} = \frac{(a; q^2)_\infty}{(q; q^2)_\infty}.$$

We will come back to (4.5.8) in section 5.4.

Exercises

1. Find the partition with distinct parts that corresponds to $21 + 19 + 15 + 15 + 15 + 9 + 7 + 7 + 3 + 1 + 1 + 1$ under Glaisher's bijection. (Answer: $30 + 21 + 19 + 15 + 14 + 9 + 3 + 2 + 1$.)

2. Find the partition with odd parts that corresponds to $22 + 18 + 16 + 14 + 11 + 10 + 9 + 6 + 5 + 3$ under Glaisher's bijection. (Answer: $11 + 11 + 11 + 9 + 9 + 9 + 7 + 7 + 5 + 5 + 5 + 3 + 3 + 3$ plus 16 1's.)
3. Find the partition with distinct parts that corresponds to $15 + 15 + 13 + 9 + 7 + 3 + 3 + 3 + 3$ under the fishhook bijection. (Answer: $16 + 15 + 14 + 9 + 7 + 6 + 3 + 1$.)
4. Find the partition with odd parts that corresponds to $21 + 20 + 14 + 12 + 9 + 6 + 5 + 4$ under the inverse fishhook bijection. (Answer: $29 + 19 + 15 + 15 + 5 + 5 + 3$.)
5. Make up your own examples of partitions with odd parts, and find the corresponding partitions into distinct parts under the fishhook bijection.
6. Make up your own examples of partitions with distinct parts, and find the corresponding partitions into odd parts under the inverse fishhook bijection.
7. How many partitions with odd parts can you think of that map to the same partition under Glaisher's bijection that they do under the fishhook bijection?
8. Same question as 7 for distinct parts and the inverse maps.
9. Take $n = 5$ in Gauss's identity (4.5.3). Write down all the partitions with odd parts generated by the right side and apply the fishhook bijection to them. Show that this gives all the partitions generated by the left side of (4.5.3).
10. Take $n = 4$ in Gauss's identity (4.5.3). Write down all the partitions with distinct parts generated by the left side and apply the inverse fishhook bijection to them. Show that this gives all the partitions generated by the right side of (4.5.3).
11. Verify (4.5.6).
12. Show that the alternating sum of the first $2n - 1$ fishhooks is
$$F_1 - F_2 + F_3 - F_4 + \cdots + F_{2n-1} = P_1 + R_n - (n-1),$$
and that the alternating sum of the first $2n$ fishhooks is
$$F_1 - F_2 + F_3 - F_4 + \cdots + F_{2n-1} - F_{2n} = P_1 - P_{2n+1} + n.$$
Explain why this implies that the alternating sum of *all* the fishhooks equals P_1.

4.6. Bibliographical Notes

In one of his provocative and entertaining blog posts [**252**], Doron Zeilberger calls Sylvester "the GREATEST Mathematician of ALL TIMES", and writes "if I had to name a *mathematician* who, all things considered (constructing a *measure* that is more concentrated on the things that really count, like vision, originality and foresight) then Sylvester has no rivals." One is reminded of Hardy's dictum (quoted by C. P. Snow on p. 46 of [**130**]) "It is never worth a first class man's time to express a majority opinion. By definition, there are plenty of others to do that." In any case, Sylvester's coming to Johns Hopkins was one of the most important events in 19th century American mathematics, a story very well documented by Karen Hunger Parshall in [**179**] and [**180**].

The first of Euler's papers to contain the pentagonal number theorem is [**93**], which was presented to the St. Petersburg Academy in the early 1740s. They did not get around to publishing [**93**] until 1751, by which time the theorem had already appeared in his book [**94**]. It is also in [**95**], [**97**], and [**99**] (which is basically the same as [**95**] except for the language). The proof in section 3.1 is sketched in

[**100**], and presented in a bit more detail much later in [**101**], but Euler never did the induction step (Lemma 3). I don't know of anyone who did before Hans Rademacher, in his masterly book [**190**]. Euler's theorem on divisor sums is in [**95**], [**99**] and [**100**]. Chapter 6 of Pólya's book [**184**] is a beautiful essay on [**95**].

Gauss's variation (4.1.5) of the pentagonal number theorem in problem 7 in section 4.1 is (29) on p. 450 of [**117**], with $x^{\frac{1}{8}}$ in place of q. Problems 11–17 come from [**96**]. Problem 18 comes from [**164**], as does the form of the pentagonal number theorem in section 4.3. Problem 19 comes from [**149**], which also has a history of the pentagonal number theorem. Jacobi proved his triple product identity in 1829 and derived the pentagonal number theorem as a special case [**148**], yet 17 years later he was still willing to work out and publish a new proof of the latter.

The material of section 4.2 is in Sylvester's monumental paper [**231**], and also in Andrews's [**16**]. Andrews has written two penetrating commentaries on [**231**], the other being [**20**], from whence the quotation and problems 12 and 15 come. The original source for problem 12 is [**73**].

Franklin's proof appeared first in [**108**], and then in [**231**]. A version of it without a diagram is in [**230**]. The fishhook bijection appears first in [**232**] in a somewhat different form, as a "Cord Rule". It was recast with fishhooks by MacMahon in the second volume of [**168**], although he does not use that word. Still another form of it is in Bressoud's wonderful book [**61**]. The remark that it proves Gauss's identity was made to me once by Krishnaswami Alladi. At the end of section 4.5 we have followed [**8**] and [**191**].

Igor Pak pointed out in [**177**] that Franklin's argument would also prove Fine's pentagonal number theorem, and that Fine's theorems in section 4.3 followed from the fishhook bijection. Fine's theorems were stated without proof in [**102**]. He published proofs of them in his book [**103**], which appeared 40 years later—just in time to be part of a renaissance in q-series that also included [**19**], [**114**], and Berndt's exegesis of Ramanujan's notebooks, for which see the notes to Chapter 6.

Lambert's identity for the divisor sum is on p. 507 of [**160**]. Clausen's identity is stated without proof at the end of [**79**]. The proof hinted at in problem 3 of section 4.4 is more or less the same as those of Scherk [**211**] and Glaisher [**122**]. Eisenstein's and Uchimura's identities for the divisor sum come from [**89**] and [**239**], respectively. Uchimura's auxiliary theorem (4.4.13) was also known to Ramanujan [**50**]. The last several problems of section 4 come from [**41**] and [**42**]. In [**41**] Bailey sketched a very involved proof of the result of problem 19 (for which see the problems in section 13.3), but he soon realized [**42**] that it could be proved as we have outlined there. The results of this group of problems have applications in number theory, but they are too complicated to get into here.

CHAPTER 5

More q-identities: Jacobi, Gauss, and Heine

5.1. Jacobi's triple product

We know from (3.6.5) in section 3.6 that if $|q| < 1$, then

(5.1.1) $$(-zq; q^2)_\infty = (1+zq)(1+zq^3)(1+zq^5)\cdots = \sum_{n=0}^\infty \frac{q^{n^2} z^n}{(q^2; q^2)_n}$$

for all z, and it follows that

(5.1.2) $$\left(-\frac{q}{z}; q^2\right)_\infty = \left(1+\frac{q}{z}\right)\left(1+\frac{q^3}{z}\right)\left(1+\frac{q^5}{z}\right)\cdots = \sum_{n=0}^\infty \frac{q^{n^2} z^{-n}}{(q^2; q^2)_n}$$

for all $z \neq 0$. The product in (5.1.1) is zero when $z = -q^{-1}, -q^{-3}, -q^{-5}, \ldots$, and the product in (5.1.2) is zero when $z = -q, -q^3, -q^5, \ldots$. If we continue either of these geometric progressions below its starting point (for example, divide $-q^5$ by q^2 to get $-q^3$, divide this by q^2 to get $-q$, and then divide by q^2 again), it runs into the other geometric progression. Since the two products fit together in this sense, we might hope that their product would have a nice expansion. Since all integer powers of z will occur, it would look like

(5.1.3) $$(-zq; q^2)_\infty \left(-\frac{q}{z}; q^2\right)_\infty = \sum_{n=-\infty}^\infty c_n(q) z^n$$

for some coefficients $c_n(q)$. Because the product is unchanged if we replace z by $\frac{1}{z}$, we must have $c_n(q) = c_{-n}(q)$.

There is a very natural way to start trying to find $c_n(q)$ that was used by Gauss. Jacobi used it too but covered his tracks when he wrote it up. If Euler had been confronted with the product (5.1.3), he would surely have tried replacing z by zq^2 there, as Gauss and Jacobi did. This gives

(5.1.4) $$(-zq^3; q^2)_\infty \left(-\frac{1}{zq}; q^2\right)_\infty = \sum_{n=-\infty}^\infty c_n(q) q^{2n} z^n,$$

and the point is that the left side of (5.1.4) is almost the same as the left side of (5.1.3). What are the differences? The product side of (5.1.3) has one factor that (5.1.4) lacks, namely $1 + zq$, and it lacks one factor that (5.1.4) has, namely $1 + \frac{1}{zq}$.

But $1 + \frac{1}{zq} = \frac{zq+1}{zq}$, so we can rewrite the product side of (5.1.4) as

$$(-zq^3; q^2)_\infty \left(-\frac{1}{zq}; q^2\right)_\infty = (-zq^3; q^2)_\infty \left(1 + \frac{1}{zq}\right)\left(-\frac{q}{z}; q^2\right)_\infty$$

$$= (-zq^3; q^2)_\infty (1 + zq) \frac{1}{zq} \left(-\frac{q}{z}; q^2\right)_\infty$$

$$= (-zq; q^2)_\infty \frac{1}{zq} \left(-\frac{q}{z}; q^2\right)_\infty$$

or in other words

(5.1.5) $\qquad zq\,(-zq^3; q^2)_\infty \left(-\frac{1}{zq}; q^2\right)_\infty = (-zq; q^2)_\infty \left(-\frac{q}{z}; q^2\right)_\infty.$

From (5.1.3), (5.1.4), and (5.1.5) it follows that

$$\sum_{n=-\infty}^{\infty} c_n(q)\, z^n = zq \sum_{n=-\infty}^{\infty} c_n(q)\, q^{2n} z^n$$

$$= \sum_{n=-\infty}^{\infty} c_n(q)\, q^{2n+1}\, z^{n+1}$$

$$= \sum_{n=-\infty}^{\infty} c_{n-1}(q)\, q^{2n-1}\, z^n,$$

where we changed n to $n-1$ in the last line. Equating coefficients of z^n it follows that

(5.1.6) $\qquad\qquad\qquad c_n(q) = q^{2n-1}\, c_{n-1}(q).$

Since $c_n(q) = c_{-n}(q)$, we only have to find $c_n(q)$ for $n \geq 0$. If $n > 0$, we can use (5.1.6) to iterate our way down to $c_0(q)$, as we have done many times before. The result is

(5.1.7) $\qquad c_n(q) = q^{(2n-1)+(2n-3)+\cdots+5+3+1}\, c_0(q) = q^{n^2} c_0(q),$

and this holds for all n since $(-n)^2$ is the same as n^2. Thus we can rewrite (5.1.3) as

(5.1.8) $\qquad\qquad (-zq; q^2)_\infty \left(-\frac{q}{z}; q^2\right)_\infty = c_0(q) \sum_{n=-\infty}^{\infty} q^{n^2} z^n,$

and we still have to find $c_0(q)$. There are many ways to do this. Perhaps the cleverest is due to Gauss, which has the additional advantage that it does not rely on any previous q-identity.

Suppose we take $z = i$ in (5.1.8). The left side becomes

$$(-iq; q^2)_\infty \left(-\frac{q}{i}; q^2\right)_\infty = (-iq; q^2)_\infty\, (iq; q^2)_\infty$$

$$= \{(1+iq)(1+iq^3)(1+iq^5)\ldots\}$$

$$\times \{(1-iq)(1-iq^3)(1-iq^5)\ldots\}$$

$$= (1+iq)(1-iq)(1+iq^3)(1-iq^3)(1+iq^5)(1-iq^5)\ldots$$

$$= (1+q^2)(1+q^6)(1+q^{10})\ldots$$

$$= (-q^2; q^4)_\infty.$$

The series becomes

$$\sum_{n=-\infty}^{\infty} q^{n^2} i^n = 1 + q\left(i + i^{-1}\right) + q^4\left(i^2 + i^{-2}\right) + q^9\left(i^3 + i^{-3}\right) + q^{16}\left(i^4 + i^{-4}\right) + \ldots$$

$$= 1 + q \cdot 0 + q^4(-2) + q^9 \cdot 0 + q^{16}(2) + q^{25} \cdot 0 + q^{36}(-2) + \ldots$$

$$= 1 - 2q^4 + 2q^{16} - 2q^{36} + 2q^{64} - + \ldots$$

$$= \sum_{n=-\infty}^{\infty} (-1)^n q^{4n^2},$$

so when $z = i$ (5.1.8) says

(5.1.9) $$(-q^2; q^4)_\infty = c_0(q) \sum_{n=-\infty}^{\infty} (-1)^n q^{4n^2}.$$

On the other hand, we can get the same series by replacing q by q^4 in (6.5.4) and setting $z = -1$. This gives

(5.1.10) $$(q^4; q^8)_\infty^2 = c_0(q^4) \sum_{n=-\infty}^{\infty} (-1)^n q^{4n^2}.$$

From (5.1.9) and (5.1.10) we have

$$\frac{c_0(q)}{(-q^2; q^4)_\infty} = \frac{1}{\sum_{n=-\infty}^{\infty} (-1)^n q^{4n^2}} = \frac{c_0(q^4)}{(q^4; q^8)_\infty^2},$$

or

(5.1.11) $$c_0(q) = c_0(q^4) \frac{(-q^2; q^4)_\infty}{(q^4; q^8)_\infty^2}.$$

Now multiply both sides of (5.1.11) by $(q^2; q^2)_\infty$. On the right side we use (exercise)

(5.1.12) $$(q^2; q^2)_\infty = (q^2; q^4)_\infty (q^4; q^4)_\infty$$
$$= (q^2; q^4)_\infty (q^4; q^8)_\infty (q^8; q^8)_\infty.$$

This cancels one of the factors of $(q^4; q^8)_\infty$ in (5.1.11), leaving

(5.1.13) $$c_0(q)(q^2; q^2)_\infty = c_0(q^4)(q^8; q^8)_\infty \frac{(q^2; q^4)_\infty (-q^2; q^4)_\infty}{(q^4; q^8)_\infty}.$$

But

$$(q^2; q^4)_\infty (-q^2; q^4)_\infty = (1 - q^2)(1 - q^6)(1 - q^{10})(1 - q^{14}) \cdots$$
$$\times (1 + q^2)(1 + q^6)(1 + q^{10})(1 + q^{14}) \cdots$$
$$= (1 - q^4)(1 - q^{12})(1 - q^{20}) \cdots = (q^4; q^8)_\infty,$$

so (5.1.13) becomes

$$c_0(q)(q^2; q^2)_\infty = c_0(q^4)(q^8; q^8)_\infty.$$

This says that the expression $c_0(q)(q^2; q^2)_\infty$ is unaltered if we replace q by q^4. It is also unchanged if we replace q by q^{16}, because this is equivalent to changing q to q^4 twice. And it is also unchanged if we replace q by q^{256}, because this is equivalent to changing q to q^{16} twice. And it is also unchanged if we replace q by q^{65536}, which is $q \mapsto q^{256}$ done twice, and so on. It is unchanged if we replace q by q^N, where

N is an arbitrarily large power of 4. Since $|q| < 1$, this means the expression is unchanged if we replace q by 0. But $(0;0)_\infty = 1$, and also $c_0(0) = 1$ since both products on the left side of (5.1.8) equal 1 when $q = 0$, and the sum on the right equals 1 (from the $n = 0$ term). Therefore

$$c_0(q)\,(q^2;q^2)_\infty = 1, \quad \text{so} \quad c_0(q) = \frac{1}{(q^2;q^2)_\infty}.$$

We have proved

THEOREM 43 (Jacobi's triple product identity). *If $|q| < 1$ and $z \neq 0$, then*

$$(5.1.14) \qquad (-zq;q^2)_\infty \left(-\frac{q}{z};q^2\right)_\infty (q^2;q^2)_\infty = \sum_{n=-\infty}^\infty q^{n^2} z^n.$$

Euler's pentagonal number theorem

$$(5.1.15) \qquad (q;q)_\infty = \sum_{n=-\infty}^\infty (-1)^n q^{\frac{n(3n-1)}{2}} = 1 - q - q^2 + q^5 + q^7 - - + + \dots$$

from section 4.1 gives us another simple way to find $c_0(q)$. Replace q by q^2 here and note that $(q^2;q^2)_\infty = (q^2;q^6)_\infty (q^4;q^6)_\infty (q^6;q^6)_\infty$. Then the pentagonal number theorem is equivalent to

$$(5.1.16) \qquad (q^2;q^6)_\infty (q^4;q^6)_\infty (q^6;q^6)_\infty = \sum_{n=-\infty}^\infty (-1)^n q^{3n^2-n}.$$

On the other hand, if we replace q by q^3 in (5.1.14) we have

$$(5.1.17) \qquad (-zq^3;q^6)_\infty \left(-\frac{q^3}{z};q^6\right)_\infty = c_0(q^3) \sum_{n=-\infty}^\infty q^{3n^2} z^n.$$

To make the series in (5.1.16) and (5.1.17) the same we need $z = -\frac{1}{q}$, which changes (5.1.17) into

$$(q^2;q^6)_\infty \left(q^4;q^6\right)_\infty = c_0(q^3) \sum_{n=-\infty}^\infty q^{3n^2} (-q)^{-n}.$$

Comparing this with (5.1.12) we see that $c_0(q^3) = 1/(q^6;q^6)_\infty$. Therefore $c_0(q) = 1/(q^2;q^2)_\infty$ as before. Since the triple product is much more general than (5.1.15) and at the same time easier to prove, this argument is more commonly used the other way around, to derive (5.1.16) from (5.1.14).

A **theta** product is a product of the form $(a;p)_\infty (b;p)_\infty$ with $ab = p$. The characteristic property of the product side of (5.1.14) is that the factors containing z are a theta product with $p = q^2$, $a = -zq$, and $b = -\frac{q}{z}$. We can rewrite (5.1.14) by replacing q^2 by q and then z by $-x\sqrt{q}$, which gives

THEOREM 44 (Jacobi's triple product identity, alternate form). *If $|q| < 1$ and $x \neq 0$, then*

$$(5.1.18) \qquad (x;q)_\infty \left(\frac{q}{x};q\right)_\infty (q;q)_\infty = \sum_{n=-\infty}^\infty (-1)^n q^{\frac{n(n-1)}{2}} x^n.$$

This again expands a theta product, now with $p = q$, $a = x$, and $b = \frac{q}{x}$.

Exercises

1. Explain why $(q^2; q^2)_\infty = (q^2; q^6)_\infty (q^4; q^6)_\infty (q^6; q^6)_\infty$.
2. Prove (5.1.12).
3. How would you prove the pentagonal number theorem (5.1.15) from (5.1.18)?
4. Show that letting $n \to \infty$ in any of the three identities in problem 10 from section 2.5 gives the pentagonal number theorem.
5. If $|q| < 1$, show that $\sum_{n=-\infty}^{\infty} q^{n^2} z^n$ converges for any z other than $z = 0$. (Use the ratio test or the root test on

$$\sum_{n=0}^{\infty} q^{n^2} z^n \quad \text{and} \quad \sum_{n=-\infty}^{-1} q^{n^2} z^n = \sum_{n=1}^{\infty} q^{n^2} z^{-n}$$

separately.)
6. If $|ab| < 1$, show that

(5.1.19) $$\sum_{n=-\infty}^{\infty} a^{\frac{n(n+1)}{2}} b^{\frac{n(n-1)}{2}} = (-a; ab)_\infty (-b; ab)_\infty (ab; ab)_\infty.$$

This is the form in which Ramanujan rediscovered Jacobi's triple product.

7. Explain why we can rewrite (5.1.14) as

$$(-zq; q^2)_\infty \left(-\frac{q}{z}; q^2\right)_\infty (q^2; q^2)_\infty = \sum_{m=-\infty}^{\infty} q^{4m^2} z^{2m} \left(1 + zq^{4m+1}\right)$$

and (5.1.18) as

$$(x; q)_\infty \left(\frac{q}{x}; q\right)_\infty (q; q)_\infty = \sum_{m=-\infty}^{\infty} x^{2m} q^{2m^2 - m} \left(1 - xq^{2m}\right).$$

8. Set $x = i$ in (5.2.7) to show that

(5.1.20) $$(-q^2; q^2)_\infty (q; q)_\infty = \sum_{k=0}^{\infty} (-1)^k \left[q^{\binom{2k+1}{2}} - q^{\binom{2k+2}{2}}\right].$$

9. Show that the left side of (5.1.20) can be rewritten as $(q; q^2)_\infty (q^4; q^4)_\infty$.
10. In view of problem 9, (5.1.20) has an interpretation similar to that of the pentagonal number theorem: the left side generates partitions with an even number of distinct parts minus partitions with an odd number of distinct parts, where the parts must either be odd or else multiples of 4. (To put it another way, the parts must be distinct and not congruent to 2 mod 4.) Explain. Write out all the partitions of this type up to $n = 10$, and check that this interpretation is consistent with the right side of (5.1.20).
11. Show that the sum in (5.1.20) can be rewritten as

$$\sum_{k=0}^{\infty} (-1)^k \left[q^{\binom{2k+1}{2}} - q^{\binom{2k+2}{2}}\right] = \sum_{k=0}^{\infty} (-1)^k q^{k(2k+1)} + \sum_{k=0}^{\infty} (-1)^{k+1} q^{(k+1)(2k+1)}$$

$$= \sum_{j=-\infty}^{\infty} (-1)^j q^{j(2j+1)} = \sum_{j=-\infty}^{\infty} q^{2j^2} (-q)^j.$$

Show that this sum equals $(q;q^4)_\infty (q^3;q^4)_\infty (q^4;q^4)_\infty$. Is this consistent with problem 9?

12. Show that
$$\sum_{n=-\infty}^{\infty} q^{\frac{n(3n+1)}{2}} = (-q;q^3)_\infty(-q^2;q^3)_\infty(q^3;q^3)_\infty = \frac{(q^3;q^3)_\infty}{(q;q^6)_\infty(q^5;q^6)_\infty}.$$
One might call this *Gauss's pentagonal number theorem*.

13. Gauss wrote down the expansion
$$1 + q + q^5 + q^8 + q^{16} + q^{21} + q^{33} + \cdots = (-q;q^6)_\infty(-q^5;q^6)_\infty(q^6;q^6)_\infty.$$
What is the series on the left side? Explain.

14. Show that the series in the previous problem also equals
$$(-q;q^2)_\infty \sum_{n=-\infty}^{\infty} (-1)^n q^{6n^2+3n}.$$

15. Use Jacobi's triple product to show that
$$q + q^9 + q^{25} + q^{49} + \cdots = \sum_{n=0}^{\infty} q^{(2n+1)^2} = q\left(-q^8;q^8\right)_\infty^2 \left(q^8;q^8\right)_\infty.$$

Hint: How are
$$\sum_{n=0}^{\infty} q^{(2n+1)^2} \quad \text{and} \quad \sum_{n=-\infty}^{\infty} q^{(2n+1)^2} \quad \text{related?}$$

16. Gauss claimed that
$$q + q^9 + q^{25} + q^{49} + \ldots$$
$$= \left[q(1+q^8)(1+q^{16})(1+q^{24})\cdots\right]\left[1 - q^{16} - q^{32} + q^{80} + q^{112} \ldots\right].$$
Use the previous problem to show this. Use Euler's pentagonal number theorem (5.1.15) to interpret the sum on the right side.

17. Gauss further claimed that
$$q^3 + q^{27} + q^{75} + \ldots$$
$$= \left[q(1+q^8)(1+q^{16})(1+q^{24})\cdots\right]\left[q^2 - q^{10} - q^{42} + q^{66} + q^{130} \ldots\right].$$
Show this. The series on the right side is
$$q^2 \sum_{n=-\infty}^{\infty} (-1)^n q^{8n(3n+2)}$$
although Gauss doesn't tell you this. Use problem 15 with q replaced by q^3 for the left side.

18. Show that
$$\left(\sum_{n=-\infty}^{\infty} q^{\frac{n(n+1)}{2}}\right)^2 = 2\left(\sum_{n=-\infty}^{\infty} q^{n^2}\right)\left(\sum_{n=-\infty}^{\infty} q^{n^2+n}\right).$$
This was observed by Cauchy.

EXERCISES

19. Either directly or by rewriting the result of the previous problem, show that

$$\left(\sum_{n=0}^{\infty} q^{\frac{n(n+1)}{2}}\right)^2 = \left(\sum_{n=-\infty}^{\infty} q^{n^2}\right)\left(\sum_{n=0}^{\infty} q^{n^2+n}\right).$$

20. Show that the identity (4.2.13)

$$\frac{1}{(xq;q)_n} = \sum_{k=0}^{n} \binom{n}{k}_q \frac{q^{k^2} x^k}{(xq;q)_k},$$

which was also problem 10 in section 2.3, becomes (5.1.22) after replacing q by q^2 and letting $n \to \infty$.

21. This problem gives another determination of $c_0(q)$ that completes the proof of the triple product.

    (i) If we multiply (5.1.1) and (5.1.2) together we get

(5.1.21) $\quad (-zq;q^2)_\infty \left(-\frac{q}{z};q^2\right)_\infty = \left(\sum_{j=0}^{\infty} \frac{q^{j^2} z^j}{(q^2;q^2)_j}\right)\left(\sum_{k=0}^{\infty} \frac{q^{k^2} z^{-k}}{(q^2;q^2)_k}\right).$

Now $c_0(q)$ is the constant term (*i.e.*, the term independent of z) in this product. Explain why we get a term independent of z in (5.1.21) exactly when $j = k$, and hence

$$c_0(q) = \sum_{k=0}^{\infty} \left(\frac{q^{k^2}}{(q^2;q^2)_k}\right)^2.$$

    (ii) If we replace q by q^2 in Jacobi's Durfee square identity (4.2.1), we get

(5.1.22) $\quad \dfrac{1}{(zq^2;q^2)_\infty} = \sum_{k=0}^{\infty} \dfrac{q^{2k^2} z^k}{(q^2;q^2)_k (zq^2;q^2)_k}.$

What value does this give for $c_0(q)$?

22. Here is an outline of Jacobi's proof of (5.1.22). He wants to find the coefficients A_n in the expansion

(5.1.23) $\quad \dfrac{1}{(zq;q)_\infty} = \sum_{n=0}^{\infty} \dfrac{A_n z^n}{(zq;q)_n}.$

    (i) Explain how we can see that $A_0 = 1$.

    (ii) Jacobi observes that the left side of (5.1.23) is unchanged if we replace z by zq and then multiply by $\frac{1}{1-zq}$. Show that this gives

(5.1.24) $\quad \dfrac{1}{(zq;q)_\infty} = \sum_{n=0}^{\infty} \dfrac{A_n z^n q^n}{(zq;q)_{n+1}}.$

    (iii) Show that $\frac{1}{1-zq^k} = 1 + \frac{zq^k}{1-zq^k}$.

(iv) Use (iii) to rewrite (5.1.24) as

$$\frac{1}{(zq;q)_\infty} = \sum_{n=0}^{\infty} \frac{A_n\, z^n\, q^n}{(zq;q)_n}\left(1 + \frac{zq^{n+1}}{1-zq^{n+1}}\right)$$

(5.1.25)
$$= \sum_{n=0}^{\infty} \frac{A_n\, z^n\, q^n}{(zq;q)_n} + \sum_{n=0}^{\infty} \frac{A_n\, z^{n+1}\, q^{2n+1}}{(zq;q)_{n+1}}.$$

(v) By comparing (5.1.25) and (5.1.23), show that

$$\sum_{n=0}^{\infty} \frac{A_n\, z^n(1-q^n)}{(zq;q)_n} = \sum_{n=0}^{\infty} \frac{A_n\, z^{n+1}\, q^{2n+1}}{(zq;q)_{n+1}}$$

and explain why this can be rewritten as

$$\sum_{n=1}^{\infty} \frac{A_n\, z^n(1-q^n)}{(zq;q)_n} = \sum_{n=1}^{\infty} \frac{A_{n-1}\, z^n\, q^{2n-1}}{(zq;q)_n}.$$

(vi) Jacobi now equates coefficients of $z^n/(zq;q)_n$ to conclude that

(5.1.26)
$$A_n = \frac{q^{2n-1}}{1-q^n}\, A_{n-1}.$$

We will discuss the legitimacy of this step in Chapter 6. Assuming it is okay, explain how we know that

$$A_n = \frac{q^{n^2}}{(q;q)_n}.$$

Hence (5.1.23) becomes

$$\frac{1}{(zq;q)_\infty} = \sum_{n=0}^{\infty} \frac{q^{n^2}\, z^n}{(q;q)_n\,(zq;q)_n},$$

and this in turn becomes (5.1.22) if we replace q by q^2.

23. Jacobi saw that (5.1.22) was powerful enough to derive the triple product directly from (5.1.3) without needing (5.1.8), and this argument was what he chose to publish.

(i) Since $c_n(q) = c_{-n}(q)$ in (5.1.3), we just have to be able to find $c_n(q)$ for $n \geq 0$. By setting $j = n+k$, explain why

$$c_n(q) = \sum_{k=0}^{\infty} \frac{q^{(n+k)^2}}{(q^2;q^2)_{n+k}} \frac{q^{k^2}}{(q^2;q^2)_k}$$

at least if $n \geq 0$.

(ii) Explain why the result of (i) can be rewritten as

$$c_n(q) = \frac{q^{n^2}}{(q^2;q^2)_n} \sum_{k=0}^{\infty} \frac{q^{2k^2+2nk}}{(q^{2n+2};q^2)_k\,(q^2;q^2)_k}$$

if $n \geq 0$. (We need this restriction because we have not yet defined $(q^2;q^2)_n$ for a negative n.)

(iii) Explain why (5.1.22) allows us to do the sum in (ii), resulting in

$$c_n(q) = \frac{q^{n^2}}{(q^2;q^2)_n}\,\frac{1}{(q^{2n+2};q^2)_\infty} = \frac{q^{n^2}}{(q^2;q^2)_\infty}.$$

Note that this is symmetric in n and $-n$, so we no longer need to assume $n \geq 0$.

(iv) Show that the triple product (5.1.14) follows from (iii).

24. Following Gauss, show that the triple product (5.1.14) can be put in the form

$$q^{\frac{1}{4}}\left(y+\frac{1}{y}\right)\left(-q^2y^2;q^2\right)_\infty \left(-\frac{q^2}{y^2};q^2\right)_\infty (q^2;q^2)_\infty = \sum_{n=0}^{\infty} q^{\left(n+\frac{1}{2}\right)^2}\left(y^{2n+1}+\frac{1}{y^{2n+1}}\right).$$

25. Bunyakowsky tried to expand the function

$$F(x,q) = \left(x-\frac{1}{x}\right)(q^2x^2;q^2)_\infty \left(\frac{q^2}{x^2};q^2\right)_\infty$$

in powers of x.

(i) Explain why if c_m is the coefficient of x^m in the expansion of $F(x,q)$, then $-c_m$ is the coefficient of x^{-m}.

(ii) Explain why the expansion of $F(x,q)$ only contains odd powers of x. Because of this and (i), Bunyakowsky takes the expansion in the form

$$F(x,q) = \sum_{n=0}^{\infty} A_{2n+1}(q)\left(x^{2n+1} - x^{-2n-1}\right).$$

(iii) Show that $F(x,q) = -x^2 q\, F(xq,q)$, and hence

$$0 = \sum_{n=0}^{\infty} A_{2n+1}(q)\left(x^{2n+1} - x^{-2n-1}\right) + x^2 q \sum_{n=0}^{\infty} A_{2n+1}(q)\left((xq)^{2n+1} - (xq)^{-2n-1}\right).$$

(iv) Show that the result of (iii) can be rewritten as

$$0 = \sum_{n=0}^{\infty} A_{2n+1}(q)x^{2n+1} - \sum_{n=0}^{\infty} A_{2n+1}(q)x^{-2n-1}$$
$$+ \sum_{n=1}^{\infty} A_{2n-1}(q)q^{2n}x^{2n+1} - \sum_{n=-1}^{\infty} A_{2n+3}(q)q^{-2n-2}x^{-2n-1},$$

and that this reduces to

$$0 = \sum_{n=1}^{\infty}\left(A_{2n+1}(q) + q^{2n} A_{2n-1}(q)\right)x^{2n+1}$$
$$- \sum_{n=0}^{\infty}\left(A_{2n+1}(q) + q^{-2n-2} A_{2n+3}(q)\right)x^{-2n-1}.$$

(v) Explain why it follows from (iv) that $A_{2n+1}(q) + q^{2n}A_{2n-1}(q) = 0$ for $n \geq 1$, and $A_{2n+1}(q) + q^{-2n-2}A_{2n+3}(q) = 0$ for $n \geq 0$, and explain why these two equations say the same thing.

(vi) Show that $A_{2n+1}(q) = (-1)^n q^{n(n+1)} A_1(q)$ for $n \geq 0$.

At this point Bunyakowsky has

(5.1.27)
$$\left(x-\frac{1}{x}\right)(q^2x^2;q^2)_\infty \left(\frac{q^2}{x^2};q^2\right)_\infty = A_1(q)\sum_{n=0}^{\infty}(-1)^n q^{n(n+1)}\left(x^{2n+1} - x^{-2n-1}\right),$$

and it remains to find $A_1(q)$. He has two ideas for doing this.

(vii) He divides (5.1.27) by $x - \frac{1}{x}$ and then lets $x \to 1$. Show that this gives

(5.1.28) $$(q^2;q^2)_\infty^2 = A_1(q) \sum_{n=0}^\infty (-1)^n (2n+1) q^{n(n+1)}.$$

(viii) He sets $x = i$ in (5.1.27). Show that this gives

(5.1.29) $$(-q^2;q^2)_\infty^2 = A_1(q) \sum_{n=0}^\infty q^{n(n+1)}.$$

(ix) Bunyakowsky never does find $A_1(q)$ (but see the next several problems). He concludes this investigation by dividing (5.1.28) by (5.1.29) and changing q^2 to q. Show that this gives

$$\frac{1 - 3q + 5q^3 - 7q^6 + 9q^{10} \cdots}{1 + q + q^3 + q^6 + q^{10} + \cdots} = \frac{(q;q)_\infty^4}{(q^2;q^2)_\infty^2}.$$

26. Use a result from this section to show that (5.1.28) implies $A_1(q) = 1/(q^2;q^2)_\infty$.

27. Use a result from this section to show that (5.1.29) implies $A_1(q) = 1/(q^2;q^2)_\infty$.

28. Show that one can derive Bunyakowsky's expansion by taking $z = -qx^2$ in (5.1.14) and multiplying by x.

29. In problem 22 in section 3.2, I asked you to verify seven quotients of infinite products that Jacobi said were equal to $(-q;q)_\infty$. You can find them in the q-shifted factorial notation below. Jacobi went on to expand the numerator and denominator of each fraction using the triple product identity. For aesthetic reasons, I'll leave the limits off of the summations below, as Jacobi did—they all go from $j = -\infty$ to ∞. Show that

$$\frac{(q^2;q^2)_\infty}{(q;q)_\infty} = \frac{\sum (-1)^j q^{j(3j+1)}}{\sum (-1)^j q^{\frac{j(3j+1)}{2}}},$$

$$\frac{(-q;q^2)_\infty (q^2;q^2)_\infty}{(q^2;q^2)_\infty} = \frac{\sum q^{j(2j+1)}}{\sum (-1)^j q^{j(3j+1)}},$$

$$\frac{(q;q)_\infty}{(q;q^2)_\infty^2 (q^2;q^2)_\infty} = \frac{\sum (-1)^j q^{\frac{j(3j+1)}{2}}}{\sum (-1)^j q^{j^2}},$$

$$\frac{(-q;q^2)_\infty (q^2;q^2)_\infty}{(q^2;q^4)_\infty^2 (q^4;q^4)_\infty} = \frac{\sum (-1)^{\frac{j(j+1)}{2}} q^{\frac{j(3j+1)}{2}}}{\sum (-1)^j q^{2j^2}},$$

$$\frac{(q^4;q^4)_\infty}{(q;q^4)_\infty (q^3;q^4)_\infty (q^4;q^4)_\infty} = \frac{\sum (-1)^j q^{2j(3j+1)}}{\sum (-1)^j q^{j(2j+1)}},$$

$$\frac{(-q;q^3)_\infty (-q^2;q^3)_\infty (q^3;q^3)_\infty}{(q^3;q^6)_\infty^2 (q^6;q^6)_\infty} = \frac{\sum q^{\frac{j(3j+1)}{2}}}{\sum (-1)^j q^{3j^2}},$$

$$\frac{(-q^3;q^{12})_\infty (-q^9;q^{12})_\infty (q^{12};q^{12})_\infty}{(q;q^6)_\infty (q^5;q^6)_\infty (q^6;q^6)_\infty} = \frac{\sum q^{3j(2j+1)}}{\sum (-1)^j q^{j(3j+2)}}.$$

The minus signs are a little tricky in the fourth expansion.

5.2. Other proofs and related results

Another proof of (5.1.14) comes from MacMahon's q-binomial theorem in section 2.7, which we rewrite with a slightly different notation here:

$$(5.2.1) \qquad \left(-\frac{q}{x};q^2\right)_m \left(-qx;q^2\right)_n = \sum_{k=-m}^{n} \binom{m+n}{m+k}_{q^2} q^{k^2} x^k.$$

It is obvious what happens to the product on the left as m and n go to infinity. On the right, we need to use Tannery's theorem from the appendix. Except for the bilateral nature of the series, the application is quite straightforward. We can write

$$\sum_{k=-m}^{n} \binom{m+n}{m+k}_{q^2} q^{k^2} x^k = \sum_{k=-m}^{-1} \binom{m+n}{m+k}_{q^2} q^{k^2} x^k + \sum_{k=0}^{n} \binom{m+n}{m+k}_{q^2} q^{k^2} x^k,$$

and we treat the latter sum first. Since k is nonnegative, it is natural to write

$$\binom{m+n}{m+k}_{q^2} = \frac{(1-q^{2m+2k+2})(1-q^{2m+2k+4})\cdots(1-q^{2m+2n})}{(q^2;q^2)_{n-k}}.$$

For $-1 < q < 1$, this increases to $1/(q^2;q^2)_\infty$ as m and n tend to ∞, and

$$\sum_{k=0}^{\infty} \frac{q^{k^2} x^k}{(q^2;q^2)_\infty}$$

converges for any x when $|q| < 1$, so we can take

$$a_k = \frac{q^{k^2} x^k}{(q^2;q^2)_\infty} = C_k$$

in Tannery's theorem if $qx > 0$. (We have to use $|q|$ and $|x|$ for C_k in general, but this does not affect the convergence of the series.) We have also to deal with the sum from $k = -m$ to -1, but we can do this painlessly by changing k to $-k$ and switching m and n, when the previous argument applies for any $x \neq 0$. Therefore

$$\lim_{m,n\to\infty} \left(-\frac{q}{x};q^2\right)_m \left(-qx;q^2\right)_n = \lim_{m,n\to\infty} \sum_{k=-m}^{n} \binom{m+n}{m+k}_{q^2} q^{k^2} x^k$$

becomes

$$\left(-\frac{q}{x};q^2\right)_\infty \left(-qx;q^2\right)_\infty = \frac{1}{(q^2;q^2)_\infty} \sum_{k=-\infty}^{\infty} q^{k^2} x^k$$

for all nonzero x, which is (5.1.14).

Frobenius devised an interesting system of notation for partitions. Let's write down the Ferrers diagram for our earlier example $7 + 6 + 6 + 3 + 3 + 1$:

```
      *  •  •  •  •  •  •
      ○  *  •  •  •  •
      ○  ○  *  •  •  •
      ○  ○  ○
      ○  ○  ○
      ○
```

Now record the lengths of the rows above the main diagonal, and of the columns below the main diagonal, in a 2-dimensional array:

$$\begin{pmatrix} 6 & 4 & 3 \\ 5 & 3 & 2 \end{pmatrix}$$

is the **Frobenius symbol** of $7+6+6+3+3+1$. Note that the number of columns of the array is the length of the main diagonal; so that, even though the $*$'s are not counted in any of the numbers in the Frobenius symbol, no information is lost. In general, a Frobenius symbol looks like

$$\begin{pmatrix} a_1 & a_2 & a_3 & \cdots & a_n \\ b_1 & b_2 & b_3 & \cdots & b_n \end{pmatrix},$$

where $a_1 > a_2 > \cdots > a_n \geq 0$ and $b_1 > b_2 > \cdots > b_n \geq 0$, the number being partitioned is $a_1 + a_2 + \cdots + a_n + b_1 + b_2 + \cdots + b_n + n$, and n is the length of the main diagonal in the Ferrers diagram. It can happen that one or both of the entries in the last column are 0.

The main point of this notation is that it reveals the conjugate of a partition immediately. For example, the conjugate of

$$\begin{pmatrix} 6 & 4 & 3 \\ 5 & 3 & 2 \end{pmatrix} \quad \text{is} \quad \begin{pmatrix} 5 & 3 & 2 \\ 6 & 4 & 3 \end{pmatrix},$$

and generally the conjugate of

$$\begin{pmatrix} a_1 & a_2 & a_3 & \cdots & a_n \\ b_1 & b_2 & b_3 & \cdots & b_n \end{pmatrix} \quad \text{is} \quad \begin{pmatrix} b_1 & b_2 & b_3 & \cdots & b_n \\ a_1 & a_2 & a_3 & \cdots & a_n \end{pmatrix}.$$

By giving up this conjugate property we can get a modified Frobenius symbol whose entries add up to the number being partitioned. Let's look at the example $8+6+6+3+2+2$:

```
      ○ ● ● ● ● ● ● ●
      ○ ○ ● ● ● ●
      ○ ○ ○ ● ●
      ○ ○ ○
      ○ ○
      ○ ○
```

This time we include the main diagonal in the lower region, rather than separating it out, so the modified symbol would be

$$\begin{pmatrix} 7 & 4 & 3 \\ 6 & 5 & 2 \end{pmatrix}$$

and the number being partitioned is $7+4+3+6+5+2 = 27$. A general modified symbol looks like

$$\begin{pmatrix} c_1 & c_2 & c_3 & \cdots & c_n \\ d_1 & d_2 & d_3 & \cdots & d_n \end{pmatrix},$$

where $c_1 > c_2 > \cdots > c_n \geq 0$ and $d_1 > d_2 > \cdots > d_n > 0$, the number being partitioned is $c_1 + c_2 + \cdots + c_n + d_1 + d_2 + \cdots + d_n$, and n is the length of the main diagonal.

5.2. OTHER PROOFS AND RELATED RESULTS

One can also complete the proof of Jacobi's triple product identity with the aid of these modified symbols. It is most convenient to apply the argument to the product

$$(5.2.2) \qquad (-z;q)_\infty \left(-\tfrac{q}{z};q\right)_\infty = \sum_{n=-\infty}^{\infty} b_n(q)\, z^n,$$

where we need to find the coefficients $b_n(q)$. By the same kind of argument as in section 5.1 we have (exercise)

$$(5.2.3) \qquad b_n(q) = q^{\frac{n(n-1)}{2}} b_0(q) \quad \text{for all } n.$$

In other words, (5.2.2) becomes

$$(5.2.4) \qquad (-z;q)_\infty \left(-\tfrac{q}{z};q\right)_\infty = b_0(q) \sum_{n=-\infty}^{\infty} q^{\frac{n(n-1)}{2}} z^n,$$

and we still have to find $b_0(q)$, which we took outside the sum since it does not depend on n.

Since $b_0(q)$ is the coefficient of z^0 in the product on the left side of (5.2.4), we get a contribution to it from any combination of factors of $(-z;q)_\infty$ and the same number of factors of $\left(-\tfrac{q}{z};q\right)_\infty$. Of course there are infinitely many such combinations, so adding them all up seems a daunting task, but let's look at a typical combination, say

$$(1+z)(1+zq^3)(1+zq^{11})(1+zq^{37})\left(1+\tfrac{q}{z}\right)\left(1+\tfrac{q^4}{z}\right)\left(1+\tfrac{q^6}{z}\right)\left(1+\tfrac{q^{19}}{z}\right).$$

If we multiply together the z terms from just these factors, taking the 1's from all the other factors, we get a term

$$z^{1+1+1+1-1-1-1-1} q^{0+3+11+37+1+4+6+19} = q^{81},$$

which is one of the many constituents of $c_0(q)$. We could encode this term as

$$\begin{pmatrix} 37 & 11 & 3 & 0 \\ 19 & 6 & 4 & 1 \end{pmatrix}$$

which is a modified Frobenius symbol. In other words, there is a 1-1 correspondence between the constituents of the constant term $b_0(q)$ and the modified Frobenius symbols. To put it another way, $b_0(q)$ is the generating function for all modified Frobenius symbols. But these symbols are in 1-1 correspondence with partitions (for example,

$$\begin{pmatrix} 37 & 11 & 3 & 0 \\ 19 & 6 & 4 & 1 \end{pmatrix}$$

corresponds to $38 + 13 + 6 + 4 + 3 + 3 + 2 +$ twelve 1's), so $b_0(q)$ is the generating function for partitions. Therefore

$$b_0(q) = \frac{1}{(q;q)_\infty}.$$

Putting this in (5.2.4) we get (5.1.18).

For several reasons, it is interesting to rewrite (5.1.18) as a sum from $n=0$ to infinity. It is convenient to start by replacing n by $-n$. Since $-n(-n-1) = n(n+1)$, and since $(-1)^{-n} = (-1)^n$, this gives

$$(x;q)_\infty \left(\frac{q}{x};q\right)_\infty (q;q)_\infty = \sum_{n=-\infty}^{\infty} (-1)^n q^{\frac{n(n+1)}{2}} x^{-n}.$$

Next, split off the terms of the series with a negative n:

$$\sum_{n=-\infty}^{\infty} (-1)^n q^{\frac{n(n+1)}{2}} x^{-n} = \sum_{n=0}^{\infty} (-1)^n q^{\frac{n(n+1)}{2}} x^{-n} + \sum_{n=-1}^{-\infty} (-1)^n q^{\frac{n(n+1)}{2}} x^{-n}$$

$$= \sum_{n=0}^{\infty} (-1)^n q^{\frac{n(n+1)}{2}} x^{-n} + \sum_{k=1}^{\infty} (-1)^k q^{\frac{k(k-1)}{2}} x^k,$$

where in the last step we replaced n by $-k$ and used $-k(-k+1) = k(k-1)$. Finally, we reindex the last sum above by letting $k-1 = n$. This gives

$$\sum_{n=-\infty}^{\infty} (-1)^n q^{\frac{n(n+1)}{2}} x^{-n} = \sum_{n=0}^{\infty} (-1)^n q^{\frac{n(n+1)}{2}} x^{-n} + \sum_{n=0}^{\infty} (-1)^{n+1} q^{\frac{n(n+1)}{2}} x^{n+1}$$

$$= \sum_{n=0}^{\infty} (-1)^n q^{\binom{n+1}{2}} x^{-n} \left(1 - x^{2n+1}\right),$$

where we can use $\binom{n+1}{2}$ for $\frac{n(n+1)}{2}$ now since n is nonnegative. Hence

(5.2.5) $$(x;q)_\infty \left(\frac{q}{x};q\right)_\infty (q;q)_\infty = \sum_{n=0}^{\infty} (-1)^n q^{\binom{n+1}{2}} x^{-n} \left(1 - x^{2n+1}\right).$$

One advantage this has over (5.1.18) is that the right side is obviously zero when $x = 1$, as it must be since $(1;q)_\infty$ has a factor of $1 - 1$. If we set $x = -1$ in (5.2.5) and divide by 2 (noting that $1 + 1$ is the first factor of $(-1;q)_\infty$), we get

(5.2.6) $$\sum_{n=0}^{\infty} q^{\binom{n+1}{2}} = (-q;q)_\infty^2 (q;q)_\infty.$$

This is essentially due to Gauss, and we will come back to it. Perhaps the most interesting thing to do with (5.2.5) is to divide by $1-x$, using the fact that $(x;q)_\infty = (1-x)(xq;q)_\infty$. This gives
(5.2.7)
$$(xq;q)_\infty \left(\frac{q}{x};q\right)_\infty (q;q)_\infty = \sum_{n=0}^{\infty} (-1)^n q^{\binom{n+1}{2}} x^{-n} \frac{1 - x^{2n+1}}{1-x}$$

$$= \sum_{n=0}^{\infty} (-1)^n q^{\binom{n+1}{2}} x^{-n} \left(1 + x + x^2 + \cdots + x^{2n}\right)$$

$$= \sum_{n=0}^{\infty} (-1)^n q^{\binom{n+1}{2}} \left(x^{-n} + \cdots + x^{-1} + 1 + x + \cdots + x^n\right).$$

A beautiful identity of Jacobi now follows on setting $x = 1$.

THEOREM 45 (Jacobi's cube identity). *If $|q| < 1$, then*

(5.2.8) $$(q;q)_\infty^3 = \sum_{n=0}^{\infty} (-1)^n (2n+1) q^{\binom{n+1}{2}}.$$

It also appears in Gauss's *Nachlass*. If we use Euler's pentagonal number theorem (5.1.15) on the left side, we can rewrite (5.2.8) (again following Jacobi) as

(5.2.9) $\left(1 - q - q^2 + q^5 + q^7 - q^{12} - q^{15} + q^{22} + q^{26} - q^{35} - q^{40} + q^{51} + q^{57} \ldots\right)^3$
$= 1 - 3q + 5q^3 - 7q^6 + 9q^{10} - 11q^{15} + 13q^{21} - 15q^{28} + 17q^{36} \ldots.$

Exercises

1. Find the Frobenius symbol of $9 + 8 + 5 + 4 + 4 + 2$.
2. Find the Frobenius symbol of $9 + 8 + 6 + 6 + 5 + 2 + 2 + 1$.
3. Find the Frobenius symbol of $9 + 8 + 6 + 6 + 6 + 5 + 2 + 2 + 1$. Any comments?
4. For a self-conjugate partition, what can you say about the Ferrers diagram? About the Frobenius symbol?
5. Explain why we have $a_1 > a_2 > \cdots > a_n$ and $b_1 > b_2 > \cdots > b_n$ in a Frobenius symbol. Can you make a stronger statement if you know the partition has distinct parts?
6. This problem outlines a derivation of (5.2.3).

    (i) Show that replacing z by zq in (5.2.2) leads to
    $$b_n(q) = q^{n-1} b_{n-1}(q).$$

    (ii) Show that iterating (i) gives
    $$b_n(q) = q^{\frac{n(n-1)}{2}} b_0(q) \quad \text{if } n \geq 0.$$

    (iii) If n is not positive, we can set $n = 1 - m$ in (i) to get
    $$b_{1-m}(q) = q^{-m} b_{-m}(q) \quad \text{or} \quad b_{-m}(q) = q^m b_{1-m}(q),$$
    where m is a positive integer. Show that iterating this gives
    $$b_{-m}(q) = q^{\frac{m(m+1)}{2}} b_0(q) \quad \text{if } m > 0,$$
    or
    $$b_n(q) = q^{\frac{n(n-1)}{2}} b_0(q) \quad \text{if } n < 0.$$

7. Show that

(5.2.10) $$(-zq;q^2)_\infty \left(-\frac{q}{z};q^2\right)_\infty (q^2;q^2)_\infty = \sum_{n=0}^{\infty} q^{n^2} z^{-n} \left(1 + (zq)^{2n+1}\right).$$

This is (5.1.14) put in a form like (5.2.5).

8. The triple product affords an easy proof of an identity that Gauss obtained differently:

(5.2.11) $$\frac{(q;q)_\infty}{(-q;q)_\infty} = 1 - 2q + 2q^4 - 2q^9 + 2q^{16} - + \ldots.$$

(i) Explain why the right side is $\sum_{n=-\infty}^{\infty} (-1)^n q^{n^2}$.

(ii) Use the triple product to sum the series in (i).

(iii) Explain why your answer to (ii) can be rewritten as

$$(q;q^2)_\infty (q;q)_\infty \quad \text{or as} \quad \frac{(q;q)_\infty}{(-q;q)_\infty}.$$

9. Gauss observed a simple but pretty corollary of (5.2.11). If
$$F(q) = 1 - 2q + 2q^4 - 2q^9 + 2q^{16} - 2q^{25} + - \ldots,$$
then (5.2.11) tells us that
$$F(q) = \frac{(q;q)_\infty}{(-q;q)_\infty} = \frac{(1-q)(1-q^2)(1-q^3)(1-q^4)\cdots}{(1+q)(1+q^2)(1+q^3)(1+q^4)\cdots}.$$
Show that $F(q)F(-q) = [F(q^2)]^2$.

10. Show that (5.2.6) can be rewritten as

(5.2.12) $$\sum_{n=0}^{\infty} q^{\binom{n+1}{2}} = \frac{(q^2;q^2)_\infty}{(q;q^2)_\infty}.$$

This is the form in which Gauss evaluated this sum, except that because of his preference for working with $(q;q)_\infty$, which he denoted as $[q]$, he would have written

$$\sum_{n=0}^{\infty} q^{\binom{n+1}{2}} = \frac{[q^2]^2}{[q]}.$$

Show that this is equivalent. (He would actually have written $[qq]^2$ instead of $[q^2]^2$, and he used x instead of q.)

11. Following Gauss, show that

$$\sum_{n=-\infty}^{\infty} (-1)^n q^{(3n-1)^2} = q - q^4 - q^{16} + q^{25} + q^{49} \ldots$$
$$= q(q^3;q^{18})_\infty (q^{15};q^{18})_\infty (q^{18};q^{18})_\infty$$
$$= q^{-\frac{1}{8}} (q^3;q^6)_\infty \sum_{n=0}^{\infty} q^{\frac{9}{8}(2n+1)^2}.$$

12. Gauss claimed that

(5.2.13) $$1 - 2q^9 + 2q^{36} - 2q^{81} + - \cdots = (1-q^3)(1-q^9)(1-q^{15})\cdots$$
$$\times q^{-\frac{1}{8}} \left[q^{\frac{1}{8}} + q^{\frac{25}{8}} + q^{\frac{49}{8}} + \ldots \right],$$

where I have written the equation almost exactly as he did. In the first place, this is an excellent illustration of the limitations of the notation ...: one might guess that

$$q^{\frac{1}{8}} + q^{\frac{25}{8}} + q^{\frac{49}{8}} + \ldots \quad \text{means} \quad \sum_{n=0}^{\infty} q^{\frac{1}{8}(24n+1)}.$$

(i) What does this series converge to?

(ii) In fact, Gauss means by $1, 25, 49, \ldots$ the odd squares that are not multiples of 3. Show that with this interpretation,

$$q^{\frac{1}{8}} + q^{\frac{25}{8}} + q^{\frac{49}{8}} + \cdots = \sum_{n=-\infty}^{\infty} q^{\frac{1}{8}(6n+1)^2},$$

and evaluate this sum using Jacobi's triple product.

(iii) What is the series on the left side of (5.2.13)? Evaluate it by Jacobi's triple product, and prove (5.2.13).

13. Just as MacMahon's q-binomial theorem is a finite form of the Jacobi triple product, Hirschhorn's identity (2.7.9) is a finite form of Jacobi's cube identity. Show that letting $n \to \infty$ in

$$(q;q)_\infty^2 = \sum_{k=0}^{n} (-1)^k (2k+1) q^{\binom{k+1}{2}} \binom{2n+1}{n-k}_q$$

gives (5.2.8). (Again, this technically requires Tannery's theorem.)

14. Weierstrass's first paper was clearly influenced by Jacobi. In it he sets

$$\phi(z) = (qz^2; q^2)_\infty \quad \text{and} \quad F(z) = \frac{\phi(z)\,\phi\left(\frac{1}{z}\right)}{\phi(1)\,\phi(1)}.$$

He expands first F and then ϕ in powers of z, but the other order makes more sense.

(i) Explain why $\phi(z) = (1 - qz^2)\phi(zq)$.

(ii) Because ϕ is a function of z^2, it has an expansion of the form

$$\phi(z) = \sum_{n=0}^{\infty} B_n(q) z^{2n}.$$

Use (i) to show that

$$B_n(q) = \frac{-q^{2n-1}}{1 - q^{2n}} B_{n-1}(q) \quad \text{for } n \geq 1.$$

(iii) Use (ii) to show that

$$\phi(z) = \sum_{n=0}^{\infty} \frac{(-1)^n q^{n^2}}{(q^2; q^2)_n} z^{2n}.$$

(iv) Similarly, F has an expansion of the form

$$F(z) = \sum_{n=-\infty}^{\infty} A_n(q) z^{2n}.$$

Explain why $A_n(q) = A_{-n}(q)$.

(v) Explain why $F(z) = -qz^2 F(zq)$.

(vi) Use (v) to show that $A_n(q) = -q^{2n-1} A_{n-1}(q)$ for all n.

(vii) Use (vi) to show that $A_n(q) = (-1)^n q^{n^2} A_0(q)$ for all n.

(viii) Weierstrass does not find $A_0(q)$. Explain why he could at least have said that
$$A_0(q) = \frac{1}{\sum_{n=-\infty}^{\infty} (-1)^n q^{n^2}}.$$

(ix) Find $A_0(q)$ and write down the final form of the expansion of $F(z)$.

15. Gauss's original proof of (5.2.11) is ingenious but rather tricky, and since it appears in his *Nachlass*, it has to be fleshed out a little. Assume throughout this problem that $|q| < 1$, as usual, and assume that m is nonnegative.

    (i) Gauss starts with the series
$$P(m) = 1 + \frac{q^m}{1+q^m} \frac{1-q^{2m+1}}{1+q^{m+1}} + \frac{q^{2m}}{1+q^m} \frac{1-q^{2m+2}}{1+q^{m+1}} \frac{1-q^{m+1}}{1+q^{m+2}}$$
$$+ \frac{q^{3m}}{1+q^m} \frac{1-q^{2m+3}}{1+q^{m+1}} \frac{1-q^{m+1}}{1+q^{m+2}} \frac{1-q^{m+2}}{1+q^{m+3}} + \cdots$$
$$= 1 + \sum_{k=1}^{\infty} q^{km}(1-q^{2m+k}) \frac{(q^{m+1};q)_{k-1}}{(-q^m;q)_{k+1}}$$

and
$$Q(m) = \frac{q^m}{1+q^m} + \frac{q^{2m}}{1+q^m} \frac{1-q^{m+1}}{1+q^{m+1}} + \frac{q^{3m}}{1+q^m} \frac{(1-q^{m+1})(1-q^{m+2})}{(1+q^{m+1})(1+q^{m+2})} + \cdots$$
$$= \sum_{k=1}^{\infty} q^{km} \frac{(q^{m+1};q)_{k-1}}{(-q^m;q)_k}.$$

Show that these converge for any $m > 0$, but not for $m = 0$.

(ii) Because of the divergence when $m = 0$, we must work with partial sums. We define $P_n(m)$ and $Q_n(m)$ to be the sums of the first $n+1$ terms of the series for $P(m)$ and $Q(m)$ respectively; that is,
$$P_n(m) = 1 + \sum_{k=1}^{n} q^{km}(1-q^{2m+k}) \frac{(q^{m+1};q)_{k-1}}{(-q^m;q)_{k+1}}$$

and
$$Q_n(m) = \sum_{k=1}^{n+1} q^{km} \frac{(q^{m+1};q)_{k-1}}{(-q^m;q)_k}.$$

Show that $Q_n(m)$ can be rewritten as

(a) $$Q_n(m) = \frac{q^m}{1+q^m} + \sum_{k=1}^{n} q^{(k+1)m} \frac{(q^{m+1};q)_k}{(-q^m;q)_{k+1}}$$

and as

(b) $$Q_n(m) = \sum_{k=1}^{n} q^{km} \frac{(q^{m+1};q)_{k-1}}{(-q^m;q)_k} + q^{m(n+1)} \frac{(q^{m+1};q)_n}{(-q^n;q)_{n+1}}.$$

(iii) Define $\phi_n(m) = P_n(m) - Q_n(m)$. Using (a), show that
$$\phi_n(m) = \sum_{k=0}^{n} q^{km} \frac{(q^m;q)_k}{(-q^m;q)_{k+1}}.$$

(iv) Using (b), show that
$$\phi_n(m) = 1 - q^{m(n+1)} \frac{(q^{m+1};q)_n}{(-q^m;q)_{n+1}} - \sum_{k=1}^{n} q^{(k+1)(m+1)-1} \frac{(q^{m+1};q)_{k-1}}{(-q^{m+1};q)_k}.$$

(v) Use (iii) and (iv) to show that
$$\sum_{k=0}^{n} q^{km} \frac{(q^m;q)_k}{(-q^m;q)_{k+1}} = 1 - q^{m(n+1)} \frac{(q^{m+1};q)_n}{(-q^m;q)_{n+1}}$$
$$- q^{2m+1} \sum_{k=0}^{n-1} q^{k(m+1)} \frac{(q^{m+1};q)_k}{(-q^{m+1};q)_{k+1}}.$$

(vi) Assuming that $m > 0$, show that letting $n \to \infty$ in (v) gives
$$\sum_{k=0}^{\infty} q^{km} \frac{(q^m;q)_k}{(-q^m;q)_{k+1}} = 1 - q^{2m+1} \sum_{k=0}^{\infty} q^{k(m+1)} \frac{(q^{m+1};q)_k}{(-q^{m+1};q)_{k+1}}.$$

Denoting the left side of this by $\phi_m(q)$, explain why we have
$$\phi_m(q) = 1 - q^{2m+1} \phi_{m+1}(q)$$
for $m > 0$.

(vii) Explain why (vi) implies that
$$\phi_m(q) = \sum_{j=0}^{k-1} (-1)^j q^{j(2m+j)} + (-1)^k q^{k(2m+k)} \phi_{m+k}(q)$$
for any $m > 0$ and any nonnegative integer k, and why this implies that
$$\phi_m(q) = \sum_{j=0}^{\infty} (-1)^j q^{j(2m+j)}$$
for $m > 0$.

(viii) If $m = 0$, the result of (vi) is false (as Gauss pointed out), but we can still let $n \to \infty$ in (v). Show that this gives
$$\frac{1}{2} = 1 - \frac{1}{2} \frac{(q;q)_\infty}{(-q;q)_\infty} - q\phi_1(q).$$

(ix) Use (vii) and (viii) to show that
$$\frac{(q;q)_\infty}{(-q;q)_\infty} = 1 - 2q + 2q^4 - 2q^9 + - \cdots = \sum_{j=-\infty}^{\infty} (-1)^j q^{j^2}.$$

This completes Gauss's original proof of (5.2.11). (This problem is dedicated to my father, Robert Shepard Johnson, who helped me translate the relevant passage in Gauss shortly before he died.)

16. Gauss's original proof of (5.2.12), the identity of problem 10, is very much like the one in problem 15. Again assume that $|q| < 1$ and that n is nonnegative.

(i) Gauss starts with the series
$$P(n) = \sum_{k=1}^{\infty} q^{(k-1)n}(1 - q^{2n+2k}) \frac{(q^{n+2};q^2)_{k-1}}{(q^{n+1};q^2)_k}$$
and
$$Q(n) = \sum_{k=1}^{\infty} q^{kn} \frac{(q^{n+2};q^2)_k}{(q^{n+1};q^2)_k}.$$
Show that these converge for any $n > 0$, but not for $n = 0$.

(ii) Define $P_m(n)$ and $Q_m(n)$ to be the sums of the first m terms of the series for $P(n)$ and $Q(n)$ respectively. Show that
$$P_m(n) = \sum_{k=1}^{m} q^{(k-1)n}(1 - q^{2n+2k}) \frac{(q^{n+2};q^2)_{k-1}}{(q^{n+1};q^2)_k}$$
$$= 1 + q^{n+1} + \sum_{k=1}^{m-1} q^{kn}(1 - q^{2n+2k+2}) \frac{(q^{n+2};q^2)_k}{(q^{n+1};q^2)_{k+1}}$$
and that
$$Q_m(n) = \sum_{k=1}^{m} q^{kn} \frac{(q^{n+2};q^2)_k}{(q^{n+1};q^2)_k}$$
$$= \sum_{k=1}^{m-1} q^{kn} \frac{(q^{n+2};q^2)_k}{(q^{n+1};q^2)_k} + q^{mn} \frac{(q^{n+2};q^2)_m}{(q^{n+1};q^2)_m}.$$

(iii) Define $\psi_m(n) = P_m(n) - Q_m(n)$. Using the first expressions for $P_m(n)$ and $Q_m(n)$ in (ii), show that
$$\psi_m(n) = \sum_{k=1}^{m} q^{(k-1)n} \frac{(q^n;q^2)_k}{(q^{n+1};q^2)_k}.$$

(iv) Using the last expressions for $P_m(n)$ and $Q_m(n)$ in (ii), show that
$$\psi_m(n) = 1 + q^{n+1} - q^{mn} \frac{(q^{n+2};q^2)_m}{(q^{n+1};q^2)_m} + q^{2n+3} \sum_{k=1}^{m-1} q^{(k-1)(n+2)} \frac{(q^{n+2};q^2)_k}{(q^{n+3};q^2)_k}.$$

Hence
$$\sum_{k=1}^{m} q^{(k-1)n} \frac{(q^n;q^2)_k}{(q^{n+1};q^2)_k} = 1 + q^{n+1} - q^{mn} \frac{(q^{n+2};q^2)_m}{(q^{n+1};q^2)_m}$$
$$+ q^{2n+3} \sum_{k=1}^{m-1} q^{(k-1)(n+2)} \frac{(q^{n+2};q^2)_k}{(q^{n+3};q^2)_k}.$$

(v) Assume that $n > 0$ and let $m \to \infty$ in (iv). Defining
$$\psi_n(q) = \sum_{k=1}^{\infty} q^{(k-1)n} \frac{(q^n;q^2)_k}{(q^{n+1};q^2)_k},$$

show that we have

$$\psi_n(q) = 1 + q^{n+1} + q^{2n+3}\psi_{n+2}(q)$$

for $n > 0$.

(vi) Explain why (v) implies that

$$\psi_n(q) = \sum_{j=0}^{2k-1} q^{nj+\binom{j+1}{2}} + q^{2nk+\binom{2k+1}{2}}\psi_{n+2k}(q)$$

for any $n > 0$ and any nonnegative integer k, and why this implies that

$$\psi_n(q) = \sum_{j=0}^{\infty} q^{nj+\binom{j+1}{2}}$$

for any nonnegative integer n.

(vii) If $n = 0$, the result of (v) is false (as Gauss again realized), but we can still let $m \to \infty$ in (iv). Show that this gives

$$0 = 1 + q - \frac{(q^2; q^2)_\infty}{(q; q^2)_\infty} + q^3 \psi_2(q),$$

and that therefore

$$\frac{(q^2; q^2)_\infty}{(q; q^2)_\infty} = \sum_{m=0}^{\infty} q^{\binom{m+1}{2}}.$$

It may help to observe that if j is a nonnegative integer, then $2j + 3 = \binom{j+1}{1} + \binom{j+2}{1}$.

17. Gauss sketched a proof of (5.2.8) that is somewhat different from Jacobi's.

(i) Assuming as usual that $|q| < 1$, he starts with

$$(5.2.14) \quad \left[\left(x - \frac{1}{x}\right)q - \left(x^3 - \frac{1}{x^3}\right)q^9 + \left(x^5 - \frac{1}{x^5}\right)q^{25} - \cdots\right]$$
$$\times \left[\left(x + \frac{1}{x}\right)q + \left(x^3 + \frac{1}{x^3}\right)q^9 + \left(x^5 + \frac{1}{x^5}\right)q^{25} + \cdots\right]$$
$$= (1 - 2q^8 + 2q^{32} - \cdots)\left[\left(x^2 - \frac{1}{x^2}\right)q^2 - \left(x^6 - \frac{1}{x^6}\right)q^{18} + \cdots\right],$$

by which he means

$$\left(\sum_{n=-\infty}^{\infty} (-1)^n q^{(2n+1)^2} x^{2n+1}\right)\left(\sum_{n=-\infty}^{\infty} q^{(2n+1)^2} x^{2n+1}\right)$$
$$= \left(\sum_{n=-\infty}^{\infty} (-1)^n q^{8n^2}\right)\left(\sum_{n=-\infty}^{\infty} (-1)^n q^{2(2n+1)^2} x^{4n+2}\right).$$

Prove this by using Jacobi's triple product on all four sums.

(ii) Next he divides both sides of (5.2.14) by

$$\left(x - \frac{1}{x}\right)\left(x + \frac{1}{x}\right) = \left(x^2 - \frac{1}{x^2}\right)$$

and then lets $x \to 1$. Show that this gives

$$(q - 3q^9 + 5q^{25} - \ldots)(q + q^9 + q^{25} + \ldots)$$
$$= (1 - 2q^8 + 2q^{32} - \ldots)(q^2 - 3q^{18} + 5q^{50} - \ldots),$$

or

$$(1 - 3q + 5q^3 - 7q^6 + \ldots)(1 + q + q^3 + q^6 + \ldots)$$
$$= (1 - 2q + 2q^4 - 2q^9 + \ldots)(1 - 3q^2 + 5q^6 - 7q^{12} + \ldots)$$

after dividing by q^2 and then replacing q^8 by q. In series form this is

$$\left(\sum_{n=0}^{\infty}(-1)^n(2n+1)q^{\binom{n+1}{2}}\right)\left(\sum_{n=0}^{\infty}q^{\binom{n+1}{2}}\right)$$
$$= \left(\sum_{n=-\infty}^{\infty}(-1)^n q^{n^2}\right)\left(\sum_{n=0}^{\infty}(-1)^n(2n+1)q^{n(n+1)}\right).$$

(iii) Define

$$f(q) = \sum_{n=0}^{\infty}(-1)^n(2n+1)q^{\binom{n+1}{2}}.$$

Using (ii) and problems 8 and 10, show that

$$\frac{f(q)}{(q;q)_{\infty}^3} = \frac{f(q^2)}{(q^2;q^2)_{\infty}^3}.$$

(iv) Show that iterating the result of (iii) gives $f(q) = (q;q)_{\infty}^3$. This is (5.2.8).

18. Gauss considered the three functions

$$P(q) = \sum_{n=-\infty}^{\infty} q^{n^2} = 1 + 2q + 2q^4 + 2q^9 + 2q^{16} + 2q^{25} + \ldots,$$

$$Q(q) = \sum_{n=-\infty}^{\infty} (-1)^n q^{n^2} = 1 - 2q + 2q^4 - 2q^9 + 2q^{16} - 2q^{25} + - \ldots,$$

$$R(q) = \sum_{n=-\infty}^{\infty} q^{(n+\frac{1}{2})^2} = q^{\frac{1}{4}} \sum_{n=-\infty}^{\infty} q^{n^2+n}$$
$$= 2q^{\frac{1}{4}} \sum_{n=0}^{\infty} q^{n^2+n} = 2q^{\frac{1}{4}}\left(1 + q^2 + q^6 + q^{12} + q^{20} + \ldots\right).$$

Show that they are related by

$$P(q^4) + R(q^4) = P(q) \quad \text{and} \quad P(q^4) - R(q^4) = Q(q).$$

19. One of the most beautiful identities from Gauss's *Nachlass* is

(5.2.15) $\quad \left(1 + 2q + 2q^4 + 2q^9 + \ldots\right)^4$
$$= \left(1 - 2q + 2q^4 - 2q^9 + \ldots\right)^4 + 16q\left(1 + q^2 + q^6 + q^{12} + \ldots\right)^4.$$

This problem sketches a proof hinted at in the *Nachlass*. See also the next problem.

(i) Show that (5.2.15) is equivalent to
$$(P(q))^4 = (Q(q))^4 + (R(q))^4$$
with P, Q, R as in the previous problem.

(ii) Use the triple product to show that
$$P(q) = (-q; q^2)_\infty^2 (q^2; q^2)_\infty,$$
$$Q(q) = (q; q^2)_\infty^2 (q^2; q^2)_\infty = (q; q)_\infty (q; q^2)_\infty,$$
$$R(q) = 2q^{\frac{1}{4}}(-q^2; q^2)_\infty^2 (q^2; q^2)_\infty = 2q^{\frac{1}{4}}(-q^2; q^2)_\infty (q^4; q^4)_\infty.$$

(iii) Show that
$$P(q)Q(q) = \left(Q(q^2)\right)^2.$$

(iv) From (iii) and the previous problem, deduce that
$$\left[P(q^4)\right]^2 - \left[R(q^4)\right]^2 = \left(Q(q^2)\right)^2.$$

(v) Show that
$$P(q^4) + iR(q^4) = P(iq) \quad \text{and} \quad P(q^4) - iR(q^4) = Q(iq),$$
and deduce that
$$\left[P(q^4)\right]^2 + \left[R(q^4)\right]^2 = P(iq)Q(iq) = \left(Q(-q^2)\right)^2 = \left(P(q^2)\right)^2.$$

(vi) Finally, deduce that
$$\left(P(q^4)\right)^4 - \left(R(q^4)\right)^4 = \left(Q(q^4)\right)^4,$$
which becomes (5.2.15) if we replace q^4 by q.

20. In a different place in the *Nachlass*, Gauss gives more details of a variation on the proof of (5.2.15) outlined in the preceding problem. It proceeds in the same way through the first three parts, so we'll start the numeration there.

(iii) In addition to
$$P(q)Q(q) = \left(Q(q^2)\right)^2, \quad \text{show also that} \quad 2P(q^2)R(q^2) = (R(q))^2.$$

(iv) Show that
$$P(q) + Q(q) = 2P(q^4), \quad \text{and} \quad P(q) - Q(q) = 2R(q^4),$$
and deduce that
$$(P(q))^2 - (Q(q))^2 = 2\left(R(q^2)\right)^2.$$

(v) Show that
$$(1+i)Q(iq) = P(q) + iQ(q)$$
$$(1-i)P(iq) = P(q) - iQ(q).$$

Hint: Separate the series for $P(q)$ and $Q(q)$ into odd and even exponents.

(vi) Deduce that
$$(P(q))^2 + (Q(q))^2 = 2\left(Q(-q^2)\right)^2 = 2\left(P(q^2)\right)^2.$$

(vii) Finally, deduce that
$$(P(q))^4 - (Q(q))^4 = (R(q))^4.$$

5.3. The quintuple product identity

We recall the Chen–Chu–Gu identity from section 2.10:

THEOREM 46 (The Chen–Chu–Gu identity). *For any nonnegative integers m and n, for all q, and for all $z \neq 0$, we have*

$$\frac{1}{\left(-\frac{q}{z};q\right)_m (-z;q)_{n+1}} = \sum_{k=-m}^{n} \binom{m+n}{m+k}_q \frac{q^{\frac{k(3k-1)}{2}} z^{3k}(1-zq^k)}{\left(\frac{q}{z^2};q\right)_{m-k} (z^2;q)_{n+k+1}}.$$

The reason why we worked so hard to get this is that it is a finite form of the quintuple product identity. If we assume $|q| < 1$ and let m and n go to infinity here, then

$$\binom{m+n}{m+k}_q = \frac{(q;q)_{m+n}}{(q;q)_{m+k}(q;q)_{n-k}} \to \frac{(q;q)_\infty}{(q;q)_\infty (q;q)_\infty} = \frac{1}{(q;q)_\infty}$$

as we have seen before, and we have (formally)

$$\frac{1}{\left(-\frac{q}{z};q\right)_\infty (-z;q)_\infty} = \sum_{k=-\infty}^{\infty} \frac{q^{\frac{k(3k-1)}{2}} z^{3k}(1-zq^k)}{(q;q)_\infty \left(\frac{q}{z^2};q\right)_\infty (z^2;q)_\infty},$$

or

$$\sum_{k=-\infty}^{\infty} q^{\frac{k(3k-1)}{2}} z^{3k}(1-zq^k) = \frac{(q;q)_\infty \left(\frac{q}{z^2};q\right)_\infty (z^2;q)_\infty}{\left(-\frac{q}{z};q\right)_\infty (-z;q)_\infty}. \tag{5.3.1}$$

As in other arguments of this type, one technically needs Tannery's theorem to justify this. We do not need any restriction on z, other than $z \neq 0$, because the series converges for all other z due to the quadratic exponent of q.

At the moment, (5.3.1) has five products, but one might prefer a "quintuple product" to have them all in the numerator. To this end we have

$$\begin{aligned}
\frac{(z^2;q)_\infty}{(-z;q)_\infty} &= \frac{(1-z^2)(1-z^2q)(1-z^2q^2)(1-z^2q^3)(1-z^2q^4)(1-z^2q^5)\cdots}{(1+z)(1+zq)(1+zq^2)(1+zq^3)\cdots} \\
&= \frac{(1-z^2)(1-z^2q^2)(1-z^2q^4)(1-z^2q^6)\cdots}{(1+z)(1+zq)(1+zq^2)(1+zq^3)\cdots} \\
&\quad \times (1-z^2q)(1-z^2q^3)(1-z^2q^5)(1-z^2q^7)\cdots \\
&= (1-z)(1-zq)(1-zq^2)\cdots \times (1-z^2q)(1-z^2q^3)(1-z^2q^5)\cdots \\
&= (z;q)_\infty (z^2q;q^2)_\infty
\end{aligned}$$

5.3. THE QUINTUPLE PRODUCT IDENTITY

and similarly

$$\frac{\left(\frac{q}{z^2};q\right)_\infty}{\left(-\frac{q}{z};q\right)_\infty} = \frac{\left(1-\frac{q}{z^2}\right)\left(1-\frac{q^2}{z^2}\right)\left(1-\frac{q^3}{z^2}\right)\left(1-\frac{q^4}{z^2}\right)\left(1-\frac{q^5}{z^2}\right)\left(1-\frac{q^6}{z^2}\right)\cdots}{\left(1+\frac{q}{z}\right)\left(1+\frac{q^2}{z}\right)\left(1+\frac{q^3}{z}\right)\left(1+\frac{q^4}{z}\right)\cdots}$$

$$= \frac{\left(1-\frac{q^2}{z^2}\right)\left(1-\frac{q^4}{z^2}\right)\left(1-\frac{q^6}{z^2}\right)\cdots}{\left(1+\frac{q}{z}\right)\left(1+\frac{q^2}{z}\right)\left(1+\frac{q^3}{z}\right)\cdots} \times \left(1-\frac{q}{z^2}\right)\left(1-\frac{q^3}{z^2}\right)\left(1-\frac{q^5}{z^2}\right)\cdots$$

$$= \left(1-\frac{q}{z}\right)\left(1-\frac{q^2}{z}\right)\left(1-\frac{q^3}{z}\right)\cdots \times \left(1-\frac{q}{z^2}\right)\left(1-\frac{q^3}{z^2}\right)\left(1-\frac{q^5}{z^2}\right)\cdots$$

$$= \left(\frac{q}{z};q\right)_\infty \left(\frac{q}{z^2};q^2\right)_\infty.$$

Therefore (5.3.1) is equivalent to

THEOREM 47 (The quintuple product identity). *If* $|q| < 1$ *and* $z \neq 0$, *then*

$$(5.3.2) \quad \sum_{k=-\infty}^{\infty} q^{\frac{k(3k-1)}{2}} z^{3k} (1-zq^k) = (q;q)_\infty (z;q)_\infty \left(\frac{q}{z};q\right)_\infty (z^2q;q^2)_\infty \left(\frac{q}{z^2};q^2\right)_\infty.$$

Note that the product side is the triple product with two more groups of factors put on. One can also give a functional equation proof of the quintuple product. Let

$$(5.3.3) \quad f(z) = (z;q)_\infty \left(\frac{q}{z};q\right)_\infty (z^2q;q^2)_\infty \left(\frac{q}{z^2};q^2\right)_\infty,$$

which has all the factors of the quintuple product that include z. Because we have infinitely many positive and negative powers of z in this product, it should have an expansion

$$(5.3.4) \quad f(z) = \sum_{n=-\infty}^{\infty} c_n z^n$$

for some coefficients c_n, which we will try to find. Look at

$$\frac{f(z)}{f(zq)} = \frac{(z;q)_\infty \left(\frac{q}{z};q\right)_\infty (z^2q;q^2)_\infty \left(\frac{q}{z^2};q^2\right)_\infty}{(zq;q)_\infty \left(\frac{1}{z};q\right)_\infty (z^2q^3;q^2)_\infty \left(\frac{1}{qz^2};q^2\right)_\infty}$$

$$= \frac{(z;q)_\infty}{(zq;q)_\infty} \times \frac{\left(\frac{q}{z};q\right)_\infty}{\left(\frac{1}{z};q\right)_\infty} \times \frac{(z^2q;q^2)_\infty}{(z^2q^3;q^2)_\infty} \times \frac{\left(\frac{q}{z^2};q^2\right)_\infty}{\left(\frac{1}{qz^2};q^2\right)_\infty}$$

$$= (1-z) \times \frac{1}{1-\frac{1}{z}} \times (1-z^2q) \times \frac{1}{1-\frac{1}{qz^2}}$$

$$= (1-z) \times \frac{-z}{1-z} \times (1-z^2q) \times \frac{-qz^2}{1-qz^2} = qz^3.$$

So $f(z) = qz^3 f(zq)$, which means that

$$\sum_{n=-\infty}^{\infty} c_n z^n = \sum_{m=-\infty}^{\infty} c_m (zq)^m (qz^3) = \sum_{m=-\infty}^{\infty} c_m q^{m+1} z^{m+3} = \sum_{n=-\infty}^{\infty} c_{n-3} q^{n-2} z^n$$

after changing the summation index m to $n-3$ to line up the powers of z. Hence $c_n = q^{n-2} c_{n-3}$. If we use this to work our way from a generic n towards zero, we will ultimately land either on $n = 1$ or on $n = 0$ or on $n = -1$, depending on where we start. Let's make another observation before we try to force these coefficients out. Consider

$$\frac{f(z)}{f\left(\frac{1}{z}\right)} = \frac{(z;q)_\infty \left(\frac{q}{z};q\right)_\infty (z^2 q;q^2)_\infty \left(\frac{q}{z^2};q^2\right)_\infty}{\left(\frac{1}{z};q\right)_\infty (zq;q)_\infty \left(\frac{q}{z^2};q^2\right)_\infty (z^2 q;q^2)_\infty}$$

$$= \frac{(z;q)_\infty}{(zq;q)_\infty} \times \frac{\left(\frac{q}{z};q\right)_\infty}{\left(\frac{1}{z};q\right)_\infty} = \frac{1-z}{1-\frac{1}{z}} = -z.$$

So $f(z) = -z\, f\left(\frac{1}{z}\right)$, which means that

$$\sum_{n=-\infty}^{\infty} c_n z^n = \sum_{m=-\infty}^{\infty} c_m z^{-m}(-z) = \sum_{m=-\infty}^{\infty} -c_m z^{1-m} = \sum_{n=-\infty}^{\infty} -c_{1-n} z^n$$

after changing the summation index from m to $1-n$ to line up the powers of z. Hence $c_n = -c_{1-n}$. In particular, taking $n = 2$ here we find that $c_2 = -c_{-1}$. But if we take $n = 2$ in our other relation $c_n = q^{n-2} c_{n-3}$, we find that $c_2 = c_{-1}$. Hence $c_{-1} = -c_{-1}$, so $c_{-1} = 0$. This means that $c_2, c_5, c_8, c_{11}, \ldots$ are all zero, as are $c_{-4}, c_{-7}, c_{-10}, \ldots$, since by repeated use of our two relations these would become c_{-1} times some power of q.

Let n be a multiple of 3, say $n = 3k$, and assume for the moment that k is nonnegative. Then we have $c_{3k} = q^{3k-2} c_{3k-3}$, and using this on itself repeatedly we get

(5.3.5)
$$c_{3k} = q^{3k-2} c_{3k-3} = c_{3k} = q^{3k-2}\left(q^{3k-5} c_{3k-6}\right) = q^{(3k-2)+(3k-5)}\left(q^{3k-8} c_{3k-9}\right) = \cdots$$
(5.3.6)
$$= q^{(3k-2)+(3k-5)+(3k-8)+\cdots+4+1} c_0 = c_0 q^{\frac{k(3k-1)}{2}},$$

at least for $k \geq 0$. Similarly, if $n \equiv 1 \pmod 3$, say $n = 3k+1$, then $c_{3k+1} = q^{3k-1} c_{3k-2}$, and using this repeatedly we get

(5.3.7)
$$c_{3k+1} = c_1 q^{\frac{k(3k+1)}{2}}$$

at least for $k \geq 0$. Note that $c_n = -c_{1-n}$ implies that $c_1 = -c_0$. Also, taking $n = 3k+1$ in this relation we have

$$c_1 q^{\frac{k(3k+1)}{2}} = c_{3k+1} = -c_{-3k}, \quad \text{or} \quad c_{-3k} = c_0 q^{\frac{k(3k+1)}{2}}$$

for $k \geq 0$. Changing k to $-k$, this says

$$c_{3k} = c_0 q^{\frac{-k(-3k+1)}{2}} = c_0 q^{\frac{k(3k-1)}{2}}$$

for $k \leq 0$, so (5.3.5) holds for all integers k, positive or not. Changing $-k$ back to k in (5.3.5) and invoking $c_n = -c_{1-n}$ again, we then have that

$$-c_0 q^{\frac{k(3k+1)}{2}} = -c_{-3k} = c_{3k+1}$$

for all integers k, or equivalently that (5.3.7) holds for all integers k.

If we split (5.3.4) into three series

$$f(z) = \sum_{k=-\infty}^{\infty} c_{3k-1} z^{3k-1} + \sum_{k=-\infty}^{\infty} c_{3k} z^{3k} + \sum_{k=-\infty}^{\infty} c_{3k+1} z^{3k+1},$$

the coefficients of the first series are all zero, and the other two combine into
(5.3.8)
$$f(z) = \sum_{k=-\infty}^{\infty} c_0 \left(q^{\frac{k(3k-1)}{2}} z^{3k} - q^{\frac{k(3k+1)}{2}} z^{3k+1} \right) = c_0 \sum_{k=-\infty}^{\infty} q^{\frac{k(3k-1)}{2}} z^{3k} \left(1 - zq^k\right).$$

We just have to find c_0 now, which can come out of the series since it does not depend on k. There are several proofs of the quintuple product in the literature that are more or less the same as this one up to this point. To find c_0 we follow Bailey: set $z = -1$, so that (5.3.8) becomes

$$f(-1) = c_0 \sum_{k=-\infty}^{\infty} (-1)^k q^{\frac{k(3k-1)}{2}} \left(1 + q^k\right) = c_0 \sum_{k=-\infty}^{\infty} (-1)^k \left(q^{\frac{k(3k-1)}{2}} + q^{\frac{k(3k+1)}{2}} \right).$$

If we split this into two sums and change k to $-k$ in one of them, we see that they are the same. In fact, this is two copies of Euler's pentagonal number theorem, so we have $f(-1) = 2c_0(q;q)_\infty$. Alternatively, we can do both sums with Jacobi's triple product, as in section 5.1.

On the other hand, according to (5.3.3) we have

$$f(-1) = (-1;q)_\infty (-q;q)_\infty (q;q^2)_\infty^2 \left(\frac{q}{z^2};q^2\right)_\infty,$$

and the very first factor is $1 + 1 = 2$, so

$$f(-1) = 2\left((-q;q)_\infty (q;q^2)_\infty\right)^2.$$

But Euler's "odd equals distinct" theorem says that

$$\frac{1}{(q;q^2)_\infty} = (-q;q)_\infty,$$

so in fact $f(-1) = 2$. Therefore $2 = 2c_0(q;q)_\infty$, so $c_0 = 1/(q;q)_\infty$. Putting this into (5.3.8) and comparing with (5.3.3), we have rederived the quintuple product identity.

There are many equivalent forms of the quintuple product identity. We content ourselves with one more that will be useful later. Replacing q by q^6 in (5.3.2) and splitting the series (as we may, since both pieces converge), we have

$$(5.3.9) \quad \sum_{k=-\infty}^{\infty} q^{3k(3k-1)} z^{3k} - \sum_{k=-\infty}^{\infty} q^{3k(3k+1)} z^{3k+1}$$
$$= (q^6;q^6)_\infty (z;q^6)_\infty \left(\frac{q^6}{z};q^6\right)_\infty (z^2 q^6;q^{12})_\infty \left(\frac{q^6}{z^2};q^{12}\right)_\infty.$$

The reason for splitting the series is that we want to reindex only one of the two sums. Replacing k by n in the first sum and by $-n-1$ in the second, the left side of (5.3.9) becomes

$$\sum_{n=-\infty}^{\infty} q^{9n^2-3n} z^{3n} - \sum_{n=-\infty}^{\infty} q^{9n^2+15n+6} z^{-3n-2}.$$

Replacing z by zq^3 equalizes the powers of q on the sum side of (5.3.9), leaving

$$\sum_{n=-\infty}^{\infty} q^{9n^2+6n}\left(z^{3n} - z^{-3n-2}\right)$$
$$= (q^6;q^6)_\infty (zq^3;q^6)_\infty \left(\frac{q^3}{z};q^6\right)_\infty (z^2 q^{12};q^{12})_\infty \left(\frac{1}{z^2};q^{12}\right)_\infty,$$

and multiplying this by zq gives the following result.

THEOREM 48 (The quintuple product identity, alternate form). *If $|q| < 1$ and $z \neq 0$, then*

(5.3.10) $$\sum_{n=-\infty}^{\infty} q^{(3n+1)^2}\left(z^{3n+1} - z^{-3n-1}\right)$$
$$= zq(q^6;q^6)_\infty (zq^3;q^6)_\infty \left(\frac{q^3}{z};q^6\right)_\infty (z^2 q^{12};q^{12})_\infty \left(\frac{1}{z^2};q^{12}\right)_\infty.$$

Since

$$\left(\frac{1}{z^2};q^{12}\right)_\infty = \left(1 - \frac{1}{z^2}\right)\left(\frac{q^{12}}{z^2};q^{12}\right)_\infty,$$

this can be made still more symmetric:

(5.3.11) $$\sum_{n=-\infty}^{\infty} q^{(3n+1)^2}\left(z^{3n+1} - z^{-3n-1}\right)$$
$$= q\left(z - z^{-1}\right)(q^6;q^6)_\infty (zq^3;q^6)_\infty \left(\frac{q^3}{z};q^6\right)_\infty (z^2 q^{12};q^{12})_\infty \left(\frac{q^{12}}{z^2};q^{12}\right)_\infty.$$

Exercises

1. Evaluate
$$\sum_{n=-\infty}^{\infty} q^{(3n+1)^2} z^{3n+1} - \sum_{n=-\infty}^{\infty} q^{(3n+1)^2} z^{-3n-1}$$
by Jacobi's triple product, and compare the result with (5.3.11).

2. The remaining exercises for this section are concerned with an identity of Gauss that is superficially similar to the quintuple product, but not as deep. For $|q| < 1$ and arbitrary nonzero x and y, let
(5.3.12)
$$F(x,y,q) = (-qxy;q^2)_\infty \left(-\frac{q}{xy};q^2\right)_\infty \left(-\frac{qx}{y};q^2\right)_\infty \left(-\frac{qy}{x};q^2\right)_\infty (q^2;q^2)_\infty^2.$$

Gauss set himself the problem of expanding $F(x,y,q)$ in powers of x and y.

(i) We first make several observations about $F(x,y,q)$. It is symmetric in x and y, it is symmetric in y and $\frac{1}{y}$, and it is symmetric in x and $\frac{1}{x}$. Explain why, in addition, any term in the expansion of (5.3.12) that has an even power of y, positive or negative, must also have an even power of x, positive or negative.

(ii) We first try to expand $F(x,y,q)$ in powers of y. Let $f_n(x,q)$ be the coefficient of y^n, so

(5.3.13) $$F(x,y,q) = \sum_{n=-\infty}^{\infty} f_n(x,q)\, y^n.$$

Explain why $f_n(x,q) = f_{-n}(x,q)$.

(iii) Show that $F(x,y,q) = y^2 q^2 F(x, yq^2, q)$.

(iv) Show that (iii) implies $f_n(x,q) = q^{2n-2} f_{n-2}(x,q)$.

(v) Show that (iv) implies

$$f_{2k}(x,q) = q^{2k^2} f_0(x,q) \quad \text{and} \quad f_{2k+1}(x,q) = q^{2k(k+1)} f_1(x,q)$$

for all integer k. (First prove these when k is positive, then explain why we can extend them to the other values.)

(vi) Because of (v), we can rewrite (5.3.13) as

$$F(x,y,q) = f_0(x,q) \sum_{k=-\infty}^{\infty} q^{2k^2} y^{2k} + f_1(x,q) \sum_{k=-\infty}^{\infty} q^{2k(k+1)} y^{2k+1}.$$

Explain why (i) implies that we can further rewrite

(5.3.14) $$F(x,y,q) = P(q) \sum_{j=-\infty}^{\infty} q^{2j^2} x^{2j} \sum_{k=-\infty}^{\infty} q^{2k^2} y^{2k}$$
$$+ Q(q) \sum_{j=-\infty}^{\infty} q^{2j(j+1)} x^{2j+1} \sum_{k=-\infty}^{\infty} q^{2k(k+1)} y^{2k+1},$$

where $P(q)$ and $Q(q)$ depend only on q.

(vii) To find $P(q)$, Gauss sets $y = i$ in (5.3.14). Explain why this eliminates the last sum, and use Jacobi's triple product on the other one to get

$$F(x,i,q) = P(q) \left(-x^2 q^2; q^4\right)_\infty \left(-\frac{q^2}{x^2}; q^4\right)_\infty (q^2; q^2)_\infty^2.$$

Show that setting $y = i$ in (5.3.12) and comparing with this gives $P(q) = 1$. This leaves us with

(5.3.15) $$F(x,y,q) = \sum_{j=-\infty}^{\infty} q^{2j^2} x^{2j} \sum_{k=-\infty}^{\infty} q^{2k^2} y^{2k}$$
$$+ Q(q) \sum_{j=-\infty}^{\infty} q^{2j(j+1)} x^{2j+1} \sum_{k=-\infty}^{\infty} q^{2k(k+1)} y^{2k+1},$$

where it remains to find $Q(q)$.

3. Gauss has two ideas for finding $Q(q)$ in (5.3.15). Show that setting $y = -xq$ makes $F(x, -xq, q) = 0$, and use Jacobi's triple product to show $Q(q) = q$, so

that we have

$$(5.3.16) \quad (-qxy; q^2)_\infty \left(-\frac{q}{xy}; q^2\right)_\infty \left(-\frac{qx}{y}; q^2\right)_\infty \left(-\frac{qy}{x}; q^2\right)_\infty (q^2; q^2)_\infty^2$$

$$= \left(\sum_{j=-\infty}^{\infty} q^{2j^2} x^{2j}\right) \left(\sum_{k=-\infty}^{\infty} q^{2k^2} y^{2k}\right)$$

$$+ q \left(\sum_{j=-\infty}^{\infty} q^{2j(j+1)} x^{2j+1}\right) \left(\sum_{k=-\infty}^{\infty} q^{2k(k+1)} y^{2k+1}\right)$$

for $|q| < 1$ and arbitrary nonzero x and y.

4. Gauss's other idea for finding $Q(q)$ is to set $y = iq$ in (5.3.15). Show that this gets rid of the other sum, and hence completes a second derivation of (5.3.16).

5. By using Jacobi's triple product, show that (5.3.16) can be rewritten as

$$(5.3.17) \quad (-qxy; q^2)_\infty \left(-\frac{q}{xy}; q^2\right)_\infty \left(-\frac{qx}{y}; q^2\right)_\infty \left(-\frac{qy}{x}; q^2\right)_\infty (q^2; q^4)_\infty^2$$

$$= (-x^2 q^2; q^4)_\infty \left(-\frac{q^2}{x^2}; q^4\right)_\infty (-y^2 q^2; q^4)_\infty \left(-\frac{q^2}{y^2}; q^4\right)_\infty$$

$$+ qxy \, (-x^2 q^4; q^4)_\infty \left(-\frac{1}{x^2}; q^4\right)_\infty (-y^2 q^4; q^4)_\infty \left(-\frac{1}{y^2}; q^4\right)_\infty.$$

6. Show that (5.3.17) can be rewritten as

$$(-qxy; q^2)_\infty \left(-\frac{q}{xy}; q^2\right)_\infty \left(-\frac{qx}{y}; q^2\right)_\infty \left(-\frac{qy}{x}; q^2\right)_\infty (q^2; q^4)_\infty^2$$

$$= (-x^2 q^2; q^4)_\infty \left(-\frac{q^2}{x^2}; q^4\right)_\infty (-y^2 q^2; q^4)_\infty \left(-\frac{q^2}{y^2}; q^4\right)_\infty$$

$$+ \frac{q}{xy} (-x^2; q^4)_\infty \left(-\frac{q^4}{x^2}; q^4\right)_\infty (-y^2; q^4)_\infty \left(-\frac{q^4}{y^2}; q^4\right)_\infty.$$

7. There is a simpler proof of (5.3.16) by series manipulation. Here is an outline:

(i) Explain why the left side of (5.3.16) is

$$\left(\sum_{r=-\infty}^{\infty} q^{r^2} (xy)^r\right) \left(\sum_{s=-\infty}^{\infty} q^{s^2} \left(\frac{x}{y}\right)^s\right) = \sum_{r=-\infty}^{\infty} \sum_{s=-\infty}^{\infty} q^{r^2 + s^2} x^{r+s} y^{r-s}.$$

(ii) The exponents $r + s$ and $r - s$ are either both even or both odd (why?). In the former case, set $r + s = 2j$ and $r - s = 2k$, and we have

$$\sum_{\substack{-\infty < r < \infty \\ -\infty < s < \infty \\ r-s \text{ even}}} q^{r^2 + s^2} x^{r+s} y^{r-s} = \left(\sum_{j=-\infty}^{\infty} q^{2j^2} x^{2j}\right) \left(\sum_{k=-\infty}^{\infty} q^{2k^2} y^{2k}\right).$$

In the latter case, set $r + s = 2j + 1$ and $r - s = 2k + 1$, and we have

$$\sum_{\substack{-\infty<r<\infty \\ -\infty<s<\infty \\ r-s \text{ odd}}} q^{r^2+s^2} x^{r+s} y^{r-s} = q \left(\sum_{j=-\infty}^{\infty} q^{2j(j+1)} x^{2j+1} \right) \left(\sum_{k=-\infty}^{\infty} q^{2k(k+1)} y^{2k+1} \right).$$

Combining the two cases we get the right side of (5.3.16), which proves Gauss's identity. Fill in the details.

8. We can make the left side of Gauss's identity more general, at a cost of making the right side less satisfying. Show that

$$(-qu; q^2)_\infty \left(-\frac{q}{u}; q^2\right)_\infty (-qv; q^2)_\infty \left(-\frac{q}{v}; q^2\right)_\infty (q^2; q^4)_\infty^2$$
$$= (-uvq^2; q^4)_\infty \left(-\frac{q^2}{uv}; q^4\right)_\infty \left(-\frac{q^2 u}{v}; q^4\right)_\infty \left(-\frac{q^2 v}{u}; q^4\right)_\infty$$
$$+ qu \, (-uvq^4; q^4)_\infty \left(-\frac{1}{uv}; q^4\right)_\infty \left(-\frac{q^4 u}{v}; q^4\right)_\infty \left(-\frac{v}{u}; q^4\right)_\infty.$$

5.4. Lebesgue's identity

Some readers will have heard of Henri Lebesgue and the Lebesgue integral, which revolutionized real analysis at the beginning of the 20th century. Several years before Cauchy published the Cauchy/Crelle series, a vaguely similar identity was given by another man named Lebesgue, who died in 1875, 18 days before Henri Lebesgue was born. (The two were apparently unrelated.) Victor-Amédée Lebesgue considered the series

$$(5.4.1) \qquad f(a) := \sum_{n=0}^{\infty} \frac{(a;q)_n}{(q;q)_n} q^{\binom{n+1}{2}},$$

though he was not the first to do so—the identity Lebesgue found for $f(a)$ appears 11 years earlier in Jacobi's 1829 paper, right after (5.2.9). Since many other results bear Jacobi's name, and Lebesgue furnished a proof but Jacobi did not, the traditional name is the best.

Let's look at $f(a) - f(aq)$, which is more or less what Euler would probably have done if faced with this series. In both cases the $n = 0$ term is 1, so we have

$$f(a) - f(aq) = \sum_{n=1}^{\infty} \frac{q^{\binom{n+1}{2}}}{(q;q)_n} \left[(a;q)_n - (aq;q)_n\right].$$

Now

$$(a;q)_n - (aq;q)_n = (aq;q)_{n-1}\left[1 - a - (1 - aq^n)\right] = -a(1-q^n)(aq;q)_{n-1},$$

so

$$f(a) - f(aq) = \sum_{n=1}^{\infty} \frac{-a(1-q^n)}{(q;q)_n} (aq;q)_{n-1} q^{\binom{n+1}{2}} = -a \sum_{n=1}^{\infty} \frac{(aq;q)_{n-1}}{(q;q)_{n-1}} q^{\binom{n+1}{2}},$$

or after reindexing

$$f(a) - f(aq) = \sum_{n=0}^{\infty} \frac{(aq;q)_n}{(q;q)_n} q^{\binom{n+2}{2}}(-a).$$

This looks similar to the series in (5.4.1), and we can make it look more similar by writing

$$\binom{n+2}{2} = \binom{n+1}{2} + \binom{n+1}{1} = \binom{n+1}{2} + n + 1.$$

Then

$$f(a) - f(aq) = \sum_{n=0}^{\infty} \frac{(aq;q)_n}{(q;q)_n} q^{\binom{n+1}{2}} \left(-aq^{n+1}\right).$$

Next we rewrite $-aq^{n+1} = 1 - aq^{n+1} - 1$, and note that (exercise)

$$(1 - aq^{n+1})(aq;q)_n = (aq;q)_{n+1} = (1-aq)(aq^2;q)_n.$$

Then

$$f(a) - f(aq) = \sum_{n=0}^{\infty} \frac{(aq;q)_{n+1}}{(q;q)_n} q^{\binom{n+1}{2}} - \sum_{n=0}^{\infty} \frac{(aq;q)_n}{(q;q)_n} q^{\binom{n+1}{2}}$$

$$= (1-aq) \sum_{n=0}^{\infty} \frac{(aq^2;q)_n}{(q;q)_n} q^{\binom{n+1}{2}} - f(aq) = (1-aq)f(aq^2) - f(aq),$$

and we finally have $f(a) = (1-aq)\, f(aq^2)$. (Now that we know this, we might try to relate $f(a)$ to $f(aq^2)$ directly, but this does not seem to be any easier.) As usual, the next step is to iterate. Replacing a by aq^2 we have $f(aq^2) = (1-aq^3)\, f(aq^4)$, which implies that $f(a) = (1-aq)(1-aq^3)\, f(aq^4)$. If we iterate n times we get

$$f(a) = (1-aq)(1-aq^3)\cdots(1-aq^{2n-1})\, f(aq^{2n})$$
$$= (aq;q^2)_n\, f(aq^{2n}).$$

Letting $n \to \infty$ under our usual assumption that $|q| < 1$, this becomes

(5.4.2) $$f(a) = (aq;q^2)_\infty\, f(0).$$

There are several ways to finish off the calculation. We give a straightforward one here; a slicker way appears in the problems. We need to find $f(0)$, which is

$$f(0) = \sum_{n=0}^{\infty} \frac{q^{\binom{n+1}{2}}}{(q;q)_n} = \sum_{n=0}^{\infty} \frac{q^{\binom{n}{2}} q^n}{(q;q)_n}.$$

But this is a series we've seen before. All we have to do is set $x = q$ in Euler's identity (3.6.1), which was

$$(-x;q)_\infty = \sum_{n=0}^{\infty} \frac{q^{\binom{n}{2}} x^n}{(q;q)_n},$$

and we see that $f(0) = (-q;q)_\infty$. This proves

THEOREM 49 (Lebesgue's identity). *If $|q| < 1$ and a is arbitrary, then*

(5.4.3) $$\sum_{n=0}^{\infty} \frac{(a;q)_n}{(q;q)_n} q^{\binom{n+1}{2}} = (aq;q^2)_\infty (-q;q)_\infty = \frac{(aq;q^2)_\infty}{(q;q^2)_\infty}.$$

The last step follows from Euler's "odd equals distinct" theorem in section 3.2. If we change a to aq and multiply both sides by $1-a$, we get

$$\sum_{n=0}^{\infty} \frac{(a;q)_{n+1}}{(q;q)_n} q^{\binom{n+1}{2}} = \frac{(a;q^2)_\infty}{(q;q^2)_\infty},$$

which was (4.5.8), the analytic equivalent of Sylvester's generalization of Euler's "odd equals distinct" theorem. Section 4.5 therefore provides a combinatorial proof of Lebesgue's identity. There is a generalization due to George Andrews.

THEOREM 50 (The double Lebesgue identity). *If $|q| < 1$ and a and b are arbitrary, then*

(5.4.4) $$\sum_{n=0}^{\infty} \frac{(a;q)_n(b;q)_n}{(q;q)_n(abq;q^2)_n} q^{\binom{n+1}{2}} = \frac{(aq;q^2)_\infty(bq;q^2)_\infty}{(q;q^2)_\infty(abq;q^2)_\infty}.$$

Note that this is symmetric in a and b, and it reduces to Lebesgue's identity if either a or b is zero. We will outline a proof in one of the problems. Andrews's proof is in section 5.6.

Exercises

1. Verify that $(1 - aq^{n+1})(aq;q)_n = (aq;q)_{n+1} = (1-aq)(aq^2;q)_n$.
2. We were able to find $f(0)$ above because we had already spent a lot of time on q-series, and $f(0)$ was a series we'd encountered before. It is fair to ask: is there a nice way to find $f(0)$ without knowing other q-series? We could set $a = 1$ in (5.4.2) to get $f(1) = (q;q^2)_\infty f(0)$. Now

$$f(1) = \sum_{n=0}^{\infty} \frac{(1;q)_n}{(q;q)_n} q^{\binom{n+1}{2}}.$$

What is $(1;q)_n$? Explain why this implies that $f(1) = 1$, and how this allows us to find $f(0)$.

3. Find $f(0)$ by setting $a = q$ in (5.4.2) and using Gauss's identity (5.2.12).
4. Lebesgue's identity was rediscovered by Atle Selberg, who considered the function

$$L_k(x) = \sum_{n=0}^{\infty} (-1)^n q^{\binom{n+1}{2} - nk} \frac{(xq;q)_n}{(q;q)_n}$$

with $|q| < 1$.

   (i) Show that the series converges for all x and all k.
   (ii) Show that
   $$q^k \left(L_k(x) - L_k(xq) \right) = L_k(xq) - \left(1 - xq^2\right) L_k(xq^2).$$

   (iii) Selberg now chooses k to satisfy $q^k = -1$ and calls the resulting function $L(x)$. Explain why this gives

   $$L(x) = \sum_{n=0}^{\infty} q^{\binom{n+1}{2}} \frac{(xq;q)_n}{(q;q)_n} \quad \text{and} \quad L(x) = (1 - xq^2) L(xq^2).$$

   (iv) Explain how Lebesgue's identity follows from (iii).

(v) Which value of k gives $q^k = -1$?

5. Prove that

(5.4.5) $$\sum_{n=0}^{\infty} \frac{(a;q)_n}{(q;q)_n} q^{\binom{n+1}{2}} z^n = (a;q)_\infty (-zq;q)_\infty \sum_{k=0}^{\infty} \frac{a^k}{(q;q)_k(-zq;q)_k}.$$

Hint: Explain why (5.4.5) can be rewritten as

$$\sum_{n=0}^{\infty} \frac{q^{\binom{n+1}{2}} z^n}{(q;q)_n (aq^n;q)_\infty} = \sum_{k=0}^{\infty} \frac{a^k}{(q;q)_k} \left(-zq^{k+1};q\right)_\infty,$$

and then use Euler's theorems (3.5.1) and (3.6.1).

6. Prove that

(5.4.6) $$\sum_{n=0}^{\infty} \frac{(a;q)_n}{(q;q)_n} q^{\binom{n+1}{2}} z^n = (-zq;q)_\infty \sum_{j=0}^{\infty} \frac{(-1)^j q^{j^2} a^j z^j}{(q;q)_j(-zq;q)_j}.$$

Hint: Rewrite the right side as in the previous problem, then use (3.6.1) and Rothe's q-binomial theorem.

7. Show that the right sides of (5.4.5) and (5.4.6) reduce to

$$(aq;q^2)_\infty (-q;q)_\infty = \frac{(aq;q^2)_\infty}{(q;q^2)_\infty}$$

when $z = 1$. This gives two more proofs of Lebesgue's identity.

8. From (5.4.5) and (5.4.6) it follows that

$$(a;q)_\infty \sum_{n=0}^{\infty} \frac{a^n}{(q;q)_n(-zq;q)_n} = \sum_{k=0}^{\infty} \frac{(-1)^k q^{k^2} a^k z^k}{(q;q)_k(-zq;q)_k},$$

or

$$\sum_{n=0}^{\infty} \frac{a^n}{(q;q)_n(-zq;q)_n} = \frac{1}{(a;q)_\infty} \sum_{k=0}^{\infty} \frac{(-1)^k q^{k^2} a^k z^k}{(q;q)_k(-zq;q)_k}.$$

Give a direct proof of the latter by using (3.5.1) and (4.2.13).

9. Recall Rowell's identity (2.5.13)

$$\sum_{k=0}^{n} \binom{n}{k}_q (-a;q)_k q^{\binom{k+1}{2}} = \sum_{j=0}^{n} \binom{n}{j}_{q^2} (-q;q)_{n-j} q^{j^2} a^j$$

from problem 13 in section 2.5. Show that letting $n \to \infty$ here (invoking Tannery's theorem again) gives Lebesgue's identity in the form

$$\sum_{k=0}^{\infty} \frac{(-a;q)_k}{(q;q)_k} q^{\binom{k+1}{2}} = (-aq;q^2)_\infty (-q;q)_\infty.$$

10. Show that

$$\sum_{k=0}^{\infty} \frac{(a;q)_k}{(q;q)_k} q^{\frac{k(k+3)}{2}} = \frac{(aq;q^2)_\infty - (a;q^2)_\infty}{a(q;q^2)_\infty}$$

$$= \frac{1}{(q;q^2)_\infty} \sum_{k=0}^{\infty} \frac{(-a)^k q^{k(k+1)}}{(q^2;q^2)_k (1+q^{k+1})}.$$

11. Jacobi proved another identity that is somewhat similar to Lebesgue's. He considered the function
$$f(z) = \sum_{k=0}^{\infty} \frac{(z^2;q^2)_k}{(q^2;q^2)_k} (-1)^k q^{k^2} + \frac{z}{q} \sum_{k=0}^{\infty} \frac{(z^2;q^2)_k}{(q^2;q^2)_k} (-1)^k q^{(k+1)^2}.$$

Here is an outline of his argument:

(i) Combining the sums above,
$$f(z) = \sum_{k=0}^{\infty} \frac{(z^2;q^2)_k}{(q^2;q^2)_k} (-1)^k q^{k^2} (1 + zq^{2k}).$$

(ii) On the other hand, we could rewrite the first sum in the definition of $f(z)$ as
$$1 + \sum_{k=0}^{\infty} \frac{(z^2;q^2)_{k+1}}{(q^2;q^2)_{k+1}} (-1)^{k+1} q^{(k+1)^2}.$$
Combine this with the other sum in the definition to get
$$f(z) = 1 + \sum_{k=0}^{\infty} \frac{(z^2;q^2)_k}{(q^2;q^2)_{k+1}} (-1)^{k+1} q^{(k+1)^2} \left(1 - \frac{z}{q}\right)(1 + zq^{2k+1}).$$

(iii) Replace z by zq in the final expression in (ii) and then multiply it by $1+z$. After reindexing, show that this gives $f(z) = (1+z)f(zq)$.

(iv) Iterate the result in (iii) to get $f(z) = (-z;q)_\infty f(0)$.

(v) There are several ways to find $f(0)$. Jacobi finds $f(1)$ first, as in problem 2. Perhaps better (as suggested to me by Sarah Schaller) is to find $f(q)$ first. Show by any method that $f(0) = 1/(-q;q)_\infty$. This proves that
$$f(z) = \frac{(-z;q)_\infty}{(-q;q)_\infty}.$$

12. Show that Jacobi's identity in problem 11 reduces to Gauss's identity (5.2.11) from problems 8 and 15 of section 5.2 if $z = -q$.

13. Show how Jacobi or Gauss could have used (5.2.11) and Euler's "odd equals distinct" theorem to finish the proof of the triple product identity in section 5.1.

14. Following Jacobi, use the result of problem 11 to show

(a) $$\frac{1}{2} \frac{(-z;q)_\infty + (z;q)_\infty}{(-q;q)_\infty} = \sum_{k=0}^{\infty} \frac{(z^2;q^2)_k}{(q^2;q^2)_k} (-1)^k q^{k^2},$$

(b) $$\frac{q}{2z} \frac{(-z;q)_\infty - (z;q)_\infty}{(-q;q)_\infty} = \sum_{k=0}^{\infty} \frac{(z^2;q^2)_k}{(q^2;q^2)_k} (-1)^k q^{(k+1)^2}.$$

15. Give an alternative derivation of (5.2.11) by setting $z = q$ or $z = -q$ in (a) or (b) of problem 14 respectively.

16. Lebesgue conceived an alternate approach to the identities of problems 11 and 14, starting from the two series in problem 14. Define
$$\theta(z) = \sum_{k=0}^{\infty} (-1)^k q^{k^2} \frac{(z^2;q^2)_k}{(q^2;q^2)_k} \quad \text{and} \quad \phi(z) = \sum_{k=0}^{\infty} (-1)^k q^{(k+1)^2} \frac{(z^2;q^2)_k}{(q^2;q^2)_k},$$

and
(5.4.7) $$f(z) = \theta(z) + \frac{z}{q}\phi(z).$$

(i) Explain why $\phi(-z) = \phi(z)$ and $\theta(-z) = \theta(z)$.

(ii) Explain why
$$\theta(z) = 1 + \sum_{k=1}^{\infty}(-1)^k q^{k^2} \frac{(z^2;q^2)_k}{(q^2;q^2)_k} \quad \text{and} \quad \phi(z) = \sum_{k=1}^{\infty}(-1)^{k-1} q^{k^2} \frac{(z^2;q^2)_{k-1}}{(q^2;q^2)_{k-1}}.$$

(iii) Starting from the forms in (ii), show that
$$\theta(zq) + \phi(zq) = \frac{1}{q}\phi(z) \quad \text{and} \quad \theta(zq) + z^2 \phi(zq) = \theta(z).$$

(iv) Use (iii) to show that $f(z) = (1+z)f(zq)$.

(v) Use (iv) to show (assuming as usual that $|q| < 1$) that
$$f(z) = (-z;q)_\infty f(0) = \frac{(-z;q)_\infty}{(-q;q)_\infty}.$$

(vi) Use (5.4.7) and (i) to show that
$$f(z) + f(-z) = 2\theta(z) \quad \text{and} \quad f(z) - f(-z) = \frac{2z}{q}\phi(z).$$

Then use these to prove the identities in problem 14.

17. This problem outlines a proof of the double Lebesgue identity. Set
$$f(a,b) = \sum_{n=0}^{\infty} \frac{(a;q)_n(b;q)_n}{(q;q)_n(abq;q^2)_n} q^{\binom{n+1}{2}} \quad \text{and} \quad g(a,b) = \sum_{n=0}^{\infty} \frac{(a;q)_n(bq;q)_n}{(q;q)_n(abq;q^2)_n} q^{\binom{n+1}{2}}.$$

The argument will succeed in evaluating $f(a,b)$, which is the left side of (5.4.4). I do not know a nice formula for $g(a,b)$, but it is a convenient thing to subtract from $f(a,b)$.

(i) Show that
$$f(a,b) - g(a,b) = -b\sum_{n=1}^{\infty} \frac{(a;q)_n}{(abq;q^2)_n} \frac{(bq;q)_{n-1}}{(q;q)_{n-1}} q^{\binom{n+1}{2}}.$$

(ii) After reindexing the series, show that (i) can be rewritten as
$$f(a,b) - g(a,b) = \sum_{n=0}^{\infty} \frac{(a;q)_n(bq;q)_n}{(q;q)_n(abq;q^2)_{n+1}} q^{\binom{n+1}{2}} \left[1 - bq^{n+1} - \left(1 - abq^{2n+1}\right)\right].$$

(iii) Show that (ii) can be rewritten as
$$f(a,b) - g(a,b) = \sum_{n=0}^{\infty} \frac{(a;q)_n(bq;q)_{n+1}}{(q;q)_n(abq;q^2)_{n+1}} q^{\binom{n+1}{2}} - \sum_{n=0}^{\infty} \frac{(a;q)_n(bq;q)_n}{(q;q)_n(abq;q^2)_n} q^{\binom{n+1}{2}}.$$

(iv) Explain why (iii) implies that
$$f(a,b) = \sum_{n=0}^{\infty} \frac{(a;q)_n(bq;q)_{n+1}}{(q;q)_n(abq;q^2)_{n+1}} q^{\binom{n+1}{2}} = \frac{1-bq}{1-abq} f(a, bq^2).$$

(v) Explain why (iv) implies that
$$f(a,b) = \frac{(bq;q^2)_\infty}{(abq;q^2)_\infty} f(a,0),$$
and complete the proof of the double Lebesgue identity.

5.5. Basic hypergeometric series

In the last several sections of this chapter we will scratch the surface of the vast subject of q-hypergeometric series. These are sometimes also called, somewhat misleadingly, basic hypergeometric series, as in the standard reference with that title by Gasper and Rahman. In this context "basic" refers to the "base" q, and is not synonymous with "elementary".

For us, a basic hypergeometric series is a series of the form

$$(5.5.1) \quad {}_{r+1}\phi_r\left(\begin{matrix} a_1, \ldots, a_{r+1} \\ b_1, \ldots, b_r \end{matrix}; q, x\right) := \sum_{n=0}^{\infty} \frac{(a_1;q)_n \cdots (a_{r+1};q)_n}{(b_1;q)_n \cdots (b_r;q)_n} \frac{x^n}{(q;q)_n}.$$

This is usually an infinite series, but it can be finite if one of the numerator parameters a_j is q to a negative integer power. When it is infinite, it converges if $|x| < 1$ (assuming as always that $|q| < 1$); we leave this as an exercise. In this section r will always be 0 or 1, but there are interesting results with r as large as 9.

We already know what happens when $r = 0$. This is the q-binomial series

$$(5.5.2) \quad {}_1\phi_0\left(\begin{matrix} s \\ - \end{matrix}; q, x\right) := \sum_{n=0}^{\infty} \frac{(s;q)_n}{(q;q)_n} x^n = \frac{(sx;q)_\infty}{(x;q)_\infty} \quad \text{if } |x| < 1$$

from section 3.7. Indeed, the main reason why this result occurs in the literature much more often than the superficially more general Cauchy/Crelle series is that it fits into the q-hypergeometric series framework.

We move on to the case $r = 1$. The development here follows Heine's epoch-making paper of 1847. A generic ${}_2\phi_1$ series looks like

$$(5.5.3) \quad {}_2\phi_1\left(\begin{matrix} a, b \\ c \end{matrix}; q, z\right) := \sum_{n=0}^{\infty} \frac{(a;q)_n (b;q)_n}{(c;q)_n (q;q)_n} z^n.$$

Unlike the q-binomial series, it is not possible to sum this for generic values of the parameters, but there are a few specific values for which we can find the sum. We start by transforming (5.5.3) into another ${}_2\phi_1$ series. Take the quotient $(b;q)_n/(c;q)_n$ in (5.5.3) and rewrite it as

$$\frac{(b;q)_n}{(c;q)_n} = \frac{(b;q)_\infty}{(bq^n;q)_\infty} \frac{(cq^n;q)_\infty}{(c;q)_\infty} = \frac{(b;q)_\infty}{(c;q)_\infty} \sum_{k=0}^{\infty} \frac{(\frac{c}{b};q)_k}{(q;q)_k} (bq^n)^k,$$

where the last equality uses (5.5.2) with $s = \frac{c}{b}$ and $x = bq^n$. Substituting this in (5.5.3) we have

$$\begin{aligned} {}_2\phi_1\left(\begin{matrix} a, b \\ c \end{matrix}; q, z\right) &= \frac{(b;q)_\infty}{(c;q)_\infty} \sum_{n=0}^{\infty} \sum_{k=0}^{\infty} \frac{(a;q)_n}{(q;q)_n} \frac{(\frac{c}{b};q)_k}{(q;q)_k} b^k z^n q^{nk} \\ &= \frac{(b;q)_\infty}{(c;q)_\infty} \sum_{k=0}^{\infty} \frac{(\frac{c}{b};q)_k}{(q;q)_k} b^k \sum_{n=0}^{\infty} \frac{(a;q)_n}{(q;q)_n} (zq^k)^n. \end{aligned}$$

Using (5.5.2) again with $x = zq^k$ and $s = a$, the inner sum is
$$\sum_{n=0}^{\infty} \frac{(a;q)_n}{(q;q)_n}(zq^k)^n = \frac{(azq^k;q)_\infty}{(zq^k;q)_\infty} = \frac{(az;q)_\infty}{(az;q)_k} \frac{(z;q)_k}{(z;q)_\infty}.$$

Using this in the previous calculation we have
$$_2\phi_1\left(\begin{matrix}a,b\\c\end{matrix};q,z\right) = \frac{(b;q)_\infty}{(c;q)_\infty} \frac{(az;q)_\infty}{(z;q)_\infty} \sum_{k=0}^{\infty} \frac{(\frac{c}{b};q)_k}{(q;q)_k} \frac{(z;q)_k}{(az;q)_k} b^k.$$

Note that the new series is again a $_2\phi_1$ with b as the power series variable, so we had better assume that $|b| < 1$ for convergence. Thus we have proved

THEOREM 51 (Heine's q-Pfaff transformation). *If q, z, b are all less than 1 in absolute value, then*

(5.5.4) $$_2\phi_1\left(\begin{matrix}a,b\\c\end{matrix};q,z\right) = \frac{(b;q)_\infty (az;q)_\infty}{(c;q)_\infty (z;q)_\infty} {_2\phi_1}\left(\begin{matrix}\frac{c}{b},z\\az\end{matrix};q,b\right).$$

One reason why this is important is that it gives us a $_2\phi_1$ that we can sum, for if we take $z = \frac{c}{ab}$, then the right side reduces to a $_1\phi_0$ that can be evaluated by (5.5.2) with $s = \frac{c}{ab}$ and $x = b$:

$$_2\phi_1\left(\begin{matrix}a,b\\c\end{matrix};q,\frac{c}{ab}\right) = \frac{(b;q)_\infty (\frac{c}{b};q)_\infty}{(c;q)_\infty (\frac{c}{ab};q)_\infty} {_1\phi_0}\left(\begin{matrix}\frac{c}{ab}\\-\end{matrix};q,b\right)$$
$$= \frac{(b;q)_\infty (\frac{c}{b};q)_\infty}{(c;q)_\infty (\frac{c}{ab};q)_\infty} \frac{(\frac{c}{a};q)_\infty}{(b;q)_\infty}.$$

After simplifying the last line we have

THEOREM 52 (Heine's q-Gauss summation formula). *If $|q| < 1$ and $|c| < |ab|$, then*

(5.5.5) $$_2\phi_1\left(\begin{matrix}a,b\\c\end{matrix};q,\frac{c}{ab}\right) = \frac{(\frac{c}{a};q)_\infty (\frac{c}{b};q)_\infty}{(c;q)_\infty (\frac{c}{ab};q)_\infty}.$$

An equivalent form of (5.5.5) was found (more or less) by Jacobi a year earlier:

THEOREM 53 (Jacobi's q-Gauss summation formula). *If $|q| < 1$ and $|t| < 1$, then*

(5.5.6) $$_2\phi_1\left(\begin{matrix}a,b\\abt\end{matrix};q,t\right) = \frac{(at;q)_\infty (bt;q)_\infty}{(t;q)_\infty (abt;q)_\infty}.$$

This follows from (5.5.5) on replacing c by abt. Ramanujan gave a similar form, which we leave as an exercise.

THEOREM 54 (Ramanujan's q-Gauss summation formula). *If $|q| < 1$ and $|abt| < 1$, then*

(5.5.7) $$_2\phi_1\left(\begin{matrix}\frac{1}{a},\frac{1}{b}\\t\end{matrix};q,abt\right) = \frac{(at;q)_\infty (bt;q)_\infty}{(t;q)_\infty (abt;q)_\infty}.$$

Next we are going to use the q-Pfaff transformation on itself. It is easy to get confused while doing this, so let's restate it first with different names for most of the parameters:

(5.5.8) $$_2\phi_1\left(\begin{matrix}u,v\\w\end{matrix};q,x\right) = \frac{(v;q)_\infty (ux;q)_\infty}{(w;q)_\infty (x;q)_\infty} {_2\phi_1}\left(\begin{matrix}\frac{w}{v},x\\ux\end{matrix};q,v\right).$$

We want to match the left side of this to
$$2\phi_1\left(\begin{matrix}\frac{c}{b},z\\az\end{matrix};q,b\right),$$
the sum on the right side of (5.5.4). We have to take $x = b$ and $w = az$, but we have two choices for the other parameters: we can either take $u = \frac{c}{b}$ and $v = z$, or the other way around. If we do the former, then we will undo the q-Pfaff transformation (we leave this as an exercise), so we take $u = z$ and $v = \frac{c}{b}$. Then $\frac{w}{v} = \frac{abz}{c}$ and $ux = bz$ and (5.5.8) becomes

$$2\phi_1\left(\begin{matrix}\frac{c}{b},z\\az\end{matrix};q,b\right) = \frac{(\frac{c}{b};q)_\infty (bz;q)_\infty}{(az;q)_\infty (b;q)_\infty} 2\phi_1\left(\begin{matrix}\frac{abz}{c},b\\bz\end{matrix};q,\frac{c}{b}\right).$$

Putting this in (5.5.4) we get

THEOREM 55 (Heine's intermediate transformation). *If $q, z, \frac{c}{b}$ are all less than 1 in absolute value, then*

(5.5.9) $$2\phi_1\left(\begin{matrix}a,b\\c\end{matrix};q,z\right) = \frac{(\frac{c}{b};q)_\infty (bz;q)_\infty}{(c;q)_\infty (z;q)_\infty} 2\phi_1\left(\begin{matrix}\frac{abz}{c},b\\bz\end{matrix};q,\frac{c}{b}\right).$$

The proof also tacitly assumes $|b| < 1$, but it is possible to relax this. As such, (5.5.9) does not seem to be a very interesting identity, but if we apply (5.5.8) to it, then we will get the formula we are really after. Again we want to make the left side of (5.5.8) match
$$2\phi_1\left(\begin{matrix}\frac{abz}{c},b\\bz\end{matrix};q,\frac{c}{b}\right),$$
the right side of (5.5.9). We have to take $w = bz$ and $x = \frac{c}{b}$. If we take $u = \frac{abz}{c}$ and $v = b$, then we'll get back to (5.5.4) (exercise), so we instead take $u = b$ and $v = \frac{abz}{c}$. Then $\frac{w}{v} = \frac{c}{a}$ and $ux = c$ and (5.5.8) becomes

$$2\phi_1\left(\begin{matrix}\frac{abz}{c},b\\bz\end{matrix};q,\frac{c}{b}\right) = \frac{(\frac{abz}{c};q)_\infty (c;q)_\infty}{(bz;q)_\infty (z;q)_\infty} 2\phi_1\left(\begin{matrix}\frac{c}{a},\frac{c}{b}\\c\end{matrix};q,\frac{abz}{c}\right).$$

Putting this into (5.5.9) we have

THEOREM 56 (Heine's q-Euler transformation). *If $q, z, \frac{abz}{c}$ are all less than 1 in absolute value, then*

(5.5.10) $$2\phi_1\left(\begin{matrix}a,b\\c\end{matrix};q,z\right) = \frac{(\frac{abz}{c};q)_\infty}{(z;q)_\infty} 2\phi_1\left(\begin{matrix}\frac{c}{a},\frac{c}{b}\\c\end{matrix};q,\frac{abz}{c}\right).$$

The proof also tacitly assumes $|\frac{c}{b}| < 1$, but it is possible to relax this. The right side of (5.5.10) has a very important feature that distinguishes it from the right sides of (5.5.4) and (5.5.9): it is visibly symmetric in a and b, as the left side is. This means that we can't get any more transformation formulas by repeated use of (5.5.8), as the reader may be relieved to hear. In the previous applications of (5.5.8) one choice of u and v reversed a previous step and one didn't, but now they both do because of the symmetry.

Exercises

1. Use the ratio test to show that, assuming $|q| < 1$, the series (5.5.1) converges if $|x| < 1$.

2. Show that the right side of (5.5.4) can also be summed if $a = 1$, but that this only leads to $1 = 1$. (What is $(1;q)_n$?)

3. Show that if we set $a = q^{-n}$ in (5.5.5) and then replace c by bt, we get

$$\sum_{k=0}^{n} \binom{n}{k}_q (-t)^k q^{\binom{k}{2}} \frac{(b;q)_k}{(bt;q)_k} = \frac{(t;q)_n}{(bt;q)_n}$$

and that this is equivalent to (2.3.12) from problem 16 in section 2.3. (2.3.11) might help.

4. Deduce Ramanujan's q-Gauss summation (5.5.7) from Heine's form (5.5.5) or from Jacobi's form (5.5.6).

5. Show that taking $u = \frac{c}{b}$ and $v = z$ along with $x = b$ and $w = az$ in (5.5.8) reverses the q-Pfaff transformation.

6. Give an alternate proof of the q-Gauss summation formula by taking $z = \frac{c}{ab}$ in (5.5.9).

7. Jacobi's form (5.5.6) of the q-Gauss summation formula is a generalization of the q-binomial series (5.5.2). Explain.

8. This problem outlines Ramanujan's proof of his form (5.5.7) of the q-Gauss summation formula. First note that $\left(\frac{1}{a};q\right)_k a^k = (a-1)(a-q)\cdots(a-q^{k-1})$, so the sum side of (5.5.7) is

$$\sum_{k=0}^{\infty} \frac{(a-1)(a-q)\cdots(a-q^{k-1})(b-1)(b-q)\cdots(b-q^{k-1})}{(t;q)_k (q;q)_k} t^k.$$

(i) Explain why $(t;q)_\infty/(bt;q)_\infty$ times this sum is

$$\sum_{k=0}^{\infty} \frac{(tq^k;q)_\infty}{(bt;q)_\infty} \times \frac{(a-1)(a-q)\cdots(a-q^{k-1})(b-1)(b-q)\cdots(b-q^{k-1})}{(q;q)_k} t^k.$$

(ii) Explain why

$$\frac{(tq^k;q)_\infty}{(bt;q)_\infty} = \sum_{j=0}^{\infty} \frac{(b-q^k)(b-q^{k+1})\cdots(b-q^{k+j-1})}{(q;q)_j} t^j.$$

(iii) Using (ii), show that the right side of (i) becomes

$$\sum_{j,k=0}^{\infty} \frac{(a-1)(a-q)\cdots(a-q^{k-1})(b-1)(b-q)\cdots(b-q^{k+j-1}) t^{k+j}}{(q;q)_{k+j}} \binom{k+j}{k}_q,$$

and explain why this is

$$\sum_{n=0}^{\infty} \frac{(b-1)(b-q)\cdots(b-q^{n-1})}{(q;q)_n} t^n \sum_{k=0}^{n} \binom{n}{k}_q (a-1)(a-q)\cdots(a-q^{k-1}).$$

(iv) The inner sum occurred as a special case of Jacobi's q-binomial theorem in section 2.6. Hence show that the sum in (iii) is

$$\sum_{n=0}^{\infty} \frac{(b-1)(b-q)\cdots(b-q^{n-1})}{(q;q)_n}(at)^n,$$

and explain why the right side is $(at;q)_\infty/(abt;q)_\infty$. Why does this prove (5.5.7)?

9. Heine's form of the q-Gauss summation formula implies a formula of Cauchy that we'll see in section 6.3. For convenience we rewrite (5.5.5) with different letters here:

(5.5.11) $$\sum_{n=0}^{\infty} \frac{(r;q)_n (s;q)_n}{(t;q)_n (q;q)_n} \left(\frac{t}{rs}\right)^n = \frac{(\frac{t}{r};q)_\infty (\frac{t}{s};q)_\infty}{(t;q)_\infty (\frac{t}{rs};q)_\infty}.$$

(i) Show that

$$\frac{(s;q)_n}{s^n} = (-1)^n \left(1 - \frac{1}{s}\right)\left(q - \frac{1}{s}\right)\cdots\left(q^{n-1} - \frac{1}{s}\right).$$

(ii) With the aid of (i), show that letting $s \to \infty$ (or, if you prefer, replacing $\frac{1}{s}$ by σ and then setting $\sigma = 0$) in (5.5.11) gives

(5.5.12) $$\sum_{n=0}^{\infty} \frac{(r;q)_n q^{\binom{n}{2}}}{(t;q)_n (q;q)_n}\left(\frac{-t}{r}\right)^n = \frac{(\frac{t}{r};q)_\infty}{(t;q)_\infty}.$$

(iii) If $t = bx$ and $\frac{t}{r} = ax$, what is r? Show that if we make these replacements in (5.5.12), we get

$$\frac{(ax;q)_\infty}{(bx;q)_\infty} = \sum_{k=0}^{\infty} \frac{q^{\binom{k}{2}} x^k}{(q;q)_k (bx;q)_k}(b-a)(bq-a)\cdots(bq^{k-1}-a),$$

which is Cauchy's formula. Ramanujan proves this from (5.5.7) in a similar way.

10. By changing q to q^2 in (5.5.12) and then setting $r = -q^2$ and $t = q^3$, show that

$$\sum_{n=0}^{\infty} \frac{(-q^2;q^2)_n}{(q;q)_{2n+1}} q^{n^2} = \frac{(-q;q^2)_\infty}{(q;q^2)_\infty}.$$

11. In Heine's intermediate transformation (5.5.9) set $a = \frac{t}{u}$, $b = w$, $c = v$, and $z = \frac{u}{w}$ and let $w \to \infty$. Show that this gives **Ramanujan's transformation**

$$(v;q)_\infty \sum_{n=0}^{\infty} \frac{(\frac{t}{u};q)_n (-1)^n q^{\binom{n}{2}} u^n}{(q;q)_n (v;q)_n} = (u;q)_\infty \sum_{n=0}^{\infty} \frac{(\frac{t}{v};q)_n (-1)^n q^{\binom{n}{2}} v^n}{(q;q)_n (u;q)_n}.$$

Thus this function is symmetric in u and v.

12. Bailey found still another derivation of (4.4.14), based on the q-Gauss sum. He used (5.5.5), but let's use Ramanujan's form, which we rewrite as

$$_2\phi_1\left(\begin{matrix}\frac{1}{a},\frac{1}{b}\\z\end{matrix};q,abz\right) = \frac{(az;q)_\infty (bz;q)_\infty}{(z;q)_\infty (abz;q)_\infty}$$

to get the same letter as in (4.4.14).

(i) Show that taking the derivative of this (use logarithmic differentiation) with respect to z gives

$$\sum_{n=1}^{\infty} \frac{\left(\frac{1}{a};q\right)_n \left(\frac{1}{b};q\right)_n (ab)^n z^{n-1}}{(z;q)_n (q;q)_n} \sum_{k=0}^{n-1} \frac{1}{1-zq^k}$$
$$= \frac{(az;q)_\infty (bz;q)_\infty}{(z;q)_\infty (abz;q)_\infty} \sum_{k=0}^{\infty} \frac{(1-a)(1-b)(1-abz^2 q^{2k}) q^k}{(1-zq^k)(1-azq^k)(1-bzq^k)(1-abzq^k)}.$$

(ii) Show that this simplifies a little, to

$$\sum_{n=1}^{\infty} \frac{\left(\frac{q}{a};q\right)_{n-1} \left(\frac{q}{b};q\right)_{n-1} (abz)^{n-1}}{(z;q)_n (q;q)_n} \sum_{k=0}^{n-1} \frac{1}{1-zq^k}$$
$$= \frac{(az;q)_\infty (bz;q)_\infty}{(z;q)_\infty (abz;q)_\infty} \sum_{k=0}^{\infty} \frac{(1-abz^2 q^{2k}) q^k}{(1-zq^k)(1-azq^k)(1-bzq^k)(1-abzq^k)}.$$

(iii) Show that multiplying (ii) by z and setting a and b equal to 1 gives

$$\sum_{n=1}^{\infty} \frac{(q;q)_{n-1} z^n}{(z;q)_n (1-q^n)} \sum_{k=0}^{n-1} \frac{1}{1-zq^k} = \sum_{k=0}^{\infty} \frac{zq^k(1+zq^k)}{(1-zq^k)^3}.$$

(iv) We now have the left side of (4.4.14), but not the right side. Use (9.1.2) from the problems in section 9.1 with $x = zq^k$ to transform the right side of (iii), and hence show that

$$\sum_{n=1}^{\infty} \frac{(q;q)_{n-1} z^n}{(z;q)_n (1-q^n)} \sum_{k=0}^{n-1} \frac{1}{1-zq^k} = \sum_{n=0}^{\infty} \frac{n^2 z^n}{1-q^n},$$

which is (4.4.14).

13. This problem outlines an alternative proof of (5.4.5) by using Heine's q-Pfaff transformation (5.5.4).

(i) Show that changing z to x, a to $-zq/x$, and b to a in (5.5.4) gives

$$\sum_{n=0}^{\infty} \frac{(a;q)_n (x+zq)(x+zq^2)\cdots(x+zq^n)}{(c;q)_n (q;q)_n}$$
$$= \frac{(a;q)_\infty (-zq;q)_\infty}{(c;q)_\infty (x;q)_\infty} \sum_{k=0}^{\infty} \frac{\left(\frac{c}{a};q\right)_k (x;q)_k}{(q;q)_k (-zq;q)_k} a^k.$$

(ii) Show that setting c and x equal to zero in (i) gives (5.4.5).

14. (a) Show that taking $a = -\frac{q}{z}$, $b = q$, and $c = xq$ in Heine's intermediate transformation (5.5.9) and letting $z \to 0$ gives

$$\sum_{n=0}^{\infty} \frac{q^{\binom{n+1}{2}}}{(xq;q)_n} = (1-x) \sum_{n=0}^{\infty} (x+q)(x+q^2)\cdots(x+q^n).$$

(b) Show that the right side of (a) can be rewritten as

$$1 + \sum_{n=0}^{\infty} (x+q)(x+q^2)\cdots(x+q^n) q^{n+1}.$$

This gives an alternative proof of the result of (ii) in problem 14 in section 3.5.

(c) Assuming as usual that $|q| < 1$, for which values of x do these series converge?

15. Imitate the proof of Heine's q-Pfaff transformation to prove **Andrews's q-Pfaff transformation**: if q, z, b are all less than 1 in absolute value and r is a positive integer, then

$$(5.5.13) \quad \sum_{n=0}^{\infty} \frac{(a;q^r)_n}{(q^r;q^r)_n} \frac{(b;q)_{rn}}{(c;q)_{rn}} z^n = \frac{(b;q)_\infty}{(c;q)_\infty} \frac{(az;q^r)_\infty}{(z;q^r)_\infty} \sum_{k=0}^{\infty} \frac{(\frac{c}{b};q)_k}{(q;q)_k} \frac{(z;q^r)_k}{(az;q^r)_k} b^k.$$

5.6. More $_2\phi_1$ identities

Another summable $_2\phi_1$ arises when we take $c = \frac{bq}{a}$ and $z = -\frac{q}{a}$ in (5.5.4). This gives

$$_2\phi_1\left(\begin{matrix} a,b \\ \frac{bq}{a} \end{matrix}; q, -\frac{q}{a}\right) = \frac{(b;q)_\infty (-q;q)_\infty}{(\frac{bq}{a};q)_\infty (-\frac{q}{a};q)_\infty} {_2\phi_1}\left(\begin{matrix} \frac{q}{a}, -\frac{q}{a} \\ -q \end{matrix}; q, b\right)$$

$$= \frac{(b;q)_\infty (-q;q)_\infty}{(\frac{bq}{a};q)_\infty (-\frac{q}{a};q)_\infty} \sum_{n=0}^{\infty} \frac{(\frac{q}{a};q)_n (-\frac{q}{a};q)_n}{(q;q)_n (-q;q)_n} b^n$$

$$= \frac{(b;q)_\infty (-q;q)_\infty}{(\frac{bq}{a};q)_\infty (-\frac{q}{a};q)_\infty} \sum_{n=0}^{\infty} \frac{(\frac{q^2}{a^2};q^2)_n}{(q^2;q^2)_n} b^n.$$

The point is that we have made the right side into the "q^2-binomial series"; that is, we can sum this series by (5.5.2), changing q to q^2 there and then taking $s = \frac{q^2}{a^2}$ and $x = b$. We get

$$(5.6.1) \quad _2\phi_1\left(\begin{matrix} a,b \\ \frac{bq}{a} \end{matrix}; q, -\frac{q}{a}\right) = \frac{(b;q)_\infty (-q;q)_\infty}{(\frac{bq}{a};q)_\infty (-\frac{q}{a};q)_\infty} \frac{(\frac{bq^2}{a^2};q^2)_\infty}{(b;q^2)_\infty},$$

and a small amount of simplification yields

THEOREM 57 (The Bailey–Daum summation formula). *If $|q| < |a|$ and $|q| < 1$, then*

$$(5.6.2) \quad _2\phi_1\left(\begin{matrix} a,b \\ \frac{bq}{a} \end{matrix}; q, -\frac{q}{a}\right) = \frac{(bq;q^2)_\infty (\frac{bq^2}{a^2};q^2)_\infty}{(\frac{bq}{a};q)_\infty (-\frac{q}{a};q)_\infty (q;q^2)_\infty}.$$

This is another of those curious instances in mathematics when two people (you guessed it, Bailey and Daum) working independently found the same theorem at almost exactly the same time. This simple proof is due to George Andrews.

The Bailey–Daum formula is an extension of Lebesgue's identity from section 5.3; see problem 7 for the details. Our main goal in this section is to rederive the double Lebesgue identity (5.4.4). We start with a $_3\phi_2$ transformation of Newman Hall. A generic $_3\phi_2$ series is

$$_3\phi_2\left(\begin{matrix} a,b,c \\ e,f \end{matrix}; q, z\right) = \sum_{n=0}^{\infty} \frac{(a;q)_n (b;q)_n (c;q)_n}{(e;q)_n (f;q)_n (q;q)_n} z^n,$$

which, using the fact that $(a;q)_\infty = (a;q)_n (aq^n;q)_\infty$, we can rewrite as

$$(5.6.3) \quad _3\phi_2\left(\begin{matrix} a,b,c \\ e,f \end{matrix}; q, z\right) = \sum_{n=0}^{\infty} \frac{(b;q)_n (c;q)_n}{(q;q)_n} z^n \frac{(a;q)_\infty (eq^n;q)_\infty (fq^n;q)_\infty}{(aq^n;q)_\infty (e;q)_\infty (f;q)_\infty}.$$

We want to use Jacobi's q-Gauss summation formula

$$(5.6.4) \qquad {}_2\phi_1\left(\begin{matrix} r,s \\ rst \end{matrix}; q, t\right) = \frac{(rt;q)_\infty (st;q)_\infty}{(t;q)_\infty (rst;q)_\infty}$$

on the infinite products containing q^n. We take $rt = eq^n$, $st = fq^n$, and $t = aq^n$, which makes $r = \frac{e}{a}$ and $s = \frac{f}{a}$, so $rst = \frac{ef}{a}q^n$. We need to put in $(rst;q)_\infty$ to have the setup for (5.6.4), so we rewrite the right side of (5.6.3) as

$$\frac{(a;q)_\infty}{(e;q)_\infty (f;q)_\infty} \sum_{n=0}^\infty \frac{(b;q)_n (c;q)_n}{(q;q)_n} z^n \left(\frac{efq^n}{a};q\right)_\infty \frac{(eq^n;q)_\infty (fq^n;q)_\infty}{(aq^n;q)_\infty (\frac{efq^n}{a};q)_\infty}$$

$$= \frac{(a;q)_\infty (\frac{ef}{a};q)_\infty}{(e;q)_\infty (f;q)_\infty} \sum_{n=0}^\infty \frac{(b;q)_n (c;q)_n}{(q;q)_n (\frac{ef}{a};q)_n} z^n \frac{(eq^n;q)_\infty (fq^n;q)_\infty}{(aq^n;q)_\infty (\frac{efq^n}{a};q)_\infty}.$$

Using (5.6.4) with r, s, t as above, this is

$$(5.6.5) \qquad {}_3\phi_2\left(\begin{matrix} a,b,c \\ e,f \end{matrix}; q, z\right) = \frac{(a;q)_\infty (\frac{ef}{a};q)_\infty}{(e;q)_\infty (f;q)_\infty}$$

$$\times \sum_{n=0}^\infty \frac{(b;q)_n (c;q)_n}{(q;q)_n (\frac{ef}{a};q)_n} z^n \sum_{k=0}^\infty \frac{(\frac{e}{a};q)_k (\frac{f}{a};q)_k}{(\frac{efq^n}{a};q)_k (q;q)_k} (aq^n)^k.$$

Note that

$$(5.6.6) \qquad \left(\frac{ef}{a};q\right)_n \left(\frac{efq^n}{a};q\right)_k = \left(\frac{ef}{a};q\right)_{n+k} = \left(\frac{ef}{a};q\right)_k \left(\frac{efq^k}{a};q\right)_n.$$

Using this in (5.6.5) and changing orders of summation we have

$$(5.6.7) \qquad {}_3\phi_2\left(\begin{matrix} a,b,c \\ e,f \end{matrix}; q, z\right) = \frac{(a;q)_\infty (\frac{ef}{a};q)_\infty}{(e;q)_\infty (f;q)_\infty}$$

$$\times \sum_{k=0}^\infty \frac{(\frac{e}{a};q)_k (\frac{f}{a};q)_k}{(\frac{ef}{a};q)_k (q;q)_k} a^k \sum_{n=0}^\infty \frac{(b;q)_n (c;q)_n}{(q;q)_n (\frac{efq^k}{a};q)_n} (zq^k)^n.$$

We now choose z so that (5.6.4) works on the inner sum. For this we need $bczq^k = \frac{efq^k}{a}$, so $z = \frac{ef}{abc}$. Then (5.6.4) gives

$${}_3\phi_2\left(\begin{matrix} a,b,c \\ e,f \end{matrix}; q, \frac{ef}{abc}\right) = \frac{(a;q)_\infty (\frac{ef}{a};q)_\infty}{(e;q)_\infty (f;q)_\infty} \sum_{k=0}^\infty \frac{(\frac{e}{a};q)_k (\frac{f}{a};q)_k}{(\frac{ef}{a};q)_k (q;q)_k} a^k \frac{(\frac{efq^k}{ab};q)_\infty (\frac{efq^k}{ac};q)_\infty}{(\frac{efq^k}{a};q)_\infty (\frac{efq^k}{abc};q)_\infty}.$$

Now

$$(5.6.8) \qquad \left(\frac{ef}{a};q\right)_k \left(\frac{efq^k}{a};q\right)_\infty = \left(\frac{ef}{a};q\right)_\infty,$$

so this simplifies a little to

$${}_3\phi_2\left(\begin{matrix} a,b,c \\ e,f \end{matrix}; q, \frac{ef}{abc}\right) = \frac{(a;q)_\infty}{(e;q)_\infty (f;q)_\infty} \sum_{k=0}^\infty \frac{(\frac{e}{a};q)_k (\frac{f}{a};q)_k}{(q;q)_k} a^k \frac{(\frac{efq^k}{ab};q)_\infty (\frac{efq^k}{ac};q)_\infty}{(\frac{efq^k}{abc};q)_\infty}.$$

Finally, we multiply top and bottom inside the sum by

$$\left(\frac{ef}{ab};q\right)_k \left(\frac{ef}{ac};q\right)_k \left(\frac{ef}{abc};q\right)_k.$$

Using (5.6.8) again, this gives

$$_3\phi_2\left(\begin{matrix}a,b,c\\e,f\end{matrix};q,\frac{ef}{abc}\right) = \frac{(a;q)_\infty(\frac{ef}{ab};q)_\infty(\frac{ef}{ac};q)_\infty}{(e;q)_\infty(f;q)_\infty(\frac{ef}{abc};q)_\infty}\sum_{k=0}^\infty \frac{(\frac{e}{a};q)_k(\frac{f}{a};q)_k(\frac{ef}{abc};q)_k}{(\frac{ef}{ab};q)_k(\frac{ef}{ac};q)_k(q;q)_k}a^k,$$

which is

THEOREM 58 (Newman Hall's $_3\phi_2$ transformation). *If $|q|<1$, then*

(5.6.9) $\quad _3\phi_2\left(\begin{matrix}a,b,c\\e,f\end{matrix};q,\frac{ef}{abc}\right) = \frac{(a;q)_\infty(\frac{ef}{ab};q)_\infty(\frac{ef}{ac};q)_\infty}{(e;q)_\infty(f;q)_\infty(\frac{ef}{abc};q)_\infty}\,_3\phi_2\left(\begin{matrix}\frac{e}{a},\frac{f}{a},\frac{ef}{abc}\\\frac{ef}{ab},\frac{ef}{ac}\end{matrix};q,a\right).$

It is interesting to observe that

$$\frac{\frac{ef}{ab}\cdot\frac{ef}{ac}}{\frac{e}{a}\cdot\frac{f}{a}\cdot\frac{ef}{abc}} = a,$$

so the new $_3\phi_2$ series is of the same type as the old one: the power series variable equals the product of the denominator parameters divided by the product of the numerator parameters. This means that (5.6.9) can be applied to itself, and the result turns out to be a slightly simpler transformation. To make this a bit less confusing, we rewrite (5.6.9) with different letters:

$$_3\phi_2\left(\begin{matrix}r,s,t\\u,v\end{matrix};q,\frac{uv}{rst}\right) = \frac{(r;q)_\infty(\frac{uv}{rs};q)_\infty(\frac{uv}{rt};q)_\infty}{(u;q)_\infty(v;q)_\infty(\frac{uv}{rst};q)_\infty}\,_3\phi_2\left(\begin{matrix}\frac{u}{r},\frac{v}{r},\frac{uv}{rst}\\\frac{uv}{rs},\frac{uv}{rt}\end{matrix};q,r\right).$$

Taking $r=\frac{f}{a}$, $s=\frac{e}{a}$, $t=\frac{ef}{abc}$, $u=\frac{ef}{ab}$, and $v=\frac{ef}{ac}$ here, we have $\frac{uv}{rst}=a$ and

$$_3\phi_2\left(\begin{matrix}\frac{e}{a},\frac{f}{a},\frac{ef}{abc}\\\frac{ef}{ab},\frac{ef}{ac}\end{matrix};q,a\right) = \frac{(\frac{f}{a};q)_\infty(\frac{ef}{bc};q)_\infty(e;q)_\infty}{(\frac{ef}{ab};q)_\infty(\frac{ef}{ac};q)_\infty(a;q)_\infty}\,_3\phi_2\left(\begin{matrix}\frac{e}{b},\frac{e}{c},a\\\frac{ef}{bc},e\end{matrix};q,\frac{f}{a}\right).$$

Using this in (5.6.9) we get

THEOREM 59 (Hall's iterated transformation). *If $|q|<1$, then*

(5.6.10) $\quad _3\phi_2\left(\begin{matrix}a,b,c\\e,f\end{matrix};q,\frac{ef}{abc}\right) = \frac{(\frac{f}{a};q)_\infty(\frac{ef}{bc};q)_\infty}{(f;q)_\infty(\frac{ef}{abc};q)_\infty}\,_3\phi_2\left(\begin{matrix}a,\frac{e}{b},\frac{e}{c}\\e,\frac{ef}{bc}\end{matrix};q,\frac{f}{a}\right).$

A key lemma of Andrews is a limiting case of (5.6.10). Note that

$$\lim_{c\to\infty}\frac{(c;q)_n}{c^n} = (-1)^n q^{\binom{n}{2}}$$

from (i) of problem 9 in the previous section. Setting $f=xa$ in (5.6.10) and letting $c\to\infty$ then gives

LEMMA 4 (Andrews's lemma). *If $|q|<1$, then*

(5.6.11) $\quad\displaystyle\sum_{n=0}^\infty \frac{(a;q)_n(b;q)_n}{(e;q)_n(q;q)_n(ax;q)_n}\left(-\frac{ex}{b}\right)^n q^{\binom{n}{2}} = \frac{(x;q)_\infty}{(ax;q)_\infty}\,_2\phi_1\left(\begin{matrix}a,\frac{e}{b}\\e\end{matrix};q,x\right).$

Andrews makes two applications of this lemma. First he takes $e=\sqrt{qab}$ and $x=-\sqrt{\frac{qb}{a}}$, so that $-\frac{ex}{b}=q$ and

$$(e;q)_n(ax;q)_n = (\sqrt{qab};q)_n(-\sqrt{qab};q)_n = (qab;q^2)_n.$$

This makes the left side of (5.6.11) into
$$\sum_{n=0}^{\infty} \frac{(a;q)_n (b;q)_n}{(q;q)_n (qab;q^2)_n}.$$

On the right side we have
$$\frac{(-\sqrt{\frac{qb}{a}};q)_\infty}{(-\sqrt{qab};q)_\infty} {}_2\phi_1\left(\begin{array}{c} a, \sqrt{\frac{qa}{b}} \\ \sqrt{qab} \end{array}; q, -\sqrt{\frac{qb}{a}}\right)$$

and the Bailey–Daum formula (rewritten with $b = r$ and $a = s$)

(5.6.12) $\quad {}_2\phi_1\left(\begin{array}{c} r, s \\ \frac{qr}{s} \end{array}; q, -\frac{q}{s}\right) = \frac{(rq;q^2)_\infty (\frac{q^2 r}{s^2};q^2)_\infty}{(\frac{qr}{s};q)_\infty (-\frac{q}{s};q)_\infty (q;q^2)_\infty}$

applies to this sum. Taking $r = a$ and $s = \sqrt{\frac{qa}{b}}$ and simplifying we have

(5.6.13) $\quad \sum_{n=0}^{\infty} \frac{(a;q)_n (b;q)_n}{(q;q)_n (qab;q^2)_n} q^{\binom{n+1}{2}} = \frac{(aq;q^2)_\infty (bq;q^2)_\infty}{(q;q^2)_\infty (qab;q^2)_\infty}.$

This is Andrews's proof of the double Lebesgue identity (5.4.4).

For a second application of Andrews's lemma, we take $a = \frac{q}{b}$ and $x = -b$, and rename e as z. Then $ax = -q$, so on the left side of (5.6.10) $(ax;q)_n (q;q)_n$ becomes $(-q;q)_n (q;q)_n = (q^2;q^2)_n$. On the right side we have

$$\frac{(-b;q)_\infty}{(-q;q)_\infty} {}_2\phi_1\left(\begin{array}{c} \frac{q}{b}, \frac{z}{b} \\ z \end{array}; q, -b\right).$$

Using (5.6.11) we get

(5.6.14) $\quad \sum_{n=0}^{\infty} \frac{(b;q)_n (\frac{q}{b};q)_n}{(z;q)_n (q^2;q^2)_n} q^{\binom{n}{2}} z^n = \frac{(zb;q)_\infty (\frac{z}{b};q)_\infty}{(z;q)_\infty}$

after a little reduction.

Exercises

1. Show that (5.6.1) reduces to the Bailey–Daum formula.
2. Verify (5.6.6).
3. Check the details of the proof of (5.6.10). How does the ${}_3\phi_2$ on the right become a ${}_2\phi_1$?
4. Check the details of the proof of (5.6.13).
5. Check the details of the proof of (5.6.14).
6. Following Hall, take $a = b = z$, $e = f = qz$, and $c = q$ in (5.6.9). Show that after multiplying the result by $z/(1-z)^2$, we have
$$\sum_{n=0}^{\infty} \frac{(q;q)_n z^{n+1}}{(1-q^{n+1})(z;q)_{n+1}} = \sum_{k=0}^{\infty} \frac{zq^k}{(1-zq^k)^2},$$

which Bailey proves in a more elementary way in problem 18 in section 4.4. As in that problem, this also implies

$$\sum_{n=0}^{\infty} \frac{(q;q)_n z^{n+1}}{(1-q^{n+1})(z;q)_{n+1}} = \sum_{n=1}^{\infty} \frac{nz^n}{1-q^n}.$$

7. In this problem we will develop the remark in the text that the Bailey–Daum formula (5.6.12) is a generalization of Lebesgue's identity.

   (i) Show that

   $$(s;q)_n \left(-\frac{q}{s}\right)^n = \left(q - \frac{q}{s}\right)\left(q^2 - \frac{q}{s}\right) \cdots \left(q^n - \frac{q}{s}\right).$$

   (ii) Show that we get Lebesgue's identity if we let $s \to \infty$ in (5.6.12) and use (i).

8. This problem outlines Fine's proof of the **Rogers–Fine identity**: if $|q| < 1$ and $|x| < 1$, then

   (5.6.15) $$\sum_{n=0}^{\infty} \frac{(a;q)_n}{(b;q)_n} x^n = \sum_{n=0}^{\infty} \frac{(a;q)_n}{(b;q)_n} \frac{q^{n^2}}{(x;q)_{n+1}} (1 - axq^{2n}) \left(\frac{axq}{b};q\right)_n \left(\frac{bx}{q}\right)^n.$$

   (i) Denoting the left side of (5.6.15) by $F(a,b,x)$, show that

   $$F(a,b,x) = 1 + \frac{1-a}{1-b} x \sum_{n=1}^{\infty} \frac{(aq;q)_{n-1}}{(bq;q)_{n-1}} x^{n-1},$$

   and explain why this implies that

   (5.6.16) $$F(a,b,x) = 1 + \frac{1-a}{1-b} x\, F(aq, bq, x).$$

   (ii) The purpose of the next several steps is to derive a relation that will allow us to change (5.6.16) into an identity connecting $F(a,b,x)$ to $F(aq, bq, xq)$ instead of $F(aq, bq, x)$. If $f_n = (a;q)_n/(b;q)_n$, show that

   $$f_n(1 - aq^n) = f_{n+1}(1 - bq^n),$$

   and hence

   $$\sum_{n=0}^{\infty} f_n(1 - aq^n) x^{n+1} = \sum_{n=0}^{\infty} f_{n+1}(1 - bq^n) x^{n+1}.$$

   (iii) By rearranging the result of (ii), show that

   (5.6.17) $$(1-x)F(a,b,x) = 1 - \frac{b}{q} + \left(\frac{b}{q} - ax\right) F(a, b, xq).$$

   (iv) Use (5.6.16) with (5.6.17) to show that

   (5.6.18) $$F(a,b,x) = \frac{1-ax}{1-x} + \frac{1-a}{1-b} \frac{1 - \frac{axq}{b}}{1-x} bx\, F(aq, bq, xq).$$

   (v) Using (5.6.18) on itself, show that

   $$F(a,b,x) = \frac{1-ax}{1-x} + \frac{1-a}{1-b} \frac{(1 - \frac{axq}{b})(1 - axq^2)}{(1-x)(1-xq)} bx$$

   $$+ \frac{(1-a)(1-aq)}{(1-b)(1-bq)} \frac{(1 - \frac{axq}{b})(1 - \frac{axq^2}{b})}{(1-x)(1-xq)} b^2 x^2 q^2\, F(aq^2, bq^2, xq^2).$$

(vi) Check that

$$(5.6.19) \quad F(a,b,x) = \sum_{n=0}^{s-1} \frac{(a;q)_n}{(b;q)_n} \frac{(\frac{axq}{b};q)_n}{(x;q)_{n+1}} \left(1 - axq^{2n}\right) \left(\frac{bx}{q}\right)^n q^{n^2}$$

$$+ \frac{(a;q)_s}{(b;q)_s} \frac{(\frac{axq}{b};q)_s}{(x;q)_s} \left(\frac{bx}{q}\right)^s q^{s^2} F(aq^s, bq^s, xq^s)$$

holds if $s = 0, 1, 2$. (Recall that an empty sum equals zero.)

(vii) Use (5.6.18) to prove (5.6.19) by induction on s.

(viii) Explain why $|q| < 1$ implies that letting $s \to \infty$ in (5.6.19) proves (5.6.15).

9. Why is it interesting to take $b = q$ in the Rogers–Fine identity?

10. One might expect that changing a to $-a$ in (5.6.15) and letting $b \to 0$ would reproduce (3.5.6). Show that it gives instead

$$(5.6.20) \quad \sum_{n=0}^{\infty}(-a;q)_n x^n = \sum_{n=0}^{\infty} \frac{(-a;q)_n}{(x;q)_{n+1}} \left(1 + axq^{2n}\right) a^n x^{2n} q^{\frac{n(3n-1)}{2}}$$

$$= \sum_{n=0}^{\infty} \frac{q^{\binom{n}{2}}(ax)^n}{(x;q)_{n+1}},$$

where the last equality is by (3.5.6).

11. Change q to q^2 in (5.6.15) and take $a = q$, $x = q$, and $b = -q^3$. Show that the result can be written as

$$\sum_{n=0}^{\infty} \frac{(q;q^2)_n}{(-q;q^2)_{n+1}} q^n = \sum_{n=0}^{\infty}(-1)^n q^{2n(n+1)}.$$

12. (a) Show that changing q to q^2 in (5.6.15) and taking $a = y = x$ and $b = yq$ gives

$$\sum_{n=0}^{\infty} \frac{(y;q^2)_n}{(yq;q^2)_n} y^n = \sum_{n=0}^{\infty} y^{2n} q^{n(2n-1)} \left(1 + yq^{2n}\right).$$

(b) Show that the right side can be rewritten as $\sum_{k=0}^{\infty} y^k q^{\binom{k}{2}}$.

13. This problem outlines a direct proof of the result of the previous problem.

(i) Assuming as usual that $|q| < 1$, show that

$$\sum_{n=0}^{\infty} \frac{(y;q^2)_n}{(yq;q^2)_n} y^n$$

converges if $|y| < 1$.

(ii) Let $f(y)$ denote the series in (i), and write

$$f(y) = 1 + \sum_{n=0}^{\infty} \frac{(y;q^2)_{n+1}}{(yq;q^2)_{n+1}} y^{n+1}$$

in the first instance of $f(y)$ below. Show that

$$f(y) - y^2 q\, f(yq^2) = 1 + y \sum_{n=0}^{\infty} \frac{(yq^2;q^2)_n\, y^n}{(yq;q^2)_{n+1}} \left[1 - y - yq^{2n+1}(1 - yq)\right].$$

(iii) By rewriting $1 - y - yq^{2n+1}(1 - yq) = 1 - yq^{2n+1} - y(1 - yq^{2n+2})$, show that (ii) simplifies to $f(y) - y^2 q f(yq^2) = 1 + y$.

(iv) Set $f(y) = \sum_{n=0}^{\infty} c_n(q) y^n$, where $c_n(q)$ is independent of y. Use the result of (iii) to show that
$$c_{n+2}(q) = q^{2n+1} c_n(q) \quad \text{for } n \geq 0, \text{ with } c_0(q) = 1 = c_1(q).$$

(v) Show that $c_n(q) = q^{\binom{n}{2}}$, and hence
$$\sum_{n=0}^{\infty} \frac{(y; q^2)_n}{(yq; q^2)_n} y^n = \sum_{n=0}^{\infty} y^n q^{\binom{n}{2}}.$$

14. Using either (5.5.2) or the result of the previous problem, show that
$$\sum_{n=0}^{\infty} \frac{(q; q^2)_n}{(q^2; q^2)_n} q^n = \frac{(q^2; q^2)_\infty}{(q; q^2)_\infty}.$$

15. Show that taking $a = \frac{z^2 q}{x}$ and $b = -zq$ in (5.6.15) and letting $x \to 0$ gives
$$\sum_{n=0}^{\infty} \frac{(-1)^n z^{2n} q^{\binom{n+1}{2}}}{(-zq; q)_n} = \sum_{n=0}^{\infty} z^{2n} q^{\frac{n(3n+1)}{2}} \left(1 - z^2 q^{2n+1}\right).$$

5.7. The q-Pfaff–Saalschütz identity

Suppose we rewrite Heine's q-Euler transformation (5.5.10) as

(5.7.1) $$\frac{(z; q)_\infty}{(\frac{abz}{c}; q)_\infty} {}_2\phi_1 \left(\begin{matrix} a, b \\ c \end{matrix}; q, z \right) = {}_2\phi_1 \left(\begin{matrix} \frac{c}{a}, \frac{c}{b} \\ c \end{matrix}; q, \frac{abz}{c} \right).$$

The fraction on the left can be expanded by the q-binomial series (5.5.2) with $x = \frac{abz}{c}$ and $s = \frac{c}{ab}$. Putting this in (5.7.1) and writing out the other series there we have

(5.7.2) $$\sum_{j=0}^{\infty} \frac{(\frac{c}{ab}; q)_j}{(q; q)_j} \left(\frac{abz}{c} \right)^j \sum_{k=0}^{\infty} \frac{(a; q)_k (b; q)_k}{(c; q)_k (q; q)_k} z^k = \sum_{n=0}^{\infty} \frac{(\frac{c}{a}; q)_n (\frac{c}{b}; q)_n}{(c; q)_n (q; q)_n} \left(\frac{abz}{c} \right)^n.$$

On the left side of (5.7.2) we set $j + k = n$. It is also convenient to insert $(q; q)_n / (q; q)_n$ there and to rewrite $z^k = (\frac{abz}{c})^k (\frac{c}{ab})^k$, so that the left side of (5.7.2) now looks like

(5.7.3) $$\sum_{n=0}^{\infty} \sum_{j+k=n} \frac{(\frac{abz}{c})^n}{(q; q)_n} \binom{n}{k}_q \left(\frac{c}{ab}; q\right)_j \frac{(a; q)_k (b; q)_k}{(c; q)_k} \left(\frac{c}{ab}\right)^k.$$

Since $j \geq 0$, the inner sum in (5.7.3) will go from $k = 0$ to $k = n$ if we take k as the summation index, as is natural, replacing the remaining j by $n - k$. The coefficient of $(\frac{abz}{c})^n / (q; q)_n$ in (5.7.3) must be the same as it is on the right side of (5.7.2), so we have

(5.7.4) $$\sum_{k=0}^{n} \binom{n}{k}_q \left(\frac{c}{ab}; q\right)_{n-k} \frac{(a; q)_k (b; q)_k}{(c; q)_k} \left(\frac{c}{ab}\right)^k = \frac{(\frac{c}{a}; q)_n (\frac{c}{b}; q)_n}{(c; q)_n}.$$

This is one of the many forms of the q-Pfaff–Saalschütz identity. We will see several others both in the text and in the exercises. The simple fact

(5.7.5) $$(s;q)_n = (s;q)_k \, (sq^k;q)_{n-k}$$

(which we leave as an exercise) is often useful in showing these equivalences. For example, (5.7.5) allows us to rewrite

(5.7.6) $$\left(\frac{c}{ab};q\right)_{n-k} = \frac{(\frac{c}{ab};q)_n}{(\frac{cq^{n-k}}{ab};q)_k}.$$

Using this in (5.7.4) we get

(5.7.7) $$\sum_{k=0}^{n} \binom{n}{k}_q \frac{(a;q)_k \, (b;q)_k}{(c;q)_k \, (\frac{cq^{n-k}}{ab};q)_k} \left(\frac{c}{ab}\right)^k = \frac{(\frac{c}{a};q)_n \, (\frac{c}{b};q)_n}{(c;q)_n \, (\frac{c}{ab};q)_n}.$$

In this guise we see that the q-Pfaff–Saalschütz identity is a finite form of the q-Gauss summation formula (5.5.5). (See problem 3.)

One can also rewrite the q-Pfaff–Saalschütz identity as a q-hypergeometric series. The factors inside the sum in (5.7.7) that need to be redone are

(5.7.8) $$\frac{(q;q)_n}{(q;q)_{n-k}} \frac{\left(\frac{c}{ab}\right)^k}{(\frac{cq^{n-k}}{ab};q)_k};$$

the other factors

(5.7.9) $$\frac{(a;q)_k \, (b;q)_k}{(c;q)_k \, (q;q)_k}$$

are fine as they are. Using (5.7.5) again we can rewrite (5.7.8) as

(5.7.10) $$\frac{\left(\frac{c}{ab}\right)^k (q^{n-k+1};q)_k}{(\frac{cq^{n-k}}{ab};q)_k} = \frac{\left(\frac{c}{ab}\right)^k (1 - q^{n-k+1})\cdots(1-q^n)}{\left(1 - \frac{cq^{n-k}}{ab}\right)\cdots\left(1 - \frac{cq^{n-1}}{ab}\right)}.$$

Now comes a bit of messy algebra: we want to turn these two products around so that we can describe them as q-shifted factorials without having to put k in the exponents, so we factor everything out of them. For example, we rewrite

$$1 - q^{n-k+1} = q^{n-k+1}\left(q^{k-1-n} - 1\right) = -q^{n-k+1}\left(1 - q^{k-1-n}\right),$$

and we do the same thing with all the other factors. Then (5.7.10) becomes

$$\frac{\left(\frac{c}{ab}\right)^k (q^{n-k+1})\cdots(q^n)(-1)^k}{\left(\frac{c}{ab}\right)^k (q^{n-k})\cdots(q^{n-1})(-1)^k} \frac{(1-q^{-n})\cdots(1-q^{k-1-n})}{(1 - \frac{ab}{c}q^{1-n})\cdots(1 - \frac{ab}{c}q^{k-n})} = q^k \frac{(q^{-n};q)_k}{(\frac{ab}{c}q^{1-n};q)_k}.$$

Putting this together with (5.7.9) and (5.7.7) we have

(5.7.11) $$\sum_{k=0}^{n} \frac{(a;q)_k \, (b;q)_k}{(c;q)_k \, (q;q)_k} \frac{(q^{-n};q)_k}{(\frac{ab}{c}q^{1-n};q)_k} q^k = \frac{(\frac{c}{a};q)_n \, (\frac{c}{b};q)_n}{(c;q)_n \, (\frac{c}{ab};q)_n}$$

as another alternate form of the q-Pfaff–Saalschütz identity. Now note that

$$(q^{-n};q)_k = (1 - q^{-n})\cdots(1 - q^{k-1-n})$$

equals zero (because it has a factor of $1 - 1$) if $k \geq n+1$. Therefore the sum in (5.7.11) would be no different if k went all the way to infinity—this would just add an infinite number of zeros to it. So we can rewrite (5.7.11) one more time, as a q-hypergeometric series:

$$\text{(5.7.12)} \qquad {}_3\phi_2\left(\begin{matrix} q^{-n}, a, b \\ c, \frac{ab}{c}q^{1-n} \end{matrix}; q, q\right) = \frac{(\frac{c}{a};q)_n (\frac{c}{b};q)_n}{(c;q)_n (\frac{c}{ab};q)_n}.$$

This is the form in which the formula was first written, by F. H. Jackson. A q-hypergeometric series is called *balanced* if the product of the numerator parameters and the power series variable is the same as the product of the denominator parameters. Here both products are abq^{1-n}, so (5.7.12) gives the sum of a terminating balanced ${}_3\phi_2$.

Exercises

1. Verify (5.7.5).

2. Verify (5.7.6), and show that it follows from (5.7.5).

3. Show that, at least formally, (5.7.7) becomes (5.5.5) when $n \to \infty$.

4. Show that the "Jacobi" form of the q-Pfaff–Saalschütz identity is

$$\sum_{k=0}^{n} \binom{n}{k}_q \frac{(a;q)_k (b;q)_k t^k}{(abt;q)_k (tq^{n-k};q)_k} = \frac{(at;q)_n (bt;q)_n}{(t;q)_n (abt;q)_n}.$$

5. Show that letting $a \to 0$ in (5.7.12) gives

$$\sum_{k=0}^{n} \binom{n}{k}_q (-1)^k q^{\binom{k+1}{2}-nk} \frac{(b;q)_k}{(c;q)_k} = \frac{(b-c)(b-cq)\cdots(b-cq^{n-1})}{(c;q)_n}$$

and that this is equivalent to (2.3.10) from problem 15 in section 2.3. Problem 16 in that section might help.

6. This problem shows that the Z-identity from section 1.7 is a consequence of the q-Pfaff–Saalschütz identity. It is convenient to start by replacing a, b, c by $\frac{1}{u}, \frac{1}{v}, w$ respectively in (5.7.4) to get

$$\text{(5.7.13)} \qquad \sum_{k=0}^{n} \binom{n}{k}_q (uvw;q)_{n-k} \frac{(\frac{1}{u};q)_k (\frac{1}{v};q)_k}{(w;q)_k} (uvw)^k = \frac{(uw;q)_n (vw;q)_n}{(w;q)_n}.$$

Rewriting the Z-identity with m replaced by $-m$ and with the q-binomial and multinomial coefficients broken down we have

$$\sum_m \frac{(q;q)_{a+b+c+d+e-m} \, q^{(d+m)(e+m)}}{(q;q)_{a-m}(q;q)_{b-m}(q;q)_{c-m}(q;q)_{d+m}(q;q)_{e+m}}$$

$$= \frac{(q;q)_{a+b+d+e}(q;q)_{a+c+d+e}(q;q)_{b+c+d+e}}{(q;q)_{a+d}(q;q)_{a+e}(q;q)_{b+d}(q;q)_{b+e}(q;q)_{c+d}(q;q)_{c+e}}.$$

(i) Since the Z-identity is symmetric in d and e and in a, b, c, we can assume without loss of generality that $d \leq e$ and $a \leq b, c$. We set $e + m = k$ and

$a + e = n$, thus replacing a and m by n and k. Show that this converts the Z-identity into

$$\sum_{k=0}^{n} \binom{n}{k}_q \frac{(q;q)_{b+c+d+e+n-k} \, q^{k(d-e+k)}}{(q;q)_{b+e-k}(q;q)_{c+e-k}(q;q)_{d-e+k}}$$
$$= \frac{(q;q)_{n+b+d}(q;q)_{n+c+d}(q;q)_{b+c+d+e}}{(q;q)_{d-e+n}(q;q)_{b+d}(q;q)_{b+e}(q;q)_{c+d}(q;q)_{c+e}}.$$

(ii) Show that

$$(q;q)_{b+e} = (q;q)_{b+e-k}(q^{b+e-k+1};q)_k \quad \text{and} \quad (q;q)_{n+b+d} = (q;q)_{b+d}(q^{b+d+1};q)_n$$

and do something similar with $(q;q)_{b+c+d+e+n-k}$. Show that this transforms the result of (i) into

$$\sum_{k=0}^{n} \binom{n}{k}_q (q^{b+c+d+e+1};q)_{n-k} \frac{(q^{b+e-k+1};q)_k (q^{c+e-k+1};q)_k}{(q^{d-e+1};q)_k} q^{k(d-e+k)}$$
$$= (q^{b+d+1};q)_n (q^{c+d+1};q)_n.$$

(iii) In order to make this match (5.7.13) we have to rewrite $(q^{b+e-k+1};q)_k$ and $(q^{c+e-k+1};q)_k$ in the form $(x;q)_k$ with an x that doesn't depend on k. By factoring out all the powers of q, show that

$$(q^{b+e-k+1};q)_k = (1 - q^{b+e-k+1})(1 - q^{b+e-k+2}) \cdots (1 - q^{b+e})$$
$$= (-1)^k q^{k(b+e) - \binom{k}{2}} (q^{-b-e};q)_k.$$

(iv) Show that using (iii) (twice) in (ii) we have

$$\sum_{k=0}^{n} \binom{n}{k}_q (q^{b+c+d+e+1};q)_{n-k} \frac{(q^{-b-e};q)_k (q^{-c-e};q)_k}{(q^{d-e+1};q)_k} q^{k(b+c+d+e+1)}$$
$$= (q^{b+d+1};q)_n (q^{c+d+1};q)_n.$$

Now choose u, v, w to make (5.7.13) match this.

7. This problem outlines a proof of the q-Pfaff–Saalschütz identity by the methods of Chapter 2. We consider the sum

$$S_n(a, b; c) := \sum_{k=0}^{n} \binom{n}{k}_q (a;q)_k (b;q)_k \left(\tfrac{c}{ab};q\right)_{n-k} (cq^k;q)_{n-k} \left(\tfrac{c}{ab}\right)^k.$$

(i) Explain why $S_0(a, b; c) = 1$, and show that

$$S_1(a, b; c) = \left(1 - \tfrac{c}{a}\right)\left(1 - \tfrac{c}{b}\right).$$

(ii) Show that

$$(1 - a)(1 - b) \frac{cq^n}{ab} S_n(aq, bq; cq)$$
$$= \sum_{j=1}^{n+1} \binom{n}{j-1}_q (a;q)_j (b;q)_j \left(\tfrac{c}{abq};q\right)_{n+1-j} (cq^j;q)_{n+1-j} \left(\tfrac{c}{ab}\right)^j q^{n-j+1}.$$

(iii) Show that

$$(1 - cq^n)\left(1 - \frac{cq^n}{ab}\right) S_n(a,b;c)$$
$$= \sum_{j=0}^{n} \binom{n}{j}_q (a;q)_j (b;q)_j \left(\frac{c}{ab};q\right)_{n-j} (cq^j;q)_{n+1-j} \left(1 - \frac{cq^n}{ab}\right) \left(\frac{c}{ab}\right)^j.$$

(iv) Using (ii) and (iii), show that

$$(1-a)(1-b)\frac{cq^n}{ab} S_n(aq,bq;cq) + (1 - cq^n)\left(1 - \frac{cq^n}{ab}\right) S_n(a,b;c)$$
$$= \sum_{j=0}^{n+1} \binom{n+1}{j}_q (a;q)_j (b;q)_j \left(\frac{c}{ab};q\right)_{n+1-j} (cq^j;q)_{n+1-j} \left(\frac{c}{ab}\right)^j.$$

(v) Explain why (iv) shows that

$$S_{n+1}(a,b;c) = (1-a)(1-b)\frac{cq^n}{ab} S_n(aq,bq;cq) + (1 - cq^n)\left(1 - \frac{cq^n}{ab}\right) S_n(a,b;c).$$

(vi) Use (v) to prove that

$$S_n(a,b;c) = \sum_{k=0}^{n} \binom{n}{k}_q (a;q)_k (b;q)_k \left(\frac{c}{ab};q\right)_{n-k} (cq^k;q)_{n-k} \left(\frac{c}{ab}\right)^k$$

(5.7.14)
$$= \left(\frac{c}{a};q\right)_n \left(\frac{c}{b};q\right)_n$$

by induction on n.

(vii) Show that multiplying (5.7.4) by $(c;q)_n$ and using (5.7.5) gives (5.7.14). Thus (5.7.14) is another equivalent form of the q-Pfaff–Saalschütz identity.

5.8. Bibliographical Notes

Jacobi's triple product is in section 64 of his *Fundamenta Nova* [148] of 1829, the work that made him a superstar at the age of 24. The steps up to (5.1.8) are sketched in a letter [147] from Jacobi to Legendre on April 12, 1828; see also Pieper's edition [183] of the Jacobi/Legendre correspondence and the nice biographical article [182].

The reasons why we use q in this subject, rather than some other letter, are (i) unlike x or z, q is infrequently used in other contexts in mathematics (exceptions are elementary probability, where it often equals $1 - p$ for some probability p; and elementary logic, where p and q are often two generic propositions); (ii) Jacobi used it in this sense in [148]; (iii) Heine, who was very much influenced by Jacobi, used it in this sense in [137], from which most of section 5.5 comes; and (iv) the Reverend F. H. Jackson, who found the q-Pfaff–Saalschütz formula [144] and made many other contributions to q-analysis, always used it.

The triple product appears in several places in volume 3 of Gauss's *Werke* [117]. The proof of section 5.1 is sketched in equations 6–9 on pp. 446–447, and the derivation from the symmetric ($m = n$) MacMahon q-binomial theorem is on p. 464. The latter is also in Cauchy [68]. In the form (5.1.19), the triple product is Entry 19 in Chapter 16 of Ramanujan's notebooks [49].

The beautiful identity (5.2.8) is (5) in the 66th and last section of [**150**]. In Gauss it is on p. 440 of [**117**] and p. 293 of [**118**]; his proof in problem 17 of section 5.2 is sketched in both places. The connection with (2.7.9) was pointed out by Hirschhorn in [**138**] and [**140**], as was the connection between problem 4 in section 5.1 and problem 10 in section 2.5. The incredible formula (5.2.15) in section 5.2 is equation 19 on p. 447 of [**117**] and equation 22 on p. 467. The proof in problem 19 is sketched on p. 447 and that of problem 20 on pp. 466–67. Problems 15 and 16 in that section come from pp. 437–440 of [**117**] (problems 8 and 9 come from p. 449), and are repeated on pp. 290–292 of [**118**]. Despite the reverence that many mathematicians have for Gauss, these arguments have received surprisingly little attention, but Ranjan Roy also discusses them in [**210**]. Problem 14 in this section comes from [**247**]. Problem 17 in section 5.1 comes from [**68**], problem 24 from [**54**], and problem 12 and several others from [**118**].

The determination of $c_0(q)$ from Frobenius symbols is more or less the same as the one in section 3 of [**18**], or in [**61**]. See pp. 67–68 of Pak's paper [**178**] for trenchant comments on the history of this proof. Frobenius symbols were introduced in [**110**].

The best reference for the quintuple product identity, which has been given in many different forms and with many different proofs, is the beautiful paper [**80**] by Shaun Cooper, a friend of mine from graduate school. In his book [**81**] he prefers the form (5.3.11). The quintuple product is often associated with G. N. Watson, who called attention to it in [**245**] and [**246**], two of his many papers on Ramanujan's notebooks, and realized that Ramanujan must have known it. For the references to Ramanujan see [**25**], p. 14; [**26**], pp. 53–54; and [**49**], pp. 80–83. An equivalent form is in Fricke's 1916 book [**109**] on elliptic functions. In some sense it dates back to the 19th century, since a result on sigma functions that implies it is in Schwarz's account [**217**] of Weierstrass's lectures on elliptic functions.

Lebesgue's identity is (8) in section 66 of [**148**]. Lebesgue published it 11 years later [**162**], with a reference to Gauss but not to Jacobi. Problems 14 and 15 in section 5.4 also come from this section of [**148**]—(a) and (b) in problem 14 are (9) and (10) in section 66—and problem 16 is from [**162**]. Jacobi does not get around to proving (a) and (b) until 1849 [**151**], when he discusses the function $f(z)$ from problem 11. The result of that problem may have a better claim to the name "Lebesgue's identity" than Lebesgue's identity does. Problem 4 in this section is essentially equation (40) in [**219**], to which we will return in Chapter 12. Problems 5–8 come from [**191**], as does problem 12 in section 5.5. Problem 9 comes from [**209**], which has some further discussion of Lebesgue's identity and partitions. The double Lebesgue identity comes from [**13**], as does section 5.6 with Andrews's simple proof of the Bailey–Daum formula. The original references for the latter are [**43**] and [**85**]. Ramanujan's transformation in problem 11 of section 5.5 comes from [**25**]. An equivalent form is used in [**56**], from which problem 10 in that section comes. Problem 14 is from [**26**]. The q-Gauss sum dates back to Jacobi [**150**] and Heine [**137**].

The Rogers–Fine identity in problem 8 in section 5.6 first appeared in [**202**], and is one of the major results in the first chapter of Fine's fine book [**103**]. It is also the centerpiece of Chapter 9 in the first volume of Andrews and Berndt's edition of Ramanujan's "lost notebook" [**25**]. Several problems in section 5.6 are from this chapter. So is (5.5.13), which originally appeared in [**7**] and may also be found in [26]. This notebook almost certainly contains the work Ramanujan did in the last year of his short life, whereas the other notebooks are of earlier vintage. The fifth and final installment of Andrews and Berndt's work on this is [29], and only the fourth volume [28] does not discuss q-analysis.

A few of the problems in this book were adapted from lectures by George Gasper at the University of Toronto in 1995, in which he made a point of trying to give different proofs than in [114], the standard reference for q-hypergeometric series. Problem 5 in section 5.7 comes from [253].

CHAPTER 6

Ramanujan's $_1\psi_1$ Summation Formula

6.1. Ramanujan's formula

This chapter is devoted to a formula found by Ramanujan around the start of the first World War, although it did not appear in print until Hardy's book on Ramanujan was published near the start of the second World War, and there was no published proof until after the war.

THEOREM 60 (Ramanujan's $_1\psi_1$ summation formula). *If $|q| < 1$ and $\left|\frac{b}{a}\right| < |x| < 1$, then*

$$(6.1.1) \qquad \frac{(ax;q)_\infty \left(\frac{q}{ax};q\right)_\infty (q;q)_\infty \left(\frac{b}{a};q\right)_\infty}{(x;q)_\infty \left(\frac{b}{ax};q\right)_\infty (b;q)_\infty \left(\frac{q}{a};q\right)_\infty} = \sum_{n=-\infty}^{\infty} \frac{(a;q)_n}{(b;q)_n} x^n.$$

We will give four proofs in section 6.2, two in the text and two more in the exercises. Section 6.3 has Schlosser's proof, which is of the finite to infinite type. A historically motivated proof, due in part to Cauchy, is in sections 6.4–6.6. Computationally, the simplest proof is that of Mourad Ismail. We will give part of it in the last exercise in this section, and sketch the rest in Chapter 13. But we need to tie up several loose ends first.

A series with both positive and negative powers of x, like (6.1.1) or Jacobi's triple product, is called a **Laurent** series. These are rarely discussed in real analysis, but are fundamental in complex analysis. Such a series obviously cannot converge if $x = 0$. If it has infinitely many positive powers, then it may diverge if $|x|$ is sufficiently large; if it has infinitely many negative powers, then it may diverge if $\left|\frac{1}{x}\right|$ is sufficiently large, so a Laurent series typically converges *between* two values of $|x|$. If we think of x as a complex variable, then the region of convergence is an **annulus** in the complex plane; in other words, a ring-shaped region with an inner circle and a concentric outer circle. Jacobi's triple product is particularly nice in that this annulus is the whole complex plane except the origin, the largest possible for a nontrivial Laurent series.

For (6.1.1), the inner radius is $\left|\frac{b}{a}\right|$ and the outer radius is 1. The ratio test shows this, as we will see presently, but it is easy to spot these two values by looking at the denominator on the left side of (6.1.1): $(x;q)_\infty$ would have zero as a factor if $x = 1$, and $\left(\frac{b}{ax};q\right)_\infty$ would be zero if $x = \frac{b}{a}$. Note that $(x;q)_\infty$ is also zero if $x = q^{-1}, q^{-2}, q^{-3}, \ldots$, but these are all outside the outer circle $|x| = 1$ since $|q| < 1$. Also $\left(\frac{b}{ax};q\right)_\infty$ is zero if $x = \frac{bq}{a}, \frac{bq^2}{a}, \frac{bq^3}{a}, \ldots$, but these are all inside the inner circle $|x| = \left|\frac{b}{a}\right|$, so the product side of (6.1.1) is a well-behaved function (an **analytic** function, in the language of complex analysis) of x inside its annulus.

Another aspect of (6.1.1) has to be explained before we go any further: what does $(a;q)_n$ mean if n is negative? If n is *not* negative, we can write

$$(a;q)_n = (1-a)(1-aq)\cdots(1-aq^{n-1})$$

$$= (1-a)(1-aq)\cdots(1-aq^{n-1})\frac{(1-aq^n)(1-aq^{n+1})\cdots}{(1-aq^n)(1-aq^{n+1})\cdots}$$

$$= \frac{(a;q)_\infty}{(aq^n;q)_\infty}.$$

This probably seems like a silly idea—why would we want to replace a simple finite product, convergent for any q, by a quotient of two infinite products that need $|q|<1$ for convergence? The answer is that the right side makes sense if n is a negative integer—in fact for any n, though we will not use the idea at this level of generality—so we can take it as the definition of the left side for a generic n. In other words, we now *define*

(6.1.2) $$(a;q)_n = \frac{(a;q)_\infty}{(aq^n;q)_\infty}.$$

Let's see what this looks like for a negative integer. Replacing n by $-n$ in (6.1.2) we have

(6.1.3) $$(a;q)_{-n} = \frac{(a;q)_\infty}{(aq^{-n};q)_\infty} = \frac{1}{(1-aq^{-n})\cdots(1-aq^{-1})} = \frac{1}{(aq^{-n};q)_n},$$

where we assume n is a positive integer. If we multiply top and bottom by $(-1)^n q^{1+2+\cdots+n}$ we have

$$(a;q)_{-n} = \frac{(-1)^n q^{\binom{n+1}{2}}}{(a-q)(a-q^2)\cdots(a-q^n)},$$

and factoring out the a's now gives

(6.1.4) $$(a;q)_{-n} = \frac{1}{(aq^{-n};q)_n} = \frac{(-1)^n q^{\binom{n+1}{2}}}{a^n \left(\frac{q}{a};q\right)_n}.$$

This will be useful below, and we will use (6.1.3) when we apply the ratio test to the terms of (6.1.1) at the end of section 6.3.

We conclude this section with a partial explanation of the name of the formula. Ramanujan's $_1\psi_1$ is the simplest member of another family of q-hypergeometric sums of a somewhat different character than the $_{r+1}\phi_r$'s. When there is a ψ instead of a ϕ, there is no factor $(q;q)_n$ built into the denominator, the number of numerator parameters a_j is typically the same as the number of denominator parameters b_j (rather than one more), and the series runs through *all* integers instead of just the nonnegative integers. One of the deepest results in q-analysis is a $_6\psi_6$ sum due to Bailey that we will discuss in our last chapter.

This is also a good place to illuminate the purpose of the denominator factor $(q;q)_n$ in a q-hypergeometric series. Why insist that such a factor be present? If we set $a=q$ in (6.1.2), then we have

(6.1.5) $$\frac{1}{(q;q)_n} = \frac{(q^{n+1};q)_\infty}{(q;q)_\infty} = \frac{(1-q^{n+1})(1-q^{n+2})\cdots(1-q^{n+k})\cdots}{(q;q)_\infty}.$$

Now suppose that n is a negative integer, say $n=-k$ for some positive integer k. Then the factor $1-q^{n+k}$ will be zero, so the whole numerator will be zero, so the whole fraction will be zero. Therefore, if we have a factor $(q;q)_n$ in the denominator of a series where n is the summation index, that causes any terms with a negative n to be zero. To put it another way, in a series $_{r+1}\phi_r$ the sum, which is supposed to go from $n=0$ to ∞, could just as well run from $n=-\infty$ to ∞, because the extra terms would all be zero due to the built-in factor of $(q;q)_n$ in the denominator. Thus the purpose of this factor is to force the sum to start at $n=0$, and this is why the ψ sums, which do not have this factor built in, run from $n=-\infty$ to ∞. The theory of these sums is less well developed than that of the $_{r+1}\phi_r$'s, but Ramanujan's evaluation of the $_1\psi_1$ is one of the most beautiful results in all of q-analysis. In this notation it reads

$$(6.1.6) \quad {}_1\psi_1\left(\begin{matrix}a\\b\end{matrix};q,x\right) = \sum_{n=-\infty}^{\infty} \frac{(a;q)_n}{(b;q)_n} x^n = \frac{(ax;q)_\infty \left(\frac{q}{ax};q\right)_\infty (q;q)_\infty \left(\frac{b}{a};q\right)_\infty}{(x;q)_\infty \left(\frac{b}{ax};q\right)_\infty (b;q)_\infty \left(\frac{q}{a};q\right)_\infty},$$

where $|q|<1$ and $|\frac{b}{a}|<|x|<1$. By the above remark, if $b=q$ it simplifies to

$$(6.1.7) \quad \sum_{n=0}^{\infty} \frac{(a;q)_n}{(q;q)_n} x^n = \frac{(ax;q)_\infty}{(x;q)_\infty},$$

the q-binomial series (3.7.4), which holds when $|q|$ and $|x|$ are both less than 1.

Exercises

1. Show that
$$\frac{(a;q)_n}{(aq;q)_n} = \frac{1-a}{1-aq^n}$$
for *all* values of n.

2. Show that
$$(6.1.8) \quad (a;q)_n (aq^n;q)_m = (a;q)_{n+m} = (a;q)_m (aq^m;q)_n$$
for all values of n and m.

3. Suppose that a sequence c_n satisfies $c_n = (1-aq^{n-1})c_{n-1}$. If n is a positive integer, we know this implies that $c_n = (a;q)_n c_0$. Show that this holds for *any* integer n. **Hint:** If n is not a positive integer, set $n = 1-m$ for a positive integer m.

4. This problem needs (3.5.6), which was

$$(6.1.9) \quad \sum_{n=0}^{\infty} \frac{q^{\binom{n}{2}}(az)^n}{(z;q)_{n+1}} = \sum_{n=0}^{\infty} (-a;q)_n z^n.$$

Evidently we have

$$\sum_{n=-\infty}^{\infty} \left(\frac{q}{s};q\right)_n t^{n+1} = \sum_{n=0}^{\infty} \left(\frac{q}{s};q\right)_n t^{n+1} + \sum_{n=1}^{\infty} \left(\frac{q}{s};q\right)_{-n} t^{1-n}.$$

Use (6.1.4) and (6.1.9) to rewrite this as

$$(6.1.10) \quad \sum_{n=-\infty}^{\infty} \left(\frac{q}{s};q\right)_n t^{n+1} = \sum_{n=0}^{\infty} \left(\frac{q}{s};q\right)_n t^{n+1} - \sum_{n=0}^{\infty} \left(\frac{q}{t};q\right)_n s^{n+1}.$$

This will be used in another exercise later in this chapter.

5. In this problem we begin to outline Ismail's proof of the $_1\psi_1$ summation formula.

(i) If $b = q^{m+1}$, where m is a nonnegative integer, explain why the right side of (6.1.1) becomes

$$\sum_{k=-m}^{\infty} \frac{(a;q)_k}{(q^{m+1};q)_k} x^k = (q;q)_m \sum_{k=-m}^{\infty} \frac{(a;q)_k}{(q;q)_{k+m}} x^k.$$

(ii) By multiplying top and bottom by $(1-aq^{-m})\cdots(1-aq^{-1})$ and setting $k+m=n$, show that (i) becomes

$$\frac{(q;q)_m}{x^m (aq^{-m};q)_m} \sum_{n=0}^{\infty} \frac{(aq^{-m};q)_n}{(q;q)_n} x^n.$$

(iii) Show that (ii) becomes

$$\frac{(ax;q)_\infty}{(x;q)_\infty} \frac{(axq^{-m};q)_m}{(aq^{-m};q)_m} \frac{(q;q)_m}{x^m}$$

after using (6.1.7).

(iv) Show that (iii) can be rewritten as

(6.1.11) $$\frac{(ax;q)_\infty}{(x;q)_\infty} \frac{\left(\frac{q}{ax};q\right)_m}{\left(\frac{q}{a};q\right)_m} (q;q)_m.$$

(v) Show that the left side of (6.1.1) reduces to (6.1.11) if $b = q^{m+1}$ for a nonnegative integer m, so (6.1.1) holds in this case. These are all the calculations that Ismail's argument needs. See the end of section 13.7 for how it concludes.

6.2. Four proofs

One of the first proofs of (6.1.1) was given around 1950 by Margaret Jackson, a student of Bailey. (We use her first name both to emphasize her gender and to distinguish her from the Reverend F. H. Jackson, a q-analyst from an earlier generation.) Assuming as usual that $|q| < 1$, consider the product

$$r(x) = \frac{(ax;q)_\infty \left(\frac{q}{ax};q\right)_\infty}{(x;q)_\infty \left(\frac{b}{ax};q\right)_\infty}.$$

We can expand each fraction by the q-binomial series (6.1.7), where we need $|x| < 1$ for the first fraction in $r(x)$ and $\left|\frac{b}{ax}\right| < 1$ for the second one. Thus we have
(6.2.1)
$$r(x) = \sum_{j=0}^{\infty} \frac{(a;q)_j}{(q;q)_j} x^j \sum_{k=0}^{\infty} \frac{\left(\frac{q}{b};q\right)_k}{(q;q)_k} \left(\frac{b}{ax}\right)^k = \sum_{j=0}^{\infty}\sum_{k=0}^{\infty} \frac{(a;q)_j \left(\frac{q}{b};q\right)_k}{(q;q)_j (q;q)_k} \left(\frac{b}{a}\right)^k x^{j-k}$$

for $\left|\frac{b}{a}\right| < |x| < 1$. Like Jacobi's triple product, this is a bilateral series, containing infinitely many positive and negative powers of x. Let's first set $j = k+n$ for a nonnegative integer n. Then the coefficient of x^n in $r(x)$ is

$$\sum_{k=0}^{\infty} \frac{(a;q)_{k+n} \left(\frac{q}{b};q\right)_k}{(q;q)_{k+n} (q;q)_k} \left(\frac{b}{a}\right)^k.$$

Writing $(a;q)_{k+n} = (a;q)_n(aq^n;q)_k$ and similarly for $(q;q)_{k+n}$, this becomes

(6.2.2)
$$\frac{(a;q)_n}{(q;q)_n} \sum_{k=0}^{\infty} \frac{(aq^n;q)_k \left(\frac{q}{b};q\right)_k}{(q^{n+1};q)_k (q;q)_k} \left(\frac{b}{a}\right)^k.$$

Now recall the q-Gauss sum (5.5.6) in the Jacobi form

(6.2.3)
$$\sum_{m=0}^{\infty} \frac{(u;q)_m(v;q)_m}{(q;q)_m(uvw;q)_m} w^m = \frac{(uw;q)_\infty (vw;q)_\infty}{(w;q)_\infty (uvw;q)_\infty},$$

where $|w| < 1$. This applies to (6.2.2) because

$$aq^n \cdot \frac{q}{b} \cdot \frac{b}{a} = q^{n+1}$$

and $\left|\frac{b}{a}\right| < 1$, so (6.2.2) becomes

$$\frac{(a;q)_n}{(q;q)_n} \frac{(bq^n;q)_\infty \left(\frac{q}{a};q\right)_\infty}{(q^{n+1};q)_\infty \left(\frac{b}{a};q\right)_\infty} \frac{(b;q)_n}{(b;q)_n} = \frac{(a;q)_n}{(b;q)_n} \frac{(b;q)_\infty \left(\frac{q}{a};q\right)_\infty}{(q;q)_\infty \left(\frac{b}{a};q\right)_\infty}.$$

This then is the coefficient of x^n in $r(x)$ when n is nonnegative. To deal with the case when n is negative, we can set $k = j + n$ in (6.2.1). Then the coefficient of x^{-n} there is

$$\sum_{j=0}^{\infty} \frac{(a;q)_j \left(\frac{q}{b};q\right)_{j+n}}{(q;q)_j (q;q)_{j+n}} \left(\frac{b}{a}\right)^{j+n}.$$

After a manipulation similar to the positive case, this becomes

(6.2.4)
$$\frac{\left(\frac{q}{b};q\right)_n}{(q;q)_n} \left(\frac{b}{a}\right)^n \sum_{j=0}^{\infty} \frac{(a;q)_j \left(\frac{q^{n+1}}{b};q\right)_j}{(q;q)_j (q^{n+1};q)_j} \left(\frac{b}{a}\right)^j.$$

The q-Gauss sum applies to this because

$$a \cdot \frac{q^{n+1}}{b} \cdot \frac{b}{a} = q^{n+1}$$

and $\left|\frac{b}{a}\right| < 1$, so (6.2.4) becomes

$$\frac{\left(\frac{q}{b};q\right)_n}{(q;q)_n} \left(\frac{b}{a}\right)^n \frac{(b;q)_\infty \left(\frac{q^{n+1}}{a};q\right)_\infty}{(q^{n+1};q)_\infty \left(\frac{b}{a};q\right)_\infty} \frac{\left(\frac{q}{a};q\right)_n}{\left(\frac{q}{a};q\right)_n} = \frac{\left(\frac{q}{b};q\right)_n}{\left(\frac{q}{a};q\right)_n} \left(\frac{b}{a}\right)^n \frac{(b;q)_\infty \left(\frac{q}{a};q\right)_\infty}{(q;q)_\infty \left(\frac{b}{a};q\right)_\infty}.$$

This then is the coefficient of x^{-n} in $r(x)$ when n is nonnegative. Note that we get the same four infinite products as in the x^n case, so we have proved that for $\left|\frac{b}{a}\right| < |x| < 1$ we have

$$\frac{(ax;q)_\infty \left(\frac{q}{ax};q\right)_\infty}{(x;q)_\infty \left(\frac{b}{ax};q\right)_\infty} = \frac{(b;q)_\infty \left(\frac{q}{a};q\right)_\infty}{(q;q)_\infty \left(\frac{b}{a};q\right)_\infty} \left(\sum_{n=0}^{\infty} \frac{(a;q)_n}{(b;q)_n} x^n + \sum_{n=1}^{\infty} \frac{\left(\frac{q}{b};q\right)_n}{\left(\frac{q}{a};q\right)_n} \left(\frac{b}{a}\right)^n x^{-n} \right).$$

Putting all the infinite products on the same side, this becomes

(6.2.5)
$$\frac{(ax;q)_\infty \left(\frac{q}{ax};q\right)_\infty (q;q)_\infty \left(\frac{b}{a};q\right)_\infty}{(x;q)_\infty \left(\frac{b}{ax};q\right)_\infty (b;q)_\infty \left(\frac{q}{a};q\right)_\infty} = \sum_{n=0}^{\infty} \frac{(a;q)_n}{(b;q)_n} x^n + \sum_{n=1}^{\infty} \frac{\left(\frac{q}{b};q\right)_n}{\left(\frac{q}{a};q\right)_n} \left(\frac{b}{a}\right)^n x^{-n}.$$

This is Ramanujan's summation formula, but it is possible to write it more neatly by using (6.1.4), which implies that

$$\frac{\left(\frac{q}{b};q\right)_n}{\left(\frac{q}{a};q\right)_n}\left(\frac{b}{a}\right)^n = \frac{(a;q)_{-n}}{(b;q)_{-n}}.$$

Therefore the second sum in (6.2.5) combines naturally with the first to give (6.1.1).

George Andrews gave an elegant variation of Jackson's proof, also based on Jacobi's q-Gauss summation (6.2.3). The argument uses the observation that we can write

(6.2.6) $\qquad {}_1\psi_1\left(\begin{matrix}a\\b\end{matrix};q,x\right) = \sum_{n=-\infty}^{\infty}\frac{(a;q)_n}{(b;q)_n}x^n = \sum_{n=-\infty}^{\infty}\frac{(a;q)_{n+m}}{(b;q)_{n+m}}x^{n+m}$

for any integer m, because as n runs through all the integers so does $n+m$.

Andrews starts with

(6.2.7) $\qquad \dfrac{(ax;q)_\infty\,(q;q)_\infty\,\left(\frac{b}{a};q\right)_\infty}{(x;q)_\infty\,(b;q)_\infty\,\left(\frac{q}{a};q\right)_\infty},$

which is most, but not all, of the right side of (6.1.6). He writes

(6.2.8) $\qquad \dfrac{(ax;q)_\infty}{(x;q)_\infty} = \sum_{n=0}^{\infty}\frac{(a;q)_n}{(q;q)_n}x^n = \sum_{n=-\infty}^{\infty}\frac{(a;q)_n}{(q;q)_n}x^n,$

using (6.1.7) and the remark at the end of section 6.1 about having $(q;q)_n$ in the denominator of a series, and he plugs this into (6.2.7) to get

$$\frac{(ax;q)_\infty\,(q;q)_\infty\,\left(\frac{b}{a};q\right)_\infty}{(x;q)_\infty\,(b;q)_\infty\,\left(\frac{q}{a};q\right)_\infty} = \frac{(q;q)_\infty\,\left(\frac{b}{a};q\right)_\infty}{(b;q)_\infty\,\left(\frac{q}{a};q\right)_\infty}\sum_{n=-\infty}^{\infty}\frac{(a;q)_n}{(q;q)_n}x^n$$

$$= \sum_{n=-\infty}^{\infty}\frac{(a;q)_n}{(b;q)_n}x^n\,\frac{(q^{n+1};q)_\infty\,\left(\frac{b}{a};q\right)_\infty}{(bq^n;q)_\infty\,\left(\frac{q}{a};q\right)_\infty}$$

after using (6.1.5) and (6.1.2) with $x=b$.

In (6.2.3) take

$$r=\frac{b}{q},\quad s=aq^n,\quad \text{and}\quad t=\frac{q}{a},$$

so that $\quad rt=\dfrac{b}{a},\quad st=q^{n+1},\quad \text{and}\quad rst=bq^n.$

Then (6.2.3) tells us that

$$\frac{(ax;q)_\infty\,(q;q)_\infty\,\left(\frac{b}{a};q\right)_\infty}{(x;q)_\infty\,(b;q)_\infty\,\left(\frac{q}{a};q\right)_\infty} = \sum_{n=-\infty}^{\infty}\frac{(a;q)_n}{(b;q)_n}x^n\,\frac{(q^{n+1};q)_\infty\,\left(\frac{b}{a};q\right)_\infty}{(bq^n;q)_\infty\,\left(\frac{q}{a};q\right)_\infty}$$

$$= \sum_{n=-\infty}^{\infty}\frac{(a;q)_n}{(b;q)_n}x^n\sum_{m=0}^{\infty}\frac{\left(\frac{b}{q};q\right)_m(aq^n;q)_m}{(bq^n;q)_m(q;q)_m}\left(\frac{q}{ax}\right)^m x^m$$

$$= \sum_{m=0}^{\infty}\frac{\left(\frac{b}{q};q\right)_m}{(q;q)_m}\left(\frac{q}{ax}\right)^m \sum_{n=-\infty}^{\infty}\frac{(a;q)_{n+m}}{(b;q)_{n+m}}x^{n+m},$$

where we used (6.1.8) to glue some of the products together. By (6.2.6) the inner sum is Ramanujan's $_1\psi_1$, which can come out of the outer sum since it does not depend on m. This gives

$$\frac{(ax;q)_\infty \, (q;q)_\infty \, \left(\frac{b}{a};q\right)_\infty}{(x;q)_\infty \, (b;q)_\infty \, \left(\frac{q}{a};q\right)_\infty} = {}_1\psi_1\left(\begin{matrix}a\\b\end{matrix};q,x\right) \sum_{m=0}^{\infty} \frac{\left(\frac{b}{q};q\right)_m}{(q;q)_m}\left(\frac{q}{ax}\right)^m$$

$$= {}_1\psi_1\left(\begin{matrix}a\\b\end{matrix};q,x\right) \frac{\left(\frac{b}{ax};q\right)_\infty}{\left(\frac{q}{ax};q\right)_\infty},$$

where we used (6.1.7) in the last step. This proves (6.1.6).

Exercises

1. Show that changing q to q^2 in (6.2.5) and then setting $x = -cqz$, $a = \frac{1}{c}$, and $b = dq^2$ gives

(6.2.9) $\quad \dfrac{(-qz;q^2)_\infty \left(-\frac{q}{z};q^2\right)_\infty (q^2;q^2)_\infty (cdq^2;q^2)_\infty}{(-cqz;q^2)_\infty \left(-\frac{dq}{z};q^2\right)_\infty (cq^2;q^2)_\infty (dq^2;q^2)_\infty}$

$$= 1 + \sum_{n=1}^{\infty} \frac{\left(\frac{1}{c};q^2\right)_n (-cq)^n}{(dq^2;q^2)_n} z^n + \sum_{n=1}^{\infty} \frac{\left(\frac{1}{d};q^2\right)_n (-dq)^n}{(cq^2;q^2)_n} z^{-n},$$

and that the convergence condition becomes $|dq| < |z| < \frac{1}{|cq|}$. This is Ramanujan's original form of the $_1\psi_1$.

2. Show that (6.2.9) is unchanged if we interchange c and d while replacing z by $\frac{1}{z}$. Some authors prefer (6.2.9) to (6.1.1) since this symmetry property is not evident in the latter, and for other reasons.

3. Andrews has given a similar proof of Jacobi's triple product from Euler's identities

$$(-x;q)_\infty = \sum_{n=0}^{\infty} \frac{q^{\binom{n}{2}} x^n}{(q;q)_n} \quad \text{and} \quad \frac{1}{(x;q)_\infty} = \sum_{n=0}^{\infty} \frac{x^n}{(q;q)_n}.$$

It is convenient to replace q by q^2 in both and x by zq in the former, so the proof actually uses

$$(-zq;q^2)_\infty = \sum_{n=0}^{\infty} \frac{q^{n^2} z^n}{(q^2;q^2)_n} \quad \text{and} \quad \frac{1}{(x;q^2)_\infty} = \sum_{n=0}^{\infty} \frac{x^n}{(q^2;q^2)_n}.$$

It starts with the former and with m in place of n.

(i) Explain why

$$(-zq;q^2)_\infty = \sum_{m=0}^{\infty} q^{m^2} z^m \frac{(q^{2m+2};q^2)_\infty}{(q^2;q^2)_\infty}.$$

(ii) Explain why we can rewrite this as

$$(-zq;q^2)_\infty = \frac{1}{(q^2;q^2)_\infty} \sum_{m=-\infty}^{\infty} q^{m^2} z^m (q^{2m+2};q^2)_\infty.$$

(iii) Explain why we can rewrite this as
$$(-zq;q^2)_\infty (q^2;q^2)_\infty = \sum_{m=-\infty}^{\infty} q^{m^2} z^m \sum_{n=0}^{\infty} \frac{(-1)^n q^{n^2} q^{n(2m+1)}}{(q^2;q^2)_n}.$$

(iv) Explain why we can rewrite this as
$$(-zq;q^2)_\infty (q^2;q^2)_\infty = \sum_{m=-\infty}^{\infty} \sum_{n=0}^{\infty} q^{n^2+2mn+m^2} z^{n+m} \left(-\frac{q}{z}\right)^n \frac{1}{(q^2;q^2)_n}.$$

(v) Explain why we can rewrite this as
$$(-zq;q^2)_\infty (q^2;q^2)_\infty = \sum_{n=0}^{\infty} \frac{\left(-\frac{q}{z}\right)^n}{(q^2;q^2)_n} \sum_{k=-\infty}^{\infty} q^{k^2} z^k$$
and how Jacobi's triple product follows from this.

4. Askey has given a beautiful functional equation proof of (6.1.1). We set
$$f(a,b,x) = \sum_{n=-\infty}^{\infty} \frac{(a;q)_n}{(b;q)_n} x^n$$
and note that, because of (6.2.8), we have
$$f(a,q,x) = \sum_{n=-\infty}^{\infty} \frac{(a;q)_n}{(q;q)_n} x^n = \sum_{n=0}^{\infty} \frac{(a;q)_n}{(q;q)_n} x^n = \frac{(ax;q)_\infty}{(x;q)_\infty}.$$

(i) Explain why we can rewrite
$$f(a,b,x) = \frac{(a;q)_\infty}{(b;q)_\infty} g(b), \quad \text{where} \quad g(b) = \sum_{n=-\infty}^{\infty} \frac{(bq^n;q)_\infty}{(aq^n;q)_\infty} x^n.$$

(ii) Explain why
$$g(q) = \frac{(ax;q)_\infty (q;q)_\infty}{(x;q)_\infty (a;q)_\infty}.$$

(iii) Explain why
$$g(b) = \sum_{n=-\infty}^{\infty} \frac{(bq^{n+1};q)_\infty}{(aq^n;q)_\infty} x^n (1-bq^n)$$
$$= \sum_{n=-\infty}^{\infty} \frac{(bq^{n+1};q)_\infty}{(aq^n;q)_\infty} x^n \left[1 - \frac{b}{a} + \frac{b}{a}(1-aq^n)\right]$$
implies that
$$g(b) = \frac{1-\frac{b}{a}}{1-\frac{b}{ax}} g(bq).$$

(iv) Explain why (iii) implies that
$$g(b) = \frac{\left(\frac{b}{a};q\right)_\infty}{\left(\frac{b}{ax};q\right)_\infty} g(0).$$

(v) Set $b=q$ in (iv) to find $g(0)$.

(vi) Find $g(b)$ and complete the proof of (6.1.1).

5. There is an earlier functional equation proof of (6.1.1) due to Andrews and Askey that works with the given sum side, rather than rewriting it as in part (i) of the previous problem. Again set

$$f(a,b,x) = \sum_{n=-\infty}^{\infty} \frac{(a;q)_n}{(b;q)_n} x^n$$

and note that, because of (6.2.8), we have

$$f(a,q,x) = \sum_{n=-\infty}^{\infty} \frac{(a;q)_n}{(q;q)_n} x^n = \sum_{n=0}^{\infty} \frac{(a;q)_n}{(q;q)_n} x^n = \frac{(ax;q)_\infty}{(x;q)_\infty}.$$

(i) Show that $\quad f(a,bq,x) - a\,f(a,bq,xq) = \dfrac{1-b}{x} f(a,b,x).$

(ii) Show that $\quad b\,f(a,bq,xq) = f(a,bq,x) - (1-b)\,f(a,b,x).$

(iii) Show that eliminating $f(a,bq,xq)$ from (i) and (ii) gives

$$(b-a)\,f(a,bq,x) = (1-b)\left(\frac{b}{x} - a\right) f(a,b,x),$$

and that this can be rewritten as

$$f(a,b,x) = \frac{1-\frac{b}{a}}{(1-b)\left(1-\frac{b}{ax}\right)} f(a,bq,x).$$

(iv) Explain why (iii) implies that

$$f(a,b,x) = \frac{\left(\frac{b}{a};q\right)_\infty}{(b;q)_\infty \left(\frac{b}{ax};q\right)_\infty} f(a,0,x).$$

(v) Set $b = q$ in (iv) to find $f(a,0,x)$ and complete the proof of (6.1.1).

6. This exercise presents a very interesting cautionary example, due to Askey. It will be convenient to have the $_1\psi_1$ with different letters: if $\left|\frac{s}{r}\right| < |z| < 1$, then

(6.2.10) $$\sum_{n=-\infty}^{\infty} \frac{(r;q)_n}{(s;q)_n} z^n = \frac{(rz;q)_\infty \left(\frac{q}{rz};q\right)_\infty (q;q)_\infty \left(\frac{s}{r};q\right)_\infty}{(z;q)_\infty \left(\frac{s}{rz};q\right)_\infty (s;q)_\infty \left(\frac{q}{r};q\right)_\infty}.$$

Askey tries to expand a similar product in a Laurent series:

(6.2.11) $$\frac{(ax;q)_\infty \left(\frac{b}{x};q\right)_\infty}{(dx;q)_\infty \left(\frac{q}{dx};q\right)_\infty} = \sum_{n=-\infty}^{\infty} c_n x^n$$

for some coefficients c_n, where we expect the expansion to be valid when $|q| < |dx| < 1$ from looking at the denominator.

(i) Show that replacing x by qx in (6.2.11) and dividing the result into (6.2.11) gives

$$q(1-ax) \sum_{n=-\infty}^{\infty} c_n x^n q^n = d(b-xq) \sum_{n=-\infty}^{\infty} c_n x^n.$$

(ii) Show that (i) implies

$$c_n = c_{n-1} \frac{q}{b} \frac{1 - \frac{aq^{n-1}}{d}}{1 - \frac{q^{n+1}}{bd}}.$$

(iii) Explain why (ii) implies

$$c_n = c_0 \left(\frac{q}{b}\right)^n \frac{\left(\frac{a}{d};q\right)_n}{\left(\frac{q^2}{bd};q\right)_n}$$

for every integer n. Exercise 3 from the previous section might help.

(iv) Using (iii) in (6.2.11) we have

$$\frac{(ax;q)_\infty \left(\frac{b}{x};q\right)_\infty}{(dx;q)_\infty \left(\frac{q}{dx};q\right)_\infty} = c_0 \sum_{n=-\infty}^{\infty} \frac{\left(\frac{a}{d};q\right)_n}{\left(\frac{q^2}{bd};q\right)_n} \left(\frac{qx}{b}\right)^n,$$

and we still need to find c_0. But (6.2.10) applies to this with $z = \frac{qx}{b}$, $r = \frac{a}{d}$, and $s = \frac{q^2}{bd}$. Show that this gives

$$\frac{(ax;q)_\infty \left(\frac{b}{x};q\right)_\infty}{(dx;q)_\infty \left(\frac{q}{dx};q\right)_\infty} = c_0 \frac{\left(\frac{aqx}{bd};q\right)_\infty \left(\frac{bd}{ax};q\right)_\infty (q;q)_\infty \left(\frac{q^2}{ab};q\right)_\infty}{\left(\frac{qx}{b};q\right)_\infty \left(\frac{q}{ax};q\right)_\infty \left(\frac{q^2}{bd};q\right)_\infty \left(\frac{qd}{a};q\right)_\infty}$$

if $\left|\frac{q}{a}\right| < |x| < \left|\frac{b}{q}\right|$.

(v) This expression for c_0 can't be correct, because c_0 is supposed to be independent of x. If we ignore that problem, the argument gives

$$\sum_{n=-\infty}^{\infty} \frac{\left(\frac{a}{d};q\right)_n}{\left(\frac{q^2}{bd};q\right)_n} \left(\frac{qx}{b}\right)^n = \frac{\left(\frac{aqx}{bd};q\right)_\infty \left(\frac{bd}{ax};q\right)_\infty (q;q)_\infty \left(\frac{q^2}{ab};q\right)_\infty}{\left(\frac{qx}{b};q\right)_\infty \left(\frac{q}{ax};q\right)_\infty \left(\frac{q^2}{bd};q\right)_\infty \left(\frac{qd}{a};q\right)_\infty},$$

which is correct if $\left|\frac{q}{a}\right| < |x| < \left|\frac{b}{q}\right|$. However, it is not an expansion of the product we started with in (6.2.11), because that cancelled in the final step.

We are left to try to explain why this argument failed to expand the product in (6.2.11). It could only have succeeded when $|q| < |dx| < 1$, which is a possible range for x, but if x is in this range, then qx is not. (Explain.) Therefore, we cannot get a functional equation for c_n by replacing x by qx. For *power* series, which contain no negative powers of x, replacing x by qx is always a good step because it moves us farther inside the circle of convergence. For Laurent series, it can move us inside the inner circle of the annulus of convergence into a region of divergence.

6.3. From the q-Pfaff–Saalschütz sum to Ramanujan's $_1\psi_1$ summation

The argument of this section is essentially due to Michael Schlosser, although our arrangement of it is a little different. Section 5.7 contained several variants of the q-Pfaff–Saalschütz identity

(6.3.1) $$\sum_{k=0}^{n} \binom{n}{k}_q \left(\frac{c}{ab};q\right)_{n-k} \frac{(a;q)_k (b;q)_k}{(c;q)_k} \left(\frac{c}{ab}\right)^k = \frac{\left(\frac{c}{a};q\right)_n \left(\frac{c}{b};q\right)_n}{(c;q)_n},$$

6.3. FROM THE q-PFAFF–SAALSCHÜTZ SUM TO RAMANUJAN'S $_1\psi_1$ SUMMATION 257

but there is yet another fruitful way to rewrite it. Start by changing n to $n+m$ and k to $k+m$ in (6.3.1). Noting that this does not change $n-k$, we then have
(6.3.2)
$$\sum_{k=-m}^{n} \binom{m+n}{m+k}_q \left(\frac{c}{ab};q\right)_{n-k} \frac{(a;q)_{m+k}(b;q)_{m+k}}{(c;q)_{m+k}} \left(\frac{c}{ab}\right)^{m+k} = \frac{(\frac{c}{a};q)_{m+n}(\frac{c}{b};q)_{m+n}}{(c;q)_{m+n}}.$$

The next step is to use (6.1.8) several times to simplify the sum, and also the related fact
$$\left(\frac{c}{ab};q\right)_{n-k} = \frac{\left(\frac{c}{ab};q\right)_n}{\left(\frac{cq^{n-k}}{ab};q\right)_k}.$$

This gives
$$\frac{(a;q)_m(b;q)_m}{(c;q)_m} \left(\frac{c}{ab};q\right)_n \left(\frac{c}{ab}\right)^m \sum_{k=-m}^{n} \binom{m+n}{m+k}_q \frac{(aq^m;q)_k(bq^m;q)_k}{(cq^m;q)_k \left(\frac{cq^{n-k}}{ab};q\right)_k} \left(\frac{c}{ab}\right)^k$$
$$= \frac{(\frac{c}{a};q)_{m+n}(\frac{c}{b};q)_{m+n}}{(c;q)_{m+n}},$$

or
$$\sum_{k=-m}^{n} \binom{m+n}{m+k}_q \frac{(aq^m;q)_k(bq^m;q)_k}{(cq^m;q)_k \left(\frac{cq^{n-k}}{ab};q\right)_k} \left(\frac{c}{ab}\right)^k$$
$$= \frac{(\frac{c}{a};q)_{m+n}(\frac{c}{b};q)_{m+n}}{(a;q)_m(b;q)_m(cq^m;q)_n\left(\frac{c}{ab};q\right)_n} \left(\frac{ab}{c}\right)^m$$

after another application of (6.1.8). Now replace a and c by aq^{-m} and cq^{-m} respectively to get
$$\sum_{k=-m}^{n} \binom{m+n}{m+k}_q \frac{(a;q)_k(bq^m;q)_k}{(c;q)_k \left(\frac{cq^{n-k}}{ab};q\right)_k} \left(\frac{c}{ab}\right)^k$$
$$= \frac{(\frac{c}{a};q)_{m+n}(\frac{c}{bq^m};q)_{m+n}}{(\frac{a}{q^m};q)_m(b;q)_m(c;q)_n\left(\frac{c}{ab};q\right)_n} \left(\frac{ab}{c}\right)^m.$$

Using (6.1.8) one more time, this is
(6.3.3) $$\sum_{k=-m}^{n} \binom{m+n}{m+k}_q \frac{(a;q)_k(bq^m;q)_k}{(c;q)_k \left(\frac{cq^{n-k}}{ab};q\right)_k} \left(\frac{c}{ab}\right)^k$$
$$= \frac{(\frac{c}{a};q)_{m+n}(\frac{c}{b};q)_n}{(b;q)_m(c;q)_n\left(\frac{c}{ab};q\right)_n} \left(\frac{ab}{c}\right)^m \frac{(\frac{cq^{-m}}{b};q)_m}{(aq^{-m};q)_m}.$$

By (6.1.4) we have
$$(aq^{-m};q)_m = (-1)^m a^m q^{-\binom{m+1}{2}} \left(\frac{q}{a};q\right)_m,$$
and using this also with a replaced by $\frac{c}{b}$ gives
$$\left(\frac{ab}{c}\right)^m \frac{(\frac{cq^{-m}}{b};q)_m}{(aq^{-m};q)_m} = \frac{\left(\frac{bq}{c};q\right)_m}{\left(\frac{q}{a};q\right)_m},$$

so (6.3.3) becomes
(6.3.4)
$$\sum_{k=-m}^{n} \binom{m+n}{m+k}_q \frac{(a;q)_k\,(bq^m;q)_k}{(c;q)_k\left(\frac{cq^{n-k}}{ab};q\right)_k} \left(\frac{c}{ab}\right)^k = \frac{(\frac{c}{a};q)_{m+n}(\frac{c}{b};q)_n(\frac{bq}{c};q)_m}{(b;q)_m(\frac{q}{a};q)_m(c;q)_n\left(\frac{c}{ab};q\right)_n}.$$

One might call (6.3.4) the MacMahon/Schlosser form of the q-Pfaff–Saalschütz identity. Schlosser gives a similar form with $m = n$, and it is analogous to MacMahon's q-binomial theorem of section 2.7, which had been given earlier by Cauchy and Gauss with the two parameters a and b both equal to n.

The point of writing the q-Pfaff–Saalschütz identity in the form of (6.3.4) is that if we formally let m and n approach infinity, assuming as usual that $|q| < 1$, we get Ramanujan's ${}_1\psi_1$ summation. Because $|q| < 1$, two of the finite products on the sum side of (6.3.4) approach $(0;q)_k = 1$ as $m, n \to \infty$; moreover, the q-binomial coefficient becomes

$$\frac{(q;q)_\infty}{(q;q)_\infty(q;q)_\infty} = \frac{1}{(q;q)_\infty}.$$

Moving this term to the other side of (6.3.4) we have

(6.3.5)
$$\sum_{k=-\infty}^{\infty} \frac{(a;q)_k}{(c;q)_k}\left(\frac{c}{ab}\right)^k = \frac{(\frac{c}{a};q)_\infty(\frac{c}{b};q)_\infty(\frac{bq}{c};q)_\infty(q;q)_\infty}{(b;q)_\infty(\frac{q}{a};q)_\infty(c;q)_\infty\left(\frac{c}{ab};q\right)_\infty}.$$

This is more commonly written with c/ab renamed as z; i.e., with $b = \frac{c}{az}$. Making this change we finally have

$$\sum_{k=-\infty}^{\infty} \frac{(a;q)_k}{(c;q)_k} z^k = \frac{(az;q)_\infty(\frac{q}{az};q)_\infty(q;q)_\infty(\frac{c}{a};q)_\infty}{(z;q)_\infty(\frac{c}{az};q)_\infty(c;q)_\infty(\frac{q}{a};q)_\infty}.$$

To justify this argument rigorously we can use Tannery's theorem. It certainly won't work unless the series converges, and we content ourselves with investigating this. Using the ratio test, we have to look at

$$\lim_{k\to\infty} \left| \frac{\frac{(a;q)_{k+1}}{(c;q)_{k+1}} z^{k+1}}{\frac{(a;q)_k}{(c;q)_k} z^k} \right| = \lim_{k\to\infty} \left| \frac{1-aq^k}{1-cq^k} z \right| = |z|,$$

since $|q| < 1$. So the series converges if $|z| < 1$. But since it is bilateral, we also have to look at what happens to the ratio of successive terms when $k \to -\infty$. If k is a positive integer, then we have

$$(a;q)_{-k} = \frac{1}{(1-aq^{-k})(1-aq^{1-k})\cdots(1-aq^{-1})}$$

from section 6.1. Then

$$\lim_{k\to\infty} \left| \frac{\frac{(a;q)_{-(k+1)}}{(c;q)_{-(k+1)}} z^{-(k+1)}}{\frac{(a;q)_{-k}}{(c;q)_{-k}} z^{-k}} \right| = \lim_{k\to\infty} \left| \frac{1-cq^{-(k+1)}}{1-aq^{-(k+1)}} \frac{1}{z} \right| = \lim_{k\to\infty} \left| \frac{q^{k+1}-c}{q^{k+1}-a} \frac{1}{z} \right| = \left| \frac{c}{az} \right|.$$

So we also need Schlosser's parameter $b = \frac{c}{az}$ to be less than one in absolute value for the series to converge. In other words, it converges if $\left|\frac{c}{a}\right| < |z| < 1$.

Exercises

1. Use Tannery's theorem to justify the above limiting process.

6.4. Another identity of Cauchy, and its finite form

To start our final proof of Ramanujan's $_1\psi_1$ summation formula, we prove another identity of Cauchy. This will also provide an occasion for some remarks on one of our standard methods. Cauchy posed to himself the following question. Consider a function $\phi(x)$ which satisfies the functional equation

$$\phi(x) = \frac{1-ax}{1-bx}\phi(xq)$$

and the initial condition $\phi(0) = 1$. By iterating the functional equation as before, under our standard assumption that $|q| < 1$, we have that $\phi(x) = (ax;q)_\infty/(bx;q)_\infty$. Cauchy asked what the coefficients c_n are if we expand $\phi(x)$ in a certain type of series:

$$\phi(x) = \sum_{n=0}^\infty \frac{c_n x^n}{(bx;q)_n}.$$

One answer to this question is that

$$c_n = (b-a)(b-aq)\cdots(b-aq^{n-1})\frac{(bx;q)_n}{(q;q)_n},$$

because this gives us back the Cauchy/Crelle series for $\phi(x)$; but this is not a good answer, because these c_n's are not independent of x. To have a series that is really of the desired type we need c_n's that are.

A principle we have often used is that if we expand a (sufficiently nice) function in a series of powers of x, then the coefficients are uniquely determined (so any legitimate method of finding them can be used; this, in a sense, is the real content of Taylor's theorem). If we have two infinite series in powers of x which are equal, then the coefficient of x^n is the same on each side for every nonnegative integer n, and we have used this to prove some of our identities. In the language of linear algebra, this comes down to the fact that the power functions $\{x^n\}$ are a basis for a certain class of functions, which we will be content with a very vague description of: "nice functions which are finite at the origin". But the power functions are far from being the only such basis—any set of polynomials $\{P_n(x)\}$ with exactly one member of each degree is nearly as good. And while the functions $x^n/(bx;q)_n$ that Cauchy wanted to use in his expansion are not polynomials, they are still a basis for the class of functions we're interested in, because no other $x^k/(bx;q)_k$ besides $x^n/(bx;q)_n$ has x^n as its *lowest* power.

Why did Cauchy ask himself this? He wanted to generalize Jacobi's Durfee square identity (4.2.1), and therefore he imitated Jacobi's argument from problem 21 in section 5.1. Since $\phi(0) = 1$, we must have $c_0 = 1$, and if

$$\phi(x) = \frac{1-ax}{1-bx}\phi(xq) \quad \text{and} \quad \phi(x) = \sum_{n=0}^\infty \frac{c_n x^n}{(bx;q)_n},$$

then
$$\sum_{n=0}^{\infty} \frac{c_n x^n}{(bx;q)_n} = \frac{1-ax}{1-bx} \sum_{n=0}^{\infty} \frac{c_n x^n q^n}{(bxq;q)_n}$$
$$= \sum_{n=0}^{\infty} \frac{c_n x^n q^n}{(bx;q)_n} \cdot \frac{1-ax}{1-bxq^n}.$$

Cauchy rewrites
$$\frac{1-ax}{1-bxq^n} = \frac{1-bxq^n+bxq^n-ax}{1-bxq^n} = 1 + \frac{x(bq^n-a)}{1-bxq^n}.$$

Then
$$\sum_{n=0}^{\infty} \frac{c_n x^n}{(bx;q)_n} = \sum_{n=0}^{\infty} \frac{c_n x^n q^n}{(bx;q)_n} + \sum_{n=0}^{\infty} \frac{c_n x^n q^n}{(bx;q)_n} \cdot \frac{x(bq^n-a)}{1-bxq^n}.$$

Moving the first sum on the right to the other side we have
$$\sum_{n=0}^{\infty} \frac{c_n x^n (1-q^n)}{(bx;q)_n} = \sum_{n=0}^{\infty} \frac{c_n q^n x^{n+1} (bq^n-a)}{(bx;q)_{n+1}}$$
$$= \sum_{n=1}^{\infty} \frac{c_{n-1} q^{n-1} x^n (bq^{n-1}-a)}{(bx;q)_n}.$$

Note that the $n = 0$ term on the left side is zero. As usual, we now equate coefficients on each side. If the c_n are independent of x we must have

(6.4.1) $\qquad c_n (1-q^n) = c_{n-1} q^{n-1} (bq^{n-1}-a).$

Then
$$c_n = c_{n-1} q^{n-1} \frac{bq^{n-1}-a}{1-q^n} \quad \text{for } n \geq 1,$$
and iterating this down to c_0 gives
$$c_n = c_0 \frac{q^{\binom{n}{2}}}{(q;q)_n} (bq^{n-1}-a)\ldots(bq-a)(b-a).$$

This proves Cauchy's identity

(6.4.2) $\qquad \phi(x) = \dfrac{(ax;q)_\infty}{(bx;q)_\infty} = \sum_{k=0}^{\infty} \dfrac{q^{\binom{k}{2}} x^k}{(q;q)_k (bx;q)_k} (b-a)(bq-a) \cdots (bq^{k-1}-a).$

When $a = 0$ and $b = q$ this reduces to (4.2.1). The case $x = 1$ of (4.2.1) has sometimes been attributed to Euler, but this seems doubtful.

Exercises

1. Show that (6.4.1) is false if
$$c_n = (b-a)(b-aq)\cdots(b-aq^{n-1}) \frac{(bx;q)_n}{(q;q)_n}.$$
   These are the c_n's in the Cauchy/Crelle series, which are not independent of x.

2. Justify the assertion in the text that no other $x^k/(bx;q)_k$ besides $x^n/(bx;q)_n$ has x^n as its lowest power.

EXERCISES

3. Jacobi found a finite form of (6.4.2). Define

$$S_n(a,b;x) := \sum_{k=0}^{n} \binom{n}{k}_q \frac{q^{\binom{k}{2}} x^k}{(bx;q)_k} (b-a)(bq-a)\cdots(bq^{k-1}-a).$$

Show that $S_0(a,b;x) = 1$ and

$$S_{n+1}(a,b;x) = \frac{1-ax}{1-bx} S_n(aq,bq;x),$$

and explain why this implies

(6.4.3) $$\sum_{k=0}^{n} \binom{n}{k}_q \frac{q^{\binom{k}{2}} x^k}{(bx;q)_k} (b-a)(bq-a)\cdots(bq^{k-1}-a) = \frac{(ax;q)_n}{(bx;q)_n}.$$

4. Show that (6.4.3) reduces to (4.2.13) if $a = 0$ and either b or x equals q.
5. Give an alternative proof of (6.4.2) by letting $n \to \infty$ in (6.4.3).
6. Show that (6.4.3) can be rewritten as

(6.4.4) $$(ax;q)_n = \sum_{k=0}^{n} \binom{n}{k}_q q^{\binom{k}{2}} x^k \left(bxq^k;q\right)_{n-k} (b-a)(bq-a)\cdots(bq^{k-1}-a).$$

7. What form of the binomial theorem does (6.4.4) reduce to when $q = 1$?
8. Alladi has given a different proof of (6.4.2), starting from the sum side. Justify each of the following steps:

   (i) By Rothe's q-binomial theorem, the right side of (6.4.2) is

   $$\sum_{k=0}^{\infty} \sum_{j=0}^{k} \frac{q^{\binom{k}{2}+\binom{j}{2}} (bx)^j (-ax)^{k-j}}{(q;q)_j (q;q)_{k-j} (bx;q)_k}.$$

   (ii) By interchanging orders of summation and setting $n = k - j$, this can be rewritten as

   $$\sum_{j=0}^{\infty} \sum_{n=0}^{\infty} \frac{q^{\binom{j}{2}+\binom{n+j}{2}} (bx)^j (-ax)^n}{(q;q)_j (q;q)_n (bx;q)_{n+j}}.$$

   (iii) This in turn can be rewritten as

   $$\sum_{n=0}^{\infty} \frac{q^{\binom{n}{2}}(-ax)^n}{(q;q)_n} \sum_{j=0}^{\infty} \frac{q^{j(j+n)} \left(\frac{bx}{q}\right)^j}{(q;q)_j (bx;q)_{n+j}}.$$

   (iv) Cauchy's theorem now follows after first using (4.2.2) and then Euler's theorem (3.6.1).

9. Recall (5.2.12) from section 5.2:

(6.4.5) $$\sum_{n=0}^{\infty} q^{\binom{n+1}{2}} = \frac{(q^2;q^2)_\infty}{(q;q^2)_\infty}.$$

For reasons connected with partitions, in the paper that engendered the previous problem Alladi finds several other q-expansions of the right side of (6.4.5). By changing q to q^2 in (6.4.2) and then setting $ax = q^2$ and $bx = q$, show that

$$\frac{(q^2;q^2)_\infty}{(q;q^2)_\infty} = \sum_{k=0}^{\infty} \frac{(-1)^k q^{k^2+k}}{(q^2;q^2)_k} \frac{1-q^{-1}}{1-q^{2k-1}}.$$

10. Alladi finds another q-expansion of (6.4.5) by using Euler's theorems from Chapter 3.

(i) By using (3.5.1) with q^2 in place of q, show that
$$\frac{(q^2;q^2)_\infty}{(q;q^2)_\infty} = (q^2;q^2)_\infty \sum_{m=0}^{\infty} \frac{q^m}{(q^2;q^2)_m} = \sum_{m=0}^{\infty} q^m \left(q^{2m+2};q^2\right)_\infty.$$

(ii) By using (3.6.1) with q^2 in place of q, show that the right side of (i) is
$$\sum_{m=0}^{\infty} \sum_{k=0}^{\infty} \frac{(-1)^k q^{k^2+k+m(2k+1)}}{(q^2;q^2)_k}.$$

(iii) By changing the order of summation in (ii), show that
$$\frac{(q^2;q^2)_\infty}{(q;q^2)_\infty} = \sum_{k=0}^{\infty} \frac{(-1)^k q^{k^2+k}}{(q^2;q^2)_k (1-q^{2k+1})}.$$

11. From the previous two problems we know that

(6.4.6) $$\sum_{k=0}^{\infty} \frac{(-1)^k q^{k^2+k}}{(q^2;q^2)_k (1-q^{2k+1})} = \sum_{k=0}^{\infty} \frac{(-1)^k q^{k^2+k}}{(q^2;q^2)_k} \frac{1-q^{-1}}{1-q^{2k-1}},$$

because both sides are equal to $(q^2;q^2)_\infty/(q;q^2)_\infty$. Alladi gives two direct proofs of (6.4.6) by working with the partial sums
$$A_n(q) = \sum_{k=0}^{n} \frac{(-1)^k q^{k^2+k}}{(q^2;q^2)_k (1-q^{2k+1})} \quad \text{and} \quad B_n(q) = \sum_{k=0}^{n} \frac{(-1)^k q^{k^2+k}}{(q^2;q^2)_k} \frac{1-q^{-1}}{1-q^{2n-1}}.$$

We outline the second proof, which was suggested by Andrews.

(i) By breaking off the $k = n$ term of $A_n(q)$ and the $k = 0$ term of $B_n(q)$, and then reindexing the sum for $B_n(q)$, show that
$$A_n(q) - B_n(q) = \frac{(-1)^n q^{n^2+n}}{(q^2;q^2)_n (1-q^{2n+1})} - 1 + \sum_{k=0}^{n-1} \frac{(-1)^k q^{k^2+k}}{(q^2;q^2)_{k+1}}.$$

(ii) The sum in (i) can be made to telescope by writing
$$\frac{(-1)^k q^{k^2+k}}{(q^2;q^2)_{k+1}} = \frac{(-1)^k q^{k^2+k} \left(1 - q^{2k+2} + q^{2k+2}\right)}{(q^2;q^2)_{k+1}}.$$

Show that this gives

(6.4.7) $$A_n(q) - B_n(q) = \frac{(-1)^n q^{n^2+3n+1}}{(q^2;q^2)_n (1-q^{2n+1})},$$

and explain why (6.4.6) follows.

12. Prove (6.4.7) by induction on n. This was Alladi's other proof of (6.4.6).

6.5. Cauchy's "mistaken identity"

Jacobi needed (4.2.1) to finish his proof of the triple product, and Cauchy wanted (6.4.2) because he was trying to generalize the triple product. He would have succeeded but for a careless error. In this section we'll work out the identity that Cauchy should have found. Suppose we try to expand the function

$$f(x) := \frac{(ax;q)_\infty \left(\frac{q}{ax};q\right)_\infty}{(bx;q)_\infty}$$

in powers of x. Because of the negative powers in the numerator, this expansion has the form

(6.5.1) $$f(x) = \frac{(ax;q)_\infty \left(\frac{q}{ax};q\right)_\infty}{(bx;q)_\infty} = \sum_{n=-\infty}^{\infty} \gamma_n x^n$$

for some coefficients γ_n, which we seek. The following argument is probably similar to what Cauchy had in mind: consider

$$\frac{f(x)}{f(xq)} = \frac{(ax;q)_\infty}{(axq;q)_\infty} \frac{\left(\frac{q}{ax};q\right)_\infty}{\left(\frac{1}{ax};q\right)_\infty} \frac{(bxq;q)_\infty}{(bx;q)_\infty} = \frac{\sum_{n=-\infty}^{\infty} \gamma_n x^n}{\sum_{n=-\infty}^{\infty} \gamma_n x^n q^n}.$$

There is a great deal of cancellation in the infinite products, and there results

$$\frac{ax}{bx-1} = \frac{\sum_{n=-\infty}^{\infty} \gamma_n x^n}{\sum_{n=-\infty}^{\infty} \gamma_n x^n q^n}.$$

Rewrite this as

$$ax \sum_{n=-\infty}^{\infty} \gamma_n x^n q^n = (bx-1) \sum_{n=-\infty}^{\infty} \gamma_n x^n$$

and rearrange to get

$$\sum_{n=-\infty}^{\infty} \gamma_n x^n = \sum_{n=-\infty}^{\infty} \gamma_n x^{n+1} (b - aq^n)$$

$$= \sum_{n=-\infty}^{\infty} \gamma_{n-1} x^n \left(b - aq^{n-1}\right).$$

By equating coefficients we have $\gamma_n = \gamma_{n-1} \left(b - aq^{n-1}\right)$. If $n > 0$ we may iterate this to get

$$\gamma_n = \gamma_0 (b-a)(b-aq) \cdots (b-aq^{n-1}),$$

which we may rewrite as

$$\gamma_n = \gamma_0 b^n \left(1 - \frac{a}{b}\right)\left(1 - \frac{aq}{b}\right) \cdots \left(1 - \frac{aq^{n-1}}{b}\right)$$

(6.5.2) $$= \gamma_0 \left(\frac{a}{b};q\right)_n b^n \quad \text{for } n \geq 0.$$

If $n < 0$ we replace n by $1-m$ in $\gamma_n = \gamma_{n-1}(b-aq^{n-1})$ to get

$$\gamma_{-m} = \frac{\gamma_{1-m}}{b - aq^{-m}} = \frac{\gamma_{1-m} q^m}{bq^m - a},$$

where m is nonnegative. Iterating this upwards gives

$$\gamma_{-m} = \frac{\gamma_0}{(b-aq^{-m})\cdots(b-aq^{-1})}$$

$$= \frac{\gamma_0\, q^{\binom{m+1}{2}}}{(bq^m-a)\cdots(bq-a)}.$$

Taking $-a$ out of each denominator factor makes this

$$\gamma_{-m} = \frac{\gamma_0\,(-1)^m q^{\binom{m+1}{2}}}{a^m\left(1-\frac{bq}{a}\right)\cdots\left(1-\frac{bq^m}{a}\right)}$$

$$= \frac{\gamma_0\,(-1)^m q^{\binom{m+1}{2}}}{a^m\left(\frac{bq}{a};q\right)_m}$$

$$= \frac{\gamma_0\,(-1)^m q^{\binom{m+1}{2}}}{b^m\left(\frac{a}{b}\right)^m \left(\frac{bq}{a};q\right)_m}.$$

Now recall (6.1.4)

$$(a;q)_{-n} = \frac{(-1)^n q^{\binom{n+1}{2}}}{a^n \left(\frac{q}{a};q\right)_n}$$

from section 6.1. This shows that

$$\gamma_{-m} = \gamma_0\, b^{-m}\left(\frac{a}{b};q\right)_{-m} \qquad \text{for } m \geq 0.$$

Combining this with (6.5.2) we have

(6.5.3) $$\gamma_n = \gamma_0\, b^n \left(\frac{a}{b};q\right)_n \qquad \text{for } all \text{ integers } n.$$

It remains to determine γ_0, and it was here that Cauchy went wrong. γ_0 is the constant term (*i.e.*, the term independent of x) in (6.5.1). One way to find it is to separate the function in (6.5.1) as

$$\frac{(ax;q)_\infty}{(bx;q)_\infty}\left(\frac{q}{ax};q\right)_\infty.$$

We can expand the fraction by the Cauchy/Crelle series, and the other infinite product by Euler's identity (3.6.1) from Chapter 3. Hence

$$\frac{(ax;q)_\infty \left(\frac{q}{ax};q\right)_\infty}{(bx;q)_\infty} = \sum_{j=0}^\infty \frac{(b-a)(b-aq)\cdots(b-aq^{j-1})}{(q;q)_j} x^j \sum_{k=0}^\infty \frac{q^{\binom{k}{2}}\left(-\frac{q}{ax}\right)^k}{(q;q)_k}.$$

The constant term in this comes from setting $j=k$:

$$\gamma_0 = \sum_{k=0}^\infty \frac{(b-a)(b-aq)\cdots(b-aq^{k-1})}{(q;q)_k (q;q)_k} q^{\binom{k}{2}}\left(-\frac{q}{a}\right)^k$$

(6.5.4) $$= \sum_{k=0}^\infty \frac{(a-b)(aq-b)\cdots(aq^{k-1}-b)}{(q;q)_k (q;q)_k} q^{\binom{k}{2}}\left(\frac{q}{a}\right)^k.$$

Here is a series that Cauchy could easily have summed, for if we interchange a and b in (6.4.2), then we have

$$\frac{(bx;q)_\infty}{(ax;q)_\infty} = \sum_{k=0}^{\infty} \frac{q^{\binom{k}{2}} x^k}{(q;q)_k (ax;q)_k} (a-b)(aq-b)\cdots(aq^{k-1}-b).$$

Setting $x = \frac{q}{a}$ in this we get the series (6.5.4), and therefore

$$\gamma_0 = \frac{\left(\frac{bq}{a};q\right)_\infty}{(q;q)_\infty}.$$

Substituting this in (6.5.3) and comparing with (6.5.1), we finally have

(6.5.5) $$\frac{(ax;q)_\infty \left(\frac{q}{ax};q\right)_\infty (q;q)_\infty}{(bx;q)_\infty \left(\frac{bq}{a};q\right)_\infty} = \sum_{n=-\infty}^{\infty} \left(\frac{a}{b};q\right)_n b^n x^n.$$

Assuming that $|q| < 1$, the series converges if $x \neq 0$ and $|bx| < 1$. We propose to call this Cauchy's mistaken identity, as the version of (6.5.5) in Cauchy's paper is missing the factor $\left(\frac{bq}{a};q\right)_\infty$ in the denominator. If $b = 0$, then (6.5.5) reduces to Jacobi's triple product identity in the form

$$(ax;q)_\infty \left(\frac{q}{ax};q\right)_\infty (q;q)_\infty = \sum_{n=-\infty}^{\infty} q^{\frac{n(n-1)}{2}} (-ax)^n.$$

Cauchy seems to have been misled by this—he apparently thought that his identity would have the same constant term as Jacobi's.

Exercises

1. Here is another way to find γ_0 above: we have

$$\frac{(ax;q)_\infty \left(\frac{q}{ax};q\right)_\infty}{(bx;q)_\infty} = \gamma_0 \sum_{n=-\infty}^{\infty} \left(\frac{a}{b};q\right)_n (bx)^n.$$

Multiply both sides by $(q;q)_\infty$ and rename $(q;q)_\infty \gamma_0$ as C. Then we need to find C in the equation

$$(ax;q)_\infty \left(\frac{q}{ax};q\right)_\infty (q;q)_\infty \frac{1}{(bx;q)_\infty} = C \sum_{n=-\infty}^{\infty} \left(\frac{a}{b};q\right)_n (bx)^n.$$

(i) Explain why

$$(ax;q)_\infty \left(\frac{q}{ax};q\right)_\infty (q;q)_\infty \frac{1}{(bx;q)_\infty} = \left(\sum_{j=-\infty}^{\infty} q^{\frac{j(j-1)}{2}} (-ax)^j\right) \left(\sum_{k=0}^{\infty} \frac{(bx)^k}{(q;q)_k}\right).$$

(ii) Explain why C is the coefficient of x^0 in

$$\left(\sum_{j=-\infty}^{\infty} q^{\frac{j(j-1)}{2}} (-ax)^j\right) \left(\sum_{k=0}^{\infty} \frac{(bx)^k}{(q;q)_k}\right).$$

(iii) Explain why
$$C = \sum_{k=0}^{\infty} \frac{q^{\frac{k(k+1)}{2}}}{(q;q)_k}\left(-\frac{b}{a}\right)^k = \sum_{k=0}^{\infty} \frac{q^{\binom{k}{2}}}{(q;q)_k}\left(-\frac{bq}{a}\right)^k.$$

(iv) Explain why $C = \left(\frac{bq}{a};q\right)_{\infty}$. This completes the proof.

2. Use Cauchy's mistaken identity (6.5.5) to show that
$$\sum_{n=-\infty}^{\infty} \left(\frac{q}{s};q\right)_n t^{n+1} - \sum_{n=-\infty}^{\infty} \left(\frac{q}{t};q\right)_n s^{n+1} = \frac{2(t-s)(q;q)_{\infty}\left(\frac{qs}{t};q\right)_{\infty}\left(\frac{qt}{s};q\right)_{\infty}}{(s;q)_{\infty}(t;q)_{\infty}}.$$

3. Use the previous problem and (6.1.10) to show that

(6.5.6) $$\sum_{n=0}^{\infty} \left(\frac{q}{s};q\right)_n t^{n+1} - \sum_{n=0}^{\infty} \left(\frac{q}{t};q\right)_n s^{n+1} = \frac{(t-s)(q;q)_{\infty}\left(\frac{qs}{t};q\right)_{\infty}\left(\frac{qt}{s};q\right)_{\infty}}{(s;q)_{\infty}(t;q)_{\infty}}.$$

4. Use (6.1.9) to rewrite (6.5.6) as

(6.5.7) $$t\sum_{n=0}^{\infty} \frac{q^{\binom{n+1}{2}}}{(t;q)_{n+1}}\left(-\frac{t}{s}\right)^n - s\sum_{n=0}^{\infty} \frac{q^{\binom{n+1}{2}}}{(s;q)_{n+1}}\left(-\frac{s}{t}\right)^n$$
$$= \frac{(t-s)(q;q)_{\infty}\left(\frac{qs}{t};q\right)_{\infty}\left(\frac{qt}{s};q\right)_{\infty}}{(s;q)_{\infty}(t;q)_{\infty}}.$$

This is **Ramanujan's reciprocity theorem**.

5. Use (5.6.20) from the exercises in section 5.6 to rewrite (6.5.7) as

(6.5.8) $$\sum_{n=0}^{\infty} \frac{\left(\frac{q}{s};q\right)_n}{(t;q)_{n+1}}\left(s-tq^{2n+1}\right)(-1)^n \frac{t^{2n+1}}{s^{n+1}} q^{\frac{n(3n+1)}{2}}$$
$$- \sum_{n=0}^{\infty} \frac{\left(\frac{q}{t};q\right)_n}{(s;q)_{n+1}}\left(t-sq^{2n+1}\right)(-1)^n \frac{s^{2n+1}}{t^{n+1}} q^{\frac{n(3n+1)}{2}}$$
$$= \frac{(t-s)(q;q)_{\infty}\left(\frac{qs}{t};q\right)_{\infty}\left(\frac{qt}{s};q\right)_{\infty}}{(s;q)_{\infty}(t;q)_{\infty}}.$$

6. Show that setting $t = -\frac{q}{z}$ and $s = -z$ in (6.5.8) gives the quintuple product identity (5.3.1), namely
$$\sum_{k=-\infty}^{\infty} \left(1-zq^k\right) z^{3k} q^{\frac{k(3k-1)}{2}} = \frac{(z^2;q)_{\infty}\left(\frac{q}{z^2};q\right)_{\infty}(q;q)_{\infty}}{(-z;q)_{\infty}\left(-\frac{q}{z};q\right)_{\infty}}.$$

You will eventually need to change n to $-k-1$ in one of the two sums in (6.5.8).

6.6. Ramanujan's formula again

Cauchy's mistaken identity is, in a sense, halfway between Jacobi's triple product and Ramanujan's $_1\psi_1$ summation. In this section we use an argument much like that of the previous section to derive the $_1\psi_1$. Since (6.5.5) arises from putting

6.6. RAMANUJAN'S FORMULA AGAIN

an infinite product in the denominator of Jacobi's triple product, it is natural to try the same thing on (6.5.5) itself. We set

$$(6.6.1) \qquad g(x) := \frac{(ax;q)_\infty \left(\frac{q}{ax};q\right)_\infty (q;q)_\infty}{(bx;q)_\infty \left(\frac{bq}{a};q\right)_\infty \left(\frac{c}{ax};q\right)_\infty} = \sum_{n=-\infty}^{\infty} r_n x^n$$

and use the same method as before to work out the coefficients r_n. We have

$$\frac{g(x)}{g(xq)} = \frac{(ax;q)_\infty \left(\frac{q}{ax};q\right)_\infty (bxq;q)_\infty \left(\frac{c}{aqx};q\right)_\infty}{(axq;q)_\infty \left(\frac{1}{ax};q\right)_\infty (bx;q)_\infty \left(\frac{c}{ax};q\right)_\infty} = \frac{\sum_{n=-\infty}^{\infty} r_n x^n}{\sum_{n=-\infty}^{\infty} r_n x^n q^n}.$$

Again there is much cancellation in the infinite products, and we get

$$\frac{\frac{c}{q} - ax}{1 - bx} = \frac{\sum_{n=-\infty}^{\infty} r_n x^n}{\sum_{n=-\infty}^{\infty} r_n x^n q^n}$$

or

$$\left(\frac{c}{q} - ax\right) \sum_{n=-\infty}^{\infty} r_n x^n q^n = (1 - bx) \sum_{n=-\infty}^{\infty} r_n x^n.$$

Rearrange this to

$$\sum_{n=-\infty}^{\infty} r_n x^n \left(1 - cq^{n-1}\right) = \sum_{n=-\infty}^{\infty} r_n x^{n+1} (b - aq^n)$$

$$= \sum_{n=-\infty}^{\infty} r_{n+1} x^n \left(b - aq^{n-1}\right)$$

and equate coefficients of x^n to get

$$r_n = r_{n-1} \frac{b - aq^{n-1}}{1 - cq^{n-1}}.$$

As before this may be iterated to get r_n in terms of r_0, and the result is

$$(6.6.2) \qquad r_n = r_0 \, b^n \, \frac{\left(\frac{a}{b};q\right)_n}{(c;q)_n}$$

for all integers n.

To find the constant term r_0, we separate the infinite products in (6.6.1) as

$$\frac{(ax;q)_\infty \left(\frac{q}{ax};q\right)_\infty (q;q)_\infty}{(bx;q)_\infty \left(\frac{bq}{a};q\right)_\infty} \cdot \frac{1}{\left(\frac{c}{ax};q\right)_\infty}.$$

Using (6.5.5) to expand the former and Euler's identity (3.5.1) from Chapter 3 for the latter we have

$$\frac{(ax;q)_\infty \left(\frac{q}{ax};q\right)_\infty (q;q)_\infty}{(bx;q)_\infty \left(\frac{bq}{a};q\right)_\infty \left(\frac{c}{ax};q\right)_\infty} = \sum_{n=-\infty}^{\infty} \left(\frac{a}{b};q\right)_n b^n x^n \sum_{k=0}^{\infty} \frac{\left(\frac{c}{ax}\right)^k}{(q;q)_k}.$$

The constant term arises when $n = k$, so

$$r_0 = \sum_{n=0}^{\infty} \frac{(\frac{a}{b};q)_n}{(q;q)_n} \left(\frac{bc}{a}\right)^n$$

$$= \sum_{n=0}^{\infty} \frac{(b-a)(b-aq)\cdots(b-aq^{n-1})}{(q;q)_n} \left(\frac{c}{a}\right)^n$$

$$= \frac{\left(a\frac{c}{a};q\right)_\infty}{\left(b\frac{c}{a};q\right)_\infty} = \frac{(c;q)_\infty}{\left(\frac{bc}{a};q\right)_\infty}$$

from the Cauchy/Crelle series. Together with (6.6.2) and (6.6.1), this proves

THEOREM 61 (Ramanujan's $_1\psi_1$ summation formula). *If* $\left|\frac{c}{a}\right| < |x| < \frac{1}{|b|}$, *so that the series in* (6.6.3) *below converges, then*

(6.6.3) $\quad \dfrac{(ax;q)_\infty \left(\frac{q}{ax};q\right)_\infty (q;q)_\infty \left(\frac{bc}{a};q\right)_\infty}{(bx;q)_\infty \left(\frac{c}{ax};q\right)_\infty (c;q)_\infty \left(\frac{bq}{a};q\right)_\infty} = \displaystyle\sum_{n=-\infty}^{\infty} \dfrac{\left(\frac{a}{b};q\right)_n}{(c;q)_n} (bx)^n.$

To get (6.1.1) from (6.6.3) we set $b = 1$ and then rename c as b. (6.6.3) is not really more general, since (exercise) we can get (6.6.3) from (6.1.1) by renaming b as c and then replacing a by $\frac{a}{b}$ and x by bx.

Exercises

1. Use the ratio test to show that the series in (6.6.3) converges if $\left|\frac{c}{a}\right| < |x| < \frac{1}{|b|}$. Again, you should consider both positive and negative values of n.
2. Show that (6.1.1) becomes (6.6.3) if we rename b as c and then replace a by $\frac{a}{b}$ and x by bx. Check the convergence conditions also.

6.7. Bibliographical Notes

The $_1\psi_1$ summation "was first brought before the mathematical world", as Bruce Berndt eloquently expressed it, by Hardy in Chapter 12 of [**133**] in 1940, in the form (6.2.9). This form is entry 17 in Chapter 16 of Ramanujan's notebooks [**49**]. The very natural proof of Jackson is in [**146**], and Andrews's variation in [**10**]. Ismail's proof appeared first in [**142**], and may also be found in [**143**] and [**24**]. There are also combinatorial proofs by Corteel and Lovejoy [**83**] and by Yee [**248**]. Andrews's proof of the triple product is in [**6**].

No account of q-series could be complete without mentioning Berndt's edition of Ramanujan's notebooks [**47**], [**48**], [**49**], [**50**], [**51**], which has the entire mathematical community in his debt. These notebooks contain thousands of wonderful facts about q-series and other topics. Some were published by Ramanujan, and others were either known before him or rediscovered since, but many were still new 75 or more years after the fact. Chapter 17 contains Ramanujan's initial development of q-series; his version of the q-Gauss sum is entry 4 there. Berndt often uses (6.2.9) instead of (6.1.1) in his work, and (6.2.9) was also preferred by Venkatachaliengar in [**241**] and by Cooper in [**81**].

6.7. BIBLIOGRAPHICAL NOTES

Askey's proof of Ramanujan's $_1\psi_1$ summation formula is in [**33**], the Andrews/Askey proof is in [**23**], and Askey's cautionary example is in [**36**]. Schlosser's proof is in [**213**].

Cauchy's identities from sections 6.4 and 6.5 are in [**67**]. He found several other special cases of the $_1\psi_1$ in [**69**]. See the next chapter for these and applications to number theory. Jacobi's identity (6.4.4) is in [**150**]. The last four problems in section 6.5 come from [**5**]. Alladi proves there a special case of (6.4.2) that appears in Ramanujan's lost notebook, but both the result and the argument extend easily to (6.4.2). Much of the material in sections 6.4–6.6 comes from my paper [**153**]. Ramanujan's reciprocity theorem (6.5.7) is in [**26**] and [**53**], and one can see these works for further references. We have more or less followed one of the proofs in [**53**], which also has the remark that it implies the quintuple product.

CHAPTER 7

Sums of Squares

7.1. Cauchy's formula

In this short chapter we give some applications of q-analysis to sums of two and four squares. The key identity is

$$(7.1.1) \qquad \frac{(ax;q)_\infty \left(\frac{q}{ax};q\right)_\infty (q;q)_\infty^2}{(x;q)_\infty \left(\frac{q}{x};q\right)_\infty (a;q)_\infty \left(\frac{q}{a};q\right)_\infty} = \sum_{n=-\infty}^{\infty} \frac{x^n}{1-aq^n},$$

where the series converges if either $|q| < |x| < 1$ or $|q| > |x| > 1$, but we will assume the former. This was stated without proof by Cauchy in 1843, though he did prove two special cases. It is itself the special case $b = aq$ of Ramanujan's ${}_1\psi_1$ summation formula.

THEOREM 62 (Ramanujan). *If $|q| < 1$ and $\left|\frac{b}{a}\right| < |x| < 1$, then*

$$(7.1.2) \qquad \frac{(ax;q)_\infty \left(\frac{q}{ax};q\right)_\infty (q;q)_\infty \left(\frac{b}{a};q\right)_\infty}{(x;q)_\infty \left(\frac{b}{ax};q\right)_\infty (b;q)_\infty \left(\frac{q}{a};q\right)_\infty} = \sum_{n=-\infty}^{\infty} \frac{(a;q)_n}{(b;q)_n} x^n.$$

Therefore we may take it for granted if we have read Chapter 6. If we haven't, then we can use a partial fractions expansion from Chapter 2 instead:

THEOREM 63. *For any nonnegative integer n, and all x except $q^{-n}, \ldots, q^n$, we have*

$$\frac{(ax;q)_n \left(\frac{q}{ax};q\right)_n}{(x;q)_{n+1} \left(\frac{q}{x};q\right)_n} = \sum_{k=-n}^{n} \frac{(a;q)_{n-k} \left(\frac{q}{a};q\right)_{n+k}}{(q;q)_{n-k}(q;q)_{n+k}} \frac{a^k}{1-xq^k}.$$

If we let $n \to \infty$ here (see the appendix for the justification of the limiting process via Tannery's theorem), we get

$$(7.1.3) \qquad \frac{(ax;q)_\infty \left(\frac{q}{ax};q\right)_\infty (q;q)_\infty^2}{(x;q)_\infty \left(\frac{q}{x};q\right)_\infty (a;q)_\infty \left(\frac{q}{a};q\right)_\infty} = \sum_{k=-\infty}^{\infty} \frac{a^k}{1-xq^k},$$

if $|q| < |a| < 1$. Assuming we also have $|q| < |x| < 1$, the left side of (7.1.3) is symmetric in x and a, so the right side must be too, and (7.1.1) follows. We will generally assume both $|q| < |x| < 1$ and $|q| < |a| < 1$ in (7.1.1) to assure the symmetry in x and a. In (7.1.1) and (7.1.3), neither a nor x can be 1 because of the denominator factors $(a;q)_\infty$ and $(x;q)_\infty$.

Cauchy's two special cases of (7.1.1) are equally beautiful, and we will need one of them in the next section. If we change q to q^2 in (7.1.1) and then replace x by xq, we get

$$(7.1.4) \qquad \frac{(aqx;q^2)_\infty \left(\frac{q}{ax};q^2\right)_\infty (q^2;q^2)_\infty^2}{(qx;q^2)_\infty \left(\frac{q}{x};q^2\right)_\infty (a;q^2)_\infty \left(\frac{q^2}{a};q^2\right)_\infty} = \sum_{n=-\infty}^{\infty} \frac{x^n q^{2n}}{1-aq^{2n}}.$$

271

We lose a little symmetry in doing so, but in the two cases of interest it is possible to restore it. If we set $a = -1$ we get

$$(7.1.5) \qquad \frac{(-qx;q^2)_\infty \left(-\frac{q}{x};q^2\right)_\infty (q^2;q^2)_\infty^2}{(qx;q^2)_\infty \left(\frac{q}{x};q^2\right)_\infty (-q^2;q^2)_\infty^2} = \sum_{n=-\infty}^{\infty} \frac{2x^n q^n}{1+q^{2n}},$$

while if we set $a = -q$ we get

$$(7.1.6) \quad \left(\sqrt{x} + \frac{1}{\sqrt{x}}\right) \frac{(-xq^2;q^2)_\infty \left(-\frac{q^2}{x};q^2\right)_\infty (q^2;q^2)_\infty^2}{(qx;q^2)_\infty \left(\frac{q}{x};q^2\right)_\infty (-q;q^2)_\infty^2}$$

$$= \sum_{n=0}^{\infty} \frac{q^n}{1+q^{2n+1}} \left((\sqrt{x})^{2n+1} + (\sqrt{x})^{-2n-1}\right).$$

We leave these two statements as exercises. The second is trickier than the first.

Exercises

1. Recall from Chapter 6 that for a general n (not necessarily a nonnegative integer), we define

   $$(a;q)_n = \frac{(a;q)_\infty}{(aq^n;q)_\infty}.$$

   Show that with this definition we have

   $$\frac{(a;q)_n}{(aq;q)_n} = \frac{1-a}{1-aq^n}$$

   for all n. Use this fact to show that (7.1.1) is the case $b = aq$ of (7.1.2).

2. Explain why neither a nor x can be q in (7.1.1) and (7.1.3).

3. Show that setting $a = -1$ in (7.1.4) gives (7.1.5).

4. Show that setting $a = -q$ in (7.1.4) gives (7.1.6). This takes some fiddling around.

5. Show that, under the usual assumption that $|q| < 1$, and also assuming $a \neq 0$, the series in (7.1.4) converges for $|q| < |x| < \frac{1}{|q|}$. Consider both positive and negative values of n.

6. By the previous problem, the series in (7.1.5) (and also (7.1.6)) converges for $|q| < |x| < \frac{1}{|q|}$ if $|q| < 1$. If $|q| > 1$, when does it converge? (**Hint:** There is a cheap way to do this.) What if $|q| = 1$?

7. Use the ratio test to show that

   $$\sum_{n=-\infty}^{\infty} \frac{x^n}{1-aq^n}$$

   converges if $|q| < |x| < 1$ or if $|q| > |x| > 1$. Consider both positive and negative values of n.

8. Suppose $|p| > |z| > 1$. Then

   $$\sum_{n=-\infty}^{\infty} \frac{z^n}{1-ap^n}$$

should converge, according to the previous problem. To see what it converges to, we can proceed as follows:

(i) Set $p = \frac{1}{q}$ and $z = \frac{1}{x}$. Show that

$$\sum_{n=-\infty}^{\infty} \frac{z^n}{1 - ap^n} = \sum_{n=-\infty}^{\infty} \frac{\frac{1}{x^n}}{1 - \frac{a}{q^n}} = \sum_{n=-\infty}^{\infty} \frac{q^n}{x^n(q^n - a)},$$

where $|q| < |x| < 1$.

(ii) Show that the sum in (i) can be rewritten as

$$-\frac{1}{a} \sum_{n=-\infty}^{\infty} \frac{\left(\frac{q}{x}\right)^n}{1 - \frac{q^n}{a}}.$$

(iii) Show that (7.1.1) can be used on the last sum, with x replaced by $\frac{q}{x}$ and with a replaced by $\frac{1}{a}$. Hence show that

$$\sum_{n=-\infty}^{\infty} \frac{q^n}{x^n(q^n - a)} = \frac{\left(\frac{q}{ax}; q\right)_\infty (ax; q)_\infty (q; q)_\infty^2}{\left(\frac{q}{x}; q\right)_\infty (x; q)_\infty \left(\frac{q}{a}; q\right)_\infty (a; q)_\infty}.$$

9. Show that

$$\sum_{k=0}^{\infty} \frac{q^{6k+1} + q^{5(6k+1)}}{1 - q^{6(6k+1)}} - \sum_{k=0}^{\infty} \frac{q^{6k+5} + q^{5(6k+5)}}{1 - q^{6(6k+5)}} = \sum_{k=-\infty}^{\infty} \frac{q^{6k+1} + q^{5(6k+1)}}{1 - q^{6(6k+1)}},$$

and use (7.1.1) to evaluate the latter sum. (See also the next two problems.)

10. You might have been surprised that half of the sum in the previous problem evaluates to zero; at least I was. Use (7.1.1) to show that

$$\sum_{k=-\infty}^{\infty} \frac{q^{rk}}{1 - q^{rk+s(k+1)}} = 0.$$

11. The previous problem is somewhat unsatisfying in that it doesn't seem to give a really good reason *why* the sum is zero. Show that

$$\sum_{k=-\infty}^{\infty} \frac{q^{rk}}{1 - q^{rk+s(k+1)}} = \sum_{k=0}^{\infty} \frac{q^{rk}}{1 - q^{rk+s(k+1)}} - \sum_{k=0}^{\infty} \frac{q^{sk}}{1 - q^{sk+r(k+1)}},$$

and (always assuming $|q| < 1$) show that everything cancels if we expand the denominators on the right into geometric series. Alternatively, expand one of the denominators and argue that the expansion is symmetric in r and s.

12. Problem 1 in section 2.8 was to show that

$$\frac{1}{(x; q)_{n+1} \left(\frac{q}{x}; q\right)_n} = \frac{1}{(1-x)(q; q)_n^2} + \sum_{k=1}^{n} \frac{(-1)^k q^{\binom{k+1}{2}}}{(q; q)_{n+k}(q; q)_{n-k}} \left(\frac{1}{1 - xq^k} + \frac{1}{q^k - x}\right).$$

Show that letting $n \to \infty$ here gives

$$\frac{(q; q)_\infty^2}{(x; q)_\infty \left(\frac{q}{x}; q\right)_\infty} = \sum_{k=-\infty}^{\infty} \frac{(-1)^k q^{\frac{k(k+1)}{2}}}{1 - xq^k}.$$

13. The Bailey–Daum formula (5.6.2) can be used to give another proof of (7.1.5).

    (i) Show that
    $$\frac{(-xq;q^2)_\infty}{(xq;q^2)_\infty} \frac{\left(-\frac{q}{x};q^2\right)_\infty}{\left(\frac{q}{x};q^2\right)_\infty} = \sum_{j=0}^\infty x^j q^j \frac{(-1;q^2)_j}{(q^2;q^2)_j} \sum_{k=0}^\infty \frac{q^k}{x^k} \frac{(-1;q^2)_k}{(q^2;q^2)_k}.$$

    (ii) We want to combine the two series on the right side in (i) into a single series $\sum_{n=-\infty}^\infty c_n x^n$. Explain how we know that c_{-n} must be the same as c_n.

    (iii) Explain why
    $$c_0 = \sum_{j=0}^\infty \left(\frac{(-1;q^2)_j}{(q^2;q^2)_j}\right)^2 q^{2j},$$
    and use the Bailey–Daum formula to show that
    $$c_0 = \left(\frac{(-q^2;q^2)_\infty}{(q^2;q^2)_\infty}\right)^2.$$

    (iv) More generally, we can evaluate c_n for a positive n by setting $j = k+n$ in (i). Show that this gives
    $$c_n = \sum_{k=0}^\infty \frac{(-1;q^2)_{k+n}(-1;q^2)_k}{(q^2;q^2)_{k+n}(q^2;q^2)_k} q^{2k+n}$$
    $$= q^n \frac{(-1;q^2)_n}{(q^2;q^2)_n} \sum_{k=0}^\infty \frac{(-q^{2n};q^2)_k(-1;q^2)_k}{(q^{2n+2};q^2)_k(q^2;q^2)_k} q^{2k}.$$

    (v) Use the Bailey–Daum formula to evaluate the sum in (iv). You should get
    $$c_n = \frac{2q^n}{1+q^{2n}} \left(\frac{(-q^2;q^2)_\infty}{(q^2;q^2)_\infty}\right)^2.$$

    (vi) Since $2q^n/(1+q^{2n})$ is symmetric in n and $-n$ (or in q and q^{-1}), we have $c_{-n} = c_n$, so the formula in (v) is correct for *all* integers n. Show how this gives (7.1.5).

14. Although he was a friend and contemporary of Abel and Jacobi, the two young men who revolutionized the subject of elliptic functions in the late 1820s, and a great mathematician in his own right, Gustav Peter Lejeune Dirichlet wrote very little about elliptic functions or q-series. But he did state one interesting theorem in passing in one of his papers:

    (7.1.7) $(q + q^9 + q^{25} + \dots)(1 + 2q^2 + 2q^8 + 2q^{18} + \dots)$
    $$= \frac{q}{1-q^2} + \frac{q^3}{1-q^6} - \frac{q^5}{1-q^{10}} - \frac{q^7}{1-q^{14}} + \dots,$$
    where the sign pattern on the right is supposed to continue $++--++--\dots$. This is the only example Dirichlet gives of a deep theorem in number theory, his favorite subject. (Cayley later gave a similar but easier example, which is in the next problem, and you might consider doing that one before this one.)

(i) Show that the two sums on the left side of (7.1.7) are

$$\frac{q}{2}\sum_{n=-\infty}^{\infty} q^{4n^2+4n} \quad \text{and} \quad \sum_{n=-\infty}^{\infty} q^{2n^2} \quad \text{respectively.}$$

(ii) Show that the sum on the right is

$$\sum_{n=0}^{\infty} \frac{(-1)^n q^{4n+1}}{1-q^{8n+2}} + \sum_{n=0}^{\infty} \frac{(-1)^n q^{4n+3}}{1-q^{8n+6}}.$$

(iii) Although this is a step in the wrong direction, show that the sum in (ii) can be rewritten as

$$\sum_{n=0}^{\infty} \frac{(-1)^n q^{4n+1}(1+q^2)(1-q^{8n+4})}{(1-q^{8n+2})(1-q^{8n+6})}.$$

(iv) For a step in the right direction, show that the sum in (ii) can be rewritten as

$$\sum_{n=-\infty}^{\infty} \frac{(-1)^n q^{4n+1}}{1-q^{8n+2}}.$$

(v) Use the Jacobi triple product to evaluate the two sums in (i).

(vi) Use (3.5.3) to evaluate the sum in (iv).

(vii) Complete the proof of (7.1.7) with the help of your answers to (v) and (vi). (This still requires a bit of work, but nothing extremely difficult. Euler's "odd equals distinct" theorem might help.) For an application of (7.1.7) see the problems in the next section.

15. Cayley read the paper of Dirichlet mentioned in the previous problem and gave another corollary of Dirichlet's theorem, which is very similar to (7.1.7) but easier:

(7.1.8) $\left(q+q^9+q^{25}+\ldots\right)\left(1+2q^4+2q^{16}+2q^{36}+\ldots\right)$

$$= \frac{q}{1-q^2} - \frac{q^3}{1-q^6} + \frac{q^5}{1-q^{10}} - \frac{q^7}{1-q^{14}} + \ldots$$

(i) Show that the two sums on the left side of (7.1.8) are

$$\frac{q}{2}\sum_{n=-\infty}^{\infty} q^{4n^2+4n} \quad \text{and} \quad \sum_{n=-\infty}^{\infty} q^{4n^2} \quad \text{respectively.}$$

(ii) The sum on the right is clearly

$$\sum_{n=0}^{\infty} \frac{(-1)^n q^{2n+1}}{1-q^{4n+2}}.$$

Show that this is

$$\frac{1}{2}\sum_{n=-\infty}^{\infty} \frac{(-1)^n q^{2n+1}}{1-q^{4n+2}} = \frac{q}{2}\sum_{n=-\infty}^{\infty} \frac{(-1)^n q^{2n}}{1-q^{4n+2}}.$$

(iii) Use the triple product and (i) to show that the left side of (7.1.8) is

$$q(-q^4;q^4)_\infty^2 (q^8;q^8)_\infty^2.$$

(iv) Use (ii) and (3.5.3) to show that the right side of (7.1.8) is
$$q \frac{(q^8;q^8)_\infty^2}{(q^4;q^8)_\infty^2},$$
and complete the proof by explaining why this is the same as the answer to (iii).

16. Show that averaging (7.1.7) and (7.1.8) gives
$$\left(\sum_{n=0}^\infty q^{(2n+1)^2}\right)\left(1+\sum_{n=1}^\infty q^{2n^2}\left(1+q^{2n^2}\right)\right) = \sum_{n=-\infty}^\infty \frac{q^{8n+1}}{1-q^{16n+2}}.$$

17. Use problem 16 to show that
$$1+\sum_{n=1}^\infty q^{2n^2}\left(1+q^{2n^2}\right) = \frac{(q^6;q^{16})_\infty (q^{10};q^{16})_\infty (q^{16};q^{16})_\infty}{(q^2;q^{16})_\infty (q^8;q^{16})_\infty (q^{14};q^{16})_\infty}.$$

7.2. Sums of two squares

Recall Cauchy's identity (7.1.5): if $|q|<1$ and $|z|$ is between $|q|$ and its reciprocal, then
$$\sum_{n=-\infty}^\infty \frac{2z^n q^n}{1+q^{2n}} = \frac{(-zq;q^2)_\infty \left(-\frac{q}{z};q^2\right)_\infty (q^2;q^2)_\infty^2}{(zq;q^2)_\infty \left(\frac{q}{z};q^2\right)_\infty (-q^2;q^2)_\infty^2}.$$
Note that $z=1$ will always work, and this special case turns out to be very interesting:

(7.2.1)
$$\sum_{n=-\infty}^\infty \frac{2q^n}{1+q^{2n}} = \left(\frac{(-q;q^2)_\infty (q^2;q^2)_\infty}{(q;q^2)_\infty (-q^2;q^2)_\infty}\right)^2.$$

Thus the sum in (7.2.1) is a perfect square. Let's see what we can say about what it's a square of. By Euler's "odd equals distinct" theorem we have
$$\frac{(-q;q^2)_\infty (q^2;q^2)_\infty}{(q;q^2)_\infty (-q^2;q^2)_\infty} = (-q;q^2)_\infty (q^2;q^2)_\infty \frac{(-q;q)_\infty}{(-q^2;q^2)_\infty}.$$
Simplifying the fraction we get
$$\frac{(-q;q^2)_\infty (q^2;q^2)_\infty}{(q;q^2)_\infty (-q^2;q^2)_\infty} = (q^2;q^2)_\infty (-q;q^2)_\infty^2,$$
and now Jacobi's triple product gives
$$\frac{(-q;q^2)_\infty (q^2;q^2)_\infty}{(q;q^2)_\infty (-q^2;q^2)_\infty} = \sum_{n=-\infty}^\infty q^{n^2}.$$
Hence (7.2.1) becomes
$$\sum_{n=-\infty}^\infty \frac{2q^n}{1+q^{2n}} = \left(\sum_{n=-\infty}^\infty q^{n^2}\right)^2.$$
To see what this is trying to tell us we rewrite it as

(7.2.2)
$$\sum_{n=-\infty}^\infty \frac{2q^n}{1+q^{2n}} = \left(\sum_{j=-\infty}^\infty q^{j^2}\right)\left(\sum_{k=-\infty}^\infty q^{k^2}\right) = \sum_{j,k=-\infty}^\infty q^{j^2+k^2}.$$

7.2. SUMS OF TWO SQUARES   277

Hence the coefficient of q^m on both sides must be the number of ways of writing m as a sum of two squares, $j^2 + k^2$, accounting for both positive and negative values of j and k and for permutations. For example, if $m = 5$, then we could have either $j = \pm 1$ and $k = \pm 2$ or vice versa, so the coefficient of q^5 should be 8. We're going to have to expand the other side soon anyway, so let's do it now to check this. Because of the symmetry in q and q^{-1}, we have

$$(7.2.3) \quad \sum_{n=-\infty}^{\infty} \frac{2q^n}{1+q^{2n}} = 1 + \sum_{n=1}^{\infty} \frac{4q^n}{1+q^{2n}}$$

$$= 1 + 4 \sum_{n=1}^{\infty} q^n \left(1 - q^{2n} + q^{4n} - q^{6n} + q^{8n} - + \ldots \right)$$

$$(7.2.4) \quad = 1 + 4 \sum_{n=1}^{\infty} \left(q^n - q^{3n} + q^{5n} - q^{7n} + q^{9n} - + \ldots \right).$$

We can get a q^5 term from (7.2.3) from q^n with $n = 5$, or from q^{5n} with $n = 1$, so the coefficient of q^5 is $4(1+1) = 8$, as hoped. Let's check a few more low-dimensional cases before we try to describe what's happening in general. The only way to get a q^4 term from (7.2.3) is from q^n with $n = 4$, and this coefficient is four, so there should be 4 ways to write 4 as a sum of squares, and (suitably interpreted) there are: $2^2 + 0^2$, $(-2)^2 + 0^2$, $0^2 + 2^2$, $0^2 + (-2)^2$. This also explains the $1 = 1q^0$ in front of the series—there is one way to write 0 as the sum of two squares, $0^2 + 0^2$.

Next let's try a number like $n = 6$, where there are (as you can check) no ways to write it as a sum of two squares. This too is consistent with (7.2.3), because we can get a q^6 term there only from q^n with $n = 6$, or from $-q^{3n}$ with $n = 2$, and they cancel.

Now let's go back to (7.2.2) and try to put it in a nice final form. We'll denote the number of ways of writing m as a sum of two squares (interpreted as above) by $\square_2(m)$. Using this notation and (7.2.3), (7.2.2) becomes

$$\sum_{m=0}^{\infty} \square_2(m)\, q^m = 1 + 4 \sum_{n=1}^{\infty} \left(q^n - q^{3n} + q^{5n} - q^{7n} + q^{9n} - + \ldots \right)$$

$$= 1 + 4 \sum_{n=1}^{\infty} \left\{ \left(q^n + q^{5n} + q^{9n} + \ldots \right) - \left(q^{3n} + q^{7n} + \ldots \right) \right\}$$

$$= 1 + 4 \sum_{n=1}^{\infty} \sum_{k=0}^{\infty} \left(q^{n(4k+1)} - q^{n(4k+3)} \right).$$

We can get a coefficient 4 for q^m from the right side whenever there is a positive integer of the form $4k + 1$ that divides m, and a coefficient -4 for q^m whenever there is a positive integer of the form $4k + 3$ that divides m. To write this neatly let's introduce some more notation: let $d_{r,4}(m)$ denote the number of divisors of m that are congruent to r mod 4 (*i.e.*, which have remainder r when divided by 4). Then we have

$$\sum_{m=0}^{\infty} \square_2(m)\, q^m = 1 + 4 \sum_{m=1}^{\infty} \left(d_{1,4}(m) - d_{3,4}(m) \right) q^m,$$

so the final form of the result is

THEOREM 64 (Jacobi's Two Square Theorem). *With the above notation,*

$$\square_2(m) = \begin{cases} 1 & \text{if } m = 0, \\ 4\left(d_{1,4}(m) - d_{3,4}(m)\right) & \text{if } m \geq 1. \end{cases}$$

A corollary of this is worth remarking. Clearly the number of ways of writing m as a sum of two squares is never negative, so we must have $d_{1,4}(m) \geq d_{3,4}(m)$ for every positive integer m. We will say a little more about this in the exercises. Let's look at a few more instances of the theorem. A relatively small number with lots of odd factors is 45, whose divisors are $1, 3, 5, 9, 15, 45$. Of these, 3 and 15 are congruent to 3 mod 4 and the others are congruent to 1 mod 4, so Jacobi's theorem predicts $4(4-2) = 8$ ways to write 45 as a sum of two squares, which are $(\pm 6)^2 + (\pm 3)^2$ and $(\pm 3)^2 + (\pm 6)^2$.

To get lots of representations as sums of two squares we need lots of prime factors congruent to 1 mod 4 (preferably distinct) and none congruent to 3 mod 4, so let's try $5 \times 13 \times 17 = 1105$. The divisors of 1105 are $1, 5, 13, 17, 65, 85, 221, 1105$, all congruent to 1 mod 4, so Jacobi's theorem predicts $4 \times 8 = 32$ ways to write 1105 as a sum of two squares. One of the two, when squared, must be at least half of 1105, and neither of the two when squared can exceed it, so there aren't so many possibilities: we just have to check everything from $24^2 = 576$ to $33^2 = 1089$, and we find that

$$24^2 + 23^2 = 1105 = 33^2 + 4^2 = 32^2 + 9^2 = 31^2 + 12^2.$$

Because we can permute each of the four pairs and there are two possible signs for each, each pair gives eight representations, just as the single pair 6 and 3 led to eight representations for 45 above.

We do one last example. The smallest number that is the hypotenuse of a Pythagorean triple in two different ways is 25, where we have $7^2 + 24^2 = 25^2 = 15^2 + 20^2$, but one could object that $(15, 20, 25)$ is not an interesting Pythagorean triple (it is not **primitive**) because all the numbers have a common factor. Jacobi's theorem implies that the next smallest number that could work is $4225 = 65^2 = 5^2 \times 13^2$. The divisors of 4225 are $1, 5, 13, 25, 65, 169, 325, 845, 4225$, all of which are congruent to 1 mod 4, so the theorem predicts $4 \times 9 = 36$ ways to write 4225 as a sum of two squares. Obviously $(\pm 65)^2 + 0^2$ and its permutations are four of them. The two primitive ones are $33^2 + 56^2 = 65^2 = 16^2 + 63^2$... wait — The two primitive ones are $33^2 + 56^2 = 65^2 = 13^2 + 64^2$, each pair again giving rise to eight representations. The others are $39^2 + 52^2 = 65^2 = 25^2 + 60^2$, which come from the primitive triples $(3, 4, 5)$ and $(5, 12, 13)$ respectively.

Exercises

1. Show that
$$\sum_{n=1}^{\infty} \frac{q^n}{1+q^{2n}} = \sum_{n=0}^{\infty} \frac{(-1)^n q^{2n+1}}{1-q^{2n+1}}.$$

2. Show that
$$\sum_{n=0}^{\infty} \frac{(-1)^n q^{2n+1}}{1-q^{4n+2}} = \sum_{n=0}^{\infty} \frac{q^{2n+1}}{1+q^{4n+2}}.$$

EXERCISES

3. Define $f(q)$ to be the function expressed by the two series in the previous problem. Show that $f(iq) = i\,f(q)$.

4. Check Jacobi's two square theorem for $m = 85$ and $m = 145$.

5. Check Jacobi's two square theorem for $m = 221$ and $m = 231$.

6. Check Jacobi's two square theorem for $m = 1885 = 5 \times 13 \times 29$.

7. What is the product of two numbers of the form $4k+1$ (say $4i+1$ and $4j+1$)? What is the product of two numbers of the form $4k+3$? What is the product of a number of the form $4k+1$ and a number of the form $4k+3$?

8. Which of the forms $4n, 4n+1, 4n+2, 4n+3$ can a sum of two even squares have? A sum of two odd squares? An even square and an odd square?

9. Can an odd prime number be the sum of two squares? If so, how? If not, why not?

10. Discuss the possibilities for writing a generic odd number as a sum of two squares, in the light of Jacobi's two square theorem and your answers to the previous three problems.

11. Recall that a primitive Pythagorean triple (a, b, c) consists of three relatively prime positive integers a, b, c (i.e., with no common factor) such that $a^2 + b^2 = c^2$. It turns out (see problems 14–16) that all the primitive Pythagorean triples have $a = m^2 - n^2$, $b = 2mn$, $c = m^2 + n^2$ for some positive integers m and n that are relatively prime, with $m > n$, and with one of m and n even. Write down all the primitive Pythagorean triples with $m < 10$.

12. As mentioned in the text, 65 is the smallest number that is the hypotenuse of two different primitive Pythagorean triples. What is the next smallest such number? What are the triples?

13. Check Jacobi's two square theorem for the square of the number from problem 12.

14. It is clear that in a primitive Pythagorean triple (a, b, c), a and b can't both be even. Show that they also can't both be odd.

15. If we have a primitive Pythagorean triple (a, b, c), then, in view of problem 14, we can assume a is odd, b is even, and c is odd. Show that in fact b is a multiple of 4. (**Hint**: Show that $(2k+1)^2 - (2j+1)^2$ must be divisible by 8.)

16. The object of this problem is to prove the characterization of primitive Pythagorean triples given in problem 11. Again we can assume a is odd, b is even, and c is odd. Fill in the following outline:

    (i) We assumed a, b, c have no common factor. Explain why this means no two of them can have a common factor.

    (ii) Explain why we have $(c+a)(c-a) = 4K^2$ for some positive integer K.

    (iii) $c+a$ and $c-a$ are not quite relatively prime, because they are both even. Explain why they can have no common factor besides 2. (**Hint**: $(c+a)+(c-a) = 2c$ and $(c+a)-(c-a) = 2a$.)

    (iv) It follows from (iii) that $\frac{c+a}{2}$ and $\frac{c-a}{2}$ are relatively prime, and from (ii) that their product is a square. Explain why this means that each of them must be a square.

(v) From (iv) we have $c + a = 2m^2$ and $c - a = 2n^2$ for some positive integers m and n, and clearly $m > n$. Show that this gives the formulas for a, b, c in problem 11.

(vi) There's still a tiny bit of work left. Explain why m and n must be relatively prime, and use problem 14 to show that one of them must be even.

17. The object of this problem is to find the "sum of two squares" type theorem that corresponds to Dirichlet's identity (7.1.7)

$$(7.2.5) \quad \left(q + q^9 + q^{25} + \ldots\right)\left(1 + 2q^2 + 2q^8 + 2q^{18} + \ldots\right)$$
$$= \frac{q}{1-q^2} + \frac{q^3}{1-q^6} - \frac{q^5}{1-q^{10}} - \frac{q^7}{1-q^{14}} + \ldots$$

from problem 14 in section 7.1.

(i) Explain why the coefficient of q^m on the left side counts the number of ways of writing m as the sum of a positive odd square plus twice another square. For example, the coefficient of q^{33} should be 4 because we can write $33 = 1 + 2(\pm 4)^2$ or $33 = 25 + 2(\pm 2)^2$.

(ii) Explain why m must be odd in (i).

(iii) Explain why the coefficient of q^m on the right side of (7.1.7) equals the number of divisors of m that are congruent to 1 or 3 mod 8, minus the number of divisors of m that are congruent to 5 or 7 mod 8. (It is probably best to start from part (ii) of problem 14 in section 7.1.)

(iv) Thus the number theory theorem equivalent to (7.1.7) is that for any positive odd number m, the number of ways to write m as a positive odd square plus twice another square equals the number of divisors of m that are congruent to 1 or 3 mod 8 minus the number of divisors of m that are congruent to 5 or 7 mod 8. For example, the divisors of 33 are $1, 3, 11, 33$, two of which are congruent to 1 mod 8 and two to 3 mod 8, so the theorem predicts four ways to write 33 as a positive odd square plus twice another square, as we found above.

(v) **Corollary:** Any positive integer m has at least as many divisors congruent to 1 or 3 mod 8 as to 5 or 7 mod 8. First explain why this holds if m is odd, and then deduce it for any positive integer m.

(vi) Write out the multiplication table for $1, 3, 5, 7$ mod 8. In other words, find every possible product of two of these numbers mod 8. For readers who know what these words mean, they form a group isomorphic to the so-called **Klein 4-group** of symmetries of a rectangle (identity, 180° rotation, horizontal reflection, vertical reflection). Discuss the corollary in (v) in the light of this multiplication table.

(vii) Check the theorem of (iv) for $m = 9, 27, 99$.

(viii) Check the theorem of (iv) for $m = 297$.

18. The "sum of two squares" type theorem that corresponds to Cayley's identity (7.1.8)

$$\left(q + q^9 + q^{25} + \ldots\right)\left(1 + 2q^4 + 2q^{16} + 2q^{36} + \ldots\right)$$
$$= \frac{q}{1-q^2} - \frac{q^3}{1-q^6} + \frac{q^5}{1-q^{10}} - \frac{q^7}{1-q^{14}} + \ldots$$

from problem 15 in section 7.1 is nearly the same as Jacobi's two square theorem. Explain why (7.1.8) implies that the number of ways of writing an odd number m as the sum of an even square and a positive odd square equals $d_{1,4}(m) - d_{3,4}(m)$, with the same notation as in Jacobi's two square theorem. Let's call this Cayley's two square theorem, even though he seems not to have noticed it.

19. Suppose m is an odd number, so that both Jacobi's and Cayley's two square theorems apply. What accounts for the extra factor of 4 in Jacobi's theorem? (It may help to try some of the examples in problems 3–5.)

7.3. Sums of four squares

Our next goal is Jacobi's four square theorem. We start with (7.1.1),

$$\sum_{n=-\infty}^{\infty} \frac{x^n}{1-aq^n} = \frac{(ax;q)_\infty \left(\frac{q}{ax};q\right)_\infty (q;q)_\infty^2}{(x;q)_\infty \left(\frac{q}{x};q\right)_\infty (a;q)_\infty \left(\frac{q}{a};q\right)_\infty},$$

which holds for $|q| < |x| < 1$, and we try to make the right side into

$$\left(\frac{(q;q)_\infty}{(-q;q)_\infty}\right)^4.$$

Taking $x = -1$ and $a = -1$ almost does this, but in the first place it makes the right side equal to zero, and in the second place we only know for sure that (7.1.1) converges for $|q| < |x| < 1$; this means it certainly diverges for $|x| > 1 > |q|$, and it might or might not converge for an x with $|x| = 1$. Then we can take $a = -1$ right away, but we have to do some work before we can set $x = -1$. When $a = -1$, (7.1.1) reduces to

$$\sum_{n=-\infty}^{\infty} \frac{x^n}{1+q^n} = \frac{(-x;q)_\infty \left(-\frac{q}{x};q\right)_\infty (q;q)_\infty^2}{(x;q)_\infty \left(\frac{q}{x};q\right)_\infty (-1;q)_\infty (-q;q)_\infty},$$

and since

$$(-1;q)_\infty = (1+1)(1+q)(1+q^2)\cdots = 2(-q;q)_\infty,$$

multiplying through by 2 we get

(7.3.1) $$\sum_{n=-\infty}^{\infty} \frac{2x^n}{1+q^n} = \frac{(-x;q)_\infty \left(-\frac{q}{x};q\right)_\infty (q;q)_\infty^2}{(x;q)_\infty \left(\frac{q}{x};q\right)_\infty (-q;q)_\infty^2}.$$

The region of convergence is still $|q| < |x| < 1$, and the right side of (7.3.1) blows up if $x = 1$ because of the factor $1 - x$ in $(x;q)_\infty$. The factor $1 + x$ in $(-x;q)_\infty$ causes it to become zero when $x = -1$, which suggests that the left side of (7.3.1) does too, so let's see if we can rewrite it to reveal this. It equals

$$1 + \sum_{n=1}^{\infty} \frac{2x^n}{1+q^n} + \sum_{n=1}^{\infty} \frac{2x^{-n}}{1+q^{-n}} \frac{(qx)^n}{(qx)^n},$$

and we need to manipulate the first sum to get something that combines nicely with the 1. We can rewrite it further as

$$1 + \sum_{n=1}^{\infty} \frac{2x^n(1+q^n-q^n)}{1+q^n} + \sum_{n=1}^{\infty} \frac{2\left(\frac{q}{x}\right)^n}{1+q^n} = 1 + 2\sum_{n=1}^{\infty} x^n + 2\sum_{n=1}^{\infty} \frac{\left(\frac{q}{x}\right)^n - q^n x^n}{1+q^n}.$$

Since we still have $|x| < 1$, we can sum the geometric series, which gives

$$1 + \frac{2x}{1-x} + 2\sum_{n=1}^{\infty} \left(\frac{q}{x}\right)^n \frac{1-x^{2n}}{1+q^n}$$

$$= \frac{1+x}{1-x} + 2\sum_{n=1}^{\infty} \frac{\left(\frac{q}{x}\right)^n}{1+q^n}(1+x^2+x^4+\cdots+x^{2n-2})(1+x)(1-x).$$

Using this on the left side of (7.3.1) and multiplying both sides by $(1-x)/(1+x)$ we get

$$1 + \sum_{n=1}^{\infty} \frac{2\left(\frac{q}{x}\right)^n}{1+q^n}(1+x^2+x^4+\cdots+x^{2n-2})(1-x)^2 = \frac{(-xq;q)_\infty \left(-\frac{q}{x};q\right)_\infty (q;q)_\infty^2}{(xq;q)_\infty \left(\frac{q}{x};q\right)_\infty (-q;q)_\infty^2}.$$

The right side of this is now symmetric in x and $\frac{1}{x}$, which is encouraging. If we let $x = -1$, then the left side is a convergent series (exercise), and we have

(7.3.2) $$1 + \sum_{n=1}^{\infty} \frac{8n(-q)^n}{1+q^n} = \left(\frac{(q;q)_\infty}{(-q;q)_\infty}\right)^4.$$

Recall Gauss's identity (5.2.11),

$$\sum_{n=-\infty}^{\infty} (-1)^n q^{n^2} = \frac{(q;q)_\infty}{(-q;q)_\infty},$$

from problem 8 in section 5.2, a special case of Jacobi's triple product. Then (7.3.2) implies

(7.3.3) $$1 + \sum_{n=1}^{\infty} \frac{8n(-q)^n}{1+q^n} = \left(\sum_{n=-\infty}^{\infty} (-1)^n q^{n^2}\right)^4.$$

Since n^2 is even if and only if n is, replacing q by $-q$ in (7.3.3) gives

(7.3.4) $$1 + \sum_{n=1}^{\infty} \frac{8nq^n}{1+(-q)^n} = \left(\sum_{n=-\infty}^{\infty} q^{n^2}\right)^4.$$

If we expand the right side of (7.3.4), the coefficient of q^m will be the number of ways of writing m as a sum of four squares, with the same interpretation as in Jacobi's two square theorem, *i.e.*, allowing positive and negative numbers and accounting for permutations. Let's denote this coefficient by $\square_4(m)$. For reasons that will soon appear, we rewrite the left side of (7.3.4) as

$$1 + \sum_{n=1}^{\infty} \frac{8nq^n}{1-q^n} - \sum_{n=1}^{\infty} 8nq^n \left(\frac{1}{1-q^n} - \frac{1}{1+(-q)^n}\right).$$

Now $1-q^n$ and $1+(-q)^n$ are the same if n is odd, so all the terms with n odd in the last sum are zero. Setting $n = 2k$ there we therefore have

$$\sum_{k=1}^{\infty} 16kq^{2k}\left(\frac{1}{1-q^{2k}} - \frac{1}{1+q^{2k}}\right) = \sum_{k=1}^{\infty} 16kq^{2k}\frac{1+q^{2k}-(1-q^{2k})}{(1+q^{2k})(1-q^{2k})}$$

$$= \sum_{k=1}^{\infty} \frac{32kq^{4k}}{1-q^{4k}}.$$

Then (7.3.4) says

(7.3.5) $$\sum_{m=0}^{\infty} \square_4(m) q^m = 1 + 8 \sum_{k=1}^{\infty} \frac{kq^k}{1-q^k} - 8 \sum_{k=1}^{\infty} \frac{4kq^{4k}}{1-q^{4k}}.$$

We now need to recall (4.4.3), which was

(7.3.6) $$\sum_{k=1}^{\infty} \frac{kq^k}{1-q^k} = \sum_{m=1}^{\infty} \sigma(m) q^m,$$

where $\sigma(m)$ is the sum of the divisors of m. Therefore the coefficient of q^m in the first sum on the right side of (7.3.5) is 8 times the sum of the divisors of m. If we replace q by q^4 in (7.3.6) and multiply by 4, we get

$$\sum_{k=1}^{\infty} \frac{4kq^{4k}}{1-q^{4k}} = \sum_{m=1}^{\infty} 4\sigma(m) q^{4m}.$$

Now $4\sigma(m)$ is the sum of the divisors of $4m$ that are multiples of 4, as there is an obvious 1-1 correspondence between these and the divisors of m. For example, the divisors of 24 that are multiples of 4 are 4, 8, 12, and 24, whose sum is $4(1 + 2 + 3 + 6) = 4\sigma(6)$. But only an integer of the form $4m$ can have a divisor that is a multiple of 4; the forms $4m+1$, $4m+2$, and $4m+3$ cannot. Therefore the coefficient of q^m in the last sum in (7.3.5) is 8 times the sum of the divisors of m that are multiples of 4, and hence, if $m \geq 1$, the coefficient of q^m on the right side of (7.3.5) must be 8 times the sum of the divisors of m that are *not* multiples of 4. We have proved

THEOREM 65 (Jacobi's four square theorem). $\square_4(0) = 1$, and if $m \geq 1$, then $\square_4(m)$ is equal to 8 times the sum of the divisors of m that are not multiples of 4.

Let's look at a few examples to get a feeling for what this theorem says, starting with $m = 4$. The divisors of 4 that aren't multiples of 4 are 1 and 2, whose sum is 3, so the theorem predicts $8 \times 3 = 24$ ways to write 4 as a sum of four squares. We can use either four 1's or one 2 and three 0's, and we just have to determine the multiplicity of each. With four 1's we just have to choose their signs; there are two choices for each sign and the choices are independent, so

$$4 = (\pm 1)^2 + (\pm 1)^2 + (\pm 1)^2 + (\pm 1)^2$$

entails 16 possibilities. As for a 2 and three 0's, there are four permutations and two choices for the sign of 2, so 8 possibilities. This makes 24, as expected.

Next we try $m = 30$. The divisors are $1, 2, 3, 5, 6, 10, 15, 30$, none of which are multiples of 4, and they add up to 72. Jacobi's theorem then predicts $8 \times 72 = 576$ ways to write 30 as a sum of four squares. There are basically two possibilities,

$$5^2 + 2^2 + 1^2 + 0^2 = 30 = 4^2 + 3^2 + 2^2 + 1^2,$$

and we just have to find the multiplicity of each. There are $4! = 24$ permutations of $4, 3, 2, 1$, and each can be either positive or negative, so $4, 3, 2, 1$ amounts to $24 \times 2^4 = 384$ possibilities. The only thing different about $5, 2, 1, 0$ is that 0 has no sign, so there are only $24 \times 2^3 = 192$ possibilities, giving $384 + 192 = 576$ possibilities in all.

Legendre and Cauchy found another corollary of (7.1.1) that has beautiful consequences in number theory. Setting $a = q$ in (7.1.4) gives

$$(7.3.7) \qquad \sum_{n=-\infty}^{\infty} \frac{x^n}{1 - q^{2n+1}} = \frac{(qx; q^2)_\infty \left(\frac{q}{x}; q^2\right)_\infty (q^2; q^2)_\infty^2}{(x; q^2)_\infty \left(\frac{q^2}{x}; q^2\right)_\infty (q; q^2)_\infty^2}.$$

We also intend to set $x = q$, but this would make the right side zero because of the first factor of $\left(\frac{q}{x}; q^2\right)_\infty$, which would become $1 - 1$. Therefore, the sum side of (7.3.7) must also become zero when $x = q$. Let's look at it and see why. The terms $n = 0, -1, 1, -2, 2, -3$ are

$$\frac{1}{1-q}, \quad \frac{x^{-1}}{1-q^{-1}}, \quad \frac{x}{1-q^3}, \quad \frac{x^{-2}}{1-q^{-3}}, \quad \frac{x^2}{1-q^5}, \quad \frac{x^{-3}}{1-q^{-5}}$$

respectively, and it is natural to combine them in pairs because they almost have the same denominators. The first two terms are

$$\frac{1}{1-q} + \frac{x^{-1}}{1-q^{-1}} \frac{q}{q} = \frac{1 - \frac{q}{x}}{1-q},$$

the next two are

$$\frac{x}{1-q^3} + \frac{x^{-2}}{1-q^{-3}} \frac{q^3}{q^3} = \frac{x - \frac{q^3}{x^2}}{1-q^3} = \frac{x\left(1 - \frac{q^3}{x^3}\right)}{1-q^3},$$

the next two are

$$\frac{x^2}{1-q^5} + \frac{x^{-3}}{1-q^{-5}} = \frac{x^2\left(1 - \left(\frac{q}{x}\right)^5\right)}{1-q^5},$$

and in general we have

$$(7.3.8) \qquad \frac{x^n}{1-q^{2n+1}} + \frac{x^{-n-1}}{1-q^{-2n-1}} \frac{q^{2n+1}}{q^{2n+1}} = \frac{x^n}{1-q^{2n+1}} \left(1 - \left(\frac{q}{x}\right)^{2n+1}\right).$$

Using (7.3.8) in (7.3.7) and dividing both sides by $1 - \frac{q}{x}$ we have

$$\sum_{n=0}^{\infty} \frac{x^n}{1-q^{2n+1}} \frac{1 - \left(\frac{q}{x}\right)^{2n+1}}{1 - \frac{q}{x}} = \frac{(qx; q^2)_\infty \left(\frac{q^3}{x}; q^2\right)_\infty (q^2; q^2)_\infty^2}{(x; q^2)_\infty \left(\frac{q^2}{x}; q^2\right)_\infty (q; q^2)_\infty^2}.$$

Taking the limit of this as $x \to q$ we finally get

$$(7.3.9) \qquad \sum_{n=0}^{\infty} \frac{(2n+1)q^n}{1-q^{2n+1}} = \left(\frac{(q^2; q^2)_\infty}{(q; q^2)_\infty}\right)^4.$$

Recall Gauss's identity (5.2.12) from problem 10 in section 5.2:

$$(7.3.10) \qquad \frac{(q^2; q^2)_\infty}{(q; q^2)_\infty} = 1 + q + q^3 + q^6 + q^{10} + \cdots = \sum_{n=0}^{\infty} q^{\binom{n+1}{2}}.$$

Therefore (7.3.9) says

$$\sum_{n=0}^{\infty} \frac{(2n+1)q^n}{1-q^{2n+1}} = \left(\sum_{n=0}^{\infty} q^{\binom{n+1}{2}}\right)^4,$$

or

$$\frac{1}{1-q} + \frac{3q}{1-q^3} + \frac{5q^2}{1-q^5} + \frac{7q^3}{1-q^7} + \cdots = \left(1 + q + q^3 + q^6 + q^{10} + q^{15} + \ldots\right)^4.$$

To see what this is trying to tell us, it helps to change q to q^2 and then multiply by q. Taking this factor of q in the form $\left(q^{\frac{1}{4}}\right)^4$ on the right side of (7.3.9), this gives

$$\sum_{n=0}^{\infty} \frac{(2n+1)q^{2n+1}}{1-q^{4n+2}} = \left(\sum_{n=0}^{\infty} q^{\left(n+\frac{1}{2}\right)^2}\right)^4.$$

Expanding the left side using the geometric series

$$\frac{x}{1-x^2} = \sum_{k=0}^{\infty} x^{2k+1} \quad \text{if } |x| < 1$$

with $x = q^{2n+1}$, we get

$$\sum_{n=0}^{\infty} \frac{(2n+1)q^{2n+1}}{1-q^{4n+2}} = \sum_{n=0}^{\infty} (2n+1) \sum_{k=0}^{\infty} q^{(2k+1)(2n+1)}.$$

Now $(2k+1)(2n+1)$ is a product of two odd numbers, so it must be another odd number, say $2m+1$, and hence

$$\sum_{n=0}^{\infty} (2n+1) \sum_{k=0}^{\infty} q^{(2k+1)(2n+1)} = \sum_{m=0}^{\infty} q^{2m+1} \sum_{n:2n+1|2m+1} (2n+1),$$

where the inner sum is of all the odd numbers $2n+1$ that divide $2m+1$. But an odd number can't have an even divisor, so these are *all* the divisors of $2m+1$. In other words, we have proved that

(7.3.11) $$\left(\sum_{n=0}^{\infty} q^{\left(n+\frac{1}{2}\right)^2}\right)^4 = \sum_{n=0}^{\infty} \frac{(2n+1)q^{2n+1}}{1-q^{4n+2}} = \sum_{m=0}^{\infty} \sigma(2m+1)q^{2m+1}.$$

Hence (7.3.9) was trying to say that

(7.3.12) $$\sum_{n=0}^{\infty} \frac{(2n+1)q^n}{1-q^{2n+1}} = \sum_{m=0}^{\infty} \sigma(2m+1)\, q^m = \left(\sum_{n=0}^{\infty} q^{\binom{n+1}{2}}\right)^4.$$

This has a nice number-theoretic interpretation of its own, but Legendre saw how to make it still more beautiful. Replacing q by q^4 in (7.3.11) we get

(7.3.13) $$\sum_{m=0}^{\infty} \sigma(2m+1)\, q^{8m+4} = \left(\sum_{n=0}^{\infty} q^{(2n+1)^2}\right)^4.$$

This implies

THEOREM 66 (Legendre's four square theorem). *The number of ways to write a number of the form $8m+4$ as a sum of four positive odd squares equals the sum of the divisors of $2m+1$.*

We still have multiplicity issues with this theorem due to permutations, but it's nicer in that we don't have to worry about signs. Let's try the example $8m+4 = 148$. Then $2m+1 = 37$, which is prime, so the sum of its divisors is $1+37 = 38$, and Legendre's theorem predicts 38 ways to write 148 as a sum of four positive odd squares. The largest odd square less than 148 is $121 = 11^2$, and we can make up the remaining 27 either with $5^2 + 1^2 + 1^2$ or with $3^2 + 3^2 + 3^2$. The next largest odd square is $81 = 9^2$, which leaves 67, which we can realize as $7^2 + 3^2 + 3^2$. If we use no odd square larger than $49 = 7^2$ we can have $7^2 + 7^2 + 7^2 + 1^2$ or $7^2 + 7^2 + 5^2 + 5^2$.

Now we have to determine the multiplicities of these five combinations. With $11^2 + 3^2 + 3^2 + 3^2$ and $7^2 + 7^2 + 7^2 + 1^2$ we have three like numbers, so the only choice we have is where the unlike number goes; thus there are four of each of these. With $7^2 + 7^2 + 5^2 + 5^2$ we have two pairs of like numbers, and we just have to choose which two of the four numbers are the 7's, which we can do in $\binom{4}{2} = 6$ ways. Finally we have $11^2 + 5^2 + 1^2 + 1^2$ and $9^2 + 7^2 + 3^2 + 3^2$. The easiest argument is that the multinomial coefficient $\binom{4}{2,1,1} = \frac{4!}{2!\,1!\,1!} = 12$ counts each of these. Alternatively, we can argue that, say for $(11,5,1,1)$, we can choose which number is the 11 in four ways, and then which number is the 5 in three ways, so there are 12 possibilities in all with two like numbers and two more unlike numbers. In total we have $12 + 12 + 6 + 4 + 4 = 38$ possibilities, as the theorem predicts.

Exercises

1. Check Jacobi's four square theorem for $m = 48$ and $m = 50$.
2. Check Jacobi's four square theorem for $m = 90$. (There are 9 different combinations of varying multiplicities.)
3. Lagrange had proved earlier that every positive integer can be written as a sum of four squares. Explain how this follows from Jacobi's four square theorem.
4. Assuming that $|q| < 1$, show that
$$\sum_{n=1}^{\infty} \frac{8n\,(-q)^n}{1+q^n}$$
converges.
5. Show that only a number of the form $8m + 4$ (for a nonnegative integer m) can be a sum of four odd squares. (Legendre's theorem implies this, but it is not difficult to show it directly.)
6. Check Legendre's theorem for $8m + 4 = 36$ and $8m + 4 = 52$.
7. Check Legendre's theorem for $8m + 4 = 180$.
8. As mentioned in Chapter 3, a number of the form $\binom{n+1}{2}$ is classically called a triangular number because
$$\binom{n+1}{2} = n + (n-1) + \cdots + 2 + 1$$
and the Ferrers diagram of $n + (n-1) + \cdots + 2 + 1$ looks like a triangle. Then (7.3.12) says that the number of ways to write m as a sum of four triangular numbers is $\sigma(2m+1)$. Check this for $m = 15$. (Note that 0 counts as a triangular number.)
9. Show that there are as many ways to write m as a sum of two triangular numbers as ways to write m as a square plus twice a triangular number. The order matters for the two triangular numbers but not for the other pair, but the other pair counts both positive and negative squares. For example, we can write $m = 6$ as $6 + 0$ or $0 + 6$ or $3 + 3$, and as either $2^2 + 2 \cdot 1$ or as $(-2)^2 + 2 \cdot 1$ or $0^2 + 2 \cdot 3$. We can write $m = 16$ as either $15 + 1$ or $1 + 15$ or $10 + 6$ or $6 + 10$, and as either $4^2 + 2 \cdot 0$ or $(-4)^2 + 2 \cdot 0$ or $2^2 + 2 \cdot 6$ or $(-2)^2 + 2 \cdot 6$.

EXERCISES

10. Are there any positive integers m that cannot be written in either of the ways described in the previous problem?

11. We can rewrite Gauss's identity (7.3.10) as

$$\frac{1}{(q;q^2)_\infty} = \frac{1}{(q^2;q^2)_\infty} \sum_{n=0}^{\infty} q^{\binom{n+1}{2}}.$$

Explain why this implies that there are as many partitions of a number m into odd parts as into even parts plus one triangular part (which could be zero).

12. Prove Gauss's identity

$$\left(1 + 2q + 2q^4 + 2q^9 + \ldots\right)^4$$
$$= \left(1 - 2q + 2q^4 - 2q^9 + \ldots\right)^4 + 16q\left(1 + q^2 + q^6 + q^{12} + \ldots\right)^4$$

from section 5.2 by replacing q by q^2 in (7.3.9) and using (7.3.3) and (7.3.4).

13. Using many of the ingredients in this section and one borrowed from Euler, Cayley proved that

(7.3.14) $$\left(1 - 2q + 2q^4 - 2q^9 + \ldots\right)^4 + 16\left(\frac{q}{1-q^2} - \frac{2q^2}{1-q^4} + \frac{3q^3}{1-q^6} - + \ldots\right)$$
$$= \frac{1 + 9q + 25q^3 + 49q^6 + 81q^{10} + \ldots}{1 + q + q^3 + q^6 + q^{10} + \ldots}.$$

(i) Using (7.3.3), show that the left side of (7.3.14) is

(a) $$1 + 8\sum_{n=1}^{\infty} \frac{n(-q)^n}{1+q^n} - 16\sum_{n=1}^{\infty} \frac{n(-q)^n}{1-q^{2n}}.$$

(ii) Show that the right side of (7.3.14) is

(b) $$\frac{\sum_{n=0}^{\infty} (2n+1)^2 q^{\binom{n+1}{2}}}{\sum_{n=0}^{\infty} q^{\binom{n+1}{2}}}.$$

(iii) We need to do something to (b) to make it look more like (a). Following Cayley we set X equal to the denominator of (b); that is, $X = 1 + q + q^3 + q^6 + q^{10} + \ldots$. If X' denotes the derivative of X with respect to q, show that the numerator of (b) is $X + 8qX'$.

(iv) To prove (7.3.14) it suffices to show that

(c) $$\sum_{n=1}^{\infty} n(-q)^n \left(\frac{1}{1+q^n} - \frac{2}{1-q^{2n}}\right) = q\frac{X'}{X},$$

with X as in part (iii). Explain.

(v) Show that (c) simplifies to

(d) $$\sum_{n=1}^{\infty} \frac{n(-q)^{n-1}}{1-q^n} = \frac{X'}{X},$$

so we just have to show this.

(vi) According to (7.3.10) we have
$$X = \frac{(q^2;q^2)_\infty}{(q;q^2)_\infty}.$$
Prove (d) by calculating the derivative with respect to q of $\log X$.

7.4. Bibliographical Notes

Cauchy's identity (7.1.1) and its special cases (7.1.5) and (7.1.6) come from [69], in which (7.3.9) also appears without proof. The two sides of (7.1.1) are called the **Jordan–Kronecker** function in the beautiful little book [241], which has more of its properties.

Dirichlet's identity (7.1.7) is on p. 468 of the first volume of his collected papers [87]. Cayley gave (7.1.7) and (7.1.8) in [70].

Jacobi's two and four square theorems are in [148]; the proofs there use q-series expansions of elliptic functions. Legendre's four square theorem is in a footnote on p. 133 of [163]. It is deduced from his identity (56) on the same page, which is more or less equivalent to our (7.3.9); (56) in turn is deduced from q-series expansions of elliptic functions. (Legendre had a great advantage here in that he had been nice to Jacobi at the beginning of the latter's career, so Jacobi was sending him his theorems. Legendre may have been the only one who understood what Jacobi was doing this early, although Abel was working in the same area from a different point of view.) Legendre restated his four square theorem in [164]. In concert with the identity that follows it, (7.3.9) is (12) in Cauchy's paper [69]. It is also in Gauss's *Nachlass* [117], as is (7.3.4).

The characterization of primitive Pythagorean triples in the exercises for section 13 is ancient. See the beginning of Chapter 4 of [86] for comments on its history, and Chapter 8 for the history of Lagrange's four square theorem, which is called Bachet's theorem there. Cayley's theorem in problem 13 of section 7.3 comes from [72]. Problem 9 in that section comes from [240].

CHAPTER 8

Ramanujan's Congruences

8.1. Ramanujan's congruences

There are five partitions of 4, namely $4, 3+1, 2+2, 2+1+1, 1+1+1+1$. There are 30 partitions of 9 and 135 partitions of 14. If we continue through the numbers congruent to 4 mod 5, here is what we find:

n	$p(n)$	n	$p(n)$	n	$p(n)$	n	$p(n)$
4	5	29	4565	54	386155	79	13848650
9	30	34	12310	59	831820	84	26543660
14	135	39	31185	64	1741630	89	49995925
19	490	44	75175	69	3554345	94	92669720
24	1575	49	173525	74	7089500	99	169229875

There are at least two patterns here: not only are all the numbers of partitions divisible by 5, but all the ones on the last line are divisible by 25. Ramanujan noticed these patterns, and a number of others, and was the first to explain some of them. He gave a proof of the first fact involving congruences, but he also found an identity that makes it obvious:

$$\sum_{n=0}^{\infty} p(5n+4)q^n = 5 \frac{(q^5;q^5)_\infty^5}{(q;q)_\infty^6}.$$

We will prove this in the next section. In this section we give a version of Ramanujan's congruence proof. It relies on two lemmas.

From Cauchy's theorem (3.5.3) we know that

$$\frac{1}{(1-x)(1-xq)(1-xq^2)(1-xq^3)(1-xq^4)} = \sum_{n=0}^{\infty} \binom{n+4}{4}_q x^n.$$

Setting $q = 1$, we have

$$\frac{1}{(1-x)^5} = \sum_{n=0}^{\infty} \binom{n+4}{4} x^n = \frac{1}{24} \sum_{n=0}^{\infty} (n+4)(n+3)(n+2)(n+1)\, x^n,$$

and it follows that

$$\frac{1-x^5}{(1-x)^5} = \frac{1}{24} \sum_{m=0}^{\infty} (m+4)(m+3)(m+2)(m+1)\, x^m$$

$$- \frac{1}{24} \sum_{m=0}^{\infty} (m+4)(m+3)(m+2)(m+1)\, x^{m+5}.$$

289

Changing m to n in the first sum and m to $n-5$ in the second, we have
$$\frac{1-x^5}{(1-x)^5} = \frac{1}{24}\sum_{n=0}^{\infty}(n+4)(n+3)(n+2)(n+1)\,x^n$$
$$-\frac{1}{24}\sum_{n=5}^{\infty}(n-1)(n-2)(n-3)(n-4)\,x^n.$$

Note that the last sum could start at $n=1$, because the $n=1,2,3,4$ terms would all be zero. Breaking off the $n=0$ term of the first sum, we have
$$\frac{1-x^5}{(1-x)^5} = 1 + \frac{1}{24}\sum_{n=1}^{\infty}\left[\begin{array}{c}(n+4)(n+3)(n+2)(n+1)\\-(n-4)(n-3)(n-2)(n-1)\end{array}\right]x^n.$$

Multiplying this out we get
$$\frac{1-x^5}{(1-x)^5} = 1 + \frac{1}{24}\sum_{n=1}^{\infty}\left(20n^3 + 100n\right)x^n = 1 + \frac{5}{6}\sum_{n=1}^{\infty}\left(n^3 + 5n\right)x^n.$$

Finally, we can write
$$\frac{n^3+5n}{6} = \frac{n^3-n}{6} + n = \binom{n+1}{3} + n,$$

so

(8.1.1) $$\frac{1-x^5}{(1-x)^5} = 1 + 5\sum_{n=1}^{\infty}\left[\binom{n+1}{3} + n\right]x^n.$$

Hence every term in this series is divisible by 5 except the first one.

For the second lemma we expand
$$q(q;q)_{\infty}^4 = q(q;q)_{\infty}(q;q)_{\infty}^3$$
using Euler's pentagonal number theorem (5.1.15) and Jacobi's cube identity (5.2.8). This gives
$$q(q;q)_{\infty}^4 = q\left(\sum_{j=-\infty}^{\infty}(-1)^j q^{\frac{j(3j+1)}{2}}\right)\left(\sum_{k=0}^{\infty}(-1)^k(2k+1)q^{\frac{k(k+1)}{2}}\right)$$
$$= \sum_{j=-\infty}^{\infty}\sum_{k=0}^{\infty}(-1)^{j+k}(2k+1)q^{1+\frac{j(3j+1)}{2}+\frac{k(k+1)}{2}}.$$

In Hardy's words, we now "consider in what circumstances the [exponent of q] is divisible by 5", and he follows Ramanujan's analysis. The exponent of q is a positive integer, so it is divisible by 5 if and only if 8 times it is divisible by 5; i.e., we need
$$8 + 4j(3j+1) + 4k(k+1) = 2j^2 + 4j + 2 + 4k^2 + 4k + 1 + 10j^2 + 5$$
to be divisible by 5, so we need $2(j+1)^2 + (2k+1)^2$ to be divisible by 5. Let's make a table of these two quantities mod 5:

j	$2(j+1)^2$	k	$(2k+1)^2$
0	2	0	1
1	3	1	4
2	3	2	0
3	0	3	1
4	2	4	4

The sum of a number in the second column and a number in the fourth column is only zero mod 5 if we take the two zeros, so the exponent of q is divisible by 5 if and only if $j \equiv 3 \pmod 5$ and $k \equiv 2 \pmod 5$. Because of the factor of $2k+1$, it follows that every coefficient of q^{5m+5} in the expansion of $q(q;q)_\infty^4$ is divisible by 5. This is the second lemma.

Now consider

(8.1.2) $$q\frac{(q^5;q^5)_\infty}{(q;q)_\infty} = q(q;q)_\infty^4 \frac{(q^5;q^5)_\infty}{(q;q)_\infty^5}.$$

We can write
$$\frac{(q^5;q^5)_\infty}{(q;q)_\infty^5} = \frac{1-q^5}{(1-q)^5}\frac{1-q^{10}}{(1-q^2)^5}\frac{1-q^{15}}{(1-q^3)^5}\frac{1-q^{20}}{(1-q^4)^5}\cdots$$

and every one of these fractions has the form $(1-x^5)/(1-x)^5$, so by (8.1.1) every term in the expansion of this product in powers of q is divisible by 5 except for an initial 1. Moreover, we know that every coefficient of q^{5m+5} in the expansion of $q(q;q)_\infty^4$ is divisible by 5, so it follows that every coefficient of q^{5m+5} in the expansion of the right side of (8.1.2) is divisible by 5. Because every term in the expansion of $(q^5;q^5)_\infty$ has the form q^{5m+5} times an integer, it further follows that every coefficient of q^{5m+5} in the expansion of $q/(q;q)_\infty$ must be divisible by 5. But

$$\frac{q}{(q;q)_\infty} = \sum_{n=0}^\infty p(n)q^{n+1},$$

and the terms with coefficients having the form q^{5m+5} are

$$\sum_{m=0}^\infty p(5m+4)q^{5m+5},$$

so finally we have that $p(5m+4)$ must be divisible by 5.

Here is another pattern that Ramanujan noticed:

n	$p(n)$	n	$p(n)$	n	$p(n)$
5	7	33	10143	61	1121505
12	77	40	37338	68	3087735
19	490	47	124754	75	8118264
26	2436	54	386155	82	20506255

This example is less obvious, but all of these partition numbers are divisible by 7, suggesting that $p(7n+5)$ is divisible by 7 for any nonnegative integer n. We will prove this in the last section.

Exercises

1. Show that we can also write
$$\frac{n^3+5n}{6} = \binom{n+2}{3} - \binom{n}{2}.$$

2. Show that
$$\frac{1-x^2}{(1-x)^2} = 1 + 2\sum_{n=1}^\infty x^n.$$

3. Show that
$$\frac{1-x^3}{(1-x)^3} = 1 + 3\sum_{n=1}^{\infty} nx^n.$$

4. Show that
$$\frac{1-x^4}{(1-x)^4} = 1 + 2\sum_{n=1}^{\infty} (n^2+1)x^n.$$
Are all of the coefficients divisible by 4?

5. Show that
$$\frac{x(1+x)}{(1-x)^3} = \sum_{n=1}^{\infty} n^2 x^n.$$

Hint: Look at problems 2 and 4.

8.2. Ramanujan's "most beautiful" identity

In this section we prove

(8.2.1) $$\sum_{n=0}^{\infty} p(5n+4) q^n = 5 \frac{(q^5; q^5)_\infty^5}{(q;q)_\infty^6},$$

which was called Ramanujan's "most beautiful" identity in a paper of Hirschhorn. This refers to Hardy's obituary notice for Ramanujan, reprinted in the *Collected Papers* of both men, which contains the sentence "It would be difficult to find more beautiful formulæ than the 'Rogers–Ramanujan' identities, proved in (19); but here Ramanujan must take second place to Prof. Rogers; and, if I had to select one formula from all Ramanujan's work, I would agree with Major MacMahon in selecting" (8.2.1). In his review of Ramanujan's *Collected Papers*, Littlewood also singles out (8.2.1) as a formula "of supreme beauty".

We start with two algebraic lemmas. It is convenient to set

(8.2.2) $\alpha = \dfrac{1+\sqrt{5}}{2}$ and $\beta = \dfrac{1-\sqrt{5}}{2}$, so that $\alpha + \beta = 1$ and $\alpha\beta = -1$.

The **Lucas** numbers are defined by

(8.2.3) $$L_n = \alpha^n + \beta^n = \left(\frac{1+\sqrt{5}}{2}\right)^n + \left(\frac{1-\sqrt{5}}{2}\right)^n,$$

where n is a nonnegative integer. We have $L_0 = 2$ and $L_1 = 1$, and if $n \geq 1$, then
$$L_{n+1} = \alpha^{n+1} + \beta^{n+1} = (\alpha^n + \beta^n)(\alpha + \beta) - \alpha^n\beta - \alpha\beta^n$$
$$= (\alpha^n + \beta^n) - \alpha\beta(\alpha^{n-1} + \beta^{n-1})$$
$$= L_n + L_{n-1},$$

so they have the same recurrence as the Fibonacci numbers, only with different starting values: the Fibonacci numbers usually start with $F_0 = 0$ and $F_1 = 1$, although in the q-Fibonacci case we started with the equivalent of two 1's. It will be convenient to have the first several values of each sequence:

n	0	1	2	3	4	5	6	7
F_n	0	1	1	2	3	5	8	13
L_n	2	1	3	4	7	11	18	29

8.2. RAMANUJAN'S "MOST BEAUTIFUL" IDENTITY

Since $L_5 = 11$, we have
$$1 - 11x^5 - x^{10} = 1 - \left(\alpha^5 + \beta^5\right)x^5 + \alpha^5\beta^5 x^{10} = \left(1 - \alpha^5 x^5\right)\left(1 - \beta^5 x^5\right).$$

Since the five solutions of $x^5 = 1$ are
$$x = 1, \quad e^{\pm\frac{2\pi i}{5}}, \quad e^{\pm\frac{4\pi i}{5}},$$

we can further factor
$$\left(1 - \alpha^5 x^5\right) = (1 - \alpha x)\left(1 - \alpha x e^{\frac{2\pi i}{5}}\right)\left(1 - \alpha x e^{-\frac{2\pi i}{5}}\right)\left(1 - \alpha x e^{\frac{4\pi i}{5}}\right)\left(1 - \alpha x e^{-\frac{4\pi i}{5}}\right)$$

and similarly for $\left(1 - \beta^5 x^5\right)$. Now
$$\left(1 - \alpha x e^{\frac{2k\pi i}{5}}\right)\left(1 - \beta x e^{\frac{2k\pi i}{5}}\right) = 1 - (\alpha + \beta)xe^{\frac{2k\pi i}{5}} + \alpha\beta x^2 e^{\frac{4k\pi i}{5}}$$
$$= 1 - xe^{\frac{2k\pi i}{5}} - x^2 e^{\frac{4k\pi i}{5}},$$

so
$$(8.2.4) \qquad 1 - 11x^5 - x^{10} = \prod_{k=-2}^{2}\left(1 - xe^{\frac{2k\pi i}{5}} - x^2 e^{\frac{4k\pi i}{5}}\right).$$

If we take $xe^{\frac{2k\pi i}{5}}$ out of each factor and divide by x^5, we finally have
$$(8.2.5) \qquad \frac{1}{x^5} - 11 - x^5 = \prod_{k=-2}^{2}\left(\frac{e^{-\frac{2k\pi i}{5}}}{x} - 1 - xe^{\frac{2k\pi i}{5}}\right).$$

We will also require a variation of this. We have
$$\left(1 - x - x^2\right)\left(F_1 + F_2 x + F_3 x^2 + F_4 x^3 + F_5 x^4 - F_4 x^5 + F_3 x^6 - F_2 x^7 + F_1 x^8\right)$$
$$= F_1 + (F_2 - F_1)x + (F_3 - F_2 - F_1)x^2 + (F_4 - F_3 - F_2)x^3$$
$$+ (F_5 - F_4 - F_3)x^4 - (2F_4 + F_5)x^5 + (F_3 + F_4 - F_5)x^6$$
$$+ (F_4 - F_3 - F_2)x^7 + (F_1 + F_2 - F_3)x^8 + (F_2 - F_1)x^9 - F_1 x^{10}$$
$$= 1 - 11x^5 - x^{10}.$$

Dividing both sides by x^5 and plugging in the Fibonacci numbers gives
$$\frac{\frac{1}{x^5} - 11 - x^5}{\frac{1}{x} - 1 - x} = \frac{1}{x^4} + \frac{1}{x^3} + \frac{2}{x^2} + \frac{3}{x} + 5 - 3x + 2x^2 - x^3 + x^4,$$

or, by (8.2.5),
$$(8.2.6)$$
$$\prod_{k=-2,-1,1,2}\left(\frac{e^{-\frac{2k\pi i}{5}}}{x} - 1 - xe^{\frac{2k\pi i}{5}}\right) = \frac{1}{x^4} + \frac{1}{x^3} + \frac{2}{x^2} + \frac{3}{x} + 5 - 3x + 2x^2 - x^3 + x^4.$$

We are ready to start the proof of (8.2.1) in earnest. The argument begins with Euler's pentagonal number theorem in the form (4.1.5) given by Gauss:
$$(8.2.7) \qquad q\left(q^{24}; q^{24}\right)_\infty = \sum_{n=-\infty}^{\infty}(-1)^n q^{(6n-1)^2} = \sum_{n=-\infty}^{\infty}(-1)^n q^{(6n+1)^2}.$$

We split the last sum into five sums $S_{-1}, S_0, S_1, S_2, S_3$, according to which of the forms $5k-1, 5k, 5k+1, 5k+2, 5k+3$ represents n. These sums are

$$(8.2.8) \qquad S_{-1} = \sum_{k=-\infty}^{\infty} (-1)^{5k-1} q^{(6(5k-1)+1)^2} = -\sum_{k=-\infty}^{\infty} (-1)^k q^{(30k-5)^2},$$

$$(8.2.9) \qquad S_0 = \sum_{k=-\infty}^{\infty} (-1)^{5k} q^{(6(5k)+1)^2} = \sum_{k=-\infty}^{\infty} (-1)^k q^{(30k+1)^2},$$

$$(8.2.10) \qquad S_1 = \sum_{k=-\infty}^{\infty} (-1)^{5k+1} q^{(6(5k+1)+1)^2} = -\sum_{k=-\infty}^{\infty} (-1)^k q^{(30k+7)^2},$$

$$(8.2.11) \qquad S_2 = \sum_{k=-\infty}^{\infty} (-1)^{5k+2} q^{(6(5k+2)+1)^2} = \sum_{k=-\infty}^{\infty} (-1)^k q^{(30k+13)^2}$$

$$(8.2.12) \qquad S_3 = \sum_{k=-\infty}^{\infty} (-1)^{5k+3} q^{(6(5k+3)+1)^2} = -\sum_{k=-\infty}^{\infty} (-1)^k q^{(30k+19)^2}.$$

We can handle S_{-1} using (8.2.7):

$$(8.2.13) \qquad S_{-1} = -\sum_{k=-\infty}^{\infty} (-1)^k q^{25(6k-1)^2} = -q^{25} \left(q^{600}; q^{600}\right)_\infty.$$

The other four sums can be combined in pairs that can be handled by the quintuple product identity from section 5.3, for we can write

$$(30k+1)^2 = 100(3k+1)^2 - 180(3k+1) + 81,$$
$$(30k+19)^2 = 100(3k+1)^2 + 180(3k+1) + 81$$

and

$$(30k+7)^2 = 100(3k+1)^2 - 60(3k+1) + 9,$$
$$(30k+13)^2 = 100(3k+1)^2 + 60(3k+1) + 9.$$

Then

$$S_0 + S_3 = q^{81} \sum_{k=-\infty}^{\infty} (-1)^k q^{100(3k+1)^2} \left(q^{-180(3k+1)} - q^{180(3k+1)}\right)$$

and

$$S_1 + S_2 = -q^9 \sum_{k=-\infty}^{\infty} (-1)^k q^{100(3k+1)^2} \left(q^{-60(3k+1)} - q^{60(3k+1)}\right).$$

Now we need (5.3.10), which was

$$\sum_{k=-\infty}^{\infty} q^{(3k+1)^2} \left(z^{3k+1} - z^{-3k-1}\right)$$

$$= zq(q^6; q^6)_\infty (zq^3; q^6)_\infty \left(\frac{q^3}{z}; q^6\right)_\infty (z^2 q^{12}; q^{12})_\infty \left(\frac{1}{z^2}; q^{12}\right)_\infty.$$

8.2. RAMANUJAN'S "MOST BEAUTIFUL" IDENTITY

If we replace z by $-z$ here, then $(-1)^{3k+1} = -(-1)^k$. Multiplying both sides by -1 and replacing q by q^{100} gives

$$(8.2.14) \quad \sum_{k=-\infty}^{\infty} (-1)^k q^{100(3k+1)^2} \left(z^{3k+1} - z^{-3k-1} \right) = zq^{100} (q^{600}; q^{600})_\infty$$

$$\times (-zq^{300}; q^{600})_\infty \left(-\frac{q^{300}}{z}; q^{600} \right)_\infty (z^2 q^{1200}; q^{1200})_\infty \left(\frac{1}{z^2}; q^{1200} \right)_\infty.$$

Taking $z = q^{-180}$ in (8.2.14), we find that $S_0 + S_3$ is $q^{81+100-180}$ times

$$\left(-q^{120}; q^{600} \right)_\infty \left(-q^{480}; q^{600} \right)_\infty \left(q^{600}; q^{600} \right)_\infty \left(q^{360}; q^{1200} \right)_\infty \left(q^{840}; q^{1200} \right)_\infty,$$

so $S_0 + S_3$ is

$$q \left(q^{600}; q^{600} \right)_\infty \frac{(q^{240}; q^{1200})_\infty (q^{840}; q^{1200})_\infty}{(q^{120}; q^{600})_\infty} \frac{(q^{360}; q^{1200})_\infty (q^{960}; q^{1200})_\infty}{(q^{480}; q^{600})_\infty}$$

$$= q \left(q^{600}; q^{600} \right)_\infty \frac{(q^{240}; q^{600})_\infty (q^{360}; q^{600})_\infty}{(q^{120}; q^{600})_\infty (q^{480}; q^{600})_\infty}.$$

Taking $z = q^{-60}$ in (8.2.14), we find that $S_1 + S_2$ is $-q^{9+100-60}$ times

$$\left(-q^{240}; q^{600} \right)_\infty \left(-q^{360}; q^{600} \right)_\infty \left(q^{600}; q^{600} \right)_\infty \left(q^{1080}; q^{1200} \right)_\infty \left(q^{120}; q^{1200} \right)_\infty,$$

so $S_1 + S_2$ is

$$-q^{49} \left(q^{600}; q^{600} \right)_\infty \frac{(q^{480}; q^{1200})_\infty (q^{1080}; q^{1200})_\infty}{(q^{240}; q^{600})_\infty} \frac{(q^{120}; q^{1200})_\infty (q^{720}; q^{1200})_\infty}{(q^{360}; q^{600})_\infty}$$

$$= -q^{49} \left(q^{600}; q^{600} \right)_\infty \frac{(q^{120}; q^{600})_\infty (q^{480}; q^{600})_\infty}{(q^{240}; q^{360})_\infty (q^{480}; q^{600})_\infty}.$$

Putting all this together, we have

$$q \left(q^{24}; q^{24} \right)_\infty = S_{-1} + S_0 + S_1 + S_2 + S_3 = (q^{600}; q^{600})_\infty$$

$$\times \left[q \frac{(q^{240}; q^{600})_\infty (q^{360}; q^{600})_\infty}{(q^{120}; q^{600})_\infty (q^{480}; q^{600})_\infty} - q^{25} - q^{49} \frac{(q^{120}; q^{600})_\infty (q^{480}; q^{600})_\infty}{(q^{240}; q^{360})_\infty (q^{480}; q^{600})_\infty} \right].$$

If we divide this by $q^{25} (q^{600}; q^{600})_\infty$ and then replace q^{24} by q, it becomes

$$\frac{1}{q} \frac{(q;q)_\infty}{(q^{25}; q^{25})_\infty} = \frac{1}{q} \frac{(q^{10}; q^{25})_\infty (q^{15}; q^{25})_\infty}{(q^5; q^{25})_\infty (q^{20}; q^{25})_\infty} - 1 - q \frac{(q^5; q^{25})_\infty (q^{20}; q^{25})_\infty}{(q^{10}; q^{25})_\infty (q^{15}; q^{25})_\infty}.$$

To write this as neatly as possible, we define

$$(8.2.15) \quad R(q) = q^{\frac{1}{5}} \frac{(q; q^5)_\infty (q^4; q^5)_\infty}{(q^2; q^5)_\infty (q^3; q^5)_\infty}.$$

Then we have proved that

$$(8.2.16) \quad \frac{1}{q} \frac{(q;q)_\infty}{(q^{25}; q^{25})_\infty} = \frac{1}{R(q^5)} - 1 - R(q^5).$$

If we replace q by $qe^{\frac{2k\pi i}{5}}$ for an integer k in

$$R(q^5) = q \frac{(q^5; q^{25})_\infty (q^{20}; q^{25})_\infty}{(q^{10}; q^{25})_\infty (q^{15}; q^{25})_\infty},$$

this only affects the factor of q in front, and we find that
$$R\left(\left(qe^{\frac{2k\pi i}{5}}\right)^5\right) = e^{\frac{2k\pi i}{5}} R(q^5).$$

Knowing this, we replace q by $qe^{\frac{2k\pi i}{5}}$ in (8.2.16) for $k = -2, -1, 0, 1, 2$ and multiply the results together. It is convenient to set $\omega = e^{\frac{2\pi i}{5}}$. Using (8.2.5) for the right side, we get

$$(8.2.17) \quad \frac{1}{q^5} \frac{\left(\omega^{-2}q; \omega^{-2}q\right)_\infty \left(\omega^{-1}q; \omega^{-1}q\right)_\infty (q;q)_\infty (\omega q; \omega q)_\infty \left(\omega^2 q; \omega^2 q\right)_\infty}{(q^{25}; q^{25})_\infty^5}$$
$$= \frac{1}{(R(q^5))^5} - 11 - \left(R(q^5)\right)^5.$$

Now
$$(8.2.18) \quad 1 - q^5 = \left(1 - \omega^{-2}q\right)\left(1 - \omega^{-1}q\right)(1-q)(1-\omega q)\left(1-\omega^2 q\right)$$

and $\omega^{-2} = \omega^3$ and $\omega^{-1} = \omega^4$, so

$$\left(\omega^{-2}q; \omega^{-2}q\right)_\infty \left(\omega^{-1}q; \omega^{-1}q\right)_\infty (q;q)_\infty (\omega q; \omega q)_\infty \left(\omega^2 q; \omega^2 q\right)_\infty$$
$$\begin{array}{cccccc}
 & (1-q) & (1-q^2) & (1-q^3) & (1-q^4) & (1-q^5) & \cdots\\
\times & (1-\omega q) & (1-\omega^2 q^2) & (1-\omega^3 q^3) & (1-\omega^4 q^4) & (1-\omega^5 q^5) & \cdots\\
\times & (1-\omega^2 q) & (1-\omega^4 q^2) & (1-\omega^6 q^3) & (1-\omega^8 q^4) & (1-\omega^{10} q^5) & \cdots\\
\times & (1-\omega^3 q) & (1-\omega^6 q^2) & (1-\omega^9 q^3) & (1-\omega^{12} q^4) & (1-\omega^{15} q^5) & \cdots\\
\times & (1-\omega^4 q) & (1-\omega^8 q^2) & (1-\omega^{12} q^3) & (1-\omega^{16} q^4) & (1-\omega^{20} q^5) & \cdots
\end{array}$$

The fifth column gives us $(1-q^5)^5$, and using (8.2.18) the first four columns are $(1-q^5)(1-q^{10})(1-q^{15})(1-q^{20})$. This pattern persists: in the fifth, tenth, fifteenth ... columns the factors are all the same, and in the other columns they are an instance of (8.2.18). Therefore the product is

$$(1-q^5)(1-q^{10})(1-q^{15})(1-q^{20})(1-q^{30})(1-q^{35})(1-q^{40})(1-q^{45})\cdots$$
$$\times (1-q^5)^5 (1-q^{10})^5 (1-q^{15})^5 (1-q^{20})^5 (1-q^{25})^5 (1-q^{30})^5 \cdots,$$

where the factors $(1-q^{25})(1-q^{50})(1-q^{75})\cdots$ are missing from the first line. If we multiply and divide by those factors we will have

$$(q^5; q^5)_\infty \frac{(q^5; q^5)_\infty^5}{(q^{25}; q^{25})_\infty},$$

and hence
(8.2.19)
$$\left(\omega^{-2}q; \omega^{-2}q\right)_\infty \left(\omega^{-1}q; \omega^{-1}q\right)_\infty (q;q)_\infty (\omega q; \omega q)_\infty \left(\omega^2 q; \omega^2 q\right)_\infty = \frac{(q^5; q^5)_\infty^6}{(q^{25}; q^{25})_\infty}.$$

Then (8.2.17) becomes
$$\frac{1}{q^5}\left(\frac{(q^5; q^5)_\infty}{(q^{25}; q^{25})_\infty}\right)^6 = \frac{1}{(R(q^5))^5} - 11 - \left(R(q^5)\right)^5,$$

and replacing q^5 by q we have

$$(8.2.20) \quad \frac{1}{q}\left(\frac{(q;q)_\infty}{(q^5; q^5)_\infty}\right)^6 = \frac{1}{(R(q))^5} - 11 - (R(q))^5.$$

8.2. RAMANUJAN'S "MOST BEAUTIFUL" IDENTITY

Now write

$$\frac{1}{(q;q)_\infty} = \frac{(\omega^{-2}q;\omega^{-2}q)_\infty (\omega^{-1}q;\omega^{-1}q)_\infty (\omega q;\omega q)_\infty (\omega^2 q;\omega^2 q)_\infty}{(\omega^{-2}q;\omega^{-2}q)_\infty (\omega^{-1}q;\omega^{-1}q)_\infty (q;q)_\infty (\omega q;\omega q)_\infty (\omega^2 q;\omega^2 q)_\infty}$$

$$= \frac{(q^{25};q^{25})_\infty}{(q^5;q^5)_\infty^6} (\omega^{-2}q;\omega^{-2}q)_\infty (\omega^{-1}q;\omega^{-1}q)_\infty (\omega q;\omega q)_\infty (\omega^2 q;\omega^2 q)_\infty.$$

We can use (8.2.16) in the form

$$(q;q)_\infty = q(q^{25};q^{25})_\infty \left(\frac{1}{R(q^5)} - 1 - R(q^5)\right)$$

for each of the factors on the right side, and (8.2.6) to multiply them together. This gives

$$\frac{1}{(q;q)_\infty} = \frac{(q^{25};q^{25})_\infty}{(q^5;q^5)_\infty^6} q^4 (q^{25};q^{25})_\infty^4$$

$$\times \left[\frac{1}{R(q^5)^4} + \frac{1}{R(q^5)^3} + \frac{2}{R(q^5)^2} + \frac{3}{R(q^5)} + 5 \right.$$
$$\left. - 3R(q^5) + 2R(q^5)^2 - R(q^5)^3 + R(q^5)^4\right]$$

$$= q^4 \frac{(q^{25};q^{25})_\infty^5}{(q^5;q^5)_\infty^6}$$

$$\times \left[\frac{1}{R(q^5)^4} + \frac{1}{R(q^5)^3} + \frac{2}{R(q^5)^2} + \frac{3}{R(q^5)} + 5 \right.$$
$$\left. - 3R(q^5) + 2R(q^5)^2 - R(q^5)^3 + R(q^5)^4\right].$$

Recall that $R(q^5)$ is actually q times a function of q^5. This means that in the expansion of the terms of

$$q^4 \left[\frac{1}{R(q^5)^4} + \frac{1}{R(q^5)^3} + \frac{2}{R(q^5)^2} + \frac{3}{R(q^5)} + 5 \right.$$
$$\left. - 3R(q^5) + 2R(q^5)^2 - R(q^5)^3 + R(q^5)^4\right]$$

in powers of q, all the exponents from each of these nine terms in turn must be congruent to $0, 1, 2, 3, 4, 0, 1, 2, 3$ respectively mod 5. In particular, only the middle term 5 has exponents congruent to 4 (mod 5). The terms of

$$\frac{1}{(q;q)_\infty} = \sum_{n=0}^\infty p(n) q^n$$

with exponents congruent to 4 (mod 5) are

$$\sum_{n=0}^\infty p(5n+4) q^{5n+4},$$

so we must have

$$\sum_{n=0}^\infty p(5n+4) q^{5n+4} = 5q^4 \frac{(q^{25};q^{25})_\infty^5}{(q^5;q^5)_\infty^6}.$$

Dividing both sides by q^4 and then replacing q^5 by q, we finally have

THEOREM 67 (Ramanujan's "most beautiful" identity). *If $|q| < 1$, then*
$$\sum_{n=0}^{\infty} p(5n+4) q^n = \frac{5\left(q^5; q^5\right)_\infty^5}{(q;q)_\infty^6}.$$

Exercises

1. Write down the 30 partitions of 9.
2. If L_n denotes the nth Lucas number for $n \geq 1$, show that
$$1 - L_n x^n + (-1)^n x^{2n} = \prod_{k=0}^{n-1} \left(1 - xe^{\frac{2k\pi i}{n}} + x^2 e^{\frac{4k\pi i}{n}}\right).$$

3. If instead of combining $1-\alpha x e^{\frac{2\pi i}{5}}$ with $1-\beta x e^{\frac{2\pi i}{5}}$ we combine it with $1-\alpha x e^{-\frac{2\pi i}{5}}$, and similarly for the other factors, we can factor
$$1 + x + 2x^2 + 3x^3 + 5x^4 - 3x^5 + 2x^6 - x^7 + x^8$$
into four *real* quadratics. Show that this gives
$$1 + x + 2x^2 + 3x^3 + 5x^4 - 3x^5 + 2x^6 - x^7 + x^8$$
$$= \left(1 - 2\alpha x \cos\frac{2\pi}{5} + \alpha^2 x^2\right)\left(1 - 2\alpha x \cos\frac{4\pi}{5} + \alpha^2 x^2\right)$$
$$\times \left(1 - 2\beta x \cos\frac{2\pi}{5} + \beta^2 x^2\right)\left(1 - 2\beta x \cos\frac{4\pi}{5} + \beta^2 x^2\right)$$
$$= \left(1 + 2\alpha x \cos\frac{\pi}{5} + \alpha^2 x^2\right)\left(1 - 2\alpha x \cos\frac{2\pi}{5} + \alpha^2 x^2\right)$$
$$\times \left(1 + 2\beta x \cos\frac{\pi}{5} + \beta^2 x^2\right)\left(1 - 2\beta x \cos\frac{2\pi}{5} + \beta^2 x^2\right).$$

4. It turns out that
$$\cos\frac{\pi}{5} = \frac{1}{\sqrt{5}-1} = \frac{\sqrt{5}+1}{4} = \frac{\alpha}{2} \quad \text{and} \quad \cos\frac{2\pi}{5} = \frac{1}{\sqrt{5}+1} = \frac{\sqrt{5}-1}{4} = -\frac{\beta}{2}.$$
One way to prove this is to draw a 36-72-72 triangle and label the sides 1, 1, and x. Then draw the line segment from one vertex to a point P on the opposite side that bisects one of the 72° angles. This creates a triangle with sides x, x, and $1-x$ that is similar to the original triangle. (Why?) This should allow you to solve for x. Now find right triangles with 36° and 72° angles in this triangle to verify the above cosines.

5. Use problem 4 to simplify the result of problem 3 to
$$1 + x + 2x^2 + 3x^3 + 5x^4 - 3x^5 + 2x^6 - x^7 + x^8$$
$$= \left(1 + \alpha^2 x + \alpha^2 x^2\right)\left(1 - x + \alpha^2 x^2\right)\left(1 - x + \beta^2 x^2\right)\left(1 + \beta^2 x + \beta^2 x^2\right).$$

6. Show that problem 5 implies
$$1 + x + 2x^2 + 3x^3 + 5x^4 - 3x^5 + 2x^6 - x^7 + x^8$$
$$= \left(1 - 2x + 4x^2 - 3x^3 + x^4\right)\left(1 + 3x + 4x^2 + 2x^3 + x^4\right).$$

7. Another evaluation of the two cosines above can be obtained from

$$x^5 - 1 = (x-1)\left(x - e^{\frac{2\pi i}{5}}\right)\left(x - e^{-\frac{2\pi i}{5}}\right)\left(x - e^{\frac{4\pi i}{5}}\right)\left(x - e^{-\frac{4\pi i}{5}}\right).$$

(i) Explain why it follows that

$$x^4 + x^3 + x^2 + x + 1 = \left(x - e^{\frac{2\pi i}{5}}\right)\left(x - e^{-\frac{2\pi i}{5}}\right)\left(x - e^{\frac{4\pi i}{5}}\right)\left(x - e^{-\frac{4\pi i}{5}}\right)$$

$$= \left(x^2 - 2x\cos\frac{2\pi}{5} + 1\right)\left(x^2 - 2x\cos\frac{4\pi}{5} + 1\right)$$

$$= \left(x^2 + 2x\cos\frac{\pi}{5} + 1\right)\left(x^2 - 2x\cos\frac{2\pi}{5} + 1\right).$$

(ii) Explain why it further follows that

$$2\left(\cos\frac{\pi}{5} - \cos\frac{2\pi}{5}\right) = 1 \quad \text{and} \quad 2 - 4\cos\frac{\pi}{5}\cos\frac{2\pi}{5} = 1,$$

or

(a) $\cos\frac{\pi}{5} - \cos\frac{2\pi}{5} = \frac{1}{2}$, (b) $4\cos\frac{\pi}{5}\cos\frac{2\pi}{5} = 1$.

(iii) One way to solve (a) and (b) is to square both sides of (a) and then add (b). Using this method or otherwise, show that they imply

$$\cos\frac{\pi}{5} = \frac{\sqrt{5}+1}{4} \quad \text{and} \quad \cos\frac{2\pi}{5} = \frac{\sqrt{5}-1}{4}.$$

8. For another algebraic derivation of these two cosines, we can construct the polynomial whose roots are $2\cos\frac{\pi}{5}$, $2\cos\frac{2\pi}{5}$, $2\cos\frac{3\pi}{5}$, and $2\cos\frac{4\pi}{5}$, namely

$$\left(x - 2\cos\frac{\pi}{5}\right)\left(x - 2\cos\frac{2\pi}{5}\right)\left(x - 2\cos\frac{3\pi}{5}\right)\left(x - 2\cos\frac{4\pi}{5}\right)$$

$$= x^4 - 2x^3\left(\cos\frac{\pi}{5} + \cos\frac{2\pi}{5} + \cos\frac{3\pi}{5} + \cos\frac{4\pi}{5}\right)$$

$$+ 4x^2\begin{pmatrix}\cos\frac{\pi}{5}\cos\frac{2\pi}{5} + \cos\frac{\pi}{5}\cos\frac{3\pi}{5} + \cos\frac{\pi}{5}\cos\frac{4\pi}{5} \\ + \cos\frac{2\pi}{5}\cos\frac{3\pi}{5} + \cos\frac{2\pi}{5}\cos\frac{4\pi}{5} + \cos\frac{3\pi}{5}\cos\frac{4\pi}{5}\end{pmatrix}$$

$$- 8x\begin{pmatrix}\cos\frac{\pi}{5}\cos\frac{2\pi}{5}\cos\frac{3\pi}{5} + \cos\frac{\pi}{5}\cos\frac{2\pi}{5}\cos\frac{4\pi}{5} \\ + \cos\frac{\pi}{5}\cos\frac{3\pi}{5}\cos\frac{4\pi}{5} + \cos\frac{2\pi}{5}\cos\frac{3\pi}{5}\cos\frac{4\pi}{5}\end{pmatrix}$$

$$+ 16\cos\frac{\pi}{5}\cos\frac{2\pi}{5}\cos\frac{3\pi}{5}\cos\frac{4\pi}{5}.$$

(i) Explain why the coefficients of x^3 and x must be zero, and why the coefficient of x^2 simplifies to $-4\left(\cos^2\frac{\pi}{5} + \cos^2\frac{2\pi}{5}\right)$, so the polynomial we are looking for is

$$x^4 - 4x^2\left(\cos^2\frac{\pi}{5} + \cos^2\frac{2\pi}{5}\right) + 16\cos\frac{\pi}{5}\cos\frac{2\pi}{5}\cos\frac{3\pi}{5}\cos\frac{4\pi}{5}.$$

(ii) To find the remaining coefficients we solve the equation $(x-1)^5 = 1$. Explain why the five roots are

$$2, \quad 2e^{\frac{\pi i}{5}}\cos\frac{\pi}{5}, \quad 2e^{\frac{2\pi i}{5}}\cos\frac{2\pi}{5}, \quad 2e^{\frac{3\pi i}{5}}\cos\frac{3\pi}{5}, \quad 2e^{\frac{4\pi i}{5}}\cos\frac{4\pi}{5}.$$

(iii) The equation we solved in (ii) is

$$x^5 - 5x^4 + 10x^3 - 10x^2 + 5x - 2 = 0,$$

and we know that one of the factors of the left side must be $x - 2$. Show that dividing out this factor leaves

$$x^4 - 3x^3 + 4x^2 - 2x + 1 = 0.$$

(iv) Explain why (iii) implies that

$$16\cos\frac{\pi}{5}\cos\frac{2\pi}{5}\cos\frac{3\pi}{5}\cos\frac{4\pi}{5} = 1.$$

(v) From (iii) we also know (why?) that

$$2e^{\frac{\pi i}{5}}\cos\frac{\pi}{5} + 2e^{\frac{2\pi i}{5}}\cos\frac{2\pi}{5} + 2e^{\frac{3\pi i}{5}}\cos\frac{3\pi}{5} + 2e^{\frac{4\pi i}{5}}\cos\frac{4\pi}{5} = 3.$$

Show that the left side simplifies to

$$4\left(\cos^2\frac{\pi}{5} + \cos^2\frac{2\pi}{5}\right).$$

9. In the previous problem we found that the polynomial equation whose roots are $2\cos\frac{\pi}{5}$, $2\cos\frac{2\pi}{5}$, $2\cos\frac{3\pi}{5}$, and $2\cos\frac{4\pi}{5}$ is $x^4 - 3x^2 + 1 = 0$. Show that algebraic expressions for the roots are

$$\pm\frac{\sqrt{5}+1}{2} \quad \text{and} \quad \pm\frac{\sqrt{5}-1}{2}.$$

(**Hint:** Rewrite the equation as $(x^2+1)^2 = 5x^2$.) Then explain how to match these up with the trigonometric expressions.

10. Show that

$$(q;q)_\infty (\omega q;q)_\infty (\omega^2 q;q)_\infty (\omega^3 q;q)_\infty (\omega^4 q;q)_\infty = (q^5;q^5)_\infty.$$

8.3. Ramanujan's congruences again

It might be asked, especially by someone who does not think that (8.2.1) is all that beautiful, why prove it when we can use the argument of section 8.1 to see that $p(5n+4)$ is divisible by 5? We can rewrite (8.2.1) as

$$(8.3.1) \qquad \sum_{n=0}^{\infty} p(5n+4)q^{n+1} = 5\frac{q}{(q;q)_\infty}\frac{(q^5;q^5)_\infty}{(q;q)_\infty^5}(q^5;q^5)_\infty^4.$$

We know from section 8.1 that every coefficient of $(q^5;q^5)_\infty/(q;q)_\infty^5$ is divisible by 5 except for an initial 1. We know that the coefficient of q^{5j+5} in $q/(q;q)_\infty$ is divisible by 5 for any nonnegative integer j, and we know that every term of the expansion of

$(q^5;q^5)_\infty^4$ is an integer times q^{5k} for some nonnegative integer k. Therefore (8.3.1) has the form

$$\sum_{n=0}^\infty p(5n+4)q^{n+1} = 5\left(\sum_{i=0}^\infty p(i)q^{i+1}\right)\left(1+5\sum_{j=1}^\infty a_j q^j\right)\left(\sum_{k=0}^\infty b_k q^{5k}\right)$$

for some integers a_j and b_k. Now consider the coefficient of q^{5m+5} on the right side of this. Every term of the last sum has this form, so we only have to look at the first two sums. If q^{i+1} has the form q^{5m+5}, then we know that $p(i)$ will be divisible by 5. If it doesn't, then multiplying it by some term $5a_j q^j$ from the second sum will give it this form, and again there is a factor of 5. Since there is also an overall factor of 5 in front, it follows that every coefficient of q^{5m+5} on the right side, and hence also on the left side, must be divisible by 25 for any nonnegative integer m. When $n+1$ has the form $5m+5$ we have $5n+4 = 5(n+1)-1 = 25(m+1)-1 = 25m+24$, so $p(25m+24)$ is divisible by 25 for any nonnegative integer m.

We can prove similarly that $p(7n+5)$ is divisible by 7. First we need a lemma about $q^2(q;q)_\infty^6$, which we can expand by Jacobi's cube identity as

$$q^2(q;q)_\infty^3(q;q)_\infty^3 = q^2\left(\sum_{j=0}^\infty (-1)^j(2j+1)q^{\binom{j+1}{2}}\right)\left(\sum_{k=0}^\infty (-1)^k(2k+1)q^{\binom{k+1}{2}}\right)$$

$$= \sum_{j=0}^\infty\sum_{k=0}^\infty (-1)^{j+k}(2j+1)(2k+1)q^{2+\frac{j(j+1)}{2}+\frac{k(k+1)}{2}}.$$

The exponent $2+\frac{j(j+1)}{2}+\frac{k(k+1)}{2}$ is a positive integer, so it is divisible by 7 if and only if 8 times it is, and we have

$$16+4j(j+1)+4k(k+1) = (2j+1)^2+(2k+1)^2+14,$$

so we need $(2j+1)^2+(2k+1)^2$ to be divisible by 7. We have

$$(2j+1)^2 \equiv 1 \pmod{7} \quad \text{if} \quad j \equiv 0 \text{ or } 6 \pmod{7},$$
$$(2j+1)^2 \equiv 2 \pmod{7} \quad \text{if} \quad j \equiv 1 \text{ or } 5 \pmod{7},$$
$$(2j+1)^2 \equiv 4 \pmod{7} \quad \text{if} \quad j \equiv 2 \text{ or } 4 \pmod{7},$$
$$(2j+1)^2 \equiv 0 \pmod{7} \quad \text{if} \quad j \equiv 3 \pmod{7}.$$

Therefore the only way for $(2j+1)^2+(2k+1)^2$ to be divisible by 7 is for both j and k to be congruent to 3 mod 7, which makes $2j+1$ and $2k+1$ both divisible by 7. Hence the coefficient of q^{7i+7} in $q^2(q;q)_\infty^6$ is actually divisible by 49.

We also need a result similar to (8.1.1). From Cauchy's theorem in section 3.5 with $q=1$ we know that

$$\frac{1}{(1-x)^7} = \sum_{n=0}^\infty \binom{n+6}{6}x^n = \frac{1}{720}\sum_{n=0}^\infty (n+6)(n+5)(n+4)(n+3)(n+2)(n+1)\,x^n,$$

and it follows that

$$\frac{1-x^7}{(1-x)^7} = \frac{1}{720}\sum_{m=0}^\infty (m+6)(m+5)(m+4)(m+3)(m+2)(m+1)\,x^m$$

$$-\frac{1}{720}\sum_{m=0}^\infty (m+6)(m+5)(m+4)(m+3)(m+2)(m+1)\,x^{m+7}.$$

Changing m to n in the first sum and m to $n-7$ in the second, we have

$$\frac{1-x^7}{(1-x)^7} = \frac{1}{720}\sum_{n=0}^{\infty}(n+6)(n+5)(n+4)(n+3)(n+2)(n+1)\,x^n$$
$$-\frac{1}{720}\sum_{n=7}^{\infty}(n-1)(n-2)(n-3)(n-4)(n-5)(n-6)\,x^n.$$

Note that the last sum could start at $n=1$, because the $n=1,2,3,4,5,6$ terms would all be zero. Breaking off the $n=0$ term of the first sum, we have

$$\frac{1-x^7}{(1-x)^7} = 1 + \frac{1}{720}\sum_{n=1}^{\infty}\left[\begin{array}{c}(n+6)(n+5)(n+4)(n+3)(n+2)(n+1)\\-(n-6)(n-5)(n-4)(n-3)(n-2)(n-1)\end{array}\right]x^n.$$

Multiplying this out we get

$$\frac{1-x^7}{(1-x)^7} = 1 + \frac{42}{720}\sum_{n=1}^{\infty}\left(n^5+35n^3+84n\right)x^n$$
$$= 1 + \frac{7}{120}\sum_{n=1}^{\infty}\left(n^5+35n^3+84n\right)x^n.$$

Finally, we can write

$$\frac{n^5+35n^3+84n}{120} = \frac{n^5-5n^3+4n+40n^3+80n}{120}$$
$$= \frac{(n-2)(n-1)n(n+1)(n+2)}{120} + \frac{2n^3+4n}{6}$$
$$= \binom{n+2}{5} + \frac{n^3+3n^2+2n+n^3-3n^2+2n}{6}$$
$$= \binom{n+2}{5} + \binom{n+2}{3} + \binom{n-2}{3},$$

so

(8.3.2) $$\frac{1-x^7}{(1-x)^7} = 1 + 7\sum_{n=1}^{\infty}\left[\binom{n+2}{5} + \binom{n+2}{3} + \binom{n-2}{3}\right]x^n.$$

Hence every term in this series is divisible by 7 except the first one.

With these two facts in hand we look at

(8.3.3) $$\sum_{n=0}^{\infty}p(n)q^{n+2} = \frac{q^2}{(q;q)_\infty} = q^2(q;q)_\infty^6\,\frac{(q^7;q^7)_\infty}{(q;q)_\infty^7}\,\frac{1}{(q^7;q^7)_\infty}.$$

Note that every factor of

$$\frac{(q^7;q^7)_\infty}{(q;q)_\infty^7} = \frac{1-q^7}{(1-q)^7}\,\frac{1-q^{14}}{(1-q^2)^7}\,\frac{1-q^{21}}{(1-q^3)^7}\cdots$$

has the form $(1-x^7)/(1-x)^7$, so every term in its expansion is divisible by 7 except for an initial 1. From the above lemmas, (8.3.3) has the form

$$\sum_{n=0}^{\infty}p(n)q^{n+2} = \left(\sum_{i=2}^{\infty}a_i q^i\right)\left(1+7\sum_{j=1}^{\infty}b_j q^j\right)\left(\sum_{k=0}^{\infty}p(k)q^{7k}\right)$$

for some integers a_i and b_j. Now consider the coefficient of q^{7m+7} on the right side of this. Every term of the last sum has this form, so we only have to look at the first two sums. If q^i has the form q^{7m+7}, then we know that a_i will be divisible by 49. If it doesn't, then multiplying it by some term $7b_j q^j$ from the second sum will give it this form with a factor of 7. It follows that every coefficient of q^{7m+7} on the right side, and hence also on the left side, must be divisible by 7 for any nonnegative integer m. When $n+2$ has the form $7m+7$ we have $n = 7m+5$, so $p(7m+5)$ is divisible by 7 for any nonnegative integer m.

8.4. Bibliographical Notes

Ramanujan's "most beautiful" identity is the subject of Hirschhorn's paper [**139**], which does not require the quintuple product but makes virtuoso use of the triple product. His book [**140**] has much more on the identity and on Ramanujan's congruences. Our proof of the identity more or less follows section 5.2 of [**81**]. Another good source for Ramanujan's identity and congruences is Chapter 2 of [**52**]. Ramanujan's "most beautiful" identity comes up again in section 13.2.

CHAPTER 9

Some Combinatorial Results

9.1. Revisiting the q-factorial

We began Chapter 1 by looking at permutations of $\{1, 2, \ldots, n\}$ and counting inversions, but there are other statistics that we can associate with permutations. We say that a permutation has a **fall** in the kth position if its kth number is larger than its $(k+1)^{\text{th}}$ number. Thus, for example, 3642175 has falls in the second, third, fourth, and sixth positions, since $6 > 4$, $4 > 2$, $2 > 1$, and $7 > 5$, respectively. Interesting things happen when we count permutations by falls, but we will reserve this subject for the exercises. We will come back to familiar ground if we record the *positions* of the falls (rather than the falls themselves), and we add the positions together. We call the sum of the positions of the falls of a permutation its **major index**. For example, 3642175 has major index $2 + 3 + 4 + 6 = 15$. The major index was first studied by MacMahon, who called it the greater index, and also looked at various other indices (lesser, equal, superior, inferior, major, minor). It is appropriate to call the most successful of MacMahon's indices the major index, since MacMahon was actually a Major in the British Army.

Let's look at the permutations of $\{1, 2, 3\}$:

permutation	falls at	q^{maj}	permutation	falls at	q^{maj}
123		q^0	231	2	q^2
132	2	q^2	312	1	q^1
213	1	q^1	321	1, 2	q^{1+2}

As with inversions, we make the major index of each permutation an exponent of q and add all the terms together to get

$$1 + q^2 + q + q^2 + q + q^3 = (1 + q^2 + q) + q(1 + q^2 + q) = (1 + q)(1 + q + q^2) = 3!_q.$$

Coincidence? Let's insert 4 into these permutations, as we did in Chapter 1. If 4 is at the beginning, this creates a new fall at the first position and moves the other falls over one spot:

permutation	q^{maj}	permutation	q^{maj}
4123	q^{1+0}	4231	q^{1+3}
4132	q^{1+3}	4312	q^{1+2}
4213	q^{1+2}	4321	q^{1+2+3}

Unlike the inversion case, the effect is not consistent—the major index usually increased by 2, but once by 1 and once by 3.

If 4 is in the second position, things are still more complicated. There is perforce a fall in the second spot. What else happens varies according to whether

4 was inserted in a fall, or in a rise:

permutation	q^{maj}	permutation	q^{maj}
1423	q^{2+0}	2431	q^{2+3}
1432	q^{2+3}	3412	q^{2+0}
2413	q^{2+0}	3421	q^{2+3}

There can't be a fall in the first position, and there must be one in the second position, so the major index for these permutations is either 5 or 2, according to whether there is or is not a fall in the third position (in other words, according to whether there was or was not a fall in the second position before 4 was inserted).

Similarly, if 4 is inserted in the third position, the major index becomes either 4 or 3, according to whether there was or was not a fall in the first position before insertion:

permutation	q^{maj}	permutation	q^{maj}
1243	q^{3+0}	2341	q^{3+0}
1342	q^{3+0}	3142	q^{3+1}
2143	q^{3+1}	3241	q^{3+1}

Finally, if 4 is put at the end, this has no effect on the major index.

If we add together all these powers of q we do indeed get $4!_q$, but why should we expect this to happen? Things become clearer if we look at these insertions from another angle. Rather than looking at each possible position for 4 one at a time, let's look at each permutation one at a time. For 123 we have

permutation	q^{maj}	permutation	q^{maj}
1234	q^{0+0}	1423	q^{0+2}
1243	q^{0+3}	4123	q^{0+1}

where 0 was the old major index, and inserting 4 added either 0, 1, 2, or 3 to it. Thus 123 contributes $1 + q + q^2 + q^3$. This is the simplest case, because 123 itself has no falls. For 132 we have:

permutation	q^{maj}	permutation	q^{maj}
1324	q^{2+0}	1432	q^{2+3}
1342	q^{2+1}	4132	q^{2+2}

Adding 4 at the end did nothing to the major index. When we put it in the third position we put it in the middle of a fall, and this pushed the fall out one unit. Putting it in the first or second position creates a new fall there and pushes an existing fall out one unit, so this adds 2 or 3 respectively to the major index. Adding this all up we get $q^2(1 + q + q^2 + q^3)$. Similarly, for 213 we have

permutation	q^{maj}	permutation	q^{maj}
2134	q^{1+0}	2413	q^{1+1}
2143	q^{1+3}	4213	q^{1+2}

and adding these we get $q(1 + q + q^2 + q^3)$. For 231 we have

permutation	q^{maj}	permutation	q^{maj}
2314	q^{2+0}	2431	q^{2+3}
2341	q^{2+1}	4231	q^{2+2}

and adding these we get $q^2(1+q+q^2+q^3)$. For 312 we have

permutation	q^{maj}	permutation	q^{maj}
3124	q^{1+0}	3412	q^{1+1}
3142	q^{1+3}	4312	q^{1+2}

and adding these we get $q(1+q+q^2+q^3)$. Finally, for 321 we have

permutation	q^{maj}	permutation	q^{maj}
3214	q^{3+0}	3421	q^{3+2}
3241	q^{3+1}	4321	q^{3+3}

and adding these we get $q^3(1+q+q^2+q^3)$. The total contribution from all the permutations is therefore

$$(1+2q+2q^2+q^3)(1+q+q^2+q^3) = (1+q)(1+q+q^2)(1+q+q^2+q^3) = 4!_q,$$

and now we can see some reason why this might happen in general. Let's do one more example before we try to prove the theorem. Consider the permutation 3642175, which has major index 15, and look at the effect of inserting 8 in all possible ways:

permutation	q^{maj}	comment
36421758	q^{15+0}	no effect
36421785	q^{15+1}	moves last fall out 1
36421875	q^{15+7}	new fall and moves last fall
36428175	q^{15+2}	moves last two falls
36482175	q^{15+3}	moves last three falls
36842175	q^{15+4}	moves all four falls
38642175	q^{15+6}	new fall and moves all falls
83642175	q^{15+5}	new fall and moves all falls

When the largest element (here 8) is inserted in a fall, it moves that fall and any subsequent falls out 1 place. When it is inserted in a rise or put at the beginning, it creates a new fall in that spot and moves any subsequent falls out 1 place. When it is put at the end it has no effect on the major index. Now let's prove the theorem that these examples suggest:

THEOREM 68. *If $\Pi(n)$ is the set of all permutations of $\{1, 2, \ldots, n\}$, then*

(9.1.1) $$n!_q = \sum_{\pi \in \Pi(n)} q^{\text{maj}\,\pi}.$$

We prove this by induction on n. We have done the cases $n = 3$ and $n = 4$ already, and we leave it to the reader to check it for smaller n. Assume it holds for n, and consider the effect of inserting $n+1$ into a permutation of $\{1, 2, \ldots, n\}$ in all possible ways. Such a permutation has $n-1$ successions, so let's suppose $k-1$ of them are falls and $n-k$ are rises. If we put $n+1$ at the end, it has no effect. If we put it in the middle of the last fall, it just pushes that fall out 1 place; if we put it in the penultimate (next-to-last) fall it moves that and the ultimate fall out 1 place, and so forth. If we put it in the jth fall from the end, it moves the last j falls out 1 place. Therefore the insertions in falls contribute $q + q^2 + \cdots + q^{k-1}$ to the sum in (9.1.1). Putting $n+1$ at the end contributes 1. Putting $n+1$ at the beginning creates a new fall in the first spot and moves each of the $k-1$ falls out

1 place, so it contributes q^k. We have $1 + q + q^2 + \cdots + q^{k-1} + q^k$ so far, and now we have to look at the rises, which are a little trickier.

Suppose the first rise is in the ith position. Then it is preceded by $i - 1$ falls and succeeded by $k - i$ falls. If we insert $n + 1$ in this rise (*i.e.*, in the $(i + 1)^{\text{th}}$ position), then we create a fall there, and we move the last $k - i$ falls over 1 place; so we add $(i + 1) + (k - i) = k + 1$ to the major index. More generally, suppose the jth rise is in the mth position. Then it is preceded by $j - 1$ rises and $m - j$ falls, and succeeded by $k - 1 - (m - j)$ falls. If we insert $n + 1$ in this rise, then we create a new fall in the $(m + 1)^{\text{th}}$ position, and we move the last $k - 1 - m + j$ falls over 1 place; so we add $(m + 1) + (k - 1 - m + j) = k + j$ to the major index. All the falls therefore contribute $q^{k+1} + q^{k+2} + \cdots + q^{k+(n-k)}$ to the sum in (9.1.1).

We have just finished considering the effect on the major index of inserting $n + 1$ in the permutations of $\{1, 2, \ldots, n\}$ in all possible ways, and we found that this contributes a factor $1 + q + q^2 + \cdots + q^k + q^{k+1} + \cdots + q^n$. It follows that

$$\sum_{\pi \in \Pi(n+1)} q^{\text{inv}\,\pi} = (1 + q + q^2 + \cdots + q^k + q^{k+1} + \cdots + q^n) \sum_{\pi \in \Pi(n)} q^{\text{maj}\,\pi},$$

and since, by induction, (9.1.1) is true for n, we have

$$\sum_{\pi \in \Pi(n+1)} q^{\text{inv}\,\pi} = (1 + q + q^2 + \cdots + q^n)\, n!_q = (n + 1)!_q,$$

and (9.1.1) is also true for $n + 1$. This proves the theorem.

There is an interesting alternative proof that foreshadows sections 9.4 and 9.5. It is essentially MacMahon's argument of section 9.4, but we follow the formulation of Chen and Xu. Let $\sigma = a_1 a_2 \ldots a_n$ be a sequence of nonnegative integers, let $S(n)$ be the set of all such sequences of length n, and define the weight of σ to be $q^{a_1 + a_2 + \cdots + a_n}$. Then the generating function of all the weighted sequences of length n is

$$\sum_{\sigma \in S(n)} q^{a_1 + \cdots + a_n} = \left(\sum_{a_1=0}^{\infty} q^{a_1}\right) \cdots \left(\sum_{a_n=0}^{\infty} q^{a_n}\right) = \frac{1}{(1-q)^n}.$$

A typical sequence of length 9 is 613684247, and we consider the array

$$\begin{pmatrix} 6 & 1 & 3 & 6 & 8 & 4 & 2 & 4 & 7 \\ 1 & 2 & 3 & 4 & 5 & 6 & 7 & 8 & 9 \end{pmatrix}.$$

We want to rearrange these columns so that the first row is a partition. The only decisions to be made are the order of the two 6's and the two 4's, and we place them so that the numbers on the second line are in increasing order whenever the ones on the first line are tied. This gives

$$\begin{pmatrix} 8 & 7 & 6 & 6 & 4 & 4 & 3 & 2 & 1 \\ 5 & \underline{9} & 1 & 4 & 6 & \underline{8} & 3 & \underline{7} & 2 \end{pmatrix},$$

where we underlined the falls on the second line. By arranging the columns this way we ensure that the first line has a fall whenever the second line has one. An array of this kind, where the first line is a partition with at most n parts (padded to n parts with zeros if necessary), the second line is a permutation of $\{1, 2, \ldots, n\}$, and the first line has a fall whenever the second line has one, is called a **standard labeled partition**, and we will see them again in section 9.5.

We now put in a third line that has the running totals of the falls from each column to the end:
$$\begin{pmatrix} 8 & 7 & 6 & 6 & 4 & 4 & 3 & 2 & 1 \\ 5 & \underline{9} & 1 & 4 & 6 & \underline{8} & 3 & \underline{7} & 2 \\ 3 & 3 & 2 & 2 & 2 & 2 & 1 & 1 & 0 \end{pmatrix}$$

The second line has three falls, but only two falls if we start in the third through sixth positions, and only one if we start in the seventh or eighth positions. Since the third line only falls one unit at a time and has exactly the same falls as the second line, if we subtract it from the first line we will still have a partition there. Doing so and writing the third line as a partition we get

$$\begin{pmatrix} 5 & 4 & 4 & 4 & 4 & 2 & 2 & 1 & 1 \\ 5 & 9 & 1 & 4 & 6 & 8 & 3 & 7 & 2 \end{pmatrix} \quad \text{and} \quad 3+3+2+2+2+2+1+1+0.$$

The new array is just a **labeled partition**. The partition line has been decoupled from the permutation in that the latter can fall when the former does not. Thus this procedure has converted the sequence 613684247 of length 9 into the permutation 591468372 of $\{1, 2, \ldots, 9\}$ and the unrelated partition $5+4+4+4+2+2+1+1$ with at most 9 parts. The conjugate of the Ferrers diagram of $3+3+2+2+2+2+1+1+0$ is

• ∗ ⊙ ⋈ ⊕ ⊖ ▽ △
• ∗ ⊙ ⋈ ⊕ ⊖
• ∗

which is the partition $8 + 6 + 2$, and the point is that this is the major index of 5<u>9</u>146<u>8</u>3<u>7</u>2. This always happens, because the position of the last fall is where the number of falls drops from 1 to 0, and more generally the position of the ith-to-last fall is where the number of falls drops from i to $i - 1$. Therefore the major index of the second line is exactly what we subtract from the original sequence to get the reduced partition on the first line of the labeled partition. Hence this procedure breaks the weight of the original sequence of length n into a partition with at most n parts and the major index of a permutation of length n. Considering all possible sequences of length n we get

$$\frac{1}{(1-q)^n} = \sum_{\sigma \in S(n)} q^{a_1 + \cdots + a_n} = \frac{1}{(q;q)_n} \sum_{\pi \in \Pi(n)} q^{\text{maj}\,\pi}.$$

It follows that

$$\sum_{\pi \in \Pi(n)} q^{\text{maj}\,\pi} = \frac{(q;q)_n}{(1-q)^n} = n!_q,$$

which is (9.1.1).

Exercises

1. Check that (9.1.1) holds in the cases $n = 0, 1, 2$.
2. Explain why the largest major index that a permutation of $\{1, 2, \ldots, n\}$ can have is $\binom{n}{2}$. Which permutation (or permutations) has (or have) this major index?
3. We can pose the analogue of Stern's problem in this setting: what is the total amount of major index in all the permutations of $\{1, 2, \ldots, n\}$?

4. Perhaps a more interesting question than the one in problem 3 is: what is the total number of falls in all the permutations of $\{1, 2, \ldots, n\}$? **Hint:** Try Rodrigues's "couples" argument.

5. A class of infinite series, which one sometimes sees examples of as "converges or diverges" questions in calculus, can be summed by the following device. Start with
$$\sum_{n=0}^{\infty} x^n = \frac{1}{1-x} \quad \text{if } |x| < 1,$$
take the derivative on both sides, and then multiply by x. This gives
$$\sum_{n=0}^{\infty} nx^n = \frac{x}{(1-x)^2} \quad \text{if } |x| < 1,$$
and we could throw away the $n = 0$ term if we want, but it does no harm to leave it in. Take the derivative again and then multiply by x again to get

(9.1.2) $$\sum_{n=0}^{\infty} n^2 x^n = \frac{x(1+x)}{(1-x)^3} \quad \text{if } |x| < 1,$$

and again to get

(9.1.3) $$\sum_{n=0}^{\infty} n^3 x^n = \frac{x(1+4x+x^2)}{(1-x)^4} \quad \text{if } |x| < 1.$$

Check these calculations. We will use (9.1.3) in Chapter 13, and (9.1.2) was needed in a problem in Chapter 5.

6. Performing the steps of the previous problem k times we'll get

(9.1.4) $$\sum_{n=0}^{\infty} n^k x^n = \frac{E_k(x)}{(1-x)^{k+1}} \quad \text{if } |x| < 1$$

for some polynomial of degree k in x, which is called an **Eulerian** polynomial. The cases worked out above are $E_0(x) = 1$, $E_1(x) = x$, $E_2(x) = x^2 + x$, and $E_3(x) = x^3 + 4x^2 + x$. By taking the derivative of both sides of (9.1.4) and then multiplying by x, show that

(9.1.5) $$E_{k+1}(x) = x\left\{(1-x)E_k'(x) + (k+1)E_k(x)\right\}.$$

7. Use (9.1.5) to show that
$$E_4(x) = x\left(x^3 + 11x^2 + 11x + 1\right).$$
Use this and (9.1.4) to show that
$$\sum_{n=0}^{\infty} \frac{n^4}{3^n} = 15.$$

8. Use (9.1.5) and the previous problem to show that
$$E_5(x) = x\left(x^4 + 26x^3 + 66x^2 + 26x + 1\right).$$
Use this and (9.1.4) to show that
$$\sum_{n=0}^{\infty} \frac{n^5}{2^n} = 1082.$$

9. Use (9.1.5) and the previous problem to show that
$$E_6(x) = x\left(x^5 + 57x^4 + 302x^3 + 302x^2 + 57x + 1\right).$$

10. Besides being useful for summing certain series, as in problems 7 and 8, the Eulerian polynomials are interesting because their coefficients count falls in permutations. Write down all the permutations of $\{1, 2, 3, 4\}$ and count the number of falls in each. You should find that there are 11 with 1 fall, 11 with 2 falls, and 1 each with 0 and 3 falls. Hence, except for the factor of x, this matches up with $E_4(x)$.

11. Problem 10 suggests the following theorem: if we define the coefficients of $E_k(x)$ by

(9.1.6) $$E_k(x) = x \sum_{j=0}^{k-1} \epsilon_{k,j}\, x^j$$

for $k \geq 1$, then $\epsilon_{k,j}$ is the number of permutations of $\{1, 2, \ldots, k\}$ with exactly j falls. (The $\epsilon_{k,j}$'s are called **Eulerian** numbers.) To prove this, first plug (9.1.6) into (9.1.5) to show that

(9.1.7) $$\epsilon_{k+1,j} = (j+1)\, \epsilon_{k,j} + (k - j + 1)\, \epsilon_{k,j-1}.$$

If we define $\epsilon_{0,0} = 1$, this holds for all nonnegative integers k and j. Now try to explain (9.1.7) combinatorially.

9.2. Revisiting the q-binomial coefficients

Since the major index on permutations of $\{1, 2, \ldots, n\}$ was so nice, let's try it with sequences of 1's and 2's, where a fall is any instance of a 2 immediately followed by a 1. For example, the sequence 12122̱11212̱ has falls in the second, fifth, and eighth positions. Adding these together, we find that the major index of 1212211212 is $2 + 5 + 8 = 15$. Given what we know from the previous section and from Chapter 1, when we look at all the sequences of a 1's and b 2's from this point of view, it is reasonable to expect the q-binomial coefficient $\binom{a+b}{a}_q$ to show up.

Let's look at all the sequences of two 1's and two 2's:

sequence	q^{maj}	sequence	q^{maj}
1122	q^0	2112	q
1212	q^2	2121	q^4
1221	q^3	2211	q^2

When we add all this up we get $1 + q + 2q^2 + q^3 + q^4$, which is $\binom{4}{2}_q$, so it looks like this is going to work. Let's do one more example before we state the theorem. This time we look at all the sequences of four 1's and two 2's:

sequence	q^{maj}	sequence	q^{maj}	sequence	q^{maj}
111122	q^0	112121	q^8	211121	q^6
111212	q^4	112211	q^4	122111	q^3
112112	q^3	121121	q^7	211211	q^5
111221	q^5	211112	q^1	212111	q^4
121112	q^2	121211	q^6	221111	q^2

When we add all this up we get

$$1 + q + 2q^2 + 2q^3 + 3q^4 + 2q^5 + 2q^6 + q^7 + q^8, \quad \text{which equals} \quad \binom{6}{2}_q.$$

The general result is

THEOREM 69. *Let $S(a,b)$ denote the set of all sequences of a 1's and b 2's. If σ is such a sequence, let maj σ denote the major index of σ. Then*

(9.2.1) $$\sum_{\sigma \in S(a,b)} q^{\mathrm{maj}\,\sigma} = \binom{a+b}{a}_q.$$

But there is an even better theorem. If we also keep track of the number of falls, we have

THEOREM 70. *Let $S(a,b)$ denote the set of all sequences of a 1's and b 2's. If σ is such a sequence, let $\mathrm{fall}\,\sigma$ and $\mathrm{maj}\,\sigma$ denote the number of falls and the major index of σ, respectively. Then*

(9.2.2) $$\sum_{\sigma \in S(a,b)} x^{\mathrm{fall}\,\sigma} q^{\mathrm{maj}\,\sigma} = \sum_{j \geq 0} x^j q^{j^2} \binom{a}{j}_q \binom{b}{j}_q.$$

The sum on the right side of (9.2.2) stops at the smaller of a and b, but it is convenient to write it unrestrictedly and let the q-binomial coefficients do the work.

Let's try to prove (9.2.2) by induction on b, the number of 2's. There is only one sequence with a 1's and no 2's. It has no falls and no major index, so (9.2.2) reduces to

$$1 = x^0 q^0 = x^0 q^{0^2} \binom{a}{0}_q \binom{0}{0}_q = 1,$$

which is a true statement. If there are a 1's and one 2, then there are no falls (and hence no major index) if the 2 is last, and otherwise there is one fall. If the 2 is in the jth (but not the last) position, then the major index is j. In this case the left side of (9.2.2) becomes

$$x^0 q^0 + x^1 \left(q + q^2 + \cdots + q^a\right) = 1 + xq[a]_q,$$

and so does the right side since the sum has only the terms $j = 0, 1$. So we assume (9.2.2) holds for a 1's and less than b 2's, and we consider sequences of a 1's and b 2's. Suppose the last 2 is in the $(b+k)^{\mathrm{th}}$ position. Then it is preceded by k 1's and $b-1$ 2's, and followed by $a-k$ 1's. If $k=a$, so that the last 2 is at the end of the sequence, then it contributes no falls and nothing to the major index, and, by induction, the rest of the sequence contributes

$$\sum_{j \geq 0} x^j q^{j^2} \binom{a}{j}_q \binom{b-1}{j}_q$$

when we sum over all the sequences with a 2 at the end, because otherwise they have a 1's and $b-1$ 2's.

If $k < a$, then the last fall is in the $(b+k)^{\mathrm{th}}$ position. It therefore contributes xq^{b+k} to the sum in (9.2.2), and the 1's that come after it contribute nothing.

9.2. REVISITING THE q-BINOMIAL COEFFICIENTS

The rest of the sequence (before the last fall) consists of k 1's and $b-1$ 2's, so by induction all such sequences contribute

$$xq^{b+k}\sum_{j\geq 0}x^j q^{j^2}\binom{k}{j}_q\binom{b-1}{j}_q$$

for each k, $0\leq k\leq a-1$. Summing over all these values of k and putting in the case $k=a$, we have the left side of (9.2.2) equal to

$$\sum_{j\geq 0}x^j q^{j^2}\binom{a}{j}_q\binom{b-1}{j}_q + \sum_{k=0}^{a-1}xq^{b+k}\sum_{j\geq 0}x^j q^{j^2}\binom{k}{j}_q\binom{b-1}{j}_q,$$

and we have to argue that this expression equals the right side of (9.2.2). Using

$$\binom{b}{j}_q = \binom{b-1}{j}_q + q^{b-j}\binom{b-1}{j-1}_q$$

takes care of one term, and leaves us having to show that

$$\sum_{j\geq 1}x^j q^{j^2+b-j}\binom{a}{j}_q\binom{b-1}{j-1}_q = \sum_{k=0}^{a-1}xq^{b+k}\sum_{j\geq 0}x^j q^{j^2}\binom{k}{j}_q\binom{b-1}{j}_q$$

$$= \sum_{j\geq 0}x^{j+1}q^{j^2+b}\binom{b-1}{j}_q\sum_{k=0}^{a-1}q^k\binom{k}{j}_q.$$

Changing j to $j+1$ on the left side, we have to show that

$$\sum_{j\geq 0}x^{j+1}q^{j^2+b+j}\binom{a}{j+1}_q\binom{b-1}{j}_q = \sum_{j\geq 0}x^{j+1}q^{j^2+b}\binom{b-1}{j}_q\sum_{k=0}^{a-1}q^k\binom{k}{j}_q.$$

This would follow if we knew that

$$\binom{a}{j+1}_q q^j = \sum_{k=0}^{a-1}q^k\binom{k}{j}_q,$$

or equivalently

(9.2.3) $$\binom{a}{j+1}_q = \sum_{k=0}^{a-1}q^{k-j}\binom{k}{j}_q,$$

and we can observe that the sum here actually has to start at $k=j$. But this is essentially (1.4.3). It follows either by using

$$\binom{a}{j+1}_q = q^{a-j-1}\binom{a-1}{j}_q + \binom{a-1}{j+1}_q$$

$$= q^{a-j-1}\binom{a-1}{j}_q + q^{a-j-2}\binom{a-2}{j}_q + \binom{a-2}{j+1}_q$$

$$= q^{a-j-1}\binom{a-1}{j}_q + q^{a-j-2}\binom{a-2}{j}_q + q^{a-j-3}\binom{a-3}{j}_q + \binom{a-3}{j+1}_q$$

repeatedly, or by considering sequences of $j+1$ 1's and $a-j-1$ 2's according to the position of the last 1 and counting inversions. The last 1 is preceded by j 1's and some number $k-j$ of 2's, and followed by $a-k-1$ 2's; the factor $\binom{k}{j}_q$ takes care of the inversions within the first k elements, q^{k-j} accounts for the inversions with the last 1, and there are no other inversions, so (9.2.3) holds, and this completes the proof of (9.2.2).

Setting $x = 1$ in (9.2.2) gives

$$\sum_{\sigma \in S(a,b)} q^{\mathrm{maj}\,\sigma} = \sum_{j \geq 0} q^{j^2} \binom{a}{j}_q \binom{b}{j}_q,$$

and to prove (9.2.1) we have to argue that the right side equals $\binom{a+b}{a}_q$. Consider sequences of a 1's and b 2's and count inversions. If j is the number of 2's among the first a elements of a sequence, then there are $a - j$ 1's among these elements, and hence j 1's among the last b elements, and hence $b - j$ 2's among the last b elements. The factor $\binom{a}{j}_q$ takes care of the inversions among the first a elements, $\binom{b}{j}_q$ takes care of the inversions among the last b elements, and q^{j^2} takes care of the inversions between the two sets. Since we know that $\binom{a+b}{a}_q$ is the generating function for all sequences of a 1's and b 2's by inversions, summing over the possible values of j we get

$$\binom{a+b}{a}_q = \sum_{j \geq 0} q^{j^2} \binom{a}{j}_q \binom{b}{j}_q,$$

which proves (9.2.1).

Exercises

1. Check (9.2.1) and (9.2.2) in the case $a = 2$, $b = 3$.
2. Explain why considering the major index in sequences of a 1's and $b + 1$ 2's by the position of the last 2 gives

$$\binom{a+b+1}{a}_q = \binom{a+b}{a}_q + \sum_{j=1}^{a} \binom{a+b-j}{a-j}_q q^{a+b-j+1},$$

and verify this identity by using the q-Pascal recurrences or inversions or both. This gives a direct proof of (9.2.1).

3. Is it obvious to you that the left sides of (9.2.1) and (9.2.2) are symmetric in a and b? They must be, because the right sides are. The symmetry of (9.2.2) in a and b suggests that there might be a 1-1 correspondence between sequences of a 1's and b 2's and sequences of b 1's and a 2's that preserves the positions of all the falls. For example, such a correspondence (if it exists) would associate 1121221 with something like 2221121 or 1221221, where the numbers of 1's and 2's have been switched but the falls are still the same. The goal of this problem is to find such a correspondence.

(a) It is pretty clear that we should just leave all the falls alone. If we change all the 1's that are *not* involved in falls to 2's and all the 2's not in falls to 1's (for example, 1121221 would become 2221121 and vice versa), explain why we have at least managed to switch the total numbers of 1's and 2's.

(b) Therefore, the procedure in (a) does what we want provided that it does not introduce any new falls. In the case 1121221 ↔ 2221121 it doesn't, but does this happen in general? If so, how do you know? If not, how could you fix the problem? **Hint:** What does a sequence of 1's and 2's look like in between two falls (or before the first fall and after the last one)?

4. Try the correspondence you constructed in problem 3 on all the sequences of two 1's and two 2's. How many sequences map to themselves, and what happens to the others?

5. Try the correspondence you constructed in problem 3 on all the sequences of three 1's and two 2's, mapping them to all the sequences of two 1's and three 2's.

6. We can also prove (9.2.2) by induction on a, the number of 1's, but the details are harder. Here is an outline:

    (i) Show that (9.2.2) holds when $a = 0$ and $a = 1$.

    (ii) Assume (9.2.2) holds for any number of 1's less than a, and divide the sequences in $S(a, b)$ into two classes: those that begin with 1, and those where the first 1 is preceded by some 2's. Explain why

$$\sum_{\substack{\sigma \in S(a,b) \\ \sigma \text{ starts with } 1}} x^{\text{fall } \sigma} q^{\text{maj } \sigma} = \sum_{\sigma \in S(a-1,b)} (xq)^{\text{fall } \sigma} q^{\text{maj } \sigma} = \sum_{j \geq 0} x^j q^{j^2+j} \binom{a-1}{j}_q \binom{b}{j}_q.$$

(iii) In the other case, suppose the first 1 is preceded by $b - k$ 2's, where $0 \leq k \leq b - 1$. For a given k, explain why this case contributes

$$xq^{b-k} \sum_{\sigma \in S(a-1,k)} \left(xq^{b-k+1}\right)^{\text{fall } \sigma} q^{\text{maj } \sigma} = xq^{b-k} \sum_{j \geq 0} \left(xq^{b-k+1}\right)^j q^{j^2} \binom{a-1}{j}_q \binom{k}{j}_q.$$

(iv) Combining (ii) with all the possible values of k in (iii) we get

$$\sum_{j \geq 0} x^j q^{j^2+j} \binom{a-1}{j}_q \binom{b}{j}_q + \sum_{k=0}^{b-1} xq^{b-k} \sum_{j \geq 0} \left(xq^{b-k+1}\right)^j q^{j^2} \binom{a-1}{j}_q \binom{k}{j}_q,$$

and we have to show that this equals

$$\sum_{j \geq 0} x^j q^{j^2} \binom{a}{j}_q \binom{b}{j}_q.$$

Explain why this reduces to showing that

$$\sum_{j \geq 1} x^j q^{j^2} \binom{a-1}{j-1}_q \binom{b}{j}_q = \sum_{j \geq 0} \sum_{k=0}^{b-1} x^{j+1} q^{j^2+j} q^{(j+1)(b-k)} \binom{a-1}{j}_q \binom{k}{j}_q.$$

(v) Explain why this further reduces to showing that

$$\binom{b}{j+1}_q = \sum_{k=j}^{b-1} \binom{k}{j}_q q^{(j+1)(b-k-1)}.$$

(vi) Prove the result of (v), either by using one of the fundamental recurrences repeatedly, or by considering sequences of $j + 1$ 1's and $b - j - 1$ 2's according to the position of the first 1, keeping track of inversions.

9.3. Foata's bijection for q-multinomial coefficients

After the first two sections, we may hope to be able to prove a similar result for q-multinomial coefficients. We use the same notation as in section 1.6: let $S_n(k_1, \ldots, k_m)$ be the set of all sequences of length n made up of k_1 1's, k_2 2's, $\ldots$, k_m m's. As usual, a fall is an instance of a smaller number coming immediately after a larger one, and the major index of such a sequence is the sum of the positions of the (first numbers in the) falls in the sequence; for example, 1$\underline{43}$1$\underline{44}$2$\underline{24}$1 has falls in the second, third, sixth, and ninth positions, so its major index is $2+3+6+9 = 20$. Then we would like to be able to prove

THEOREM 71 (MacMahon's theorem). *With the above notation,*

(9.3.1) $$\binom{n}{k_1, \ldots, k_m}_q = \sum_{\sigma \in S_n(k_1, \ldots, k_m)} q^{\operatorname{maj} \sigma}.$$

We will give two proofs of MacMahon's theorem. A third is outlined in the problems for section 9.4. The first proof is due to Foata, whose strategy was to find a way to take a given permutation of k_1 1's, k_2 2's, $\ldots$, k_m m's with major index M and convert it into another permutation of k_1 1's, k_2 2's, $\ldots$, k_m m's with M inversions. If we can find a method for doing this which is a bijection (in other words, we have to be able to undo it), then we can rely on the corresponding result for inversions. The problem of finding this type of proof of MacMahon's theorem was suggested to Foata by Schützenberger.

Foata's idea is to move through the permutation from left to right, "correcting" the inversions as needed by rearranging the permutation. Let's look at 1431442241, which has major index 20 as we noted above. We don't have to do anything until we get to 143, because 1 and 14 have 0 inversions and 0 major index. But 143 has major index 2 and only one inversion, so we have to do something to create a new inversion. We have $4 > 3$, so we split 143 after everything > 3: 14|3. Then we move the last element to the front in each component: 41|3. So 143 is replaced by 413, which has the same elements as 143 and has the same number of inversions as 143 has major index. Having dealt with the first three elements of 1431442241, we move on to the fourth element 1. Again the element we are trying to tack on (1) is smaller than the last element of the current sequence (3), so we split 4131 after everything bigger than the new element 1: 4|13|1. In each component we move the last element to the front: 4|31|1. So the first four numbers in the transformed sequence are (at the moment; they may change) 4311. This was necessary because adding 1 to 413 produced a fall at the third position, so it added three to the major index, but it only caused two new inversions; changing 4131 to 4311 created a third inversion.

Next we add 4, and no action is necessary, because this creates no new inversions and no new major index. Technically, when the new number is at least as big as the old one (here $1 \leq 4$) we split after everything ≤ 4 and move the last element to the front in each component. In this case, though, all the components have length one and nothing moves. Adding the next 4 also causes no change. This brings us to the first 2: 4311442. Here the new last number (2) is smaller than the previous last number (4); $4 > 2$, so we split after everything > 2: 4|3|114|4|2. We move the last element to the front in each component: 4|3|411|4|2; so now we have 4341142. The 2 created six more units of major index, but only four new inversions, and moving the second 4 forward two places gave us two more inversions, making up the difference.

Now we append the next 2: 43411422. This time the new element is not smaller than the previous last element; $2 \leq 2$, so we split after everything ≤ 2: 4341|1|42|2. Move the last element to the front in each component to get 14341242. This has removed four inversions, which was necessary because the new 2 did not create any more major index, but it did cause four new inversions.

Next add the last 4 in the original sequence; $2 \leq 4$, so we split after everything ≤ 4. All components have length one, so nothing moves, and we have 143412424. Finally we add the last 1. Now $4 > 1$, so we split after everything > 1: 14|3|4|12|4|2|4|1. The last shall be first in each component, and we finally reach 4134214241 as the sequence that corresponds to the original sequence 1431442241 under Foata's map. 4134214241 has the same numbers as 1431442241, and it has as many inversions (20) as 1431442241 has major index.

Let's prove that this works in general, for which it is enough to show that it works at any one stage. Suppose we are using Foata's procedure on a sequence, that the current last element of the sequence is a, and that we want to append b. We have to consider two cases: $a > b$ and $a \leq b$.

$a > b$: Suppose a and $j-1$ numbers before a are $> b$, and suppose that k numbers before a are $\leq b$. Let the j numbers which exceed b be $\omega_1, \omega_2, \ldots, \omega_j$, in that order from left to right (so $\omega_j = a$). Let k_1 be the number of numbers before ω_1 in the current sequence; if there are any such numbers, by construction they are all $\leq b$. Let k_2 be the number of numbers strictly between ω_1 and ω_2 in the current sequence; k_2 could be zero, but again all such numbers are $\leq b$. Generally, let k_i be the number of numbers strictly between ω_{i-1} and ω_i in the current sequence; all such numbers are $\leq b$. Then we have $k_1 + k_2 + \cdots + k_j = k$.

When $a > b$ we split after everything $> b$, so in this case we split after each of the ω_i, but nowhere else. In each split component the last shall be first, so ω_i is moved ahead of the other k_i elements in the ith component for each i, which creates k_i new inversions for each i. The total number of new inversions is $k_1 + k_2 + \cdots + k_j = k$. This was necessary because, since $a > b$, appending b to the sequence created a fall in the $(k+j)^{\text{th}}$ position, but the new b was only inverted with the j ω_i's. Thus Foata's map works in the case $a > b$.

$a \leq b$: We use the same system of notation as above. Let j be the number of things before a which are $> b$, and let there be $k-1$ things besides a which precede b and are $\leq b$. In this case appending b does not add to the major index, but it still creates j new inversions. In the previous case we had to create k additional inversions, but in this case we have to get rid of j inversions. Let the k things which precede b and are $\leq b$ be $\omega_1, \omega_2, \ldots, \omega_k$, in that order from left to right (so $\omega_k = a$). Let the number of things which precede ω_1 in the current sequence (all of which are $> b$) be j_1; let the number of things strictly between ω_1 and ω_2 (all $> b$) be j_2; and so forth, so that there are j_i things (all $> b$) strictly between ω_{i-1} and ω_i for each i. Then we have $j_1 + j_2 + \cdots + j_k = j$.

When $a \leq b$ we split after everything $\leq b$, so we split after each of the ω_i's and nowhere else. In each split component the last shall be first, so we move ω_i in front of j_i things (all of which are $> \omega_i$) for each i. This removes j_i inversions for each i, so it removes $j_1 + j_2 + \cdots + j_k = j$ inversions in all, which is exactly what we wanted to accomplish. Thus Foata's map also works in the case $a \leq b$.

Next we describe the inverse map—how to get back from the number of inversions to the major index. Let's consider the example 3141223142, which has 17 inversions. To reverse the above procedure we strip off the permutation one element at a time, starting at the right, and again there are two cases. There is one subtlety: it might seem that to undo what we were doing above, we should compare the sizes of the last two elements (here 4 and 2). This does not work in general, for the following reason: in case $a > b$ in the above proof, $\omega_j = a$ was moved in front of k_j things that were $\leq b$. After this the last element was at least as big as the penultimate element, unless $k_j = 0$, in which case the last element is smaller than the penultimate element. On the other hand, in case $a \leq b$, $\omega_k = a$ was moved in front of j_k things that were $> b$. After this the last element was smaller than the penultimate element, unless $j_k = 0$, in which case the last element is at least as big as the penultimate element. So we can't tell which of the two moves to undo by looking at the last two elements.

Fortunately, we *can* tell by comparing the last element with the *first* element. In case $a < b$, the element ω_1, which was the first thing in the sequence $> b$, was moved to the beginning of the sequence. If it was not moved (*i.e.*, if $k_1 = 0$), then it was already at the beginning of the sequence. Thus when Foata's map is applied in the case $a < b$, the result is always a sequence with the first element bigger than the last. In case $a \leq b$, again ω_1 was moved to the front of the sequence, if it was not there already. In this case ω_1 was the first element $\leq b$, so the result of Foata's map in this case is invariably a sequence where the last element is at least as big as the first.

Now that we know this, let's do our example 3141223142. We'll use α as a generic name for the first element in the sequence, and ω as a generic name for the last element, so we start with $\alpha = 3$ and $\omega = 2$. When $\alpha > \omega$, we remove ω and split before every element $> \omega$: |31|4122|31|4. In each component we move the first element to the end: |13|1224|13|4. So the new permutation is 131224134, and 2 has been removed.

Now $\alpha = 1$ and $\omega = 4$. When $\alpha \leq \omega$, we remove ω and split before every element $\leq \omega$. In this case $\omega = 4$ is at least as big as every other element in the sequence, so all components have length 1 and nothing moves. Now we have 13122413 and we have removed 42. Then $\alpha = 1$ and $\omega = 3$, so we remove 3 and split before every element ≤ 3: |1|3|1|2|24|1. Moving the first element to the end in each component we get 1312421 and we have removed 342.

Now $\alpha = 1$ and $\omega = 1$; $\alpha \leq \omega$, so we remove 1 (on the right) and split before everything ≤ 1: |13|1242. Moving the first element to the end in each component we get 312421 and we have removed 1342. Now $\alpha = 3$ and $\omega = 1$; $\alpha > \omega$, so we remove 1 and split before everything > 1: |31|2|4|2. The first shall be last in each component, so we get 13242 and we have removed 11342. Now $\alpha = 1$ and $\omega = 2$; $\alpha \leq \omega$, so we remove 2 and split before everything ≤ 2: |13|24. Make the first last in each component: we have 3142, and we have removed 211342.

Now $\alpha = 3$ and $\omega = 2$; $\alpha > \omega$, so we remove 2 and split before everything > 2: |31|4. Making the first last in each component we get 134 and we have removed 2211342. We have three steps left, but you can check that the remaining elements come off one at a time in the current (right to left) order, so we finally wind up with 1342211342. It has falls in the third, fifth, and ninth spots, so its major index is $3+5+9 = 17$, which is the same as the inversion number of the original sequence 3141223142.

Exercises

1. Check that if we apply Foata's map to 1342211342, we undo the above steps one at a time and the result is 3141223142.
2. Check that if we apply the inverse map to 4134214241, we undo the steps in our first example one at a time, and the result is 1431442241.
3. Show that 214313431132 has major index 34, and that applying Foata's map to it gives 412343311312, with 34 inversions.
4. Show that applying the inverse map to 412343311312 gives 214313431132.
5. Show that 1323224341431 has major index 45, and that applying Foata's map to it gives 4433122234311, with 45 inversions.
6. Show that applying the inverse map to 4433122234311 gives 1323224341431.
7. Make up your own examples and solve them.

9.4. MacMahon's proof

MacMahon's proof of his theorem is much different, and is also quite interesting. We saw a special case of it already in the alternative proof of section 9.1. It relies on a fact we know well: $1/(q;q)_k$ is the generating function for partitions with at most k parts; or with exactly k parts, some of which might be zero. Then we can think of
$$\frac{1}{(q;q)_{k_1}(q;q)_{k_2}\cdots(q;q)_{k_m}}$$
as the generating function for m-line arrays, where the first line is a partition into exactly k_1 parts, some of which might be zero; the second line is a partition into exactly k_2 parts, some of which might be zero, and so on. (In section 9.1, every k_i was 1.) Let's write down an example at this point:
$$\frac{1}{(q;q)_5(q;q)_3(q;q)_6(q;q)_4}$$
generates arrays such as:

$$\begin{array}{cccccc} 8 & 4 & 3 & 1 & 0 \\ 6 & 6 & 0 \\ 9 & 8 & 6 & 5 & 5 & 1 \\ 6 & 5 & 3 & 2 \end{array}$$

We combine this into one big partition, recording which line each part came from and breaking ties in favor of earlier lines:

Parts	9	8	8	6	6	6	6	5	5	5	4	3	3	2	1	1	0	0
Lines	3	1	3	2	2	3	4	3	3	4	1	1	4	4	1	3	1	2

The first line is a partition with exactly $5+3+6+4 = 18$ parts, some of which might be zero, and the second line is a permutation of five 1's, three 2's, six 3's, and four 4's. In general, at this stage we would have a partition with exactly $k_1+k_2+\cdots+k_m$ parts, some of which might be zero, on the first line; and a permutation of k_1 1's,

k_2 2's, and so forth on the second line. The generating function for the former is $1/(q;q)_{k_1+\cdots+k_m}$, and we need to figure out what the latter contributes to the overall generating function.

The problem is that the partition and the permutation are not completely general; rather, they are related in that the permutation has a fall everywhere that the partition has one, because of our policy of breaking ties in favor of earlier lines. The partition may have more falls than the permutation, as happens here in going from 4 to 3 and from 3 to 2. Let's put a third line in our array:

Parts	9	8	8	6	6	6	6	5	5	5	4	3	3	2	1	1	0	0
Lines	<u>3</u>	1	<u>3</u>	2	2	3	<u>4</u>	3	3	<u>4</u>	1	1	<u>4</u>	<u>4</u>	1	<u>3</u>	1	2
Falls	6	5	5	4	4	4	4	3	3	3	2	2	2	2	1	1	0	0

We underlined all the falls in the permutation on line 2, and on line 3 we listed the numbers of falls in the permutation when we start at the corresponding positions on line 2. (That is, there are 6 falls in total on line 2, but only 5 falls if we start in the second or third positions on line 2, and only 4 falls if we start in the fourth through seventh positions, and so on.) Note that the third line has *exactly* the same falls as the second one.

We now create another new array, by subtracting the third line from the first in the old array to get the first line of the new one, and otherwise recopying the second and third lines:

3	3	3	2	2	2	2	2	2	2	1	1	1	0	0	0	0
<u>3</u>	1	<u>3</u>	2	2	3	<u>4</u>	3	3	<u>4</u>	1	1	<u>4</u>	<u>4</u>	1	<u>3</u>	1	2
6	5	5	4	4	4	4	3	3	3	2	2	2	2	1	1	0	0

This decouples the partition from the permutation, in that the new partition on the first line need not have falls in the same places that the permutation has them. The first line is still a partition into $k_1 + k_2 + \cdots + k_m$ parts, some of which might be zero, so the generating function for the first lines of the new arrays is still $1/(q;q)_{k_1+\cdots+k_m}$. What about the second lines?

We have to put in a power of q to compensate for the subtractions we did above, and that power must be the sum of the numbers on the third line, because that's what we subtracted. We claim that this sum is exactly the major index of the permutation on the second line. We can check that this happens in this example, but why would we expect it to happen? To see this, we write the third line as the Ferrers diagram of a partition. To save space (and since we're going to look at the conjugate anyway) let's write the parts as columns:

```
●  *  ⊙  ◁  ★  ⊕  ▽  ×  ⊖  △  ⊗  ⋈  ⊘  ▷  o  ◇
●  *  ⊙  ◁  ★  ⊕  ▽  ×  ⊖  △  ⊗  ⋈  ⊘  ▷
●  *  ⊙  ◁  ★  ⊕  ▽  ×  ⊖  △
●  *  ⊙  ◁  ★  ⊕  ▽
●  *  ⊙
●
```

The partition represented by the rows is $16 + 14 + 10 + 7 + 3 + 1$, and this is the major index of the permutation in the array since the underlined numbers are the first, third, seventh, tenth, fourteenth, and sixteenth ones. That is, the numbers in the conjugate partition of the third line are the positions of the falls on the second line. This is because the last fall on the second line occurs at the same place that the third line drops from 1 to 0, so its position equals the number of things on the third line that are at least 1; the next-to-last fall on the second line occurs where the third line drops from 2 to 1, so its position equals the number of things on the third line that are at least 2; and so forth.

If $n = k_1 + k_2 + \cdots + k_m$, then it follows that the second lines in the arrays are generated by

$$\sum_{\sigma \in S_n(k_1,\ldots,k_m)} q^{\mathrm{maj}\,\sigma},$$

where we use the same notation as the previous section—the sum is over all permutations ω of k_1 1's, k_2 2's, ..., k_m m's. It follows that

$$\frac{1}{(q;q)_{k_1}(q;q)_{k_2}\cdots(q;q)_{k_m}} = \left(\sum_{\sigma \in S_n(k_1,\ldots,k_m)} q^{\mathrm{maj}\,\sigma}\right) \frac{1}{(q;q)_{k_1+\cdots+k_m}},$$

and this becomes MacMahon's theorem when we multiply both sides by

$$(q;q)_{k_1+\cdots+k_m}.$$

Exercises

1. Explain why (9.4.2) is obvious if $n = 0$ or $n = 1$, and verify it when $n = 2$.

2. MacMahon's theorem can also be proved by induction using the the recurrence relation

(9.4.1)
$$\binom{n+1}{k_1,\ldots,k_m}_q = \binom{n}{k_1-1,k_2,\ldots,k_m}_q + q^{k_1}\binom{n}{k_1,k_2-1,k_3,\ldots,k_m}_q$$
$$+ q^{k_1+k_2}\binom{n}{k_1,k_2,k_3-1,\ldots,k_m}_q + \cdots$$
$$+ q^{k_1+\cdots+k_{m-1}}\binom{n}{k_1,\ldots,k_{m-1},k_m-1}_q$$

for the q-multinomial coefficients (which was (1.6.3) in section 1.6), where $k_1 + \cdots + k_m = n+1$ and each k_i is a nonnegative integer, by exploiting the fact that (9.4.1) is symmetric in $k_1,\ldots,k_m$. After the previous problem, we may assume (9.4.2) holds for n and try to show

(9.4.2)
$$\binom{n+1}{k_1,\ldots,k_m}_q = \sum_{\sigma \in S_{n+1}(k_1,\ldots,k_m)} q^{\mathrm{maj}\,\sigma}.$$

We need to evaluate the last sum, which we break up according to which number is at the end of the sequence:

(9.4.3)
$$\sum_{\sigma \in S_{n+1}(k_1,\ldots,k_m)} q^{\mathrm{maj}\,\sigma} = \sum_{j=1}^{m} \sum_{\substack{\sigma \in S_{n+1}(k_1,\ldots,k_m) \\ j \text{ at end}}} q^{\mathrm{maj}\,\sigma}.$$

(i) Explain why the sequences where m is last contribute $\binom{n}{k_1,\ldots,k_{m-1},k_m-1}_q$.

(ii) Next suppose $m-1$ is last. By considering the two cases (a) m is next to last, (b) m is not next to last, show that these sequences contribute

$$(q^n - 1)\binom{n-1}{k_1,\ldots,k_{m-1}-1,k_m-1}_q + \binom{n}{k_1,\ldots,k_{m-1}-1,k_m}_q$$
$$= q^{k_m}\binom{n}{k_1,\ldots,k_{m-1}-1,k_m}_q.$$

(iii) We claim that the sequences with j at the end contribute a term

(9.4.4) $$q^{k_m+k_{m-1}+\cdots+k_{j+1}}\binom{n}{k_1,\ldots,k_{j-1},k_j-1,k_{j+1},\ldots,k_m}_q.$$

We have seen this already when j is m or $m-1$. We prove it by induction on increasing n and decreasing j; i.e., for n and $j-1$ by using the fact that it holds for n and $j, j+1, \ldots, m$, and also for $n-1$ and $1, 2, \ldots, j-1$.

Suppose $j-1$ is at the end of a sequence. There is a fall at the nth position if this position is occupied by one of $j, j+1, \ldots, m$. Explain why this case gives

(9.4.5) $$q^n \sum_{i=j}^{m} q^{k_m+k_{m-1}+\cdots+k_{i+1}}\binom{n-1}{k_1,\ldots,k_{j-1}-1,\ldots,k_i-1,\ldots,k_m}_q.$$

(iv) Explain why if the nth position is occupied by one of $1, 2, \ldots, j-1$ instead, we get

(9.4.6) $$\sum_{i=1}^{j-2} q^{k_m+k_{m-1}+\cdots+k_{j-1}-1+\cdots+k_{i+1}}\binom{n-1}{k_1,\ldots,k_i-1,\ldots,k_{j-1}-1,\ldots,k_m}_q$$
$$+ q^{k_m+k_{m-1}+\cdots+k_j}\binom{n-1}{k_1,\ldots,k_{j-1}-2,\ldots,k_m}_q.$$

(v) Show that combining (9.4.5) and (9.4.6) gives

(9.4.7) $$\binom{n-1}{k_1,\ldots,k_{j-1}-2,\ldots,k_m}_q$$
$$+ \sum_{i=1}^{j-2} q^{k_{j-1}-1+\cdots+k_{i+1}}\binom{n-1}{k_1\ldots,k_i-1,\ldots,k_{j-1}-1,\ldots,k_m}_q$$
$$+ \sum_{i=j}^{m} q^{k_m+k_{m-1}+\cdots+k_{i+1}+k_{j-1}+\cdots+k_1-1}\binom{n-1}{k_1,\ldots,k_{j-1}-1,\ldots,k_i-1,\ldots,k_m}_q.$$

(vi) Use (9.4.1) with parameters in the order $k_{j-1}, k_{j-2}, \ldots, k_1, k_m, \ldots, k_j$ to get

$$\binom{n}{k_1, \ldots, k_{j-1}-1, \ldots, k_m}_q = \binom{n-1}{k_1, \ldots, k_{j-1}-2, \ldots, k_m}_q$$
$$+ q^{k_{j-1}-1} \binom{n-1}{k_1, \ldots, k_{j-2}-1, k_{j-1}-1, \ldots, k_m}_q$$
$$+ q^{k_{j-1}+k_{j-2}-1} \binom{n-1}{k_1, \ldots, k_{j-3}-1, k_{j-2}, k_{j-1}-1, \ldots, k_m}_q$$
$$+ \cdots + q^{k_1+\cdots+k_{j-1}-1} \binom{n-1}{k_1, \ldots, k_{j-1}-1, k_j, \ldots, k_m-1}_q$$
$$+ q^{k_1+\cdots+k_{j-1}-1+k_m} \binom{n-1}{k_1, \ldots, k_{j-1}-1, k_j, \ldots, k_{m-1}-1, k_m}_q$$
$$+ \cdots + q^{k_1+\cdots+k_{j-1}-1+k_m+\cdots+k_{j+1}} \binom{n-1}{k_1, \ldots, k_{j-1}-1, k_j-1, k_{j+1}, \ldots, k_m}_q$$
$$= q^{k_m+\cdots+k_j} \binom{n}{k_1, \ldots, k_{j-1}-1, \ldots, k_m}_q.$$

This proves (9.4.4).

(vii) Show that using (9.4.4) in (9.4.3) gives

$$\sum_{j=1}^{m} \sum_{\substack{\sigma \in S_{n+1}(k_1,\ldots,k_m) \\ j \text{ at end}}} q^{\operatorname{maj}\sigma}$$
$$= \sum_{j=1}^{m} q^{k_m+k_{m-1}+\cdots+k_{j+1}} \binom{n}{k_1, \ldots, k_{j-1}, k_j-1, k_{j+1}, \ldots, k_m}_q,$$

and explain why (9.4.1) implies this is

$$\sum_{\sigma \in S_{n+1}(k_1,\ldots,k_m)} q^{\operatorname{maj}\sigma} = \binom{n+1}{k_1, \ldots, k_m}_q.$$

9.5. q-derangement numbers

In a typical permutation it often happens that at least one number is in its "proper" place. In 6137524, for example, 3 is in the third position and 5 is in the fifth position. We call these the **fixed points** of the permutation, and we call the five other numbers its **deranged points**. If a permutation has no fixed points, such as 6317254, we call it a **derangement**.

Suppose a professor has n students and has never bothered to learn their names. When homework is handed back, each student receives a paper at random. What is the probability that no student gets her own paper back? How different is it for large classes than for small ones? This is a famous problem in the history of probability, often called the *problème des rencontres*, and a moment's thought shows that it is intimately bound up with derangements. Suppose for example that there are five students, and number them 1–5. If the professor gives student 1 the paper of student 3, student 2 the paper of student 4, student 3 the paper of student 1, student 4 the paper of student 5, and student 5 the paper of student 2, then no student got the correct paper, and we can think of this as the derangement 34152. If D_n is the number of derangements of $\{1, 2, 3, \ldots, n\}$, then the problem is asking for $D_n/n!$, and we can call it the derangement problem.

For a q-analogue, let Δ_n be the set of derangements of $\{1, 2, 3, \ldots, n\}$ and set

$$(9.5.1) \qquad D_n(q) = \sum_{\delta \in \Delta_n} q^{\mathrm{maj}\,\delta}, \quad \text{where} \quad D_0(q) = 1.$$

Even the most inept professor couldn't mess up the case $n = 1$, so Δ_1 is empty and $D_1(q) = 0$. If $n = 2$, then 21 is a derangement (with major index 1) and 12 is not, so $D_2(q) = q$. There are two derangements of $\{1, 2, 3\}$, namely 231 and 312, with major index 2 and 1 respectively, so $D_3(q) = q + q^2$. Let's make a table for $n = 4$:

derangement	q^{maj}	derangement	q^{maj}	derangement	q^{maj}
2143	q^4	3142	q^4	4123	q
2341	q^3	3412	q^2	4312	q^3
2413	q^2	3421	q^5	4321	q^6

Thus $D_4(q) = q + 2q^2 + 2q^3 + 2q^4 + q^5 + q^6$. The goal of this section is to prove

THEOREM 72 (Gessel's theorem). *For all nonnegative integers n,*

$$(9.5.2) \qquad D_n(q) = n!_q \sum_{k=0}^{n} \frac{(-1)^k q^{\binom{k}{2}}}{k!_q}.$$

The reason for defining $D_0(q) = 1$ is to make this hold for $n = 0$. Note that if $n \geq 1$, then the $k = 0$ and $k = 1$ terms always cancel, so the sum can start at $k = 2$ except in degenerate cases. Let's check Gessel's theorem for $n = 4$. The right side of (9.5.2) is

$$4!_q \left(\frac{q}{2!_q} - \frac{q^3}{3!_q} + \frac{q^6}{4!_q} \right) = q[3]_q[4]_q - q^3[4]_q + q^6 = q[2]_q[4]_q + q^6$$

$$= q + 2q^2 + 2q^3 + 2q^4 + q^5 + q^6,$$

as desired.

To prove Gessel's theorem we follow the approach of Chen and Xu. They use MacMahon's argument from section 9.4 to simplify the proof of a theorem of Michelle Wachs, who deduced Gessel's theorem from hers. We need to go back to an example like 6137524, which was not a derangement because of the fixed points 3 and 5. The deranged points are 61724, and we can make these into a permutation by simply relabeling them with $\{1, 2, 3, 4, 5\}$ preserving the order; so 1 and 2 stay as they are, 4 is relabeled down to 3, 6 is relabeled down to 4, and 7 is relabeled down to 5. Thus the **reduced derangement** of 6137524 is 41523. Note that if a permutation is a derangement, then its reduced derangement is itself. For the identity permutation $123 \cdots n$ where every element is a fixed point, the reduced derangement is empty.

We need to show that the "reduced derangement" really is a derangement. Let π be a permutation of $\{1, 2, \ldots, n\}$ and suppose f is a fixed point. (If it hasn't got a fixed point, then it is already a derangement.) Suppose we remove f, move everything that came after it forward one unit, and relabel every number $> f$ down one unit. Now we have a permutation π' of $\{1, 2, \ldots, n-1\}$, and there are four types of numbers to consider: (i) numbers $< f$ that appear before the fth position don't move and are not relabeled, so they are only fixed after removing f if they were fixed before; (ii) numbers $< f$ that appear after the fth position are not relabeled but move forward one unit, leaving them still smaller than the position they occupy; (iii) numbers $> f$ that appear before the fth position don't move but are relabeled down one unit, leaving them still larger than the position they occupy; (iv) numbers $> f$ that appear after the fth position move forward one unit and are relabeled down one unit, so they are only fixed after removing f if they were fixed before. Therefore π' has exactly one less fixed point than π had. Thus the fixed points may be removed in any order without creating any new ones, so once we remove them all, we have a derangement.

It is easy to reconstruct a permutation from its reduced derangement and its fixed points: just insert each fixed point where it naturally belongs and relabel as needed. For the example 41523 with fixed points 3 and 5, we first insert 3 in the third position and relabel all the numbers ≥ 3 up one unit to get 513624. Then we insert 5 in the fifth position and relabel all the numbers ≥ 5 up one unit to get 6137524. (The insertions must be done in increasing order so that we know what the fifth position actually is.) We are now ready to state

THEOREM 73 (Wachs's theorem). *If $\Pi(n; \delta)$ is the set of all permutations of $\{1, 2, \ldots, n\}$ whose reduced derangement is δ, then*

$$(9.5.3) \qquad \sum_{\pi \in \Pi(n;\delta)} q^{\text{maj}\,\pi} = q^{\text{maj}\,\delta} \binom{n}{k}_q,$$

where δ is a derangement of $\{1, 2, \ldots, k\}$.

The Chen/Xu proof relies on the idea of a standard labeled partition, which we saw already in section 9.1. Recall that a **labeled** partition with n columns is a 2-line array where the first line (or partition line) is a partition with at most n parts, padded to n parts by adding zeros at the end if necessary, and the second line (or permutation line) is a permutation of $\{1, 2, \ldots, n\}$. For example,

$$\begin{pmatrix} 11 & 8 & 4 & 4 & 4 & 3 & 1 & 1 & 0 \\ 3 & 6 & 1 & \underline{8} & 5 & 4 & \underline{9} & 2 & 7 \end{pmatrix}$$

is a labeled partition with 9 columns. In a **standard** labeled partition, the partition line has a fall everywhere that the permutation line has one. This fails in the example above in the fourth and seventh positions, where the permutation falls but the partition does not. We would have a standard labeled partition if we exchanged the fourth and fifth columns, and also the seventh and eighth columns, to get

$$\begin{pmatrix} 11 & 8 & 4 & 4 & 3 & 1 & 1 & 0 \\ 3 & 6 & 1 & 5 & 8 & 4 & 2 & 9 & 7 \end{pmatrix}.$$

Note that the partition may have more falls than the permutation, as in the first column here. We can convert a labeled partition into a standard labeled one without changing the permutation line by first putting in a third line with the running totals of the falls

$$\begin{array}{ccccccccc} 11 & 8 & 4 & 4 & 4 & 3 & 1 & 1 & 0 \\ 3 & \underline{6} & 1 & \underline{8} & \underline{5} & 4 & \underline{9} & 2 & 7 \\ 4 & 4 & 3 & 3 & 2 & 1 & 1 & 0 & 0 \end{array}$$

and then adding the third line to the first to get
$$\begin{pmatrix} 15 & 12 & 7 & 7 & 6 & 4 & 2 & 1 & 0 \\ 3 & 6 & 1 & 8 & 5 & 4 & 9 & 2 & 7 \end{pmatrix}.$$

Another example of a standard labeled partition with 9 columns is
$$P = \begin{pmatrix} 13 & 11 & 10 & 8 & 8 & 6 & 4 & 4 & 1 \\ 5 & 9 & 3 & 4 & 8 & 6 & 1 & 7 & 2 \end{pmatrix}.$$

The fixed points of the permutation line are $3, 4, 6$ and the deranged points are $5, 9, 8, 1, 7, 2$. The Chen/Xu bijection decomposes the partition into two pieces as follows: the **fixed** partition is $10 + 8 + 6$, since these are the parts in the columns corresponding to the fixed points; and the **broken** partition is $13 + 11 + 8 + 4 + 4 + 1$, since these are the parts in the columns corresponding to the deranged points. The reduced derangement of 598172 is 365142, so we can think of this as a decomposition of P into the fixed partition $F = 10 + 8 + 6$ and the labeled partition
$$p = \begin{pmatrix} 13 & 11 & 8 & 4 & 4 & 1 \\ 3 & 6 & 5 & 1 & 4 & 2 \end{pmatrix},$$

where the broken partition is on the first line and the reduced derangement is on the second. Note that p is in fact a standard labeled partition. This is clear in the case of the last three numbers 142 on the second line: they came from 172, which were contiguous in the standard labeled partition P, so the standardness must carry over. It is less clear for 65, which came from 98, which originally had the fixed points 3 and 4 between them.

Suppose then that a and b are two contiguous numbers in the reduced derangement of p, that A and B are the corresponding numbers on the second line of P, and that at least one fixed point was originally between A and B. If $a < b$, then there is nothing to check, so suppose $a > b$, and hence $A > B$. If there is no fall corresponding to ab on the partition line of p, then the two corresponding parts must be the same number, say N, in which case the parts corresponding to the fixed points between them must also be N. But this implies that the fixed point right after A is larger than A, and the fixed point right before B is smaller than B. Since the fixed points are increasing and $A > B$, this is impossible. Therefore this decomposition must give a standard labeled partition p.

It is not hard to reconstruct the original standard labeled partition P from the fixed partition F and the reduced standard labeled partition p; we may seem to have lost information about the positions of the fixed points, but it is easily recovered. In the above example, the part 10 has to go between 11 and 8, so we add the column $\binom{10}{3}$ to p and relabel everything ≥ 3 on the second line up one unit to get
$$p' = \begin{pmatrix} 13 & 11 & 10 & 8 & 4 & 4 & 1 \\ 4 & 7 & 3 & 6 & 1 & 5 & 2 \end{pmatrix}.$$

The 8 in F is a little trickier, but it has to go before the 8 in p' and not after. Otherwise we would have
$$\begin{pmatrix} 13 & 11 & 10 & 8 & 8 & 4 & 4 & 1 \\ 4 & 8 & 3 & 7 & 5 & 1 & 6 & 2 \end{pmatrix}$$

which is not standard since there is a fall in the fourth position on the second line but not the first. Instead we insert the column $\binom{8}{4}$ and relabel the second line to

9.5. q-DERANGEMENT NUMBERS

get
$$p'' = \begin{pmatrix} 13 & 11 & 10 & 8 & 8 & 4 & 4 & 1 \\ 5 & 8 & 3 & 4 & 7 & 1 & 6 & 2 \end{pmatrix}.$$

Finally the 6 must go between the last 8 and the first 4, so we insert the column $\binom{6}{6}$ and relabel the second line to get
$$P = \begin{pmatrix} 13 & 11 & 10 & 8 & 8 & 6 & 4 & 4 & 1 \\ 5 & 9 & 3 & 4 & 8 & 6 & 1 & 7 & 2 \end{pmatrix}.$$

Suppose then that we start with a partition λ with at most n parts, say $\lambda = 9+7+7+5+5+4+3+3+1$, and a permutation π of $\{1,2,\ldots,n\}$, say 593486172. Assemble these into a labeled partition L and write down the running totals of the falls of π. In this example this gives:

$$L = \begin{matrix} & 9 & 7 & 7 & 5 & 5 & 4 & 3 & 3 & 1 \\ & 5 & 9 & 3 & 4 & \underline{8} & \underline{6} & 1 & \underline{7} & 2 \\ & 4 & 4 & 3 & 3 & 3 & 2 & 1 & 1 & 0 \end{matrix}$$

By adding the third line to the first we create a standard labeled partition P, in this case
$$P = \begin{pmatrix} 13 & 11 & 10 & 8 & 8 & 6 & 4 & 4 & 1 \\ 5 & 9 & 3 & 4 & 8 & 6 & 1 & 7 & 2 \end{pmatrix}.$$

Let the partition on the first line be μ, and let $|\mu|$ be the number that μ is a partition of. Using the same notation for λ, we have $|\mu| = |\lambda| + \mathrm{maj}\,\pi$ by MacMahon's argument. Now run the Chen/Xu bijection described above on P. As above, in this example this produces the fixed partition $10 + 8 + 6$ and the standard labeled partition
$$p = \begin{pmatrix} 13 & 11 & 8 & 4 & 4 & 1 \\ 3 & 6 & 5 & 1 & 4 & 2 \end{pmatrix}$$

with the broken partition B on the first line and the reduced derangement δ on the second. Finally, put in the running totals of the falls of the reduced derangement to get

$$\begin{matrix} 13 & 11 & 8 & 4 & 4 & 1 \\ 3 & \underline{6} & \underline{5} & 1 & \underline{4} & 2 \\ 3 & 3 & 2 & 1 & 1 & 0 \end{matrix}$$

and subtract the third line from the first to get a labeled partition ℓ, in this case
$$\ell = \begin{pmatrix} 10 & 8 & 6 & 3 & 3 & 1 \\ 3 & 6 & 5 & 1 & 4 & 2 \end{pmatrix}$$

with the **deranged** partition D on the first line and the reduced derangement δ on the second. Since $|\mu| = |B| + |F|$ and $|B| = |D| + \mathrm{maj}\,\delta$, we have

(9.5.4) $$|\lambda| + \mathrm{maj}\,\pi = |\mu| = |F| + |D| + \mathrm{maj}\,\delta,$$

where λ is a partition with at most n parts, π is a permutation of $\{1,2,\ldots,n\}$, $|F|$ is a partition with at most $n-k$ parts, D is a partition with at most k parts, and δ is a derangement of $\{1,2,\ldots,k\}$ for some k with $0 \leq k \leq n$. Making the beginning and end of (9.5.4) an exponent of q and summing over all the partitions with at most n parts and permutations of $\{1,2,\ldots,n\}$ whose reduced derangement is δ, we get

$$\frac{\sum_{\pi \in \Pi(n;\delta)} q^{\mathrm{maj}\,\pi}}{(q;q)_n} = \frac{q^{\mathrm{maj}\,\delta}}{(q;q)_k (q;q)_{n-k}}.$$

This proves Wachs's theorem.

The rest of the argument is due to Wachs. If we sum (9.5.3) over all the derangements of $\{1, 2, \ldots, k\}$ we get

$$\sum_{\delta \in \Delta_k} \sum_{\pi \in \Pi(n;\delta)} q^{\mathrm{maj}\,\pi} = \binom{n}{k}_q D_k(q).$$

If we further sum this from $k = 0$ to n, then every permutation of $\{1, 2, \ldots, n\}$ will be counted exactly once on the left side, so by (9.1.1) we get

(9.5.5) $$n!_q = \sum_{k=0}^{n} \binom{n}{k}_q D_k(q).$$

Now recall (2.3.6) from section 2.3, which says

$$A_n = \sum_{k=0}^{n} \binom{n}{k}_q B_k \iff B_n = \sum_{k=0}^{n} \binom{n}{k}_q (-1)^k q^{\binom{k}{2}} A_{n-k}$$

after changing k to $n-k$ in the formula for B_n. Taking $B_k = D_k(q)$ here, (9.5.5) implies that $A_{n-k} = (n-k)!_q$ and we have

$$D_n(q) = \sum_{k=0}^{n} \binom{n}{k}_q (-1)^k q^{\binom{k}{2}} (n-k)!_q = n!_q \sum_{k=0}^{n} \frac{(-1)^k q^{\binom{k}{2}}}{k!_q}.$$

This proves Gessel's theorem.

Let's collect some values of $D_n(q)/n!_q$. The first few are

$$\frac{D_0(q)}{0!_q} = 1, \quad \frac{D_1(q)}{1!_q} = 0, \quad \frac{D_2(q)}{2!_q} = \frac{q}{1+q}, \quad \frac{D_3(q)}{3!_q} = \frac{q}{1+q+q^2}$$

and the next two are

$$\frac{D_4(q)}{4!_q} = \frac{q+q^2+q^4}{1+2q+2q^2+2q^3+q^4},$$

$$\frac{D_5(q)}{5!_q} = \frac{q+2q^2+2q^3+2q^4+2q^5+2q^6}{1+3q+5q^2+6q^3+6q^4+5q^5+3q^6+q^7}.$$

It is not too illuminating to write out more values of $D_n(q)/n!_q$, but the values of $D_n/n!$ are interesting:

n	$\frac{D_n}{n!}$	which is
4	$\frac{3}{8}$	.375
5	$\frac{11}{30}$	.366666666666666...
6	$\frac{53}{144}$	.368055555555555...
7	$\frac{103}{280}$	.367857142857142...

So the answer to the probability question is that the size of the class is almost irrelevant. As long as there are at least five students, the probability that none gets their own paper back is converging rapidly to a number slightly below .368. It is a little bigger for even class sizes than for odd.

If we recall the q-exponential function

$$E_q(x) = \sum_{k=0}^{\infty} \frac{q^{\binom{k}{2}} x^k}{k!_q}$$

from section 3.8, we see that $\frac{D_n(q)}{n!_q}$ is a partial sum of the series for $E_q(-1) = (1-q;q)_\infty$. It follows that

$$\frac{D_n}{n!} \to \frac{1}{e} \approx .367879441171442321595523 8\ldots.$$

Exercises

1. Show that
$$\sum_{n=0}^{\infty} D_n(q) \frac{x^n}{n!_q} = \frac{E_q(-x)}{1-x}.$$

2. For the partition $\lambda = 12 + 9 + 8 + 8 + 8 + 6 + 5 + 3$ and the permutation $\pi = 73142658$, find:

    (i) the corresponding standard labeled partition P;

    (ii) the corresponding fixed partition F, and the standard labeled partition p containing the broken partition B and the reduced derangement δ;

    (iii) the corresponding labeled partition ℓ containing δ and the deranged partition D. Also check (9.5.4) for these objects.

3. In the previous problem you should find $F = 10 + 7 + 3$, $\delta = 53124$, and $D = 14 + 11 + 10 + 9 + 5$. Reconstruct the original partition λ and permutation π from these.

4. Make up your own examples like the previous two problems and solve them.

5. Explain why the highest degree term of $D_n(q)$ is the same as that of $n!_q$ if n is even, but not if n is odd.

6. The derangement problem is often solved as an application of the inclusion-exclusion principle. We outline the argument in this problem and the next.

    (i) Let $|S|$ denote the size of the set S. Explain why
    $$|A \cup B| = |A| + |B| - |A \cap B|.$$

    (ii) Explain why
    $$|A \cup B \cup C| = |A| + |B| + |C| - |A \cap B| - |A \cap C| - |B \cap C| + |A \cap B \cap C|.$$

    (iii) The generalization of (i) and (ii) to n sets is

(9.5.6) $\quad |A_1 \cup \cdots \cup A_n| = |A_1| + \cdots + |A_n| - \text{all 2-way intersections}$
$\qquad + \text{all 3-way intersections} - + \cdots + (-1)^{n-1}|A_1 \cap \cdots \cap A_n|.$

To prove this, assume x is in all of these sets. (Otherwise we could drop out any terms on the right side that x is not in, which gives the same expression with a smaller n.) The left side of (9.5.6) counts x once, so we have to show that the right side does too. Explain why the right side counts x

$$\sum_{k=1}^{n} \binom{n}{k}(-1)^{k-1}$$

times, and show that this sum equals 1 if $n \geq 1$.

7. To apply (9.5.6) to the derangement problem, let A_j be the set of all permutations of n papers in which the jth student *does* get her own paper back.

   (i) Explain why $|A_j| = (n-1)!$, and more generally why every k-way intersection of these sets has size $(n-k)!$.

   (ii) Explain why (i) and (9.5.6) imply that the number of permutations of the papers in which at least one student gets her own paper back is

$$n! \sum_{k=1}^{n} \frac{(-1)^{k-1}}{k!},$$

and why this implies

$$D_n = n! \sum_{k=0}^{n} \frac{(-1)^k}{k!}.$$

8. Show that

(9.5.7) $\qquad D_{n+1}(q) = [n]_q \left(D_n(q) + q^n D_{n-1}(q) \right) \quad \text{for } n \geq 1.$

This factor of $[n]_q$ explains the simplification in the expressions for $D_n(q)/n!_q$ given above. **Note:** I do not know how to derive (9.5.7) directly from (9.5.1). If one could do that, it would lead to an alternative proof of Gessel's theorem that I will outline in the next several problems.

9. Use (9.5.7) to show that

(9.5.8) $\quad D_n(q) - [n]_q D_{n-1}(q) = -q^{n-1}\left(D_{n-1}(q) - [n-1]_q D_{n-2}(q) \right) \quad \text{for } n \geq 2.$

10. Use (9.5.8) to show that

(9.5.9) $\qquad D_n(q) = [n]_q D_{n-1}(q) + (-1)^n q^{\binom{n}{2}} \quad \text{for } n \geq 1.$

11. Use (9.5.9) to prove Gessel's theorem.

12. It is not difficult to prove the $q=1$ case of (9.5.7) combinatorially. Consider all the derangements of $\{1, 2, \ldots, n+1\}$, and let k be the number in the last position (*i.e.*, the paper that the last student gets), where we know k must be one of $\{1, 2, \ldots, n\}$ since we have a derangement. There are two cases.

    (i) In case 1, $n+1$ is not in the kth position. Then erase $n+1$ and move k where $n+1$ was. Explain why this gives a derangement of $\{1, 2, \ldots, n\}$.

    (ii) In case 2, $n+1$ is in the kth position, so the procedure of case 1 does not give a derangement. In this case, erase both $n+1$ and k, move everything after the kth position forward one unit, and relabel every number $>k$ down one unit. Explain why this gives a derangement of $\{1, 2, \ldots, n-1\}$. (The argument is similar to the proof that the reduced derangement is a derangement.)

    (iii) Explain why (i) and (ii) imply the $q=1$ case of (9.5.7). Unfortunately this argument does not track the major index well.

9.6. q-Eulerian numbers and polynomials

In sections 2.4 and 3.8 we looked at the q-**derivative** of a function $f(x)$, which we defined by

(9.6.1) $$\mathbf{D}_q f(x) = \frac{f(x) - f(qx)}{x(1-q)},$$

where this means the ordinary derivative if $q = 1$. In this section we only need its simplest property:

(9.6.2) $$\mathbf{D}_q x^n = \frac{x^n - x^n q^n}{x(1-q)} = \frac{x^n(1-q^n)}{x(1-q)} = [n]_q x^{n-1}.$$

We'll be doing the q-analogue of what we were doing in problems 5–11 in section 9.1. If we calculate the q-derivative of both sides of the equation

(9.6.3) $$\sum_{n=0}^{\infty} x^n = \frac{1}{1-x},$$

assuming that $|x| < 1$, we get

$$\sum_{n=0}^{\infty} [n]_q x^{n-1} = \frac{\frac{1}{1-x} - \frac{1}{1-xq}}{x(1-q)}$$

$$= \frac{1 - xq - (1-x)}{x(1-q)(1-x)(1-xq)}$$

$$= \frac{x - xq}{x(1-q)(1-x)(1-xq)}$$

$$= \frac{1}{(1-x)(1-xq)}.$$

Multiplying both sides by x this becomes

(9.6.4) $$\sum_{n=0}^{\infty} [n]_q x^n = \frac{x}{(1-x)(1-xq)}.$$

Next, take the q-derivative of both sides of (9.6.4):

$$\sum_{n=0}^{\infty} [n]_q^2 x^{n-1} = \frac{\frac{x}{(1-x)(1-xq)} - \frac{xq}{(1-xq)(1-xq^2)}}{x(1-q)}$$

$$= \frac{1 - xq^2 - q(1-x)}{(1-q)(1-x)(1-xq)(1-xq^2)}$$

$$= \frac{1 - q + xq(1-q)}{(1-q)(1-x)(1-xq)(1-xq^2)}$$

$$= \frac{1 + xq}{(1-x)(1-xq)(1-xq^2)}.$$

Multiplying both sides by x this becomes

(9.6.5) $$\sum_{n=0}^{\infty} [n]_q^2 x^n = \frac{x(1+xq)}{(1-x)(1-xq)(1-xq^2)}.$$

We play this game one more time. Take the q-derivative of both sides of (9.6.5):

$$\sum_{n=0}^{\infty}[n]_q^3 x^{n-1} = \frac{\frac{x(1+xq)}{(1-x)(1-xq)(1-xq^2)} - \frac{xq(1+xq^2)}{(1-xq)(1-xq^2)(1-xq^3)}}{x(1-q)}$$

$$= \frac{(1+xq)(1-xq^3) - q(1+xq^2)(1-x)}{(1-q)(1-x)(1-xq)(1-xq^2)(1-xq^3)}$$

$$= \frac{1 - q + 2xq(1-q^2) + x^2 q^3(1-q)}{(1-q)(1-x)(1-xq)(1-xq^2)(1-xq^3)}$$

$$= \frac{1 + 2[2]_q xq + x^2 q^3}{(1-x)(1-xq)(1-xq^2)(1-xq^3)},$$

and multiplying through by x gives

(9.6.6) $$\sum_{n=0}^{\infty}[n]_q^3 x^n = \frac{x(1 + 2[2]_q xq + x^2 q^3)}{(1-x)(1-xq)(1-xq^2)(1-xq^3)}.$$

Let's look again at the permutations of $\{1, 2, 3\}$ and count falls and major index:

permutation	falls	maj	permutation	falls	maj
123	0	0	231	1	2
132	1	2	312	1	1
213	1	1	321	2	3

If we write down $x^{\text{falls}} q^{\text{maj}}$ for each of these and add the results, we get

$$x^0 q^0 + x^1 q^2 + x^1 q^1 + x^1 q^2 + x^1 q^1 + x^2 q^3 = 1 + 2xq + 2xq^2 + x^2 q^3$$
$$= 1 + 2[2]_q xq + x^2 q^3,$$

the same polynomial that occurs in the numerator of (9.6.6). This is the result we would like to prove in general. For nonnegative integer values of k, define functions $E_k(x, q)$ by

(9.6.7) $$\frac{E_k(x, q)}{(x; q)_{k+1}} = \sum_{n=0}^{\infty} [n]_q^k x^n.$$

Then (9.6.3)–(9.6.6) say that $E_0(x, q) = 1$, $E_1(x, q) = x$, $E_2(x, q) = x(1 + xq)$, and $E_3(x, q) = x(1 + 2[2]_q xq + x^2 q^3)$. $E_k(x, q)$ is called a q-Eulerian polynomial. In general it will have degree k in x and leading coefficient $q^{\binom{k}{2}}$, and it will have x as a factor if $k \geq 1$. The latter follows from the fact that $[0]_q^k = 0$ if $k > 0$, so that the $n = 0$ term in the sum in (9.6.7) is zero unless $k = 0$.

We will prove the former by obtaining a recurrence relation for the functions $E_k(x, q)$, and this comes via the same type of calculation that we were doing above, namely taking the q-derivative of both sides of (9.6.7) and then multiplying through by x. Applying $\mathbf{D}_q$ to (9.6.7) gives

$$\sum_{n=0}^{\infty} [n]_q^{k+1} x^{n-1} = \frac{\frac{E_k(x,q)}{(x;q)_{k+1}} - \frac{E_k(xq,q)}{(xq;q)_{k+1}}}{x(1-q)}$$

$$= \frac{\left(1 - xq^{k+1}\right) E_k(x, q) - (1-x) E_k(xq, q)}{x(1-q)(x; q)_{k+2}},$$

and multiplying through by x we get
$$\sum_{n=0}^{\infty} [n]_q^{k+1} x^n = \frac{\left(1 - xq^{k+1}\right) E_k(x,q) - (1-x) E_k(xq,q)}{(1-q)(x;q)_{k+2}}.$$
But according to (9.6.7)
$$\sum_{n=0}^{\infty} [n]_q^{k+1} x^n = \frac{E_{k+1}(x,q)}{(x;q)_{k+2}},$$
so we must have
$$\frac{E_{k+1}(x,q)}{(x;q)_{k+2}} = \frac{\left(1 - xq^{k+1}\right) E_k(x,q) - (1-x) E_k(xq,q)}{(1-q)(x;q)_{k+2}},$$
which implies that

(9.6.8) $$E_{k+1}(x,q) = \frac{\left(1 - xq^{k+1}\right) E_k(x,q) - (1-x) E_k(xq,q)}{1-q}.$$

This allows us to prove by induction that
$$E_k(x,q) = x^k q^{\binom{k}{2}} + \text{terms of lower degree in } x.$$
We have seen this already when $k \leq 3$. Assuming it is true for k, (9.6.8) implies that
$$E_{k+1}(x,q) = \frac{1}{1-q} \left\{ -xq^{k+1} \cdot x^k q^{\binom{k}{2}} + x \cdot (xq)^k q^{\binom{k}{2}} \right\} + \text{lower terms}$$
$$= \frac{1}{1-q} \left\{ x^{k+1} q^k q^{\binom{k}{2}} (1-q) \right\} + \text{terms of lower degree in } x$$
$$= x^{k+1} q^{\binom{k+1}{2}} + \text{terms of lower degree in } x.$$
Let's use (9.6.8) to work out $E_4(x,q)$. Taking $k=3$ there, we have
$$E_4(x,q) = \frac{\left(1 - xq^4\right) E_3(x,q) - (1-x) E_3(xq,q)}{1-q}$$
$$= \frac{\left(1 - xq^4\right) x \left(1 + 2[2]_q xq + x^2 q^3\right) - (1-x) xq \left(1 + 2[2]_q xq^2 + x^2 q^5\right)}{1-q}$$
$$= \frac{x}{1-q} \left\{ \begin{array}{l} 1 - q + xq(1-q^3) + 2[2]_q xq(1-q^2) \\ + x^2 q^3(1-q^3) + 2[2]_q x^2 q^3 (1-q^2) + x^3 q^6 (1-q) \end{array} \right\}$$
$$= x \left\{ 1 + [3]_q xq + 2[2]_q^2 xq + [3]_q x^2 q^3 + 2[2]_q^2 x^2 q^3 + x^3 q^6 \right\}$$
$$= x \left\{ 1 + xq \left([3]_q + 2[2]_q^2\right) + x^2 q^3 \left([3]_q + 2[2]_q^2\right) + x^3 q^6 \right\}.$$

Let's look at the permutations of $\{1, 2, 3, 4\}$ with one fall. There are 11 such permutations, and we look at the major index of each:

permutation	q^{maj}	permutation	q^{maj}
1243	q^3	2134	q^1
1324	q^2	2314	q^2
1342	q^3	2341	q^3
1423	q^2	2413	q^2
3124	q^1	3412	q^2
4123	q^1		

When we add all this up we get
$$3q + 5q^2 + 3q^3, \quad \text{which equals} \quad q\left([3]_q + 2[2]_q^2\right),$$
which was the coefficient of x^2 in $E_4(x, q)$.

If we look at the permutations of $\{1, 2, 3, 4\}$ with two falls, and compute the major index of each, we get the following table:

permutation	q^{maj}	permutation	q^{maj}
4123	q^3	3421	q^5
4231	q^4	3241	q^4
4213	q^3	3214	q^3
4132	q^4	3142	q^4
2431	q^5	2143	q^4
1432	q^5		

Adding all this up we get
$$3q^3 + 5q^4 + 3q^5, \quad \text{which equals} \quad q^3\left([3]_q + 2[2]_q^2\right),$$
which was the coefficient of x^3 in $E_4(x, q)$. There is also a certain symmetry in the results of these two calculations, but we hold off on discussing it until we've stated the theorem we're after, which was found by Carlitz. For $k \geq 1$, define the coefficients of $E_k(x, q)$ as follows:

$$(9.6.9) \qquad E_k(x, q) = x \sum_{j=0}^{k-1} \epsilon_{k,j}(q)\, x^j.$$

In addition, define $\epsilon_{k,j}(q) = 0$ whenever k and j are not integers such that $0 \leq j \leq k-1$, with the single exception that $\epsilon_{0,0}(q)$ is defined to be 1. The $\epsilon_{k,j}(q)$'s are called q-Eulerian numbers, although they are actually polynomials in q. Our main goal in this section is to prove

THEOREM 74 (Carlitz's theorem). *If $k \geq 1$ and $F(k, j)$ is the set of all permutations of $\{1, 2, \ldots, k\}$ with exactly j falls, then*
$$\epsilon_{k,j}(q) = \sum_{\pi \in F(k,j)} q^{\text{maj}\, \pi}.$$

We verified this above for $\epsilon_{4,1}(q) = 3q + 5q^2 + 3q^3$ and $\epsilon_{4,2}(q) = 3q^3 + 5q^4 + 3q^5$. Note that $\epsilon_{4,1}(q)$ and $\epsilon_{4,2}(q)$ are practically the same polynomial, except that $\epsilon_{4,2}(q)$ has a larger power of q as a common factor. We will return to this point after we prove Carlitz's theorem. There is another symmetry property of the $\epsilon_{k,j}$'s: like the q-factorials and the q-binomial coefficients, they are reciprocal polynomials—recall that this means that the coefficients of each one come in the same order (3-5-3 in the above examples) when read backwards as when read forwards. Before we try to prove this, let's do a calculation which will allow us to generate more examples of it. Substituting (9.6.9) into (9.6.8) we get

$$x \sum_{j=0}^{k} \epsilon_{k+1,j}(q)\, x^j$$

$$= \frac{1}{1-q} \left\{ (1 - xq^{k+1})\, x \sum_{j=0}^{k-1} \epsilon_{k,j}(q)\, x^j - (1-x)xq \sum_{j=0}^{k-1} \epsilon_{k,j}(q)\, x^j\, q^j \right\}.$$

9.6. q-EULERIAN NUMBERS AND POLYNOMIALS

We cancel the factors of x and distribute the sums:

$$\sum_{j=0}^{k} \epsilon_{k+1,j}(q)\, x^j = \frac{1}{1-q} \left\{ \begin{array}{l} \displaystyle\sum_{j=0}^{k-1} \epsilon_{k,j}(q)\, x^j - q^{k+1} \sum_{j=0}^{k-1} \epsilon_{k,j}(q)\, x^{j+1} \\ \displaystyle -\sum_{j=0}^{k-1} \epsilon_{k,j}(q)\, x^j\, q^{j+1} + \sum_{j=0}^{k-1} \epsilon_{k,j}(q)\, x^{j+1}\, q^{j+1} \end{array} \right\}.$$

Rearrange this to

$$\sum_{j=0}^{k} \epsilon_{k+1,j}(q)\, x^j = \sum_{j=0}^{k-1} \epsilon_{k,j}(q)\, x^j\, \frac{1 - q^{j+1}}{1 - q} + \sum_{j=0}^{k-1} \epsilon_{k,j}(q)\, x^{j+1}\, \frac{q^{j+1} - q^{k+1}}{1 - q}.$$

Now replace $j+1$ by j in the last sum, leaving the others alone:

$$\sum_{j=0}^{k} \epsilon_{k+1,j}(q)\, x^j = \sum_{j=0}^{k-1} \epsilon_{k,j}(q)\, x^j\, \frac{1 - q^{j+1}}{1 - q} + \sum_{j=1}^{k} \epsilon_{k,j-1}(q)\, x^j\, q^j\, \frac{1 - q^{k-j+1}}{1 - q}.$$

Both sums on the right can go over the range $0 \leq j \leq k$, since this just adds a zero term to each of them. Thus we have

$$\sum_{j=0}^{k} \epsilon_{k+1,j}(q)\, x^j = \sum_{j=0}^{k} \epsilon_{k,j}(q)\, x^j\, [j+1]_q + \sum_{j=0}^{k} \epsilon_{k,j-1}(q)\, x^j\, q^j\, [k-j+1]_q,$$

and equating coefficients of x^j gives us the recurrence

(9.6.10) $\qquad \epsilon_{k+1,j}(q) = [j+1]_q\, \epsilon_{k,j}(q) + [k-j+1]_q\, q^j\, \epsilon_{k,j-1}(q),$

which holds for all nonnegative integers k and j.

If we take $k = 4$ and $j = 2$ in (9.6.10), it says

$$\epsilon_{5,2}(q) = [3]_q\, \epsilon_{4,2}(q) + [3]_q\, q^2\, \epsilon_{4,1}(q),$$

and using the expressions for $\epsilon_{4,2}(q)$ and $\epsilon_{4,1}(q)$ that we worked out above, this becomes

$$\epsilon_{5,2}(q) = [3]_q \left\{ 3q^5 + 5q^4 + 3q^3 + q^2(3q^3 + 5q^2 + 3q) \right\}$$
$$= 2[3]_q q^3 \left(3q^2 + 5q + 3 \right)$$
$$= 6q^7 + 16q^6 + 22q^5 + 16q^4 + 6q^3$$

and again the coefficients are the same forwards as backwards. To prove that this happens in general, we start by replacing q by $\frac{1}{q}$ in (9.6.7).

$$[n]_q = \frac{1 - q^n}{1 - q} \quad \text{becomes} \quad \frac{1 - \frac{1}{q^n}}{1 - \frac{1}{q}} \frac{q^n}{q^n} = \frac{q^n - 1}{q - 1} q^{1-n} = q^{1-n}[n]_q,$$

and

$$\left(x; \frac{1}{q} \right)_{k+1} = (1-x)\left(1 - \frac{x}{q}\right)\left(1 - \frac{x}{q^2}\right) \cdots \left(1 - \frac{x}{q^k}\right) = \left(\frac{x}{q^k}; q \right)_{k+1}$$

so (9.6.7) becomes

(9.6.11) $\qquad \displaystyle\frac{E_k\left(x, \frac{1}{q}\right)}{\left(\frac{x}{q^k}; q\right)_{k+1}} = \sum_{n=0}^{\infty} [n]_q^k\, q^{k-kn}\, x^n = q^k \sum_{n=0}^{\infty} [n]_q^k \left(\frac{x}{q^k}\right)^n.$

On the other hand, if we replace x by $\frac{x}{q^k}$ in (9.6.7) then we get

$$\frac{E_k\left(\frac{x}{q^k}, q\right)}{\left(\frac{x}{q^k}; q\right)_{k+1}} = \sum_{n=0}^{\infty} [n]_q^k \left(\frac{x}{q^k}\right)^n,$$

and comparing this with (9.6.11) there results

$$E_k\left(x, \frac{1}{q}\right) = q^k E_k\left(\frac{x}{q^k}, q\right).$$

Now plug (9.6.9) into this to get

$$x \sum_{j=0}^{k-1} \epsilon_{k,j}(\tfrac{1}{q}) x^j = q^k \left(\frac{x}{q^k}\right) \sum_{j=0}^{k-1} \epsilon_{k,j}(q) \left(\frac{x}{q^k}\right)^j.$$

The factors outside the summation signs cancel, and equating coefficients of x^j and rearranging we get

(9.6.12) $\qquad q^{kj} \epsilon_{k,j}(\tfrac{1}{q}) = \epsilon_{k,j}(q).$

With (9.6.12) in hand we can see that $\epsilon_{k,j}(q)$ is a reciprocal polynomial. Changing q to $\frac{1}{q}$ in a polynomial in q makes all the exponents negative, and also the highest power becomes the lowest power and vice versa. To make the result back into a polynomial in q again we need to multiply by a power of q large enough to make all the exponents nonnegative. Thus what we have to argue is that kj is the right power of q to multiply by. If a permutation of $\{1, 2, \ldots, k\}$ has j falls, then the smallest major index it can have is

$$1 + 2 + 3 + \cdots + j = \binom{j+1}{2},$$

and the largest major index it can have is

$$(k-1) + (k-2) + (k-3) + \cdots + (k-j) = kj - \binom{j+1}{2}.$$

Therefore $\epsilon_{k,j}(q)$ has the form

$$A q^{kj - \binom{j+1}{2}} + \cdots + \Omega q^{\binom{j+1}{2}}$$

for some coefficients A and Ω. Changing q to $\frac{1}{q}$ and then multiplying by q^{kj} this becomes

$$q^{kj} \left(A q^{\binom{j+1}{2} - kj} + \cdots + \Omega q^{-\binom{j+1}{2}}\right) = A q^{\binom{j+1}{2}} + \cdots + \Omega q^{kj - \binom{j+1}{2}}.$$

If this equals $\epsilon_{k,j}(q)$, then we must have $A = \Omega$, and similarly for all the other coefficients.

Now we use (9.6.10) to prove Carlitz's theorem by induction on k. We verified it above for $\epsilon_{4,1}(q)$ and $\epsilon_{4,2}(q)$, and it is easy to see that it holds generally for $k = 4$, since $\epsilon_{4,0}(q) = 1$ and $\epsilon_{4,3}(q) = q^6$: there is only one permutation of $\{1, 2, 3, 4\}$ with 0 falls, namely 1234, and it has 0 major index; there is also only one permutation of $\{1, 2, 3, 4\}$ with 3 falls, namely 4321, and it has major index 6. Thus Carlitz's theorem holds if $k = 4$. Assuming it holds for k, (9.6.10) will allow us to conclude that it also holds for $k + 1$.

The argument is much the same as the one in section 9.1. We can get a permutation of $\{1, 2, \ldots, k, k+1\}$ with j falls from a permutation of $\{1, 2, \ldots, k\}$

with j falls by putting $k+1$ at the end, or by inserting $k+1$ in the middle of a fall. Putting $k+1$ at the end adds nothing to the major index. Inserting it in the last fall just pushes the last fall out one place, so it adds 1 to the major index. Inserting it in the next-to-last fall pushes the last two falls out one place, hence adds 2 to the major index, and so on. Inserting $k+1$ in the first fall pushes all j falls out one place, so adds j to the major index.

Let $F(k+1,j)$ denote the set of permutations of $\{1,2,\ldots,k+1\}$ with exactly j falls, as in Carlitz's theorem. Divide these permutations into two classes: class j consists of those that still have j falls when $k+1$ is erased, and class $j-1$ consists of those that lose one fall when we erase $k+1$. Then

$$(9.6.13) \qquad \sum_{\pi \in F(k+1,j)} q^{\mathrm{maj}\,\pi} = \sum_{\pi \in \text{class } j} q^{\mathrm{maj}\,\pi} + \sum_{\pi \in \text{class } j-1} q^{\mathrm{maj}\,\pi},$$

and the argument above shows that

$$(9.6.14) \qquad \sum_{\pi \in \text{class } j} q^{\mathrm{maj}\,\pi} = \left(1 + q + q^2 + \cdots + q^j\right) \sum_{\pi \in F(k,j)} q^{\mathrm{maj}\,\pi} = [j+1]_q \sum_{\pi \in F(k,j)} q^{\mathrm{maj}\,\pi}.$$

We can also get a permutation of $\{1,2,\ldots,k,k+1\}$ with j falls from a permutation of $\{1,2,\ldots,k\}$ with $j-1$ falls by putting $k+1$ at the beginning, or by inserting $k+1$ in the middle of a rise. Putting $k+1$ at the beginning adds 1 fall with major index 1, and pushes the other $j-1$ falls out one place, so it adds j in total to the major index.

Suppose the first rise is in the ith position. Then it is preceded by $i-1$ falls and succeeded by $j-i$ falls. If we insert $k+1$ in this rise (i.e., in the $(i+1)^{\text{th}}$ position), then we create a fall there, and we move the last $j-i$ falls out 1 place; so we add $(i+1)+(j-i) = j+1$ to the major index. More generally, suppose the tth rise is in the mth position. Then it is preceded by $t-1$ rises and $m-t$ falls, and succeeded by $j-1-(m-t)$ falls. If we insert $k+1$ in this rise, then we create a new fall in the $(m+1)^{\text{th}}$ position, and we move the last $j-1-m+t$ falls out 1 place; so we add $(m+1)+(j-1-m+t) = j+t$ to the major index. Since there are $k-j$ falls, the largest value of t is $k-j$, and therefore

$$\sum_{\pi \in \text{class } j-1} q^{\mathrm{maj}\,\pi} = \left(q^j + q^{j+1} + q^{j+2} + \cdots + q^k\right) \sum_{\pi \in F(k,j-1)} q^{\mathrm{maj}\,\pi}$$

$$= q^j \left(1 + q + q^2 + \cdots + q^{k-j}\right) \sum_{\pi \in F(k,j-1)} q^{\mathrm{maj}\,\pi}$$

$$= q^j [k-j+1]_q \sum_{\pi \in F(k,j-1)} q^{\mathrm{maj}\,\pi}.$$

Using this and (9.6.14) in (9.6.13), we have

$$\sum_{\pi \in F(k+1,j)} q^{\mathrm{maj}\,\pi} = \sum_{\pi \in \text{class } j} q^{\mathrm{maj}\,\pi} + \sum_{\pi \in \text{class } j-1} q^{\mathrm{maj}\,\pi}$$

$$= [j+1]_q \sum_{\pi \in F(k,j)} q^{\mathrm{maj}\,\pi} + q^j [k-j+1]_q \sum_{\pi \in F(k,j-1)} q^{\mathrm{maj}\,\pi}.$$

Therefore

$$\sum_{\pi \in F(k+1,j)} q^{\mathrm{maj}\,\pi} = [j+1]_q\, \epsilon_{k,j}(q) + [k-j+1]_q\, q^j\, \epsilon_{k,j-1}(q) = \epsilon_{k+1,j}(q),$$

where the first equality is by the induction assumption and the second equality is (9.6.10). This proves Carlitz's theorem.

We conclude this section by using Carlitz's theorem to pin down the relationship between $\epsilon_{k,j}(q)$ and $\epsilon_{k,k-j-1}(q)$. We have done several examples in which these came out to be the same polynomial except for some power of q. Suppose we have a permutation π of $\{1, 2, \ldots, k\}$ with j falls, and suppose the positions of the falls are $f_1, f_2, \ldots, f_j$. Then the major index of π is $f_1 + f_2 + \cdots + f_j$, which we abbreviate as m. If we read π backwards (i.e., if π was 18356742 we now read it as 24765381), then it has j rises, which are at $k - f_j, \ldots, k - f_2, k - f_1$, so the sum of the positions of the rises is $kj - m$. But if we add up the positions of the rises and the falls in a permutation of $\{1, 2, \ldots, k\}$ we get

$$1 + 2 + 3 + \cdots + (k-1) = \binom{k}{2}.$$

Hence π when read backwards has $k - j - 1$ falls and its major index is $\binom{k}{2} - kj + m$. This proves that

$$\epsilon_{k,k-j-1}(q) = q^{\binom{k}{2}-kj}\epsilon_{k,j}(q).$$

Exercises

1. The material of this section allows us to find q-analogues of facts like

$$\sum_{n=1}^{\infty} \frac{n^2}{2^n} = 6 \quad \text{and} \quad \sum_{n=1}^{\infty} \frac{n^2}{3^n} = \frac{3}{2}.$$

Show that

$$\sum_{n=1}^{\infty} \frac{[n]^2}{[2]_q^n} = \frac{(1+q)(1+2q)}{q(1+q-q^2)} \quad \text{and} \quad \sum_{n=1}^{\infty} \frac{[n]^2}{[3]_q^n} = \frac{[3]_q}{q(1+q^2)}.$$

9.7. q-trigonometric functions

There are two obvious ways to define q-analogues of the sine and cosine. One is by power series analogous to those for the ordinary sine and cosine:

(9.7.1) $$\sin_q x = \sum_{n=0}^{\infty} (-1)^n \frac{x^{2n+1}}{(2n+1)!_q},$$

(9.7.2) $$\cos_q x = \sum_{n=0}^{\infty} (-1)^n \frac{x^{2n}}{(2n)!_q}.$$

For the other, recall the very important relation between the ordinary exponential function e^x and the sine and cosine,

$$e^{ix} = \cos x + i \sin x,$$

9.7. q-TRIGONOMETRIC FUNCTIONS

where $i^2 = -1$ as usual. If we have read section 3.8 on the q-exponential functions $e_q(x)$ and $E_q(x)$, it is logical to define q-sines and cosines by

(9.7.3) $$e_q(ix) = \cos_q x + i \sin_q x,$$
(9.7.4) $$E_q(ix) = \text{Cos}_q x + i \, \text{Sin}_q x.$$

For a real x, these are perfectly good definitions, because then $\cos_q x$ is the real part of $e_q(ix)$ and $\sin_q x$ is the imaginary part, and similarly for the capital letter functions. Fortunately, it is easy to see that (9.7.1) and (9.7.2) are defining the same functions as (9.7.3). Since the powers of i are

$$i = i = i^5 = i^9 = i^{13} = \cdots = i^{4k+1},$$
$$i^2 = -1 = i^6 = i^{10} = i^{14} = \cdots = i^{4k+2},$$
$$i^3 = -i = i^7 = i^{11} = i^{15} = \cdots = i^{4k+3},$$
$$i^4 = 1 = i^0 = i^8 = i^{12} = \cdots = i^{4k},$$

we have

$$e_q(ix) = \sum_{n=0}^{\infty} \frac{i^n x^n}{n!_q}$$

$$= \sum_{k=0}^{\infty} \frac{i^{4k} x^{4k}}{(4k)!_q} + \sum_{k=0}^{\infty} \frac{i^{4k+1} x^{4k+1}}{(4k+1)!_q} + \sum_{k=0}^{\infty} \frac{i^{4k+2} x^{4k+2}}{(4k+2)!_q} + \sum_{k=0}^{\infty} \frac{i^{4k+3} x^{4k+3}}{(4k+3)!_q}$$

$$= \sum_{k=0}^{\infty} \frac{x^{4k}}{(4k)!_q} + i \sum_{k=0}^{\infty} \frac{x^{4k+1}}{(4k+1)!_q} - \sum_{k=0}^{\infty} \frac{x^{4k+2}}{(4k+2)!_q} - i \sum_{k=0}^{\infty} \frac{x^{4k+3}}{(4k+3)!_q}$$

$$= \sum_{k=0}^{\infty} (-1)^k \frac{x^{2k}}{(2k)!_q} + i \sum_{k=0}^{\infty} (-1)^k \frac{x^{2k+1}}{(2k+1)!_q}$$

$$= \cos_q x + i \sin_q x,$$

which shows the equivalence. By a similar calculation, the q-trig functions defined by (9.7.4) have the series expansions

(9.7.5) $$\text{Sin}_q x = \sum_{n=0}^{\infty} (-1)^n q^{\binom{2n+1}{2}} \frac{x^{2n+1}}{(2n+1)!_q},$$

(9.7.6) $$\text{Cos}_q x = \sum_{n=0}^{\infty} (-1)^n q^{\binom{2n}{2}} \frac{x^{2n}}{(2n)!_q}.$$

We can also see right away that it is worth having both sets of functions. We know from section 3.8 that

$$e_q(ix) E_q(-ix) = 1,$$

so it follows that

$$1 = (\cos_q x + i \sin_q x)(\text{Cos}_q x - i \, \text{Sin}_q x)$$
$$= \cos_q x \, \text{Cos}_q x + \sin_q x \, \text{Sin}_q x + i(\sin_q x \, \text{Cos}_q x - \cos_q x \, \text{Sin}_q x).$$

Assuming x is real, we can equate real and imaginary parts of this to get

(9.7.7) $$\cos_q x \, \text{Cos}_q x + \sin_q x \, \text{Sin}_q x = 1$$

and

$$\sin_q x \, \text{Cos}_q x = \cos_q x \, \text{Sin}_q x,$$

which it is more interesting to write as

(9.7.8) $$\frac{\sin_q x}{\cos_q x} = \frac{\text{Sin}_q x}{\text{Cos}_q x}.$$

(9.7.7) is the best q-analogue of $\cos^2 x + \sin^2 x = 1$ that we are going to be able to get (although not the only one), and we will come back to (9.7.8) presently.

It is easy to see from (9.7.1) and (9.7.2) that

(9.7.9) $$\mathbf{D}_q \sin_q x = \cos_q x,$$
(9.7.10) $$\mathbf{D}_q \cos_q x = -\sin_q x.$$

The capital letter functions are only slightly harder. We have

$$\mathbf{D}_q \text{Sin}_q x = \sum_{n=0}^{\infty} (-1)^n q^{\binom{2n+1}{2}} \frac{x^{2n}}{(2n)!_q}$$
$$= \sum_{n=0}^{\infty} (-1)^n q^{\binom{2n}{2}+2n} \frac{x^{2n}}{(2n)!_q}$$
$$= \sum_{n=0}^{\infty} (-1)^n q^{\binom{2n}{2}} \frac{(qx)^{2n}}{(2n)!_q}$$

and it follows that

(9.7.11) $$\mathbf{D}_q \text{Sin}_q x = \text{Cos}_q qx.$$

Similarly

(9.7.12) $$\mathbf{D}_q \text{Cos}_q x = -\text{Sin}_q qx.$$

If we define q-analogues of the secant and tangent in the most obvious way, by

(9.7.13) $$\sec_q x = \frac{1}{\cos_q x},$$
(9.7.14) $$\tan_q x = \frac{\sin_q x}{\cos_q x}$$

and

(9.7.15) $$\text{Sec}_q x = \frac{1}{\text{Cos}_q x},$$
(9.7.16) $$\text{Tan}_q x = \frac{\text{Sin}_q x}{\text{Cos}_q x},$$

then we already have an interesting theorem from (9.7.8):

(9.7.17) $$\tan_q x = \text{Tan}_q x.$$

The remaining calculations in this section are mostly routine applications of the q-product or q-quotient or q-reciprocal rules from the problems in section 2.4, and many of them will be left as exercises. If we apply the q-reciprocal rule (2.4.13) to (9.7.13) and (9.7.15) we get

(9.7.18) $$\mathbf{D}_q \sec_q x = \sec_q qx \tan_q x,$$
(9.7.19) $$\mathbf{D}_q \text{Sec}_q x = \text{Sec}_q x \, \text{Tan}_q qx.$$

9.7. q-TRIGONOMETRIC FUNCTIONS

Because $\tan_q x = \text{Tan}_q x$ and we have two superficially different q-quotient rules, we can expect as many as four different-looking results for the q-derivative of the q-tangent. If we apply (2.4.15) to (9.7.14) we get

$$\mathbf{D}_q \tan_q x = \mathbf{D}_q \frac{\sin_q x}{\cos_q x} = \frac{\cos_q qx \cos_q x - \sin_q qx(-\sin_q x)}{\cos_q x \cos_q qx}$$

(9.7.20)
$$= 1 + \tan_q x \tan_q qx.$$

Similarly, if we apply (2.4.14) to (9.7.16) we get

(9.7.21) $$\mathbf{D}_q \text{Tan}_q x = 1 + \text{Tan}_q x \text{Tan}_q qx,$$

which is clearly the same as (9.7.20). If instead we use (2.4.14) with (9.7.14) we have

(9.7.22) $$\mathbf{D}_q \tan_q x = \frac{\cos_q^2 x + \sin_q^2 x}{\cos_q x \cos_q qx},$$

which is not as nice as (9.7.20) since we don't have a nice identity for $\cos_q^2 x + \sin_q^2 x$, a point we will return to presently. Actually, this calculation and (9.7.20) imply the curious fact

(9.7.23) $$\cos_q^2 x + \sin_q^2 x = \cos_q x \cos_q qx + \sin_q x \sin_q qx.$$

Similarly, if we use (2.4.15) with (9.7.16) we get

(9.7.24) $$\mathbf{D}_q \text{Tan}_q x = \frac{\text{Cos}_q^2 qx + \text{Sin}_q^2 qx}{\text{Cos}_q x \text{Cos}_q qx},$$

which together with (9.7.21) implies

(9.7.25) $$\text{Cos}_q^2 qx + \text{Sin}_q^2 qx = \text{Cos}_q x \text{Cos}_q qx + \text{Sin}_q x \text{Sin}_q qx.$$

Let's see what, if anything, we can say about $\cos_q^2 x + \sin_q^2 x$. We have

$$\cos_q^2 x + \sin_q^2 x = \sum_{j=0}^{\infty} (-1)^j \frac{x^{2j}}{(2j)!_q} \sum_{k=0}^{\infty} (-1)^k \frac{x^{2k}}{(2k)!_q}$$
$$+ \sum_{j=0}^{\infty} (-1)^j \frac{x^{2j+1}}{(2j+1)!_q} \sum_{k=1}^{\infty} (-1)^{k-1} \frac{x^{2k-1}}{(2k-1)!_q}$$
$$= \sum_{n=0}^{\infty} \sum_{j+k=n} (-1)^n \frac{x^{2n}}{(2n)!_q} \binom{2n}{2j}_q$$
$$+ \sum_{n=1}^{\infty} \sum_{j+k=n} (-1)^{n-1} \frac{x^{2n}}{(2n)!_q} \binom{2n}{2j+1}_q$$
$$= \sum_{n=0}^{\infty} (-1)^n \frac{x^{2n}}{(2n)!_q} \sum_{j=0}^{n} \left[\binom{2n}{2j}_q - \binom{2n}{2j+1}_q \right]$$
$$= \sum_{n=0}^{\infty} (-1)^n \frac{x^{2n}}{(2n)!_q} \sum_{j=0}^{2n} (-1)^j \binom{2n}{j}_q .$$

A similar calculation shows that

$$(9.7.26) \quad \cos_q x \cos_q qx + \sin_q x \sin_q qx = \sum_{n=0}^{\infty}(-1)^n \frac{x^{2n}}{(2n)!_q} \sum_{j=0}^{2n}(-q)^j \binom{2n}{j}_q.$$

These series do simplify a little, but not all the way to 1. Recall Gauss's identity (2.5.2) and its companion (2.5.3) from section 2.5:

$$\sum_{j=0}^{2n}(-1)^j \binom{2n}{j}_q = (q;q^2)_n = \sum_{j=0}^{2n}(-q)^j \binom{2n}{j}_q.$$

This means that

$$\cos_q^2 x + \sin_q^2 x = \cos_q x \cos_q qx + \sin_q x \sin_q qx = \sum_{n=0}^{\infty}(-1)^n (q;q^2)_n \frac{x^{2n}}{(2n)!_q}.$$

This does reduce to $\cos^2 x + \sin^2 x = 1$ when $q = 1$, since $(q;q^2)_n$ has the factor $1-q$ for all $n \geq 1$, but as a q-analogue of $\cos^2 x + \sin^2 x = 1$ it is a little disappointing.

Exercises

1. Use (3.8.7) and (9.7.4) to derive (9.7.5) and (9.7.6).
2. Prove (9.7.9) and (9.7.10).
3. Prove (9.7.12).
4. Prove (9.7.18) and (9.7.19).
5. Use the q-product rule (2.4.11) with $\sec_q x \cos_q x = 1$ and $\text{Sec}_q x \, \text{Cos}_q x = 1$ to give alternate proofs of (9.7.18) and (9.7.19).
6. Use the q-product rule (2.4.12) with $\sec_q x \cos_q x = 1$ and $\text{Sec}_q x \, \text{Cos}_q x = 1$ to give alternate proofs of (9.7.18) and (9.7.19).
7. Use the q-product rule (2.4.12) with $\tan_q x \cos_q x = \sin_q x$ to give an alternate proof of (9.7.20).
8. Verify (9.7.21), (9.7.22), and (9.7.24).
9. Verify (9.7.26).
10. Another possible definition of $\sin_q x$ and $\cos_q x$ is (9.7.9) and (9.7.10) together with $\sin_q 0 = 0$ and $\cos_q 0 = 1$. Use (2.4.9) to show that this implies (9.7.1) and (9.7.2).
11. Similarly to problem 10, we could have defined $\text{Sin}_q x$ and $\text{Cos}_q x$ by (9.7.11) and (9.7.12) together with $\text{Sin}_q 0 = 0$ and $\text{Cos}_q 0 = 1$. Use (2.4.9) and (2.4.5) to show that this implies (9.7.5) and (9.7.6).
12. Show that $1 + \tan_q^2 x = \sec_q x \, \text{Sec}_q x$. (Don't forget (9.7.17).)
13. Obvious definitions of the q-cotangent and q-cosecant are

$$\cot_q x = \frac{1}{\tan_q x} = \frac{\cos_q x}{\sin_q x} \quad \text{and} \quad \csc_q x = \frac{1}{\sin_q x},$$

and similarly for the capital letter functions. Note that $\cot_q x = \text{Cot}_q x$, since $\tan_q x = \text{Tan}_q x$. Use the q-reciprocal rule (2.4.13) to show that

$$\mathbf{D}_q \csc_q x = -\cot_q x \csc_q qx \quad \text{and} \quad \mathbf{D}_q \text{Csc}_q x = -\text{Cot}_q qx \, \text{Csc}_q x$$

and that

$$\mathbf{D}_q \cot_q x = -1 - \cot_q x \cot_q qx = -1 - \operatorname{Cot}_q x \operatorname{Cot}_q qx = \mathbf{D}_q \operatorname{Cot}_q x.$$

14. Show that $1 + \cot_q^2 x = \csc_q x \operatorname{Csc}_q x$.

9.8. Combinatorics of q-tangents and secants

In combinatorics the fundamental trigonometric functions are not the sine and cosine, but the tangent and secant. We begin this section with a brief sketch of this theory, so the reader will know what we are attempting a q-analogue of. Start by defining the Euler numbers E_n by

$$\sec x + \tan x = \sum_{n=0}^{\infty} E_n \frac{x^n}{n!},$$

where the series converges if $|x| < \frac{\pi}{2}$, because that's the distance from the origin to the closest singularity of the function. Since $\sec x$ is even and $\tan x$ is odd, this decouples into

$$\sec x = \sum_{n=0}^{\infty} E_{2n} \frac{x^{2n}}{(2n)!} \quad \text{and} \quad \tan x = \sum_{n=0}^{\infty} E_{2n+1} \frac{x^{2n+1}}{(2n+1)!}.$$

There are various relations between tangent and secant that one can use to find recurrences for E_n. For example, if $f(x) = \sec x + \tan x$, then $f'(x) = f(x) \sec x$ and $2f'(x) = 1 + (f(x))^2$ and $f''(x) = f(x) f'(x)$, and these respectively imply

$$E_{n+1} = \sum_{k=0}^{\lfloor n/2 \rfloor} \binom{n}{2k} E_{2k} E_{n-2k},$$

$$2 E_{n+1} = \sum_{k=0}^{n} \binom{n}{k} E_k E_{n-k} \quad \text{if } n \geq 1,$$

$$E_{n+2} = \sum_{k=0}^{n} \binom{n}{k} E_{k+1} E_{n-k},$$

where $E_0 = 1$, since $\sec 0 = 1$ and $\tan 0 = 0$. Any of these can be used to calculate the Euler numbers recursively. The first fourteen of them are:

$$E_0 = 1 = E_1 = E_2 \quad E_3 = 2 \quad E_4 = 5 \quad E_5 = 16$$
$$E_6 = 61 \quad E_7 = 272 \quad E_8 = 1385 \quad E_9 = 7936$$
$$E_{10} = 50521 \quad E_{11} = 353792$$
$$E_{12} = 2702765 \quad E_{13} = 22368256$$

Starting with E_2, the last digits repeat the pattern $1, 2, 5, 6$. The Euler numbers count what are called **up-down permutations**. Let's look at all the permutations of $\{1, 2, 3\}$:

permutation	description
123	up, up
132	up, down
213	down, up
231	up, down
312	down, up
321	down, down

We read the permutations from left to right, and write "up" whenever they increase, and "down" whenever they decrease. In general, an up-down permutation will be any permutation that alternately increases and decreases, starting with an increase, and a down-up permutation will be the same thing only starting with a decrease, where 1 and the empty permutation are considered both up-down and down-up. Thus the up-down permutations of $\{1, 2, 3\}$ are 132 and 231, and the down-up permutations are 213 and 312. Let's classify the permutations of $\{1, 2, 3, 4\}$:

up-down	down-up	up	down	uud	udd	duu	ddu
1324	2143	1234	4321	1243	1432	2134	3214
1423	3142			1342	2431	3124	4213
2314	3241			2341	3421	4123	4312
2413	4132						
3412	4231						

where uud stands for "up-up-down" and so forth. As expected, since $E_4 = 5$, we found five up-down permutations of length 4. We also found five down-up permutations of length 4, which could lead us to suspect that E_n also counts these. It is not too hard to see that the number of down-up permutations of a given length equals the number of up-down ones, and we will come back to this shortly.

Once one has proved these facts about the Euler numbers, one can define cosine as the reciprocal of secant, and sine as tangent divided by secant, and derive the power series for sine and cosine by the inclusion-exclusion principle, but this is as far as we will develop the theory in the $q = 1$ case.

Since $\cos_q x$ and $\text{Cos}_q x$ are even functions, and $\sin_q x$ and $\text{Sin}_q x$ are odd functions, $\sec_q x$ and $\text{Sec}_q x$ will be even functions, and $\tan_q x$ and $\text{Tan}_q x$ will be odd functions. Define coefficients $E_n(q)$ and $\varepsilon_n(q)$ by

$$(9.8.1) \qquad \sec_q x + \tan_q x = \sum_{n=0}^{\infty} E_n(q) \frac{x^n}{n!_q},$$

$$(9.8.2) \qquad \text{Sec}_q x + \text{Tan}_q x = \sum_{n=0}^{\infty} \varepsilon_n(q) \frac{x^n}{n!_q}.$$

Then, because of the even/oddness, these decouple into

(9.8.3) $$\sec_q x = \sum_{n=0}^{\infty} E_{2n}(q) \frac{x^{2n}}{(2n)!_q},$$

(9.8.4) $$\tan_q x = \sum_{n=0}^{\infty} E_{2n+1}(q) \frac{x^{2n+1}}{(2n+1)!_q},$$

(9.8.5) $$\text{Sec}_q x = \sum_{n=0}^{\infty} \varepsilon_{2n}(q) \frac{x^{2n}}{(2n)!_q},$$

(9.8.6) $$\text{Tan}_q x = \sum_{n=0}^{\infty} \varepsilon_{2n+1}(q) \frac{x^{2n+1}}{(2n+1)!_q}.$$

Now (9.7.18) and (9.7.20) imply that

(9.8.7) $$\mathbf{D}_q \sec_q x + \tan_q x = 1 + \tan_q x \left(\sec_q qx + \tan_q qx\right)$$

and (9.7.19) and (9.7.21) that

(9.8.8) $$\mathbf{D}_q \text{Sec}_q x + \text{Tan}_q x = 1 + \text{Tan}_q qx \left(\text{Sec}_q x + \text{Tan}_q x\right).$$

With the above definitions in terms of infinite series, (9.8.7) translates into

$$\sum_{n=0}^{\infty} E_{n+1}(q) \frac{x^n}{n!_q} = 1 + \left(\sum_{j=0}^{\infty} E_{2j+1}(q) \frac{x^{2j+1}}{(2j+1)!_q}\right)\left(\sum_{k=0}^{\infty} E_k(q) \frac{x^k q^k}{k!_q}\right)$$

$$= 1 + \sum_{n=1}^{\infty} \frac{x^n}{n!_q} \sum_{2j+1+k=n} \binom{n}{2j+1}_q E_{2j+1}(q) q^k E_k(q)$$

which implies that

(9.8.9) $$E_{n+1}(q) = \sum_j \binom{n}{2j+1}_q q^{n-2j-1} E_{2j+1}(q) E_{n-2j-1}(q) \quad \text{if } n \geq 1; \; E_1(q) = 1.$$

In a similar way, (9.8.8) implies

(9.8.10) $$\varepsilon_{n+1}(q) = \sum_j \binom{n}{2j+1}_q q^{2j+1} \varepsilon_{2j+1}(q) \varepsilon_{n-2j-1}(q) \quad \text{if } n \geq 1; \; \varepsilon_1(q) = 1.$$

We also have $\sec_q 0 = 1 = \text{Sec}_q 0$ and $\tan_q 0 = 0 = \text{Tan}_q 0$, since $\cos_q 0 = 1 = \text{Cos}_q 0$ and $\sin_q 0 = 0 = \text{Sin}_q 0$, and therefore $E_0(q) = 1 = \varepsilon_0(q)$. Then we can use (9.8.9) and (9.8.9) to calculate the first several values of $E_n(q)$ and $\varepsilon_n(q)$. We find $E_2(q) = 1$, $\varepsilon_2(q) = q$, $E_3(q) = q + q^2 = \varepsilon_3(q)$, $E_4(q) = q + 2q^2 + q^3 + q^4$, $\varepsilon_4(q) = q^2 + q^3 + 2q^4 + q^5$. As a sample calculation we work out $E_5(q)$:

$$E_5(q) = \sum_{j=0}^{1} \binom{4}{2j+1}_q q^{3-2j} E_{2j+1}(q) E_{3-2j}(q)$$

$$= \binom{4}{1}_q q E_1(q) E_3(q) + \binom{4}{3}_q q^3 E_3(q) E_1(q)$$

$$= [4]_q q(1+q^2)(1)(q+q^2) = [4]_q q^2(1+q)(1+q^2) = ([4]_q q)^2$$

$$= q^2 + 2q^3 + 3q^4 + 4q^5 + 3q^6 + 2q^7 + q^8.$$

$\varepsilon_5(q)$ also equals this.

We know that in the $q = 1$ case, $E_n(q)$ and $\varepsilon_n(q)$ would count up-down permutations, and the presence of q could lead us to look at inversions in these permutations. Let's look at the up-down permutations of $\{1, 2, 3, 4\}$:

up-down permutation	inversions	non-inversions
1324	1	5
1423	2	4
2314	2	4
2413	3	3
3412	4	2

In the third column we have counted all the pairs of numbers that are *not* inverted, which will be $\binom{4}{2} = 6$ minus the number of pairs that are inverted. Note that the 1-2-2-3-4 pattern precisely matches $E_4(q)$ and the 2-3-4-4-5 pattern precisely matches $\varepsilon_4(q)$. This suggests a theorem, but let's make another table before we state it. We know that E_n also counts down-up permutations, so let's look at the down-up permutations of $\{1, 2, 3, 4\}$:

down-up permutation	inversions	non-inversions
2143	2	4
3142	3	3
3241	4	2
4132	4	2
4231	5	1

This time the second column matches $\varepsilon_4(q)$, and the third column matches $E_4(q)$. Of course, this is no accident, since if we read an up-down permutation of even length backwards, we get a down-up permutation, and any pair of numbers that is an inversion in one direction is a non-inversion in the other direction.

However, if we read an up-down permutation of odd length backwards, we don't get a down-up permutation but rather a different up-down permutation, unless the length is 1. (Among other things, this implies that the number of up-down permutations of odd length greater than 1 is always even.) In this case, when we have up-down permutations of some odd length $2n + 1$, we can convert them into down-up permutations of the same length by subtracting every element from $2n + 2$. For example, subtracting every element of the up-down permutation 25341 from 6 gives the down-up permutation 41325. Moreover, this converts every inversion in the up-down permutation to a non-inversion in the down-up permutation, and conversely; because if $a < b$, then $-a > -b$, and so anything minus a is bigger than the same thing minus b.

This subtraction procedure will work in both the even length and odd length cases, so we can conclude from this that there are just as many up-down permutations of length n with k non-inversions as there are down-up permutations of length n with k inversions, and conversely.

The result we have been heading toward is

THEOREM 75 (The Stanley–Gessel theorem). *Let U_n and D_n denote the sets of up-down permutations and down-up permutations of length n, respectively, and*

9.8. COMBINATORICS OF q-TANGENTS AND SECANTS

let $\operatorname{inv}\pi$ denote the number of inversions in the permutation π. Then

(a) $$E_n(q) = \sum_{\pi \in U_n} q^{\operatorname{inv}\pi},$$

(b) $$\varepsilon_n(q) = \sum_{\pi \in D_n} q^{\operatorname{inv}\pi}.$$

Let's make one more table before we try to prove this. We'll list each up-down permutation of $\{1,2,3,4,5\}$, its number of inversions, the corresponding down-up permutation of $\{1,2,3,4,5\}$ that comes from the subtraction procedure, and its number of inversions (which will be the same as the number of non-inversions in the original up-down permutation):

up-down permutation	inversions	down-up permutation	inversions
13254	2	53412	8
14253	3	52413	7
14352	4	52314	6
15243	4	51423	6
15342	5	51324	5
23154	3	43512	7
24153	4	42513	6
24351	5	42315	5
25143	5	41523	5
25341	6	41325	4
34152	5	32514	5
34251	6	32415	4
35142	6	31524	4
35241	7	31425	3
45132	7	21534	3
45231	8	21435	2

Our proof of the Stanley–Gessel theorem will be by induction on n, using the recurrence relations (9.8.9) and (9.8.10), which we repeat here for easy reference:

$$E_{n+1}(q) = \sum_j \binom{n}{2j+1}_q q^{n-2j-1} E_{2j+1}(q) E_{n-2j-1}(q) \quad \text{if } n \geq 1;\ E_1(q) = 1,$$

$$\varepsilon_{n+1}(q) = \sum_j \binom{n}{2j+1}_q q^{2j+1} \varepsilon_{2j+1}(q) \varepsilon_{n-2j-1}(q) \quad \text{if } n \geq 1;\ \varepsilon_1(q) = 1.$$

In the above tables we verified the theorem in the cases $n = 4, 5$, and it is easily checked for smaller values of n. Suppose we have an up-down permutation of length $n + 1$. Then the element $n + 1$ will be in an even position, so some odd number $2j + 1$ of elements will precede it, and $n - 2j - 1$ elements will follow it. Note that none of the elements that precede $n + 1$ will be inverted with it, but that all of the ones that follow it will be. Therefore there will be four kinds of inversions in the sequence:

(1) Inversions between two elements that precede $n + 1$. By the induction hypothesis, $E_{2j+1}(q)$ counts these.
(2) Inversions between two elements that succeed $n + 1$. Again by induction these are counted by $E_{n-2j-1}(q)$.

(3) Inversions between one element that precedes $n+1$, and one element that succeeds it. These are between-set inversions with a first set of size $2j+1$ and a second set of size $n-2j-1$, so $\binom{n}{2j+1}_q$ counts these.

(4) Inversions between $n+1$ and an element that succeeds it. Every pair of this type is an inversion, so q^{n-2j-1} counts these.

Thus all the possible inversions are accounted for by the various terms in the recurrence (9.8.9), and so (a) is true by induction.

The proof of (b) is very similar. Now we are looking at down-up permutations of length $n+1$. The element 1 must be in an even position, and it will be inverted with each of the odd number $2j+1$ of elements that precede it. The other details of the proof are exactly the same as before, and so (b) is true by induction.

COROLLARY 2. $E_{2j+1}(q) = \varepsilon_{2j+1}(q)$ for all nonnegative integers j, and therefore $\tan_q x = \text{Tan}_q x$.

We already knew the second part of the corollary, but the first part affords a combinatorial proof of it. We have seen that if we read an up-down permutation of odd length backwards, we get another up-down permutation where all the inversions have been converted to non-inversions. If we now perform the subtraction procedure on this, it will become a down-up permutation, and all the non-inversions will be converted back to inversions. Thus we arrive at a down-up permutation that has the same length and the same number of inversions as the original up-down permutation. Since this procedure is easily reversed, we get a bijection between up-down and down-up permutations of odd length that preserves the number of inversions. This implies the corollary.

If we try this with up-down permutations of even length, it doesn't quite work—rather it brings us to another up-down permutation of even length with the same number of inversions, which may or may not be the same as the one you started with. For example, 1324, 2413 and 3412 get mapped to themselves, while 1423 and 2314 are mapped to each other. The fact that the epsilons count non-inversions in up-down permutations while the E's count inversions can be translated into

(9.8.11) $$\varepsilon_n(q) = q^{\binom{n}{2}} E_n(q^{-1})$$

since the number of non-inversions in a sequence of length n equals $\binom{n}{2}$ minus the number of inversions. This implies

THEOREM 76. *The capital letter q-trigonometric functions are obtained from the small letter ones by replacing q by q^{-1}. In other words,*

$$\text{Sin}_{q^{-1}} x = \sin_q x,$$
$$\text{Cos}_{q^{-1}} x = \cos_q x,$$
$$\text{Sec}_{q^{-1}} x = \sec_q x,$$
$$\text{Tan}_{q^{-1}} x = \tan_q x.$$

In effect, we already knew this from the corresponding fact $E_{q^{-1}}(x) = e_q(x)$ for the q-exponential functions, but it is more interesting to have a combinatorial explanation of it. One can also give a direct analytic proof similar to the exponential function case, using (9.8.11) to do the q-tangent and q-secant.

9.9. Bibliographical Notes

The major index was introduced by MacMahon in [**166**]. As noted above, he actually called it the greater index there and in his book [**168**]. His argument of section 9.4 can be found in both, but the version in [**166**] is clearer. (Andrews's version [**15**] is better than either of them.) Foata's proof of MacMahon's theorem is in [**104**], [**107**] and [**105**]. Our argument in the special case of the q-factorial is similar to that in [**236**].

The *problème des rencontres* dates back to the first edition of [**171**] in 1708. The solution via inclusion-exclusion is due to Nicolas Bernoulli and Abraham de Moivre, independently. The early history has been discussed by Todhunter in sections 160–162 of [**238**]. Gessel's theorem comes from the manuscript [**121**], which was cited by Wachs in [**242**], from whence her eponymous theorem comes. The proof of Chen and Xu is in [**77**]. For a q-analogue of the inclusion-exclusion principle see [**76**].

The first half-dozen Eulerian polynomials are on p. 373 of Euler's differential calculus book [**98**]. The combinatorial interpretation in the text was apparently not known before a paper of Carlitz and Riordan in 1953 [**66**], but was well known after Riordan's classic book [**195**] appeared in 1958. Carlitz [**65**] did the q-case in 1975.

The Stanley–Gessel theorem is one of my favorites. It is due to Ira Gessel [**120**] and Richard Stanley [**223**], and is generalized in section 3.19 of [**224**]. The two volumes [**224**] and [**225**] are the current state of the art in combinatorics, and, as Gian-Carlo Rota wrote in the Foreword to [**225**], "I find it impossible to predict when [they] may be superseded." Now that a second edition of [**224**] has appeared, this may be even more true today than when Rota said it in 1998. They are an alternative source for many of the topics presented here, and many other beautiful results within q-analysis and without.

CHAPTER 10

The Rogers–Ramanujan Identities I: Schur

10.1. Schur's extension of Franklin's argument

We met Georg Frobenius in section 5.2. His name may be familiar in connection with series solutions of differential equations, but he was also one of the people who changed "algebra" into "abstract algebra" in the late 1800s and early 1900s. His best Ph.D. student was Issai Schur, another excellent algebraist who also made important contributions to several other areas of mathematics before being forced out of his professorship in Berlin when the Nazis came to power. He was, in particular, an expert on infinite series, which (without meaning to offend anybody) would be quite unusual for an algebraist today (as, for that matter, would expertise on differential equations).

Schur's greatest contribution to q-analysis was his independent discovery and combinatorial proof of the so-called Rogers–Ramanujan identities. Schur adapted Franklin's proof of the pentagonal number theorem to show the following identities:

(10.1.1) $$(q;q)_\infty \sum_{k=0}^{\infty} \frac{q^{k^2}}{(q;q)_k} = \sum_{m=-\infty}^{\infty} (-1)^m q^{\frac{m(5m-1)}{2}},$$

(10.1.2) $$(q;q)_\infty \sum_{k=0}^{\infty} \frac{q^{k^2+k}}{(q;q)_k} = \sum_{m=-\infty}^{\infty} (-1)^m q^{\frac{m(5m+3)}{2}}.$$

These become more interesting if we use the Jacobi triple product on the right sides, and we will return to this point after we prove them. For now we will focus on (10.1.1); only one tiny modification is needed to adapt the argument to (10.1.2). As in Franklin's argument, the factor $(q;q)_\infty$ generates partitions with distinct parts, where a partition gets a $+$ sign if it has an even number of parts and a $-$ sign if it has an odd number. We know that $1/(q;q)_k$ generates partitions with at most k parts, and that $1+3+5+\cdots+(2k-1) = k^2$. Suppose we have a partition with at most k parts. First, make it have exactly k "parts" by adding a sufficient number of zeros. Then add 1 to the smallest part, 3 to the next smallest, 5 to the next smallest, and so on, finally adding $2k-1$ to the largest. This creates a partition with k nonzero parts, which are not only distinct but differ by at least two from one another, and these are the partitions generated by the sum on the left side of (10.1.1). Thus the whole left side of (10.1.1) generates ordered pairs (S_1, S_2) of partitions, where S_1 has distinct parts, S_2 has parts that differ by at least two, and a pair counts positively if S_1 has an even number of parts and negatively if it has an odd number of parts. As with Franklin's proof, we want an argument that cancels most of these pairs, leaving only a few to be counted by the right side of (10.1.1). Let $|S|$ denote the number that S is a partition of, so, for example,

$|7+6+3+2+2| = 20$. Schur's algorithm changes (S_1, S_2) into a new pair of partitions (S_1', S_2') such that S_1' has distinct parts just as S_1 does, S_2' has gaps of at least 2 between parts as S_2 does, $|S_1'| + |S_2'| = |S_1| + |S_2|$, S_1' has either one more part than S_1 or one less, and such that the same algorithm changes (S_1', S_2') back into (S_1, S_2). If we could do this for every possible pair (S_1, S_2), then the right side of (10.1.1) would be zero. What will happen instead, as with Franklin's near-bijection, is that there will be a class of pairs to which the algorithm does not apply, and these will give the sum on the right side of (10.1.1).

Let the largest parts of S_1 and S_2 be P_1 and P_2 respectively. There are two easy cases and several trickier ones. Case 1(a) is when $P_1 \geq P_2 + 2$. Then we remove P_1 from S_1 and make it the new largest part of S_2. In other words, S_1' is S_1 with P_1 deleted, and S_2' is S_2 with P_1 added, so S_2' still has gaps ≥ 2 between its parts. Case 1(b) occurs when $P_1 < P_2$, when we make P_2 the new largest part in S_1. Clearly these two cases satisfy all the above conditions, and they undo each other.

In all the remaining cases we have either $P_1 = P_2$ or $P_1 = P_2 + 1$. For these pairs (S_1, S_2) we put in the Durfee square and the Franklin triangle for S_1, and what we'll call a **Schur shape** for S_2. Let's write down an example at this point: let S_1 be $9+7+6+3+1$ (distinct parts) and let S_2 be $9+7+4+1$ (gaps at least 2 between parts). The Ferrers diagrams are

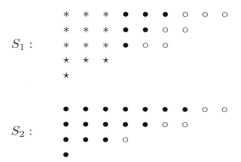

and

respectively, and you can see the Schur shape in S_2. Since we know this type of partition has gaps of at least 2 between parts, we can always take 1 out of the smallest part, 3 out of the next smallest, 5 out of the next smallest, and so on, and that's the Schur shape (in this case; the only change in the proof of (10.1.2) is a different Schur shape). Taking these three figures out of (S_1, S_2) leaves three smaller partitions, which we will call π_1, π_2, and π_3. π_1 is the stuff below the Durfee square in S_1, so in this case π_1 is $3+1$, the $\star$'s in S_1. π_2 is the o's to the right of the Franklin triangle in S_1, but read by columns (or diagonals). In other words, in this example π_2 is the conjugate of $3+2+2$, which is $3+3+1$. Similarly, π_3 is the o's to the right of the Schur shape in S_2 but again read by columns, so in this case π_3 is the conjugate of $2+2+1$, which is $3+2$.

Denote the smallest part in π_1, π_2, π_3 by p_1, p_2, p_3 respectively. It could happen that one or more of $\pi_1, \pi_2,$ and π_3 is empty. In fact, the case where the bijection fails is when all three are empty. We deal below with the cases in which at least one of the three is not empty, and if π_i is empty, then we set $p_i = \infty$. In the above example we have $p_1 = 1 = p_2$ and $p_3 = 2$.

Case 2(a) has $P_1 = P_2$, $p_1 \leq p_2$, and $p_1 \leq p_3$, as in the example above. In this case we make p_1 the new smallest part in π_2. Then $S_2' = S_2$, S_1' has one part less than S_1 had, and its largest part is one greater than that of S_1, so this will satisfy all of our conditions once we describe how to reverse it. In our example S_1' is $10 + 7 + 6 + 2$.

Case 2(b) has $P_1 = P_2 + 1$, $p_2 \leq p_3$, and $p_2 < p_1$, which is precisely the situation we find ourselves in if we apply the algorithm to Case 2(a). In this case we make p_2 the new smallest part in π_1, which clearly undoes Case 2(a).

Case 3(a) has $P_1 = P_2$, $p_3 < p_1$, and $p_3 \leq p_2$. An example of this case is $S_1 = 8 + 6 + 5 + 3 + 2$ and $S_2 = 8 + 5 + 3$. The Ferrers diagrams with the shapes put in are:

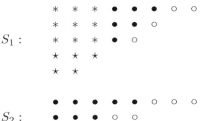

and

We have $\pi_1 = 3 + 2$, so $p_1 = 2$. The o's in S_1 are $2 + 1 + 1$, and π_2 is the conjugate of this, namely $3 + 1$, so $p_2 = 1$. The o's in S_2 are $3 + 2 + 2$, and π_3 is the conjugate of this, which is $3 + 3 + 1$, so $p_3 = 1$. Note that we can't move p_1 in this example, because it's too big. We also can't move p_2 to π_1 or to π_3, because this would put us in Case 1(b).

In this case we make p_3 the new smallest part in π_1. Since this cuts one off of P_2, we have $P_1 = P_2 + 1$ in (S_1', S_2'), and S_1' has one more part than S_1 had. It still has distinct parts, and S_2' still has gaps at least 2 between parts. In the example π_1 changes to $3 + 2 + 1$ and π_2 doesn't change, so $S_1' = 8 + 6 + 5 + 3 + 2 + 1$; we took the new 1 in S_1' off the end of P_2, so $S_2' = 7 + 5 + 3$. To undo Case 3(a) we have Case 3(b), when $P_1 = P_2 + 1$, $p_1 \leq p_2$, and $p_1 \leq p_3$. Here we make p_1 the new smallest part in π_3, which makes $P_1 = P_2$. We couldn't make p_1 the new smallest part in π_2, because this would put us in Case 1(a).

The trickiest cases are the remaining two. In Case 4(a) we have $P_1 = P_2$, $p_2 < p_1$, and $p_2 < p_3$. An example of this case has $S_1 = 10 + 6 + 5 + 4 + 2$ and $S_2 = 10 + 8 + 3$. Here are the Ferrers diagrams with the appropriate shapes put in:

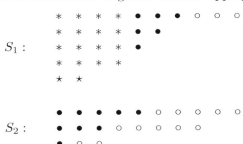

and

We have $\pi_1 = 2 = p_1$. π_2 is the conjugate of 3, which is $1 + 1 + 1$, so $p_2 = 1$, and π_3 is the conjugate of $5 + 5 + 2$, which is $3 + 3 + 2 + 2 + 2$, so $p_3 = 2$. Moving p_2 to either π_1 or π_3 right away wouldn't work, because this would put us in Case 1(b). Instead we start to form (S_1', S_2') by putting P_2 above P_1. Clearly we have to do something more, because P_1 and P_2 have the same size and S_1' must have distinct parts, so then we make p_2 the new smallest part in π_3. In our example we temporarily have $S_1' = 10 + 10 + 6 + 5 + 4 + 2$ and $S_2' = 8 + 3$, but after moving p_2 to π_3 we ultimately have $S_1' = 10 + 9 + 6 + 5 + 4 + 2$ and $S_2' = 9 + 3$. In this example we have $P_1 = P_2 + 1$ after applying the algorithm, but it is not so obvious that this always happens. But if $p_3 > p_2$, then $p_3 \geq 2$, so the largest part in π_3 is at least 2, which means that the second largest part in S_2 must have been $P_2 - 2$; note that this is even true if π_3 is empty. This part has 1 added to it when p_2 comes over, so it now has size $P_2 - 1$, whereas the new P_1 in S_1' is the old P_2, so we do indeed always have $P_1 = P_2 + 1$ after Case 4(a).

To undo Case 4(a) we have Case 4(b), in which $P_1 = P_2 + 1$, $p_3 < p_1$, and $p_3 < p_2$. Here we first put P_1 above P_2 and then make p_3 the smallest part in the new π_2. This evidently undoes Case 4(a).

We leave it to the reader to check that these 8 cases cover every possible situation except when either $P_1 = P_2$ or $P_1 = P_2 + 1$, and in addition π_1, π_2, π_3 are all empty, meaning that (S_1, S_2) consist of a Durfee square, a Franklin triangle, a Schur shape, and nothing else. Note that all the parts in a Schur shape (in the case of (10.1.1)) are odd.

Suppose the Durfee square has side m. Then the Franklin triangle could have side length m or $m-1$. If the length is $m-1$, then $P_1 = 2m-1$, and since P_2 must be odd we have $P_2 = 2m-1$ also. Then the Durfee square contributes m^2, the Franklin triangle $1 + 2 + \cdots + (m-1) = \binom{m}{2}$, and the Schur shape $1 + 3 + \cdots + (2m-1) = m^2$, so in total we have $m^2 + \binom{m}{2} + m^2 = \frac{m(5m-1)}{2}$, and this counts positively if m is even and negatively if m is odd.

The other possibility is that the Franklin triangle also has side m. Then $P_1 = m + m = 2m$, but since P_2 must be odd we must have $P_2 = 2m - 1$, which means that the Schur shape is the same as before. The only difference in this case is that the Franklin triangle is now $1 + 2 + \cdots + m = \binom{m+1}{2}$, so in total we have $m^2 + \binom{m+1}{2} + m^2 = \frac{m(5m+1)}{2}$, counted positively if m is even and negatively if m is odd. Thus Schur's argument proves that

$$(q;q)_\infty \sum_{k=0}^{\infty} \frac{q^{k^2}}{(q;q)_k} = 1 + \sum_{m=1}^{\infty} (-1)^m \left(q^{\frac{m(5m-1)}{2}} + q^{\frac{m(5m+1)}{2}} \right),$$

where the 1 counts the case where S_1 and S_2 are both empty. But

$$\frac{m(5m+1)}{2} = \frac{-m(5(-m)-1)}{2},$$

so this is equivalent to

$$(q;q)_\infty \sum_{k=0}^{\infty} \frac{q^{k^2}}{(q;q)_k} = \sum_{m=-\infty}^{\infty} (-1)^m q^{\frac{m(5m-1)}{2}},$$

which is (10.1.1).

As we mentioned earlier, this theorem becomes much more interesting if we apply the Jacobi triple product

(10.1.3) $\qquad (z;q)_\infty (\frac{q}{z};q)_\infty (q;q)_\infty = \sum_{m=-\infty}^{\infty} (-1)^m q^{\frac{m(m-1)}{2}} z^m$

to it. Replacing q by q^5 and then taking $z = -q^2$, (10.1.1) becomes

$$(q;q)_\infty \sum_{k=0}^\infty \frac{q^{k^2}}{(q;q)_k} = (q^2;q^5)_\infty (q^3;q^5)_\infty (q^5;q^5)_\infty,$$

or

(10.1.4) $$\sum_{k=0}^\infty \frac{q^{k^2}}{(q;q)_k} = \frac{(q^2;q^5)_\infty (q^3;q^5)_\infty (q^5;q^5)_\infty}{(q;q)_\infty}.$$

Finally, this simplifies (exercise) to

THEOREM 77 (The first Rogers–Ramanujan identity). *If $|q| < 1$, then*

(10.1.5) $$\sum_{k=0}^\infty \frac{q^{k^2}}{(q;q)_k} = \frac{1}{(q;q^5)_\infty (q^4;q^5)_\infty}.$$

We will say more about Rogers and Ramanujan in Chapter 11. Schur was the third person to discover the Rogers–Ramanujan identities. He gave two proofs, which are not only completely unlike the proofs of Rogers and Ramanujan, but also completely unlike each other. He was the second person to interpret them in terms of partitions, after MacMahon, and the first person to do both.

We know that the left side of (10.1.5) generates partitions with gaps of at least 2 between parts. The right side is

$$\frac{1}{(1-q)(1-q^6)(1-q^{11})\cdots(1-q^4)(1-q^9)(1-q^{14})\cdots}$$

which generates partitions where the allowed part sizes are $1, 6, 11, 16, \ldots$ and $4, 9, 14, 19, \ldots$. In other words, the parts are either 1 more or 1 less than a multiple of 5; *i.e.*, congruent to 1 or 4 mod 5 (or to ± 1 mod 5). So the theorem says that these two types of partitions are equinumerous, a remarkable fact. We explore this further in Chapter 11.

To prove (10.1.2) by Schur's method, the only change on the left side is that the exponent of q in the sum is now $k^2 + k = 2 + 4 + 6 + \cdots + 2k$ instead of $k^2 = 1 + 3 + 5 + \cdots + (2k-1)$, so the new Schur shape looks like

Otherwise the proof goes through word for word with all the same cases. The partitions that don't cancel again have a Durfee square, a Franklin triangle, a Schur shape, and nothing else, and either $P_1 = P_2$ or $P_1 = P_2 + 1$. If the Durfee square has side m and the Franklin triangle does too, then $P_1 = 2m$, which must equal P_2 since P_2 is even. Then the square contributes m^2, the triangle $\binom{m+1}{2}$, and the Schur shape $m^2 + m$, and this adds up to $\frac{m(5m+3)}{2}$, counted positively if m is even and negatively if m is odd. If the square has side m and the triangle has side $m-1$, then $P_1 = 2m - 1$. This can't equal P_2 since P_2 is even, so we must have $P_1 = P_2 + 1$, which implies $P_2 = 2m - 2$. Then the square contributes m^2, the triangle $\binom{m}{2}$, and the Schur shape $2+4+\cdots+(2m-2) = m^2 - m$, and this adds up to $\frac{m(5m-3)}{2}$. Therefore we have

$$(10.1.6) \qquad (q;q)_\infty \sum_{k=0}^\infty \frac{q^{k^2+k}}{(q;q)_k} = 1 + \sum_{m=1}^\infty (-1)^m \left(1 + q^{\frac{m(5m+3)}{2}} + q^{\frac{m(5m-3)}{2}}\right)$$

$$(10.1.7) \qquad\qquad = \sum_{m=-\infty}^\infty (-1)^m q^{\frac{m(5m+3)}{2}},$$

which is (10.1.1). Again, this gets better if we apply Jacobi's triple product to the sum on the right. To match (10.1.3) we replace q by q^5 and then set $z = -q^4$. If we also divide through by $(q;q)_\infty$ then we get

$$(10.1.8) \qquad \sum_{k=0}^\infty \frac{q^{k^2+k}}{(q;q)_k} = \frac{(q;q^5)_\infty (q^4;q^5)_\infty (q^5;q^5)_\infty}{(q;q)_\infty}.$$

Simplifying this gives

THEOREM 78 (The second Rogers–Ramanujan identity). *If $|q| < 1$, then*

$$(10.1.9) \qquad \sum_{k=0}^\infty \frac{q^{k^2+k}}{(q;q)_k} = \frac{1}{(q^2;q^5)_\infty (q^3;q^5)_\infty}.$$

By the same reasoning as before, the right side of (10.1.9) generates partitions where the allowable parts are $2, 7, 12, 17, \ldots$ and $3, 8, 13, 18, \ldots$, i.e., congruent to 2 or 3 mod 5 (or to ± 2 mod 5). On the left side of (12.2.7) the partitions look like the Schur shape $2 + 4 + 6 + \ldots$ with a partition to the right, so again we have gaps of at least 2 between parts, and now (since we add 2 to the smallest part, 4 to the next smallest, and so on), we also don't have any 1's. Thus the second Rogers–Ramanujan identity says that these types of partitions are equinumerous.

Exercises

1. Check that cases 2(a), 3(a), and 4(a) cover every possibility with $P_1 = P_2$, except $p_1 = p_2 = p_3 = \infty$ (i.e., π_1, π_2, π_3 all empty), and that no more than one of them applies at any given time.
2. Check that cases 2(b), 3(b), and 4(b) cover every possibility with $P_1 = P_2 + 1$, except $p_1 = p_2 = p_3 = \infty$ (i.e., π_1, π_2, π_3 all empty), and that no more than one of them applies at any given time.
3. Explain why (10.1.4) simplifies to (10.1.5) and (10.1.8) simplifies to (10.1.9).
4. Apply Schur's algorithm to the case $S_1 = 9 + 8 + 5 + 4 + 3$, $S_2 = 9 + 7 + 5 + 2$.
5. Apply Schur's algorithm to the case $S_1 = 9 + 7 + 4 + 3$, $S_2 = 8 + 6 + 3 + 1$.
6. Apply Schur's algorithm to the case $S_1 = 10 + 9 + 6 + 5 + 3$, $S_2 = 10 + 8 + 5 + 2$.
7. Apply Schur's algorithm to the case $S_1 = 10 + 9 + 8 + 7 + 6 + 3 + 2$, $S_2 = 9 + 5 + 2$.
8. In the example $S_1 = 9 + 7 + 6 + 3 + 1$, $S_2 = 9 + 7 + 4 + 1$ that was used to illustrate Case 2(a), why can't we move p_2 to π_1? Why can't we move p_2 to π_3?

10.2. The Bressoud–Chapman proof

Schur's second proof of the Rogers–Ramanujan identities is in the same spirit as some of the arguments in Chapter 2: he proves a finite identity by recurrence and induction, and then takes a limit. The simplest known proof of this type is a fairly recent one by Robin Chapman that modifies an earlier argument of David Bressoud. It relies heavily on the fundamental recurrences

(10.2.1) $$\binom{m+1}{k}_q = \binom{m}{k-1}_q + q^k \binom{m}{k}_q$$

(10.2.2) $$= \binom{m}{k}_q + q^{m-k+1} \binom{m}{k-1}_q.$$

We define four sets of polynomials:

(10.2.3) $$b_n(q) = \sum_{j=0}^n \binom{n}{j}_q q^{j^2}, \qquad B_n(q) = \sum_j (-1)^j q^{\frac{j(5j+1)}{2}} \binom{2n}{n+2j}_q,$$

(10.2.4) $$c_n(q) = \sum_{j=0}^n \binom{n}{j}_q q^{j^2+j}, \qquad C_n(q) = \sum_j (-1)^j q^{\frac{j(5j-3)}{2}} \binom{2n+1}{n+2j}_q.$$

In the expressions for $B_n(q)$ and $C_n(q)$ the sums are over all integer values of j; i.e., from $j = -\infty$ to ∞, but there are only finitely many nonzero terms, so they really are polynomials. For $B_n(q)$, we have $\binom{2n}{n+2j}_q = 0$ unless $n + 2j$ and $2n - (n+2j) = n - 2j$ are both nonnegative. Thus we must have $n \geq 2j \geq -n$, or $\frac{n}{2} \geq j \geq -\frac{n}{2}$, or finally, since j must be an integer,

$$\lfloor \tfrac{n}{2} \rfloor \geq j \geq \lceil -\tfrac{n}{2} \rceil,$$

where $\lfloor x \rfloor$ and $\lceil x \rceil$ are respectively the floor and ceiling of x: the former is the greatest integer less than or equal to x, and the latter is the smallest integer greater than or equal to x. So really we have

$$B_n(q) = \sum_{j=\lceil -\frac{n}{2} \rceil}^{\lfloor \frac{n}{2} \rfloor} (-1)^j q^{\frac{j(5j+1)}{2}} \binom{2n}{n+2j}_q,$$

but it is convenient to write the sums unrestrictedly and let the q-binomial coefficients do the restriction. Similarly,

$$C_n(q) = \sum_{j=\lceil -\frac{n}{2} \rceil}^{\lfloor \frac{n+1}{2} \rfloor} (-1)^j q^{\frac{j(5j-3)}{2}} \binom{2n+1}{n+2j}_q,$$

and the same comment applies to the first two lemmas below.

We first observe that when $n = 0$ we have $b_0(q) = \binom{0}{0}_q = 1 = c_0(q) = B_0(q)$ and $C_0(q) = \binom{1}{0}_q = 1$. Next we prove a sequence of lemmas.

LEMMA 5. *For any nonnegative integer n we have*

$$\sum_j (-1)^j q^{\frac{5}{2} j(j+1)} \binom{2n}{n+1+2j}_q = 0.$$

To see this, denote the sum in the lemma by S and change j to $-j-1$. Then $j(j+1)$ becomes $(-j-1)(-j) = j(j+1)$, and
$$\binom{2n}{n+1+2j}_q \text{ becomes } \binom{2n}{n+1-2j-2}_q = \binom{2n}{n-1-2j}_q = \binom{2n}{n+1+2j}_q$$
by the symmetry of the q-binomial coefficients. Since $(-1)^k = (-1)^{-k}$, we have
$$S = \sum_j (-1)^{j+1} q^{\frac{5}{2} j(j+1)} \binom{2n}{n+1+2j}_q = -S,$$
so $S = 0$.

LEMMA 6. *For any nonnegative integer n we have*
$$B_n(q) = \sum_j (-1)^j q^{\frac{j(5j+1)}{2}} \binom{2n+1}{n+1+2j}_q.$$

Applying (10.2.1) to this sum we get
$$\sum_j (-1)^j q^{\frac{j(5j+1)}{2}} \binom{2n+1}{n+1+2j}_q = \sum_j (-1)^j q^{\frac{j(5j+1)}{2}} \binom{2n}{n+2j}_q$$
$$+ \sum_j (-1)^j q^{\frac{j(5j+1)}{2} + n+1+2j} \binom{2n}{n+1+2j}_q$$
$$= B_n(q) + q^{n+1} \sum_j (-1)^j q^{\frac{j(5j+1+4)}{2}} \binom{2n}{n+1+2j}_q$$
and the last sum is zero by Lemma 5.

LEMMA 7. *For any nonnegative integer n we have*
$$B_{n+1}(q) = B_n(q) + q^{n+1} C_n(q).$$

To see this, start with
$$B_{n+1}(q) = \sum_j (-1)^j q^{\frac{j(5j+1)}{2}} \binom{2n+2}{n+1+2j}_q$$
and apply (10.2.2) to get
$$B_{n+1}(q) = \sum_j (-1)^j q^{\frac{j(5j+1)}{2}} \binom{2n+1}{n+1+2j}_q + \sum_j (-1)^j q^{\frac{j(5j+1)}{2}} q^{n+1-2j} \binom{2n+1}{n+2j}_q.$$
According to Lemma 6, the first sum on the right is $B_n(q)$. Rewriting the other one, we have
$$B_{n+1}(q) = B_n(q) + q^{n+1} \sum_j (-1)^j q^{\frac{j(5j+1-4)}{2}} \binom{2n+1}{n+2j}_q$$
and we see that the remaining sum is $C_n(q)$.

LEMMA 8. *For any positive integer n we have*
$$C_n(q) - q^n B_n(q) = (1 - q^n) C_{n-1}(q).$$

This even holds for $n = 0$, because both sides are zero, as long as $C_{-1}(q)$ is not infinite. (The most natural definition, if we need one, is $C_{-1}(q) = 0$.) For positive n we have

$$C_n(q) - q^n B_n(q) = \sum_j (-1)^j \left[\binom{2n+1}{n+2j}_q q^{\frac{j(5j-3)}{2}} - \binom{2n}{n+2j}_q q^{n+\frac{j(5j+1)}{2}} \right]$$

$$= \sum_j (-1)^j q^{\frac{j(5j-3)}{2}} \left[\binom{2n+1}{n+2j}_q - q^{n+2j} \binom{2n}{n+2j}_q \right],$$

and by (10.2.1) this is just

$$C_n(q) - q^n B_n(q) = \sum_j (-1)^j q^{\frac{j(5j-3)}{2}} \binom{2n}{n+2j-1}_q.$$

Now use (10.2.1) on this:

$$C_n(q) - q^n B_n(q) = \sum_j (-1)^j q^{\frac{j(5j-3)}{2}} \binom{2n-1}{n+2j-1}_q$$

$$+ \sum_j (-1)^j q^{\frac{j(5j-3)}{2}} q^{n-2j+1} \binom{2n-1}{n+2j-2}_q$$

$$= C_{n-1}(q) + q^n \sum_j (-1)^j q^{1+\frac{j(5(j-1)-2)}{2}} \binom{2n-1}{n+2j-2}_q.$$

Now change $j-1$ to $-j$ in the last sum, so that j becomes $1-j$. This gives

$$C_n(q) - q^n B_n(q) - C_{n-1}(q) = q^n \sum_j (-1)^{-j-1} q^{1+\frac{(1-j)(-5j-2)}{2}} \binom{2n-1}{n-2j}_q$$

$$= -q^n \sum_j (-1)^j q^{\frac{2+(5j+2)(j-1)}{2}} \binom{2n-1}{n+2j-1}_q$$

$$= -q^n \sum_j (-1)^j q^{\frac{j(5j-3)}{2}} \binom{2n-1}{n+2j-1}_q$$

$$= -q^n C_{n-1}(q).$$

Lemmas 7 and 8 tell us what we need to know about $B_n(q)$ and $C_n(q)$. We also need two (very) similar lemmas about $b_n(q)$ and $c_n(q)$, whose proofs we leave as exercises.

LEMMA 9. *For any nonnegative integer n we have*

$$b_{n+1}(q) = b_n(q) + q^{n+1} c_n(q).$$

LEMMA 10. *For any positive integer n we have*

$$c_n(q) - q^n b_n(q) = (1-q^n) c_{n-1}(q).$$

The similarity of Lemmas 7 and 9 and of Lemmas 8 and 10 is not a coincidence. The result we have been leading up to is

THEOREM 79. *For any nonnegative integer n we have $b_n(q) = B_n(q)$ and $c_n(q) = C_n(q)$.*

As we noted above, all four of these equal 1 when $n = 0$, so the theorem holds in that case. The short version of the rest of the proof is that since the recurrences are the same, the theorem must then hold for all n by induction.

Let's do this a little more slowly (but not much more). Taking $n = 0$ in Lemmas 7 and 9 we see that $B_1(q) = 1 + 1 \cdot q = 1 + q = b_1(q)$. Taking $n = 1$ in Lemmas 8 and 10 we further see that $C_1(q) = q(1+q) + (1-q)(1) = 1 + q^2 = c_1(q)$. Assuming that the theorem holds for n, Lemmas 7 and 9 say

$$B_{n+1}(q) = B_n(q) + q^{n+1}C_n(q) = b_n(q) + q^{n+1}c_n(q) = b_{n+1}(q),$$

so half of the theorem holds for $n+1$. Replacing n by $n+1$ in Lemmas 8 and 10 we also have

$$C_{n+1}(q) = q^{n+1}B_{n+1}(q) + (1 - q^{n+1})C_n(q)$$
$$= q^{n+1}b_{n+1}(q) + (1 - q^{n+1})c_n(q) = c_{n+1}(q),$$

so the other half of the theorem holds for $n+1$ too.

The point of this theorem is that it is a finite form of the Rogers–Ramanujan identities. Assuming $|q| < 1$ and letting $n \to \infty$ we have

$$\binom{n}{j}_q = \frac{(q^{n+1-j};q)_j}{(q;q)_j} \to \frac{(0;q)_j}{(q;q)_j} = \frac{1}{(q;q)_j},$$

so

$$b(q) := \lim_{n \to \infty} b_n(q) = \lim_{n \to \infty} \sum_{j=0}^n \binom{n}{j}_q q^{j^2} = \sum_{j=0}^\infty \frac{q^{j^2}}{(q;q)_j}$$

and

$$c(q) := \lim_{n \to \infty} c_n(q) = \lim_{n \to \infty} \sum_{j=0}^n \binom{n}{j}_q q^{j^2+j} = \sum_{j=0}^\infty \frac{q^{j^2+j}}{(q;q)_j}.$$

As $n \to \infty$ we further have

$$\binom{2n}{n+2j}_q = \frac{(q^{n+2j+1};q)_{n-2j}}{(q;q)_{n-2j}} \to \frac{(0;q)_\infty}{(q;q)_\infty} = \frac{1}{(q;q)_\infty}$$

and similarly

$$\binom{2n+1}{n+2j}_q \to \frac{1}{(q;q)_\infty},$$

so

$$B(q) := \lim_{n \to \infty} B_n(q) = \lim_{n \to \infty} \sum_{j=\lceil -\frac{n}{2} \rceil}^{\lfloor \frac{n}{2} \rfloor} (-1)^j q^{\frac{j(5j+1)}{2}} \binom{2n}{n+2j}_q$$

$$= \frac{1}{(q;q)_\infty} \sum_{j=-\infty}^\infty (-1)^j q^{\frac{j(5j+1)}{2}}$$

and

$$C(q) := \lim_{n \to \infty} C_n(q) = \lim_{n \to \infty} \sum_{j=\lceil -\frac{n}{2} \rceil}^{\lfloor \frac{n+1}{2} \rfloor} (-1)^j q^{\frac{j(5j-3)}{2}} \binom{2n+1}{n+2j}_q$$

$$= \frac{1}{(q;q)_\infty} \sum_{j=-\infty}^\infty (-1)^j q^{\frac{j(5j-3)}{2}}$$

and we have $b(q) = B(q)$ and $c(q) = C(q)$; that is,

$$\sum_{j=0}^{\infty} \frac{q^{j^2}}{(q;q)_j} = \frac{1}{(q;q)_{\infty}} \sum_{j=-\infty}^{\infty} (-1)^j q^{\frac{j(5j+1)}{2}}$$

and

$$\sum_{j=0}^{\infty} \frac{q^{j^2+j}}{(q;q)_j} = \frac{1}{(q;q)_{\infty}} \sum_{j=-\infty}^{\infty} (-1)^j q^{\frac{j(5j-3)}{2}}.$$

The Rogers–Ramanujan identities follow as in section 10.1 by using the Jacobi triple product and simplifying. It is not difficult to justify the limiting process using Tannery's theorem.

Exercises

1. Use (10.2.2) to prove Lemma 9.
2. Prove Lemma 10. (You should be able to do this without (10.2.1) or (10.2.2).)
3. Justify taking the limit as $n \to \infty$ above via Tannery's theorem. You may assume $0 < q < 1$ for convenience.
4. Define r^{th} order q-Fibonacci numbers by $F_0^{(r)}(q) = 1 = F_1^{(r)}(q)$ and

(10.2.5) $$F_{n+1}^{(r)}(q) = F_n^{(r)}(q) + q^{n+r} F_{n-1}^{(r)}(q).$$

Show that

(10.2.6) $$F_n^{(r)}(q) = \sum_k \binom{n-k}{k}_q q^{k^2+rk}$$

by showing that this sum satisfies (10.2.5) and the initial conditions. The sum goes from $k = 0$ to $\lfloor \frac{n}{2} \rfloor$, but since the q-binomial coefficient is 0 for other values of k it is convenient to leave it unrestricted.

5. Schur's second proof of the Rogers–Ramanujan identities used the polynomials

(10.2.7) $$S_n(a,q) := \sum_k (-1)^k q^{\frac{k(5k+1)}{2} - 2ak} \binom{n}{\lfloor \frac{n-5k}{2} \rfloor + a}_q.$$

As in section 10.2, the sum runs from $k = -\infty$ to ∞, but these are still polynomials since all but finitely many of the q-binomial coefficients will be zero. The goal of this problem is to show that they too have a q-Fibonacci type recurrence, namely

(10.2.8) $$S_{n+1}(a,q) = S_n(a,q) + q^n S_{n-1}(a,q) \quad \text{if } n \geq 1.$$

(i) By setting $n = 2m$ in (10.2.7) and considering even and odd k separately ($k = 2j$ and $k = 2j+1$), show that

(10.2.9) $$S_{2m}(a,q) = \sum_j q^{j(10j+1-4a)} \binom{2m}{m-5j+a}_q$$
$$- \sum_j q^{(2j+1)(5j+3-2a)} \binom{2m}{m-5j-3+a}_q.$$

(ii) Show similarly that

$$(10.2.10) \quad S_{2m+1}(a,q) = \sum_j q^{j(10j+1-4a)} \binom{2m+1}{m-5j+a}_q$$

$$- \sum_j q^{(2j+1)(5j+3-2a)} \binom{2m+1}{m-5j-2+a}_q.$$

(iii) Use the q-Pascal recurrences

$$(10.2.11) \quad \binom{n+1}{k}_q = \binom{n}{k-1}_q + q^k \binom{n}{k}_q$$

$$(10.2.12) \quad = \binom{n}{k}_q + q^{n-k+1} \binom{n}{k-1}_q$$

on (10.2.10) to show that

$$(10.2.13) \quad S_{2m+1}(a,q) = S_{2m}(a,q) + q^{m-a+1} D_{2m}(a,q),$$

where

$$(10.2.14)$$

$$D_{2m}(a,q) := \sum_j q^{2j(5j+3-2a)} \left\{ \binom{2m}{m-5j+a-1}_q - \binom{2m}{m-5j+a-2}_q \right\}.$$

(iv) Use (10.2.11) and (10.2.12) on (10.2.14) to show that

$$D_{2m}(a,q) = q^{m+a-1} S_{2m-1}(a,q),$$

and hence

$$(10.2.15) \quad S_{2m+1}(a,q) = S_{2m}(a,q) + q^{2m} S_{2m-1}(a,q).$$

(v) Use (10.2.11) and (10.2.12) on (10.2.9) to show that

$$(10.2.16) \quad S_{2m}(a,q) = S_{2m-1}(a,q) + q^{m+a} D_{2m-1}(a,q),$$

where

$$(10.2.17)$$

$$D_{2m-1}(a,q) = \sum_k q^{2k(5k-2-2a)} \left\{ \binom{2m-1}{m-5k+a}_q - \binom{2m-1}{m-5k+a+1}_q \right\}.$$

(vi) Use (10.2.11) and (10.2.12) on (10.2.17) to show that

$$D_{2m-1}(a,q) = q^{m-a-1} S_{2m-2}(a,q),$$

and hence

$$S_{2m}(a,q) = S_{2m-1}(a,q) + q^{2m-1} S_{2m-2}(a,q).$$

This completes the proof of (10.2.8).

6. Show that $S_n(0,q) = F_n^{(0)}(q)$, by showing that they satisfy the same recurrence and the same initial conditions.
7. Show that $S_n(1,q) = F_{n-1}^{(1)}(q)$, where both sides are zero if $n = 0$, by showing that they satisfy the same recurrence and the same initial conditions.
8. Show that the Rogers–Ramanujan identities follow by letting $n \to \infty$ in the previous two problems.

9. Schur used not the q-Fibonacci numbers but the $(n+1) \times (n+1)$ tridiagonal determinant

$$D_{n+1}(xq, xq^2, \ldots, xq^n) = \begin{vmatrix} 1 & xq & 0 & \cdots & 0 & 0 \\ -1 & 1 & xq^2 & \cdots & 0 & 0 \\ 0 & -1 & 1 & \ddots & 0 & 0 \\ \vdots & \vdots & \ddots & \ddots & \ddots & \vdots \\ 0 & 0 & 0 & \ddots & 1 & xq^n \\ 0 & 0 & 0 & \cdots & -1 & 1 \end{vmatrix}$$

where all the entries on the main diagonal are 1, all the ones on the diagonal below it are -1, the one on the diagonal above it in the kth row is xq^k, and every other entry is zero.

(i) It is convenient to define $D_0 = 1$. Show that $D_1 = 1$ and $D_2(xq) = 1 + xq$.

(ii) By expanding on the last row and then the last column, or vice versa, show that

$$D_{n+1}(xq, xq^2, \ldots, xq^n) = D_n(xq, xq^2, \ldots, xq^{n-1}) + xq^n \, D_{n-1}(xq, xq^2, \ldots, xq^{n-2}).$$

(iii) Show that $S_n(0, q) = D_n(q, q^2, \ldots, q^{n-1})$, by showing that they satisfy the same recurrence and the same initial conditions.

(iv) Show that $S_n(1, q) = D_{n-1}(q^2, q^3, \ldots, q^{n-1})$, where both sides are zero if $n = 0$, by showing that they satisfy the same recurrence and the same initial conditions.

(v) Define $\Delta(x, q) = \lim_{n \to \infty} D_{n+1}(xq, xq^2, \ldots, xq^n)$. By expanding on the first row and then the first column, or vice versa, show that

$$\Delta(x, q) = \Delta(xq, q) + xq \, \Delta(xq^2, q).$$

(vi) Set $\Delta(x, q) = \sum_{n=0}^{\infty} s_n(q) x^n$. By substituting this in the recurrence in (v), show that

$$s_n(q) = \frac{q^{n^2}}{(q; q)_n} s_0(q).$$

(vii) What is $\Delta(0, q)$? Explain why this implies that

$$\Delta(x, q) = \sum_{n=0}^{\infty} \frac{q^{n^2} x^n}{(q; q)_n}.$$

(viii) Explain how the Rogers–Ramanujan identities follow by letting $n \to \infty$ in (iii) and (iv) and using (v) and (vii). This was Schur's second proof.

10.3. The AKP and GIS identities

Recall the rth order q-Fibonacci numbers from problem 4 of the previous section. If

(10.3.1) $$F_n^{(r)}(q) = \sum_k \binom{n-k}{k}_q q^{k^2 + rk},$$

then $F_0^{(r)}(q) = 1 = F_1^{(r)}(q)$ and

(10.3.2) $$F_{n+1}^{(r)}(q) = F_n^{(r)}(q) + q^{n+r} F_{n-1}^{(r)}(q),$$

and we also proved that

(10.3.3) $$\lim_{n\to\infty} F_n^{(r)}(q) = \sum_{k=0}^{\infty} \frac{q^{k^2+rk}}{(q;q)_k}.$$

In this section we will establish a relationship between these numbers that is essentially due to Andrews, Knopfmacher, and Paule, namely

THEOREM 80 (The AKP identity). *If m and k are integers with $m \geq 0$ and $k \geq -1$, then*

(10.3.4) $$F_{m+k}^{(1)}(q) F_m^{(0)}(q) - F_{m+k+1}^{(0)}(q) F_{m-1}^{(1)}(q) = (-1)^m q^{\binom{m+1}{2}} F_k^{(m+1)}(q),$$

where $F_{-1}^{(r)}(q) = 0$ for any r.

The proof will be an induction on k. If $k = -1$, then the terms on the left side of the AKP identity cancel, and the right side is also zero. Let's set $f_{m,k}$ equal to the left side of the AKP identity. If $k = 0$ and $m > 0$, then we have, using (10.3.2),

$$\begin{aligned}
f_{m,0} &= F_m^{(1)}(q) F_m^{(0)}(q) - F_{m+1}^{(0)}(q) F_{m-1}^{(1)}(q) \\
&= \left(F_{m-1}^{(1)}(q) + q^m F_{m-2}^{(1)}(q)\right) F_m^{(0)}(q) - \left(F_m^{(0)}(q) + q^m F_{m-1}^{(0)}(q)\right) F_{m-1}^{(1)}(q) \\
&= -q^m \left(F_{m-1}^{(1)}(q) F_{m-1}^{(0)}(q) - F_m^{(0)}(q) F_{m-2}^{(1)}(q)\right) = -q^m f_{m-1,0}.
\end{aligned}$$

Therefore, by iteration,

$$\begin{aligned}
f_{m,0} &= -q^m f_{m-1,0} = -q^m \left(-q^{m-1} f_{m-2,0}\right) = \cdots \\
&= (-1)^j q^{m+(m-1)+\cdots+(m-j+1)} f_{m-j,0} = \cdots \\
&= (-1)^m q^{m+(m-1)+\cdots+1} f_{0,0} \\
&= (-1)^m q^{\binom{m+1}{2}} (1 \cdot 1 - 1 \cdot 0) = (-1)^m q^{\binom{m+1}{2}},
\end{aligned}$$

and this is what the right side of the AKP identity becomes if $k = 0$.

By a similar argument, again using (10.3.2), we can find a recurrence for $f_{m,k}$. If $k \geq 0$, then

$$\begin{aligned}
f_{m,k+1} &= F_{m+k+1}^{(1)}(q) F_m^{(0)}(q) - F_{m+k+2}^{(0)}(q) F_{m-1}^{(1)}(q) \\
&= \left(F_{m+k}^{(1)}(q) + q^{m+k+1} F_{m+k-1}^{(1)}(q)\right) F_m^{(0)}(q) \\
&\quad - \left(F_{m+k+1}^{(0)}(q) + q^{m+k+1} F_{m+k}^{(0)}(q)\right) F_{m-1}^{(1)}(q) \\
&= \left(F_{m+k}^{(1)}(q) F_m^{(0)}(q) - F_{m+k+1}^{(0)}(q) F_{m-1}^{(1)}(q)\right) \\
&\quad + q^{m+k+1} \left(F_{m+k-1}^{(1)}(q) F_m^{(0)}(q) - F_{m+k}^{(0)}(q) F_{m-1}^{(1)}(q)\right) \\
&= f_{m,k} + q^{m+k+1} f_{m,k-1}.
\end{aligned}$$

Finally, we use this to prove by induction on k that $f_{m,k}$ equals the right side of the AKP identity. We have proved already that this is true for $k = -1$ and for

$k = 0$. Assuming that it is true for k, we have

$$\begin{aligned} f_{m,k+1} &= f_{m,k} + q^{m+k+1} f_{m,k-1} \\ &= (-1)^m q^{\binom{m+1}{2}} F_k^{(m+1)}(q) + q^{m+k+1} (-1)^m q^{\binom{m+1}{2}} F_{k-1}^{(m+1)}(q) \\ &= (-1)^m q^{\binom{m+1}{2}} \left(F_k^{(m+1)}(q) + q^{m+k+1} F_{k-1}^{(m+1)}(q) \right) \\ &= (-1)^m q^{\binom{m+1}{2}} F_{k+1}^{(m+1)}(q), \end{aligned}$$

so by induction $f_{m,k}$ equals the right side of the AKP identity for all $k \geq -1$. Since $f_{m,k}$ was defined to be the left side of the AKP identity, this completes the proof.

Next we let $k \to \infty$ in the AKP identity, as Andrews, Knopfmacher, and Paule did. According to (10.3.3) we get

$$F_m^{(0)}(q) \sum_{j=0}^{\infty} \frac{q^{j^2+j}}{(q;q)_j} - F_{m-1}^{(1)}(q) \sum_{j=0}^{\infty} \frac{q^{j^2}}{(q;q)_j} = (-1)^m q^{\binom{m+1}{2}} \sum_{j=0}^{\infty} \frac{q^{j^2+(m+1)j}}{(q;q)_j}.$$

Using the Rogers–Ramanujan identities on the left side, this becomes an identity of Garrett, Ismail, and Stanton.

THEOREM 81 (The GIS identity). *Assume (as usual) that $|q| < 1$, and define the rth order q-Fibonacci numbers as above. In addition, define $F_{-2}^{(1)}(q) = 1$. Then if m is an integer ≥ -1, we have*

$$\frac{F_m^{(0)}(q)}{(q^2;q^5)_\infty (q^3;q^5)_\infty} - \frac{F_{m-1}^{(1)}(q)}{(q;q^5)_\infty (q^4;q^5)_\infty} = (-1)^m q^{\binom{m+1}{2}} \sum_{j=0}^{\infty} \frac{q^{j^2+(m+1)j}}{(q;q)_j}.$$

Note that (10.3.2) would say

$$F_0^{(r)}(q) = F_{-1}^{(r)}(q) + q^{r-1} F_{-2}^{(r)}(q)$$

if it held for $n = -1$, and this reduces to

$$1 = 0 + q^{r-1} F_{-2}^{(r)}(q) \quad \text{or} \quad F_{-2}^{(r)}(q) = q^{1-r}.$$

Thus the definition $F_{-2}^{(1)}(q) = 1$ is in fact consistent with (10.3.2), although not with (10.3.1). When $m = -1$ the GIS identity reduces to the first Rogers–Ramanujan identity, and it becomes the second one if $m = 0$.

10.4. Schur's second partition theorem

Schur made a second major contribution to q-analysis. In this section we discuss this second paper, which contains a surprising partition theorem that has a beautiful bijective proof, due to David Bressoud. We state Schur's theorem in the notation of George Andrews. Let $A(n)$ be the number of partitions of n using only odd parts not divisible by 3, and let $B(n)$ be the number of partitions of n with distinct parts not divisible by 3. It is easy to see that $A(n) = B(n)$, either by Glaisher's bijection or by a generating function argument. The generating function for $B(n)$ is $(-q;q^3)_\infty(-q^2;q^3)_\infty$, and

$$\begin{aligned}(-q;q^3)_\infty(-q^2;q^3)_\infty &= (1+q)(1+q^2)(1+q^4)(1+q^5)(1+q^7)(1+q^8)\cdots \\ &= \frac{1-q^2}{1-q}\frac{1-q^4}{1-q^2}\frac{1-q^8}{1-q^4}\frac{1-q^{10}}{1-q^5}\frac{1-q^{14}}{1-q^7}\frac{1-q^{16}}{1-q^8}\cdots \\ &= \frac{(q^2;q^6)_\infty(q^4;q^6)_\infty}{(q;q^3)_\infty(q^2;q^3)_\infty} \\ &= \frac{(q^2;q^6)_\infty(q^4;q^6)_\infty}{(q;q^6)_\infty(q^4;q^6)_\infty(q^2;q^6)_\infty(q^5;q^6)_\infty} \\ &= \frac{1}{(q;q^6)_\infty(q^5;q^6)_\infty},\end{aligned}$$

which is the generating function for $A(n)$, since an odd number not divisible by 3 must be congruent to 1 or 5 mod 6. Schur finds a third class of partitions that is equinumerous with these two. Let $C(n)$ be the number of partitions of n where all the parts differ by at least 3 from each other, and in addition any multiples of 3 must differ by at least 6, and let's call these **Schur partitions**. Schur proved that $C(n) = B(n) = A(n)$. For example, when $n = 17$ we have:

$A(17)$	$B(17)$	$C(17)$
17	17	17
$13+1+1+1+1$	$16+1$	$16+1$
$11+5+1$	$14+2+1$	$15+2$
$11+$ six 1's	$13+4$	$14+3$
$7+7+1+1+1$	$11+5+1$	$13+4$
$7+5+5$	$11+4+2$	$12+5$
$7+5+$ five 1's	$10+7$	$12+4+1$
$7+$ ten 1's	$10+5+2$	$11+6$
$5+5+5+1+1$	$10+4+2+1$	$11+5+1$
$5+5+$ seven 1's	$8+7+2$	$10+7$
$5+$ twelve 1's	$8+5+4$	$10+6+1$
seventeen 1's	$7+5+4+1$	$10+5+2$

Note that several partitions appear in both the $B(17)$ and $C(17)$ columns, which suggests trying to construct a bijection between $B(n)$ and $C(n)$. A gap of less than 3 between successive parts in a partition of $B(n)$ type can only occur when one of the parts is congruent to 1 mod 3, and the other to 2 mod 3. If we add these two parts together, we always get a multiple of 3. Moreover, if we do this whenever we have a gap less than 3, we can never get two consecutive multiples of 3, because each number in the larger pair must be at least 3 more than each number in the smaller pair.

Take for example $34 + 23 + 22 + 19 + 16 + 14 + 10 + 8 + 4 + 2 + 1$, a partition of 153 into distinct parts that aren't divisible by 3. Starting with the smallest parts and working up, add two parts together whenever they differ by less than 3. Thus we add 1 and 2, leave 4 alone, add 8 and 10, add 14 and 16, leave 19 alone, add 22 and 23, and leave 34 alone, which gives $34 + 45 + 19 + 30 + 18 + 4 + 3$. Now just reordering this doesn't work because we have 19 and 18, and also 4 and 3. To ensure the gaps are at least 3 after reordering, we build them in as in the right column below:

$$\begin{array}{ccc} 34 & 16 & 18 \\ 45 & 30 & 15 \\ 19 & 7 & 12 \\ 30 & 21 & 9 \\ 18 & 12 & 6 \\ 4 & 1 & 3 \\ 3 & 3 & 0 \end{array}$$

Because the difference between multiples of 3 in the left column must be at least 6, the difference between multiples of 3 in the middle column must be at least 3. Therefore, if we reorder the middle column and add it to the right column, not only will we have gaps of at least 3 between parts, but any multiples of 3 will have gaps of at least 6. The reordering gives

$$\begin{array}{ccc} 30 & 18 & 48 \\ 21 & 15 & 36 \\ 16 & 12 & 28 \\ 12 & 9 & 21 \\ 7 & 6 & 13 \\ 3 & 3 & 6 \\ 1 & 0 & 1 \end{array}$$

so the Schur partition of 153 that corresponds to $34+23+22+19+16+14+10+8+4+2+1$ under Bressoud's bijection is $48+36+28+21+13+6+1$.

To prove that this is really a bijection we describe the inverse map, which is somewhat harder. The first step is clear: split $48+36+28+21+13+6+1$ into two columns as above. Next, write each multiple of 3 in the left column as the number in the middle column plus two remainders, one congruent to 1 mod 3 and the other to 2 mod 3, and differing from each other by either 1 or 2. For example, we need to write 30 as 18 plus 12, and then split up 12. The only way to write 12 as a sum of two numbers differing by at most 2 without using a multiple of 3 is $7+5$, so we write $30 = 18+7+5$. Similarly we write 21 as $15+6$, and the only way to split up 6 is $4+2$, so $21 = 15+4+2$. We have $12 = 9+3 = 9+2+1$, and for $3 = 3+0$ we have to use a negative remainder: $3 = 3+1+(-1)$. To distinguish the remainders from the other numbers let's write all of them in parentheses, so we now have:

$$\begin{array}{cc} 18+(7)+(5) & 18 \\ 15+(4)+(2) & 15 \\ 16 & 12 \\ 9+(2)+(1) & 9 \\ 7 & 6 \\ 3+(1)+(-1) & 3 \\ 1 & 0 \end{array}$$

Working from the bottom up in the left column, we now perform the following operation repeatedly: if there is a single number on the line below a remainder and the remainder is smaller, switch the single number with the sum of three numbers, and also transfer 3 units from the first number to the small remainder, so that the first number still matches the corresponding number in the right column. In this example we have $1 > -1$, so we switch 1 with $3 + (1) + (-1)$ and rewrite $3 + (1) + (-1) = 0 + (2) + (1)$, which gives:

$$\begin{array}{cc} 18 + (7) + (5) & 18 \\ 15 + (4) + (2) & 15 \\ 16 & 12 \\ 9 + (2) + (1) & 9 \\ 7 & 6 \\ 1 & 3 \\ 0 + (2) + (1) & 0 \end{array}$$

The next instance of a single number exceeding a remainder right above it is with 7 and (1) on lines 4 and 5, so we rewrite:

$$\begin{array}{cc} 18 + (7) + (5) & 18 \\ 15 + (4) + (2) & 15 \\ 16 & 12 \\ 7 & 9 \\ 6 + (4) + (2) & 6 \\ 1 & 3 \\ 0 + (2) + (1) & 0 \end{array}$$

Next we have $16 > (2)$ on lines 2 and 3, so we rewrite:

$$\begin{array}{cc} 18 + (7) + (5) & 18 \\ 16 & 15 \\ 12 + (5) + (4) & 12 \\ 7 & 9 \\ 6 + (4) + (2) & 6 \\ 1 & 3 \\ 0 + (2) + (1) & 0 \end{array}$$

Now $7 > (4)$ on lines 3 and 4, so we rewrite:

$$\begin{array}{cc} 18 + (7) + (5) & 18 \\ 16 & 15 \\ 7 & 12 \\ 9 + (7) + (5) & 9 \\ 6 + (4) + (2) & 6 \\ 1 & 3 \\ 0 + (2) + (1) & 0 \end{array}$$

Finally $16 > (5)$ on lines 1 and 2, so we rewrite:

$$\begin{array}{cc} 16 & 18 \\ 15 + (8) + (7) & 15 \\ 7 & 12 \\ 9 + (7) + (5) & 9 \\ 6 + (4) + (2) & 6 \\ 1 & 3 \\ 0 + (2) + (1) & 0 \end{array}$$

Once all the single numbers are less than or equal to the remainders on the line above them, on any line without remainders we just add the two numbers together. The other lines have two copies of some multiple of 3 and two remainders, and we just add one remainder to each multiple of 3, which cannot create any multiples of 3. In this example, this gives $34 + 23 + 22 + 19 + 16 + 14 + 10 + 8 + 4 + 2 + 1$, which is the partition of type $B(n)$ that we started with.

While it is possible to have negative remainders during this procedure, there can't be any at the end of it. If there was one, there must be a lowest line with a negative remainder. It can't be the bottom line, because any number of the form $a + (r_1) + (r_2)$ is at least 3 initially, and this sum never changes throughout the algorithm. If it winds up at the bottom, then it has the form $0 + (r_3) + (r_4)$, where $r_3 + r_4 \geq 3$ and they differ by at most 2, so neither of them can be negative.

The lowest line with negative remainders also can't be directly above a number without remainders, because if it was then the algorithm wouldn't be finished. Therefore, if it exists, it has to be directly above a line with positive remainders, so that we have

$$a + 3 + (r_1) + (r_2) \quad a + 3$$
$$a + (r_3) + (r_4) \quad a$$

with r_2 negative and $r_3 > r_4 > 0$. But $a + 3 + r_1 + r_2$ exceeds $a + r_3 + r_4$ by at least 3, because they're both multiples of 3, so we must have $r_1 + r_2 \geq r_3 + r_4$. But this is impossible with r_2 negative and r_3 and r_4 positive because r_1 exceeds r_2 by no more than 2.

The algorithm cannot create any multiples of 3, because the remainders are never multiples of 3 and neither are the single numbers in the left column. Can we be sure that it creates distinct parts? Whenever two parts come from the same line, as with 16 and 14 above, they must be distinct, because the two remainders are never the same. To prove that parts coming from different lines are never the same, it is enough to show that a part coming from a given line is always larger than a part coming from the line right below it. This would also be good to know since it implies no reordering is necessary after the parts are created.

There are several cases. If neither line has remainders, then this is clear, because the order of the single numbers in the left column never changes and they are weakly decreasing to begin with. In this case two parts coming from consecutive lines must differ by at least 3. If both lines have remainders, then as above we must have $r_1 + r_2 \geq r_3 + r_4$. We can assume $r_1 > r_2 > 0$ and $r_3 > r_4 > 0$, and we claim that $a + 3 + r_2 > a + r_3$. If not, then $r_3 \geq r_2 + 3$, so $r_3 \geq r_1 + 1$. This makes $r_4 \geq r_1 - 1$, so $r_3 + r_4 \geq 2r_1 > r_1 + r_2$, a contradiction.

If the top line has remainders and the bottom line does not, then we have

$$a + 3 + (r_1) + (r_2) \quad a + 3$$
$$b \quad a$$

with $r_1 > r_2 > 0$. We must have $b \leq r_2$, because otherwise the algorithm wouldn't have terminated, so $a + 3 + r_2$ exceeds $a + b$ by at least 3. Finally, if the bottom line has remainders and the top line does not, say

$$b \quad a + 3$$
$$a + (r_1) + (r_2) \quad a$$

with $r_1 > r_2 > 0$, there are two possibilities. If b was always above $a + (r_1) + (r_2)$, then $b > a + r_1 + r_2$ (the inequality must be strict because $a + r_1 + r_2$ is divisible by 3 and b is not), so it is clear that $a + b + 3 > a + r_1 > a + r_2$. If b was moved above $a + (r_1) + (r_2)$ by the algorithm, then before this move we had

$$a + 3 + (r_2) + (r_1 - 3) \quad a + 3$$
$$b \qquad\qquad\qquad\qquad a$$

so $b > r_1 - 3$, so again $a + b + 3 > a + r_1 > a + r_2$.

Exercises

1. Explain why Glaisher's bijection proves that $A(n) = B(n)$.
2. Does Sylvester's fishhook bijection prove that $A(n) = B(n)$?
3. Apply the bijections of Glaisher and Bressoud to the partitions counted by $A(17)$, $B(17)$, and $C(17)$, to see which ones match up with one another.
4. Here is a result from real analysis that is sometimes useful in q-analysis. In particular, it is needed in one of the parts of the next problem.

    THEOREM 82 (Abel's limit theorem). *If $a_n \to a$ as $n \to \infty$ and a_n is not infinite for $n \geq 0$, then*

    $$\lim_{x \to 1^-} (1-x) \sum_{n=0}^{\infty} a_n x^n = a.$$

    The intuition is that the series must behave essentially like the geometric series

    $$\sum_{n=0}^{\infty} a x^n = \frac{a}{1-x} \quad \text{for } |x| < 1.$$

    The limit is taken from below because (as Abel proved) the series does not converge for $x > 1$. We outline a typical real analysis argument: let ϵ be a very small positive number. Since $a_n \to a$ as $n \to \infty$, there must be a positive integer N such that $|a_n - a| < \epsilon$ for all $n \geq N$, no matter how small an ϵ we chose. Now write

    $$\sum_{n=0}^{\infty} a_n x^n = \sum_{n=0}^{N-1} a_n x^n + \sum_{n=N}^{\infty} (a_n - a) x^n + \sum_{n=N}^{\infty} a x^n = S_1 + S_2 + S_3.$$

    (i) Explain why $\lim_{x \to 1^-} (1-x) S_1 = 0$.

    (ii) Explain why $\lim_{x \to 1^-} (1-x) S_3 = a$.

    (iii) Explain why $\lim_{x \to 1^-} (1-x) |S_2| < \epsilon$, and why this together with (i) and (ii) proves Abel's limit theorem.

5. In this problem we outline Andrews's analytic proof of Schur's theorem, which is shorter than Schur's proof. Let $s_j(n, k)$ denote the number of Schur partitions of n with exactly k parts and with all parts larger than j, where we define $s_j(0,0) = 1$. Note that $s_j(n,k) = 0$ if either n or k is nonpositive, unless they are both zero. Then $C(n) = \sum_{k \geq 0} s_0(n,k)$.

(i) Andrews first observes that

$$s_0(n,k) - s_1(n,k) = s_0(n - 3k + 2, k - 1),$$
$$s_1(n,k) - s_2(n,k) = s_1(n - 3k + 1, k - 1),$$
$$s_2(n,k) - s_3(n,k) = s_3(n - 3k, k - 1),$$
$$s_3(n,k) = s_0(n - 3k, k).$$

For the first equality, the left side is the number of Schur partitions of n with k parts whose smallest part is 1, which means that the next smallest part is at least 4. If we delete the 1 and subtract 3 from the other $k-1$ parts, we get a Schur partition of $n - 3(k-1) - 1 = n - 3k + 2$ that has exactly $k-1$ parts. Give similar proofs of the other three equalities. Be sure to explain why we have 3 and not 2 as the subscript on the right side of the third equality.

(ii) Next Andrews sets

$$S_j(x) = \sum_{n,k \geq 0} s_j(n,k) x^k q^n.$$

Explain why $S_j(0) = 1$, and use (i) to show the four equalities

$$S_0(x) - S_1(x) = xq S_0(xq^3),$$
$$S_1(x) - S_2(x) = xq^2 S_1(xq^3),$$
$$S_2(x) - S_3(x) = xq^3 S_3(xq^3),$$
$$S_3(x) = S_0(xq^3).$$

(iii) Use (ii) to show that

$$S_0(x) = \left(1 + xq + xq^2\right) S_0(xq^3) + xq^3(1 - xq^3) S_0(xq^6).$$

(iv) To simplify (iii) slightly set $g(x) = S_0(x)/(x; q^3)_\infty$. Show that $g(0) = 1$ and that

$$(1-x) g(x) = \left(1 + xq + xq^2\right) g(xq^3) + xq^3 g(xq^6).$$

(v) Andrews now looks for a power series solution of the recurrence in (iv). If we set $g(x) = \sum_{n=0}^{\infty} a_n(q) x^n$, show that $a_0(q) = 1$ and that

$$a_n(q) = a_{n-1}(q) \frac{(1 + q^{3n-2})(1 + q^{3n-1})}{1 - q^{3n}},$$

and explain why this implies that

$$a_n(q) = \frac{(-q; q^3)_n (-q^2; q^3)_n}{(q^3; q^3)_n}.$$

(vi) Andrews now has

(10.4.1) $$\sum_{n,k \geq 0} s_0(n,k) x^k q^n = S_0(x) = (x; q^3)_\infty \sum_{n=0}^{\infty} \frac{(-q; q^3)_n (-q^2; q^3)_n}{(q^3; q^3)_n} x^n,$$

and he is trying to say something about $C(n) = \sum_{k \geq 0} s_0(n,k)$. Explain why

$$S_0(1) = \sum_{n=0}^{\infty} C(n) q^n.$$

(vii) It is not quite straightforward to get $S_0(1)$ from the right side of (10.4.1) because $(1; q^3)_\infty = 0$. Andrews is able to think of several ways to surmount this obstacle. One is to apply Abel's limit theorem from the previous problem, since

$$S_0(1) = \lim_{x \to 1^-} (xq^3; q^3)_\infty (1-x) \sum_{n=0}^{\infty} \frac{(-q; q^3)_n (-q^2; q^3)_n}{(q^3; q^3)_n} x^n.$$

Show that this gives $S_0(1) = (-q; q^3)_\infty (-q^2; q^3)_\infty$, and explain why this proves that $C(n) = B(n)$.

6. In the previous problem we wanted to find $S_0(1)$, when direct substitution of $x = 1$ gives zero times infinity. Andrews also uses Heine's intermediate transformation (5.5.9), which was

(10.4.2) $$\quad {}_2\phi_1\left(\begin{matrix} a, b \\ c \end{matrix}; q, z\right) = \frac{(\frac{c}{b}; q)_\infty (bz; q)_\infty}{(c; q)_\infty (z; q)_\infty} {}_2\phi_1\left(\begin{matrix} \frac{abz}{c}, b \\ bz \end{matrix}; q, \frac{c}{b}\right),$$

to get around this, where $q, z, \frac{c}{b}$ are all less than 1 in absolute value.

(i) Explain why

$$S_0(x) = (x; q^3)_\infty \, {}_2\phi_1\left(\begin{matrix} -q, -q^2 \\ 0 \end{matrix}; q^3, x\right).$$

(ii) Explain why

$$\left(\frac{abx}{c}; q^3\right)_n \left(\frac{c}{b}\right)^n = \left(\frac{c}{b} - ax\right)\left(\frac{c}{b} - axq^3\right) \cdots \left(\frac{c}{b} - axq^{3n-3}\right).$$

(iii) By changing q to q^3 in (10.4.2) and using (ii), show that

$$S_0(x) = (-xq^2; q^3)_\infty \sum_{n=0}^{\infty} \frac{(-q^2; q^3)_n \, x^n q^{\frac{n(3n-1)}{2}}}{(-xq^2; q^3)_n (q^3; q^3)_n},$$

and hence

$$S_0(1) = (-q^2; q^3)_\infty \sum_{n=0}^{\infty} \frac{q^{\frac{n(3n-1)}{2}}}{(q^3; q^3)_n}.$$

(iv) We know how to do the remaining sum in (iii) (why?), so

$$S_0(1) = (-q; q^3)_\infty (-q^2; q^3)_\infty$$

as in the previous problem.

7. Schur's proof of his theorem is quite indirect. It starts with the tridiagonal determinant

$$D_n = \begin{vmatrix} 1+aq & q^3-q^4 & 0 & \cdots & 0 & 0 \\ -1 & 1+aq^2 & q^4-q^6 & \cdots & 0 & 0 \\ 0 & -1 & 1+aq^3 & \cdots & 0 & 0 \\ \vdots & \vdots & \vdots & \ddots & \vdots & \vdots \\ 0 & 0 & 0 & \cdots & 1+aq^{n-1} & q^{n+1}-q^{2n} \\ 0 & 0 & 0 & \cdots & -1 & 1+aq^n \end{vmatrix}$$

where the ii entry is $1+aq^i$ for $1 \leq i \leq n$, the $i, i-1$ entry is -1 for $2 \leq i \leq n$, the $i-1, i$ entry is $q^{i+1} - q^{2i}$ for $2 \leq i \leq n$, all the other entries are zero, and we define $D_0 = 1$.

(i) By expanding on the last row, show that
$$D_n = (1 + aq^n)D_{n-1} + (q^{n+1} - q^{2n})D_{n-2} \quad \text{for } n \geq 2.$$

(ii) Next we have the product
$$P_n(a,q) = \prod_{j=1}^{n} \left(1 + aq^j + q^{2j+1}\right), \quad \text{where } P_0(a,q) = 1,$$

and the sum
$$S_n = \sum_{k=0}^{n} (-1)^k \binom{n}{k}_q q^{k(n+2) - \binom{k}{2}} P_{n-k}(a,q).$$

Show that $S_n = D_n$ for $n = 0, 1, 2$. For future use, also explain why
$$P_{n-k}(a,q) = \left(1 + aq^{n-k} + q^{2n-2k+1}\right) P_{n-k-1}(a,q).$$

(iii) Using
$$\binom{n}{k}_q = \binom{n-1}{k}_q + q^{n-k} \binom{n-1}{k-1}_q$$

in the definition of S_n, show that
$$S_n = \sum_{k=0}^{n-1} (-1)^k \binom{n-1}{k}_q q^{k(n+1) - \binom{k}{2}} P_{n-1-k}(a,q) \left(q^k + aq^n\right) \quad \text{for } n \geq 1.$$

(iv) By writing $q^k + aq^n = 1 + aq^n - (1 - q^k)$, show that (iii) implies
$$S_n = (1 + aq^n)S_{n-1} + (q^{n+1} - q^{2n})S_{n-2} \quad \text{for } n \geq 2,$$

and explain why this proves that $S_n = D_n$ for $n \geq 0$. The proof continues in the next problem.

8. It is now convenient to use a different letter in Schur's determinant. Following Schur we use t, so that
$$D_n = \begin{vmatrix} 1 + at & t^3 - t^4 & 0 & \cdots & 0 & 0 \\ -1 & 1 + at^2 & t^4 - t^6 & \cdots & 0 & 0 \\ 0 & -1 & 1 + at^3 & \cdots & 0 & 0 \\ \vdots & \vdots & \vdots & \ddots & \vdots & \vdots \\ 0 & 0 & 0 & \cdots & 1 + at^{n-1} & t^{n+1} - t^{2n} \\ 0 & 0 & 0 & \cdots & -1 & 1 + at^n \end{vmatrix}$$

and we have $D_0 = 1$, $D_1 = 1 + at$, and
$$D_n = (1 + at^n)D_{n-1} + (t^{n+1} - t^{2n})D_{n-2} \quad \text{for } n \geq 2.$$

Schur now sets $a = x + y$ and $t = xy$ for new variables x and y, and then defines three sets of functions $\phi_n, \phi'_n, \phi''_n$ by $\phi_0 = 1 + x$, $\phi'_0 = y$, $\phi''_0 = xy = t$, $\phi_1 = xt(1+x)$, $\phi'_1 = yt(1+x+y)$, $\phi''_1 = t^2(1+x+y)$, and for $n \geq 2$
$$\phi_n = xt^n \left(\phi_0 + \phi'_0 + \phi''_0 + \cdots + \phi_{n-2} + \phi'_{n-2} + \phi''_{n-2} + \phi_{n-1}\right),$$
$$\phi'_n = yt^n \left(\phi_0 + \phi'_0 + \phi''_0 + \cdots + \phi_{n-2} + \phi'_{n-2} + \phi''_{n-2} + \phi_{n-1} + \phi'_{n-1}\right),$$
$$\phi''_n = t^{n+1} \left(\phi_0 + \phi'_0 + \phi''_0 + \cdots + \phi_{n-2} + \phi'_{n-2} + \phi''_{n-2} + \phi_{n-1} + \phi'_{n-1}\right).$$

(Note that the last equality does not include ϕ''_{n-1}.) He also sets

$$T_n = \sum_{k=0}^{n} (\phi_k + \phi'_k + \phi''_k).$$

The goal of this problem is to show that $T_n = (1+x)(1+y)D_n$ for $n \geq 0$.

(i) Show that $T_0 = (1+x)(1+y)$ and $T_1 = (1+x)(1+y)(1+at)$.

(ii) Next, Schur sets $\Delta_n = T_n - (1 + (x+y)t^n) T_{n-1}$ for $n \geq 1$. Explain why our goal will be achieved if we can show that $\Delta_n = (t^{n+1} - t^{2n}) T_{n-2}$ for $n \geq 2$.

(iii) Show that
$$\phi_n = xt^n (T_{n-2} + \phi_{n-1}),$$
$$\phi'_n = yt^n (T_{n-2} + \phi_{n-1} + \phi'_{n-1}),$$
$$\phi''_n = t^{n+1} (T_{n-2} + \phi_{n-1} + \phi'_{n-1}).$$

(iv) Show that
$$\phi'_n = \frac{y}{x} \phi_n + yt^n \phi'_{n-1},$$
$$\phi''_n = x\phi'_n = y\phi_n + t^{n+1} \phi'_{n-1}.$$

(v) Show that
$$\phi_n = xt^n (T_{n-1} - (1+x)\phi'_{n-1})$$
$$\phi'_n = yt^n (T_{n-1} - x\phi'_{n-1}).$$

(vi) On one hand we have $\phi_n + \phi'_n = T_n - T_{n-1} - \phi''_n$, and on the other hand we can get an expression for $\phi_n + \phi'_n$ from (v). By setting these two expressions equal, show that
$$\Delta_n = \phi''_n - xt^n(1+x+y)\phi'_{n-1}.$$

(vii) Show that the result of (vi) can be rewritten as
$$\Delta_n = y\phi'_n - x(1+x)t^n \phi'_{n-1}.$$

(viii) Using (v) with $n-1$ in place of n, show that the result of (vii) can be rewritten as
$$\Delta_n = t^{n+1} T_{n-2} - t^{2n} T_{n-2}.$$

This proves (ii). The proof of Schur's theorem continues in the next problem.

9. Schur finally brings in the Schur partitions at this point. He defines $\psi_1 = 1 + q$, $\psi_2 = q^2$, $\psi_3 = q^3$, $\psi_4 = q^5 + q^4 = q^{4+1} + q^4$, $\psi_5 = q^7 + q^6 + q^5 = q^{5+2} + q^{5+1} + q^5$, $\psi_6 = q^8 + q^7 + q^6 = q^{6+2} + q^{6+1} + q^6$, and so on, so that in general ψ_n is the generating function for Schur partitions with largest part n (except that ψ_1 includes the empty partition of zero).

(i) Explain why
$$\psi_{3k+1} = q^{3k+1} (\psi_1 + \psi_2 + \cdots + \psi_{3k-2}),$$
$$\psi_{3k+2} = q^{3k+2} (\psi_1 + \psi_2 + \cdots + \psi_{3k-1}),$$
$$\psi_{3k+3} = q^{3k+3} (\psi_1 + \psi_2 + \cdots + \psi_{3k-1})$$

for $k \geq 1$. (Note the similarity to the ϕ functions of the previous problem.)

(ii) Explain why taking $x = q$ and $y = q^2$ (so $t = q^3$) makes $\psi_{3k+1} = \phi_k$, $\psi_{3k+2} = \phi'_k$, and $\psi_{3k+3} = \phi''_k$, and hence makes

$$T_n = \psi_1 + \psi_2 + \psi_3 + \cdots + \psi_{3n+1} + \psi_{3n+2} + \psi_{3n+3}$$

the generating function for Schur partitions with largest part less than or equal to $3n + 3$.

(iii) Explain why the previous two problems imply that

$$T_n = (1+q)(1+q^2) \sum_{k=0}^{n} (-1)^k q^{3k(n+2) - 3\binom{k}{2}} \binom{n}{k}_{q^3} P_{n-k}(1+q, q^3).$$

(iv) By simplifying $(1+q)(1+q^2)P_{n-k}(1+q, q^3)$, show that the generating function for Schur partitions with largest part at most $3n$ is

$$T_{n-1} = \sum_{k=0}^{n-1} (-1)^k q^{3k(n+1) - 3\binom{k}{2}} \binom{n-1}{k}_{q^3} (-q; q^3)_{n-k} (-q^2; q^3)_{n-k}.$$

(v) If we let $n \to \infty$ in (iv), assuming as usual that $|q| < 1$, we must get the generating function for all Schur partitions, which is $\sum_{n=0}^{\infty} C(n) q^n$ in Andrews's notation. But if $n \to \infty$, then all the terms tend rapidly to zero in (iv) because of the factor $q^{3k(n+1)}$, except for the $k = 0$ term. This is a bit glib, but explain why it implies

$$\sum_{n=0}^{\infty} C(n) q^n = (-q; q^3)_\infty (-q^2; q^3)_\infty,$$

and why this implies Schur's theorem.

10.5. Bibliographical Notes

Schur's two proofs of the Rogers–Ramanujan identities are in [**214**], but for his bijective proof I have followed Pak's epic paper [**178**]. Pak's comments on the issue of pictures in proofs near the end of [**178**] are very much worth reading. Problems 4–8 in section 10.2 come from [**11**]. The GIS identity comes from [**112**], and is also discussed in [**31**], from whence comes the AKP identity.

The sources for the Bressoud–Chapman proof of section 10.2 are [**59**] and [**74**]. Before Chapman's work, a different argument of Bressoud [**60**] was sometimes said to be the simplest proof of the Rogers–Ramanujan identities.

Schur's second partition theorem is in [**215**]. Andrews's analytic proof is in [**19**], as is Bressoud's bijection of section 10.5. The latter appears in greater generality in [**57**]. Abel's limit theorem is essentially Theorem 4 in his great paper [**1**]; the hard part, which we skipped since we did not need it, is to show that the series does not converge for $x > 1$. For further references on the Rogers–Ramanujan identities see the notes for the next chapter.

CHAPTER 11

The Rogers–Ramanujan Identities II: Rogers

11.1. Ramanujan's proof

The Rogers–Ramanujan identities, which we proved by Schur's methods in Chapter 10, have one of the most interesting backstories of any mathematical result. Ramanujan discovered them in India, but could not prove them. When he came to England he showed them to Hardy, the leading British analyst of the time. Hardy could not prove them either, nor could any of the other mathematicians he showed them to. (It is not recorded who these were, except that MacMahon and Perron were two of them. One would expect Littlewood to be another, but he was away from Cambridge doing war work for most of the time that Ramanujan was there.) MacMahon was so impressed with their implications for partitions, which we mentioned in Chapter 10 and will discuss in more detail in the next section, that he published them in the second volume of his book *Combinatory Analysis* in 1916 even though he had no proof. Ramanujan was often criticized by British mathematicians for not giving proofs of his theorems, but in this case it was he who continued to seek the proof. Shortly thereafter, he was browsing through old issues of the *Proceedings of the London Mathematical Society*, and to his amazement he found the identities stated and proved by L. J. Rogers in 1894. Rogers was still alive and still working in Britain; he had been ignored throughout his career, but suddenly he was famous.

After looking at Rogers's work, Ramanujan was able to find a simpler proof of the identities; and, his interest in them rekindled, Rogers was also able to simplify his proof in a manner similar to Ramanujan's. These proofs were published together in 1917, at just about the same time as Schur's paper. Since Germany and England were on opposite sides of the first World War at the time, British mathematicians got used to the name "Rogers–Ramanujan identities" for several years before they learned what Schur had done.

Ramanujan's proof begins with the function $G(x)$ from problem 13 in section 4.2:

$$(11.1.1) \qquad G(x) = 1 + \sum_{n=1}^{\infty} (-1)^n x^{2n} q^{\frac{n(5n-1)}{2}} \frac{(xq;q)_{n-1}}{(q;q)_n} (1 - xq^{2n})$$

$$(11.1.2) \qquad = \sum_{n=0}^{\infty} (-1)^n x^{2n} q^{\frac{n(5n+1)}{2}} \frac{(xq;q)_n}{(q;q)_n} (1 - x^2 q^{4n+2}).$$

After he showed that (11.1.1) and (11.1.2) are the same function, Ramanujan defined a new function $H(x)$ by

$$(11.1.3) \qquad H(x) = \frac{G(x)}{1 - xq} - G(xq).$$

Rewrite (11.1.2) as

(11.1.4) $$G(x) = 1 - x^2 q^2 + \sum_{n=1}^{\infty} (-1)^n x^{2n} q^{\frac{n(5n+1)}{2}} \frac{(xq;q)_n}{(q;q)_n} (1 - x^2 q^{4n+2})$$

and substitute it in for $G(x)$ in its occurrence in (11.1.3), while substituting (11.1.1) in for the occurrence of $G(xq)$. This gives

$$H(x) = 1 + xq + \sum_{n=1}^{\infty} (-1)^n x^{2n} q^{\frac{n(5n+1)}{2}} \frac{(xq^2;q)_{n-1}}{(q;q)_n} (1 - x^2 q^{4n+2})$$

$$- 1 - \sum_{n=1}^{\infty} (-1)^n (xq)^{2n} q^{\frac{n(5n-1)}{2}} \frac{(xq^2;q)_{n-1}}{(q;q)_n} (1 - xq^{2n+1})$$

$$= xq + \sum_{n=1}^{\infty} (-1)^n x^{2n} q^{\frac{n(5n+1)}{2}} \frac{(xq^2;q)_{n-1}}{(q;q)_n} \left[1 - x^2 q^{4n+2} - q^n (1 - xq^{2n+1}) \right].$$

Ramanujan resists the temptation to factor $1 - xq^{2n+1}$ out of the term in brackets, and instead rewrites

$$1 - x^2 q^{4n+2} - q^n(1 - xq^{2n+1}) = 1 - q^n + xq^{3n+1}(1 - xq^{n+1}).$$

Then we have

$$H(x) = xq + \sum_{n=1}^{\infty} (-1)^n x^{2n} q^{\frac{n(5n+1)}{2}} \frac{(xq^2;q)_{n-1}}{(q;q)_n} \left[(1 - q^n) + xq^{3n+1}(1 - xq^{n+1}) \right]$$

$$= xq + \sum_{n=1}^{\infty} (-1)^n x^{2n} q^{\frac{n(5n+1)}{2}} \frac{(xq^2;q)_{n-1}}{(q;q)_{n-1}}$$

$$+ \sum_{n=1}^{\infty} (-1)^n x^{2n+1} q^{3n+1+\frac{n(5n+1)}{2}} \frac{(xq^2;q)_n}{(q;q)_n}.$$

Note that xq is the $n = 0$ term of the latter sum. If we also replace n by $n + 1$ in the first sum, then we have (writing the second sum first)

$$H(x) = \sum_{n=0}^{\infty} (-1)^n x^{2n+1} q^{\frac{(n+1)(5n+2)}{2}} \frac{(xq^2;q)_n}{(q;q)_n}$$

$$- \sum_{n=0}^{\infty} (-1)^n x^{2n+2} q^{\frac{(n+1)(5n+6)}{2}} \frac{(xq^2;q)_n}{(q;q)_n}$$

$$= \sum_{n=0}^{\infty} (-1)^n x^{2n+1} q^{\frac{(n+1)(5n+2)}{2}} \frac{(xq^2;q)_n}{(q;q)_n} (1 - xq^{2n+2}).$$

Since

$$x^{2n+1} q^{\frac{(n+1)(5n+2)}{2}} = xq \cdot x^{2n} q^{\frac{5n^2+7n}{2}} = xq \cdot (xq^2)^{2n} q^{\frac{5n^2-n}{2}},$$

$H(x)$ has a factor of xq. It also has a factor of $1 - xq^2$, for the $n = 0$ term is $xq(1 - xq^2)$ and otherwise $1 - xq^2$ is the first factor of $(xq^2;q)_n$. Therefore

$$H(x) = xq(1 - xq^2) \left[1 + \sum_{n=1}^{\infty} (-1)^n (xq^2)^{2n} q^{\frac{n(5n-1)}{2}} \frac{(xq^3;q)_{n-1}}{(q;q)_n} (1 - xq^{2n+2}) \right].$$

Comparing this with (11.1.1) we finally see that

$$H(x) = xq(1 - xq^2) G(xq^2),$$

and using this in (11.1.3) gives

(11.1.5) $$\frac{G(x)}{1-xq} - G(xq) = xq(1-xq^2)\,G(xq^2).$$

The argument gets easier from here. To simplify (11.1.5) Ramanujan set

(11.1.6) $$G(x) = R(x)\,(xq;q)_\infty$$

for a new function $R(x)$ (which Ramanujan called $F(x)$). Substituting (11.1.6) in (11.1.5) gives

(11.1.7) $$R(x) = R(xq) + xq\,R(xq^2).$$

We leave this as an exercise. Since $1 = G(0) = R(0)\,(0;q)_\infty = R(0)$, either of two previous exercises, problem 11 of section 3.6 or parts (v)–(vii) of problem 9 in section 10.2, shows that

(11.1.8) $$R(x) = \sum_{n=0}^{\infty} \frac{q^{n^2} x^n}{(q;q)_n}.$$

Putting all of this together, we have proved that

(11.1.9) $$(xq;q)_\infty \sum_{n=0}^{\infty} \frac{q^{n^2} x^n}{(q;q)_n} = 1 + \sum_{n=1}^{\infty} (-1)^n\, x^{2n}\, q^{\frac{n(5n-1)}{2}}\, \frac{(xq;q)_{n-1}}{(q;q)_n}\,(1-xq^{2n})$$

(11.1.10) $$= \sum_{n=0}^{\infty} (-1)^n\, x^{2n}\, q^{\frac{n(5n+1)}{2}}\, \frac{(xq;q)_n}{(q;q)_n}\,(1-x^2 q^{4n+2}).$$

Now set $x = 1$ in (11.1.9). This gives

$$(q;q)_\infty \sum_{n=0}^{\infty} \frac{q^{n^2}}{(q;q)_n} = 1 + \sum_{n=1}^{\infty} (-1)^n\, q^{\frac{n(5n-1)}{2}}\, \frac{(q;q)_{n-1}}{(q;q)_n}\,(1-q^{2n})$$

$$= 1 + \sum_{n=1}^{\infty} (-1)^n\, q^{\frac{n(5n-1)}{2}}\, \frac{1-q^{2n}}{1-q^n}$$

$$= 1 + \sum_{n=1}^{\infty} (-1)^n\, q^{\frac{n(5n-1)}{2}}\,(1+q^n)$$

$$= 1 + \sum_{n=1}^{\infty} (-1)^n\, \left\{ q^{\frac{n(5n-1)}{2}} + q^{\frac{n(5n+1)}{2}} \right\}.$$

Since

$$\frac{(-n)(5(-n)+1)}{2} = \frac{n(5n-1)}{2},$$

we can rewrite this as

(11.1.11) $$(q;q)_\infty \sum_{n=0}^{\infty} \frac{q^{n^2}}{(q;q)_n} = \sum_{n=-\infty}^{\infty} (-1)^n\, q^{\frac{n(5n+1)}{2}}.$$

Now we need the Jacobi triple product with base q^5:

(11.1.12) $$(x;q^5)_\infty \left(\frac{q^5}{x};q^5\right)_\infty (q^5;q^5)_\infty = \sum_{n=-\infty}^{\infty} (-1)^n\, q^{\frac{5n(n-1)}{2}}\, x^n.$$

Taking $x = q^3$ here gives

$$(q^3; q^5)_\infty (q^2; q^5)_\infty (q^5; q^5)_\infty = \sum_{n=-\infty}^{\infty} (-1)^n q^{\frac{n(5n+1)}{2}},$$

and comparing this with (11.1.11) we have

(11.1.13) $$\sum_{n=0}^{\infty} \frac{q^{n^2}}{(q;q)_n} = \frac{(q^3; q^5)_\infty (q^2; q^5)_\infty (q^5; q^5)_\infty}{(q;q)_\infty}.$$

Simplifying the right side of (11.1.13) we finally reach

THEOREM 83 (The first Rogers–Ramanujan identity). *If* $|q| < 1$, *then*

$$\sum_{n=0}^{\infty} \frac{q^{n^2}}{(q;q)_n} = \frac{1}{(q; q^5)_\infty (q^4; q^5)_\infty}.$$

If instead we set $x = q$ in (11.1.8) we get

$$(q^2; q)_\infty \sum_{n=0}^{\infty} \frac{q^{n^2+n}}{(q;q)_n} = 1 + \sum_{n=1}^{\infty} (-1)^n q^{2n} q^{\frac{n(5n-1)}{2}} \frac{(q^2;q)_{n-1}}{(q;q)_n} (1 - q^{2n+1})$$

$$= 1 + \sum_{n=1}^{\infty} (-1)^n q^{\frac{n(5n+3)}{2}} \frac{1 - q^{2n+1}}{1 - q},$$

or, multiplying through by $1 - q$,

$$(q; q)_\infty \sum_{n=0}^{\infty} \frac{q^{n^2+n}}{(q;q)_n} = 1 - q + \sum_{n=1}^{\infty} (-1)^n q^{\frac{n(5n+3)}{2}} (1 - q^{2n+1})$$

$$= \sum_{n=0}^{\infty} (-1)^n q^{\frac{n(5n+3)}{2}} (1 - q^{2n+1})$$

$$= \sum_{n=0}^{\infty} (-1)^n q^{\frac{n(5n+3)}{2}} + \sum_{n=0}^{\infty} (-1)^{n+1} q^{\frac{(n+1)(5n+2)}{2}}.$$

In the last sum we replace $n + 1$ by $-m$; and for consistency we rename n as m in the first sum on the right. Since

$$\frac{(n+1)(5n+2)}{2} \quad \text{becomes} \quad \frac{(-m)(-5m - 5 + 2)}{2}, \quad \text{which equals} \quad \frac{m(5m+3)}{2},$$

this gives

$$(q;q)_\infty \sum_{n=0}^{\infty} \frac{q^{n^2+n}}{(q;q)_n} = \sum_{m=0}^{\infty} (-1)^m q^{\frac{m(5m+3)}{2}} + \sum_{m=-1}^{-\infty} (-1)^m q^{\frac{m(5m+3)}{2}}$$

(11.1.14) $$= \sum_{m=-\infty}^{\infty} (-1)^m q^{\frac{m(5m+3)}{2}}.$$

Taking $x = q^4$ in (11.1.12) gives

$$(q^4; q^5)_\infty (q; q^5)_\infty (q^5; q^5)_\infty = \sum_{n=-\infty}^{\infty} (-1)^n q^{\frac{n(5n+3)}{2}},$$

and using this in (11.1.14) we get
$$\sum_{n=0}^{\infty} \frac{q^{n^2+n}}{(q;q)_n} = \frac{(q^4;q^5)_\infty (q;q^5)_\infty (q^5;q^5)_\infty}{(q;q)_\infty}.$$

Simplifying the right side we finally reach

THEOREM 84 (The second Rogers–Ramanujan identity). *If $|q| < 1$, then*
$$\sum_{n=0}^{\infty} \frac{q^{n^2+n}}{(q;q)_n} = \frac{1}{(q^2;q^5)_\infty (q^3;q^5)_\infty}.$$

Exercises

1. Verify (11.1.7).

2. Explain why
$$\frac{(q^3;q^5)_\infty (q^2;q^5)_\infty (q^5;q^5)_\infty}{(q;q)_\infty} = \frac{1}{(q;q^5)_\infty (q^4;q^5)_\infty}$$
and why
$$\frac{(q^4;q^5)_\infty (q;q^5)_\infty (q^5;q^5)_\infty}{(q;q)_\infty} = \frac{1}{(q^2;q^5)_\infty (q^3;q^5)_\infty}.$$

3. The text of this section has the entire contents of Ramanujan's paper except for one thing: the Rogers–Ramanujan continued fraction. Ramanujan considered a function $K(x)$ defined by

(11.1.15) $$K(x) = \frac{G(x)}{(1-xq)\,G(xq)},$$

with $G(x)$ as above.

(i) Use (11.1.5) to show that
$$K(x) = 1 + \frac{xq(1-xq^2)\,G(xq^2)}{G(xq)}.$$

(ii) Use (11.1.6) to show that $K(x) = R(x)/R(xq)$.

(iii) Show that the result of (i) can be rewritten as

(11.1.16) $$K(x) = 1 + \frac{xq}{K(xq)}.$$

This leads to a **continued fraction** for $K(x)$.

(iv) Write down the result of replacing x by xq in (11.1.16).

(v) Substitute the result of (iv) into (11.1.16) for $K(xq)$ to show that

$$K(x) = 1 + \cfrac{xq}{1 + \cfrac{xq^2}{K(xq^2)}}.$$

(vi) Write down the result of replacing x by xq^2 in (11.1.16), and substitute it into the result of (v) to show that

$$K(x) = 1 + \cfrac{xq}{1 + \cfrac{xq^2}{1 + \cfrac{xq^3}{K(xq^3)}}}.$$

Continuing in this way we have

$$K(x) = 1 + \cfrac{xq}{1 + \cfrac{xq^2}{1 + \cfrac{xq^3}{1 + \cfrac{xq^4}{\ddots}}}} \quad \text{and} \quad \frac{1}{K(x)} = \cfrac{1}{1 + \cfrac{xq}{1 + \cfrac{xq^2}{1 + \cfrac{xq^3}{1 + \cfrac{xq^4}{\ddots}}}}}.$$

(vii) By setting $x = 1$ in the last continued fraction in (vi) and comparing with (ii), show that

$$\frac{R(q)}{R(1)} = \cfrac{1}{1 + \cfrac{q}{1 + \cfrac{q^2}{1 + \cfrac{q^3}{1 + \cfrac{q^4}{\ddots}}}}}.$$

(viii) Use (vii) and the Rogers–Ramanujan identities to show that

(11.1.17) $$\cfrac{1}{1 + \cfrac{q}{1 + \cfrac{q^2}{1 + \cfrac{q^3}{1 + \cfrac{q^4}{\ddots}}}}} = \frac{(q;q^5)_\infty \, (q^4;q^5)_\infty}{(q^2;q^5)_\infty \, (q^3;q^5)_\infty}.$$

(11.1.17) is the Rogers–Ramanujan continued fraction, which was also found by Schur.

4. By taking the limit of (11.1.17) as $q \to 1$, show that

$$\cfrac{1}{1 + \cfrac{1}{1 + \cfrac{1}{1 + \cfrac{1}{1 + \cfrac{1}{\ddots}}}}} = \prod_{n=1}^{\infty} \frac{(5n-1)(5n-4)}{(5n-2)(5n-3)}.$$

5. By using gamma functions it is possible to evaluate the infinite product in problem 4: it equals
$$\frac{\Gamma\left(\frac{2}{5}\right)\Gamma\left(\frac{3}{5}\right)}{\Gamma\left(\frac{1}{5}\right)\Gamma\left(\frac{4}{5}\right)}, \quad \text{which equals} \quad \frac{\sin\frac{2\pi}{5}}{\sin\frac{\pi}{5}} = \frac{1}{2\cos\frac{\pi}{5}} = \frac{\sqrt{5}-1}{2},$$
using a result from the problems in section 8.2. But the continued fraction in problem 4 can be evaluated much more easily. If
$$R = \cfrac{1}{1 + \cfrac{1}{1 + \cfrac{1}{1 + \cfrac{1}{1 + \cfrac{1}{\ddots}}}}}$$
evaluate R by explaining why it must be the positive solution of $R = 1/(1+R)$. As Dick Askey once observed, this together with the previous two problems constitutes the most complicated proof ever given that
$$\cos\frac{\pi}{5} = \frac{1}{\sqrt{5}-1} = \frac{\sqrt{5}+1}{4}.$$

11.2. The Rogers–Ramanujan identities and partitions

We have proved the Rogers–Ramanujan identities

(11.2.1) $$\sum_{n=0}^{\infty} \frac{q^{n^2}}{(q;q)_n} = \frac{1}{(q;q^5)_\infty (q^4;q^5)_\infty},$$

(11.2.2) $$\sum_{n=0}^{\infty} \frac{q^{n^2+n}}{(q;q)_n} = \frac{1}{(q^2;q^5)_\infty (q^3;q^5)_\infty}$$

where $|q| < 1$. As we mentioned in Chapter 10, MacMahon and Schur realized that the product side of (11.2.1) is the generating function for partitions using only the parts $1, 4, 6, 9, 11, 14, 16, 19, \ldots$; i.e., using only parts which are either one less or one more than a multiple of 5. In other words, the product side of (11.2.1) is the generating function for partitions with parts congruent to 1 or 4 mod 5, or to ± 1 (mod 5). Similarly, the product side of (11.2.2) is the generating function for parts congruent to 2 or 3 mod 5, or to ± 2 (mod 5). In other words, if $\rho_1(m)$ denotes the number of partitions of m with parts congruent to 1 or 4 mod 5, and $\varrho_1(m)$ denotes the number of partitions of m with parts congruent to 2 or 3 mod 5, then

$$\frac{1}{(q;q^5)_\infty (q^4;q^5)_\infty} = \sum_{m=0}^{\infty} \rho_1(m) q^m \quad \text{and} \quad \frac{1}{(q^2;q^5)_\infty (q^3;q^5)_\infty} = \sum_{m=0}^{\infty} \varrho_1(m) q^m.$$

MacMahon was fascinated by (11.2.1) and (11.2.2) because the sum sides also have partition interpretations. Let's look at (11.2.1) first. We know that $1/(q;q)_n$ generates partitions into at most n parts, or into exactly n parts where some parts might be zero. Recall that
$$n^2 = 1 + 3 + 5 + \cdots + (2n-1).$$

If we take a partition with exactly n parts, some of which might be zero, and add 1 to the smallest part, 3 to the next smallest, 5 to the next smallest, and so on, adding $2n-1$ to the largest part, then we get a partition with exactly n (nonzero) parts, which are not only distinct but have a difference of at least two between any two parts. In other words, if $\rho_2(m)$ denotes the number of partitions of m into parts that all differ by at least 2, then the sum side of (11.2.1) is the generating function for $\rho_2(m)$:

$$\sum_{n=0}^{\infty} \frac{q^{n^2}}{(q;q)_n} = \sum_{m=0}^{\infty} \rho_2(m)\, q^m.$$

Therefore we have

THEOREM 85 (The first Rogers–Ramanujan identity, classical partition version). *For any nonnegative integer m we have $\rho_1(m) = \rho_2(m)$. In other words, there are exactly as many partitions of m using only parts $\equiv \pm 1 \pmod{5}$ as there are partitions of m using parts that differ by at least 2 (in other words, the gaps between consecutive parts are ≥ 2).*

For example we consider partitions of 12:

$\pm 1 \pmod 5$	gaps ≥ 2
$11 + 1$	12
$9 + 1 + 1 + 1$	$11 + 1$
$6 + 6$	$10 + 2$
$6 + 4 + 1 + 1$	$9 + 3$
$6 + 1 + 1 + 1 + 1 + 1 + 1$	$8 + 4$
$4 + 4 + 4$	$8 + 3 + 1$
$4 + 4 + 1 + 1 + 1 + 1$	$7 + 4 + 1$
$4 + 1 + 1 + 1 + 1 + 1 + 1 + 1 + 1$	$7 + 5$
$1+1+1+1+1+1+1+1+1+1+1+1$	$6 + 4 + 2$

We can play the same game with the sum side of (11.2.2). Here we use

$$n^2 + n = 2 + 4 + 6 + \cdots + 2n.$$

Starting with a partition with exactly n parts, some of which might be zero, we add 2 to the smallest part, 4 to the next smallest, 6 to the next smallest, and so on, finally adding $2n$ to the largest part. As before this gives partitions with exactly n nonzero parts that differ by at least 2, with one extra condition: 1 is not a part. If $\varrho_2(m)$ denotes the number of partitions of m into parts at least 2 that differ by at least 2, then

$$\sum_{n=0}^{\infty} \frac{q^{n^2+n}}{(q;q)_n} = \sum_{m=0}^{\infty} \varrho_2(m)\, q^m,$$

and therefore we have

THEOREM 86 (The second Rogers–Ramanujan identity, classical partition version). *For any nonnegative integer m we have $\varrho_1(m) = \varrho_2(m)$. In other words, there are exactly as many partitions of m using only parts $\equiv \pm 2 \pmod{5}$ as there are partitions of m using parts that are at least 2 and differ by at least 2.*

For example we again consider partitions of 12:

$$
\begin{array}{cc}
\pm 2 \pmod 5 & \text{parts} \geq 2,\ \text{gaps} \geq 2 \\
12 & 12 \\
8+2+2 & 10+2 \\
7+3+2 & 9+3 \\
3+3+3+3 & 8+4 \\
3+3+2+2+2 & 7+5 \\
2+2+2+2+2+2 & 6+4+2
\end{array}
$$

A simple bijective proof of the Rogers–Ramanujan identities is still not known, and is one of the holy grails of mathematics. It may seem that it should be possible to adapt Schur's proof to this task, and it is—a very complicated bijective proof along these lines was found by Garsia and Milne. It was simplified somewhat by Zeilberger and Bressoud, but their argument still uses the Garsia–Milne "involution principle", a powerful idea that was brought into being by this problem. Unfortunately it does not usually lead to simple bijections. In their paper, Zeilberger and Bressoud expressed doubt that a simple bijection for the Rogers–Ramanujan identities would ever be found.

There are, however, simple bijections between the partitions counted by $\rho_2(m)$ and $\varrho_2(m)$ and some other classes of partitions. Let's go back to the sum side of (11.2.1), and write

$$n^2 = n + n + n + \cdots + n \quad \text{instead of} \quad n^2 = 1 + 3 + 5 + \cdots + (2n-1).$$

If we start with a partition with exactly n parts, some of which might be zero, and we add n to each part, we wind up with a partition with exactly n parts each of which is at least n. In other words, if we define $\rho_3(m)$ to be the number of partitions of m such that each part is at least as large as the number of parts, then

$$\sum_{n=0}^{\infty} \frac{q^{n^2}}{(q;q)_n} = \sum_{m=0}^{\infty} \rho_3(m)\, q^m,$$

and therefore $\rho_1(m) = \rho_2(m) = \rho_3(m)$ for any nonnegative integer m. Let's add a column to a previous table:

$$
\begin{array}{ccc}
\pm 1 \pmod 5 & \text{gaps} \geq 2 & \text{parts} \geq \#\ \text{of parts} \\
11+1 & 12 & 12 \\
9+1+1+1 & 11+1 & 10+2 \\
6+6 & 10+2 & 9+3 \\
6+4+1+1 & 9+3 & 8+4 \\
6\ \text{and six 1's} & 8+4 & 7+5 \\
4+4+4 & 8+3+1 & 6+6 \\
4+4+1+1+1+1 & 7+4+1 & 4+4+4 \\
4\ \text{and eight 1's} & 7+5 & 5+4+3 \\
\text{twelve 1's} & 6+4+2 & 6+3+3
\end{array}
$$

Similarly, on the sum side of (11.2.2) we can write

$$n^2 + n = (n+1) + (n+1) + (n+1) + \cdots + (n+1).$$

If we start with a partition with exactly n parts, some of which might be zero, and add $n+1$ to each part, we wind up with a partition with exactly n parts each of which is at least $n+1$. In other words, if $\varrho_3(m)$ is the number of partitions of m in which each part is larger than the number of parts, then

$$\sum_{n=0}^{\infty} \frac{q^{n^2+n}}{(q;q)_n} = \sum_{m=0}^{\infty} \varrho_3(m)\, q^m,$$

and therefore $\varrho_1(m) = \varrho_2(m) = \varrho_3(m)$ for any nonnegative integer m. Adding this column to a previous table we have

$\pm 2 \pmod 5$	≥ 2, gaps ≥ 2	parts $>$ # of parts
12	12	12
$8+2+2$	$10+2$	$9+3$
$7+3+2$	$9+3$	$8+4$
$3+3+3+3$	$8+4$	$7+5$
$3+3+2+2+2$	$7+5$	$6+6$
$2+2+2+2+2+2$	$6+4+2$	$4+4+4$

Next, let $\rho_4(m)$ denote the number of partitions of m with distinct parts and with each even part being larger than twice the number of odd parts; and let $\varrho_4(m)$ be those partitions counted by $\rho_4(m)$ which also have no 1's. In the previous two tables the middle columns (counted by $\rho_2(12)$ and $\varrho_2(12)$ respectively) correspond to these almost exactly. In the first one every partition in the second column is part of $\rho_4(12)$ except $7+4+1$, which needs to be replaced by $6+5+1$. The second table shows no difference between $\varrho_2(12)$ and $\varrho_4(12)$.

There is a pretty bijection of Bressoud that shows that $\rho_2(m) = \rho_4(m)$ and $\varrho_2(m) = \varrho_4(m)$ for all nonnegative integers m. Suppose we start with a partition of the type counted by $\rho_2(m)$, i.e., with gaps ≥ 2 between parts, for example $19+16+13+11+8+5+2$ (which, since it has no 1's, is also counted by $\varrho_2(m)$ for $m=74$). This has four odd parts and three even ones, and only one of the even ones exceeds 8, so we have to transform the partition by fixing the parts 8 and 2 somehow. Since the partition has gaps ≥ 2, the parts are guaranteed to be at least as large as $1, 3, 5, \ldots$, so let's write the Ferrers diagram in a form that reflects this:

Next, rearrange the • rows so that the ones of odd length come first, in descending order, and then the even-length rows, in descending order. In this example the • rows are in the order $6, 5, 4, 4, 3, 2, 1$, and we rearrange them in the order

5, 3, 1, 6, 4, 4, 2:

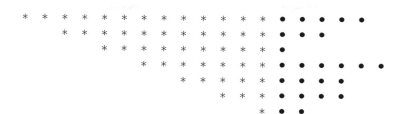

This gives the partition $18+14+10+13+9+7+3$ (or $18+14+13+10+9+7+3$), and now each even part is greater than twice the number of odd parts, and the parts are distinct although they no longer always have gaps of 2. Thus this is one of the partitions counted by $p_4(m)$ (and also by $\varrho_4(m)$, since it has no 1's) when $m = 74$.

The general tendency of Bressoud's bijection is to make the even parts larger and the odd parts smaller, because an odd number of •'s is added to the longer ∗ rows (which always have odd length, so this creates relatively large even parts), and an even number of •'s is added to the shorter ∗ rows (creating relatively small odd parts). The even parts maintain a gap size of at least 2, and so do the odd ones, so the parts stay distinct although an odd part may now differ by only one from an even part. The numbers of odd parts and even parts also stay the same.

Suppose we start (and hence finish) with k parts, and we have j even parts after (and hence before) Bressoud's bijection. Then the largest even part (after the bijection) is at least $(2k-1)+1 = 2k$, and the second largest even part is at least $(2k-3)+1 = 2k-2$, and so on; generally the ith largest even part is at least $[2k-(2i-1)]+1 = 2k-2i+2$. Then the smallest even part is at least $2k-2j+2$, and the number of odd parts is $k-j$, so the bijection does exactly what we want.

Moreover, if we start without a 1 then we always finish without a 1, because the last • row is no longer than any other • row, but this last • row has length at least 1 since the smallest part is at least 2.

What about the other direction? Let's find the partition with gaps ≥ 2 that corresponds to $20 + 17 + 16 + 15 + 9 + 8$, which has distinct parts with each even part greater than twice the number of odd parts. Start by listing the even parts in descending order, followed by the odd parts in descending order: $20, 16, 8, 17, 15, 9$. Then make an indented Ferrers diagram out of these with $11, 9, 7, 5, 3, 1$ highlighted (it is not possible that any of these could be larger than the corresponding number in the partition, since the even and odd parts are distinct and each even part is larger than twice the number of odd parts):

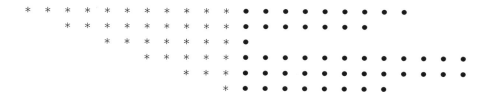

The lengths of the • rows here are 9, 7, 1, 12, 12, 8, and we rearrange them in descending order 12, 12, 9, 8, 7, 1:

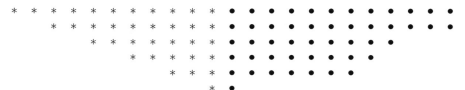

This gives the partition 23+21+16+13+10+2, which has gap size ≥ 2 between all parts. Clearly a partition formed in this way will always have these gaps, because they are built in by the *'s. And again we will not wind up with 1 as a part if we did not have a 1 initially, because each part has at least one * and at least one •. This shows that $\rho_2(m) = \rho_4(m)$ and $\varrho_2(m) = \varrho_4(m)$ for all nonnegative integers m.

Exercises

1. For each of the partitions of 12 where each part is at least as large as the number of parts, draw the Ferrers diagram and the Durfee square. What do you notice? Is it true in general?

2. Write down the conjugate of each partition in problem 1. What property do the Ferrers diagram of these partitions have?

3. Let $\rho_5(m)$ denote the number of partitions of m in which the largest part is less than or equal to the number of copies of the largest part. Explain why $\rho_5(m) = \rho_3(m)$ (and hence all the other ρ's).

4. True or false: the sum side of (11.2.1) is the generating function for partitions where each part is less than or equal to the square root of the sum of the parts. Explain.

5. Does the property you found in problem 1 have an analogue for partitions where each part is larger than the sum of the parts? In other words (with reference to problem 3), is there a natural way to define $\varrho_5(m)$? Explain.

6. Find a simple bijective proof of the Rogers–Ramanujan identities. (If you succeed, you will be famous.)

7. Although problem 6 seems to be very difficult, if not impossible, it is not too hard to construct a bijection between partitions with gaps ≥ 2 between parts and partitions where each part is at least as big as the number of parts, in other words a bijective proof that $\rho_2(m) = \rho_3(m)$ for all nonnegative integers m, by recalling how each type is generated by $q^{k^2}/(q;q)_k$. See if you can do this.

8. Does your bijection in problem 7 also show that $\varrho_2(m) = \varrho_3(m)$ for all nonnegative integers m? Explain.

11.3. Rogers's second proof

Rogers published three different proofs of the Rogers–Ramanujan identities. We will see the third proof in the next chapter. The first two are similar in spirit, but the second is much shorter. We will follow it in part and outline the rest of it in the problems. Problem 12 in section 2.5 showed that

$$\frac{(q;q)_{2n}}{(q;q)_n} = \sum_k (-1)^k q^{\frac{k(3k-1)}{2}} \binom{2n}{n+k}_q,$$

$$\frac{(q;q)_{2n+1}}{(q;q)_n} = \sum_k (-1)^k q^{\frac{k(3k-1)}{2}} \binom{2n+1}{n+k}_q.$$

If we change k to $-k$ and cancel the denominator with the first n factors of the numerator, we get

(11.3.1) $$(q^{n+1};q)_n = \sum_{k=-n}^{n} \binom{2n}{n-k}_q (-1)^k q^{\frac{k(3k+1)}{2}}$$

and

(11.3.2) $$(q^{n+1};q)_{n+1} = \sum_{k=-n-1}^{n} \binom{2n+1}{n-k}_q (-1)^k q^{\frac{k(3k+1)}{2}},$$

respectively. Besides these, we need the identity

(11.3.3) $$\frac{1}{(q;q)_\infty} = \sum_{m=0}^{\infty} \frac{q^{m(m+r)}}{(q;q)_m (q;q)_{m+r}},$$

which holds for an arbitrary nonnegative integer r. This is a special case of (4.2.2); it comes from looking at the largest $m \times (m+r)$ rectangle that fits in the upper left corner of the Ferrers diagram of a partition. The rectangle accounts for the factor of $q^{m(m+r)}$. The region below the rectangle accounts for the factor $1/(q;q)_{m+r}$, since it is a partition whose parts are no larger than $m+r$, and the region to the right of the rectangle accounts for the factor $1/(q;q)_m$ since it is a partition with at most m parts. It is convenient to replace m by $n-k$ for a generic integer k, which gives

(11.3.4) $$\frac{1}{(q;q)_\infty} = \sum_{n=k}^{\infty} \frac{q^{(n-k)(n-k+r)}}{(q;q)_{n-k}(q;q)_{n-k+r}}.$$

If we take $r = 2k$ in (11.3.4) and multiply both sides by q^{k^2}, there results

$$\frac{q^{k^2}}{(q;q)_\infty} = \sum_{n=k}^{\infty} \frac{q^{n^2}}{(q;q)_{n-k}(q;q)_{n+k}}$$

(11.3.5) $$= \sum_{n=k}^{\infty} \frac{q^{n^2}}{(q;q)_{2n}} \binom{2n}{n-k}_q.$$

If we take $r = 2k+1$ in (11.3.4) and multiply both sides by q^{k^2+k}, there results

$$\frac{q^{k^2+k}}{(q;q)_\infty} = \sum_{n=k}^{\infty} \frac{q^{n^2+n}}{(q;q)_{n-k}(q;q)_{n+k+1}}$$

(11.3.6) $$= \sum_{n=k}^{\infty} \frac{q^{n^2}}{(q;q)_{2n+1}} \binom{2n+1}{n-k}_q.$$

Now we multiply both sides of (11.3.5) by

$$(-1)^k q^{k^2 + \binom{k}{2}} (1 + q^k)$$

for every $k \geq 1$, and we sum the result over all $k \geq 0$. This gives

$$(11.3.7) \quad \frac{1}{(q;q)_\infty}\left[1+\sum_{k=1}^\infty (-1)^k q^{2k^2+\binom{k}{2}}(1+q^k)\right]$$

$$=\sum_{n=0}^\infty \frac{q^{n^2}}{(q;q)_{2n}}\binom{2n}{n}_q + \sum_{k=1}^\infty \sum_{n=k}^\infty \frac{q^{n^2}}{(q;q)_{2n}}\binom{2n}{n-k}_q (-1)^k q^{k^2+\binom{k}{2}}(1+q^k).$$

The left side of (11.3.7) is

$$\frac{1}{(q;q)_\infty}\left[1+\sum_{k=1}^\infty (-1)^k\left(q^{\frac{k(5k-1)}{2}}+q^{\frac{k(5k+1)}{2}}\right)\right]=\frac{1}{(q;q)_\infty}\sum_{k=-\infty}^\infty (-1)^k q^{\frac{k(5k+1)}{2}}.$$

Changing the order of summation on the right side of (11.3.7) we have

$$\sum_{n=0}^\infty \frac{q^{n^2}}{(q;q)_{2n}}\binom{2n}{n}_q + \sum_{n=1}^\infty \frac{q^{n^2}}{(q;q)_{2n}}\sum_{k=1}^n \binom{2n}{n-k}_q (-1)^k q^{k^2+\binom{k}{2}}(1+q^k).$$

We can extend the last sum down to $n=0$ because the inner sum is empty in that case, so we have

$$\sum_{n=0}^\infty \frac{q^{n^2}}{(q;q)_{2n}}\left[\binom{2n}{n}_q + \sum_{k=1}^n \binom{2n}{n-k}_q (-1)^k q^{k^2+\binom{k}{2}}(1+q^k)\right].$$

Now

$$\sum_{k=1}^n \binom{2n}{n-k}_q (-1)^k q^{k^2+\binom{k}{2}}(1+q^k) = \sum_{k=1}^n \binom{2n}{n-k}_q (-1)^k \left(q^{\frac{k(3k-1)}{2}}+q^{\frac{k(3k+1)}{2}}\right),$$

and hence

$$\binom{2n}{n}_q + \sum_{k=1}^n \binom{2n}{n-k}_q (-1)^k q^{k^2+\binom{k}{2}}(1+q^k) = \sum_{k=-n}^n \binom{2n}{n-k}_q (-1)^k q^{\frac{k(3k+1)}{2}}$$

$$= (q^{n+1};q)_n$$

by (11.3.1). Therefore (11.3.7) takes the form

$$\frac{1}{(q;q)_\infty}\sum_{k=-\infty}^\infty (-1)^k q^{\frac{k(5k+1)}{2}} = \sum_{n=0}^\infty \frac{q^{n^2}}{(q;q)_{2n}}(q^{n+1};q)_n = \sum_{n=0}^\infty \frac{q^{n^2}}{(q;q)_n},$$

which becomes the first Rogers–Ramanujan identity when we apply Jacobi's triple product to the left side.

To get the second Rogers–Ramanujan identity we multiply both sides of (11.3.6) by

$$(-1)^k q^{k^2+k+\binom{k}{2}}(1-q^{2k+1})$$

for each $k \geq 0$ and then sum over all such k. This gives

$$(11.3.8) \quad \frac{1}{(q;q)_\infty} \sum_{k=0}^{\infty} (-1)^k q^{2k^2+2k+\binom{k}{2}} (1-q^{2k+1})$$

$$= \sum_{k=0}^{\infty} \sum_{n=k}^{\infty} \frac{q^{n^2+n}}{(q;q)_{2n+1}} \binom{2n+1}{n-k}_q (-1)^k q^{k^2+k+\binom{k}{2}} (1-q^{2k+1})$$

$$= \sum_{n=0}^{\infty} \frac{q^{n^2+n}}{(q;q)_{2n+1}} \sum_{k=0}^{n} \binom{2n+1}{n-k}_q (-1)^k q^{k^2+k+\binom{k}{2}} (1-q^{2k+1}).$$

The left side of (11.3.8) is

$$\frac{1}{(q;q)_\infty} \left[\sum_{k=0}^{\infty} (-1)^k q^{\frac{k(5k+3)}{2}} + \sum_{k=0}^{\infty} (-1)^{k+1} q^{\frac{(5k+2)(k+1)}{2}} \right],$$

which becomes

$$\frac{1}{(q;q)_\infty} \sum_{k=-\infty}^{\infty} (-1)^k q^{\frac{k(5k+3)}{2}}$$

if we replace $k+1$ by $-k$ in the latter sum. The inner sum on the right side of (11.3.8) is

$$\sum_{k=0}^{n} \binom{2n+1}{n-k}_q (-1)^k q^{\frac{k(3k+1)}{2}} + \sum_{k=0}^{n} \binom{2n+1}{n-k}_q (-1)^{k+1} q^{\frac{(k+1)(3k+2)}{2}},$$

which becomes

$$\sum_{k=-n-1}^{n} \binom{2n+1}{n-k}_q (-1)^k q^{\frac{k(3k+1)}{2}}$$

if we replace $k+1$ by $-k$ in the latter sum. Using (11.3.2), (11.3.8) takes the form

$$\frac{1}{(q;q)_\infty} \sum_{k=-\infty}^{\infty} (-1)^k q^{\frac{k(5k+3)}{2}} = \sum_{n=0}^{\infty} \frac{q^{n^2+n}}{(q;q)_{2n+1}} (q^{n+1};q)_{n+1} = \sum_{n=0}^{\infty} \frac{q^{n^2+n}}{(q;q)_n},$$

which becomes the second Rogers–Ramanujan identity when we apply Jacobi's triple product to the left side.

Exercises

1. Change q to q^2 in (11.3.5), then multiply both sides by $(x^k + x^{-k})q^{k^2}$ for $k \geq 1$, and then sum over $k \geq 0$. Show that this gives

$$\sum_{n=0}^{\infty} \frac{q^{2n^2}}{(q^2;q^2)_n} (-qx;q^2)_n \left(-\tfrac{q}{x};q^2\right)_n = \frac{(-q^3x;q^6)_\infty \left(-\tfrac{q^3}{x};q^6\right)_\infty}{(q^2;q^6)_\infty (q^4;q^6)_\infty}.$$

2. What happens in the previous problem if we take $x = -q$?

3. Rogers's second proof uses (11.3.1) and (11.3.2) as above, but (11.3.3) is replaced in a way that we will now outline.

  (i) Show that taking $z = xe^{i\theta}$ and $z = xe^{-i\theta}$ in Euler's identity

$$(-z; q)_\infty = \sum_{k=0}^{\infty} \frac{q^{\binom{k}{2}} z^k}{(q; q)_k}$$

and multiplying the results together gives

$$\left(-xe^{i\theta}; q\right)_\infty \left(-xe^{-i\theta}; q\right)_\infty = \sum_{n=0}^{\infty} \frac{q^{\binom{n}{2}} x^n}{(q; q)_n} \sum_{k=0}^{n} \binom{n}{k}_q q^{-k(n-k)} e^{i(n-2k)\theta}.$$

  (ii) Either by changing θ to $-\theta$ in (i) and averaging the result with (i), or by assuming θ and q are real and taking the real part of both sides of (i), show that

$$\left(-xe^{i\theta}; q\right)_\infty \left(-xe^{-i\theta}; q\right)_\infty = \sum_{n=0}^{\infty} \frac{q^{\binom{n}{2}} x^n}{(q; q)_n} \sum_{k=0}^{n} \binom{n}{k}_q q^{-k(n-k)} \cos(n-2k)\theta.$$

  (iii) Rogers now asks himself whether Jacobi's triple product can be used on the product in (ii). Explain why this is only possible for $x = q^{\frac{1}{2}}$.

  (iv) By taking $x = q^{\frac{1}{2}}$ in (ii) and using Jacobi's triple product, show that

$$\frac{1}{(q;q)_\infty} \sum_{k=-\infty}^{\infty} q^{\frac{k^2}{2}} e^{ik\theta} = \sum_{n=0}^{\infty} \frac{q^{\frac{n^2}{2}}}{(q;q)_n} \sum_{k=0}^{n} \binom{n}{k}_q q^{-k(n-k)} \cos(n-2k)\theta.$$

  (v) By using one of the ideas from (ii), show that the result of (iv) can be rewritten as

$$\frac{1}{(q;q)_\infty}\left[1 + 2\sum_{k=1}^{\infty} q^{\frac{k^2}{2}} \cos k\theta\right] = \sum_{n=0}^{\infty} \frac{q^{\frac{n^2}{2}}}{(q;q)_n} \sum_{k=0}^{n} \binom{n}{k}_q q^{-k(n-k)} \cos(n-2k)\theta.$$

  (vi) Rogers's plan is to exploit the fact that the coefficients of $\cos m\theta$ on each side of (v) must be the same for every nonnegative integer m, which is the reason for grouping the $\cos n\theta$ and $\cos(-n\theta)$ terms together in (v). Eventually we will have to do the same thing on the right side of (v), but first Rogers separates the even terms from the odd terms there. Show that the even terms give

$$\frac{1}{(q;q)_\infty}\left[1 + 2\sum_{k=1}^{\infty} q^{2k^2} \cos 2k\theta\right] = \sum_{n=0}^{\infty} \frac{q^{n^2}}{(q;q)_{2n}} \sum_{k=0}^{2n} \binom{2n}{k}_q q^{(n-k)^2} \cos 2(n-k)\theta$$

and that the odd terms give

$$\frac{2}{(q;q)_\infty} \sum_{k=0}^{\infty} q^{2(k+\frac{1}{2})^2} \cos(2n+1)\theta$$

$$= \sum_{n=0}^{\infty} \frac{q^{(n+\frac{1}{2})^2}}{(q;q)_{2n+1}} \sum_{k=0}^{2n+1} \binom{2n+1}{k}_q q^{(n-k+\frac{1}{2})^2} \cos(2n-2k+1)\theta.$$

(vii) By splitting the inner sum in the even case into $0 \leq k \leq n-1$, $k = n$, and $n+1 \leq k \leq 2n$, and changing k to $n-k$ in the first case and to $n+k$ in the third, show that

$$\sum_{k=0}^{2n} \binom{2n}{k}_q q^{(n-k)^2} \cos 2(n-k)\theta = \binom{2n}{n}_q + 2\sum_{k=1}^{n} \binom{2n}{n-k}_q q^{k^2} \cos 2k\theta.$$

(viii) By splitting the inner sum in the odd case into $0 \leq k \leq n$ and $n+1 \leq k \leq 2n+1$, and changing k to $n-k$ in the first case and to $n+k+1$ in the second, show that

$$\sum_{k=0}^{2n+1} \binom{2n+1}{k}_q q^{(n-k+\frac{1}{2})^2} \cos(2n-2k+1)\theta$$

$$= 2\sum_{k=0}^{n} \binom{2n+1}{n-k}_q q^{(k+\frac{1}{2})^2} \cos(2k+1)\theta.$$

(ix) Using (vii) and (viii) in (vi), Rogers now has

$$\frac{1}{(q;q)_\infty} \left[1 + 2\sum_{k=1}^{\infty} q^{2k^2} \cos 2k\theta\right]$$

$$= \sum_{n=0}^{\infty} \frac{q^{n^2}}{(q;q)_{2n}} \left[\binom{2n}{n}_q + 2\sum_{k=1}^{n} \binom{2n}{n-k}_q q^{k^2} \cos 2k\theta\right]$$

in the even case and

$$\frac{2}{(q;q)_\infty} \sum_{k=0}^{\infty} q^{2(k+\frac{1}{2})^2} \cos(2k+1)\theta$$

$$= 2\sum_{n=0}^{\infty} \frac{q^{(n+\frac{1}{2})^2}}{(q;q)_{2n+1}} \sum_{k=0}^{n} \binom{2n+1}{n-k}_q q^{(k+\frac{1}{2})^2} \cos(2k+1)\theta$$

in the odd case. Explain why we can rewrite the odd case as

$$\frac{1}{(q;q)_\infty} \sum_{k=0}^{\infty} q^{2k^2+2k} \cos(2k+1)\theta$$

$$= \sum_{n=0}^{\infty} \frac{q^{n^2+n}}{(q;q)_{2n+1}} \sum_{k=0}^{n} \binom{2n+1}{n-k}_q q^{k^2+k} \cos(2k+1)\theta.$$

(x) Each of the identities in (ix) contains two Fourier series for the same function, so, quoting Rogers, "the coefficients of the several cosines of multiples of θ are equal. This being so, we may replace the cosines by any quantities we please consistent with the convergency of the resulting series."

Show that replacing $2\cos 2k\theta$ by

$$(-1)^k q^{\binom{k}{2}}(1+q^k)$$

in the even case gives (11.3.7), and that replacing $\cos(2k+1)\theta$ by

$$(-1)^k q^{\binom{k}{2}}(1-q^{2k+1})$$

in the odd case gives (11.3.8). Rogers's proof now concludes as in the text.

11.4. More identities of Rogers

In addition to the Rogers–Ramanujan identities, L. J. Rogers proved many other q-series identities. This, as much as anything else, explains why his discovery of the Rogers–Ramanujan identities was overlooked; they were buried in a mass of other material. If he had proved less he might have received more credit, at least in the short run. It is because of the work of Dick Askey and George Andrews in the last quarter of the 20th century that we now see Rogers as a nearly great mathematician unrecognized in his own time. In this section we derive some more of his identities and recast them as statements about partitions. We first establish several lemmas.

Consider the expression

$$(-aq;q^2)_\infty \sum_{k=0}^{\infty} \frac{q^{k^2+k} a^k}{(q^2;q^2)_k (-aq;q^2)_k}.$$

The first k factors of the infinite product outside the sum are cancelled by the $(-aq;q^2)_k$ inside it, so we have

$$(11.4.1) \qquad (-aq;q^2)_\infty \sum_{k=0}^{\infty} \frac{q^{k^2+k} a^k}{(q^2;q^2)_k (-aq;q^2)_k} = \sum_{k=0}^{\infty} \frac{q^{k^2+k} a^k}{(q^2;q^2)_k} (-aq^{2k+1};q^2)_\infty.$$

Now recall Euler's identity

$$(11.4.2) \qquad (-x;q)_\infty = \sum_{k=0}^{\infty} \frac{q^{\binom{k}{2}} x^k}{(q;q)_k}$$

from section 3.6. Replacing q by q^2 in (11.4.2) and then taking $x = aq^{2k+1}$, we can expand the factor $(-aq^{2k+1};q^2)_\infty$ in (11.4.1):

$$(-aq^{2k+1};q^2)_\infty = \sum_{j=0}^{\infty} \frac{q^{2\binom{j}{2}}}{(q^2;q^2)_j} \left(aq^{2k+1}\right)^j$$

$$= \sum_{j=0}^{\infty} \frac{a^j q^{j^2+2kj}}{(q^2;q^2)_j}.$$

Substituting this in (11.4.1) we get

$$(-aq;q^2)_\infty \sum_{k=0}^{\infty} \frac{q^{k^2+k} a^k}{(q^2;q^2)_k (-aq;q^2)_k} = \sum_{k=0}^{\infty} \frac{q^{k^2+k} a^k}{(q^2;q^2)_k} \sum_{j=0}^{\infty} \frac{a^j q^{j^2+2kj}}{(q^2;q^2)_j}$$

$$= \sum_{j=0}^{\infty} \sum_{k=0}^{\infty} \frac{a^{k+j} q^{k^2+2kj+j^2} q^k}{(q^2;q^2)_k (q^2;q^2)_j}.$$

If we set $j+k=n$ here, this becomes

$$(-aq;q^2)_\infty \sum_{k=0}^{\infty} \frac{q^{k^2+k} a^k}{(q^2;q^2)_k (-aq;q^2)_k} = \sum_{n=0}^{\infty} \frac{q^{n^2} a^n}{(q^2;q^2)_n} \sum_{k=0}^{n} \binom{n}{k}_{q^2} q^k.$$

The inner sum equals $(-q;q)_n$ by (2.5.8). Also

$$(11.4.3) \qquad (q^2;q^2)_n = (q;q)_n (-q;q)_n,$$

so we finally have

11.4. MORE IDENTITIES OF ROGERS

LEMMA 11.
$$(-aq;q^2)_\infty \sum_{k=0}^{\infty} \frac{q^{k^2+k} a^k}{(q^2;q^2)_k (-aq;q^2)_k} = \sum_{n=0}^{\infty} \frac{q^{n^2} a^n}{(q;q)_n}.$$

Let's prove another identity of this type. Consider

(11.4.4) $$(aq;q^2)_\infty \sum_{k=0}^{\infty} \frac{q^{k^2} a^k}{(q^2;q^2)_k (aq;q^2)_k} = \sum_{k=0}^{\infty} \frac{q^{k^2} a^k}{(q^2;q^2)_k} (aq^{2k+1};q^2)_\infty,$$

and expand $(aq^{2k+1};q^2)_\infty$ using (11.4.2) as before:

$$(aq^{2k+1};q^2)_\infty = \sum_{j=0}^{\infty} \frac{q^{2\binom{j}{2}}}{(q^2;q^2)_j} \left(-aq^{2k+1}\right)^j$$

$$= \sum_{j=0}^{\infty} \frac{(-1)^j a^j q^{j^2+2kj}}{(q^2;q^2)_j}.$$

Substituting this in (11.4.4) we get

$$(aq;q^2)_\infty \sum_{k=0}^{\infty} \frac{q^{k^2} a^k}{(q^2;q^2)_k (aq;q^2)_k} = \sum_{k=0}^{\infty} \frac{q^{k^2} a^k}{(q^2;q^2)_k} \sum_{j=0}^{\infty} \frac{(-1)^j a^j q^{j^2+2kj}}{(q^2;q^2)_j}$$

$$= \sum_{j=0}^{\infty} \sum_{k=0}^{\infty} \frac{a^{k+j} q^{k^2+2kj+j^2} q^k}{(q^2;q^2)_k (q^2;q^2)_j}$$

$$= \sum_{n=0}^{\infty} \frac{q^{n^2} a^n}{(q^2;q^2)_n} \sum_{j=0}^{n} \binom{n}{j}_{q^2} (-1)^j,$$

where we set $j+k=n$. We have seen the inner sum before. It equals zero if n is odd, because the terms cancel in pairs, so we may as well set $n=2m$ above:

$$(aq;q^2)_\infty \sum_{k=0}^{\infty} \frac{q^{k^2} a^k}{(q^2;q^2)_k (aq;q^2)_k} = \sum_{m=0}^{\infty} \frac{q^{4m^2} a^{2m}}{(q^2;q^2)_{2m}} \sum_{j=0}^{2m} \binom{2m}{j}_{q^2} (-1)^j.$$

If we replace q by q^2 in Gauss's identity (2.5.2), we see that the inner sum equals $(q^2;q^4)_m$. Since we also have $(q^2;q^2)_{2m} = (q^2;q^4)_m (q^4;q^4)_m$ (why?), the final result is

LEMMA 12.
$$(aq;q^2)_\infty \sum_{k=0}^{\infty} \frac{q^{k^2} a^k}{(q^2;q^2)_k (aq;q^2)_k} = \sum_{m=0}^{\infty} \frac{q^{4m^2} a^{2m}}{(q^4;q^4)_m}.$$

We have established two of the four lemmas we will need for Rogers's identities. The other two are

LEMMA 13.
$$(-aq^2;q^2)_\infty \sum_{k=0}^{\infty} \frac{q^{k^2} a^k}{(q^2;q^2)_k (-aq^2;q^2)_k} = \sum_{n=0}^{\infty} \frac{q^{n^2} a^n}{(q;q)_n}.$$

LEMMA 14.
$$(aq;q^2)_\infty \sum_{k=0}^{\infty} \frac{q^{\frac{k(3k-1)}{2}} a^k}{(q;q)_k (aq;q^2)_k} = \sum_{n=0}^{\infty} \frac{q^{2n^2} a^n}{(q^2;q^2)_n}.$$

We leave these as exercises. See problem 12 in section 3.6 for alternative proofs of Lemmas 11 and 13.

For convenience we recall the Rogers–Ramanujan identities:

(11.4.5) $$\sum_{n=0}^{\infty} \frac{q^{n^2}}{(q;q)_n} = \frac{1}{(q;q^5)_\infty (q^4;q^5)_\infty},$$

(11.4.6) $$\sum_{n=0}^{\infty} \frac{q^{n^2+n}}{(q;q)_n} = \frac{1}{(q^2;q^5)_\infty (q^3;q^5)_\infty}.$$

Let's find some other identities of a similar form. We can rewrite (why?)

$$\sum_{n=0}^{\infty} \frac{q^{n^2}}{(q^4;q^4)_n} = \sum_{n=0}^{\infty} \frac{q^{n^2}}{(q^2;q^2)_n (-q^2;q^2)_n}$$

and then transform the right side by taking $a=1$ in Lemma 13. This gives

(11.4.7) $$\sum_{n=0}^{\infty} \frac{q^{n^2}}{(q^4;q^4)_n} = \frac{1}{(-q^2;q^2)_\infty} \sum_{n=0}^{\infty} \frac{q^{n^2}}{(q;q)_n},$$

and hence (11.4.5) gives

$$\sum_{n=0}^{\infty} \frac{q^{n^2}}{(q^4;q^4)_n} = \frac{1}{(-q^2;q^2)_\infty (q;q^5)_\infty (q^4;q^5)_\infty}.$$

Similarly,

$$\sum_{n=0}^{\infty} \frac{q^{n^2+2n}}{(q^4;q^4)_n} = \sum_{n=0}^{\infty} \frac{q^{n^2+2n}}{(q^2;q^2)_n (-q^2;q^2)_n}$$

$$= \frac{1}{(-q^2;q^2)_\infty} \sum_{n=0}^{\infty} \frac{q^{n^2+n}}{(q;q)_n} \quad \text{taking } a=q \text{ in Lemma 11}$$

$$= \frac{1}{(-q^2;q^2)_\infty (q^2;q^5)_\infty (q^3;q^5)_\infty} \quad \text{by (11.4.6)}.$$

We can also rewrite

$$\sum_{n=0}^{\infty} \frac{q^{n^2}}{(q;q)_{2n}} = \sum_{n=0}^{\infty} \frac{q^{n^2}}{(q;q^2)_n (q^2;q^2)_n}$$

and then apply Lemma 12 with $a=1$ to get

$$\sum_{n=0}^{\infty} \frac{q^{n^2}}{(q;q)_{2n}} = \frac{1}{(q;q^2)_\infty} \sum_{n=0}^{\infty} \frac{q^{4n^2}}{(q^4;q^4)_n}.$$

If we now apply (11.4.5) with q replaced by q^4 there results

(11.4.8) $$\sum_{n=0}^{\infty} \frac{q^{n^2}}{(q;q)_{2n}} = \frac{1}{(q;q^2)_\infty (q^4;q^{20})_\infty (q^{16};q^{20})_\infty}.$$

Similarly we can rewrite

$$\sum_{n=0}^{\infty} \frac{q^{n^2+2n}}{(q;q)_{2n+1}} = \sum_{n=0}^{\infty} \frac{q^{n^2+2n}}{(q;q^2)_{n+1} (q^2;q^2)_n} = \frac{1}{1-q} \sum_{n=0}^{\infty} \frac{q^{n^2+2n}}{(q^2;q^2)_n (q^3;q^2)_n}.$$

11.4. MORE IDENTITIES OF ROGERS

Using Lemma 12 with $a = q^2$ we get

$$\sum_{n=0}^{\infty} \frac{q^{n^2+2n}}{(q;q)_{2n+1}} = \frac{1}{1-q} \frac{1}{(q^3;q^2)_\infty} \sum_{n=0}^{\infty} \frac{q^{4n^2+4n}}{(q^4;q^4)_n} = \frac{1}{(q;q^2)_\infty} \sum_{n=0}^{\infty} \frac{q^{4n^2+4n}}{(q^4;q^4)_n}.$$

Finally, replacing q by q^4 in (11.4.6) this becomes

(11.4.9) $$\sum_{n=0}^{\infty} \frac{q^{n^2+2n}}{(q;q)_{2n+1}} = \frac{1}{(q;q^2)_\infty \, (q^8;q^{20})_\infty \, (q^{12};q^{20})_\infty}.$$

An identity less natural in appearance than the above, but easy to prove using Lemma 14, is

$$\sum_{n=0}^{\infty} \frac{q^{\frac{n(3n-1)}{2}}}{(q;q)_n \, (q;q^2)_n} = \frac{1}{(q;q^2)_\infty} \sum_{n=0}^{\infty} \frac{q^{2n^2}}{(q^2;q^2)_n} \quad \text{taking } a = 1 \text{ in Lemma 14}$$

$$= \frac{1}{(q;q^2)_\infty \, (q^2;q^{10})_\infty \, (q^8;q^{10})_\infty} \quad \text{replacing } q \text{ by } q^2 \text{ in (11.4.5)}.$$

This has a companion which is slightly trickier:

$$\sum_{n=0}^{\infty} \frac{q^{\frac{3n(n+1)}{2}}}{(q;q)_n \, (q;q^2)_{n+1}} = \frac{1}{1-q} \sum_{n=0}^{\infty} \frac{q^{\frac{n(3n-1)}{2}+2n}}{(q;q)_n \, (q^3;q^2)_n}$$

$$= \frac{1}{1-q} \frac{1}{(q^3;q^2)_\infty} \sum_{n=0}^{\infty} \frac{q^{2n^2+2n}}{(q^2;q^2)_n} \quad \text{taking } a = q^2 \text{ in Lemma 14}$$

$$= \frac{1}{(q;q^2)_\infty \, (q^4;q^{10})_\infty \, (q^6;q^{10})_\infty} \quad \text{replacing } q \text{ by } q^2 \text{ in (11.4.6)}.$$

And here are two more:

$$\sum_{n=0}^{\infty} \frac{q^{n^2+n}}{(q^2;q^2)_n \, (-q;q^2)_n} = \frac{1}{(-q;q^2)_\infty} \sum_{n=0}^{\infty} \frac{q^{n^2}}{(q;q)_n} \quad \text{taking } a = 1 \text{ in Lemma 11}$$

$$= \frac{1}{(-q;q^2)_\infty \, (q;q^5)_\infty \, (q^4;q^5)_\infty} \quad \text{by (11.4.5)},$$

and

$$\sum_{n=0}^{\infty} \frac{q^{n^2+n}}{(q^2;q^2)_n \, (-q;q^2)_{n+1}} = \frac{1}{1+q} \sum_{n=0}^{\infty} \frac{q^{n^2+n}}{(q^2;q^2)_n \, (-q^3;q^2)_n}$$

$$= \frac{1}{1+q} \frac{1}{(-q^3;q^2)_\infty} \sum_{n=0}^{\infty} \frac{q^{n^2+n}}{(q;q)_n} \quad a = q \text{ in Lemma 13}$$

$$= \frac{1}{(-q;q^2)_\infty \, (q^2;q^5)_\infty \, (q^3;q^5)_\infty} \quad \text{by (11.4.6)}.$$

These last two identities look more natural with q replaced by $-q$. We give the details for the first one. To do the replacement expeditiously let's note that

$$(q;q^5)_\infty = (1-q)(1-q^6)(1-q^{11})(1-q^{16})(1-q^{21})(1-q^{26})\cdots$$
$$= \{(1-q)(1-q^{11})(1-q^{21})\cdots\}\{(1-q^6)(1-q^{16})(1-q^{26})\cdots\}$$
$$= (q;q^{10})_\infty \, (q^6;q^{10})_\infty$$

and similarly

$$(q^4;q^5)_\infty = (q^4;q^{10})_\infty \, (q^9;q^{10})_\infty.$$

Hence
$$\sum_{n=0}^{\infty} \frac{q^{n^2+n}}{(q^2;q^2)_n \, (-q;q^2)_n} = \frac{1}{(-q;q^2)_\infty \, (q;q^{10})_\infty \, (q^4;q^{10})_\infty \, (q^6;q^{10})_\infty \, (q^9;q^{10})_\infty},$$

and if we now replace q by $-q$ this becomes

(11.4.10)
$$\sum_{n=0}^{\infty} \frac{q^{n^2+n}}{(q^2;q^2)_n \, (q;q^2)_n} = \frac{1}{(q;q^2)_\infty \, (-q;q^{10})_\infty \, (q^4;q^{10})_\infty \, (q^6;q^{10})_\infty \, (-q^9;q^{10})_\infty}$$

since $(-1)^{n^2+n} = 1$ for any integer n. The advantage of this is that the denominator of the left side simplifies to $(q;q)_{2n}$, which will help us interpret this as a partition identity. To make further progress toward that we need to rewrite the $(-q;q^{10})_\infty$ and $(-q^9;q^{10})_\infty$ terms. We have

$$\begin{aligned}
(-q;q^{10})_\infty &= (1+q)(1+q^{11})(1+q^{21})(1+q^{31}) \cdots \\
&= \frac{1-q^2}{1-q} \frac{1-q^{22}}{1-q^{11}} \frac{1-q^{42}}{1-q^{21}} \frac{1-q^{62}}{1-q^{31}} \cdots \\
&= \frac{(q^2;q^{20})_\infty}{(q;q^{10})_\infty}
\end{aligned}$$

and similarly

$$(-q^9;q^{10})_\infty = \frac{(q^{18};q^{20})_\infty}{(q^9;q^{10})_\infty}.$$

Then (11.4.10) becomes

$$\sum_{n=0}^{\infty} \frac{q^{n^2+n}}{(q;q)_{2n}} = \frac{(q;q^{10})_\infty \, (q^9;q^{10})_\infty}{(q;q^2)_\infty \, (q^2;q^{20})_\infty \, (q^4;q^{10})_\infty \, (q^6;q^{10})_\infty \, (q^{18};q^{20})_\infty}.$$

One last thing needs to be done to this—we have to cancel the terms in the numerator. We can do this by splitting up $(q;q^2)_\infty$. Since any odd number is congruent to either 1, 3, 5, 7, or 9 mod 10, we have

$$\begin{aligned}
\frac{(q;q^{10})_\infty \, (q^9;q^{10})_\infty}{(q;q^2)_\infty} &= \frac{(q;q^{10})_\infty \, (q^9;q^{10})_\infty}{(q;q^{10})_\infty \, (q^3;q^{10})_\infty \, (q^5;q^{10})_\infty \, (q^7;q^{10})_\infty \, (q^9;q^{10})_\infty} \\
&= \frac{1}{(q^3;q^{10})_\infty \, (q^5;q^{10})_\infty \, (q^7;q^{10})_\infty}.
\end{aligned}$$

This is a good time to introduce a common abbreviation in this subject:

$$(a_1;q)_n \, (a_2;q)_n \cdots (a_k;q)_n =: (a_1, a_2, \ldots, a_k; q)_n,$$

so that the above becomes

$$\frac{(q;q^{10})_\infty \, (q^9;q^{10})_\infty}{(q;q^2)_\infty} = \frac{1}{(q^3, q^5, q^7; q^{10})_\infty}.$$

Then we finally have

(11.4.11)
$$\sum_{n=0}^{\infty} \frac{q^{n^2+n}}{(q;q)_{2n}} = \frac{1}{(q^2, q^{18}; q^{20})_\infty \, (q^3, q^4, q^5, q^6, q^7; q^{10})_\infty}.$$

Similarly, if we replace q by $-q$ in the identity

(11.4.12) $$\sum_{n=0}^{\infty} \frac{q^{n^2+n}}{(q^2;q^2)_n \, (-q;q^2)_{n+1}} = \frac{1}{(-q;q^2)_\infty \, (q^2;q^5)_\infty \, (q^3;q^5)_\infty}$$

and then try to write the result so that the right side looks like a partition generating function, we ultimately reach

(11.4.13) $$\sum_{n=0}^{\infty} \frac{q^{n^2+n}}{(q;q)_{2n+1}} = \frac{1}{(q^6, q^{14}; q^{20})_\infty \, (q, q^2, q^5, q^8, q^9; q^{10})_\infty}.$$

Exercises

1. Verify (11.4.3).
2. Check that $(q^2; q^2)_{2m} = (q^2; q^4)_m \, (q^4; q^4)_m$.
3. Prove Lemma 13 by the same method that we used to prove Lemma 11.
4. Prove Lemma 14 by the same method as Lemmas 11, 12, and 13. You may need the result of problem 5 in section 2.5.
5. Show that replacing q by $-q$ in (11.4.12) leads to (11.4.13).

11.5. Rogers's identities and partitions

The right side of (11.4.11) is the generating function for partitions with parts congruent to 3, 4, 5, 6, or 7 mod 10, or to 2 or 18 mod 20. Since any number congruent to 3 mod 10 is congruent to either 3 or 13 mod 20, we could also write (11.4.11) as

$$\sum_{n=0}^{\infty} \frac{q^{n^2+n}}{(q;q)_{2n}} = \frac{1}{(q^2, q^3, q^4, q^5, q^6, q^7, q^{13}, q^{14}, q^{15}, q^{16}, q^{17}, q^{18}; q^{20})_\infty}$$

and think of the right side as generating partitions with parts congruent to ± 2, ± 3, ± 4, ± 5, ± 6, or ± 7 mod 20. What about the left side? We know that the denominator generates partitions with at most $2n$ parts. In the numerator we can say

$$n^2 + n = 2\binom{n+1}{2} = 2(1 + 2 + 3 + \cdots + n)$$
$$= 1 + 1 + 2 + 2 + 3 + 3 + \cdots + n + n.$$

If we start with a partition with at most $2n$ parts (or exactly $2n$ parts some of which might be 0) and add 1 to the two smallest parts, 2 to the next two smallest, 3 to the next two smallest, and so on, finally adding n to the two largest parts, we get partitions with exactly $2n$ parts (*i.e.*, with an even number of parts) where the second largest part is definitely bigger than the third largest, the fourth largest part is definitely bigger than the fifth largest, and so on. For example, there are sixteen partitions of 12 of each type. For the partitions of 12 with parts congruent to 3, 4, 5, 6, or 7 mod 10, or to 2 or 18 mod 20, the possible parts are 2,3,4,5,6,7, and the partitions are:

$$
\begin{array}{cccc}
7+5 & 7+3+2 & 6+6 & 6+4+2 \\
6+3+3 & 6+2+2+2 & & 5+5+2 \\
5+4+3 & 5+3+2+2 & & 4+4+4 \\
& 4+4+2+2 & & 4+3+3+2 \\
& 4+2+2+2+2 & & 3+3+3+3 \\
& 3+3+2+2+2 & & 2+2+2+2+2+2
\end{array}
$$

The partitions of 12 with an even number of parts where the $2k^{\text{th}}$ part is always bigger than the $2k+1^{\text{st}}$ part for any k are:

$$
\begin{array}{cccccc}
11+1 & 10+2 & 9+3 & 8+4 & 7+5 & 6+6 \\
5+5+1+1 & & 6+4+1+1 & & 7+3+1+1 & \\
8+2+1+1 & & 6+3+2+1 & & 5+4+2+1 & \\
5+3+2+2 & & 4+4+2+2 & & 4+4+3+1 & \\
& & 3+3+2+2+1+1 & & &
\end{array}
$$

Similarly, we can rewrite the right side of (11.4.13) as

$$\sum_{n=0}^{\infty} \frac{q^{n^2+n}}{(q;q)_{2n+1}} = \frac{1}{(q, q^2, q^5, q^6, q^8, q^9, q^{11}, q^{12}, q^{14}, q^{15}, q^{18}, q^{19}; q^{20})_{\infty}}$$

and the right side generates partitions with parts congruent to ± 1, ± 2, ± 5, ± 6, ± 8, or ± 9 mod 20. On the left side, the denominator generates partitions with at most $2n-1$ parts, and we can think of $n^2 + n$ as

$$n^2 + n = 0 + 1 + 1 + 2 + 2 + 3 + 3 + \cdots + n + n$$

in much the same way as before. Now we add 0 to the smallest part (which might already have been 0), 1 to the next two smallest, 2 to the next two smallest, and so on, finally adding n to the two largest parts, and we arrive at the same kind of partitions as before (with the $2k^{\text{th}}$ part always bigger than the $2k+1^{\text{st}}$ part for any k) except that now the number of parts could be either even or odd.

The identity (11.4.8)

$$\sum_{n=0}^{\infty} \frac{q^{n^2}}{(q;q)_{2n}} = \frac{1}{(q;q^2)_{\infty} (q^4;q^{20})_{\infty} (q^{16};q^{20})_{\infty}}$$

has a very similar interpretation. The right side generates partitions where the parts are either odd or congruent to ± 4 mod 20. If desired, we could rewrite (11.4.8) as

$$\sum_{n=0}^{\infty} \frac{q^{n^2}}{(q;q)_{2n}} = \frac{1}{(q, q^3, q^4, q^5, q^7, q^9, q^{11}, q^{13}, q^{14}, q^{15}, q^{17}, q^{19}; q^{20})_{\infty}}.$$

The denominator on the left side generates partitions with at most $2n$ parts. We can rewrite the exponent in the numerator as

$$n^2 = \binom{n+1}{2} + \binom{n}{2}$$
$$= (1 + 2 + \cdots + (n-1) + n) + (0 + 1 + \cdots + (n-2) + (n-1))$$
$$= 0 + 1 + 1 + 2 + 2 + \cdots + (n-1) + (n-1) + n.$$

Then start with a partition with exactly $2n$ parts, some of which might be 0, and add 0 to the smallest part, 1 to the next two smallest, 2 to the next two smallest, and so on, finally adding n to the largest and $n-1$ to the next two largest. This creates a partition with

largest part $>$ 2nd largest part $\geq$ 3rd largest part $>$ 4th largest part $\geq \cdots$,

in other words a partition where the $2k-1^{\text{st}}$ part is larger than the $2k^{\text{th}}$ part for every k. Thus these partitions are equinumerous with the partitions whose parts are either odd or congruent to ± 4 mod 20. Here are the partitions of 8 generated by (11.4.8):

left side	right side
8	7 + 1
7 + 1	5 + 3
6 + 2	5 + 1 + 1 + 1
6 + 1 + 1	4 + 4
5 + 3	4 + 3 + 1
5 + 2 + 1	4 + 1 + 1 + 1 + 1
4 + 3 + 1	3 + 3 + 1 + 1
4 + 2 + 2	3 + 1 + 1 + 1 + 1 + 1
3 + 2 + 2 + 1	1 + 1 + 1 + 1 + 1 + 1 + 1 + 1

The identity (11.4.9)

$$\sum_{n=0}^{\infty} \frac{q^{n^2+2n}}{(q;q)_{2n+1}} = \frac{1}{(q;q^2)_\infty (q^8;q^{20})_\infty (q^{12};q^{20})_\infty}$$

has a similar interpretation. The denominator on the left side generates partitions with at most $2n+1$ parts. We need to rewrite the exponent in the numerator in the same sort of way as before. Since we know already that

$$n^2 = 0 + 0 + 1 + 1 + 2 + 2 + \cdots + (n-1) + (n-1) + n,$$

we can get $n^2 + 2n$ as a sum of $2n+1$ numbers by adding 1 to the first $2n$ of these:

$$n^2 + 2n = 1 + 1 + 2 + 2 + 3 + 3 + \cdots + (n-1) + (n-1) + n + n + n.$$

Then start with a partition with exactly $2n+1$ parts, some of which might be 0, and add 1 to the smallest two parts, 2 to the next two smallest, 3 to the next two smallest, and so on, finally adding n to the three largest parts and $n-1$ to the fourth and fifth largest. This creates a partition with an odd number of parts and with the $2k-1^{\text{st}}$ part larger than the $2k^{\text{th}}$ part for every k except possibly for $k=1$ (the three largest parts could be the same size). The right side of (4.11) generates partitions whose parts are either odd or congruent to ± 8 mod 20; if desired we could rewrite it as

$$\sum_{n=0}^{\infty} \frac{q^{n^2+2n}}{(q;q)_{2n+1}} = \frac{1}{(q,q^3,q^5,q^7,q^8,q^9,q^{11},q^{12},q^{13},q^{15},q^{17},q^{19};q^{20})_\infty}.$$

Thus these two types of partitions are equinumerous. Here are the partitions of 9 generated by (11.4.9):

left side	right side
9	9
$7+1+1$	$8+1$
$6+2+1$	$7+1+1$
$5+3+1$	$5+3+1$
$5+2+2$	$5+1+1+1+1$
$4+4+1$	$3+3+3$
$4+3+2$	$3+3+1+1+1$
$3+3+3$	$3+1+1+1+1+1+1$
$3+2+2+2+1$	$1+1+1+1+1+1+1+1+1$

The bijection of Bressoud at the end of section 11.2 is really a combinatorial proof of Rogers's identity

$$(11.5.1) \qquad \sum_{n=0}^{\infty} \frac{q^{n^2}}{(q^4;q^4)_n} = \frac{1}{(-q^2;q^2)_\infty} \sum_{n=0}^{\infty} \frac{q^{n^2}}{(q;q)_n}.$$

To see this, rewrite (11.5.1) as

$$\sum_{n=0}^{\infty} \frac{q^{n^2}}{(q;q)_n} = (-q^2;q^2)_\infty \sum_{n=0}^{\infty} \frac{q^{n^2}}{(q^2;q^2)_n (-q^2;q^2)_n}$$

$$(11.5.2) \qquad\qquad = \sum_{n=0}^{\infty} \frac{q^{n^2}}{(q^2;q^2)_n} (-q^{2n+2};q^2)_\infty.$$

The denominator on the right side of (11.5.2) generates partitions with at most n parts, all of which are even. Writing

$$n^2 = 1 + 3 + 5 + \cdots + (2n-1)$$

and adding 1 to the smallest (even) part, 3 to the next smallest, and so on, finally adding $2n-1$ to the largest, we get a partition with exactly n parts, which are odd and distinct; these are the parts generated by $q^{n^2}/(q^2;q^2)_n$. Moreover, $(-q^{2n+2};q^2)_\infty$ is the generating function for partitions whose parts are even, distinct, and at least $2n+2$; so each of these parts are more than twice the number of odd parts generated by $q^{n^2}/(q^2;q^2)_n$. If as before $p_4(m)$ denotes the number of partitions of m with distinct parts and with each even part being larger than twice the number of odd parts, then this shows that the right side of (11.5.2) is the generating function for $p_4(m)$; i.e., we have

$$(11.5.3) \qquad \sum_{n=0}^{\infty} \frac{q^{n^2}}{(q^2;q^2)_n} (-q^{2n+2};q^2)_\infty = \sum_{m=0}^{\infty} p_4(m) q^m.$$

11.6. The Göllnitz–Gordon identities

But we saw in section 11.2 that, if $\rho_2(m)$ denotes the number of partitions of m into parts which all differ by at least 2, then

(11.5.4) $$\sum_{n=0}^{\infty} \frac{q^{n^2}}{(q;q)_n} = \sum_{m=0}^{\infty} \rho_2(m)\, q^m.$$

Thus we may use (11.5.3), (11.5.4), and (11.5.2) to prove that $\rho_2(m) = \rho_4(m)$ for all nonnegative integers m. Alternatively, if we combine Bressoud's proof that $\rho_2(m) = \rho_4(m)$ with (11.5.3) and (11.5.4), then we have a partition counting proof of (11.5.2), and hence of (11.5.1).

11.6. The Göllnitz–Gordon identities

In this section we present Krishnaswami Alladi's beautiful proof of a pair of partition identities found almost simultaneously around 1960 by Heinz Göllnitz and Basil Gordon. Alladi's idea is to split Gauss's identity (5.2.12), which was

(11.6.1) $$\sum_{n=0}^{\infty} q^{\binom{n+1}{2}} = \frac{(q^2;q^2)_\infty}{(q;q^2)_\infty},$$

into its even and odd parts. For a function $f(q)$, these are

$$\frac{f(q)+f(-q)}{2} \quad \text{and} \quad \frac{f(q)-f(-q)}{2}$$

respectively, so on the series side of (11.6.1) we have to consider

$$\sum_{n=0}^{\infty} q^{\binom{n+1}{2}} \frac{1+(-1)^{\binom{n+1}{2}}}{2} \quad \text{and} \quad \sum_{n=0}^{\infty} q^{\binom{n+1}{2}} \frac{1-(-1)^{\binom{n+1}{2}}}{2}.$$

Starting at $n = 0$, $(-1)^{\binom{n+1}{2}}$ has the pattern $1, -1, -1, 1, 1, -1, -1, 1, 1, -1, -1, \ldots$, which suggests that we should look at residue classes mod 4.

If $n = 4k$, then
$$\binom{n+1}{2} = \frac{(4k+1)4k}{2} = 2k(4k+1), \quad \text{which is even, so} \quad (-1)^{\binom{n+1}{2}} = 1.$$

If $n = 4k+1$, then
$$\binom{n+1}{2} = \frac{(4k+2)(4k+1)}{2} = (2k+1)(4k+1), \quad \text{which is odd, so} \quad (-1)^{\binom{n+1}{2}} = -1.$$

If $n = 4k+2$, then
$$\binom{n+1}{2} = \frac{(4k+3)(4k+2)}{2} = (2k+1)(4k+3), \quad \text{which is odd, so} \quad (-1)^{\binom{n+1}{2}} = -1.$$

If $n = 4k+3$, then
$$\binom{n+1}{2} = \frac{(4k+4)(4k+3)}{2} = (2k+2)(4k+3), \quad \text{which is even, so} \quad (-1)^{\binom{n+1}{2}} = 1.$$

It follows that

$$\sum_{n=0}^{\infty} q^{\binom{n+1}{2}} \frac{1+(-1)^{\binom{n+1}{2}}}{2} = \sum_{k=0}^{\infty} q^{2k(4k+1)} + \sum_{k=0}^{\infty} q^{(2k+2)(4k+3)}$$
(11.6.2)
$$= \sum_{k=0}^{\infty} q^{8k^2+2k}\left(1+q^{12k+6}\right)$$

and

(11.6.3)
$$\sum_{n=0}^{\infty} q^{\binom{n+1}{2}} \frac{1-(-1)^{\binom{n+1}{2}}}{2} = \sum_{k=0}^{\infty} q^{(2k+1)(4k+1)} + \sum_{k=0}^{\infty} q^{(2k+1)(4k+3)}$$
$$= \sum_{k=0}^{\infty} q^{8k^2+6k+1}\left(1+q^{4k+2}\right).$$

But it is better to combine the series together in a different way. Since

$$\sum_{j=0}^{\infty} q^{(2j+2)(4j+3)} = \sum_{k=-1}^{-\infty} q^{(-2k)(-4k-1)} = \sum_{k=-\infty}^{-1} q^{2k(4k+1)}$$

(by replacing j by $-k-1$), (11.6.2) becomes

$$\sum_{n=0}^{\infty} q^{\binom{n+1}{2}} \frac{1+(-1)^{\binom{n+1}{2}}}{2} = \sum_{k=0}^{\infty} q^{2k(4k+1)} + \sum_{k=-\infty}^{-1} q^{2k(4k+1)}$$
$$= \sum_{k=-\infty}^{\infty} q^{2k(4k+1)}.$$

Similarly, we have

$$\sum_{j=0}^{\infty} q^{(2j+1)(4j+3)} = \sum_{k=-1}^{-\infty} q^{(-2k-1)(-4k-1)} = \sum_{k=-\infty}^{-1} q^{(2k+1)(4k+1)},$$

so (11.6.3) becomes

$$\sum_{n=0}^{\infty} q^{\binom{n+1}{2}} \frac{1-(-1)^{\binom{n+1}{2}}}{2} = \sum_{k=0}^{\infty} q^{(2k+1)(4k+1)} + \sum_{k=-\infty}^{-1} q^{(2k+1)(4k+1)}$$
$$= \sum_{k=-\infty}^{\infty} q^{(2k+1)(4k+1)}.$$

Now we bring in Jacobi's triple product with base q^8:

(11.6.4)
$$(-zq^8; q^{16})_\infty \left(-\frac{q^8}{z}; q^{16}\right)_\infty (q^{16}; q^{16})_\infty = \sum_{k=-\infty}^{\infty} q^{8k^2} z^k.$$

Taking $z = q^2$ in (11.6.4) we have

(11.6.5)
$$\sum_{k=-\infty}^{\infty} q^{2k(4k+1)} = (-q^6; q^{16})_\infty (-q^{10}; q^{16})_\infty (q^{16}; q^{16})_\infty.$$

11.6. THE GÖLLNITZ–GORDON IDENTITIES

Taking $z = q^6$ in (11.6.4) we have

(11.6.6) $$\sum_{k=-\infty}^{\infty} q^{(2k+1)(4k+1)} = q(-q^2; q^{16})_\infty (-q^{14}; q^{16})_\infty (q^{16}; q^{16})_\infty.$$

Thus the products (11.6.5) and (11.6.6) are the even and odd parts respectively of the series side of (11.6.1). The even part of the product side is

(11.6.7) $$\frac{1}{2} \left[\frac{(q^2; q^2)_\infty}{(q; q^2)_\infty} + \frac{(q^2; q^2)_\infty}{(-q; q^2)_\infty} \right],$$

which is

(11.6.8) $$(q^4; q^4)_\infty \frac{1}{2} \left[(-q; q^2)_\infty + (q; q^2)_\infty \right],$$

and similarly the odd part is

(11.6.9) $$(q^4; q^4)_\infty \frac{1}{2} \left[(-q; q^2)_\infty - (q; q^2)_\infty \right].$$

We recall Euler's identity (3.6.1), in Jacobi's form with q replaced by q^2 and x replaced by zq:

(11.6.10) $$(-zq; q^2)_\infty = \sum_{n=0}^{\infty} \frac{q^{n^2} z^n}{(q^2; q^2)_n}.$$

Taking $z = \pm 1$ here we have

$$\frac{1}{2} \left[(-q; q^2)_\infty + (q; q^2)_\infty \right] = \sum_{n=0}^{\infty} \frac{q^{n^2}}{(q^2; q^2)_n} \frac{1 + (-1)^n}{2}$$

(11.6.11) $$= \sum_{k=0}^{\infty} \frac{q^{4k^2}}{(q^2; q^2)_{2k}}$$

and

$$\frac{1}{2} \left[(-q; q^2)_\infty - (q; q^2)_\infty \right] = \sum_{n=0}^{\infty} \frac{q^{n^2}}{(q^2; q^2)_n} \frac{1 - (-1)^n}{2}$$

(11.6.12) $$= \sum_{k=0}^{\infty} \frac{q^{4k^2 + 4k + 1}}{(q^2; q^2)_{2k+1}}.$$

Using (11.6.11) and (11.6.12) in (11.6.8) and (11.6.9) respectively, the even and odd parts of the product side of (11.6.1) are

$$(q^4; q^4)_\infty \sum_{k=0}^{\infty} \frac{q^{4k^2}}{(q^2; q^2)_{2k}}$$

and

$$q(q^4; q^4)_\infty \sum_{k=0}^{\infty} \frac{q^{4k^2 + 4k}}{(q^2; q^2)_{2k+1}}$$

respectively. Comparing these with (11.6.5) and (11.6.6), we have proved that

$$\sum_{k=0}^{\infty} \frac{q^{4k^2}}{(q^2; q^2)_{2k}} = \frac{(-q^6; q^{16})_\infty (-q^{10}; q^{16})_\infty (q^{16}; q^{16})_\infty}{(q^4; q^4)_\infty}$$

and
$$\sum_{k=0}^{\infty} \frac{q^{4k^2+4k}}{(q^2;q^2)_{2k+1}} = \frac{(-q^2;q^{16})_\infty (-q^{14};q^{16})_\infty (q^{16};q^{16})_\infty}{(q^4;q^4)_\infty}.$$

In both of these we can cancel $(q^{16};q^{16})_\infty$ into the denominator. Moreover, since all the powers of q are even in both of them, it makes sense to replace q^2 by q. Making these changes we get

$$(11.6.13) \qquad \sum_{k=0}^{\infty} \frac{q^{2k^2}}{(q;q)_{2k}} = \frac{(-q^3;q^8)_\infty (-q^5;q^8)_\infty}{(q^2;q^4)_\infty (q^4;q^8)_\infty}$$

and

$$(11.6.14) \qquad \sum_{k=0}^{\infty} \frac{q^{2k^2+2k}}{(q;q)_{2k+1}} = \frac{(-q;q^8)_\infty (-q^7;q^8)_\infty}{(q^2;q^4)_\infty (q^4;q^8)_\infty}.$$

These are both quite interesting in their own right, and we will return to this. We can transform them into the Göllnitz–Gordon identities by using Alladi's lemma below. It is convenient to observe first that

$$(q;q)_{2k} = (q;q^2)_k (q^2;q^2)_k \quad \text{and} \quad (q;q)_{2k+1} = (q;q^2)_{k+1} (q^2;q^2)_k,$$

and hence replacing q by $-q$ in (11.6.13) and (11.6.14) gives

$$(11.6.15) \qquad \sum_{k=0}^{\infty} \frac{q^{2k^2}}{(-q;q^2)_k (q^2;q^2)_k} = \frac{(q^3;q^8)_\infty (q^5;q^8)_\infty}{(q^2;q^4)_\infty (q^4;q^8)_\infty}$$

and

$$(11.6.16) \qquad \sum_{k=0}^{\infty} \frac{q^{2k^2+2k}}{(-q;q^2)_{k+1} (q^2;q^2)_k} = \frac{(q;q^8)_\infty (q^7;q^8)_\infty}{(q^2;q^4)_\infty (q^4;q^8)_\infty}.$$

Now we are ready for

LEMMA 15 (Alladi's lemma). *If $|q| < 1$, then*

$$(11.6.17) \qquad \sum_{n=0}^{\infty} \frac{q^{n^2} a^n (-bq;q^2)_n}{(q^2;q^2)_n} = \sum_{k=0}^{\infty} \frac{q^{2k^2}(ab)^k (-aq^{2k+1};q^2)_\infty}{(q^2;q^2)_k}.$$

To see this, expand $(-aq^{2k+1};q^2)_\infty$ using (11.6.10) with $z = aq^{2k}$. This makes the right side of (11.6.17) into

$$\sum_{k=0}^{\infty} \frac{q^{2k^2}(ab)^k}{(q^2;q^2)_k} \sum_{j=0}^{\infty} \frac{q^{j^2} q^{2kj} a^j}{(q^2;q^2)_j} = \sum_{j=0}^{\infty} \sum_{k=0}^{\infty} \frac{q^{k^2+2kj+j^2} a^{k+j} q^{k^2} b^k}{(q^2;q^2)_k (q^2;q^2)_j} \frac{(q^2;q^2)_{k+j}}{(q^2;q^2)_{k+j}}.$$

Setting $j+k = n$, this becomes

$$\sum_{n=0}^{\infty} \frac{q^{n^2} a^n}{(q^2;q^2)_n} \sum_{k=0}^{n} \binom{n}{k}_{q^2} q^{k^2} b^k,$$

which, by the form (2.5.10) of Rothe's q-binomial theorem, equals the left side of (11.6.17).

When $a = b = 1$, Alladi's lemma says

$$\sum_{n=0}^{\infty} \frac{q^{n^2}(-q;q^2)_n}{(q^2;q^2)_n} = \sum_{k=0}^{\infty} \frac{q^{2k^2}(-q^{2k+1};q^2)_\infty}{(q^2;q^2)_k} \frac{(-q;q^2)_k}{(-q;q^2)_k}$$

$$= (-q;q^2)_\infty \sum_{k=0}^{\infty} \frac{q^{2k^2}}{(-q;q^2)_k (q^2;q^2)_k}$$

$$= \frac{(-q;q^2)_\infty (q^3;q^8)_\infty (q^5;q^8)_\infty}{(q^2;q^4)_\infty (q^4;q^8)_\infty},$$

where we used (11.6.15) in the last step. Similarly, when $b = 1$ and $a = q^2$, Alladi's lemma says

$$\sum_{n=0}^{\infty} \frac{q^{n^2+2n}(-q;q^2)_n}{(q^2;q^2)_n} = \sum_{k=0}^{\infty} \frac{q^{2k^2+2k}(-q^{2k+3};q^2)_\infty}{(q^2;q^2)_k} \frac{(-q;q^2)_{k+1}}{(-q;q^2)_{k+1}}$$

$$= (-q;q^2)_\infty \sum_{k=0}^{\infty} \frac{q^{2k^2+2k}}{(-q;q^2)_{k+1}(q^2;q^2)_k}$$

$$= \frac{(-q;q^2)_\infty (q;q^8)_\infty (q^7;q^8)_\infty}{(q^2;q^4)_\infty (q^4;q^8)_\infty},$$

where we used (11.6.16) in the last step. Now

(11.6.18) $\qquad \dfrac{(-q;q^2)_\infty}{(q^2;q^4)_\infty} = \dfrac{1}{(q;q^2)_\infty} = \dfrac{1}{(q,q^3,q^5,q^7;q^8)_\infty},$

so we finally have

THEOREM 87 (The Göllnitz–Gordon identities). *If $|q| < 1$, then*

(11.6.19) $\qquad \displaystyle\sum_{n=0}^{\infty} \frac{q^{n^2}(-q;q^2)_n}{(q^2;q^2)_n} = \frac{1}{(q;q^8)_\infty (q^4;q^8)_\infty (q^7;q^8)_\infty},$

(11.6.20) $\qquad \displaystyle\sum_{n=0}^{\infty} \frac{q^{n^2+2n}(-q;q^2)_n}{(q^2;q^2)_n} = \frac{1}{(q^3;q^8)_\infty (q^4;q^8)_\infty (q^5;q^8)_\infty}.$

Exercises

1. Prove (11.6.2).
2. Prove (11.6.3).
3. Prove (11.6.8) and (11.6.9).
4. Explain why Lebesgue's identity (5.4.3) can be rewritten as

$$\sum_{n=0}^{\infty} q^{n^2+n} \frac{(a;q^2)_n}{(q^2;q^2)_n} = \frac{(aq^2;q^4)_\infty}{(q^2;q^4)_\infty}.$$

5. Alladi has pointed out that the result of problem 4 can be used to derive two theorems very similar in appearance to the Göllnitz–Gordon identities, which he has christened the **little Göllnitz identities**. Show that

(11.6.21) $\qquad \displaystyle\sum_{n=0}^{\infty} q^{n^2+n} \frac{(-q;q^2)_n}{(q^2;q^2)_n} = \frac{1}{(q^2;q^8)_\infty (q^3;q^8)_\infty (q^7;q^8)_\infty}$

and

(11.6.22) $$\sum_{n=0}^{\infty} q^{n^2+n} \frac{(-\frac{1}{q};q^2)_n}{(q^2;q^2)_n} = \frac{1}{(q;q^8)_\infty (q^5;q^8)_\infty (q^6;q^8)_\infty}.$$

6. Andrews had earlier given a derivation of the Göllnitz–Gordon identities that uses a similar splitting idea. It is convenient to begin with

(11.6.23) $$\sum_{n=0}^{\infty} \frac{(-1)^n q^{n^2} (q;q^2)_n}{(q^2;q^2)_n},$$

which is the left side of the first Göllnitz–Gordon identity with q replaced by $-q$ (since n^2 is even if and only if n is).

(i) Explain why (11.6.23) can be rewritten as
$$(q;q^2)_\infty \sum_{n=0}^{\infty} \frac{(-1)^n q^{n^2}}{(q^2;q^2)_n} \sum_{k=0}^{\infty} \frac{(q^{2n+1})^k}{(q^2;q^2)_k}.$$

(ii) Explain why the series in (i) can be rewritten as
$$(q;q^2)_\infty \sum_{k=0}^{\infty} \frac{q^k}{(q^2;q^2)_k} (q^{2k+1};q^2)_\infty.$$

(iii) Explain why the series in (ii) can be rewritten as
$$(q;q^2)_\infty^2 \sum_{k=0}^{\infty} \frac{q^k}{(q;q)_{2k}}.$$

The simplification in the denominator is the reason for changing q to $-q$ at the start. The splitting (or perhaps unsplitting) idea comes next.

(iv) Explain why we can rewrite
$$\sum_{k=0}^{\infty} \frac{q^k}{(q;q)_{2k}} = \sum_{m=0}^{\infty} \frac{q^{\frac{m}{2}}}{(q;q)_m} \frac{1+(-1)^m}{2}.$$

(v) Explain why we can evaluate (iv) as
$$\frac{1}{2}\left[\frac{1}{(q^{\frac{1}{2}};q)_\infty} + \frac{1}{(-q^{\frac{1}{2}};q)_\infty}\right] = \frac{(q^{\frac{1}{2}};q)_\infty + (-q^{\frac{1}{2}};q)_\infty}{2(q;q^2)_\infty}$$
$$= \frac{(q^{\frac{1}{2}};q^2)_\infty (q^{\frac{3}{2}};q^2)_\infty + (-q^{\frac{1}{2}};q^2)_\infty (-q^{\frac{3}{2}};q^2)_\infty}{2(q;q^2)_\infty}.$$

(vi) Combining (v) with (iii), we have that (11.6.23) equals
$$\frac{(q;q^2)_\infty}{2}\left[(q^{\frac{1}{2}};q^2)_\infty (q^{\frac{3}{2}};q^2)_\infty + (-q^{\frac{1}{2}};q^2)_\infty (-q^{\frac{3}{2}};q^2)_\infty\right]$$
$$= \frac{(q;q^2)_\infty}{2(q^2;q^2)_\infty}\left[(q^{\frac{1}{2}};q^2)_\infty (q^{\frac{3}{2}};q^2)_\infty (q^2;q^2)_\infty + (-q^{\frac{1}{2}};q^2)_\infty (-q^{\frac{3}{2}};q^2)_\infty (q^2;q^2)_\infty\right].$$

Use Jacobi's triple product to show that this is
$$\frac{(q;q^2)_\infty}{(q^2;q^2)_\infty} \sum_{n=-\infty}^{\infty} q^{4n^2+n} = \frac{(q;q^2)_\infty}{(q^2;q^2)_\infty} (-q^3;q^8)_\infty (-q^5;q^8)_\infty (q^8;q^8)_\infty.$$

(vii) Changing q to $-q$ again, we have proved that

$$\sum_{n=0}^{\infty} \frac{q^{n^2}(-q;q^2)_n}{(q^2;q^2)_n} = \frac{(-q;q^2)_\infty}{(q^2;q^2)_\infty} (q^3;q^8)_\infty (q^5;q^8)_\infty (q^8;q^8)_\infty.$$

Show that this simplifies to (11.6.19).

7. Give a similar proof of (11.6.20).
8. Göllnitz's proof of the Göllnitz–Gordon identities is intricate, but of a type that we have seen before. He considers the function

(11.6.24) $\quad G(x) = 1 + \sum_{n=1}^{\infty} (-1)^n x^{2n} q^{n(4n-1)} (1 - xq^{4n}) \frac{(-q;q^2)_n (xq^2;q^2)_{n-1}}{(-xq;q^2)_n (q^2;q^2)_n}.$

Note that $G(0) = 1$.

(i) Assuming $|q| < 1$, show that this series converges for all x.

(ii) Show that

$$(1-x)G(x) = \sum_{n=0}^{\infty} (-1)^n x^{2n} q^{n(4n-1)} (1 - xq^{4n}) \frac{(-q;q^2)_n (x;q^2)_n}{(-xq;q^2)_n (q^2;q^2)_n}.$$

This will be useful later.

(iii) By writing $1 - xq^{4n} = 1 - q^{2n} + q^{2n}(1 - xq^{2n})$ in (11.6.24), show that

$$G(x) = 1 + \sum_{n=1}^{\infty} (-1)^n x^{2n} q^{n(4n-1)} \frac{(-q;q^2)_n (xq^2;q^2)_{n-1}}{(-xq;q^2)_n (q^2;q^2)_{n-1}}$$
$$+ \sum_{n=1}^{\infty} (-1)^n x^{2n} q^{n(4n+1)} \frac{(-q;q^2)_n (xq^2;q^2)_n}{(-xq;q^2)_n (q^2;q^2)_n}.$$

(iv) Note that 1 is the $n = 0$ term of the last series in (iii). By changing n to $n+1$ in the first series in (iii) and combining the two series, show that $G(x)$ equals

$$\sum_{n=0}^{\infty} (-1)^n x^{2n} q^{n(4n+1)} (1 - xq^{4n+2})(1 + xq^{2n+1} + xq^{4n+2}) \frac{(-q;q^2)_n (xq^2;q^2)_n}{(-xq;q^2)_{n+1} (q^2;q^2)_n}.$$

(v) Show that

$$\frac{G(x)}{1-xq^2} - G(xq^2) = \frac{xq^2}{1+xq}$$
$$+ \sum_{n=1}^{\infty} (-1)^n x^{2n} q^{n(4n+1)} (1 - xq^{4n+2})(1 - q^{2n} + xq^{4n+2}) \frac{(-q;q^2)_n (xq^4;q^2)_{n-1}}{(-xq;q^2)_{n+1} (q^2;q^2)_n}.$$

Use (iv) (with the $n = 0$ term broken off) for the first term and (11.6.24) for the second.

(vi) Show that

$$(1 - xq^{4n+2})(1 - q^{2n} + xq^{4n+2}) = 1 - q^{2n} + xq^{6n+2}(1 - xq^{2n+2}),$$

and hence
$$\frac{G(x)}{1-xq^2} - G(xq^2) = \frac{xq^2}{1+xq} + \sum_{n=1}^{\infty}(-1)^n x^{2n} q^{n(4n+1)} \frac{(-q;q^2)_n (xq^4;q^2)_{n-1}}{(-xq;q^2)_{n+1}(q^2;q^2)_{n-1}}$$
$$+ \sum_{n=1}^{\infty}(-1)^n x^{2n+1} q^{4n^2+7n+2} \frac{(-q;q^2)_n (xq^4;q^2)_n}{(-xq;q^2)_{n+1}(q^2;q^2)_n}.$$

(vii) Note that $xq^2/(1+xq)$ is the $n=0$ term of the last series in (vi). By changing n to $n+1$ in the first series in (vi) and combining the two series, show that
$$\frac{G(x)}{1-xq^2} - G(xq^2) = \sum_{n=0}^{\infty}(-1)^n x^{2n+1} q^{4n^2+7n+2}(1-xq^{4n+4}) \frac{(-q;q^2)_n (xq^4;q^2)_n}{(-xq;q^2)_{n+2}(q^2;q^2)_n}.$$

(viii) By pulling out factors and comparing with (ii), show that
(11.6.25)
$$\frac{G(x)}{1-xq^2} - G(xq^2) = \frac{xq^2(1-xq^4)}{(1+xq)(1+xq^3)} G(xq^4).$$

(ix) To simplify (11.6.25), Göllnitz sets
$$G(x) = \frac{(xq^2;q^2)_\infty}{(-xq;q^2)_\infty} H(x).$$
Show that this gives
$$H(x) = (1+xq)H(xq^2) + xq^2 H(xq^4).$$
Note that $H(0) = 1$, since $G(0) = 1$.

(x) If $H(x) = \sum_{n=0}^{\infty} g_n(q) x^n$, show that (ix) implies
$$g_n(q) = q^{2n-1} \frac{1+q^{2n-1}}{1-q^{2n}} g_{n-1}(q) \quad \text{for } n \geq 1.$$

(xi) Explain why (x) implies
$$g_n(q) = q^{n^2} \frac{(-q;q^2)_n}{(q^2;q^2)_n},$$
and hence
$$G(x) = \frac{(xq^2;q^2)_\infty}{(-xq;q^2)_\infty} \sum_{n=0}^{\infty} \frac{(-q;q^2)_n}{(q^2;q^2)_n} q^{n^2} x^n.$$

(xii) Show that (11.6.24) implies
$$G(1) = 1 + \sum_{n=1}^{\infty}(-1)^n q^{n(4n-1)}(1+q^{2n}),$$
and that this is equivalent to
$$G(1) = \sum_{n=-\infty}^{\infty}(-1)^n q^{n(4n-1)}.$$

(xiii) Use Jacobi's triple product to show that the sum in (xii) is
$$(q^3;q^8)_\infty (q^5;q^8)_\infty (q^8;q^8)_\infty.$$

(xiv) Comparing with (xi), we therefore have
$$\sum_{n=0}^{\infty} \frac{(-q;q^2)_n}{(q^2;q^2)_n} q^{n^2} = \frac{(-q;q^2)_\infty}{(q^2;q^2)_\infty} (q^3;q^8)_\infty (q^5;q^8)_\infty (q^8;q^8)_\infty.$$
Show that this simplifies to (11.6.19).

(xv) Show that (11.6.24) implies
$$G(q^2) = 1 + \sum_{n=1}^{\infty} (-1)^n q^{n(4n+3)} \frac{1-q^{2n+1}}{1-q},$$
and this is equivalent to
$$(1-q)G(q^2) = \sum_{n=0}^{\infty} (-1)^n q^{n(4n+3)}(1-q^{2n+1}) = \sum_{n=-\infty}^{\infty} (-1)^n q^{n(4n+3)}.$$

(xvi) Use Jacobi's triple product to show that the sum in (xv) is
$$(q;q^8)_\infty (q^7;q^8)_\infty (q^8;q^8)_\infty.$$

(xiv) On the other hand, (xi) says
$$\sum_{n=0}^{\infty} \frac{(-q;q^2)_n}{(q^2;q^2)_n} q^{n^2+2n} = \frac{(-q^3;q^2)_\infty}{(q^4;q^2)_\infty} G(q^2).$$
Explain why (xv) and (xvi) imply that
$$\sum_{n=0}^{\infty} \frac{(-q;q^2)_n}{(q^2;q^2)_n} q^{n^2+2n} = \frac{(-q;q^2)_\infty}{(q^2;q^2)_\infty} (q;q^8)_\infty (q^7;q^8)_\infty (q^8;q^8)_\infty,$$
and show that this simplifies to (11.6.20).

9. By taking $x = q$ in problem 8 and comparing with (11.6.21), show that
$$\sum_{n=0}^{\infty} (-1)^n q^{n(4n+1)}(1-q^{4n+1}) \frac{(q^2;q^4)_n}{(q^4;q^4)_n} = (q;q^4)_\infty (q^6;q^8)_\infty.$$

10. (If you have not done problem 3 in section 11.1, you should at least look at it before you try this one.) As with the Rogers–Ramanujan identities, there is also a continued fraction associated with the Göllnitz–Gordon identities; in fact, Gordon's paper is primarily concerned with it. Define a function $K(x)$ by

(11.6.26)
$$K(x) = \frac{G(x)(1+xq)}{(1-xq^2)\,G(xq^2)},$$

with $G(x)$ as in problem 8.

(i) Use (11.6.25) to show that

(11.6.27)
$$K(x) = 1 + xq + \frac{xq^2}{K(xq^2)}.$$

This leads to a continued fraction for $K(x)$.

(ii) Write down the result of replacing x by xq^2 in (11.6.27), and substitute it into (11.6.27) to show that
$$K(x) = 1 + xq + \cfrac{xq^2}{1 + xq^3 + \cfrac{xq^4}{K(xq^4)}}.$$

(iii) Write down the result of replacing x by xq^4 in (11.6.27), and substitute it into the result of (v) to show that

$$K(x) = 1 + xq + \cfrac{xq^2}{1 + xq^3 + \cfrac{xq^4}{1 + xq^5 + \cfrac{xq^6}{K(xq^6)}}}.$$

Continuing in this way we have

$$K(x) = 1 + xq + \cfrac{xq^2}{1 + xq^3 + \cfrac{xq^4}{1 + xq^5 + \cfrac{xq^6}{1 + xq^7 + \cfrac{xq^8}{\ddots}}}}.$$

(iv) By setting $x = 1$ in the last continued fraction in (iii) and using (ii) and the Göllnitz–Gordon identities, show that

$$1 + q + \cfrac{q^2}{1 + q^3 + \cfrac{q^4}{1 + q^5 + \cfrac{q^6}{1 + q^7 + \cfrac{q^8}{\ddots}}}} = \frac{G(1)(1+q)}{G(q^2)(1-q^2)} = \frac{(q^3;q^8)_\infty \, (q^5;q^8)_\infty}{(q;q^8)_\infty \, (q^7;q^8)_\infty}.$$

This is the **Göllnitz–Gordon continued fraction**.

11.7. The Göllnitz–Gordon identities and partitions

What do the identities of section 11.6 tell us about partitions? Let's go back to (11.6.13) first. There are two natural interpretations of the product side, one of which we reserve for the problems. In its present form the numerator generates partitions into distinct parts congruent to 3 or 5 mod 8 (or to ± 3 mod 8), and the denominator generates even parts that are not multiples of 8. Since there is no overlap, the product side of (11.6.13) generates partitions into parts congruent to 2, 3, 4, 5, or 6 mod 8, where the odd parts must be distinct. Let's let $G_1(n)$ denote the set of partitions of n of this type.

On the sum side, the denominator for a generic k generates partitions with at most $2k$ parts. We can decompose the $2k^2$ in the exponent of the numerator as

$$2k^2 = 1 + 1 + 3 + 3 + 5 + 5 + \cdots + (2k-1) + (2k-1),$$

and then add these $2k$ numbers to the parts generated by the denominator in increasing order. This gives us exactly $2k$ parts (*i.e.*, an even number of parts), where the second largest part exceeds the third largest by at least 2, the fourth largest part exceeds the fifth largest by at least 2, and so on. Let's let $G_2(n)$ denote the set of partitions of n of this type.

Alternatively, we can decompose $2k^2 = k + k + k + k + \cdots + k$, and add these $2k$ numbers to the parts generated by the denominator. This again gives us an even number of parts, and now each part is at least half as big as the number of parts.

11.7. THE GÖLLNITZ–GORDON IDENTITIES AND PARTITIONS

Let's let $G_3(n)$ denote the set of partitions of n of this type. Here are all of these partitions for $n = 12$:

$G_1(12)$	$G_2(12)$	$G_3(12)$
12	11 + 1	11 + 1
10 + 2	10 + 2	10 + 2
6 + 6	9 + 3	9 + 3
6 + 4 + 2	8 + 4	8 + 4
6 + 2 + 2 + 2	7 + 5	7 + 5
5 + 4 + 3	6 + 6	6 + 6
5 + 3 + 2 + 2	7 + 3 + 1 + 1	6 + 2 + 2 + 2
4 + 4 + 4	6 + 4 + 1 + 1	5 + 3 + 2 + 2
4 + 4 + 2 + 2	5 + 5 + 1 + 1	4 + 4 + 2 + 2
4 + 2 + 2 + 2 + 2	5 + 4 + 2 + 1	4 + 3 + 3 + 2
2 + 2 + 2 + 2 + 2 + 2	4 + 4 + 2 + 2	3 + 3 + 3 + 3

As for the Göllnitz–Gordon identities (11.6.19) and (11.6.20), let's define a **GG partition** to be a partition with distinct, nonconsecutive parts, and also with nonconsecutive even parts. Thus $8 + 5 + 3$ is a GG partition, but $8 + 6 + 3$ is not because 8 and 6 are consecutive even parts. Let's also call a GG partition with no 1's or 2's a **GGG partition**. These two classes of partitions are closely related to the ee partitions of section 3.3, in which the even parts may be repeated but the odd parts must be distinct. If we start with a GG partition with exactly n parts and subtract 1 from the smallest part, 3 from the next smallest, and so on, finally subtracting $2n - 1$ from the largest, we will still have a partition because of the gaps between parts, and it will have at most n parts. It could have a repeated even part, if the GG partition had two consecutive odd parts, but it could not have a repeated odd part. Therefore, it is precisely an ee partition with at most n parts. For a GGG partition with exactly n parts, we simply subtract 3 from the smallest part, 5 from the next smallest, and so on, finally subtracting $2n+1$ from the largest part, and we again get an ee partition with at most n parts. Since the generating function for ee partitions with at most n parts is $(-q;q^2)_n/(q^2;q^2)_n$, it follows that the generating function for GG partitions with exactly n parts is

$$(11.7.1) \qquad q^{1+3+5+\cdots+2n-1} \frac{(-q;q^2)_\infty}{(q^2;q^2)_\infty} = q^{n^2} \frac{(-q;q^2)_\infty}{(q^2;q^2)_\infty},$$

and that the generating function for GGG partitions with exactly n parts is

$$(11.7.2) \qquad q^{3+5+7+\cdots+2n+1} \frac{(-q;q^2)_\infty}{(q^2;q^2)_\infty} = q^{n^2+2n} \frac{(-q;q^2)_\infty}{(q^2;q^2)_\infty}.$$

It follows that the generating function for *all* GG partitions is

$$\sum_{n=0}^{\infty} q^{n^2} \frac{(-q;q^2)_n}{(q^2;q^2)_n},$$

and that the generating function for all GGG partitions is

$$\sum_{n=0}^{\infty} q^{n^2+2n} \frac{(-q;q^2)_n}{(q^2;q^2)_n}.$$

These are precisely the left sides of (11.6.19) and (11.6.20). Therefore (11.6.19) is trying to tell us that there are just as many GG partitions of n as there are partitions of n into parts congruent to 1, 4, or 7 mod 8; and (11.6.20) is trying to tell us that there are just as many GGG partitions of n as there are partitions of n into parts congruent to 3, 4, or 5 mod 8.

Here are all the partitions of 14 relevant to (11.6.19):

GG partitions	$1, 4, 7$ (mod 8)
14	$12 + 1 + 1$
$13 + 1$	$9 + 4 + 1$
$12 + 2$	$9 + 1 + 1 + 1 + 1 + 1$
$11 + 3$	$7 + 7$
$10 + 4$	$7 + 4 + 1 + 1 + 1$
$10 + 3 + 1$	$7 + 1 + 1 + 1 + 1 + 1 + 1 + 1$
$9 + 5$	$4 + 4 + 4 + 1 + 1$
$9 + 4 + 1$	$4 + 4 + 1 + 1 + 1 + 1 + 1 + 1$
$8 + 5 + 1$	4 + ten 1's
$7 + 5 + 2$	fourteen 1's

There are only a few partitions of 14 relevant to (11.6.20):

GGG partitions	$3, 4, 5$ (mod 8)
14	$11 + 3$
$11 + 3$	$5 + 5 + 4$
$10 + 4$	$5 + 3 + 3 + 3$
$9 + 5$	$4 + 4 + 3 + 3$

There is also a nice partition theorem associated with (11.6.21). If we have an ee partition with at most n parts, first add some zeros if necessary to get exactly n parts, and then add 2 to the smallest part, 4 to the next smallest, and so on, finally adding $2n$ to the largest. This again creates gaps at least 2 between parts, and this time there are no consecutive odd parts because an ee partition can't have a repeated odd part. Also, there are no 1's. If we call these AGG partitions (for Alladi), then the partitions of 14 relevant to (11.6.21) are:

AGG partitions	$2, 3, 7$ (mod 8)
14	$11 + 3$
$12 + 2$	$10 + 2 + 2$
$11 + 3$	$7 + 7$
$10 + 4$	$7 + 3 + 2 + 2$
$9 + 5$	$3 + 3 + 3 + 3 + 2$
$8 + 6$	$3 + 3 + 2 + 2 + 2 + 2$
$8 + 4 + 2$	$2 + 2 + 2 + 2 + 2 + 2 + 2$

Exercises

1. Show that (11.6.13) can be rewritten as

$$\sum_{k=0}^{\infty} \frac{q^{2k^2}}{(q;q)_{2k}} = \frac{1}{(q^2, q^3, q^4, q^5, q^{11}, q^{12}, q^{13}, q^{14}; q^{16})_{\infty}}.$$

What kind of partitions does the right side generate?

2. Show that (11.6.14) can be rewritten as
$$\sum_{k=0}^{\infty} \frac{q^{2k^2+2k}}{(q;q)_{2k+1}} = \frac{1}{(q,q^4,q^6,q^7,q^9,q^{10},q^{12},q^{15};q^{16})_\infty}.$$
What kind of partitions does the right side generate?

3. Write down all the GG partitions of 16, and all the partitions of 16 with parts congruent to 1, 4, or 7 mod 8. Do the same for the GGG partitions of 16, and the partitions of 16 with parts congruent to 3, 4, or 5 mod 8; and for the AGG partitions of 16 and those with parts congruent to 2, 3, or 7 mod 8.

4. Suppose we write $n^2 = n+n+n+\cdots+n$ instead of $n^2 = 1+3+5+\cdots+(2n-1)$. What partition theorem does (11.6.19) imply then?

5. Give alternate partition interpretations of (11.6.20) and (11.6.21), similarly to the previous problem.

6. This problem outlines a direct proof of (11.7.1), without reducing (explicitly) to the ee partitions.

(i) Explain why the generating function for GG partitions with exactly one part is
$$q + q^2 + q^3 + q^4 + \cdots = \frac{q}{1-q} = q\frac{1+q}{1-q^2}.$$

(ii) Here are the GG partitions with exactly two parts, listed according to the size of the smallest part:

3+1	5+2	5+3	7+4	7+5	9+6	9+7	...
4+1	6+2	6+3	8+4	8+5	10+6	10+7	...
5+1	7+2	7+3	9+4	9+5	11+6	11+7	...
6+1	8+2	8+3	10+4	10+5	12+6	12+7	...
⋮	⋮	⋮	⋮	⋮	⋮	⋮	...

By constructing the generating function for each column and adding them all together, or otherwise, show that the generating function for GG partitions with exactly two parts is
$$\frac{q^4}{1-q} + \frac{q^7(1+q)}{(1-q)(1-q^4)} = q^4 \frac{(1+q)(1+q^3)}{(1-q^2)(1-q^4)}.$$

(iii) Explain why every GG partition with three parts and smallest part $2k$ or $2k+1$ corresponds uniquely to a GG partition with two parts by deleting the $2k$ or $2k+1$ and subtracting $2k+2$ from the other parts, and why this implies that the generating function for such GG partitions is
$$q^{6k+8} \frac{(1+q)^2(1+q^3)}{(1-q^2)(1-q^4)}.$$

(iv) Use (iii) to show that the generating function for GG partitions with exactly three parts is
$$q^9 \frac{(1+q)(1+q^3)(1+q^5)}{(1-q^2)(1-q^4)(1-q^6)}.$$

(v) Prove (11.7.1) by induction on n by the same type of argument as in (iii) and (iv).

11.8. Bibliographical Notes

In Chapter 3 of the second volume of [**168**], MacMahon wrote that (11.2.1) "has been verified as far as the coefficient of $[q^{89}]$ by actual expansion so that there is practically no reason to doubt its truth; but it has not yet been established." At that time no one knew about Rogers's work, but that would all change a year later.

Rogers's first proof of the Rogers–Ramanujan identities is in [**200**]. His work on this subject is vaguely reminiscent of *Star Wars*, in that there is a trilogy [**199**], [**200**], [**201**], a prequel [**198**], and a prequel [**197**] to the prequel. The sequel, his second proof of the Rogers–Ramanujan identities [**202**], came out many years later. His third proof [**203**] appeared simultaneously with Ramanujan's proof of section 11.1. The same proof was given later by Selberg [**219**], as we shall see in the next chapter. For a modern treatment of Rogers's work see Bowman's paper [**55**].

The bijective proof of Garsia and Milne is in [**113**], and the shorter version by Zeilberger and Bressoud in [**62**]. Rogers's identities of section 11.4 are in [**200**] and [**202**]; see also [**19**]. In that section I have followed some of the exercises in [**24**]. (For the longest time I could not remember where that section came from, but I think this is right.) Bressoud's bijection of section 11.2 is in [**58**] and [**19**]. The latter also has a more readable account of some of Rogers's best work than the original.

The Göllnitz–Gordon identities were known for several years before they were published. They are already in Göllnitz's undergraduate thesis(!) of 1960 [**124**], and again in his Ph.D thesis [**125**] of 1963. His paper [**126**], based on the latter, finally appeared in 1967. Gordon's paper [**127**] appeared in 1965, but he knew the identities at least four years earlier. Gordon spent some time in Göttingen on sabbatical in the early 1960s, when Göllnitz was a student there, but neither of them knew that the other knew the identities and they never spoke. We have followed Alladi's beautiful paper [**4**] for this topic. Andrews's proof from problem 6 is in [**12**]. The recent book [**220**] by Andrew Sills has much more on Rogers–Ramanujan type identities.

CHAPTER 12

The Rogers–Selberg Function

12.1. The Rogers–Selberg function

Atle Selberg was one of the greatest mathematicians of the 20th century. He came from a mathematical family and was reading his father's copy of Ramanujan's collected papers at a young age, before getting his own copy as a present. While he did most of his work on analytic number theory, his first paper was on q-analysis. It was a thorough study of a function depending on x, q, and two other parameters that, following Selberg, we will call k and i, namely

(12.1.1) $$S_{k,i}(x) = \sum_{j=0}^{\infty} (-1)^j x^{jk} q^{(2k+1)\binom{j+1}{2} - ij} (1 - x^i q^{i(2j+1)}) \frac{(xq;q)_j}{(q;q)_j},$$

where as usual $|q| < 1$. We also assume $k > -\frac{1}{2}$ for the convergence of the series. Among other things, as we shall see, Selberg was able to rederive the Rogers–Ramanujan identities from this function, which reduces to the one Ramanujan used in his proof when $k = 2 = i$. In fact, Selberg's argument is almost identical with Rogers's third proof, which used the same function with a different name. Rogers called the function V_m, though it also depended on n, x, and q. Selberg's k and i are Rogers's n and m respectively, and Selberg had xq in the place of Rogers's x. It is not clear whether Selberg knew this. He may have, because this short paper of Rogers was included in an appendix to Ramanujan's collected papers.

We start by noting that $S_{k,i}(0) = 1$, from the $j = 0$ term, and that $S_{k,0}(x) = 0$, from the factor $1 - x^i q^{i(2j+1)}$. It will save us trouble later to show that $S_{k,i}(1)$ can be summed by Jacobi's triple product. We have

$$S_{k,i}(1) = \sum_{j=0}^{\infty} (-1)^j q^{(2k+1)\binom{j+1}{2} - ij} (1 - q^{i(2j+1)})$$

$$= \sum_{j=0}^{\infty} (-1)^j q^{\frac{j(j+1)(2k+1)}{2} - ij} + \sum_{j=0}^{\infty} (-1)^{j+1} q^{\frac{j(j+1)(2k+1)}{2} + i(j+1)}.$$

Changing j to m in the first sum, and $j+1$ to $-m$ in the second, this becomes

$$S_{k,i}(1) = \sum_{m=0}^{\infty} (-1)^m q^{\frac{m(m+1)(2k+1)}{2} - im} + \sum_{m=-1}^{-\infty} (-1)^m q^{\frac{-m(-m-1)(2k+1)}{2} + i(-m)}$$

$$= \sum_{m=-\infty}^{\infty} (-1)^m q^{\frac{m(m+1)(2k+1)}{2} - im}.$$

417

Jacobi's triple product says that

$$(x;q)_\infty \left(\tfrac{q}{x};q\right)_\infty (q;q)_\infty = \sum_{m=-\infty}^{\infty} (-1)^m q^{\frac{m(m-1)}{2}} x^m,$$

or, if we replace q by q^{2k+1}, that
(12.1.2)

$$(x;q^{2k+1})_\infty \left(\tfrac{q^{2k+1}}{x};q^{2k+1}\right)_\infty (q^{2k+1};q^{2k+1})_\infty = \sum_{m=-\infty}^{\infty} (-1)^m q^{\frac{m(m-1)(2k+1)}{2}} x^m.$$

If we choose $x = q^{2k-i+1}$, this becomes

(12.1.3) $\qquad S_{k,i}(1) = (q^{2k-i+1};q^{2k+1})_\infty (q^i;q^{2k+1})_\infty (q^{2k+1};q^{2k+1})_\infty.$

Selberg began his study of $S_{k,i}(x)$ by observing that

$$S_{k,-i}(x) = \sum_{j=0}^{\infty} (-1)^j x^{jk} q^{(2k+1)\binom{j+1}{2}+ij} (1 - x^{-i}q^{-i(2j+1)}) \frac{(xq;q)_j}{(q;q)_j}$$

$$= \sum_{j=0}^{\infty} (-1)^j x^{jk} q^{(2k+1)\binom{j+1}{2}-ij} \left(q^{2ij} - x^{-i}q^{-i}\right) \frac{(xq;q)_j}{(q;q)_j}$$

$$= -x^{-i}q^{-i} \sum_{j=0}^{\infty} (-1)^j x^{jk} q^{(2k+1)\binom{j+1}{2}-ij} \left(1 - x^i q^i q^{2ij}\right) \frac{(xq;q)_j}{(q;q)_j},$$

and consequently

(12.1.4) $\qquad S_{k,-i}(x) = -x^{-i}q^{-i} S_{k,i}(x).$

Next Selberg derives the fundamental recurrence for his function, which was also given by Rogers. Observe that $S_{k,i}(x) - S_{k,i-1}(x)$ is

$$\sum_{j=0}^{\infty} (-1)^j x^{jk} q^{(2k+1)\binom{j+1}{2}-ij} \frac{(xq;q)_j}{(q;q)_j} \left[1 - x^i q^{i(2j+1)} - q^j \left(1 - x^{i-1} q^{(i-1)(2j+1)}\right)\right]$$

and that the quantity in brackets is

$$1 - q^j + x^{i-1} q^{(i-1)(2j+1)+j} \left(1 - xq^{j+1}\right).$$

Therefore

$$S_{k,i}(x) - S_{k,i-1}(x) = \sum_{j=0}^{\infty} (-1)^j x^{jk} q^{(2k+1)\binom{j+1}{2}-ij} \frac{(xq;q)_j}{(q;q)_j} (1 - q^j)$$

$$+ \sum_{j=0}^{\infty} (-1)^j x^{jk} q^{(2k+1)\binom{j+1}{2}-ij} \frac{(xq;q)_j}{(q;q)_j} x^{i-1} q^{(i-1)(2j+1)+j} \left(1 - xq^{j+1}\right)$$

$$= \sum_{j=1}^{\infty} (-1)^j x^{jk} q^{(2k+1)\binom{j+1}{2}-ij} \frac{(xq;q)_j}{(q;q)_{j-1}}$$

$$+ \sum_{j=0}^{\infty} (-1)^j x^{jk+i-1} q^{(2k+1)\binom{j+1}{2}+ij-j+i-1} \frac{(xq;q)_{j+1}}{(q;q)_j}.$$

Replacing j by $j+1$ in the penultimate sum, we get

$$S_{k,i}(x) - S_{k,i-1}(x) = \sum_{j=0}^{\infty} (-1)^{j+1} x^{jk+k} q^{(2k+1)\binom{j+2}{2}-ij-i} \frac{(xq;q)_{j+1}}{(q;q)_j}$$

$$+ \sum_{j=0}^{\infty} (-1)^j x^{jk+i-1} q^{(2k+1)\binom{j+1}{2}+ij-j+i-1} \frac{(xq;q)_{j+1}}{(q;q)_j}$$

$$= \sum_{j=0}^{\infty} (-1)^j x^{jk} q^{(2k+1)\binom{j+1}{2}-ij} \frac{(xq;q)_{j+1}}{(q;q)_j} \left[x^{i-1} q^{(i-1)(2j+1)+j} - x^k q^{(2k+1)(j+1)-i} \right].$$

The quantity in brackets is

$$(xq)^{i-1} \left[q^{j(2i-1)} - x^{k-i+1} q^{(2k+1)(j+1)-(2i-1)} \right],$$

which is

$$(xq)^{i-1} q^{j(2i-1)} \left[1 - (xq)^{k-i+1} q^{(2j+1)(k-i+1)} \right].$$

Putting in a factor of q^{jk-jk} we have

$$S_{k,i}(x) - S_{k,i-1}(x) = (xq)^{i-1}$$

$$\times \sum_{j=0}^{\infty} (-1)^j (xq)^{jk} q^{(2k+1)\binom{j+1}{2}-j(k-i+1)} \frac{(xq;q)_{j+1}}{(q;q)_j} \left[1 - (xq)^{k-i+1} q^{(2j+1)(k-i+1)} \right].$$

Note also that $(xq;q)_{j+1} = (1-xq)(xq^2;q)_j$, which finally implies

THEOREM 88 (Fundamental recurrence for the Rogers–Selberg function).

(12.1.5) $\quad S_{k,i}(x) = S_{k,i-1}(x) + (xq)^{i-1}(1-xq)S_{k,k-i+1}(xq).$

Exercises

1. Show that the series in (12.1.1) converges if $k > -\frac{1}{2}$.
2. What happens if $k = -\frac{1}{2}$?
3. Show that $S_{k,i}(q^{-m}) = 0$ for a positive integer m.
4. Show that (12.1.5) implies

    $$S_{k,i+1}(x) = S_{k,i}(x) + (xq)^i (1-xq) S_{k,k-i}(xq)$$

    and

    $$S_{k,k-i}(xq) = S_{k,k-i+1}(xq) - (xq^2)^{k-i}(1-xq^2) S_{k,i}(xq^2).$$

5. Explain why the previous problem implies that $S_{k,i+1}(x)$ can be expressed in terms of $S_{k,i}(x)$, $S_{k,i}(xq^2)$, and $S_{k,i-1}(x)$.
6. Show that

    $$S_{k,i}(x) = (xq)^{i-1} \left[(1-xq) S_{k,k-i+1}(xq) - S_{k,k-i}(x) \right].$$

12.2. Some applications

When $k = 1$, the fundamental recurrence (12.1.5) becomes

(12.2.1) $\qquad S_{1,i}(x) = S_{1,i-1}(x) + (xq)^{i-1}(1-xq)S_{1,2-i}(xq).$

Taking $i = 2$ in (12.2.1) we get

$$S_{1,2}(x) = S_{1,1}(x) + (xq)(1-xq)S_{1,0}(xq) = S_{1,1}(x).$$

Taking $i = 1$, (12.2.1) says

$$S_{1,1}(x) = S_{1,0}(x) + (1-xq)S_{1,1}(xq) = (1-xq)S_{1,1}(xq).$$

Iterating this and using $S_{k,i}(0) = 1$ we have immediately

$$S_{1,1}(x) = (xq;q)_\infty = S_{1,2}(x).$$

Taking $i = 3$ in (12.2.1) we get

$$S_{1,3}(x) = S_{1,2}(x) + (xq)^2(1-xq)S_{1,-1}(xq).$$

But (12.1.4) tells us that

$$S_{1,-1}(x) = -(xq)^{-1}S_{1,1}(x),$$

so

$$\begin{aligned}S_{1,3}(x) &= S_{1,2}(x) - (xq)^2(1-xq)(xq^2)^{-1}S_{1,1}(xq)\\ &= (xq;q)_\infty - x(1-xq)(xq^2;q)_\infty \\ &= (xq;q)_\infty(1-x) = (x;q)_\infty.\end{aligned}$$

Comparing with (12.1.1), we have proved that

(12.2.2) $\qquad (xq;q)_\infty = \sum_{j=0}^\infty (-1)^j x^j q^{\frac{j(3j+1)}{2}} \left(1 - xq^{2j+1}\right) \frac{(xq;q)_j}{(q;q)_j},$

(12.2.3) $\qquad (xq;q)_\infty = \sum_{j=0}^\infty (-1)^j x^j q^{\frac{j(3j-1)}{2}} \left(1 - x^2 q^{4j+2}\right) \frac{(xq;q)_j}{(q;q)_j},$

(12.2.4) $\qquad (x;q)_\infty = \sum_{j=0}^\infty (-1)^j x^j q^{3\binom{j}{2}} \left(1 - x^3 q^{6j+3}\right) \frac{(xq;q)_j}{(q;q)_j}.$

Selberg only wrote down the first of these. He seems not to have known Sylvester's identity (4.2.5) (to which it is equivalent), but he did point out that (12.2.2) reduces to Euler's pentagonal number theorem when $x = 1$.

When $k = 2$, the fundamental recurrence (12.1.5) becomes

$$S_{2,i}(x) = S_{2,i-1}(x) + (xq)^{i-1}(1-xq)S_{2,3-i}(xq).$$

Taking $i = 1$, this says

(12.2.5) $\qquad S_{2,1}(x) = S_{2,0}(x) + (1-xq)S_{2,2}(xq) = (1-xq)S_{2,2}(xq)$

since $S_{2,0}(x) = 0$. Similarly, when $i = 3$ it says

$$S_{2,3}(x) = S_{2,2}(x) + (xq)^2(1-xq)S_{2,0}(xq) = S_{2,2}(x).$$

When $i = 2$ it says

$$S_{2,2}(x) = S_{2,1}(x) + xq(1-xq)S_{2,1}(xq),$$

12.2. SOME APPLICATIONS

and using (12.2.5) here we have

$$S_{2,2}(x) = (1-xq)S_{2,2}(xq) + xq(1-xq)(1-xq^2)S_{2,2}(xq^2).$$

To simplify this Selberg sets

$$S_{2,2}(x) = (xq;q)_\infty R(x),$$

which gives

(12.2.6) $$R(x) = R(xq) + xq\, R(xq^2)$$

with $R(0) = 1$. Rogers also reaches (12.2.6) in his third proof of the Rogers–Ramanujan identities, with a different order of operations. Hardy said of this proof that it is "in principle the same [as Ramanujan's], though the details differ." This is clear from section 11.1 of the previous chapter, where Ramanujan also arrived at (12.2.6). As in that section, the power series solution of (12.2.6) is

$$R(x) = \sum_{n=0}^\infty \frac{q^{n^2} x^n}{(q;q)_n}.$$

Hence

$$S_{2,2}(x) = (xq;q)_\infty \sum_{n=0}^\infty \frac{q^{n^2} x^n}{(q;q)_n} = S_{2,3}(x),$$

and it follows from (12.2.5) that

$$S_{2,1}(x) = (xq;q)_\infty \sum_{n=0}^\infty \frac{q^{n^2+n} x^n}{(q;q)_n}.$$

Therefore we have
(12.2.7)
$$(xq;q)_\infty \sum_{n=0}^\infty \frac{q^{n^2+n} x^n}{(q;q)_n} = S_{2,1}(x) = \sum_{j=0}^\infty (-1)^j x^{2j} q^{\frac{j(5j+3)}{2}} \left(1 - xq^{2j+1}\right) \frac{(xq;q)_j}{(q;q)_j}$$

and also
(12.2.8)
$$(xq;q)_\infty \sum_{n=0}^\infty \frac{q^{n^2} x^n}{(q;q)_n} = S_{2,2}(x) = \sum_{j=0}^\infty (-1)^j x^{2j} q^{\frac{j(5j+1)}{2}} \left(1 - x^2 q^{4j+2}\right) \frac{(xq;q)_j}{(q;q)_j}$$

$$= S_{2,3}(x) = \sum_{j=0}^\infty (-1)^j x^{2j} q^{\frac{j(5j-1)}{2}} \left(1 - x^3 q^{6j+3}\right) \frac{(xq;q)_j}{(q;q)_j}.$$

The first half of (12.2.8) is also a key equation in Ramanujan's proof. Was Selberg trying to generalize Ramanujan's argument, or was he trying to see what else could be got out of Rogers's function V_m?

The second Rogers–Ramanujan identity follows by taking $x = 1$ in (12.2.7), and the first by taking $x = 1$ in either half of (12.2.8). Because of (12.1.3), this gives

$$(12.2.9) \qquad (q;q)_\infty \sum_{n=0}^{\infty} \frac{q^{n^2+n}}{(q;q)_n} = S_{2,1}(1) = (q;q^5)_\infty (q^4;q^5)_\infty (q^5;q^5)_\infty,$$

$$(12.2.10) \qquad (q;q)_\infty \sum_{n=0}^{\infty} \frac{q^{n^2}}{(q;q)_n} = S_{2,2}(1) = (q^2;q^5)_\infty (q^3;q^5)_\infty (q^5;q^5)_\infty,$$

$$(12.2.11) \qquad (q;q)_\infty \sum_{n=0}^{\infty} \frac{q^{n^2}}{(q;q)_n} = S_{2,3}(1) = (q^2;q^5)_\infty (q^3;q^5)_\infty (q^5;q^5)_\infty.$$

Selberg also attacks

$$S_{0,\frac{1}{2}}(x) = \sum_{j=0}^{\infty} (-1)^j q^{\frac{j^2}{2}} \left(1 - x^{\frac{1}{2}} q^{\frac{2j+1}{2}}\right) \frac{(xq;q)_j}{(q;q)_j}.$$

The fundamental recurrence (12.1.5) gives

$$S_{0,\frac{1}{2}}(x) = S_{0,-\frac{1}{2}}(x) + (xq)^{-\frac{1}{2}}(1-xq) S_{0,\frac{1}{2}}(xq),$$

and (12.1.4) tells us that

$$S_{0,-\frac{1}{2}}(x) = -(xq)^{-\frac{1}{2}} S_{0,\frac{1}{2}}(x),$$

so

$$S_{0,\frac{1}{2}}(x) \left[1 + (xq)^{-\frac{1}{2}}\right] = (xq)^{-\frac{1}{2}} (1-xq) S_{0,\frac{1}{2}}(xq)$$
$$= (xq)^{-\frac{1}{2}} \left[1 + (xq)^{\frac{1}{2}}\right] \left[1 - (xq)^{\frac{1}{2}}\right] S_{0,\frac{1}{2}}(xq),$$

and this simplifies to

$$S_{0,\frac{1}{2}}(x) = \left[1 - (xq)^{\frac{1}{2}}\right] S_{0,\frac{1}{2}}(xq).$$

Iterating this we get

$$S_{0,\frac{1}{2}}(x) = \left[1 - (xq)^{\frac{1}{2}}\right] \left[1 - (xq^2)^{\frac{1}{2}}\right] S_{0,\frac{1}{2}}(xq^2)$$
$$= \left[1 - (xq)^{\frac{1}{2}}\right] \left[1 - (xq^2)^{\frac{1}{2}}\right] \left[1 - (xq^3)^{\frac{1}{2}}\right] S_{0,\frac{1}{2}}(xq^3)$$
$$= \ldots$$
$$= \left((xq)^{\frac{1}{2}}; q^{\frac{1}{2}}\right)_\infty S_{0,\frac{1}{2}}(0)$$

since $|q| < 1$. Now

$$S_{0,\frac{1}{2}}(0) = \sum_{j=0}^{\infty} \frac{(-1)^j q^{\frac{j^2}{2}}}{(q;q)_j},$$

so using Euler's identity (3.5.1)

$$(-x;q)_\infty = \sum_{j=0}^{\infty} \frac{q^{\binom{j}{2}} x^j}{(q;q)_j}$$

we have

$$S_{0,\frac{1}{2}}(0) = \left(q^{\frac{1}{2}};q\right)_\infty.$$

Thus we have proved that

$$\sum_{j=0}^{\infty}(-1)^j q^{\frac{j^2}{2}}\left(1-x^{\frac{1}{2}}q^{\frac{2j+1}{2}}\right)\frac{(xq;q)_j}{(q;q)_j} = S_{0,\frac{1}{2}}(x) = \left((xq)^{\frac{1}{2}};q^{\frac{1}{2}}\right)_{\infty}\left(q^{\frac{1}{2}};q\right)_{\infty}.$$

To make this look nicer we replace q by q^2 and x by x^2, which gives

$$(12.2.12) \quad \sum_{j=0}^{\infty}(-1)^j q^{j^2}\left(1-xq^{2j+1}\right)\frac{(x^2q^2;q^2)_j}{(q^2;q^2)_j} = (xq;q)_{\infty}(q;q^2)_{\infty} = \frac{(xq;q)_{\infty}}{(-q;q)_{\infty}},$$

where the last equality uses Euler's "odd equals distinct" theorem. We saw (12.2.12) in problem 11 in section 5.4, with xq replaced by $-z$. As Selberg points out, it reduces to Gauss's theorem (5.2.11), namely

$$(12.2.13) \quad \sum_{j=-\infty}^{\infty}(-1)^j q^{j^2} = \frac{(q;q)_{\infty}}{(-q;q)_{\infty}},$$

if $x = 1$.

Exercises

1. Is there a nice formula for $S_{1,4}(x)$? Show that $S_{1,4}(q) = -q(q;q)_{\infty}$.
2. Show that (12.2.10) and (12.2.11) simplify to

$$\sum_{n=0}^{\infty}\frac{q^{n^2}}{(q;q)_n} = \frac{1}{(q;q^5)_{\infty}(q^4;q^5)_{\infty}},$$

the first Rogers–Ramanujan identity.

3. Show that (12.2.9) simplifies to

$$\sum_{n=0}^{\infty}\frac{q^{n^2+n}}{(q;q)_n} = \frac{1}{(q^2;q^5)_{\infty}(q^3;q^5)_{\infty}},$$

the second Rogers–Ramanujan identity.

4. Show that (12.2.12) reduces to (12.2.13) when $x = 1$.

12.3. The Selberg coefficients

For the further development of the theory of the Rogers–Selberg function, it is convenient to have two more functions

$$(12.3.1) \qquad a_{m,j}(x) = \sum_{i=0}^{m-j-1} x^i q^{i(j+1)}\binom{i+j}{i}_q \binom{m-i-1}{j}_q$$

and

$$(12.3.2) \qquad b_{m,j}(x) = \sum_{i=0}^{m-j} x^i q^{ij}\binom{i+j}{i}_q \binom{m-i-1}{j-1}_q,$$

where m and j are nonnegative integers, and we define $b_{m,0}(x) = x^m$.

We require two lemmas about these functions that follow from the q-Pascal relation

(12.3.3) $$\binom{n+1}{k}_q = \binom{n}{k-1}_q + q^k \binom{n}{k}_q.$$

Using (12.3.3), we have
$$a_{m+1,j}(x) = \sum_{i=0}^{m-j} x^i q^{i(j+1)} \binom{i+j}{i}_q \left[\binom{m-i-1}{j-1}_q + q^j \binom{m-i-1}{j}_q\right]$$
$$= \sum_{i=0}^{m-j} (xq)^i q^{ij} \binom{i+j}{i}_q \binom{m-i-1}{j-1}_q$$
$$+ q^j \sum_{i=0}^{m-j} x^i q^{i(j+1)} \binom{i+j}{i}_q \binom{m-i-1}{j}_q.$$

The first sum on the last line is $b_{m,j}(xq)$, and, since the $i = m - j$ term may be discarded from the second sum, we have

LEMMA 16.
$$a_{m+1,j}(x) = b_{m,j}(xq) + q^j a_{m,j}(x).$$

Using (12.3.3) again, we have
$$b_{m+1,j}(x) = \sum_{i=0}^{m+1-j} x^i q^{ij} \binom{m-i}{j-1}_q \left[\binom{i+j-1}{i-1}_q + q^i \binom{i+j-1}{i}_q\right]$$
$$= \sum_{i=1}^{m+1-j} x^i q^{ij} \binom{i+j-1}{i-1}_q \binom{m-i}{j-1}_q$$
$$+ \sum_{i=0}^{m+1-j} (xq)^i q^{ij} \binom{i+j-1}{i}_q \binom{m-i}{j-1}_q \binom{i+j-1}{i}_q.$$

The last sum is $a_{m+1,j-1}(xq)$. Reindexing the other one (replacing i by $i+1$) we have
$$\sum_{i=0}^{m-j} x^{i+1} q^{(i+1)j} \binom{i+j}{i}_q \binom{m-i-1}{j-1}_q = xq^j \sum_{i=0}^{m-j} x^i q^{ij} \binom{i+j}{i}_q \binom{m-i-1}{j-1}_q.$$

This proves

LEMMA 17.
$$b_{m+1,j}(x) = a_{m+1,j-1}(xq) + xq^j b_{m,j}(x).$$

Selberg introduced these functions because he wanted to express $S_{k,m}(x)$ as a sum of functions of the form $S_{k,k}(xq^s)$. It turns out to be convenient to have such an expression for $S_{k,k-m}(x)$ as well. From the fundamental recurrence (12.1.5) we have
$$S_{k,1}(x) = S_{k,0}(x) + (1 - xq)S_{k,k}(xq) = (1 - xq)S_{k,k}(xq),$$
and then
$$S_{k,k-1}(x) = S_{k,k}(x) - (xq)^{k-1}(1 - xq)S_{k,1}(xq),$$
which, from the previous line, gives
$$S_{k,k-1}(x) = S_{k,k}(x) - (xq)^{k-1}(1 - xq)(1 - xq^2)S_{k,k}(xq^2).$$

In general we have

$$(12.3.4) \quad S_{k,m}(x) = \sum_{j=0}^{m-1} (-1)^j x^{jk} q^{(2k+1)\binom{j+1}{2} - mj} (xq;q)_{2j+1} a_{m,j}(x) S_{k,k}(xq^{2j+1})$$

and

$$(12.3.5) \quad S_{k,k-m}(x) = \sum_{j=0}^{m} (-1)^j (xq^j)^{jk-m} q^{\binom{j}{2}} (xq;q)_{2j} b_{m,j}(x) S_{k,k}(xq^{2j}).$$

Assume these both hold up to m. By the fundamental recurrence (12.1.5) we have

$$S_{k,m+1}(x) = S_{k,m}(x) + (xq)^m (1 - xq) S_{k,k-m}(xq),$$

and plugging in (12.3.4) and (12.3.5) we get

$$S_{k,m+1}(x) = \sum_{j=0}^{m-1} (-1)^j x^{jk} q^{(2k+1)\binom{j+1}{2} - mj} (xq;q)_{2j+1} a_{m,j}(x) S_{k,k}(xq^{2j+1})$$

$$+ \sum_{j=0}^{m} (-1)^j (xq)^m (xq^{j+1})^{jk-m} q^{\binom{j}{2}} (1 - xq)(xq^2;q)_{2j} b_{m,j}(xq) S_{k,k}(xq^{2j+1}).$$

We can change $m-1$ to m in the first sum since $a_{m,m}(x) = 0$. The second sum is

$$\sum_{j=0}^{m} (-1)^j x^{jk} q^{j(j+1)k - jm + \binom{j}{2}} (xq;q)_{2j+1} b_{m,j}(xq) S_{k,k}(xq^{2j+1}),$$

and it is convenient to rewrite the exponent of q as

$$(2k+1)\binom{j+1}{2} - (m+1)j.$$

Making this change and combining the second sum with the first we have

$$S_{k,m+1}(x) =$$

$$\sum_{j=0}^{m} (-1)^j x^{jk} (xq;q)_{2j+1} S_{k,k}(xq^{2j+1}) q^{(2k+1)\binom{j+1}{2} - (m+1)j} \left[q^j a_{m,j}(x) + b_{m,j}(xq) \right].$$

Invoking Lemma 16 we have (12.3.4) with $m+1$ in place of m. We are half done: we still need (12.3.5) with $m+1$ in place of m. Using the fundamental recurrence (12.1.5) again we have

$$S_{k,k-m-1}(x) = S_{k,k-m}(x) - (xq)^{k-m-1} S_{k,m+1}(xq).$$

Using (12.3.5) and what we just proved, this says

$$S_{k,k-m-1}(x) = \sum_{j=0}^{m} (-1)^j (xq^j)^{jk-m} q^{\binom{j}{2}} (xq;q)_{2j} b_{m,j}(x) S_{k,k}(xq^{2j})$$

$$+ \sum_{j=1}^{m+1} (-1)^j (xq)^{jk-m-1} (xq;q)_{2j} q^{(2k+1)\binom{j}{2} - (m+1)(j-1)} a_{m+1,j-1}(xq) S_{k,k}(xq^{2j}),$$

where we changed $j+1$ to j in the last sum. The other piece of $S_{k,k-m-1}(x)$ is

$$x^{-m}b_{m,0}(x)S_{k,k}(x) + \sum_{j=1}^{m}(-1)^j(xq^j)^{jk-m}q^{\binom{j}{2}}(xq;q)_{2j}b_{m,j}(x)S_{k,k}(xq^{2j}),$$

and we can rewrite this as

$$S_{k,k}(x) + \sum_{j=1}^{m+1}(-1)^j(xq^j)^{jk-m}q^{\binom{j}{2}}(xq;q)_{2j}b_{m,j}(x)S_{k,k}(xq^{2j})$$

since $b_{m,0}(x) = x^m$ and $b_{m,m+1}(x) = 0$. Hence

$$S_{k,k-m-1}(x) = S_{k,k}(x) + \sum_{j=1}^{m+1}(-1)^j(xq^j)^{jk-m}q^{\binom{j}{2}}(xq;q)_{2j}b_{m,j}(x)S_{k,k}(xq^{2j})$$

$$+ \sum_{j=1}^{m+1}(-1)^j(xq)^{jk-m-1}(xq;q)_{2j}q^{(2k+1)\binom{j}{2}-(m+1)(j-1)}a_{m+1,j-1}(xq)S_{k,k}(xq^{2j})$$

and we hope that the two sums combine nicely. We can see some common factors, so we write

$$S_{k,k-m-1}(x) = S_{k,k}(x) +$$

$$\sum_{j=1}^{m+1}(-1)^j x^{jk-m-1}q^{\binom{j}{2}-mj}(xq;q)_{2j}S_{k,k}(xq^{2j})\begin{bmatrix}xq^{j^2k}b_{m,j}(x)+\\ q^{jk+2k\binom{j}{2}-j}a_{m+1,j-1}(xq)\end{bmatrix}.$$

Now $jk + 2k\binom{j}{2} = j^2k$, so this becomes

$$S_{k,k-m-1}(x) = S_{k,k}(x) +$$

$$\sum_{j=1}^{m+1}(-1)^j x^{jk-m-1}q^{\binom{j}{2}-mj+j^2k-j}(xq;q)_{2j}S_{k,k}(xq^{2j})\left[xq^j b_{m,j}(x) + a_{m+1,j-1}(xq)\right]$$

which, by Lemma 17, is

$$S_{k,k-m-1}(x) = S_{k,k}(x)$$

$$+ \sum_{j=1}^{m+1}(-1)^j x^{jk-m-1}q^{\binom{j}{2}-(m+1)j+j^2k}(xq;q)_{2j}b_{m+1,j}(x)S_{k,k}(xq^{2j}).$$

Moreover, $S_{k,k}(x) = x^{-m-1}b_{m+1,0}(x)S_{k,k}(x)$ is the $j = 0$ term of the sum, so

$$S_{k,k-m-1}(x) = \sum_{j=0}^{m+1}(-1)^j x^{jk-m-1}q^{\binom{j}{2}-(m+1)j+j^2k}(xq;q)_{2j}b_{m+1,j}(x)S_{k,k}(xq^{2j}),$$

and this is (12.3.5) with $m+1$ in place of m since $j^2k - (m+1)j = j(jk-m-1)$. This proves (12.3.4) and (12.3.5) by induction on m.

Exercises

1. Show directly that
$$S_{k,2}(x) = (1+xq)(1-xq)S_{k,k}(xq) - x^k q^{2k-1}(xq;q)_3 S_{k,k}(xq^3).$$

2. Show directly that
$$S_{k,k-2}(x) = S_{k,k}(x) - (xq)^{k-2}(1+xq+xq^2)(xq;q)_2 S_{k,k}(xq^2)$$
$$+ x^{2k-2} q^{4k-3}(xq;q)_4 S_{k,k}(xq^4).$$

12.4. The case $k=3$

When $k=3$, the fundamental recurrence (12.1.5) gives

(12.4.1) $\quad S_{3,4}(x) = S_{3,3}(x) + (xq)^3(1-xq)S_{3,0}(xq) = S_{3,3}(x),$

(12.4.2) $\quad S_{3,3}(x) = S_{3,2}(x) + (xq)^2(1-xq)S_{3,1}(xq),$

(12.4.3) $\quad S_{3,2}(x) = S_{3,1}(x) + xq(1-xq)S_{3,2}(xq),$

(12.4.4) $\quad S_{3,1}(x) = S_{3,0}(x) + (1-xq)S_{3,3}(xq) = (1-xq)S_{3,3}(xq).$

These will ultimately be useful, but Selberg starts instead with (12.3.4) and (12.3.5). Taking $k=3$ in both, $m=2$ in (12.3.4), and $m=1$ in (12.3.5), we have
$$S_{3,2}(x) = (1-xq)a_{2,0}(x)S_{3,3}(xq) - x^3 q^5 (xq;q)_3 a_{2,1}(x) S_{3,3}(xq^3)$$
$$= x^{-1} b_{1,0}(x) S_{3,3}(x) - (xq)^2 (xq;q)_2 b_{1,1}(x) S_{3,3}(xq^2).$$

Using $a_{2,0}(x) = 1+xq$, $b_{1,0}(x) = x$, $a_{2,1}(x) = 1 = b_{1,1}(x)$, and setting the two expressions for $S_{3,2}(x)$ equal to each other, we get
$$S_{3,3}(x) = (1+xq)(1-xq)S_{3,3}(xq) + x^2 q^2 (1-xq)(1-xq^2)S_{3,3}(xq^2)$$
$$- x^3 q^5 (1-xq)(1-xq^2)(1-xq^3)S_{3,3}(xq^3).$$

To simplify this Selberg sets

(12.4.5) $\quad S_{3,3}(x) = (xq;q)_\infty Q_0(x)$

for a new function $Q_0(x)$. Since $S_{3,3}(0) = 1$, we have $Q_0(0) = 1$. Plugging in (12.4.5), the recurrence becomes

(12.4.6) $\quad Q_0(x) = (1+xq)Q_0(xq) + x^2 q^2 Q_0(xq^2) - x^3 q^5 Q_0(xq^3).$

Then Selberg defines a sequence of functions $Q_n(x)$ by

(12.4.7) $\quad Q_n(x) = Q_{n-1}(x) - x^2 q^{2n} Q_{n-1}(xq^2) \quad$ for $n \geq 1$.

Note that $Q_n(0) = Q_{n-1}(0)$, so $Q_n(0) = 1$ for every n. These functions satisfy an extension of (12.4.6), namely

(12.4.8) $\quad Q_n(x) = (1+xq)Q_n(xq) + x^2 q^{2n+2} Q_n(xq^2) - x^3 q^{2n+5} Q_n(xq^3).$

When $n=0$, (12.4.8) reduces to (12.4.6). To see it in general, we define

(12.4.9) $\quad R_n(x) = Q_n(x) - (1+xq)Q_n(xq) - x^2 q^{2n+2} Q_n(xq^2) + x^3 q^{2n+5} Q_n(xq^3).$

Then $R_0(x) = 0$ because of (12.4.6), and we hope to prove that $R_n(x) = 0$. This would follow if we could show that

(12.4.10) $\quad R_{n+1}(x) = R_n(x) - x^2 q^{2n+4} R_n(xq^2).$

But by definition
$$R_{n+1}(x) = Q_{n+1}(x) - (1+xq)Q_{n+1}(xq) - x^2 q^{2n+4} Q_{n+1}(xq^2) + x^3 q^{2n+7} Q_{n+1}(xq^3),$$
and using (12.4.7) this becomes
$$R_{n+1}(x) = Q_n(x) - x^2 q^{2n+2} Q_n(xq^2) - (1+xq)\left[Q_n(xq) - x^2 q^{2n+4} Q_n(xq^3)\right]$$
$$- x^2 q^{2n+4} \left[Q_n(xq^2) - x^2 q^{2n+6} Q_n(xq^4)\right] + x^3 q^{2n+7} \left[Q_n(xq^3) - x^2 q^{2n+8} Q_n(xq^5)\right].$$
Rearranging this we have
$$R_{n+1}(x) = Q_n(x) - (1+xq)Q_n(xq) - x^2 q^{2n+2} Q_n(xq^2) + x^3 q^{2n+5} Q_n(xq^3)$$
$$- x^2 q^{2n+4} \left[Q_n(xq^2) - (1+xq^3)Q_n(xq^3) - x^2 q^{2n+6} Q_n(xq^4) + x^3 q^{2n+11} Q_n(xq^5)\right]$$
which is (12.4.10). This establishes (12.4.8).

Selberg now proves that $Q_n(x)$ approaches a function $Q(x)$ as $n \to \infty$. Taking this for granted and using the fact that $|q| < 1$, (12.4.8) simplifies to $Q(x) = (1+xq)Q(xq)$. Iterating this and using $Q(0) = 1$ we have
$$Q(x) = (-xq; q)_\infty.$$

For completeness we sketch a proof that the sequence $\{Q_n(x)\}$ converges. We will follow Selberg for the most part, but for simplicity we assume q is real. Recall that $Q_0(x)$ is a convergent infinite series divided by a convergent infinite product, and choose a positive number r closer to zero than q^{-1} to avoid the zeros of the product. Then there is a positive constant A such that $|Q_0(x)| \leq A$ whenever $|x| \leq r$. By the triangle inequality we have
$$|Q_1(x)| \leq |Q_0(x)| + |x^2 q^2 Q_0(xq^2)| \leq A(1 + r^2 q^2)$$
and
$$|Q_2(x)| \leq |Q_1(x)| + |x^2 q^4 Q_1(xq^2)| \leq A(1 + r^2 q^2)(1 + r^2 q^4)$$
and so on. Continuing in this way we get
$$|Q_n(x)| \leq A(1 + r^2 q^2) \cdots (1 + r^2 q^{2n}) = A(-r^2 q^2; q^2)_n,$$
and since every factor of $(-r^2 q^2; q^2)_n$ is at least 1, we can replace this by the uniform bound
$$(12.4.11) \qquad |Q_n(x)| \leq A(-r^2 q^2; q^2)_\infty,$$
a convergent infinite product. Next, observe that
$$Q_{n+m}(x) - Q_n(x) = \sum_{k=1}^{m} (Q_{n+k}(x) - Q_{n+k-1}(x)),$$
and hence, by the triangle inequality and the definition of $Q_{n+k}(x)$,
$$|Q_{n+m}(x) - Q_n(x)| \leq \sum_{k=1}^{m} |Q_{n+k}(x) - Q_{n+k-1}(x)|$$
$$= \sum_{k=1}^{m} |x^2 q^{2n+2k} Q_{n+k-1}(xq^2)|.$$
Using (12.4.11) and the fact that q is real, this becomes
$$|Q_{n+m}(x) - Q_n(x)| \leq Ar^2(-r^2 q^2; q^2)_\infty \sum_{k=1}^{m} q^{2n+2k} \leq \frac{Ar^2 q^{2n+2}}{1 - q^2} (-r^2 q^2; q^2)_\infty,$$

12.5. Explicit formulas for the Q functions

which approaches zero as $n \to \infty$, independently of m and x. Given this, only a tiny bit of knowledge of real analysis is necessary to see that $\{Q_n(x)\}$ converges for $|x| \leq r$: if the functions $Q_n(x)$ become arbitrarily close to one another, there should be some function $Q(x)$ that they become close to. The technical term is that $\{Q_n(x)\}$ is a *Cauchy sequence*, and therefore a convergent one. This may, or may not, seem intuitive. A proof is not too hard, but it requires the completeness of the real number system. One can find it in any undergraduate real analysis book.

12.5. Explicit formulas for the Q functions

Selberg is able to obtain explicit formulas for the functions $Q_n(x)$ from the previous section. The argument starts with (12.4.7), which we rewrite as

$$Q_n(x) = Q_{n+1}(x) + x^2 q^{2n+2} Q_n(xq^2).$$

It is easy to prove that

$$(12.5.1) \quad Q_n(x) = x^{2m} q^{2m(m+n)} Q_n(xq^{2m}) + \sum_{j=0}^{m-1} x^{2j} q^{2j(j+n)} Q_{n+1}(xq^{2j})$$

for any nonnegative integer m, by induction on m. (When $m = 0$ the sum is empty and (12.5.1) just says $Q_n(x) = Q_n(x)$. When $m = 1$ it is equivalent to (12.4.7).) Letting $m \to \infty$ we get

$$(12.5.2) \quad Q_n(x) = \sum_{j=0}^{\infty} x^{2j} q^{2j(j+n)} Q_{n+1}(xq^{2j})$$

since $|q| < 1$. This too can be iterated: for any nonnegative integer m we have

$$(12.5.3) \quad Q_n(x) = \sum_{k=0}^{\infty} x^{2k} q^{2k(k+n)} \binom{m+k}{k}_{q^2} Q_{n+m+1}(xq^{2k}).$$

When $m = 0$ this is (12.5.2). Let's assume (12.5.3) holds for $m - 1$; in other words, that

$$Q_n(x) = \sum_{i=0}^{\infty} x^{2i} q^{2i(i+n)} \binom{m+i-1}{i}_{q^2} Q_{n+m}(xq^{2i}).$$

Using (12.5.2) with n replaced by $n + m$ and x replaced by xq^{2i}, we get the double sum

$$Q_n(x) = \sum_{i=0}^{\infty} x^{2i} q^{2i(i+n)} \binom{m+i-1}{i}_{q^2} \sum_{j=0}^{\infty} (xq^{2i})^{2j} q^{2j(j+n+m)} Q_{n+m+1}(xq^{2i+2j})$$

$$= \sum_{i=0}^{\infty} \sum_{j=0}^{\infty} x^{2i+2j} q^{2i^2+4ij+2j^2+2in+2jn+2jm} \binom{m+i-1}{i}_{q^2} Q_{n+m+1}(xq^{2i+2j}).$$

Setting $i + j = k$ (replacing j), this becomes

$$Q_n(x) = \sum_{k=0}^{\infty} \sum_{i=0}^{k} x^{2k} q^{2k^2+2nk+2m(k-i)} \binom{m+i-1}{i}_{q^2} Q_{n+m+1}(xq^{2k})$$

$$= \sum_{k=0}^{\infty} x^{2k} q^{2k(k+n)} Q_{n+m+1}(xq^{2k}) \sum_{i=0}^{k} q^{2m(k-i)} \binom{m+i-1}{i}_{q^2},$$

and all we need to complete the proof is
$$\binom{m+k}{k}_{q^2} = \sum_{i=0}^{k} q^{2m(k-i)} \binom{m+i-1}{i}_{q^2}.$$
But we've seen this in section 1.4; it's a q-analogue of the diagonal property of Pascal's triangle. Combinatorially, we can look at sequences of m 0's and k 1's and count inversions. $\binom{m+k}{m}_q$ is the generating function for all such sequences, and $q^{mi}\binom{m+k-i-1}{m-1}_q$ is the generating function for the sequences in which i 1's precede the first 0, so
$$\binom{m+k}{m}_q = \sum_{i=0}^{k} q^{mi} \binom{m+k-i-1}{m-1}_q.$$
Changing i to $k-i$ and q to q^2 we complete the proof of (12.5.3). Next, letting $m \to \infty$ there we get
$$Q_n(x) = \sum_{k=0}^{\infty} \frac{x^{2k} q^{2k(k+n)}}{(q^2; q^2)_k} Q\left(xq^{2k}\right).$$
But $Q(x) = (-xq; q)_\infty$, so
$$Q_n(x) = \sum_{k=0}^{\infty} \frac{x^{2k} q^{2k(k+n)}}{(q^2; q^2)_k} \left(-xq^{2k+1}; q\right)_\infty.$$
Multiplying inside the sum by $(-xq; q)_{2k}$ over itself we finally have

(12.5.4) $$Q_n(x) = (-xq; q)_\infty \sum_{k=0}^{\infty} \frac{q^{2k(k+n)} x^{2k}}{(-xq; q)_{2k} (q^2; q^2)_k},$$

and in particular

(12.5.5) $$Q_0(x) = (-xq; q)_\infty \sum_{k=0}^{\infty} \frac{q^{2k^2} x^{2k}}{(-xq; q)_{2k} (q^2; q^2)_k}.$$

Exercises

1. If $Q_n(x)$ were *defined* by (12.5.4), would $Q_n(x) \to (-xq; q)_\infty$ as $n \to \infty$? Explain.
2. Prove (12.5.1) by induction on m.

12.6. Explicit formulas for $S_{3,i}(x)$

From (12.5.5) and (12.4.5) we have immediately

(12.6.1) $$S_{3,3}(x) = (xq; q)_\infty Q_0(x) = (x^2 q^2; q^2)_\infty \sum_{k=0}^{\infty} \frac{q^{2k^2} x^{2k}}{(-xq; q)_{2k} (q^2; q^2)_k},$$

and because of (12.4.1) we also know that

(12.6.2) $$S_{3,4}(x) = (x^2 q^2; q^2)_\infty \sum_{k=0}^{\infty} \frac{q^{2k^2} x^{2k}}{(-xq; q)_{2k} (q^2; q^2)_k}.$$

Moreover, from (12.4.4) we have $S_{3,1}(x) = (1-xq)S_{3,3}(xq)$. Combining this with (12.6.1) we get

$$S_{3,1}(x) = (1-xq)(x^2q^4;q^2)_\infty \sum_{k=0}^\infty \frac{q^{2k^2+2k}x^{2k}}{(-xq^2;q)_{2k}(q^2;q^2)_k}.$$

If we multiply the right side by $1+xq$ over itself, we get a formula as nice as (12.6.1), namely

(12.6.3) $$S_{3,1}(x) = (x^2q^2;q^2)_\infty \sum_{k=0}^\infty \frac{q^{2k^2+2k}x^{2k}}{(-xq;q)_{2k+1}(q^2;q^2)_k}.$$

With a bit more work, we can get a similar expression for $S_{3,2}(x)$. From (12.4.2) we have

$$S_{3,2}(x) = S_{3,3}(x) - (xq)^2(1-xq)S_{3,1}(xq).$$

Using (12.6.1) and (12.6.3) here we get

$$S_{3,2}(x) = (x^2q^2;q^2)_\infty \sum_{k=0}^\infty \frac{q^{2k^2}x^{2k}}{(-xq;q)_{2k}(q^2;q^2)_k}$$
$$- (xq)^2(1-xq)(x^2q^4;q^2)_\infty \sum_{k=0}^\infty \frac{q^{2k^2+4k}x^{2k}}{(-xq^2;q)_{2k+1}(q^2;q^2)_k}.$$

It is again helpful to multiply the last sum by $1+xq$ over itself. If we also multiply it by $1-q^{2k+2}$ over itself, we get

$$S_{3,2}(x) = (x^2q^2;q^2)_\infty \sum_{k=0}^\infty \frac{q^{2k^2}x^{2k}}{(-xq;q)_{2k}(q^2;q^2)_k}$$
$$- (x^2q^2;q^2)_\infty \sum_{k=0}^\infty \frac{q^{2(k+1)^2}x^{2(k+1)}(1-q^{2k+2})}{(-xq^2;q)_{2(k+1)}(q^2;q^2)_{k+1}},$$

and changing $k+1$ to k in the last sum leaves us with

$$S_{3,2}(x) = (x^2q^2;q^2)_\infty \left[\sum_{k=0}^\infty \frac{q^{2k^2}x^{2k}}{(-xq;q)_{2k}(q^2;q^2)_k} - \sum_{k=1}^\infty \frac{q^{2k^2}x^{2k}(1-q^{2k})}{(-xq^2;q)_{2k}(q^2;q^2)_k} \right].$$

We can change the initial value for the last sum from $k=1$ back to $k=0$, since the factor $1-q^{2k}$ is zero when $k=0$. If we do so, the two sums combine into

(12.6.4) $$S_{3,2}(x) = (x^2q^2;q^2)_\infty \sum_{k=0}^\infty \frac{q^{2k^2+2k}x^{2k}}{(-xq;q)_{2k}(q^2;q^2)_k},$$

a formula just as nice as (12.6.1) and (12.6.3).

Exercises

1. Show that (12.6.3) and (12.6.4) satisfy (12.4.3).
2. Show that

$$S_{3,5}(x) = (x^2q^2;q^2)_\infty \sum_{k=0}^\infty \frac{q^{2k^2}x^{2k}\left[1-x(1-q^{2k})\right]}{(-xq;q)_{2k}(q^2;q^2)_k}.$$

12.7. The payoff for $k = 3$

Following Selberg, we are now in a position to prove several identities similar to, but more complicated than, the Rogers–Ramanujan identities. By (12.6.1) and (12.1.3) we have

$$(q^2; q^2)_\infty \sum_{k=0}^{\infty} \frac{q^{2k^2}}{(-q; q)_{2k}(q^2; q^2)_k} = S_{3,3}(1) = (q^3; q^7)_\infty (q^4; q^7)_\infty (q^7; q^7)_\infty,$$

or

(12.7.1) $$\sum_{k=0}^{\infty} \frac{q^{2k^2}}{(-q; q)_{2k}(q^2; q^2)_k} = \frac{(q^3; q^7)_\infty (q^4; q^7)_\infty (q^7; q^7)_\infty}{(q^2; q^2)_\infty}.$$

Selberg rewrites both sides of this. He prefers to have the denominator on the sum side as $(-q; q^2)_k (q^4; q^4)_k$; we leave this as an exercise. On the product side the manipulation is a bit more complicated. We start by rewriting everything to the base 14. Using the notation for multiple q-shifted factorials from the previous chapter, we have

$$\frac{(q^3; q^7)_\infty (q^4; q^7)_\infty (q^7; q^7)_\infty}{(q^2; q^2)_\infty} = \frac{(q^3, q^4, q^7, q^{10}, q^{11}, q^{14}; q^{14})_\infty}{(q^2, q^4, q^6, q^8, q^{10}, q^{12}, q^{14}; q^{14})_\infty}$$
$$= \frac{(q^3, q^7, q^{11}; q^{14})_\infty}{(q^2, q^6, q^8, q^{12}; q^{14})_\infty}.$$

He also uses

(12.7.2) $$(q^7; q^{14})_\infty = \frac{1}{(-q^7; q^7)_\infty},$$

which is just Euler's "odd equals distinct" theorem. He further writes

$$\frac{(q^3; q^{14})_\infty}{(q^6; q^{14})_\infty} = \frac{1 - q^3}{1 - q^6} \frac{1}{1 - q^{20}} \frac{1 - q^{17}}{1 - q^{34}} \frac{1}{1 - q^{48}} \frac{1 - q^{31}}{1 - q^{62}} \cdots$$
$$= \frac{1}{(1 + q^3)(1 - q^{20})(1 + q^{17})(1 - q^{48})(1 + q^{31}) \cdots}$$
$$= \frac{1}{(-q^3; q^{14})_\infty (q^{20}; q^{28})_\infty}.$$

and similarly

$$\frac{(q^{11}; q^{14})_\infty}{(q^8; q^{14})_\infty} = \frac{1}{(-q^{11}; q^{14})_\infty (q^8; q^{28})_\infty}.$$

After all these changes, Selberg's final form of (12.7.1) is

(12.7.3) $$\sum_{k=0}^{\infty} \frac{q^{2k^2}}{(-q; q^2)_k (q^4; q^4)_k} = \frac{1}{(-q^7; q^7)_\infty (q^2, -q^3, -q^{11}, q^{12}; q^{14})_\infty (q^8, q^{20}; q^{28})_\infty}.$$

You can argue about whether this is an improvement or not.

Next, setting $x = 1$ in (12.6.3) and recalling (12.1.3), we have

(12.7.4) $$\sum_{k=0}^{\infty} \frac{q^{2k^2 + 2k}}{(-q; q)_{2k+1}(q^2; q^2)_k} = \frac{(q; q^7)_\infty (q^6; q^7)_\infty (q^7; q^7)_\infty}{(q^2; q^2)_\infty}.$$

Again Selberg rewrites both sides. He prefers $(-q;q^2)_{k+1}(q^4;q^4)_k$ as the denominator on the sum side. On the product side we again rewrite everything to base 14:

$$\frac{(q;q^7)_\infty(q^6;q^7)_\infty(q^7;q^7)_\infty}{(q^2;q^2)_\infty} = \frac{(q^3,q^6,q^7,q^8,q^{13},q^{14};q^{14})_\infty}{(q^2,q^4,q^6,q^8,q^{10},q^{12},q^{14};q^{14})_\infty}$$

$$= \frac{(q^3,q^7,q^{13};q^{14})_\infty}{(q^2,q^4,q^{10},q^{12};q^{14})_\infty}.$$

We use (12.7.2) again, and also write

$$\frac{(q^{13};q^{14})_\infty}{(q^{12};q^{14})_\infty} = \frac{1}{1-q^{12}} \frac{1-q^{13}}{1-q^{26}} \frac{1}{1-q^{40}} \frac{1-q^{27}}{1-q^{54}} \frac{1}{1-q^{68}} \frac{1-q^{41}}{1-q^{82}} \cdots$$

$$= \frac{1}{(1-q^{12})(1+q^{13})(1-q^{40})(1+q^{27})(1-q^{68})(1+q^{41})\cdots}$$

$$= \frac{1}{(-q^{13};q^{14})_\infty(q^{12};q^{28})_\infty}$$

and similarly

$$\frac{(q;q^{14})_\infty}{(q^2;q^{14})_\infty} = \frac{1}{(-q;q^{14})_\infty(q^{16};q^{28})_\infty}.$$

After all these changes, Selberg's final form of (12.7.4) is
(12.7.5)
$$\sum_{k=0}^\infty \frac{q^{2k^2+2k}}{(-q;q^2)_{k+1}(q^4;q^4)_k} = \frac{1}{(-q^7;q^7)_\infty(-q,q^4,q^{10},-q^{13};q^{14})_\infty(q^{12},q^{16};q^{28})_\infty}.$$

Finally, setting $x = 1$ in (12.6.4) and recalling (12.1.3) we have

(12.7.6) $$\sum_{k=0}^\infty \frac{q^{2k^2+2k}}{(-q;q)_{2k}(q^2;q^2)_k} = \frac{(q^2;q^7)_\infty(q^5;q^7)_\infty(q^7;q^7)_\infty}{(q^2;q^2)_\infty}.$$

As before, Selberg prefers $(-q;q^2)_k(q^4;q^4)_k$ as the denominator on the sum side. On the product side we again rewrite everything to base 14:

$$\frac{(q^2;q^7)_\infty(q^5;q^7)_\infty(q^7;q^7)_\infty}{(q^2;q^2)_\infty} = \frac{(q^2,q^5,q^7,q^9,q^{12},q^{14};q^{14})_\infty}{(q^2,q^4,q^6,q^8,q^{10},q^{12},q^{14};q^{14})_\infty}$$

$$= \frac{(q^5,q^7,q^9;q^{14})_\infty}{(q^4,q^6,q^8,q^{10};q^{14})_\infty}.$$

We use (12.7.2) again, and also write

$$\frac{(q^5;q^{10})_\infty}{(q^{10};q^{14})_\infty} = \frac{1}{(-q^5;q^{14})_\infty(q^{24};q^{28})_\infty}$$

and

$$\frac{(q^9;q^{14})_\infty}{(q^4;q^{14})_\infty} = \frac{1}{(-q^9;q^{14})_\infty(q^4;q^{28})_\infty}.$$

Making all these changes, Selberg's final form of (12.7.6) is
(12.7.7)
$$\sum_{k=0}^\infty \frac{q^{2k^2+2k}}{(-q;q^2)_k(q^4;q^4)_k} = \frac{1}{(-q^7;q^7)_\infty(-q^5,q^6,q^8,-q^9;q^{14})_\infty(q^4,q^{24};q^{28})_\infty}.$$

Exercises

1. Show that $(-q;q)_{2k}(q^2;q^2)_k = (-q;q^2)_k(q^4;q^4)_k$.
2. Show that $(-q;q)_{2k+1}(q^2;q^2)_k = (-q;q^2)_{k+1}(q^4;q^4)_k$.
3. Show that
$$\frac{(q^{11};q^{14})_\infty}{(q^8;q^{14})_\infty} = \frac{1}{(-q^{11};q^{14})_\infty(q^8;q^{28})_\infty}.$$
4. Show that
$$\frac{(q;q^{14})_\infty}{(q^2;q^{14})_\infty} = \frac{1}{(-q;q^{14})_\infty(q^{16};q^{28})_\infty}.$$
5. Show that
$$\frac{(q^5;q^{14})_\infty}{(q^{10};q^{14})_\infty} = \frac{1}{(-q^5;q^{14})_\infty(q^{24};q^{28})_\infty}.$$
6. Show that
$$\frac{(q^9;q^{14})_\infty}{(q^4;q^{14})_\infty} = \frac{1}{(-q^9;q^{14})_\infty(q^4;q^{28})_\infty}.$$
7. Explain why looking at $S_{3,4}(1)$ gives (12.7.3) again.
8. Do we get anything interesting from $S_{3,5}(1)$?

12.8. Gordon's theorem

Mathematicians divide into two types in many different ways. One such dichotomy is special versus general. Mathematics consists in large part of generalization, as we have seen many times in these pages, so all mathematicians are generalizers to a greater or lesser degree. But some of us find the greatest beauty in generality, and others find it in well-chosen specific cases.

As mathematical subjects go, q-analysis tends to be special rather more than general, but more general partition theorems than the ones we have seen began to appear in the second half of the 20th century. We will discuss one such theorem, a generalization of the Rogers–Ramanujan identities due to Basil Gordon. George Andrews observed that Gordon's theorem could be obtained from the Rogers–Selberg function, and this is the path we shall follow.

If we set
$$S_{k,i}(x) = (xq;q)_\infty C_{k,i}(x)$$
and plug this into the fundamental recurrence (12.1.5), we get

(12.8.1) $\qquad C_{k,i}(x) = C_{k,i-1}(x) + (xq)^{i-1} C_{k,k-i+1}(xq).$

Suppose now that k and i are nonnegative integers with $i < 2k+1$. Then $S_{k,i}(x)$ has no negative powers of x or q, and
$$\frac{1}{(xq;q)_\infty} = \sum_{n=0}^\infty \frac{(xq)^n}{(q;q)_n}$$
has none either. It follows that $C_{k,i}(x)$ has none, so we can write

(12.8.2) $\qquad C_{k,i}(x) = \sum_{m=0}^\infty \sum_{N=0}^\infty c_{k,i}(m,N) x^m q^N$

12.8. GORDON'S THEOREM

for some coefficients $c_{k,i}(m, N)$. Since $S_{k,0}(x) = 0$, we have $c_{k,0}(m, N) = 0$ for all k, m, N, and having observed this we now assume i is a positive integer. Since $S_{k,i}(0) = 1$, we have $c_{k,i}(0,0) = 1$. Plugging (12.8.1) into (12.8.2) we get

$$\sum_{m=0}^{\infty}\sum_{N=0}^{\infty} c_{k,i}(m,N)x^m q^N - \sum_{m=0}^{\infty}\sum_{N=0}^{\infty} c_{k,i-1}(m,N)x^m q^N$$
$$= \sum_{r=0}^{\infty}\sum_{S=0}^{\infty} c_{k,k-i+1}(r,S)(xq)^{r+i-1}q^S = \sum_{r=0}^{\infty}\sum_{S=0}^{\infty} c_{k,k-i+1}(r,S)x^{r+i-1}q^{r+S+i-1}.$$

The coefficient of $x^m q^N$ must be the same on both sides. If we set $m = r + i - 1$ and $N = r + S + i - 1$, then $S = N - m$ and $r = m - i + 1$, so

(12.8.3) $$c_{k,i}(m, N) - c_{k,i-1}(m, N) = c_{k,k-i+1}(m - i + 1, N - m).$$

We claim that $c_{k,i}(m, N)$ is the number of partitions of N with exactly m parts, say $N = b_1 + b_2 + \cdots + b_m$, where as usual the parts b_r are weakly decreasing, and they satisfy two additional conditions:

- For each r, $b_r - b_{r+k-1} \geq 2$.
- There are at most $i - 1$ 1's.

These conditions will look more natural if we observe that when $k = 2$ and $i = 1, 2$, they are the conditions for one side of the Rogers–Ramanujan identities. Let's call these AG partitions.

Clearly there are no AG partitions if $i = 0$, and the empty partition of zero, with no parts, gives us $c_{k,i}(0, 0) = 1$; these are the base cases for an induction argument. The point is that AG partitions satisfy (12.8.3). Assume the claim is true for m, N, k, and $i - 1$, and for all smaller values of m and N. If the claim were also true for i, m, N, k, then we would be done by induction on i, m, N.

If the claim were also true for i, then $c_{k,i}(m, N) - c_{k,i-1}(m, N)$ would be the number of AG partitions where 1 is a part *exactly* $i - 1$ times. We need to make an observation in this case. Suppose an AG partition has $i - 1$ 1's and j 2's. Since there are m parts in all, there are $m - i - j + 1$ parts larger than 2. This means that the first 2 is in the $(m - i - j + 2)^{\text{th}}$ position, and the last 1 is in the mth position; i.e., $b_{m-i-j+2} = 2$ and $b_m = 1$, so that $b_{m-i-j+2} - b_m = 1$. This violates the gap condition unless $i + j - 2 \leq k - 2$, so we must have $j \leq k - i$. In other words, an AG partition with exactly $i - 1$ 1's can have at most $k - i$ 2's.

If we now subtract 1 from all the parts, the gap condition remains in place and we will have an AG partition of $N - m$ with exactly $m - i + 1$ parts, of which at most $k - i$ are 1's. By induction, $c_{k,k-i+1}(m - i + 1, N - m)$ counts these. Reading the argument backwards, we see that if the claim is true for $c_{k,k-i+1}(m - i + 1, N - m)$ and for $c_{k,i-1}(m, N)$, then it must also be true for $c_{k,i}(m, N)$. This proves the claim.

If we set $x = 1$ in (12.8.2) and define $G_{k,i}(n)$ to be the number of AG partitions of n with any number of parts, then (12.8.2) becomes

$$C_{k,i}(1) = \sum_{n=0}^{\infty} G_{k,i}(n) q^n.$$

But by (12.1.3) and the definition of $C_{k,i}(x)$ we also have
$$C_{k,i}(1) = \frac{(q^{2k-i+1};q^{2k+1})_\infty (q^i;q^{2k+1})_\infty (q^{2k+1};q^{2k+1})_\infty}{(q;q)_\infty}.$$
This implies

THEOREM 89 (Gordon's theorem). *Let $G_{k,i}(n)$ be the number of AG partitions of n, and let $A_{k,i}(n)$ be the number of partitions of n with parts not congruent to 0 or i or $2k-i+1$ modulo $2k+1$. Then*
$$A_{k,i}(n) = G_{k,i}(n).$$

12.9. Bibliographical Notes

In this chapter we have largely followed Selberg's paper [**219**], except in the last section where we largely followed [**9**]. Gordon's theorem first appeared in [**128**].

Selberg attributed (1.4.2) to [**148**], which was published well before [**117**] but probably written later; (12.2.12) appeared earlier in [**162**] and [**149**]; see the notes for Chapter 5. The identities (12.7.3), (12.7.5), and (12.7.7) were found earlier by Rogers [**202**], and they also appear in Ramanujan's lost notebook [**25**].

Rogers's version of the proof of the Rogers–Ramanujan identities in section 12.2 is in [**203**]. In the classic book [**135**], Hardy chose it instead of Ramanujan's proof, which is a bit curious in that Hardy spent much of his career promoting Ramanujan's work. He wrote there that "no proof is really easy (and it would perhaps be unreasonable to expect an easy proof)." As Igor Pak has pointed out (with Schur's combinatorial proof from Chapter 10 in mind), Hardy might not have said this if he had been a better combinatorialist. He was a great analyst and had many other interests within mathematics and without. Outstanding examples in each category are number theory and cricket.

Hardy is often regarded as the best English prose stylist among mathematicians, and the recent book [**3**] was published largely on this theory. As a writer he is most famous for *A Mathematician's Apology* [**134**]. I am supposed to say at this point that every young mathematician should read [**134**], which I have used in several courses, but I think it is somewhat overrated, and better advice—unless you see yourself more as an intellectual who happens to be a mathematician—would be to read [**130**] and [**135**]. If you do read [**134**], make sure you get the edition with C. P. Snow's Foreword and read that too. In my opinion, Hardy's best writing is about Ramanujan, in his obituary [**131**] and in [**132**]. The latter has been reprinted as the first chapter of [**133**] and in [**3**], and the former in the *Collected Papers* of both men. The very best paragraph Hardy wrote was the last one in [**131**]. He returns to it in [**132**] and partially contradicts it, but he was right the first time.

CHAPTER 13

Bailey's $_6\psi_6$ Sum

13.1. Bailey's formula

We begin this chapter with a beautiful identity of Bailey, though it is only a special case of the one that the chapter title refers to.

THEOREM 90 (Bailey's formula). *If $|q| < 1$ and $xy \neq 0$, then*

$$(13.1.1) \quad \sum_{n=-\infty}^{\infty} \left(\frac{xq^n}{(1-xq^n)^2} - \frac{yq^n}{(1-yq^n)^2} \right)$$

$$= \frac{(x-y)(xy;q)_\infty \left(\frac{q}{xy};q\right)_\infty \left(\frac{qx}{y};q\right)_\infty \left(\frac{qy}{x};q\right)_\infty (q;q)_\infty^4}{(x;q)_\infty^2 \left(\frac{q}{x};q\right)_\infty^2 (y;q)_\infty^2 \left(\frac{q}{y};q\right)_\infty^2}.$$

We follow the proof of Dobbie, who denotes the right side of (13.1.1) by $F(x,y)$ and begins by trying to determine its behavior near $x = 1$. This is a double root of the denominator because of the factor $(x;q)_\infty^2$, so we define

$$\phi(x) = (1-x)^2 F(x,y) = \frac{(x-y)(xy;q)_\infty \left(\frac{q}{xy};q\right)_\infty \left(\frac{qx}{y};q\right)_\infty \left(\frac{qy}{x};q\right)_\infty (q;q)_\infty^4}{(xq;q)_\infty^2 \left(\frac{q}{x};q\right)_\infty^2 (y;q)_\infty^2 \left(\frac{q}{y};q\right)_\infty^2}$$

(evidently $\phi(x)$ also depends on y, but we focus on the x dependence). Then

$$\phi(1) = \frac{(1-y)(y;q)_\infty \left(\frac{q}{y};q\right)_\infty \left(\frac{q}{y};q\right)_\infty (qy;q)_\infty (q;q)_\infty^4}{(q;q)_\infty^2 (q;q)_\infty^2 (y;q)_\infty^2 \left(\frac{q}{y};q\right)_\infty^2} = 1$$

since $(1-y)(qy;q)_\infty = (y;q)_\infty$. Therefore $F(x,y)$ behaves like $\frac{1}{(1-x)^2}$ near $x = 1$, but we might need to add a $\frac{1}{1-x}$ term to capture the behavior precisely. To find the right one, we determine the constant A that makes

$$\lim_{x \to 1} \left(F(x,y) - \frac{1}{(1-x)^2} - \frac{A}{1-x} \right)$$

exist. Note that we can rewrite this as

$$\lim_{x \to 1} \frac{\phi(x) - 1 - A(1-x)}{(1-x)^2},$$

which has the form $\frac{0}{0}$ since $\phi(1) = 1$. By L'Hopital's rule it equals

$$\lim_{x \to 1} \frac{\phi'(x) + A}{2(x-1)},$$

437

so we need $A = -\phi'(1)$ for the limit to exist. To get this we first calculate

$$\log \phi(x) = \log(x-y) + \sum_{n=0}^{\infty} \log(1-xyq^n) + \sum_{n=1}^{\infty} \log\left(1 - \frac{q^n}{xy}\right)$$

$$+ \sum_{n=1}^{\infty} \log\left(1 - \frac{q^n x}{y}\right) + \sum_{n=1}^{\infty} \log\left(1 - \frac{q^n y}{x}\right) - 2\sum_{n=1}^{\infty} \log(1-xq^n)$$

$$- 2\sum_{n=1}^{\infty} \log\left(1 - \frac{q^n}{x}\right) + 4\log(q;q)_\infty - 2\log(y;q)_\infty - 2\log\left(\frac{q}{y};q\right)_\infty$$

and then take the derivative with respect to x to get

$$\frac{\phi'(x)}{\phi(x)} = \frac{1}{x-y} - \sum_{n=0}^{\infty} \frac{yq^n}{1-xyq^n} + \sum_{n=1}^{\infty} \frac{\frac{q^n}{x^2 y}}{1 - \frac{q^n}{xy}} - \sum_{n=1}^{\infty} \frac{\frac{q^n}{y}}{1 - \frac{q^n x}{y}}$$

$$+ \sum_{n=1}^{\infty} \frac{\frac{q^n y}{x^2}}{1 - \frac{q^n y}{x}} + 2\sum_{n=1}^{\infty} \frac{q^n}{1-xq^n} - 2\sum_{n=1}^{\infty} \frac{\frac{q^n}{x^2}}{1 - \frac{q^n}{x}}.$$

Setting $x = 1$ and using $\phi(1) = 1$, we have

$$\phi'(1) = \frac{1}{1-y} - \sum_{n=0}^{\infty} \frac{yq^n}{1-yq^n} + \sum_{n=1}^{\infty} \frac{\frac{q^n}{y}}{1 - \frac{q^n}{y}} - \sum_{n=1}^{\infty} \frac{\frac{q^n}{y}}{1 - \frac{q^n}{y}}$$

$$+ \sum_{n=1}^{\infty} \frac{q^n y}{1-q^n y} + 2\sum_{n=1}^{\infty} \frac{q^n}{1-q^n} - 2\sum_{n=1}^{\infty} \frac{q^n}{1-q^n}.$$

Almost everything cancels here, and we are left with

$$\phi'(1) = \frac{1}{1-y} - \frac{y}{1-y} = 1.$$

It follows that the constant A should equal -1, so that $F(x,y)$ behaves like

$$\frac{1}{(1-x)^2} - \frac{1}{1-x} = \frac{x}{(1-x)^2}$$

near $x = 1$. Since (exercise) $F(x,y) = F(xq, y)$, F has the same behavior near any power of q. This suggests constructing the series

$$\sum_{n=-\infty}^{\infty} \frac{xq^n}{(1-xq^n)^2},$$

which, following Dobbie, we call $G(x)$. Since $G(x) = G(xq)$, if we write $F(x,y) = G(x) + H(x,y)$ for some function $H(x,y)$, then we will also have $H(x,y) = H(xq, y)$. Because $G(x)$ matches the behavior of $F(x,y)$ at every singularity of F (as a function of x) in the finite plane except possibly the origin, we can write

$$H(x,y) = \sum_{n=-\infty}^{\infty} a_n(y) x^n,$$

where this converges for all finite x except possibly $x = 0$. But since $H(x,y) = H(xq, y)$, we have $a_n(y) = q^n a_n(y)$ for all n and an arbitrary q with absolute value less than 1. This is impossible unless $a_n(y) = 0$ for all $n \neq 0$. In other words,

$H(x, y)$ reduces to $a_0(y)$, so we have $F(x, y) = G(x) + a_0(y)$. But $F(y, y) = 0$, so $a_0(y) = -G(y)$, and therefore

$$F(x, y) = G(x) - G(y) = \sum_{n=-\infty}^{\infty} \left(\frac{xq^n}{(1 - xq^n)^2} - \frac{yq^n}{(1 - yq^n)^2} \right)$$

which is the left side of (13.1.1).

Exercises

1. Show that $\sum_{n=-\infty}^{\infty} \frac{xq^n}{(1-xq^n)^2}$ converges for any x if $|q| < 1$.

2. Denoting the right side of (13.1.1) by $F(x, y)$ as above, show that $F(x, y) = F(xq, y)$. Does $F(x, y) = F(x, yq)$?

3. Show that (13.1.1) can be rewritten as
(13.1.2)
$$\sum_{n=-\infty}^{\infty} \frac{q^n(1 - xyq^{2n})}{[(1 - xq^n)(1 - yq^n)]^2} = \frac{(xy; q)_\infty \left(\frac{q}{xy}; q\right)_\infty \left(\frac{qx}{y}; q\right)_\infty \left(\frac{qy}{x}; q\right)_\infty (q; q)_\infty^4}{(x; q)_\infty^2 \left(\frac{q}{x}; q\right)_\infty^2 (y; q)_\infty^2 \left(\frac{q}{y}; q\right)_\infty^2}.$$

4. This problem outlines a proof of the **Andrews–Warnaar identity**

(13.1.3) $\quad (q; q)_\infty (a; q)_\infty (b; q)_\infty \sum_{n=0}^{\infty} \frac{(abq^{n-1}; q)_n \, q^n}{(q; q)_n (a; q)_n (b; q)_n}$

$$= \left(\sum_{r=0}^{\infty} (-1)^r q^{\binom{r}{2}} a^r \right) \left(\sum_{s=0}^{\infty} (-1)^s q^{\binom{s}{2}} b^s \right).$$

We have to show that the coefficient of $a^r b^s$ on the left side of (13.1.3) is

$$(-1)^{r+s} q^{\binom{r}{2} + \binom{s}{2}}.$$

(i) Show that rewriting the left side as

$$(q; q)_\infty \sum_{n=0}^{\infty} \frac{(abq^{n-1}; q)_n \, q^n}{(q; q)_n} (aq^n; q)_\infty (bq^n; q)_\infty$$

and expanding the three numerator factors gives the quadruple sum

$$(q; q)_\infty \sum_{n=0}^{\infty} \sum_{i=0}^{n} \sum_{j=0}^{\infty} \sum_{k=0}^{\infty} \frac{(-1)^{i+j+k} a^{i+j} b^{i+k} q^{n(i+j+k+1) - i + \binom{i}{2} + \binom{j}{2} + \binom{k}{2}}}{(q; q)_i (q; q)_j (q; q)_k (q; q)_{n-i}}.$$

(ii) Fortunately, we do not really have to deal with a quadruple sum. For a given value of i, the only j we want is $r - i$ and the only k is $s - i$. This means the coefficient we want is

$$(q; q)_\infty (-1)^{r+s} \sum_{n=0}^{\infty} \sum_{i=0}^{n} \frac{(-1)^i q^{n(r+s-i+1) - i + \binom{i}{2} + \binom{r-i}{2} + \binom{s-i}{2}}}{(q; q)_i (q; q)_{r-i} (q; q)_{s-i} (q; q)_{n-i}}.$$

Note that the sum on i actually goes to the smallest of the three numbers r, s, n. Show that switching the order of summation and then setting $m = n - i$ makes this into

$$(q;q)_\infty (-1)^{r+s} \sum_{i=0}^\infty \frac{(-1)^i q^{i(r+s-i)+\binom{i}{2}+\binom{r-i}{2}+\binom{s-i}{2}}}{(q;q)_i (q;q)_{r-i}(q;q)_{s-i}} \sum_{m=0}^\infty \frac{q^{m(r+s-i+1)}}{(q;q)_m},$$

where again the outer sum is actually finite.

(iii) Show that doing the inner sum and simplifying gives

$$(-1)^{r+s} q^{\binom{r}{2}+\binom{s}{2}} \sum_i (-1)^i q^{\binom{i+1}{2}} \binom{r+s-i}{i, r-i, s-i}_q.$$

Then use problem 6 in section 1.6 to complete the proof.

5. Andrews and Warnaar derived (13.1.3) to simplify the proof of

THEOREM 91 (Warnaar's formula). *For $|q| < 1$ and all a and b, we have*

(13.1.4) $\quad (q;q)_\infty (a;q)_\infty (b;q)_\infty \sum_{n=0}^\infty \frac{\left(\frac{ab}{q};q\right)_{2n} q^n}{(q;q)_n (a;q)_n (b;q)_n (ab;q)_n}$

$$= 1 + \sum_{r=1}^\infty (-1)^r q^{\binom{r}{2}} a^r + \sum_{s=1}^\infty (-1)^s q^{\binom{s}{2}} b^s.$$

It is convenient to denote the left side of (13.1.3) by $L(a,b)$ and that of (13.1.4) by $W(a,b)$.

(i) Show that $W(a,b)$ equals

$$(q;q)_\infty (a;q)_\infty (b;q)_\infty \left[1 + \sum_{n=1}^\infty \frac{\left[1 - abq^{n-1} - \frac{ab}{q}(1-q^n)\right](abq^n;q)_{n-1} q^n}{(q;q)_n (a;q)_n (b;q)_n} \right].$$

Was it necessary to split off the $n = 0$ term?

(ii) By splitting the numerator, show that $W(a,b) = L(a,b) - ab\, L(aq, bq)$.

(iii) Show that using the right side of (13.1.3) for $L(a,b)$ in (ii) gives the right side of (13.1.4). This proves Warnaar's formula. It has some affinity with Bailey's formula in that it splits a complicated bivariate expression into two pieces that each involve only one of the variables.

6. Show that Warnaar's formula reduces to Jacobi's triple product if $b = \frac{q}{a}$. This may be its most remarkable feature.

7. Show that taking $a = \sqrt{xq}$ and $b = -\sqrt{xq}$ in Warnaar's formula gives

$$(q;q)_\infty (xq;q^2)_\infty \sum_{n=0}^\infty \frac{(-x;q)_{2n} q^n}{(q;q)_n (xq;q^2)_n (-xq;q)_n} = 1 + 2 \sum_{k=1}^\infty q^{2k^2} x^k.$$

For which values of x can we use Jacobi's triple product on the right side? What happens when we do?

8. An identity similar to (13.1.4) was found by Schilling and Warnaar, and independently by Berkovich. If $x \neq y$, then

$$(13.1.5) \quad (q;q)_\infty (qx;q)_\infty (qy;q)_\infty \sum_{n=0}^{\infty} \frac{(xy;q)_{2n}\, q^n}{(q;q)_n (qx;q)_n (qy;q)_n (xy;q)_n}$$

$$= \sum_{k=1}^{\infty} (-1)^{k+1} q^{\binom{k}{2}} \frac{x^k - y^k}{x - y}.$$

(i) Explain why the right side of (13.1.5) can be rewritten as

$$\sum_{k=1}^{\infty} (-1)^{k+1} q^{\binom{k}{2}} \sum_{s=0}^{k-1} x^{k-1-s} y^s = \sum_{s=0}^{\infty} \sum_{k=s+1}^{\infty} (-1)^{k+1} q^{\binom{k}{2}} x^{k-1-s} y^s$$

$$= \sum_{r=0}^{\infty} \sum_{s=0}^{\infty} (-1)^{r+s} q^{\binom{r+s+1}{2}} x^r y^s.$$

Therefore we have to show that the coefficient of $x^r y^s$ on the left side of (13.1.5) is

$$(-1)^{r+s} q^{\binom{r+s+1}{2}}.$$

(ii) Show that rewriting the left side as

$$(q;q)_\infty \sum_{n=0}^{\infty} \frac{(xyq^n;q)_n\, q^n}{(q;q)_n} (xq^{n+1};q)_\infty (yq^{n+1};q)_\infty$$

and expanding the three numerator factors gives the quadruple sum

$$(q;q)_\infty \sum_{n=0}^{\infty} \sum_{i=0}^{n} \sum_{j=0}^{\infty} \sum_{k=0}^{\infty} \frac{(-1)^{i+j+k} x^{i+j} y^{i+k} q^{n(i+j+k+1)+\binom{i}{2}+\binom{j+1}{2}+\binom{k+1}{2}}}{(q;q)_i (q;q)_j (q;q)_k (q;q)_{n-i}}.$$

(iii) Again, we do not really have to deal with a quadruple sum. For a given value of i, the only j we want is $r - i$ and the only k is $s - i$. This means the coefficient we want is

$$(q;q)_\infty (-1)^{r+s} \sum_{n=0}^{\infty} \sum_{i=0}^{n} \frac{(-1)^i q^{n(r+s-i+1)+\binom{i}{2}+\binom{r-i+1}{2}+\binom{s-i+1}{2}}}{(q;q)_i (q;q)_{r-i} (q;q)_{s-i} (q;q)_{n-i}}.$$

Note that the sum on i actually goes to the smallest of the three numbers r, s, n. Show that switching the order of summation and then setting $m = n - i$ makes this into

$$(q;q)_\infty (-1)^{r+s} \sum_{i=0}^{\infty} \frac{(-1)^i q^{i(r+s-i+1)+\binom{i}{2}+\binom{r-i+1}{2}+\binom{s-i+1}{2}}}{(q;q)_i (q;q)_{r-i} (q;q)_{s-i}} \sum_{m=0}^{\infty} \frac{q^{m(r+s-i+1)}}{(q;q)_m},$$

where again the outer sum is actually finite.

(iv) Show that doing the inner sum and simplifying gives

$$(-1)^{r+s} q^{\binom{r+1}{2}+\binom{s+1}{2}} \sum_{i} (-1)^i q^{\binom{i}{2}} \binom{r+s-i}{i, r-i, s-i}_q.$$

Then use problem 6 in section 1.6 to complete the proof.

9. Not only are the proofs of (13.1.4) and (13.1.5) quite similar, but Berkovich has shown how to obtain each identity from the other. In this problem we follow his method for deducing (13.1.4) from (13.1.5). (As he points out, "it is a bit more of a challenge" to go the other way.)

(i) Show that $\quad x^{n+1} - y^{n+1} = (x^n + y^n)(x-y) + xy(x^{n-1} - y^{n-1}).$

(ii) Denote the right side of (13.1.5) by $B(x,y)$ and write the right side of (13.1.4) as

$$W(x,y) = 1 + \sum_{n=1}^{\infty} (-1)^n q^{\binom{n}{2}} (x^n + y^n).$$

Use (i) to show that $\quad W(x,y) = B\left(\dfrac{x}{q}, \dfrac{y}{q}\right) - xyq\, B(xq, yq).$

(iii) Show that

$$\frac{\left(\frac{xy}{q};q\right)_{2n}}{(xy;q)_n} = \frac{\left(\frac{xy}{q^2};q\right)_{2n}}{\left(\frac{xy}{q^2};q\right)_n} - \frac{xy}{q}(1-q^n)(1-q^{n-1})(xyq^n;q)_{n-2},$$

where the last term is zero if $n < 2$.

(iv) If $S(x,y)$ denotes the left side of (13.1.5), use (iii) to show that

$$(q;q)_\infty (x;q)_\infty (y;q)_\infty \sum_{n=0}^{\infty} \frac{\left(\frac{xy}{q};q\right)_{2n} q^n}{(q;q)_n (x;q)_n (y;q)_n (xy;q)_n} = S\left(\frac{x}{q},\frac{y}{q}\right) - xyq\, S(xq, yq).$$

(v) Explain why (ii), (iv), and (13.1.5) together imply (13.1.4).

13.2. Another proof of Ramanujan's "most beautiful" identity

If we change q to q^5 in Bailey's formula (13.1.1) and set $x = q$ and $y = q^2$, we get

$$\sum_{n=-\infty}^{\infty} \left(\frac{q^{5n+1}}{(1-q^{5n+1})^2} - \frac{q^{5n+2}}{(1-q^{5n+2})^2}\right)$$
$$= \frac{q(1-q)(q^3;q^5)_\infty (q^2;q^5)_\infty (q^4;q^5)_\infty (q^6;q^5)_\infty (q^5;q^5)_\infty^4}{(q;q^5)_\infty^2 (q^4;q^5)_\infty^2 (q^2;q^5)_\infty^2 (q^3;q^5)_\infty^2}.$$

Noting that $(1-q)(q^6;q^5)_\infty = (q;q^5)_\infty$, we can make a number of cancellations here and be left with

$$\sum_{n=-\infty}^{\infty} \left(\frac{q^{5n+1}}{(1-q^{5n+1})^2} - \frac{q^{5n+2}}{(1-q^{5n+2})^2}\right) = \frac{q(q^5;q^5)_\infty^4}{(q;q^5)_\infty (q^2;q^5)_\infty (q^3;q^5)_\infty (q^4;q^5)_\infty}.$$

Multiplying top and bottom of the right side by $(q^5;q^5)_\infty$ we get

(13.2.1) $$\sum_{n=-\infty}^{\infty} \left(\frac{q^{5n+1}}{(1-q^{5n+1})^2} - \frac{q^{5n+2}}{(1-q^{5n+2})^2}\right) = \frac{q(q^5;q^5)_\infty^5}{(q;q)_\infty}.$$

13.2. ANOTHER PROOF OF RAMANUJAN'S "MOST BEAUTIFUL" IDENTITY

Ramanujan states this formula without proof in one of his notebooks. He then asks himself which terms of the series in (13.2.1) will have exponents that are multiples of 5. For this we just have to look at

(13.2.2) $$\frac{x}{(1-x)^2} = x + 2x^2 + 3x^3 + 4x^4 + 5x^5 + \ldots \quad \text{if } |x| < 1.$$

The terms which have exponents divisible by 5 are

$$5x^5 + 10x^{10} + 15x^{15} + \ldots,$$

and by (13.2.2) we have

$$5x^5 + 10x^{10} + 15x^{15} + \cdots = \frac{5x^5}{(1-x^5)^2}.$$

Therefore the terms on the series side of (13.2.1) with exponents divisible by 5 are

$$\sum_{n=-\infty}^{\infty} \left(\frac{5q^{25n+5}}{(1-q^{25n+5})^2} - \frac{5q^{25n+10}}{(1-q^{25n+10})^2} \right).$$

On the other side of (13.2.1) we have $q(q^5;q^5)_\infty^5$, and every term in the expansion of this will have an exponent congruent to 1 mod 5. It follows that the only terms of $1/(q;q)_\infty$ that can give us an exponent divisible by 5 are those with exponents congruent to 4 mod 5, namely

$$\sum_{n=0}^{\infty} p(5n+4) q^{5n+4},$$

and so we have

$$q(q^5;q^5)_\infty^5 \sum_{n=0}^{\infty} p(5n+4) q^{5n+4} = \sum_{n=-\infty}^{\infty} \left(\frac{5q^{25n+5}}{(1-q^{25n+5})^2} - \frac{5q^{25n+10}}{(1-q^{25n+10})^2} \right).$$

Moving the factor of q inside the sum and changing q^5 to q we get

$$(q;q)_\infty^5 \sum_{n=0}^{\infty} p(5n+4) q^{n+1} = \sum_{n=-\infty}^{\infty} \left(\frac{5q^{5n+1}}{(1-q^{5n+1})^2} - \frac{5q^{5n+2}}{(1-q^{5n+2})^2} \right).$$

We can use (13.2.1) to sum the right side, which gives

$$(q;q)_\infty^5 \sum_{n=0}^{\infty} p(5n+4) q^{n+1} = \frac{5q(q^5;q^5)_\infty^5}{(q;q)_\infty},$$

and finally

$$\sum_{n=0}^{\infty} p(5n+4) q^n = \frac{5(q^5;q^5)_\infty^5}{(q;q)_\infty^6}.$$

This is Ramanujan's "most beautiful" identity again.

13.3. Sums of eight squares and of eight triangular numbers

In this section we use Bailey's formula to derive Jacobi's theorem on sums of eight squares. If we set $y = x$ in (13.1.2), we get

$$(13.3.1) \qquad \sum_{n=-\infty}^{\infty} \frac{q^n(1 + xq^n)}{(1 - xq^n)^3} = \frac{(x^2; q)_\infty \left(\frac{q}{x^2}; q\right)_\infty (q; q)_\infty^6}{(x; q)_\infty^4 \left(\frac{q}{x}; q\right)_\infty^4}.$$

Multiplying (13.3.1) by

$$\frac{(1-x)^3}{1+x} = \frac{(1-x)^4}{1-x^2}$$

we have

$$(13.3.2) \qquad \sum_{n=-\infty}^{\infty} \frac{q^n(1 + xq^n)(1-x)^3}{(1 - xq^n)^3(1+x)} = \frac{(x^2q; q)_\infty \left(\frac{q}{x^2}; q\right)_\infty (q; q)_\infty^6}{(xq; q)_\infty^4 \left(\frac{q}{x}; q\right)_\infty^4}.$$

The $n = 0$ term of the sum in (13.3.2) is 1, and the terms with n negative are

$$\sum_{n=1}^{\infty} \frac{q^{-n}(1 + xq^{-n})(1-x)^3}{(1 - xq^{-n})^3(1+x)} \frac{q^{3n}}{q^{3n}} = \sum_{n=1}^{\infty} \frac{q^n(q^n + x)(1-x)^3}{(q^n - x)^3(1+x)},$$

so (13.3.2) becomes

$$(13.3.3) \qquad \frac{(x^2q; q)_\infty \left(\frac{q}{x^2}; q\right)_\infty (q; q)_\infty^6}{(xq; q)_\infty^4 \left(\frac{q}{x}; q\right)_\infty^4} = 1 + (1-x)^3 \sum_{n=1}^{\infty} q^n \frac{\frac{1+xq^n}{(1-xq^n)^3} + \frac{q^n+x}{(q^n-x)^3}}{1+x}.$$

We now let $x \to -1$, using L'Hopital's rule on the fraction inside the sum. Since the derivative of the denominator is 1, we have

$$\lim_{x \to -1} \frac{\frac{1+xq^n}{(1-xq^n)^3} + \frac{q^n+x}{(q^n-x)^3}}{1+x} = \lim_{x \to -1} \frac{d}{dx}\left[\frac{1+xq^n}{(1-xq^n)^3} + \frac{q^n+x}{(q^n-x)^3}\right].$$

Calculating the derivatives we get

$$\frac{(1-xq^n)q^n + (1+xq^n)3q^n}{(1-xq^n)^4} = \frac{2q^n(2+xq^n)}{(1-xq^n)^4}$$

and

$$\frac{q^n - x + 3(q^n + x)}{(q^n - x)^4} = \frac{2(2q^n + x)}{(q^n - x)^4}$$

so

$$\lim_{x \to -1} \frac{\frac{1+xq^n}{(1-xq^n)^3} + \frac{q^n+x}{(q^n-x)^3}}{1+x} = \lim_{x \to -1} \left[\frac{2q^n(2+xq^n)}{(1-xq^n)^4} + \frac{2(2q^n+x)}{(q^n-x)^4}\right]$$

$$= \frac{2(2q^n - q^{2n} + 2q^n - 1)}{(1+q^n)^4}$$

$$(13.3.4) \qquad = \frac{-2(1 - 4q^n + q^{2n})}{(1+q^n)^4}.$$

Hence the result of letting $x \to -1$ in (13.3.3) is

$$(13.3.5) \qquad \left(\frac{(q;q)_\infty}{(-q;q)_\infty}\right)^8 = 1 - 16 \sum_{n=1}^{\infty} \frac{q^n(1 - 4q^n + q^{2n})}{(1+q^n)^4}.$$

13.3. SUMS OF EIGHT SQUARES AND OF EIGHT TRIANGULAR NUMBERS

Equation (9.1.3) was

(13.3.6) $$\sum_{k=0}^{\infty} k^3 x^k = \frac{x(1+4x+x^2)}{(1-x)^4} \quad \text{for } |x|<1,$$

where the $k=0$ term can be discarded. Taking $x = -q^n$ here we have

$$\left[\frac{(q;q)_\infty}{(-q;q)_\infty}\right]^8 = 1 + 16 \sum_{n=1}^{\infty} \sum_{k=1}^{\infty} k^3 (-1)^k q^{nk}.$$

If we set $nk = m$ here, then k has to be a divisor of m, and we have

(13.3.7) $$\left[\frac{(q;q)_\infty}{(-q;q)_\infty}\right]^8 = 1 + 16 \sum_{m=1}^{\infty} \left(\sum_{k|m} (-1)^k k^3\right) q^m.$$

On the left side we use Gauss's identity (5.2.11)

$$\frac{(q;q)_\infty}{(-q;q)_\infty} = \sum_{j=-\infty}^{\infty} (-1)^j q^{j^2}$$

to get

$$\left[\frac{(q;q)_\infty}{(-q;q)_\infty}\right]^8 = \sum_{j_1,\ldots,j_8=-\infty}^{\infty} (-1)^{j_1+\cdots+j_8} q^{j_1^2+\cdots+j_8^2}.$$

Observe that

$$(-1)^{j_1+\cdots+j_8} = (-1)^{j_1^2+\cdots+j_8^2}$$

since $j_1^2 - j_1 + \cdots + j_8^2 - j_8 = j_1(j_1-1) + \cdots + j_8(j_8-1)$ is a sum of eight even numbers. Then

$$\left[\frac{(q;q)_\infty}{(-q;q)_\infty}\right]^8 = \sum_{j_1,\ldots,j_8=-\infty}^{\infty} (-q)^{j_1^2+\cdots+j_8^2}.$$

The coefficient of $(-q)^m$ here is clearly the number of ways that m can arise as a sum of eight squares $j_1^2 + \cdots + j_8^2$. Using the notation $\square_8(m)$ for it, (13.3.7) becomes

$$\sum_{m=0}^{\infty} (-1)^m \square_8(m) q^m = 1 + 16 \sum_{m=1}^{\infty} \left(\sum_{k|m} (-1)^k k^3\right) q^m$$

and we have proved

THEOREM 92 (Jacobi's eight square theorem). *If $\square_8(m)$ denotes the number of ways to write the nonnegative integer m as a sum of eight squares (counting signs and permutations), then*

$$\square_8(m) = \begin{cases} 1 & \text{if } m = 0, \\ 16 \sum_{k|m} (-1)^{m-k} k^3 & \text{if } m \geq 1. \end{cases}$$

A similar argument gives a theorem about sums of eight triangular numbers. We start by replacing q by q^2 in (13.3.1) and then replacing x by xq. This gives

$$\sum_{n=-\infty}^{\infty} \frac{q^{2n}(1+xq^{2n+1})}{(1-xq^{2n+1})^3} = \frac{(x^2q^2;q^2)_\infty \left(\frac{1}{x^2};q^2\right)_\infty (q^2;q^2)_\infty^6}{(xq;q^2)_\infty^4 \left(\frac{q}{x};q^2\right)_\infty^4}.$$

Note that the right side becomes zero if $x = 1$ because of the first factor of $\left(\frac{1}{x^2}; q^2\right)_\infty$. Dividing both sides by this factor we have

(13.3.8) $\qquad \dfrac{x^2}{x^2 - 1} \displaystyle\sum_{n=-\infty}^{\infty} \dfrac{q^{2n}(1 + xq^{2n+1})}{(1 - xq^{2n+1})^3} = \dfrac{(x^2q^2; q^2)_\infty \left(\frac{q^2}{x^2}; q^2\right)_\infty (q^2; q^2)_\infty^6}{(xq; q^2)_\infty^4 \left(\frac{q}{x}; q^2\right)_\infty^4}.$

If we let $x \to 1$, then the right side becomes

$$\left(\dfrac{(q^2; q^2)_\infty}{(q; q^2)_\infty}\right)^8 = \left(\sum_{n=0}^{\infty} q^{\binom{n+1}{2}}\right)^8$$

by Gauss's identity (5.2.12). To be able to take this limit on the left side, we rewrite

$$\sum_{n=-\infty}^{\infty} \dfrac{q^{2n}(1 + xq^{2n+1})}{(1 - xq^{2n+1})^3} = \sum_{m=0}^{\infty} \dfrac{q^{2m}(1 + xq^{2m+1})}{(1 - xq^{2m+1})^3} + \sum_{m=-1}^{-\infty} \dfrac{q^{2m}(1 + xq^{2m+1})}{(1 - xq^{2m+1})^3}.$$

Setting $m = -n - 1$ in the last sum and $m = n$ in the one before it, we get

$$\sum_{n=-\infty}^{\infty} \dfrac{q^{2n}(1 + xq^{2n+1})}{(1 - xq^{2n+1})^3} = \sum_{n=0}^{\infty} \dfrac{q^{2n}(1 + xq^{2n+1})}{(1 - xq^{2n+1})^3} + \sum_{n=0}^{\infty} \dfrac{q^{-2n-2}(1 + xq^{-2n-1})}{(1 - xq^{-2n-1})^3} \dfrac{q^{6n+3}}{q^{6n+3}}$$

$$= \sum_{n=0}^{\infty} \dfrac{q^{2n}(1 + xq^{2n+1})}{(1 - xq^{2n+1})^3} + \sum_{n=0}^{\infty} \dfrac{q^{2n}(q^{2n+1} + x)}{(q^{2n+1} - x)^3},$$

so (13.3.8) becomes

(13.3.9) $\qquad \dfrac{x^2}{x^2 - 1} \displaystyle\sum_{n=0}^{\infty} q^{2n} \left[\dfrac{1 + xq^{2n+1}}{(1 - xq^{2n+1})^3} - \dfrac{x + q^{2n+1}}{(x - q^{2n+1})^3}\right]$

$$= \dfrac{(x^2q^2; q^2)_\infty \left(\frac{q^2}{x^2}; q^2\right)_\infty (q^2; q^2)_\infty^6}{(xq; q^2)_\infty^4 \left(\frac{q}{x}; q^2\right)_\infty^4}.$$

We now need to calculate

$$\lim_{x \to 1} \dfrac{x^2}{x^2 - 1} \left[\dfrac{1 + xq^{2n+1}}{(1 - xq^{2n+1})^3} - \dfrac{x + q^{2n+1}}{(x - q^{2n+1})^3}\right],$$

and by L'Hopital's rule this is

$$\dfrac{1}{2} \lim_{x \to 1} \dfrac{d}{dx} \left[\dfrac{1 + xq^{2n+1}}{(1 - xq^{2n+1})^3} - \dfrac{x + q^{2n+1}}{(x - q^{2n+1})^3}\right]$$

$$= \dfrac{1}{2} \lim_{x \to 1} \left[\dfrac{2q^{2n+1}(2 + xq^{2n+1})}{(1 - xq^{2n+1})^4} + \dfrac{2(x + 2q^{2n+1})}{(x - q^{2n+1})^4}\right]$$

$$= \dfrac{q^{2n+1}(2 + q^{2n+1}) + 1 + q^{2n+1}}{(1 - q^{2n+1})^4} = \dfrac{1 + 4q^{2n+1} + q^{4n+2}}{(1 - q^{2n+1})^4}.$$

Therefore the result of letting $x \to 1$ in (13.3.9) is

(13.3.10) $\qquad \displaystyle\sum_{n=0}^{\infty} \dfrac{q^{2n}\left(1 + 4q^{2n+1} + q^{4n+2}\right)}{(1 - q^{2n+1})^4} = \left(\dfrac{(q^2; q^2)_\infty}{(q; q^2)_\infty}\right)^8 = \left(\sum_{n=0}^{\infty} q^{\binom{n+1}{2}}\right)^8.$

If we multiply the left side by $\frac{q}{q}$, then (13.3.6) applies to it with $x = q^{2n+1}$. Denoting by $\triangle_8(m)$ the number of ways to write the nonnegative integer m as a sum of eight triangular numbers $\binom{n+1}{2}$, we have

$$\sum_{n=0}^{\infty} \sum_{k=1}^{\infty} k^3 q^{(2n+1)k-1} = \sum_{m=0}^{\infty} \triangle_8(m)\, q^m.$$

If we now set $(2n+1)k = m+1$, then we get a contribution of k^3 to the coefficient of q^m on the left side exactly when $k(2n+1) = m+1$ for some $n \geq 0$, i.e., exactly when $\frac{m+1}{k}$ is a positive odd number. We have proved the following remarkable theorem.

THEOREM 93. *If $\triangle_8(m)$ denotes the number of ways to write the nonnegative integer m as a sum of eight triangular numbers, then*

(13.3.11) $$\triangle_8(m) = \sum_k k^3,$$

where the sum is over all positive integers k such that $\frac{m+1}{k}$ is odd.

We give one illustration of this theorem. If $m = 14$, we have to look at the divisors of $m+1 = 15$, namely $1, 3, 5, 15$. Since they are all odd, 15 divided by any of them is odd, so the admissible values of k in (13.3.11) are $k = 1, 3, 5, 15$, and therefore the right side of (13.3.11) is $1^3 + 3^3 + 5^3 + 15^3 = 1 + 27 + 125 + 3375 = 3528$. The triangular numbers $\leq m = 14$ are $0, 1, 3, 6, 10$. If we write $14 = 10 + 3 + 1$ plus five 0's, there are 8 ways to place the 10, then 7 ways to place the 3, then 6 ways to place the 1, so this gives $8 \cdot 7 \cdot 6 = 336$ possibilities, and by the same reasoning $14 = 6 + 3 + 1 + 1 + 1 + 1 + 1 + 0$ gives 336 possibilities. If we write $14 = 10 + 1 + 1 + 1 + 1 + 0 + 0 + 0$, then there are 8 ways to place the 10 and then $\binom{7}{4} = 35$ ways to place the 1's, so this gives $8 \cdot 35 = 280$ possibilities. If we write $14 = 6 + 6 + 1 + 1$ plus four 0's, there are $\binom{8}{2} = 28$ ways to place the 6's and then $\binom{6}{2} = 15$ ways to place the 1's, so this gives $28 \cdot 15 = 420$ possibilities, and by the same reasoning $14 = 3 + 3 + 3 + 3 + 1 + 1 + 0 + 0$ gives 420 possibilities. The largest number of options comes from $14 = 6 + 3 + 3 + 1 + 1 + 0 + 0 + 0$, where we have 8 ways to place the 6, then $\binom{7}{2} = 21$ ways to place the 3's, then $\binom{5}{2} = 10$ ways to place the 1's, so $8 \cdot 21 \cdot 10 = 1680$ possibilities. The only other way to write 14 with eight triangular numbers is as a sum of three 3's and five 1's, which gives $\binom{8}{3} = 56$ possibilities. Since $336 + 336 + 280 + 420 + 420 + 1680 + 56 = 3528$, (13.3.11) is verified. Can you explain why all these numbers are divisible by 7? (In fact, they are all divisible by 28.)

Exercises

1. Explain why Jacobi's eight square theorem is particularly nice when m is odd.

2. How many ways are there to write 1 as a sum of eight squares? (In some sense this explains where the 16 comes from.)

3. How many ways are there to write 2 as a sum of eight squares? Describe them.

4. If p is an odd prime number, how many ways are there to write p as a sum of eight squares? Explain.

5. Check Jacobi's eight square theorem when $m = 8$. Both sides should equal 16×583.

6. Check Jacobi's eight square theorem when $m = 9$. Both sides should equal 16×757.

7. Check Jacobi's eight square theorem when $m = 10$. Both sides should equal 16×882.

8. If m is an odd prime power, say $m = p^k$ for an odd prime p and a nonnegative integer k, show that there are

$$16 \frac{p^{3k+3} - 1}{p^3 - 1}$$

ways to write m as a sum of eight squares.

9. If m is a power of 2, say $m = 2^k$ for a positive integer k, show that there are

$$\frac{16}{7} \left(8^{k+1} - 15 \right)$$

ways to write m as a sum of eight squares.

10. With some algebra we can avoid using L'Hopital's rule in (13.3.3). Show that

$$\frac{\frac{1+xq^n}{(1-xq^n)^3} + \frac{q^n+x}{(q^n-x)^3}}{1+x} = \frac{(1-x)\left[x(1 - 6q^{2n} + q^{4n}) + q^n(1+x^2)(1+q^{2n})\right]}{(1-xq^n)^3 (q^n - x)^3}$$

if $x \neq -1$, and that the right side reduces to (13.3.4) when $x = -1$. This gives an alternate derivation of (13.3.5).

11. We can also avoid L'Hopital's rule in (13.3.9). Show that

$$\frac{1 + xq^{2n+1}}{(1 - xq^{2n+1})^3} - \frac{x + q^{2n+1}}{(x - q^{2n+1})^3}$$
$$= \frac{(x^2 - 1)\left[q^{2n+1}\left(1 + q^{4n+2}\right)\left(x^2 + 1\right) + x\left(1 - 6q^{4n+2} + q^{8n+4}\right)\right]}{(1 - xq^{2n+1})^3 (x - q^{2n+1})^3}$$

and that this implies that (13.3.9) becomes (13.3.10) when $x \to 1$.

12. Check (13.3.11) for $m = 0, 1, 2, 3, 4, 5$. Both sides should equal $1, 8, 28, 64, 126$, and 224 respectively.

13. Check (13.3.11) for $m = 6, 7, 8, 9$. Both sides should equal $344, 512, 757, 1008$ respectively.

14. Check (13.3.11) for $m = 10$ and $m = 11$. Both sides should equal 1332 and 1792 respectively.

15. Check (13.3.11) for $m = 12$ and $m = 13$. Both sides should equal 2198 and 2752 respectively.

16. Check (13.3.11) for $m = 15$ and $m = 16$. Both sides should equal 4096 and 4914 respectively.

13.4. Bailey's $_6\psi_6$ summation formula

In 1936, Bailey proved a very general series identity, which we state in the form given by Askey. It is convenient to use the abbreviation

$$(a_1, a_2, \ldots, a_k; q)_\infty = (a_1; q)_\infty (a_2; q)_\infty \cdots (a_k; q)_\infty$$

introduced in Chapter 11, and similarly for finite products.

THEOREM 94 (Bailey's very well poised $_6\psi_6$ sum). *If $|q| < |bcde|$ and $|q| < 1$, then*

$$(13.4.1) \quad \sum_{n=-\infty}^{\infty} \frac{(aq, -aq, ab, ac, ad, ae; q)_n}{(a, -a, \frac{aq}{b}, \frac{aq}{c}, \frac{aq}{d}, \frac{aq}{e}; q)_n} \left(\frac{q}{bcde}\right)^n$$
$$= \frac{(a^2 q, q, \frac{q}{a^2}, \frac{q}{bc}, \frac{q}{bd}, \frac{q}{be}, \frac{q}{cd}, \frac{q}{ce}, \frac{q}{de}; q)_\infty}{(\frac{q}{ab}, \frac{q}{ac}, \frac{q}{ad}, \frac{q}{ae}, \frac{aq}{b}, \frac{aq}{c}, \frac{aq}{d}, \frac{aq}{e}, \frac{q}{bcde}; q)_\infty}.$$

Note that this means none of b, c, d, e can be zero. As in Chapter 6, "$_6\psi_6$" means the series runs from $-\infty$ to ∞ with 6 q-shifted factorials in the numerator and 6 in the denominator. Beyond the symmetry in b, c, d, e, "well poised" means that every denominator parameter can be paired with a numerator parameter to give a constant product. In this case

$$a^2 q = a(aq) = -a(-aq) = \frac{aq}{b}(ab) = \frac{aq}{c}(ac) = \frac{aq}{d}(ad) = \frac{aq}{e}(ae).$$

"Very" well poised means that, in addition, two numerator parameters are q times the corresponding denominator parameters, meaning that these four parameters contribute only a factor of

$$(13.4.2) \quad \frac{(aq; q)_n (-aq; q)_n}{(a; q)_n (-a; q)_n} = \frac{1 - a^2 q^{2n}}{1 - a^2}$$

to the summands. This means we can rewrite (13.4.1) as

$$(13.4.3) \quad \sum_{n=-\infty}^{\infty} (1 - a^2 q^{2n}) \frac{(ab, ac, ad, ae; q)_n}{(\frac{aq}{b}, \frac{aq}{c}, \frac{aq}{d}, \frac{aq}{e}; q)_n} \left(\frac{q}{bcde}\right)^n$$
$$= \frac{(a^2, q, \frac{q}{a^2}, \frac{q}{bc}, \frac{q}{bd}, \frac{q}{be}, \frac{q}{cd}, \frac{q}{ce}, \frac{q}{de}; q)_\infty}{(\frac{q}{ab}, \frac{q}{ac}, \frac{q}{ad}, \frac{q}{ae}, \frac{aq}{b}, \frac{aq}{c}, \frac{aq}{d}, \frac{aq}{e}, \frac{q}{bcde}; q)_\infty}.$$

It is also possible to motivate the power series variable $q/bcde$, as we shall see in the next section, but as Askey says "the real reason [for it] is that it is the choice that allows the series to be summed."

Before we try to prove this, we derive Bailey's formula (13.1.1) from it. Restoring the factor $x - y$ to (13.1.2) we have

$$\sum_{n=-\infty}^{\infty} \left(\frac{xq^n}{(1 - xq^n)^2} - \frac{yq^n}{(1 - yq^n)^2}\right) = (x - y) \sum_{n=-\infty}^{\infty} \frac{(1 - \sqrt{xy}q^n)(1 + \sqrt{xy}q^n)}{(1 - xq^n)^2 (1 - yq^n)^2} q^n,$$

where we also factored the numerator so we can use (13.4.1). Next recall that

$$(13.4.4) \quad \frac{(z; q)_n}{(zq; q)_n} = \frac{1 - z}{1 - zq^n},$$

and that we observed in Chapter 6 that this holds even if n is negative or zero. It follows that
$$1 - zq^n = (1-z)\frac{(zq;q)_n}{(z;q)_n}.$$
Using this six times on the sum we get
$$\sum_{n=-\infty}^{\infty}\left(\frac{xq^n}{(1-xq^n)^2} - \frac{yq^n}{(1-yq^n)^2}\right)$$
$$= \frac{(x-y)(1-xy)}{(1-x)^2(1-y)^2}\sum_{n=-\infty}^{\infty}\frac{\left(\sqrt{xy}q, -\sqrt{xy}q, x, x, y, y; q\right)_n}{\left(\sqrt{xy}, -\sqrt{xy}, xq, xq, yq, yq; q\right)_n}q^n.$$
This fits (13.4.1) if we take
$$a = \sqrt{xy}, \quad b = \sqrt{\frac{x}{y}} = c, \quad d = \sqrt{\frac{y}{x}} = e,$$
which makes $bcde = 1$ and
$$\frac{aq}{b} = yq = \frac{aq}{c} \quad \text{and} \quad \frac{aq}{d} = xq = \frac{aq}{e}.$$
Using these values in (13.4.1) we get
$$\sum_{n=-\infty}^{\infty}\left(\frac{xq^n}{(1-xq^n)^2} - \frac{yq^n}{(1-yq^n)^2}\right)$$
$$= \frac{(x-y)(1-xy)}{(1-x)^2(1-y)^2}\frac{\left(xyq, q, \frac{q}{xy}, \frac{qy}{x}, q, q, q, q, \frac{qx}{y}; q\right)_\infty}{\left(\frac{q}{x}, \frac{q}{x}, \frac{q}{y}, \frac{q}{y}, yq, yq, xq, xq, q; q\right)_\infty}.$$
Cancelling one $(q;q)_\infty$ from numerator and denominator and gluing a few factors onto the infinite products, we finally have
$$\sum_{n=-\infty}^{\infty}\left(\frac{xq^n}{(1-xq^n)^2} - \frac{yq^n}{(1-yq^n)^2}\right) = \frac{(x-y)\left(xy, \frac{q}{xy}, \frac{qy}{x}, \frac{qx}{y}, q, q, q, q; q\right)_\infty}{\left(\frac{q}{x}, \frac{q}{x}, \frac{q}{y}, \frac{q}{y}, x, x, y, y; q\right)_\infty},$$
which is Bailey's formula (13.1.1).

Exercises

1. Explain why we can rewrite (13.4.1) as (13.4.3).
2. Prove (13.4.2). Does your proof work if n is negative?
3. This problem establishes a formula that is useful in the next problem.
    (i) Show that
    $$\sum_{n=-\infty}^{\infty}\frac{aq^n}{(1-aq^n)^2} = \frac{a}{(1-a)^2} + \sum_{n=1}^{\infty}\left(\frac{aq^n}{(1-aq^n)^2} + \frac{\frac{q^n}{a}}{\left(1-\frac{q^n}{a}\right)^2}\right).$$
    (ii) Using (13.2.2), show that the result of (i) can be transformed to
    $$\sum_{n=-\infty}^{\infty}\frac{aq^n}{(1-aq^n)^2} = \frac{a}{(1-a)^2} + \sum_{m=1}^{\infty}\frac{m\left(a^m + a^{-m}\right)q^m}{1-q^m}.$$

4. In this problem we outline Bailey's proof, based on (13.1.1), of the formula

(13.4.5) $$\frac{(q;q)_\infty^5}{(q^5;q^5)_\infty} = 1 - 5\sum_{n=0}^{\infty}\left(\begin{array}{c}\frac{(5n+1)q^{5n+1}}{1-q^{5n+1}} - \frac{(5n+2)q^{5n+2}}{1-q^{5n+2}} \\ -\frac{(5n+3)q^{5n+3}}{1-q^{5n+3}} + \frac{(5n+4)q^{5n+4}}{1-q^{5n+4}}\end{array}\right),$$

which was stated by Ramanujan. Bailey proves this by taking $x = \omega$ and $y = \omega^2$ in (13.1.1), where $\omega = e^{\frac{2\pi i}{5}}$.

(i) First we deal with the product side of (13.1.1). Show that it becomes

$$\frac{\omega(1-\omega^3)}{(1-\omega)(1-\omega^2)^2} \frac{(q;q)_\infty^5}{(q^5;q^5)_\infty}.$$

Hint: The infinite products other than $(q;q)_\infty^4$ in the numerator should largely cancel half of the ones in the denominator. After accomplishing this, try to rewrite the denominator to be able to use problem 10 in section 8.2.

(ii) On the series side of (13.1.1), use the previous problem to show that

$$\sum_{n=-\infty}^{\infty}\left(\frac{\omega q^n}{(1-\omega q^n)^2} - \frac{\omega^2 q^n}{(1-\omega^2 q^n)^2}\right)$$

$$= \frac{\omega}{(1-\omega)^2} - \frac{\omega^2}{(1-\omega^2)^2} + \sum_{m=1}^{\infty}\frac{mq^m}{1-q^m}\omega^m(1-\omega^m)(1-\omega^{2m}).$$

(iii) Show that

$$\omega^m(1-\omega^m)(1-\omega^{2m}) = \begin{cases} 0 & \text{if } m = 0, \\ \omega(1-\omega)(1-\omega^2) & \text{if } m = 1 \text{ or } m = 4, \\ -\omega(1-\omega)(1-\omega^2) & \text{if } m = 2 \text{ or } m = 3. \end{cases}$$

Since $\omega^5 = 1$, it follows that

$$\omega^m(1-\omega^m)(1-\omega^{2m}) = \begin{cases} 0 & \text{if } m \equiv 0 \pmod{5}, \\ \omega(1-\omega)(1-\omega^2) & \text{if } m \equiv 1 \text{ or } 4 \pmod{5}, \\ -\omega(1-\omega)(1-\omega^2) & \text{if } m \equiv 2 \text{ or } 3 \pmod{5}. \end{cases}$$

(iv) Explain why (ii) and (iii) imply that

$$\sum_{m=1}^{\infty}\frac{mq^m}{1-q^m}\omega^m(1-\omega^m)(1-\omega^{2m})$$

$$= \omega(1-\omega)(1-\omega^2)\sum_{n=0}^{\infty}\left(\begin{array}{c}\frac{(5n+1)q^{5n+1}}{1-q^{5n+1}} - \frac{(5n+2)q^{5n+2}}{1-q^{5n+2}} \\ -\frac{(5n+3)q^{5n+3}}{1-q^{5n+3}} + \frac{(5n+4)q^{5n+4}}{1-q^{5n+4}}\end{array}\right).$$

(v) From (i) and (iv) it follows that

$$\frac{(q;q)_\infty^5}{(q^5;q^5)_\infty} = \frac{(1-\omega)(1-\omega^2)^2}{\omega(1-\omega^3)}\left(\frac{\omega}{(1-\omega)^2} - \frac{\omega^2}{(1-\omega^2)^2}\right)$$

$$+ \frac{(1-\omega)^2(1-\omega^2)^3}{1-\omega^3}\sum_{n=0}^\infty \left(\begin{array}{c}\frac{(5n+1)q^{5n+1}}{1-q^{5n+1}} - \frac{(5n+2)q^{5n+2}}{1-q^{5n+2}} \\ -\frac{(5n+3)q^{5n+3}}{1-q^{5n+3}} + \frac{(5n+4)q^{5n+4}}{1-q^{5n+4}}\end{array}\right).$$

Show that this simplifies to (13.4.5). **Hint:** For the factor that multiplies the series, show that

$$\omega + \omega^2 + \omega^3 + \omega^4 = -1 = (\omega+\omega^4)(\omega^2+\omega^3)$$

and

$$(\omega+\omega^4)^2 + (\omega^2+\omega^3)^2 = 3.$$

5. Show that setting $e = a$ in (13.4.1) gives Bailey's very well poised $_6\phi_5$ sum

(13.4.6) $$\sum_{n=0}^\infty \frac{(aq,-aq,a^2,ab,ac,ad;q)_n}{(a,-a,\frac{aq}{b},\frac{aq}{c},\frac{aq}{d},q;q)_n}\left(\frac{q}{abcd}\right)^n = \frac{(a^2q,\frac{q}{bc},\frac{q}{bd},\frac{q}{cd};q)_\infty}{(\frac{aq}{b},\frac{aq}{c},\frac{aq}{d},\frac{q}{abcd};q)_\infty}$$

or

$$\sum_{n=0}^\infty (1-a^2q^{2n})\frac{(a^2,ab,ac,ad;q)_n}{(\frac{aq}{b},\frac{aq}{c},\frac{aq}{d},q;q)_n}\left(\frac{q}{abcd}\right)^n = \frac{(a^2,\frac{q}{bc},\frac{q}{bd},\frac{q}{cd};q)_\infty}{(\frac{aq}{b},\frac{aq}{c},\frac{aq}{d},\frac{q}{abcd};q)_\infty},$$

where $|q/abcd| < 1$. It may help to see the comment about (6.1.5).

6. Bailey's original form of (13.4.6) was

(13.4.7) $$\sum_{n=0}^\infty \frac{(q\sqrt{a},-q\sqrt{a},a,b,c,d;q)_n}{(\sqrt{a},-\sqrt{a},\frac{aq}{b},\frac{aq}{c},\frac{aq}{d},q;q)_n}\left(\frac{aq}{bcd}\right)^n = \frac{(aq,\frac{aq}{bc},\frac{aq}{bd},\frac{aq}{cd};q)_\infty}{(\frac{aq}{b},\frac{aq}{c},\frac{aq}{d},\frac{aq}{bcd};q)_\infty}.$$

Derive this from (13.4.6) by changing a to $\sqrt{a}$, b to $b/\sqrt{a}$, c to $c/\sqrt{a}$, and d to $d/\sqrt{a}$.

7. Bailey's original proof of the identity

(13.4.8) $$\sum_{n=1}^\infty \left(\frac{q^n}{(1-q^n)^2} - \frac{aq^n}{(1-aq^n)^2}\right)\left(\begin{array}{c}\frac{1}{1-q}+\frac{1}{1-q^2}+\cdots+\frac{1}{1-q^n} \\ +\frac{a}{1-a}+\frac{aq}{1-aq}+\cdots+\frac{aq^{n-1}}{1-aq^{n-1}}\end{array}\right)$$
$$= \sum_{n=1}^\infty \frac{n^2q^n}{1-q^n}$$

from problem 19 in section 4.4 (rewritten here with a in place of z) used his $_6\phi_5$ sum, which we rewrite as

(13.4.9) $$\sum_{n=0}^\infty \frac{1-aq^{2n}}{1-a}\frac{(a,b,c,z;q)_n}{(\frac{aq}{b},\frac{aq}{c},\frac{aq}{z},q;q)_n}\left(\frac{aq}{bcz}\right)^n = \frac{(aq,\frac{aq}{bc},\frac{aq}{bz},\frac{aq}{cz};q)_\infty}{(\frac{aq}{b},\frac{aq}{c},\frac{aq}{z},\frac{aq}{bcz};q)_\infty}.$$

The argument is long, and he omits all details.

(i) Show that the derivative with respect to z of the left side of (13.4.9) is

$$-\sum_{n=1}^{\infty} \frac{1-aq^{2n}}{1-a} \frac{(a,b,c,z;q)_n}{\left(\frac{aq}{b}, \frac{aq}{c}, \frac{aq}{z}, q; q\right)_n} \left(\frac{aq}{bcz}\right)^n \sum_{k=1}^{n} \left(\frac{q^{k-1}}{1-zq^{k-1}} + \frac{1}{z-aq^k}\right).$$

Why does the sum start at $n = 1$?

(ii) Show that the derivative with respect to z of the right side of (13.4.9) is

$$\frac{\left(aq, \frac{aq}{bc}, \frac{aq}{bz}, \frac{aq}{cz}; q\right)_\infty}{\left(\frac{aq}{b}, \frac{aq}{c}, \frac{aq}{z}, \frac{aq}{bcz}; q\right)_\infty}$$

$$\times \sum_{k=1}^{\infty} \left[\frac{\frac{a}{b}q^k}{z\left(z-\frac{a}{b}q^k\right)} + \frac{\frac{a}{c}q^k}{z\left(z-\frac{a}{c}q^k\right)} - \frac{aq^k}{z(z-aq^k)} - \frac{\frac{a}{bc}q^k}{z\left(z-\frac{a}{bc}q^k\right)}\right].$$

(iii) After (i) and (ii) we have

$$\sum_{n=1}^{\infty} \frac{1-aq^{2n}}{1-a} \frac{(a,b,c,z;q)_n}{\left(\frac{aq}{b}, \frac{aq}{c}, \frac{aq}{z}, q; q\right)_n} \left(\frac{aq}{bcz}\right)^n \sum_{k=1}^{n} \left(\frac{q^{k-1}}{1-zq^{k-1}} + \frac{1}{z-aq^k}\right)$$

$$= \frac{\left(aq, \frac{aq}{bc}, \frac{aq}{bz}, \frac{aq}{cz}; q\right)_\infty}{\left(\frac{aq}{b}, \frac{aq}{c}, \frac{aq}{z}, \frac{aq}{bcz}; q\right)_\infty} \sum_{k=1}^{\infty} \frac{aq^k}{z} \left[\frac{1}{z-aq^k} + \frac{1}{bcz-aq^k} - \frac{1}{bz-aq^k} - \frac{1}{cz-aq^k}\right].$$

Show that this can be rewritten as

$$\sum_{n=1}^{\infty} \frac{1-aq^{2n}}{1-a} \frac{(a,b,c,z;q)_n}{\left(\frac{aq}{b}, \frac{aq}{c}, \frac{aq}{z}, q; q\right)_n} \left(\frac{aq}{bcz}\right)^n \sum_{k=1}^{n} \left(\frac{q^{k-1}}{1-zq^{k-1}} + \frac{1}{z-aq^k}\right)$$

$$= \frac{\left(aq, \frac{aq}{bc}, \frac{aq}{bz}, \frac{aq}{cz}; q\right)_\infty}{\left(\frac{aq}{b}, \frac{aq}{c}, \frac{aq}{z}, \frac{aq}{bcz}; q\right)_\infty} \sum_{k=1}^{\infty} \frac{aq^k(1-b)(1-c)(bcz^2 - a^2q^{2k})}{(z-aq^k)(bcz-aq^k)(bz-aq^k)(cz-aq^k)}.$$

(iv) Show that dividing both sides by $(1-b)(1-c)$ and then setting $b = 1 = c$ gives

$$\sum_{n=1}^{\infty} \frac{1-aq^{2n}}{1-a} \frac{(a;q)_n(z;q)_n(q;q)_{n-1}^2}{(aq, aq, \frac{aq}{z}, q; q)_n} \left(\frac{aq}{z}\right)^n \sum_{k=1}^{n} \left(\frac{q^{k-1}}{1-zq^{k-1}} + \frac{1}{z-aq^k}\right)$$

$$= \sum_{k=1}^{\infty} \frac{aq^k(z+aq^k)}{(z-aq^k)^3}.$$

(v) Show that taking $z = a$ and multiplying both sides by a gives

$$\sum_{n=1}^{\infty} \frac{(1-a)(1-aq^{2n})}{(1-q^n)^2(1-aq^n)^2} q^n \sum_{k=1}^{n} \left(\frac{aq^{k-1}}{1-zq^{k-1}} + \frac{1}{1-q^k}\right) = \sum_{k=1}^{\infty} \frac{q^k(1+q^k)}{(1-q^k)^3}.$$

(vi) By using (9.1.2) with $x = q^k$ and changing orders of summation, show that (v) is equivalent to

$$\sum_{n=1}^{\infty} \frac{(1-a)(1-aq^{2n})}{(1-q^n)^2(1-aq^n)^2} q^n \sum_{k=1}^{n} \left(\frac{aq^{k-1}}{1-zq^{k-1}} + \frac{1}{1-q^k}\right) = \sum_{n=1}^{\infty} \frac{n^2 q^n}{1-q^n}$$

and that this is equivalent to (13.4.8).

13.5. Askey's proof: Phase 1

We begin this section by noting a curious symmetry property of (13.4.1) that in some sense explains the power series variable $q/bcde$. Note that we could just as well write the series side of (13.4.1) as

$$\sum_{n=-\infty}^{\infty} \frac{(aq, -aq, ab, ac, ad, ae; q)_{-n}}{\left(a, -a, \frac{aq}{b}, \frac{aq}{c}, \frac{aq}{d}, \frac{aq}{e}; q\right)_{-n}} \left(\frac{q}{bcde}\right)^{-n},$$

and we can use

(13.5.1) $$(x;q)_{-n} = \frac{(-1)^n q^{\binom{n+1}{2}}}{x^n \left(\frac{q}{x}; q\right)_n}$$

(which was (6.1.4)) to rewrite this. All of the factors of $(-1)^n q^{\binom{n+1}{2}}$ will cancel, as will many factors of a^n and two more factors of $(-1)^n$, leaving

$$\sum_{n=-\infty}^{\infty} \frac{\left(\frac{q}{a}, -\frac{q}{a}, \frac{b}{a}, \frac{c}{a}, \frac{d}{a}, \frac{e}{a}; q\right)_n}{\left(\frac{1}{a}, -\frac{1}{a}, \frac{q}{ab}, \frac{q}{ac}, \frac{q}{ad}, \frac{q}{ae}; q\right)_n} \left(\frac{bcde}{q}\right)^n \frac{\left(\frac{q}{b} \frac{q}{c} \frac{q}{d} \frac{q}{e}\right)^n}{q^{2n}(bcde)^n}$$

$$= \sum_{n=-\infty}^{\infty} \frac{\left(\frac{q}{a}, -\frac{q}{a}, \frac{b}{a}, \frac{c}{a}, \frac{d}{a}, \frac{e}{a}; q\right)_n}{\left(\frac{1}{a}, -\frac{1}{a}, \frac{q}{ab}, \frac{q}{ac}, \frac{q}{ad}, \frac{q}{ae}; q\right)_n} \left(\frac{q}{bcde}\right)^n.$$

In other words, we have

(13.5.2) $$\sum_{n=-\infty}^{\infty} \frac{(aq, -aq, ab, ac, ad, ae; q)_n}{\left(a, -a, \frac{aq}{b}, \frac{aq}{c}, \frac{aq}{d}, \frac{aq}{e}; q\right)_n} \left(\frac{q}{bcde}\right)^n$$

$$= \sum_{n=-\infty}^{\infty} \frac{\left(\frac{q}{a}, -\frac{q}{a}, \frac{b}{a}, \frac{c}{a}, \frac{d}{a}, \frac{e}{a}; q\right)_n}{\left(\frac{1}{a}, -\frac{1}{a}, \frac{q}{ab}, \frac{q}{ac}, \frac{q}{ad}, \frac{q}{ae}; q\right)_n} \left(\frac{q}{bcde}\right)^n,$$

and the right side is the same as the left with a replaced by $\frac{1}{a}$, so it is also very well poised. This shows that $q/bcde$ is natural in the sense that any other choice of the power series variable would have changed when we summed the series in the other direction. It also explains the symmetry of the right side of (13.4.1) in a and $\frac{1}{a}$.

Askey starts to prove (13.4.1) by using (13.4.2) for the first two numerator and denominator parameters and transforming the others using

(13.5.3) $$(x;q)_n = \frac{(x;q)_\infty}{(xq^n;q)_\infty}.$$

We will work instead with (13.4.3), the sum side of which becomes

$$\frac{(ab, ac, ad, ae; q)_\infty}{\left(\frac{aq}{b}, \frac{aq}{c}, \frac{aq}{d}, \frac{aq}{e}; q\right)_\infty} \sum_{n=-\infty}^{\infty} (1 - a^2 q^{2n}) \frac{\left(\frac{aq^{n+1}}{b}, \frac{aq^{n+1}}{c}, \frac{aq^{n+1}}{d}, \frac{aq^{n+1}}{e}; q\right)_\infty}{(abq^n, acq^n, adq^n, aeq^n; q)_\infty} \left(\frac{q}{bcde}\right)^n.$$

Putting the infinite products outside the sum on the other side and introducing the notation

(13.5.4) $$h_n(b) = \frac{\left(\frac{aq^{n+1}}{b}; q\right)_\infty}{(abq^n; q)_\infty b^n},$$

we see that (13.4.3) is equivalent to

$$(13.5.5) \quad \sum_{n=-\infty}^{\infty} \left(1 - a^2 q^{2n}\right) h_n(b) h_n(c) h_n(d) h_n(e) \, q^n$$

$$= \frac{\left(a^2, q, \frac{q}{a^2}, \frac{q}{bc}, \frac{q}{bd}, \frac{q}{be}, \frac{q}{cd}, \frac{q}{ce}, \frac{q}{de}; q\right)_\infty}{\left(\frac{q}{ab}, \frac{q}{ac}, \frac{q}{ad}, \frac{q}{ae}, ab, ac, ad, ae, \frac{q}{bcde}; q\right)_\infty}.$$

Next, observe that

$$h_n(bq) = \frac{\left(\frac{aq^n}{b}; q\right)_\infty}{(abq^{n+1}; q)_\infty (bq)^n} = \frac{\left(1 - \frac{aq^n}{b}\right)\left(\frac{aq^{n+1}}{b}; q\right)_\infty}{(abq^{n+1}; q)_\infty b^n q^n} \cdot \frac{1 - abq^n}{1 - abq^n}$$

$$= \frac{\left(1 - \frac{aq^n}{b}\right)(1 - abq^n)}{q^n} \cdot \frac{\left(\frac{aq^{n+1}}{b}; q\right)_\infty}{(abq^n; q)_\infty b^n} = \frac{1 - aq^n\left(b + \frac{1}{b}\right) + a^2 q^{2n}}{q^n} h_n(b)$$

$$(13.5.6) \quad = \left(q^{-n} - a\left(b + \frac{1}{b}\right) + a^2 q^n\right) h_n(b).$$

Denote the left side of (13.5.5) by $f(b,c,d,e)$ and look at

$$(13.5.7) \quad f(bq, c, d, e) - f(b, cq, d, e)$$

$$= \sum_{n=-\infty}^{\infty} \left(1 - a^2 q^{2n}\right) \left[h_n(bq) h_n(c) - h_n(b) h_n(cq)\right] h_n(d) h_n(e) \, q^n.$$

Using (13.5.6) we have

$$h_n(bq) h_n(c) - h_n(b) h_n(cq) = h_n(b) h_n(c) \left[q^{-n} - a\left(b + \frac{1}{b}\right) + a^2 q^n \right.$$
$$\left. - \left(q^{-n} - a\left(c + \frac{1}{c}\right) + a^2 q^n\right) \right]$$

$$= a h_n(b) h_n(c) \left(c + \frac{1}{c} - b - \frac{1}{b}\right)$$

$$= a h_n(b) h_n(c)(c - b)\left(1 - \frac{1}{bc}\right),$$

and using this in (13.5.7) we get the functional equation

$$(13.5.8) \quad f(bq, c, d, e) - f(b, cq, d, e) = a(c - b)\left(1 - \frac{1}{bc}\right) f(b, c, d, e).$$

Of course, we could get a similar equation involving any two of b, c, d, e. We now show that the right side of (13.5.5) satisfies the same set of functional equations. Denote it by $k(b, c, d, e)$ and look at

$$(13.5.9) \quad k(bq, c, d, e) - k(b, cq, d, e) = \frac{\left(a^2, q, \frac{q}{a^2}, \frac{q}{de}; q\right)_\infty}{\left(ad, \frac{q}{ad}, ae, \frac{q}{ae}; q\right)_\infty} \frac{\left(\frac{1}{bc}; q\right)_\infty}{\left(\frac{1}{bcde}; q\right)_\infty}$$

$$\times \left[\frac{\left(\frac{1}{bd}, \frac{1}{be}, \frac{q}{cd}, \frac{q}{ce}; q\right)_\infty}{\left(abq, \frac{1}{ab}, ac, \frac{q}{ac}; q\right)_\infty} - \frac{\left(\frac{q}{bd}, \frac{q}{be}, \frac{1}{cd}, \frac{1}{ce}; q\right)_\infty}{\left(ab, \frac{q}{ab}, acq, \frac{1}{ac}; q\right)_\infty} \right].$$

We want to find a factor of $k(b,c,d,e)$ on the right side of (13.5.9), and there are three things to fix. First note that

$$(abq;q)_\infty \left(\frac{1}{ab};q\right)_\infty = \left(1 - \frac{1}{ab}\right)(abq;q)_\infty \left(\frac{q}{ab};q\right)_\infty$$

$$= -\frac{1}{ab}(1-ab)(abq;q)_\infty \left(\frac{q}{ab};q\right)_\infty = -\frac{1}{ab}(ab;q)_\infty \left(\frac{q}{ab};q\right)_\infty,$$

and hence

$$(acq;q)_\infty \left(\frac{1}{ac};q\right)_\infty = -\frac{1}{ac}(ac;q)_\infty \left(\frac{q}{ac};q\right)_\infty.$$

We can also pull the factors

$$\left(\frac{q}{bd},\frac{q}{be},\frac{q}{cd},\frac{q}{ce};q\right)_\infty$$

out of the numerators in the brackets, and we can write

$$\frac{\left(\frac{1}{bc};q\right)_\infty}{\left(\frac{1}{bcde};q\right)_\infty} = \frac{1-\frac{1}{bc}}{1-\frac{1}{bcde}}\frac{\left(\frac{q}{bc};q\right)_\infty}{\left(\frac{q}{bcde};q\right)_\infty}.$$

This makes (13.5.9) into

$$k(bq,c,d,e) - k(b,cq,d,e) = \frac{\left(a^2, q, \frac{q}{a^2}, \frac{q}{bc}, \frac{q}{bd}, \frac{q}{be}, \frac{q}{cd}, \frac{q}{ce}, \frac{q}{de};q\right)_\infty}{\left(ab, \frac{q}{ab}, ac, \frac{q}{ac}, ad, \frac{q}{ad}, ae, \frac{q}{ae}, \frac{q}{bcde};q\right)_\infty} \frac{1-\frac{1}{bc}}{1-\frac{1}{bcde}}$$

$$\times \left[\frac{\left(1-\frac{1}{bd}\right)\left(1-\frac{1}{be}\right)}{-\frac{1}{ab}} - \frac{\left(1-\frac{1}{cd}\right)\left(1-\frac{1}{ce}\right)}{-\frac{1}{ac}}\right],$$

which is

(13.5.10) $\quad k(bq,c,d,e) - k(b,cq,d,e) = k(b,c,d,e)\frac{1-\frac{1}{bc}}{1-\frac{1}{bcde}}a$

$$\times \left[c\left(1-\frac{1}{cd}\right)\left(1-\frac{1}{ce}\right) - b\left(1-\frac{1}{bd}\right)\left(1-\frac{1}{be}\right)\right].$$

But

$$c\left(1-\frac{1}{cd}\right)\left(1-\frac{1}{ce}\right) - b\left(1-\frac{1}{bd}\right)\left(1-\frac{1}{be}\right)$$

$$= c\left[1 - \frac{1}{c}\left(\frac{1}{d}+\frac{1}{e}\right) + \frac{1}{c^2 de}\right] - b\left[1 - \frac{1}{b}\left(\frac{1}{d}+\frac{1}{e}\right) + \frac{1}{b^2 de}\right]$$

$$= c - b + \frac{1}{cde} - \frac{1}{bde} = c - b + \frac{b-c}{bcde} = (c-b)\left(1-\frac{1}{bcde}\right),$$

so (13.5.10) becomes

$$k(bq,c,d,e) - k(b,cq,d,e) = a(c-b)\left(1-\frac{1}{bc}\right)k(b,c,d,e),$$

which has the same form as (13.5.8). Because of the symmetry in b,c,d,e, $f(b,c,d,e)$ and $k(b,c,d,e)$ satisfy $\binom{4}{2} = 6$ functional equations of this type.

Exercises

1. Use the ratio test to show that (13.4.1) converges for $|q| < |bcde|$. **Hint:** Since it is a bilateral series, you have to look at what happens when $n \to \infty$ and also when $n \to -\infty$, but for the latter we can use the symmetry in a and $\frac{1}{a}$.

2. Show that

$$(13.5.11) \qquad \lim_{x \to \infty} \frac{(ax; q)_n}{x^n} = (-1)^n a^n q^{\frac{n(n-1)}{2}}$$

for any integer n. Use (13.5.1) for the negative n's.

3. Show that (13.5.8) is unchanged if we replace b by $\frac{1}{b}$ and c by $\frac{1}{c}$. In other words, if we set either $f(\frac{1}{b}, \frac{1}{c}, \frac{1}{d}, \frac{1}{e})$ or $k(\frac{1}{b}, \frac{1}{c}, \frac{1}{d}, \frac{1}{e})$ equal to $g(b, c, d, e)$, then g satisfies the same set of functional equations as f and k.

13.6. Askey's proof: Phase 2

Does the functional equation argument of the previous section convince you that f and k are the same? So far we don't know they're the same for *any* values of b, c, d, e, let alone all admissible values. Askey next proves that f and k are the same in four special cases using Ramanujan's ${}_1\psi_1$ sum

$$\sum_{n=-\infty}^{\infty} \frac{(u; q)_n}{(v; q)_n} x^n = \frac{\left(ux, \frac{q}{ux}, q, \frac{v}{u}; q\right)_\infty}{\left(x, \frac{v}{ux}, v, \frac{q}{u}; q\right)_\infty}.$$

Multiplying this by $(v; q)_\infty / (u; q)_\infty$ and using (13.5.3) we have

$$\sum_{n=-\infty}^{\infty} \frac{(vq^n; q)_\infty}{(uq^n; q)_\infty} x^n = \frac{\left(ux, \frac{q}{ux}, q, \frac{v}{u}; q\right)_\infty}{\left(x, \frac{v}{ux}, u, \frac{q}{u}; q\right)_\infty}.$$

We will need this with the specific values $v = \frac{aq}{b}$ and $u = ab$, which gives

$$(13.6.1) \qquad \sum_{n=-\infty}^{\infty} \frac{\left(\frac{aq^{n+1}}{b}; q\right)_\infty}{(abq^n; q)_\infty} x^n = \frac{\left(abx, \frac{q}{abx}, q, \frac{q}{b^2}; q\right)_\infty}{\left(x, \frac{q}{b^2 x}, ab, \frac{q}{ab}; q\right)_\infty}.$$

We will also need a few specific values of (13.5.4), which we leave as exercises:

$$(13.6.2) \qquad h_n\left(q^{\frac{1}{2}}\right) = q^{-\frac{n}{2}}, \quad h_n\left(-q^{\frac{1}{2}}\right) = (-1)^n q^{-\frac{n}{2}},$$

$$h_n(1) = \frac{1}{1 - aq^n}, \quad h_n(-1) = \frac{(-1)^n}{1 + aq^n}.$$

Again using the notation $f(b, c, d, e)$ for the left side of (13.5.5), (13.6.2) tells us that

$$f\left(b, q^{\frac{1}{2}}, 1, -1\right) = \sum_{n=-\infty}^{\infty} (1 - a^2 q^{2n}) \frac{\left(\frac{aq^{n+1}}{b}; q\right)_\infty}{(abq^n; q)_\infty b^n} q^{-\frac{n}{2}} \frac{1}{1 - aq^n} \frac{(-1)^n}{1 + aq^n} q^n$$

$$= \sum_{n=-\infty}^{\infty} \frac{\left(\frac{aq^{n+1}}{b}; q\right)_\infty}{(abq^n; q)_\infty} \left(-\frac{q^{\frac{1}{2}}}{b}\right)^n.$$

Using (13.6.1) with $x = -q^{\frac{1}{2}}/b$ we have

(13.6.3) $$f\left(b, q^{\frac{1}{2}}, 1, -1\right) = \frac{\left(-aq^{\frac{1}{2}}, -\frac{q^{\frac{1}{2}}}{a}, q, \frac{q}{b^2}; q\right)_\infty}{\left(-\frac{q^{\frac{1}{2}}}{b}, -\frac{q^{\frac{1}{2}}}{b}, ab, \frac{q}{ab}; q\right)_\infty}.$$

On the other hand, again using the notation $k(b, c, d, e)$ for the right side of (13.5.5), we have

$$k\left(b, q^{\frac{1}{2}}, 1, -1\right) = \frac{\left(a^2, q, \frac{q}{a^2}, \frac{q^{\frac{1}{2}}}{b}, \frac{q}{b}, -\frac{q}{b}, q^{\frac{1}{2}}, -q^{\frac{1}{2}}, -q; q\right)_\infty}{\left(ab, \frac{q}{ab}, aq^{\frac{1}{2}}, \frac{q^{\frac{1}{2}}}{a}, a, \frac{q}{a}, -a, -\frac{q}{a}, -\frac{q^{\frac{1}{2}}}{b}; q\right)_\infty}.$$

Not content to have 9 infinite products divided by 9 others, we multiply top and bottom of this by

$$\left(-\frac{q^{\frac{1}{2}}}{b}, -\frac{q^{\frac{1}{2}}}{a}, -aq^{\frac{1}{2}}; q\right)_\infty.$$

The point is that in the numerator we now have

$$\left(\frac{q^{\frac{1}{2}}}{b}, \frac{-q^{\frac{1}{2}}}{b}, \frac{q}{b}, -\frac{q}{b}; q\right)_\infty = \left(1 - \frac{q^{\frac{1}{2}}}{b}\right)\left(1 + \frac{q^{\frac{1}{2}}}{b}\right)\left(1 - \frac{q}{b}\right)\left(1 + \frac{q}{b}\right)$$
$$\times \left(1 - \frac{q^{\frac{3}{2}}}{b}\right)\left(1 + \frac{q^{\frac{3}{2}}}{b}\right)\left(1 - \frac{q^2}{b}\right)\left(1 + \frac{q^2}{b}\right)\cdots$$
$$= \left(1 - \frac{q}{b^2}\right)\left(1 - \frac{q^2}{b^2}\right)\left(1 - \frac{q^3}{b^2}\right)\left(1 - \frac{q^4}{b^2}\right)\cdots$$
$$= \left(\frac{q}{b^2}; q\right)_\infty.$$

Similarly

$$\left(q^{\frac{1}{2}}, -q^{\frac{1}{2}}, q, -q; q\right)_\infty = (q; q)_\infty$$

and in the denominator

$$\left(a, -a, aq^{\frac{1}{2}}, -aq^{\frac{1}{2}}; q\right)_\infty = (a^2; q)_\infty \quad \text{and} \quad \left(\frac{q^{\frac{1}{2}}}{a}, \frac{-q^{\frac{1}{2}}}{a}, \frac{q}{a}, -\frac{q}{a}; q\right)_\infty = \left(\frac{q}{a^2}; q\right)_\infty.$$

Hence

$$k\left(b, q^{\frac{1}{2}}, 1, -1\right) = \frac{\left(a^2, \frac{q}{a^2}, q, \frac{q}{b^2}, -aq^{\frac{1}{2}}, -\frac{q^{\frac{1}{2}}}{a}; q\right)_\infty}{\left(ab, \frac{q}{ab}, a^2, \frac{q}{a^2}, -\frac{q^{\frac{1}{2}}}{b}, -\frac{q^{\frac{1}{2}}}{b}; q\right)_\infty} = \frac{\left(q, \frac{q}{b^2}, -aq^{\frac{1}{2}}, -\frac{q^{\frac{1}{2}}}{a}; q\right)_\infty}{\left(ab, \frac{q}{ab}, -\frac{q^{\frac{1}{2}}}{b}, -\frac{q^{\frac{1}{2}}}{b}; q\right)_\infty},$$

in agreement with (13.6.3). Hence (13.5.5) is true in the special case where $c = q^{\frac{1}{2}}$, $d = 1$, and $e = -1$. By a very similar calculation, which we leave as an exercise,

we have

$$(13.6.4) \quad f(b, -q^{\frac{1}{2}}, 1, -1) = \frac{\left(aq^{\frac{1}{2}}, \frac{q^{\frac{1}{2}}}{a}, q, \frac{q}{b^2}; q\right)_\infty}{\left(\frac{q^{\frac{1}{2}}}{b}, \frac{q^{\frac{1}{2}}}{b}, ab, \frac{q}{ab}; q\right)_\infty} = k(b, -q^{\frac{1}{2}}, 1, -1),$$

so (13.5.5) is also true in the special case where $c = -q^{\frac{1}{2}}$, $d = 1$, and $e = -1$.

Askey's third special case is $c = 1$, $d = q^{\frac{1}{2}}$, $e = -q^{\frac{1}{2}}$, in which case the left side of (13.5.5) becomes

$$f\left(b, 1, q^{\frac{1}{2}}, -q^{\frac{1}{2}}\right) = \sum_{n=-\infty}^{\infty} (1 - a^2 q^{2n}) \frac{\left(\frac{aq^{n+1}}{b}; q\right)_\infty}{(abq^n; q)_\infty b^n} \frac{1}{1 - aq^n} q^{-\frac{n}{2}} (-1)^n q^{-\frac{n}{2}} q^n$$

$$= \sum_{n=-\infty}^{\infty} (1 + aq^n) \frac{\left(\frac{aq^{n+1}}{b}; q\right)_\infty}{(abq^n; q)_\infty} \left(-\frac{1}{b}\right)^n$$

after using (13.6.2) again. This is a little trickier than the previous cases because we have to split the sum into

$$f\left(b, 1, q^{\frac{1}{2}}, -q^{\frac{1}{2}}\right) = \sum_{n=-\infty}^{\infty} \frac{\left(\frac{aq^{n+1}}{b}; q\right)_\infty}{(abq^n; q)_\infty} \left(-\frac{1}{b}\right)^n + a \sum_{n=-\infty}^{\infty} \frac{\left(\frac{aq^{n+1}}{b}; q\right)_\infty}{(abq^n; q)_\infty} \left(-\frac{q}{b}\right)^n.$$

Using (13.6.1) with $x = -\frac{1}{b}$ and with $x = -\frac{q}{b}$, we have

$$f\left(b, 1, q^{\frac{1}{2}}, -q^{\frac{1}{2}}\right) = \frac{\left(-a, -\frac{q}{a}, q, \frac{q}{b^2}; q\right)_\infty}{\left(-\frac{1}{b}, -\frac{q}{b}, ab, \frac{q}{ab}; q\right)_\infty} + a \frac{\left(-aq, -\frac{1}{a}, q, \frac{q}{b^2}; q\right)_\infty}{\left(-\frac{q}{b}, -\frac{1}{b}, ab, \frac{q}{ab}; q\right)_\infty}.$$

These two products are actually the same. The denominators are visibly the same, and in the numerator we have

$$a(-aq; q)_\infty \left(-\frac{1}{a}; q\right)_\infty = a\left(1 + \frac{1}{a}\right)(-aq; q)_\infty \left(-\frac{q}{a}; q\right)_\infty = (-a; q)_\infty \left(-\frac{q}{a}; q\right)_\infty,$$

so

$$(13.6.5) \quad f\left(b, 1, q^{\frac{1}{2}}, -q^{\frac{1}{2}}\right) = 2 \frac{\left(-a, -\frac{q}{a}, q, \frac{q}{b^2}; q\right)_\infty}{\left(-\frac{1}{b}, -\frac{q}{b}, ab, \frac{q}{ab}; q\right)_\infty}.$$

On the other hand, when $c = 1$, $d = q^{\frac{1}{2}}$, and $e = -q^{\frac{1}{2}}$ the right side of (13.5.5) becomes

$$k\left(b, 1, q^{\frac{1}{2}}, -q^{\frac{1}{2}}\right) = \frac{\left(a^2, q, \frac{q}{a^2}, \frac{q}{b}, \frac{q^{\frac{1}{2}}}{b}, -\frac{q^{\frac{1}{2}}}{b}, q^{\frac{1}{2}}, -q^{\frac{1}{2}}, -1; q\right)_\infty}{\left(ab, \frac{q}{ab}, a, \frac{q}{a}, aq^{\frac{1}{2}}, \frac{q^{\frac{1}{2}}}{a}, -aq^{\frac{1}{2}}, -\frac{q^{\frac{1}{2}}}{a}, -\frac{1}{b}; q\right)_\infty}.$$

If we multiply top and bottom of this by

$$\left(-a, -\frac{q}{a}, -\frac{q}{b}; q\right)_\infty$$

and use $(-1; q)_\infty = (1+1)(-q; q)_\infty = 2(-q; q)_\infty$, then the same calculations as before reduce this to an expression that matches (13.6.5). Hence (13.5.5) is also true when $c = 1$, $d = q^{\frac{1}{2}}$, and $e = -q^{\frac{1}{2}}$. Askey's fourth special case is $c = -1$, $d = q^{\frac{1}{2}}$, and $e = -q^{\frac{1}{2}}$. By very similar calculations this case gives

$$(13.6.6) \quad f\left(b, -1, q^{\frac{1}{2}}, -q^{\frac{1}{2}}\right) = 2 \frac{\left(a, \frac{q}{a}, q, \frac{q}{b^2}; q\right)_\infty}{\left(\frac{1}{b}, \frac{q}{b}, ab, \frac{q}{ab}; q\right)_\infty} = k\left(b, -1, q^{\frac{1}{2}}, -q^{\frac{1}{2}}\right).$$

Exercises

1. Prove (13.6.2).
2. Prove (13.6.4).
3. Verify that $k\left(b, 1, q^{\frac{1}{2}}, -q^{\frac{1}{2}}\right)$ reduces to the expression in (13.6.5).
4. Prove (13.6.6).

13.7. Askey's proof: Phase 3

To avoid a technical problem, we begin this section by using the remark from problem 3 in section 13.5 that (13.5.8) still holds if b, c, d, e are replaced by their reciprocals. If we set $g(b, c, d, e) = f(\frac{1}{b}, \frac{1}{c}, \frac{1}{d}, \frac{1}{e})$ and $r(b, c, d, e) = k(\frac{1}{b}, \frac{1}{c}, \frac{1}{d}, \frac{1}{e})$, then we have

$$(13.7.1) \quad g(b, cq, d, e) = g(bq, c, d, e) - a(c - b)\left(1 - \frac{1}{bc}\right) g(b, c, d, e)$$

and similarly for r. The reason for doing this is that we have convergence in (13.4.1) for $|bcde| > |q|$, so we do not want to make b, c, d, e smaller there, as using (13.5.8) would tend to do. When rephrased in terms of g, the convergence condition becomes $|bcde| < \frac{1}{|q|}$, and this is compatible with (13.7.1). In the previous section we showed results equivalent to

$$(13.7.2) \quad g\left(b, q^{-\frac{1}{2}}, 1, -1\right) = r\left(b, q^{-\frac{1}{2}}, 1, -1\right),$$

$$(13.7.3) \quad g\left(b, -q^{-\frac{1}{2}}, 1, -1\right) = r\left(b, -q^{-\frac{1}{2}}, 1, -1\right),$$

$$(13.7.4) \quad g\left(b, 1, q^{-\frac{1}{2}}, -q^{-\frac{1}{2}}\right) = r\left(b, 1, q^{-\frac{1}{2}}, -q^{-\frac{1}{2}}\right),$$

$$(13.7.5) \quad g\left(b, -1, q^{-\frac{1}{2}}, -q^{-\frac{1}{2}}\right) = r\left(b, -1, q^{-\frac{1}{2}}, -q^{-\frac{1}{2}}\right)$$

with convergence when $|bcde| < \frac{1}{|q|}$. We will try to establish

$$(13.7.6) \quad g\left(q^{i-\frac{1}{2}}, -q^{j-\frac{1}{2}}, q^m, -q^n\right) = r\left(q^{i-\frac{1}{2}}, -q^{j-\frac{1}{2}}, q^m, -q^n\right)$$

for as many values of i, j, m, n as possible, where we need $i + j + m + n > 0$ for convergence. By (13.7.3), (13.7.6) holds for any positive i as long as j, m, n are all zero. By (13.7.2) and symmetry we have

$$g\left(q^{\frac{1}{2}}, c, 1, -1\right) = r\left(q^{\frac{1}{2}}, c, 1, -1\right),$$

so (13.7.6) holds for arbitrary positive j when i, m, n are all zero. By (13.7.5) and symmetry we have

$$g\left(q^{\frac{1}{2}}, -q^{\frac{1}{2}}, d, -1\right) = r\left(q^{\frac{1}{2}}, -q^{\frac{1}{2}}, d, -1\right),$$

so (13.7.6) holds for arbitrary positive m when i, j, n are all zero, and by (13.7.4) and symmetry we have

$$g\left(q^{\frac{1}{2}}, -q^{\frac{1}{2}}, 1, e\right) = r\left(q^{\frac{1}{2}}, -q^{\frac{1}{2}}, 1, e\right),$$

so (13.7.6) holds for arbitrary positive n when i, j, m are all zero.

Moreover, (13.7.5) also implies that (13.7.6) holds for arbitrary $i > \frac{1}{2}$ when $j = \frac{1}{2}$ and $m = -\frac{1}{2} = n$. By (13.7.4) and symmetry we have

$$g\left(1, c, q^{\frac{1}{2}}, -q^{\frac{1}{2}}\right) = r\left(1, c, q^{\frac{1}{2}}, -q^{\frac{1}{2}}\right),$$

so (13.7.6) holds for arbitrary $j > \frac{1}{2}$ when $i = \frac{1}{2}$ and $m = -\frac{1}{2} = n$. By (13.7.3) and symmetry we have

$$g\left(1, -1, d, -q^{\frac{1}{2}}\right) = r\left(1, -1, d, -q^{\frac{1}{2}}\right),$$

so (13.7.6) holds for arbitrary $m > -\frac{1}{2}$ when $n = -\frac{1}{2}$ and $i = \frac{1}{2} = j$. By (13.7.2) and symmetry we have

$$g\left(1, -1, q^{\frac{1}{2}}, e\right) = r\left(1, -1, q^{\frac{1}{2}}, e\right),$$

so (13.7.6) holds for arbitrary $n > -\frac{1}{2}$ when $m = -\frac{1}{2}$ and $i = \frac{1}{2} = j$.

Therefore we can go from the cases $j = 0$ and $j = \frac{1}{2}$ of (13.7.6), known for $m = n = 0$ and $m = n = -\frac{1}{2}$ respectively, to $j = 1, 2, 3, \ldots$ and $j = \frac{3}{2}, \frac{5}{2}, \frac{7}{2}, \ldots$ for the same values of m and n, and by symmetry we can do the same with $m = 0$ and $m = \frac{1}{2}$ and with $n = 0$ and $n = \frac{1}{2}$. Therefore we can establish (13.7.6) for arbitrary nonnegative integers i, j, m, n whose sum is positive, and also for many sets of half-integers with a positive sum. Askey's final argument is that this is enough to prove that $g = r$, and hence that $f = k$, for all admissible b, c, d, e since g and r are analytic functions of b, c, d, e agreeing for infinitely many values with a limit point, namely $(b, c, d, e) \to (0, 0, 0, 0)$. This rests on the identity theorem for analytic functions, which is in any good complex analysis book, if not always under that name. Note that because we replaced b, c, d, e by their reciprocals, $(b, c, d, e) \to (0, 0, 0, 0)$ here amounts to b, c, d, e all going to infinity in (13.4.1), which we could actually calculate using (13.5.11).

This argument generalizes Ismail's proof of Ramanujan's ${}_1\psi_1$ summation, which we sketched in problem 5 in section 6.1. We had shown that if $|q| < 1$ and $|x| < 1$, then

$$(13.7.7) \quad \frac{(ax;q)_\infty \left(\frac{q}{ax};q\right)_\infty (q;q)_\infty \left(\frac{b}{a};q\right)_\infty}{(x;q)_\infty \left(\frac{b}{ax};q\right)_\infty (b;q)_\infty \left(\frac{q}{a};q\right)_\infty} = \sum_{n=-\infty}^{\infty} \frac{(a;q)_n}{(b;q)_n} x^n$$

holds if $b = q^{m+1}$ for any nonnegative integer m. Both sides are analytic functions of b if $|b| < |ax|$ (which ensures that $\left(\frac{b}{ax};q\right)_\infty$ is not zero), and they agree for infinitely many values of b with a limit point, namely $b = q^{m+1} \to 0$ as $m \to \infty$.

Exercises

1. This problem outlines a derivation of the quintuple product identity (5.3.2) from (13.4.3).

(i) Using (13.5.11), show that if we let $b, c, d \to \infty$ in (13.1.1) and set $e = q^{\frac{1}{2}}$, we get (at least formally)

$$\sum_{n=-\infty}^{\infty} \left(1 - a^2 q^{2n}\right) \left((-1)^n a^n q^{\frac{n(n-1)}{2}}\right)^3 q^{\frac{n}{2}} = \frac{(a^2; q)_\infty \left(\frac{q}{a^2}; q\right)_\infty (q; q)_\infty}{\left(aq^{\frac{1}{2}}; q\right)_\infty \left(\frac{q^{\frac{1}{2}}}{a}; q\right)_\infty}.$$

(ii) Show that (i) simplifies to

(13.7.8) $\displaystyle\sum_{n=-\infty}^{\infty} \left(1 - a^2 q^{2n}\right)(-1)^n a^{3n} q^{\frac{3n^2}{2} - n}$

$$= (a^2; q^2)_\infty \left(\frac{q^2}{a^2}; q^2\right)_\infty (q; q)_\infty \left(-aq^{\frac{1}{2}}; q\right)_\infty \left(-\frac{q^{\frac{1}{2}}}{a}; q\right)_\infty.$$

(iii) This is already the quintuple product identity, but we have to fiddle with it to give it the same form as in section 5.3. Show that if we set $a = -q^{\frac{1}{2}}/z$ and split the sum, (13.7.8) becomes

(13.7.9) $\displaystyle\sum_{n=-\infty}^{\infty} z^{-3n} q^{\frac{3n^2+n}{2}} - \sum_{n=-\infty}^{\infty} z^{-3n-2} q^{\frac{(n+1)(3n+2)}{2}}$

$$= (qz^2; q^2)_\infty \left(\frac{q}{z^2}; q^2\right)_\infty (q; q)_\infty (z; q)_\infty \left(\frac{q}{z}; q\right)_\infty.$$

(iv) Show that if we replace n by $-k$ in the first sum in (13.7.9) and n by $-k - 1$ in the second one, we get

$$\sum_{k=-\infty}^{\infty} z^{3k} q^{\frac{3k^2-k}{2}}(1 - zq^k) = (qz^2; q^2)_\infty \left(\frac{q}{z^2}; q^2\right)_\infty (q; q)_\infty (z; q)_\infty \left(\frac{q}{z}; q\right)_\infty,$$

which is (5.3.2).

2. Show that if we let b, c, d, e all tend to ∞ in (13.4.3) and replace a^2 by z, we get

$$\sum_{n=-\infty}^{\infty} z^{2n} q^{2n^2 - n}(1 - zq^{2n}) = (q; q)_\infty (z; q)_\infty \left(\frac{q}{z}; q\right)_\infty,$$

and that this is equivalent to Jacobi's triple product identity.

3. In the next several problems we outline Andrews's derivation of Jacobi's eight square theorem from Bailey's $_6\psi_6$ sum. First, show that if we set b, c, d, e all equal to $-\frac{1}{a}$ in (13.4.1) and replace a^2 by z, we get

(13.7.10) $\displaystyle\sum_{n=-\infty}^{\infty} \frac{1 - zq^{2n}}{1 - z} \left[\frac{(-1; q)_n}{(-q; q)_n}\right]^4 (qz^2)^n = \frac{(q; q)_\infty \left(\frac{q}{z}; q\right)_\infty (zq; q)_\infty^7}{(-q; q)_\infty^4 (-zq; q)_\infty^4 (qz^2; q)_\infty}.$

4. We want to set $z = 1$ in (13.7.10), but there is an obvious problem in the denominator on the sum side, which is

$$\sum_{n=-\infty}^{\infty} \frac{q^n z^{2n}}{1 - z} \left[\frac{(-1; q)_n}{(-zq; q)_n}\right]^4 - \sum_{n=-\infty}^{\infty} \frac{q^{3n} z^{2n+1}}{1 - z} \left[\frac{(-1; q)_n}{(-zq; q)_n}\right]^4.$$

Show that changing n to $-n$ in the second sum and using (13.5.1) makes this

(13.7.11)
$$\sum_{n=-\infty}^{\infty} \frac{q^n z^{2n}}{1-z} \left[\frac{(-1;q)_n}{(-zq;q)_n}\right]^4 - \sum_{n=-\infty}^{\infty} \frac{q^n z^{2n+1}}{1-z} \left[\frac{(-\frac{1}{z};q)_n}{(-q;q)_n}\right]^4,$$

and that combining these two sums makes the left side of (13.7.10) into

(13.7.12)
$$\sum_{n=-\infty}^{\infty} \frac{q^n z^{2n}}{[(-q;q)_n(-zq;q)_n]^4} \frac{z\left[(-\frac{1}{z};q)_n(-zq;q)_n\right]^4 - [(-1;q)_n(-q;q)_n]^4}{z-1}.$$

5. Using logarithmic differentiation or otherwise, show that

$$\frac{d}{dz}(-zq;q)_n = (-zq;q)_n \left(\frac{q}{1+zq} + \cdots + \frac{q^n}{1+zq^n}\right)$$

and

$$\frac{d}{dz}\left(-\frac{1}{z};q\right)_n = -\left(-\frac{1}{z};q\right)_n \left(\frac{1}{z(z+1)} + \frac{q}{z(z+q)} + \cdots + \frac{q^{n-1}}{z(z+q^{n-1})}\right).$$

6. Using L'Hopital's rule and the derivatives from the previous problem, or otherwise, show that

$$\lim_{z \to 1} \frac{z\left[(-\frac{1}{z};q)_n(-zq;q)_n\right]^4 - [(-1;q)_n(-q;q)_n]^4}{z-1} = [(-1;q)_n(-q;q)_n]^4 \frac{3q^n - 1}{1+q^n}.$$

7. Using the previous problem and (13.7.12), show that the result of letting $z \to 1$ in (13.7.10) is

$$\sum_{n=-\infty}^{\infty} \frac{q^n(3q^n - 1)}{1+q^n} \left[\frac{(-1;q)_n}{(-q;q)_n}\right]^4 = \left[\frac{(q;q)_\infty}{(-q;q)_\infty}\right]^8,$$

and that this can be rewritten as

$$\left[\frac{(q;q)_\infty}{(-q;q)_\infty}\right]^8 = 1 - 16 \sum_{n=1}^{\infty} \frac{q^n(q^{2n} - 4q^n + 1)}{(1+q^n)^4}.$$

This was (13.3.5), the key identity for the eight square theorem in section 13.3.

8. Show that setting $d = 1$ and $e = -1$ in (13.4.1) gives

(13.7.13)
$$\sum_{n=-\infty}^{\infty} \frac{(ab;q)_n (ac;q)_n}{\left(\frac{aq}{b};q\right)_n \left(\frac{aq}{c};q\right)_n} \left(-\frac{q}{bc}\right)^n = \frac{\left(q^2, a^2q, \frac{q}{a^2}, \frac{q^2}{b^2}, \frac{q^2}{c^2}; q^2\right)_\infty \left(\frac{q}{bc};q\right)_\infty}{\left(\frac{aq}{b}, \frac{aq}{c}, \frac{q}{ab}, \frac{q}{ac}; q\right)_\infty \left(-\frac{q}{bc};q\right)_\infty},$$

where $|q/bc| < 1$.

9. Show that setting $a = 1$ and $b = -1$ in (13.7.13) gives

(13.7.14)
$$\sum_{n=-\infty}^{\infty} \frac{2}{1+q^n} \frac{(c;q)_n}{\left(\frac{q}{c};q\right)_n} \left(\frac{q}{c}\right)^n = \left[\frac{(q;q)_\infty \left(-\frac{q}{c};q\right)_\infty}{(-q;q)_\infty \left(\frac{q}{c};q\right)_\infty}\right]^2,$$

where $|q/c| < 1$.

10. Use (13.5.11) to take the limit of (13.7.14) as $c \to \infty$, and hence show that

(13.7.15) $$1 + 4\sum_{n=1}^{\infty} \frac{(-1)^n q^{\binom{n+1}{2}}}{1+q^n} = \left[\frac{(q;q)_\infty}{(-q;q)_\infty}\right]^2.$$

11. The result (13.7.15) of the previous problem can be used to derive Jacobi's two square theorem from section 7.2.

(i) Explain why we have

$$\left[\frac{(q;q)_\infty}{(-q;q)_\infty}\right]^2 = \left(\sum_{j=-\infty}^{\infty}(-1)^j q^{j^2}\right)\left(\sum_{k=-\infty}^{\infty}(-1)^k q^{k^2}\right),$$

and why we can rewrite the right side as

$$\sum_{n=0}^{\infty} \Box_2(n)(-q)^n,$$

with the same notation as in section 7.2: $\Box_2(n)$ is the number of ways to write the nonnegative integer n as a sum of two squares, accounting for signs and permutations.

(ii) By splitting into even and odd n, show that

$$\sum_{n=1}^{\infty} \frac{(-1)^n q^{\binom{n+1}{2}}}{1+q^n} = \sum_{k=1}^{\infty} \frac{q^{k(2k+1)}}{1+q^{2k}} - \sum_{m=0}^{\infty} \frac{q^{(m+1)(2m+1)}}{1+q^{2m+1}}.$$

It is convenient to use two different letters for the summation indices here, for reasons that will soon appear.

(iii) By expanding the denominator into a geometric series, using j as the summation index, and then setting $j+k = m$, show that

$$\sum_{k=1}^{\infty} \frac{q^{k(2k+1)}}{1+q^{2k}} = \sum_{m=1}^{\infty} \sum_{k=1}^{m} (-1)^{m-k} q^{k(2m+1)}.$$

Note that we could extend the sum on the right side down to $m = 0$, because the inner sum would be empty.

(iv) By expanding the denominator into a geometric series, using j as the summation index, and then setting $j+m+1 = k$, show that

$$-\sum_{m=0}^{\infty} \frac{q^{(m+1)(2m+1)}}{1+q^{2m+1}} = \sum_{m=0}^{\infty} \sum_{k=m+1}^{\infty} (-1)^{k-m} q^{k(2m+1)}.$$

(v) Explain why (ii)–(iv) imply that

$$\sum_{n=1}^{\infty} \frac{(-1)^n q^{\binom{n+1}{2}}}{1+q^n} = \sum_{m=0}^{\infty} \sum_{k=1}^{\infty} (-1)^{m+k} q^{k(2m+1)}.$$

(vi) By splitting into even and odd values of m, show that the right side of (v) can be rewritten as

$$\sum_{j=0}^{\infty} \sum_{k=1}^{\infty} (-1)^k \left[q^{k(4j+1)} - q^{k(4j+3)}\right] = \sum_{j=0}^{\infty} \sum_{k=1}^{\infty} \left[(-q)^{k(4j+1)} - (-q)^{k(4j+3)}\right].$$

Combining (13.7.15) with (i), (v), and (vi), we have

$$\sum_{n=0}^{\infty} \square_2(n)(-q)^n = 1 + 4 \sum_{j=0}^{\infty} \sum_{k=1}^{\infty} \left[(-q)^{k(4j+1)} - (-q)^{k(4j+3)} \right].$$

If we change $-q$ to q here, we have the identity that gave us Jacobi's two square theorem in section 7.2.

12. Show that setting $b = a$ and $c = d = e = -\frac{1}{a}$ in (13.4.1) and then setting $a^2 = z$ gives

(13.7.16) $\quad 1 + 8 \sum_{n=1}^{\infty} \frac{(1-zq^{2n})(zq;q)_{n-1}(-q;q)_{n-1}^3(-zq)^n}{(q;q)_n(-zq;q)_n^3} = \left[\frac{(zq;q)_\infty}{(-zq;q)_\infty}\right]^4.$

Again the comment about (6.1.5) in section 6.1 might help.

13. Show that setting $z = 1$ in (13.7.16) gives

$$1 + 8 \sum_{n=1}^{\infty} \frac{(-q)^n}{(1+q^n)^2} = \left[\frac{(q;q)_\infty}{(-q;q)_\infty}\right]^4.$$

Then use (13.2.2) with $x = -q^n$ and interchange orders of summation to transform this into

$$1 + 8 \sum_{k=1}^{\infty} \frac{k(-q)^k}{1+q^k} = \left[\frac{(q;q)_\infty}{(-q;q)_\infty}\right]^4.$$

This was the key identity for Jacobi's four square theorem in section 7.3.

13.8. An integral

Askey used a similar functional equation argument a few years earlier to evaluate a beautiful integral depending on 5 parameters a_1, a_2, a_3, a_4, a_5. We will generally assume these are less than 1 in absolute value, but it is convenient to allow one of them to be 1 and one to be -1. For a generic a with $|a| \leq 1$, define

(13.8.1) $\quad r(x,a) = \prod_{n=0}^{\infty} \left(1 - 2axq^n + a^2 q^{2n} \right) = (1 - 2ax + a^2) r(x, aq).$

Then Askey's integral is

(13.8.2) $I(a_1, a_2, a_3, a_4, a_5)$
$$= \int_{-1}^{1} \frac{r(x,1)\, r\left(x, q^{\frac{1}{2}}\right) r(x,-1)\, r\left(x, -q^{\frac{1}{2}}\right) r(x, a_1 a_2 a_3 a_4 a_5)}{r(x, a_1)\, r(x, a_2)\, r(x, a_3)\, r(x, a_4)\, r(x, a_5)} \frac{dx}{\sqrt{1-x^2}}.$$

The case $a_5 = 0$ is the Askey–Wilson integral, which is of fundamental importance in the theory of orthogonal polynomials. We will follow Askey's method of systematically trying to guess (and ultimately to prove) what this integral must be equal to.

We first seek three quantities A, B, C independent of x such that

(13.8.3) $\quad \dfrac{A\, r(x, a_1 a_2 a_3 a_4 a_5 q)}{r(x, a_1 q)\, r(x, a_2)} - \dfrac{B\, r(x, a_1 a_2 a_3 a_4 a_5 q)}{r(x, a_1)\, r(x, a_2 q)} = \dfrac{C\, r(x, a_1 a_2 a_3 a_4 a_5)}{r(x, a_1)\, r(x, a_2)}.$

Using (13.8.1) and clearing fractions we get
$$A\left(1 - 2xa_1 + a_1^2\right) - B\left(1 - 2xa_2 + a_2^2\right) = C\left[1 - 2xa_1a_2a_3a_4a_5 + (a_1a_2a_3a_4a_5)^2\right],$$
which forces

(13.8.4) $$Aa_1 - Ba_2 = Ca_1a_2a_3a_4a_5,$$

(13.8.5) $$A(1 + a_1^2) - B(1 + a_2^2) = C\left[1 + (a_1a_2a_3a_4a_5)^2\right]$$

since A, B, C are to be independent of x. Solving (13.8.4) for C and substituting in (13.8.5) we get
$$A(1 + a_1^2) - \frac{Aa_1}{a_1a_2a_3a_4a_5}\left[1 + (a_1a_2a_3a_4a_5)^2\right]$$
$$= B(1 + a_2^2) - \frac{Ba_2}{a_1a_2a_3a_4a_5}\left[1 + (a_1a_2a_3a_4a_5)^2\right].$$

Clearing fractions and temporarily setting $u = a_2a_3a_4a_5$ and $v = a_1a_3a_4a_5$, this is
$$Aa_1\left[u + a_1^2 u - 1 - a_1^2 u^2\right] = Ba_2\left[v + a_2^2 v - 1 - a_2^2 v^2\right],$$
or

(13.8.6) $$Aa_1(1 - u)(a_1^2 u - 1) = Ba_2(1 - v)(a_2^2 v - 1).$$

Note that if we find values A, B, C satisfying (13.8.4) and (13.8.5), then any constant times the same values will also work. Therefore we can choose one of them arbitrarily, so we take
$$A = a_2(1 - v)(a_2^2 v - 1) = a_2(1 - a_1a_3a_4a_5)(a_1a_2^2a_3a_4a_5 - 1),$$
in which case (13.8.6) gives
$$B = a_1(1 - u)(a_1^2 u - 1) = a_1(1 - a_2a_3a_4a_5)(a_1^2a_2a_3a_4a_5 - 1).$$

Using these values in (13.8.4) gives
$$C = \frac{a_1a_2}{a_1a_2a_3a_4a_5}\left[\begin{array}{l}a_1a_3a_4a_5 - 1 + a_1a_2^2a_3a_4a_5 - (a_1a_2a_3a_4a_5)^2 \\ - a_2a_3a_4a_5 + 1 - a_1^2a_2a_3a_4a_5 + (a_1a_2a_3a_4a_5)^2\end{array}\right]$$
$$= \frac{1}{a_3a_4a_5}\left[a_3a_4a_5(a_2 - a_1) - a_1a_2a_3a_4a_5(a_2 - a_1)\right] = (a_2 - a_1)(1 - a_1a_2).$$

Therefore we have (13.8.3) in the more precise form
$$\frac{A\,r(x, a_1a_2a_3a_4a_5 q)}{r(x, a_1 q)\,r(x, a_2)} - \frac{B\,r(x, a_1a_2a_3a_4a_5 q)}{r(x, a_1)\,r(x, a_2 q)} = \frac{(a_2 - a_1)(1 - a_1a_2)\,r(x, a_1a_2a_3a_4a_5)}{r(x, a_1)\,r(x, a_2)}$$
with
$$A = a_2(1 - a_1a_3a_4a_5)(a_1a_2^2a_3a_4a_5 - 1)$$
$$B = a_1(1 - a_2a_3a_4a_5)(a_1^2a_2a_3a_4a_5 - 1),$$
and it follows that

(13.8.7) $$A\,I(a_1 q, a_2, a_3, a_4, a_5) - B\,I(a_1, a_2 q, a_3, a_4, a_5),$$
$$= (a_2 - a_1)(1 - a_1a_2)I(a_1, a_2, a_3, a_4, a_5).$$

13.8. AN INTEGRAL

The factor $1-a_1a_2$ suggests that I might have a factor of $f(a_1a_2)$ for some function f, in which case, by symmetry, it would have to have 9 similar factors, so we try setting

$I(a_1, a_2, a_3, a_4, a_5)$
$= f(a_1a_2)f(a_1a_3)f(a_1a_4)f(a_1a_5)f(a_2a_3)f(a_2a_4)f(a_2a_5)f(a_3a_4)f(a_3a_5)f(a_4a_5).$

Plugging this into (13.8.7) and setting $a_3, a_4, a_5 = 0$, A reduces to a_2 and B to a_1 and we have

$$a_2 f(a_1a_2q)\,[f(0)]^9 - a_1 f(a_1a_2q)\,[f(0)]^9 = (a_2 - a_1)(1 - a_1a_2)f(a_1a_2)\,[f(0)]^9.$$

Now $f(0)$ cannot be zero, or else I would be zero when any of a_1, a_2, a_3, a_4, a_5 were zero. Therefore

$(a_2-a_1)f(a_1a_2q) = (a_2-a_1)(1-a_1a_2)f(a_1a_2),$ or $f(a_1a_2q) = (1-a_1a_2)f(a_1a_2).$

Iterating this we get

$$f(a_1a_2) = \frac{f(0)}{(a_1a_2;q)_\infty},$$

so our current guess is

$$I(a_1, a_2, a_3, a_4, a_5) = \frac{[f(0)]^{10}}{(a_1a_2, a_1a_3, a_1a_4, a_1a_5, a_2a_3, a_2a_4, a_2a_5, a_3a_4, a_3a_5, a_4a_5; q)_\infty}.$$

This does not work in (13.8.7), but (Askey continues) the presence of the factors $1 - a_1a_3a_4a_5$ and $1 - a_2a_3a_4a_5$ there suggests the improved guess

$$I(a_1, a_2, a_3, a_4, a_5) = \frac{g(a_1a_2a_3a_4)g(a_1a_2a_3a_5)g(a_1a_2a_4a_5)g(a_1a_3a_4a_5)g(a_2a_3a_4a_5)}{(a_1a_2, a_1a_3, a_1a_4, a_1a_5, a_2a_3, a_2a_4, a_2a_5, a_3a_4, a_3a_5, a_4a_5; q)_\infty}$$

for some function g. Putting this in (13.8.7) and setting $a_5 = 0$ gives

$$\frac{a_2\, g(a_1a_2a_3a_4q)\,[g(0)]^4}{(a_1a_2q, a_1a_3q, a_1a_4q, a_2a_3, a_2a_4, a_3a_4; q)_\infty}$$
$$- \frac{a_1\, g(a_1a_2a_3a_4q)\,[g(0)]^4}{(a_1a_2q, a_1a_3, a_1a_4, a_2a_3q, a_2a_4q, a_3a_4; q)_\infty}$$
$$= \frac{(a_2 - a_1)(1 - a_1a_2)g(a_1a_2a_3a_4)\,[g(0)]^4}{(a_1a_2, a_1a_3, a_1a_4, a_2a_3, a_2a_4, a_3a_4; q)_\infty}$$
$$= \frac{(a_2 - a_1)g(a_1a_2a_3a_4)\,[g(0)]^4}{(a_1a_2q, a_1a_3, a_1a_4, a_2a_3, a_2a_4, a_3a_4; q)_\infty}.$$

Again $g(0)$ cannot be zero, or else I would be zero if any of a_1, a_2, a_3, a_4, a_5 were zero. Cancelling $[g(0)]^4$ and clearing fractions we get

$$g(a_1a_2a_3a_4q)\,[a_2(1 - a_1a_3)(1 - a_1a_4) - a_1(1 - a_2a_3)(1 - a_2a_4)]$$
$$= (a_2 - a_1)g(a_1a_2a_3a_4).$$

The quantity in brackets is

$$a_2\left[1 - a_1(a_3 + a_4) + a_1^2 a_3 a_4\right] - a_1\left[1 - a_2(a_3 + a_4) + a_2^2 a_3 a_4\right]$$
$$= a_2 - a_1 + a_1a_2a_3a_4(a_1 - a_2) = (a_2 - a_1)(1 - a_1a_2a_3a_4),$$

and it follows that

$$g(a_1a_2a_3a_4) = (1 - a_1a_2a_3a_4)g(a_1a_2a_3a_4q).$$

Iterating this we get
$$g(a_1a_2a_3a_4) = g(0)(a_1a_2a_3a_4;q)_\infty,$$
so now the simplest thing that I could be is

(13.8.8) $\quad I(a_1,a_2,a_3,a_4,a_5)$
$$\stackrel{?}{=} \frac{(a_1a_2a_3a_4, a_1a_2a_3a_5, a_1a_2a_4a_5, a_1a_3a_4a_5, a_2a_3a_4a_5; q)_\infty}{(a_1a_2, a_1a_3, a_1a_4, a_1a_5, a_2a_3, a_2a_4, a_2a_5, a_3a_4, a_3a_5, a_4a_5; q)_\infty}.$$

Substituting this in (13.8.3) and clearing fractions we get

(13.8.9) $\quad a_2 \left(1 - a_1 a_2^2 a_3 a_4 a_5\right)(1 - a_1 a_3)(1 - a_1 a_4)(1 - a_1 a_5)$
$$- a_1 \left(1 - a_1^2 a_2 a_3 a_4 a_5\right)(1 - a_2 a_3)(1 - a_2 a_4)(1 - a_2 a_5)$$

on the left side and

(13.8.10) $\qquad (a_2 - a_1)(1 - a_1 a_2 a_3 a_4)(1 - a_1 a_2 a_3 a_5)(1 - a_1 a_2 a_4 a_5)$

on the right. We will show that (13.8.9) equals (13.8.10). First,
$$(1 - a_1 a_3)(1 - a_1 a_4)(1 - a_1 a_5)$$
$$= 1 - a_1(a_3 + a_4 + a_5) + a_1^2(a_3 a_4 + a_3 a_5 + a_4 a_5) - a_1^3 a_3 a_4 a_5$$
$$= 1 - a_1 X + a_1^2 Y - \frac{a_1^2}{a_2} Z,$$

where
$$X = a_3 + a_4 + a_5, \quad Y = a_3 a_4 + a_3 a_5 + a_4 a_5, \quad Z = a_1 a_2^2 a_3 a_4 a_5,$$

and similarly
$$(1 - a_2 a_3)(1 - a_2 a_4)(1 - a_2 a_5) = 1 - a_2 X + a_2^2 Y - \frac{a_2^2}{a_1} Z.$$

Then (13.8.9) takes the form
$$(1 - a_2 Z)\left(a_2 - a_1 a_2 X + a_1^2 a_2 Y - a_1^2 Z\right) - (1 - a_1 Z)\left(a_1 - a_1 a_2 X + a_1 a_2^2 Y - a_2^2 Z\right).$$

The first product is
$$a_2 - a_1 a_2 X + a_1^2 a_2 \left(Y + Z^2\right) - \left(a_1^2 + a_2^2\right) Z + a_1 a_2^2 X Z - (a_1 a_2)^2 Y Z$$

and the second is the same with a_1 and a_2 switched. Therefore all the terms that are symmetric in a_1 and a_2 cancel when we subtract, which leaves
$$a_2 - a_1 + a_1^2 a_2 \left(Y - XZ + Z^2\right) - a_1 a_2^2 \left(Y - XZ + Z^2\right)$$
$$= (a_2 - a_1)\left[1 - a_1 a_2 \left(Y - XZ + Z^2\right)\right].$$

We leave the proof that (13.8.10) also equals this as an exercise.

We have more or less completed the analogue of Phase 1 of Askey's evaluation of Bailey's $_6\psi_6$ sum in this context, by finding in (13.8.8) an expression that satisfies the same functional equation as (13.8.2). While Phase 1 was longer here (and it isn't quite over yet), Phases 2 and 3 will be shorter. If we set $A(a_1, a_2, a_3, a_4, a_5)$ equal to the right side of (13.8.8), then we need to find values of a_1, a_2, a_3, a_4, a_5 for which we can actually evaluate I, and then see if A is the same for those values. We take $a_1 = 1$, $a_2 = q^{\frac{1}{2}}$, $a_3 = -1$, $a_4 = -q^{\frac{1}{2}}$, and a_5 equal to an arbitrary real

13.8. AN INTEGRAL

(for simplicity) number a with $|a| < 1$, in which case $a_1 a_2 a_3 a_4 a_5 = aq$. With these choices, most of the integrand cancels and I reduces to

$$I\left(1, q^{\frac{1}{2}}, -1, -q^{\frac{1}{2}}, a\right) = \int_{-1}^{1} \frac{r(x, aq)}{r(x, a)} \frac{dx}{\sqrt{1-x^2}} = \int_{-1}^{1} \frac{dx}{1 - 2ax + a^2} \frac{1}{\sqrt{1-x^2}},$$

or

$$I\left(1, q^{\frac{1}{2}}, -1, -q^{\frac{1}{2}}, a\right) = \int_{-\frac{\pi}{2}}^{\frac{\pi}{2}} \frac{d\theta}{1 + a^2 - 2a \sin\theta}$$

after substituting $x = \sin\theta$. There is a nonobvious substitution that works beautifully on this type of integral, which is to let

(13.8.11) $$\sin\phi = \frac{(1 + a^2)\sin\theta - 2a}{1 + a^2 - 2a \sin\theta}$$

for a new angle ϕ. When $\theta = -\frac{\pi}{2}$ we have

$$\sin\phi = \frac{-1 - 2a - a^2}{1 + 2a + a^2} = -1, \quad \text{so} \quad \phi = -\frac{\pi}{2},$$

and when $\theta = \frac{\pi}{2}$ we have

$$\sin\phi = \frac{1 + 2a + a^2}{1 + 2a + a^2} = 1, \quad \text{so} \quad \phi = \frac{\pi}{2}.$$

To see that this expression is really the sine of something we compute

$$(1 + a^2 - 2a \sin\theta)^2 - ((1 + a^2)\sin\theta - 2a)^2$$
$$= [1 + a^2 - 2a \sin\theta + (1 + a^2)\sin\theta - 2a][1 + a^2 - 2a \sin\theta - (1 + a^2)\sin\theta + 2a]$$
$$= (1 - 2a + a^2)(1 + \sin\theta)(1 + 2a + a^2)(1 - \sin\theta)$$
$$= (1 - a)^2 (1 + a)^2 (1 - \sin^2\theta)$$
$$= (1 - a^2)^2 \cos^2\theta.$$

Since $|a| < 1$ and $-\frac{\pi}{2} < \theta < \frac{\pi}{2}$, $(1 - a^2)\cos\theta$ is positive, so $1 + a^2 - 2a\sin\theta$ is the hypotenuse of a right triangle with legs $(1 + a^2)\sin\theta - 2a$ and $(1 - a^2)\cos\theta$, and we have

(13.8.12) $$\cos\phi = \frac{(1 - a^2)\cos\theta}{1 + a^2 - 2a \sin\theta}.$$

Taking derivatives of (13.8.11) we get

$$\cos\phi \frac{d\phi}{d\theta} = \frac{(1 + a^2 - 2a\sin\theta)(1 + a^2)\cos\theta - ((1 + a^2)\sin\theta - 2a)(-2a\cos\theta)}{(1 + a^2 - 2a\sin\theta)^2}$$
$$= \frac{\cos\theta}{(1 + a^2 - 2a\sin\theta)^2} \left[(1 + a^2)^2 - 2a(1 + a^2)\sin\theta + 2a(1 + a^2)\sin\theta - 4a^2\right],$$

so

$$\cos\phi \, d\phi = \frac{(1 - a^2)^2 \cos\theta \, d\theta}{(1 + a^2 - 2a\sin\theta)^2} = \frac{(1 - a^2)\cos\theta}{1 + a^2 - 2a\sin\theta} \frac{(1 - a^2) d\theta}{1 + a^2 - 2a\sin\theta},$$

which simplifies to

$$d\phi = \frac{(1 - a^2) d\theta}{1 + a^2 - 2a\sin\theta}$$

because of (13.8.12). Therefore

$$I\left(1, q^{\frac{1}{2}}, -1, -q^{\frac{1}{2}}, a\right) = \int_{-\frac{\pi}{2}}^{\frac{\pi}{2}} \frac{d\theta}{1 + a^2 - 2a \sin \theta} = \int_{-\frac{\pi}{2}}^{\frac{\pi}{2}} \frac{d\phi}{1 - a^2} = \frac{\pi}{1 - a^2}.$$

On the other hand,

$$A\left(1, q^{\frac{1}{2}}, -1, -q^{\frac{1}{2}}, a\right) = \frac{\left(q, -aq^{\frac{1}{2}}, -aq, aq^{\frac{1}{2}}, aq; q\right)_\infty}{\left(q^{\frac{1}{2}}, -1, -q^{\frac{1}{2}}, a, -q^{\frac{1}{2}}, -q, aq^{\frac{1}{2}}, q^{\frac{1}{2}}, -a, -aq^{\frac{1}{2}}; q\right)_\infty}$$

$$= \frac{(q;q)_\infty (aq;q)_\infty (-aq;q)_\infty}{2\left(q^{\frac{1}{2}}, -q^{\frac{1}{2}}, -q; q\right)_\infty^2 (a;q)_\infty (-a;q)_\infty}.$$

Using Euler's "odd equals distinct" theorem we have

$$\left(q^{\frac{1}{2}}; q\right)_\infty \left(-q^{\frac{1}{2}}; q\right)_\infty (-q;q)_\infty = (q;q^2)_\infty (-q;q)_\infty = 1,$$

so this reduces to

$$A\left(1, q^{\frac{1}{2}}, -1, -q^{\frac{1}{2}}, a\right) = \frac{(q;q)_\infty}{2} \frac{(aq;q)_\infty}{(a;q)_\infty} \frac{(-aq;q)_\infty}{(-a;q)_\infty} = \frac{(q;q)_\infty}{2(1-a^2)}.$$

So in fact (13.8.8) is wrong, but introducing a factor of $2\pi/(q;q)_\infty$ makes it correct: actually

(13.8.13) $\quad I(a_1, a_2, a_3, a_4, a_5)$

$$= \frac{2\pi \left(a_1 a_2 a_3 a_4, a_1 a_2 a_3 a_5, a_1 a_2 a_4 a_5, a_1 a_3 a_4 a_5, a_2 a_3 a_4 a_5; q\right)_\infty}{(q, a_1 a_2, a_1 a_3, a_1 a_4, a_1 a_5, a_2 a_3, a_2 a_4, a_2 a_5, a_3 a_4, a_3 a_5, a_4 a_5; q)_\infty},$$

and we have done Phases 1 and 2 of the proof. By symmetry in a_1, a_2, a_3, a_4, a_5 we have proved (13.8.13) for arbitrary a_1 with $|a_1| < 1$ and $a_2 = 1$, $a_3 = q^{\frac{1}{2}}$, $a_4 = -1$, $a_5 = -q^{\frac{1}{2}}$. The functional equation (13.8.7) then allows us to multiply any value of a_2 for which (13.8.13) is known by q, and hence by any positive integer power of q. By symmetry in a_2, a_3, a_4, a_5, we can do the same for a_3, a_4, a_5, so we can claim to have proved (13.8.13) for arbitrary a_1 with $a_2 = q^j$, $a_3 = q^{k+\frac{1}{2}}$, $a_4 = -q^\ell$, and $a_5 = -q^{m+\frac{1}{2}}$ for any nonnegative integers j, k, ℓ, m. Both (13.8.2) and (13.8.13) are analytic in a_1, a_2, a_3, a_4, a_5 when all their absolute values are less than 1, and they agree for infinitely many values with limit point $(a_1, 0, 0, 0, 0)$, so by the identity theorem for analytic functions, they must agree on the whole domain of analyticity.

Exercises

1. Show that the product of the last three factors in (13.8.10) is

$$1 - a_1 a_2 \left(Y - XZ + Z^2\right)$$

with X, Y, Z as above. This completes the proof that (13.8.9) equals (13.8.10).

2. Give an alternative proof that

$$\int_{-\frac{\pi}{2}}^{\frac{\pi}{2}} \frac{d\theta}{1 + a^2 - 2a \sin \theta} = \frac{\pi}{1 - a^2}$$

for $-1 < a < 1$, by multiplying top and bottom by $1 + a^2 + 2a\sin\theta$ inside the integral or otherwise.

13.9. Bailey's lemma

In the next section we will conclude with one of the central results in q-hypergeometric series, Watson's transformation. We follow Bailey's approach as refined by Andrews. It starts with the following simple result.

LEMMA 18 (Bailey's transform). *Let $U_n, V_n, \alpha_n, \delta_n$ be four sequences, and define two more sequences by*

$$\text{(13.9.1)} \qquad \beta_n = \sum_{m=0}^{n} \alpha_m U_{n-m} V_{n+m} \quad \text{and} \quad \gamma_m = \sum_{n=m}^{\infty} \delta_n U_{n-m} V_{n+m}.$$

Then

$$\text{(13.9.2)} \qquad \sum_{m=0}^{\infty} \alpha_m \gamma_m = \sum_{n=0}^{\infty} \beta_n \delta_n$$

subject to convergence conditions.

To see this we just substitute the definition of γ_m into the left side of (13.9.2), which gives

$$\sum_{m=0}^{\infty} \alpha_m \gamma_m = \sum_{m=0}^{\infty} \sum_{n=m}^{\infty} \alpha_m \delta_n U_{n-m} V_{n+m}.$$

The range of the double sum is $0 \leq m \leq n < \infty$, so we can just as well write it as

$$\sum_{m=0}^{\infty} \alpha_m \gamma_m = \sum_{n=0}^{\infty} \delta_n \sum_{m=0}^{n} \alpha_m U_{n-m} V_{n+m} = \sum_{n=0}^{\infty} \delta_n \beta_n$$

as desired. We will apply Bailey's transform to a finite sum, so there will be no convergence issues.

Two sequences α_n and β_n are called a *Bailey pair* if they are related by

$$\text{(13.9.3)} \qquad \beta_n = \sum_{k=0}^{n} \frac{\alpha_k}{(q;q)_{n-k}(aq;q)_{n+k}},$$

where a is arbitrary.

If we substitute

$$\text{(13.9.4)} \qquad \alpha_k = \frac{(-1)^k q^{\binom{k}{2}} (a;q)_k (1 - aq^{2k})}{(1-a)(q;q)_k}$$

into (13.9.3), we get

$$\beta_n = \sum_{k=0}^{n} \frac{(-1)^k q^{\binom{k}{2}} (a;q)_k (1 - aq^{2k})}{(q;q)_k (q;q)_{n-k} (a;q)_{n+k+1}} \frac{(q;q)_n}{(q;q)_n}$$

$$= \frac{1}{(q;q)_n (a;q)_{n+1}} \sum_{k=0}^{n} \binom{n}{k}_q (-1)^k q^{\binom{k}{2}} \frac{(1-aq^{2k})(a;q)_k}{(aq^{n+1};q)_k}.$$

If $n = 0$, then the right side is 1. Otherwise we can use Agarwal's q-binomial theorem (2.3.15) from problem 19 in section 2.3, which was

$$\sum_{k=0}^{m} \binom{n}{k}_q (-1)^k q^{\binom{k}{2}} (1-aq^{2k}) \frac{(a;q)_k}{(aq^{n+1};q)_k} = \binom{n-1}{m}_q (-1)^m q^{\binom{m+1}{2}} \frac{(a;q)_{m+1}}{(aq^{n+1};q)_m}.$$

Taking $m = n$ here we find that

$$\sum_{k=0}^{n} \binom{n}{k}_q (-1)^k q^{\binom{k}{2}} \frac{(1-aq^{2k})(a;q)_k}{(aq^{n+1};q)_k} = 0 \quad \text{if } n \geq 1.$$

Therefore one example of a Bailey pair is $\beta_n = \delta_{n0}$ (the Kronecker delta) with α_k as above.

To prepare for our next result, we set

(13.9.5) $$\delta_k = \frac{(\rho_1;q)_k (\rho_2;q)_k (q^{-N};q)_k q^k}{\left(\frac{\rho_1 \rho_2}{aq^N};q\right)_k}$$

in (13.9.1) for a positive integer N. Note that $\delta_k = 0$ if $k > N$ because of the factor $(q^{-N};q)_n$. Using (13.9.5) in (13.9.1) we get

$$\gamma_m = \sum_{r=m}^{\infty} \frac{(\rho_1;q)_r (\rho_2;q)_r (q^{-N};q)_r q^r}{\left(\frac{\rho_1 \rho_2}{aq^N};q\right)_r (q;q)_{r-m} (aq;q)_{r+m}}$$

$$= \sum_{n=0}^{\infty} \frac{(\rho_1;q)_{n+m} (\rho_2;q)_{n+m} (q^{-N};q)_{n+m} q^{n+m}}{\left(\frac{\rho_1 \rho_2}{aq^N};q\right)_{n+m} (q;q)_n (aq;q)_{n+2m}}.$$

Writing $(\rho_1;q)_{n+m} = (\rho_1;q)_m (\rho_1 q^m;q)_n$ and similarly for most of the other terms, and taking what we can out of the sum, we get

$$\gamma_m = \frac{(\rho_1;q)_m (\rho_2;q)_m (q^{-N};q)_m q^m}{\left(\frac{\rho_1 \rho_2}{aq^N};q\right)_m (aq;q)_{2m}} \sum_{n=0}^{\infty} \frac{(\rho_1 q^m;q)_n (\rho_2 q^m;q)_n (q^{-(N-m)};q)_n q^n}{\left(\frac{\rho_1 \rho_2 q^m}{aq^N};q\right)_n (q;q)_n (aq^{2m+1};q)_n},$$

where the sum is finite since the factor $(q^{-(N-m)};q)_n$ becomes zero when $n > N-m$. In fact, the sum is a terminating balanced $_3\phi_2$ since the product of the numerator parameters and the power series variable is

$$\rho_1 q^m \rho_2 q^m q^{-N+m+1} = \rho_1 \rho_2 q^{3m+1-N},$$

which is also the product of the denominator parameters. Therefore we can use the q-Pfaff–Saalschütz identity

(13.9.6) $$_3\phi_2 \left(\begin{matrix} q^{-n}, u, v \\ w, \frac{uv}{w} q^{1-n} \end{matrix} ; q, q \right) = \frac{\left(\frac{w}{u};q\right)_n \left(\frac{w}{v};q\right)_n}{(w;q)_n \left(\frac{w}{uv};q\right)_n}$$

from section 5.7 to evaluate it. With $w = aq^{2m+1}$ and $n = N - m$, this gives

$$\gamma_m = \frac{(\rho_1;q)_m (\rho_2;q)_m (q^{-N};q)_m q^m}{\left(\frac{\rho_1 \rho_2}{aq^N};q\right)_m (aq;q)_{2m}} \cdot \frac{\left(\frac{aq^{m+1}}{\rho_1};q\right)_{N-m} \left(\frac{aq^{m+1}}{\rho_2};q\right)_{N-m}}{(aq^{2m+1};q)_{N-m} \left(\frac{aq}{\rho_1 \rho_2};q\right)_{N-m}}.$$

We want to rewrite this so that all the subscripts are either N or m. The denominator factors $(aq;q)_{2m} (aq^{2m+1};q)_{N-m}$ combine nicely into $(aq;q)_{N+m}$, which is

$(aq;q)_N(aq^{N+1};q)_m$. We can rewrite another denominator factor as

$$\left(\frac{\rho_1\rho_2}{aq^N};q\right)_m = \left(1-\frac{\rho_1\rho_2}{aq^N}\right)\left(1-\frac{\rho_1\rho_2 q}{aq^N}\right)\cdots\left(1-\frac{\rho_1\rho_2 q^{m-1}}{aq^N}\right)$$

$$= \left(-\frac{\rho_1\rho_2}{aq^N}\right)\left(1-\frac{aq^N}{\rho_1\rho_2}\right)\cdots\left(-\frac{\rho_1\rho_2 q^{m-1}}{aq^N}\right)\left(1-\frac{aq^{N-m+1}}{\rho_1\rho_2}\right)$$

$$= \left(-\frac{\rho_1\rho_2}{aq^N}\right)^m q^{\binom{m}{2}}\left(1-\frac{aq^{N-m+1}}{\rho_1\rho_2}\right)\cdots\left(1-\frac{aq^N}{\rho_1\rho_2}\right).$$

This now combines nicely with

$$\left(\frac{aq}{\rho_1\rho_2};q\right)_{N-m} = \left(1-\frac{aq}{\rho_1\rho_2}\right)\left(1-\frac{aq^2}{\rho_1\rho_2}\right)\cdots\left(1-\frac{aq^{N-m}}{\rho_1\rho_2}\right)$$

to produce

$$\left(-\frac{\rho_1\rho_2}{aq^N}\right)^m q^{\binom{m}{2}}\left(\frac{aq}{\rho_1\rho_2};q\right)_N.$$

We also have

$$\left(\frac{aq^{m+1}}{\rho_1};q\right)_{N-m} = \frac{\left(\frac{aq}{\rho_1};q\right)_N}{\left(\frac{aq}{\rho_1};q\right)_m}$$

and similarly for ρ_2, so finally
(13.9.7)
$$\gamma_m = \frac{\left(\frac{aq}{\rho_1};q\right)_N\left(\frac{aq}{\rho_2};q\right)_N}{(aq;q)_N\left(\frac{aq}{\rho_1\rho_2};q\right)_N}\frac{(\rho_1;q)_m(\rho_2;q)_m(q^{-N};q)_m}{\left(\frac{aq}{\rho_1};q\right)_m\left(\frac{aq}{\rho_2};q\right)_m(aq^{N+1};q)_m}\left(-\frac{aq}{\rho_1\rho_2}\right)^m q^{Nm-\binom{m}{2}}.$$

Note that $\gamma_m = 0$ if $m > N$ because of the factor $(q^{-N};q)_m$. It is convenient to make one more observation before stating the major result of this section. Note that

$$\frac{1}{(q;q)_{N-k}} = \frac{(1-q^{N-k+1})\cdots(1-q^N)}{(q;q)_N},$$

and factoring $-q^m$ out of each numerator factor $1-q^m$, as we have done several times before, this becomes

$$\frac{1}{(q;q)_{N-k}} = (-q^N)^k q^{-(1+2+\cdots+(k-1))}\frac{(1-q^{-N})\cdots(1-q^{k-N-1})}{(q;q)_N},$$

which is

(13.9.8) $$\frac{1}{(q;q)_{N-k}} = (-1)^k q^{Nk-\binom{k}{2}}\frac{(q^{-N};q)_k}{(q;q)_N}.$$

We are now ready for

THEOREM 95 (Bailey's lemma). *If α_n and β_n are a Bailey pair, then so are α'_n and β'_n, where*

(13.9.9) $$\alpha'_n = \frac{(\rho_1;q)_n(\rho_2;q)_n}{\left(\frac{aq}{\rho_1};q\right)_n\left(\frac{aq}{\rho_2};q\right)_n}\left(\frac{aq}{\rho_1\rho_2}\right)^n \alpha_n$$

and

(13.9.10) $$\beta'_n = \sum_{k=0}^{n} \frac{(\rho_1;q)_k(\rho_2;q)_k \left(\frac{aq}{\rho_1\rho_2};q\right)_{n-k}}{\left(\frac{aq}{\rho_1};q\right)_n \left(\frac{aq}{\rho_2};q\right)_n (q;q)_{n-k}} \left(\frac{aq}{\rho_1\rho_2}\right)^k \beta_k.$$

We need to show that inserting (13.9.9) into (13.9.3) produces (13.9.10). We have

$$\sum_{m=0}^{N} \frac{\alpha'_m}{(q;q)_{N-m}(aq;q)_{N+m}}$$

$$= \sum_{m=0}^{N} \frac{(\rho_1;q)_m(\rho_2;q)_m}{\left(\frac{aq}{\rho_1};q\right)_m \left(\frac{aq}{\rho_2};q\right)_m} \left(\frac{aq}{\rho_1\rho_2}\right)^m \frac{\alpha_m}{(q;q)_{N-m}(aq;q)_{N+m}}$$

$$= \sum_{m=0}^{N} \frac{(\rho_1;q)_m(\rho_2;q)_m(q^{-N};q)_m}{\left(\frac{aq}{\rho_1};q\right)_m \left(\frac{aq}{\rho_2};q\right)_m (aq^{N+1};q)_m} \left(-\frac{aq}{\rho_1\rho_2}\right)^m q^{Nm-\binom{m}{2}} \frac{\alpha_m}{(q;q)_N(aq;q)_N}$$

by (13.9.8). Most of (13.9.7) is present, so we rewrite the last line as

$$\frac{\left(\frac{aq}{\rho_1\rho_2};q\right)_N}{\left(\frac{aq}{\rho_1};q\right)_N \left(\frac{aq}{\rho_2};q\right)_N (q;q)_N (aq;q)_N} \frac{\left(\frac{aq}{\rho_1};q\right)_N \left(\frac{aq}{\rho_2};q\right)_N}{\left(\frac{aq}{\rho_1\rho_2};q\right)_N}$$

$$\times \sum_{m=0}^{N} \frac{(\rho_1;q)_m(\rho_2;q)_m(q^{-N};q)_m}{\left(\frac{aq}{\rho_1};q\right)_m \left(\frac{aq}{\rho_2};q\right)_m (aq^{N+1};q)_m} \left(-\frac{aq}{\rho_1\rho_2}\right)^m q^{Nm-\binom{m}{2}} \alpha_m$$

$$= \frac{\left(\frac{aq}{\rho_1\rho_2};q\right)_N}{\left(\frac{aq}{\rho_1};q\right)_N \left(\frac{aq}{\rho_2};q\right)_N (q;q)_N} \sum_{m=0}^{N} \gamma_m \alpha_m.$$

Since $\gamma_m = 0$ if $m > N$, we can extend this sum out to infinity without change, so Bailey's transform (13.9.2) applies to it and we get

$$\frac{\left(\frac{aq}{\rho_1\rho_2};q\right)_N}{\left(\frac{aq}{\rho_1};q\right)_N \left(\frac{aq}{\rho_2};q\right)_N (q;q)_N} \sum_{k=0}^{\infty} \beta_k \delta_k$$

with δ_k as in (13.9.5). Since $\delta_k = 0$ for $k > N$, this too is a finite sum, namely

(13.9.11) $$\frac{\left(\frac{aq}{\rho_1\rho_2};q\right)_N}{\left(\frac{aq}{\rho_1};q\right)_N \left(\frac{aq}{\rho_2};q\right)_N (q;q)_N} \sum_{k=0}^{N} \frac{(\rho_1;q)_k(\rho_2;q)_k(q^{-N};q)_k q^k}{\left(\frac{\rho_1\rho_2}{aq^N};q\right)_k} \beta_k.$$

By the same manipulations as above we have

$$\frac{(q^{-N};q)_k}{(q;q)_N} \frac{\left(\frac{aq}{\rho_1\rho_2};q\right)_N}{\left(\frac{\rho_1\rho_2}{aq^N};q\right)_k} = \frac{\left(\frac{aq}{\rho_1\rho_2};q\right)_{N-k} \left(\frac{aq^{N-k+1}}{\rho_1\rho_2};q\right)_k}{\left(\frac{\rho_1\rho_2}{a}\right)^k (-1)^k q^{\binom{k}{2}-Nk} \left(\frac{aq^{N-k+1}}{\rho_1\rho_2};q\right)_k} \frac{(-1)^k q^{\binom{k}{2}-Nk}}{(q;q)_{N-k}}$$

$$= \left(\frac{a}{\rho_1\rho_2}\right)^k \frac{\left(\frac{aq}{\rho_1\rho_2};q\right)_{N-k}}{(q;q)_{N-k}},$$

so (13.9.11) becomes

$$\frac{1}{\left(\frac{aq}{\rho_1};q\right)_N \left(\frac{aq}{\rho_2};q\right)_N} \sum_{k=0}^{N} \frac{(\rho_1;q)_k (\rho_2;q)_k \left(\frac{aq}{\rho_1\rho_2};q\right)_{N-k}}{(q;q)_{N-k}} \left(\frac{aq}{\rho_1\rho_2}\right)^k \beta_k,$$

which is β'_N as given in (13.9.10). This proves Bailey's lemma.

13.10. Watson's transformation

In this concluding section we get the reward for the hard work of the preceding section. The only example of a Bailey pair we know so far is $\beta_n = \delta_{n0}$ and α_n as given by (13.9.4), but we can use these expressions in Bailey's lemma to get another example. Using $\beta_k = \delta_{k0}$ in (13.9.10) just picks out the $k=0$ term, so

$$\beta'_n = \frac{\left(\frac{aq}{\rho_1\rho_2};q\right)_n}{\left(\frac{aq}{\rho_1};q\right)_n \left(\frac{aq}{\rho_2};q\right)_n (q;q)_n}.$$

Using (13.9.4) in (13.9.9) we get

$$\alpha'_n = \frac{(\rho_1;q)_n (\rho_2;q)_n (a;q)_n (1-aq^{2n})(-1)^n q^{\binom{n}{2}}}{\left(\frac{aq}{\rho_1};q\right)_n \left(\frac{aq}{\rho_2};q\right)_n (q;q)_n (1-a)} \left(\frac{aq}{\rho_1\rho_2}\right)^n.$$

Since this is a Bailey pair, (13.9.3) tells us that

$$\sum_{k=0}^{n} \frac{(\rho_1;q)_k (\rho_2;q)_k (a;q)_k (1-aq^{2k})(-1)^k q^{\binom{k}{2}}}{\left(\frac{aq}{\rho_1};q\right)_k \left(\frac{aq}{\rho_2};q\right)_k (q;q)_k (q;q)_{n-k}(1-a)(aq;q)_{n+k}} \left(\frac{aq}{\rho_1\rho_2}\right)^k$$

$$= \frac{\left(\frac{aq}{\rho_1\rho_2};q\right)_n}{\left(\frac{aq}{\rho_1};q\right)_n \left(\frac{aq}{\rho_2};q\right)_n (q;q)_n}.$$

Writing $(aq;q)_{n+k} = (aq;q)_n (aq^{n+1};q)_k$ again and assembling a q-binomial coefficient, this becomes

$$(13.10.1) \quad \sum_{k=0}^{n} \binom{n}{k}_q (-1)^k q^{\binom{k}{2}} \frac{1-aq^{2k}}{1-a} \frac{(\rho_1;q)_k (\rho_2;q)_k (a;q)_k}{\left(\frac{aq}{\rho_1};q\right)_k \left(\frac{aq}{\rho_2};q\right)_k (aq^{n+1};q)_k} \left(\frac{aq}{\rho_1\rho_2}\right)^k$$

$$= \frac{(aq;q)_n \left(\frac{aq}{\rho_1\rho_2};q\right)_n}{\left(\frac{aq}{\rho_1};q\right)_n \left(\frac{aq}{\rho_2};q\right)_n}.$$

But the factor $(1-aq^{2k})/(1-a)$ might also remind us of Bailey's very well poised $_6\psi_6$ sum (13.4.1). As in (13.4.2), we can write

$$(13.10.2) \quad \frac{1-aq^{2k}}{1-a} = \frac{(\sqrt{a}q;q)_k (-\sqrt{a}q;q)_k}{(\sqrt{a};q)_k (-\sqrt{a};q)_k}.$$

Using (13.9.8) we also have

$$(13.10.3) \quad \binom{n}{k}_q (-1)^k q^{\binom{k}{2}} = \frac{(q^{-n};q)_k}{(q;q)_k} q^{nk},$$

so (13.10.1) becomes

$$(13.10.4) \quad \sum_{k=0}^{n} \frac{(q^{-n}, \rho_1, \rho_2, a, \sqrt{aq}, -\sqrt{aq}; q)_k}{\left(q, \frac{aq}{\rho_1}, \frac{aq}{\rho_2}, aq^{n+1}, \sqrt{a}, -\sqrt{a}; q\right)_k} \left(\frac{aq^{n+1}}{\rho_1 \rho_2}\right)^k = \frac{(aq;q)_n \left(\frac{aq}{\rho_1 \rho_2}; q\right)_n}{\left(\frac{aq}{\rho_1}; q\right)_n \left(\frac{aq}{\rho_2}; q\right)_n}.$$

In the series, each numerator parameter may be paired with a denominator parameter to produce a product of aq, so this is a terminating very well poised $_6\phi_5$ series, which in q-hypergeometric notation reads

$$(13.10.5) \quad {}_6\phi_5\left(\begin{matrix} q^{-n}, \rho_1, \rho_2, a, \sqrt{aq}, -\sqrt{aq} \\ \frac{aq}{\rho_1}, \frac{aq}{\rho_2}, aq^{n+1}, \sqrt{a}, -\sqrt{a} \end{matrix}; q, \frac{aq^{n+1}}{\rho_1 \rho_2}\right) = \frac{(aq;q)_n \left(\frac{aq}{\rho_1 \rho_2}; q\right)_n}{\left(\frac{aq}{\rho_1}; q\right)_n \left(\frac{aq}{\rho_2}; q\right)_n}.$$

We can now use

$$\alpha'_n = \frac{(\rho_3;q)_n (\rho_4;q)_n (a;q)_n (1 - aq^{2n})(-1)^n q^{\binom{n}{2}}}{\left(\frac{aq}{\rho_3};q\right)_n \left(\frac{aq}{\rho_4};q\right)_n (q;q)_n (1-a)} \left(\frac{aq}{\rho_3 \rho_4}\right)^n$$

and

$$\beta'_k = \frac{\left(\frac{aq}{\rho_3 \rho_4};q\right)_k}{\left(\frac{aq}{\rho_3};q\right)_k \left(\frac{aq}{\rho_4};q\right)_k (q;q)_k}$$

in (13.9.9) and (13.9.10) respectively to get a third example of a Bailey pair. (Andrews refers to this iteration of Bailey's lemma as a *Bailey chain*.) This gives

$$\alpha''_n = \frac{(\rho_1, \rho_2, \rho_3, \rho_4, a;q)_n}{\left(\frac{aq}{\rho_1}, \frac{aq}{\rho_2}, \frac{aq}{\rho_3}, \frac{aq}{\rho_4}, q;q\right)_n} \frac{1 - aq^{2n}}{1-a} q^{\binom{n}{2}} \left(-\frac{a^2 q^2}{\rho_1 \rho_2 \rho_3 \rho_4}\right)^n$$

and

$$\beta''_n = \frac{1}{\left(\frac{aq}{\rho_1};q\right)_n \left(\frac{aq}{\rho_2};q\right)_n} \sum_{k=0}^{n} \frac{(\rho_1;q)_k (\rho_2;q)_k \left(\frac{aq}{\rho_3 \rho_4};q\right)_k \left(\frac{aq}{\rho_1 \rho_2};q\right)_{n-k}}{\left(\frac{aq}{\rho_3};q\right)_k \left(\frac{aq}{\rho_4};q\right)_k (q;q)_k (q;q)_{n-k}} \left(\frac{aq}{\rho_1 \rho_2}\right)^k,$$

and by Bailey's lemma we also have

$$\beta''_n = \sum_{k=0}^{n} \frac{\alpha''_k}{(q;q)_{n-k}(aq;q)_{n+k}}$$

$$= \sum_{k=0}^{n} \frac{(\rho_1, \rho_2, \rho_3, \rho_4, a;q)_k}{\left(\frac{aq}{\rho_1}, \frac{aq}{\rho_2}, \frac{aq}{\rho_3}, \frac{aq}{\rho_4}, q;q\right)_k (q;q)_{n-k}(aq;q)_{n+k}} \frac{1 - aq^{2k}}{1-a} q^{\binom{k}{2}} \left(-\frac{a^2 q^2}{\rho_1 \rho_2 \rho_3 \rho_4}\right)^k.$$

Multiplying both expressions for β''_n by $(q;q)_n (aq;q)_n$ we get the q-binomial identity

$$\frac{(aq;q)_n}{\left(\frac{aq}{\rho_1};q\right)_n \left(\frac{aq}{\rho_2};q\right)_n} \sum_{k=0}^{n} \binom{n}{k}_q \frac{(\rho_1;q)_k (\rho_2;q)_k \left(\frac{aq}{\rho_3 \rho_4};q\right)_k \left(\frac{aq}{\rho_1 \rho_2};q\right)_{n-k}}{\left(\frac{aq}{\rho_3};q\right)_k \left(\frac{aq}{\rho_4};q\right)_k} \left(\frac{aq}{\rho_1 \rho_2}\right)^k$$

$$= \sum_{k=0}^{n} \binom{n}{k}_q \frac{(\rho_1, \rho_2, \rho_3, \rho_4, a;q)_k}{\left(\frac{aq}{\rho_1}, \frac{aq}{\rho_2}, \frac{aq}{\rho_3}, \frac{aq}{\rho_4}, aq^{n+1};q\right)_k} \frac{1 - aq^{2k}}{1-a} q^{\binom{k}{2}} \left(-\frac{a^2 q^2}{\rho_1 \rho_2 \rho_3 \rho_4}\right)^k.$$

13.10. WATSON'S TRANSFORMATION

In the last section we showed that

$$\left(\frac{\rho_1\rho_2}{aq^n};q\right)_k \left(\frac{aq}{\rho_1\rho_2};q\right)_{n-k} = \left(-\frac{\rho_1\rho_2}{aq^n}\right)^k q^{\binom{k}{2}} \left(\frac{aq}{\rho_1\rho_2};q\right)_n,$$

so

$$\left(\frac{aq}{\rho_1\rho_2};q\right)_{n-k} \left(\frac{aq}{\rho_1\rho_2}\right)^k = (-1)^k q^{\binom{k+1}{2}-nk} \frac{\left(\frac{aq}{\rho_1\rho_2};q\right)_n}{\left(\frac{\rho_1\rho_2}{aq^n};q\right)_k},$$

and making this replacement on the left side above we get

$$\frac{(aq;q)_n \left(\frac{aq}{\rho_1\rho_2};q\right)_n}{\left(\frac{aq}{\rho_1};q\right)_n \left(\frac{aq}{\rho_2};q\right)_n} \sum_{k=0}^n \binom{n}{k}_q (-1)^k q^{\binom{k+1}{2}-nk} \frac{(\rho_1;q)_k(\rho_2;q)_k \left(\frac{aq}{\rho_3\rho_4};q\right)_k}{\left(\frac{aq}{\rho_3};q\right)_k \left(\frac{aq}{\rho_4};q\right)_k \left(\frac{\rho_1\rho_2}{aq^n};q\right)_k}$$

$$= \sum_{k=0}^n \binom{n}{k}_q \frac{(\rho_1,\rho_2,\rho_3,\rho_4,a;q)_k}{\left(\frac{aq}{\rho_1},\frac{aq}{\rho_2},\frac{aq}{\rho_3},\frac{aq}{\rho_4},aq^{n+1};q\right)_k} \frac{1-aq^{2k}}{1-a} q^{\binom{k}{2}} \left(-\frac{a^2q^2}{\rho_1\rho_2\rho_3\rho_4}\right)^k.$$

This is Watson's transformation, but it isn't usually written in q-binomial form. Using (13.10.3) to replace the q-binomial coefficients on both sides, it becomes

$$\frac{(aq;q)_n \left(\frac{aq}{\rho_1\rho_2};q\right)_n}{\left(\frac{aq}{\rho_1};q\right)_n \left(\frac{aq}{\rho_2};q\right)_n} \sum_{k=0}^n \frac{\left(q^{-n},\rho_1,\rho_2,\frac{aq}{\rho_3\rho_4};q\right)_k}{\left(q,\frac{aq}{\rho_3},\frac{aq}{\rho_4},\frac{\rho_1\rho_2}{aq^n};q\right)_k} q^k$$

$$= \sum_{k=0}^n \frac{(q^{-n},\rho_1,\rho_2,\rho_3,\rho_4,a;q)_k}{\left(q,\frac{aq}{\rho_1},\frac{aq}{\rho_2},\frac{aq}{\rho_3},\frac{aq}{\rho_4},aq^{n+1};q\right)_k} \frac{1-aq^{2k}}{1-a} \left(\frac{a^2q^{n+2}}{\rho_1\rho_2\rho_3\rho_4}\right)^k.$$

On the right side, $(1-aq^{2k})/(1-a)$ can be rewritten as in (13.10.2), and then each numerator can be paired with a denominator parameter to give the product aq, so it is a terminating very well poised $_8\phi_7$. On the left side, the product of the four numerator parameters and the power series variable is the same as the product of the three denominator parameters, so it is a terminating balanced $_4\phi_3$. Thus Watson's transformation gives a remarkable connection between these two different classes of q-hypergeometric series. In q-hypergeometric form it appears as

THEOREM 96 (Watson's transformation). *If n is a nonnegative integer, then*

(13.10.6)
$$_8\phi_7\left(\begin{matrix} q^{-n},\rho_1,\rho_2,\rho_3,\rho_4,a,\sqrt{aq},-\sqrt{aq} \\ \frac{aq}{\rho_1},\frac{aq}{\rho_2},\frac{aq}{\rho_3},\frac{aq}{\rho_4},aq^{n+1},\sqrt{a},-\sqrt{a} \end{matrix};q,\frac{a^2q^{n+2}}{\rho_1\rho_2\rho_3\rho_4}\right)$$

$$= \frac{(aq;q)_n \left(\frac{aq}{\rho_1\rho_2};q\right)_n}{\left(\frac{aq}{\rho_1};q\right)_n \left(\frac{aq}{\rho_2};q\right)_n} {}_4\phi_3\left(\begin{matrix} q^{-n},\rho_1,\rho_2,\frac{aq}{\rho_3\rho_4} \\ \frac{aq}{\rho_3},\frac{aq}{\rho_4},\frac{\rho_1\rho_2}{aq^n} \end{matrix};q,q\right)$$

$$= \frac{(aq;q)_n \left(\frac{aq}{\rho_3\rho_4};q\right)_n}{\left(\frac{aq}{\rho_3};q\right)_n \left(\frac{aq}{\rho_4};q\right)_n} {}_4\phi_3\left(\begin{matrix} q^{-n},\rho_3,\rho_4,\frac{aq}{\rho_1\rho_2} \\ \frac{aq}{\rho_1},\frac{aq}{\rho_2},\frac{\rho_3\rho_4}{aq^n} \end{matrix};q,q\right).$$

The left side of (13.10.6) is symmetric in $\rho_1,\rho_2,\rho_3,\rho_4$, so the right side must be too, and hence we get the last equality in (13.10.6).

If we set $\rho_1\rho_2\rho_3\rho_4 = a^2q^{n+1}$, then the power series variable on the $_8\phi_7$ side reduces to q, and one of the numerator parameters on the $_4\phi_3$ side equals one of the denominator parameters, leaving a balanced $_3\phi_2$ that can be summed by the q-Pfaff–Saalschütz identity (13.9.6). Taking $u = \rho_1$, $v = \rho_2$, and $w = \frac{aq}{\rho_3}$ there, we get

THEOREM 97 (Jackson's q-Dougall identity). *If $a, \rho_1, \rho_2, \rho_3, \rho_4$ are related by $\rho_1\rho_2\rho_3\rho_4 = a^2 q^{n+1}$ and n is a nonnegative integer, then*
(13.10.7)
$$_8\phi_7\left(\begin{matrix} q^{-n}, \rho_1, \rho_2, \rho_3, \rho_4, a, \sqrt{aq}, -\sqrt{aq} \\ \frac{aq}{\rho_1}, \frac{aq}{\rho_2}, \frac{aq}{\rho_3}, \frac{aq}{\rho_4}, aq^{n+1}, \sqrt{a}, -\sqrt{a} \end{matrix}; q, q\right) = \frac{\left(aq, \frac{aq}{\rho_1\rho_2}, \frac{aq}{\rho_1\rho_3}, \frac{aq}{\rho_2\rho_3}; q\right)_n}{\left(\frac{aq}{\rho_1}, \frac{aq}{\rho_2}, \frac{aq}{\rho_3}, \frac{aq}{\rho_1\rho_2\rho_3}; q\right)_n}.$$

If we use $\rho_1\rho_2\rho_3\rho_4 = a^2 q^{n+1}$ to eliminate ρ_4 from the left side, this takes the somewhat uglier form
(13.10.8)
$$_8\phi_7\left(\begin{matrix} q^{-n}, \rho_1, \rho_2, \rho_3, a, \sqrt{aq}, -\sqrt{aq}, \frac{a^2 q^{n+1}}{\rho_1\rho_2\rho_3} \\ \frac{aq}{\rho_1}, \frac{aq}{\rho_2}, \frac{aq}{\rho_3}, aq^{n+1}, \sqrt{a}, -\sqrt{a}, \frac{\rho_1\rho_2\rho_3}{aq^n} \end{matrix}; q, q\right) = \frac{\left(aq, \frac{aq}{\rho_1\rho_2}, \frac{aq}{\rho_1\rho_3}, \frac{aq}{\rho_2\rho_3}; q\right)_n}{\left(\frac{aq}{\rho_1}, \frac{aq}{\rho_2}, \frac{aq}{\rho_3}, \frac{aq}{\rho_1\rho_2\rho_3}; q\right)_n}.$$

Note that
$$\frac{(q^{-n}; q)_k}{\left(\frac{\rho_1\rho_2\rho_3}{aq^n}; q\right)_k} = \frac{(1-q^{-n})\cdots(1-q^{-n+k-1})}{\left(1-\frac{\rho_1\rho_2\rho_3 q^{-n}}{a}\right)\cdots\left(1-\frac{\rho_1\rho_2\rho_3 q^{-n+k-1}}{a}\right)}$$
$$= \frac{(q^n - 1)\cdots(q^{n-k+1} - 1)}{\left(q^n - \frac{\rho_1\rho_2\rho_3}{a}\right)\cdots\left(q^{n-k+1} - \frac{\rho_1\rho_2\rho_3}{a}\right)}.$$

If we assume $|q| < 1$ and let $n \to \infty$ in (13.10.8), these two factors will tend to
$$\left(\frac{a}{\rho_1\rho_2\rho_3}\right)^k$$
and the two factors containing q^{n+1} will tend to zero, leaving

(13.10.9) $\quad _6\phi_5\left(\begin{matrix} \rho_1, \rho_2, \rho_3, a, \sqrt{aq}, -\sqrt{aq} \\ \frac{aq}{\rho_1}, \frac{aq}{\rho_2}, \frac{aq}{\rho_3}, \sqrt{a}, -\sqrt{a} \end{matrix}; q, \frac{aq}{\rho_1\rho_2\rho_3}\right) = \frac{\left(aq, \frac{aq}{\rho_1\rho_2}, \frac{aq}{\rho_1\rho_3}, \frac{aq}{\rho_2\rho_3}; q\right)_\infty}{\left(\frac{aq}{\rho_1}, \frac{aq}{\rho_2}, \frac{aq}{\rho_3}, \frac{aq}{\rho_1\rho_2\rho_3}; q\right)_\infty}.$

This is Bailey's very well poised $_6\phi_5$ sum (13.4.7) from the problems in section 13.4.

Watson used his transformation to give another proof of the Rogers–Ramanujan identities, and this is an appropriate place to end. Note that
$$\frac{(x;q)_k}{x^k} = \frac{1-x}{x}\frac{1-xq}{x}\cdots\frac{1-xq^{k-1}}{x}$$
$$= \left(\frac{1}{x} - 1\right)\left(\frac{1}{x} - q\right)\cdots\left(\frac{1}{x} - q^{k-1}\right) \to (-1)^k q^{\binom{k}{2}} \quad \text{as } x \to \infty.$$

If we let all of $\rho_1, \rho_2, \rho_3, \rho_4$ tend to infinity in (13.10.6), the $_8\phi_7$ side becomes
$$\sum_{k=0}^n \frac{(q^{-n}, a, \sqrt{aq}, -\sqrt{aq}; q)_k}{(q, aq^{n+1}, \sqrt{a}, -\sqrt{a}; q)_k} \left((-1)^k q^{\binom{k}{2}}\right)^4 a^{2k} q^{(n+2)k}$$
$$= \sum_{k=0}^n \frac{(q^{-n}; q)_k (a; q)_k}{(q; q)_k (aq^{n+1}; q)_k} \frac{1 - aq^{2k}}{1-a} a^{2k} q^{2k^2 + nk},$$

or, using (13.10.3),

(13.10.10) $$\sum_{k=0}^{n} \binom{n}{k}_q \frac{(a;q)_k}{(aq^{n+1};q)_k} \frac{1-aq^{2k}}{1-a} (-1)^k a^{2k} q^{2k^2+\binom{k}{2}}.$$

On the $_4\phi_3$ side we need to look at

$$\frac{(\rho_1;q)_k (\rho_2;q)_k}{\left(\frac{\rho_1 \rho_2}{aq^n};q\right)_k}.$$

This consists of k factors having the form

$$\frac{(1-\rho_1 q^j)(1-\rho_2 q^j)}{1-\frac{\rho_1 \rho_2 q^j}{aq^n}} = \frac{\left(\frac{1}{\rho_1}-q^j\right)\left(\frac{1}{\rho_2}-q^j\right)}{\frac{1}{\rho_1 \rho_2}-\frac{q^j}{aq^n}}$$

for $j = 0, 1, 2, \ldots, k-1$. As ρ_1 and ρ_2 tend to infinity, this quotient tends to

$$\frac{(-q^j)(-q^j)}{\frac{-q^j}{aq^n}} = -aq^{n+j},$$

so the product of all these factors tends to

$$(-1)^k a^k q^{nk} q^{0+1+\cdots+(k-1)} = (-1)^k a^k q^{nk} q^{\binom{k}{2}},$$

and the $_4\phi_3$ side of (13.10.6) becomes

$$(aq;q)_n \sum_{k=0}^{n} \frac{(q^{-n};q)_k}{(q;q)_k} (-1)^k (aq)^k q^{nk} q^{\binom{k}{2}} = (aq;q)_n \sum_{k=0}^{n} \binom{n}{k}_q (aq)^k \left((-1)^k q^{\binom{k}{2}}\right)^2$$

using (13.10.3) again. Equating this to (13.10.10), we have

$$\sum_{k=0}^{n} \binom{n}{k}_q \frac{(a;q)_k}{(aq^{n+1};q)_k} \frac{1-aq^{2k}}{1-a} (-1)^k a^{2k} q^{\frac{k(5k-1)}{2}} = (aq;q)_n \sum_{k=0}^{n} \binom{n}{k}_q a^k q^{k^2}.$$

Letting $n \to \infty$ here (assuming $|q| < 1$), we get

$$\sum_{k=0}^{\infty} \frac{(a;q)_k}{(q;q)_k} \frac{1-aq^{2k}}{1-a} (-1)^k a^{2k} q^{\frac{k(5k-1)}{2}} = (aq;q)_\infty \sum_{k=0}^{\infty} \frac{q^{k^2} a^k}{(q;q)_k},$$

using Tannery's theorem and the fact that $\binom{n}{k}_q \to \frac{1}{(q;q)_k}$ as $n \to \infty$. This is the result (11.1.9) used by Ramanujan in his proof of the Rogers–Ramanujan identities in section 11.1, so the argument concludes as it did there.

Exercises

1. Show that (13.10.9) reduces to (13.10.5) if $\rho_3 = q^{-n}$.
2. F. H. Jackson's original proof of (13.10.7) is quite interesting, and we outline it in this problem and the next. Instead of using the condition $\rho_1 \rho_2 \rho_3 \rho_4 = a^2 q^{n+1}$ to eliminate ρ_4 from (13.10.7), which gives the less symmetric (13.10.8), we can use the condition that n must be a nonnegative integer to make (13.10.7) appear

more symmetric. It is convenient to replace ρ_i by $\frac{1}{x_i}$ for $1 \leq i \leq 4$. If we also set $x_5 = q^n$, then (13.10.7) becomes

(13.10.11) $\quad {}_8\phi_7 \left(\begin{array}{c} \frac{1}{x_1}, \frac{1}{x_2}, \frac{1}{x_3}, \frac{1}{x_4}, \frac{1}{x_5}, a, \sqrt{a}q, -\sqrt{a}q \\ aqx_1, aqx_2, aqx_3, aqx_4, aqx_5, \sqrt{a}, -\sqrt{a} \end{array}; q, q \right)$

$$= \frac{(aq, aqx_1x_2, aqx_1x_3, aqx_2x_3; q)_n}{(aqx_1, aqx_2, aqx_3, aqx_1x_2x_3; q)_n},$$

where $x_5 = q^n$ and $a^2qx_1x_2x_3x_4x_5 = 1$. This problem checks (13.10.11) in a specific case, where (assuming $n \geq 1$) we multiply both sides by $1 - ax_3q^n$ and then set $ax_3 = q^{-n}$. Since $x_5 = q^n$, the condition $a^2qx_1x_2x_3x_4x_5 = 1$ becomes $aqx_1x_2x_4 = 1$.

(i) Show that on the right side of (13.10.11), this gives

(13.10.12) $\quad \dfrac{(aq;q)_n(aqx_1x_2;q)_n(x_1q^{1-n};q)_n(x_2q^{1-n};q)_n}{(aqx_1;q)_n(aqx_2;q)_n(x_1x_2q^{1-n};q)_n(q^{1-n};q)_{n-1}}.$

(ii) The only thing on the left side of (13.10.11) that contains the factor $1 - ax_3q^n$ is $(aqx_3;q)_n$ in the denominator of the $k = n$ term

(13.10.13) $\quad \dfrac{\left(\frac{1}{x_1}, \frac{1}{x_2}, \frac{1}{x_3}, \frac{1}{x_4}, \frac{1}{x_5}, a; q\right)_n}{(aqx_1, aqx_2, aqx_3, aqx_4, aqx_5, q; q)_n} \dfrac{1-aq^{2n}}{1-a} q^n,$

so all the other terms become zero when we multiply by this factor and set $ax_3 = q^{-n}$. Keeping in mind that $x_5 = q^n$, show that (13.10.13) becomes

(13.10.14)
$$\dfrac{\left(\frac{1}{x_1};q\right)_n \left(\frac{1}{x_2};q\right)_n (aqx_1x_2;q)_n (a;q)_n (aq^n;q)_n (q^{-n};q)_n}{(aqx_1;q)_n(aqx_2;q)_n \left(\frac{1}{x_1x_2};q\right)_n (q^{1-n};q)_{n-1}(aq^{n+1};q)_n(q;q)_n} \dfrac{1-aq^{2n}}{1-a} q^n$$

after using $aqx_1x_2x_4 = 1$ to eliminate x_4.

(iii) We just have to convince ourselves that (13.10.12) equals (13.10.14). Show that

$$(zq^{1-n};q)_n = (-1)^n q^{-\binom{n}{2}} z^n \left(\frac{1}{z};q\right)_n.$$

(iv) Explain why (iii) allows us to rewrite

$$\dfrac{(q^{-n};q)_n}{(q;q)_n} q^n = (-1)^n q^{-\binom{n}{2}}$$

in (13.10.14), and to rewrite (13.10.12) as

(13.10.15) $\quad \dfrac{(aq;q)_n(aqx_1x_2;q)_n \left(\frac{1}{x_1};q\right)_n \left(\frac{1}{x_2};q\right)_n}{(aqx_1;q)_n(aqx_2;q)_n \left(\frac{1}{x_1x_2};q\right)_n (q^{1-n};q)_{n-1}} (-1)^n q^{-\binom{n}{2}}.$

(v) Show that (13.10.14) equals (13.10.15).

3. The advantage of rewriting (13.10.7) as (13.10.11) is that, since the left side is symmetric in $x_1, \ldots, x_5$, (13.10.11) must still hold if *any* of $x_1, \ldots, x_5$ is equal to q^n. Jackson's proof is by induction on n.

(i) Explain why (13.10.11) holds if $x_5 = 1$. Since $x_5 = q^n$, this is the case $n = 0$.

(ii) Next, Jackson assumes (13.10.11) holds if $x_5 = 1, q, q^2, \ldots, q^{n-1}$, and tries to prove it for $x_5 = q^n$. Explain why the induction hypothesis implies that at least it holds for $x_5 = q^n$ if x_3 is one of $1, q, q^2, \ldots, q^{n-1}$.

(iii) Explain why the induction hypothesis also implies that (13.10.11) holds for $x_5 = q^n$ if

$$x_3 = \frac{1}{ax_1 x_2 q^{n+k+1}} \quad \text{for } k = 0, 1, 2, \ldots, n-1.$$

(iv) If we multiply (13.10.11) by $(aqx_3; q)_n (aqx_1 x_2 x_3; q)_n$ and rewrite

$$\left(\frac{1}{x_3}; q\right)_k = \left(a^2 q^{n+1} x_1 x_2 x_4; q\right)_k,$$

explain why both sides of (13.10.11) become polynomials in x_3 of degree $2n$.

(v) The two polynomials in (iv) are equal if x_3 has one of the values in (ii), and also if x_3 has one of the values in (iii). The previous problem gives us one more value where they agree, namely $x_3 = 1/aq^n$, so they agree for $2n+1$ different values of x_3. Explain why this makes them the same polynomial. Therefore (13.10.11) holds for n if it holds for $0, 1, 2, \ldots, n-1$, so it holds for all nonnegative integers n.

13.11. Bibliographical Notes

Wilfrid Norman Bailey was the leading q-analyst for perhaps 20 years, from the early 1930s to the early 1950s. His beautiful little book [**39**] on hypergeometric series, which has one chapter on q-hypergeometric series, is still worth reading today. Among his Ph.D. students were two women who did excellent work on q-hypergeometric series, Lucy Slater and Margaret Jackson. Slater's book [**221**] is an updated and expanded version of [**39**] and was the best book on q-hypergeometric series before [**114**].

Bailey evidently got Ramanujan's "lost notebook" from Watson sometime in 1950, and found Ramanujan's sketch of a proof of his "most beautiful" identity based on (13.2.1) there. In [**44**] Bailey observed that (13.2.1) is a special case of (13.1.1), which in turn was a special case of his summation (13.4.1) of a very well poised $_6\psi_6$ series from [**40**]. In the subsequent note [**45**] he derived (13.1.1) out of elliptic functions instead, on the theory that this would have been more familiar to Ramanujan. A few years later Dobbie [**88**] published the relatively simple proof of (13.1.1) that we presented in section 13.1. For a thorough account of the relevant part of Ramanujan's lost notebook see [**27**].

Bailey's formula (13.1.1) is equivalent to an identity from the Weierstrass theory of elliptic functions, a subject I have tried to avoid in spite of its importance. A penetrating study of Ramanujan's work on elliptic functions has been made by Venkatachaliengar, whose manuscript was itself studied intensively and revised by Shaun Cooper [**241**]. One can see section 3.3 of [**241**] for another proof of (13.2.1). Cooper's larger work [**81**] is another natural continuation of this book into elliptic functions.

The references for the last six problems in section 13.1 are [**32**], [**26**], [**243**], [**212**], and [**46**]. Problem 7 in section 13.4 comes from [**41**]. The problems in section 13.7 all come from Andrews's beautiful survey paper [**14**].

As Askey observed in [**35**], most of the proofs of (13.4.1) prove (13.4.6) first. He remarked "it is still annoying that a sum that is this important has not been obtained from a more elementary special case", and set himself to satisfying the want. His evaluation of the integral of section 13.8 is in [**34**]. For the connection with orthogonal polynomials, the best sources are the original paper [**38**] and the more recent book [**143**]. The latter also has Ismail's simple proof of Ramanujan's $_1\psi_1$ sum using the identity theorem for analytic functions. In [**37**], Askey and Ismail use (13.4.6) and the identity theorem to derive (13.4.1). There are many sources for the identity theorem, for example [**237**], p. 89, or [**157**], p. 87.

Although it is implicit in Jacobi's *Fundamenta Nova* [**148**], the first really clear statement of the eight square theorem seems to be in article 127 of Henry John Stephen Smith's report on number theory [**222**] from 1865, a remarkable work of scholarship that is still worth reading today. The odd m case was stated by Eisenstein in [**91**]. The two, four, and eight square theorems, corresponding theorems about triangular numbers, and other kindred results are all part of Theorem 3.44 in [**81**]. They can also be found in [**82**] and in Berndt's book [**52**], an excellent introduction to the connections between Ramanujan's work and number theory.

The q-Dougall sum was published by F. H. Jackson in [**145**]. His proof, and the ones in [**133**] and [**39**], are (as one might guess) q-versions of an argument given by Dougall for a $q = 1$ hypergeometric series identity. Our proof more or less follows [**133**]. Watson published his eponymous transformation and proof of the Rogers–Ramanujan identities in [**244**]. One can also find it in [**39**], [**221**], [**19**], and [**24**]. The latter two works also have more on Bailey chains. The name comes from [**17**], and anyone with a serious interest in them should look at [**22**].

APPENDIX A

A Brief Guide to Notation

If you glanced at a few random pages of this book, you might have been intimidated, or repulsed, by the notation. This subject makes extensive use of a system of notation that is not used in other parts of mathematics. But it isn't as complicated as it may look at first, because it is just a handful of special symbols that are used over and over and over again. Moreover, although many years ago George Andrews could speak of a one-to-one correspondence between authors and systems of notation, nowadays everyone working on q-analysis uses pretty much the same system. The notation is always explained wherever it is introduced in the text, but if you ever need to remind yourself what a symbol means, you can look here.

$[n]_q$ denotes the q-analogue of the number n. If n is a positive integer, then $[n]_q = 1 + q + q^2 + \cdots + q^{n-1}$, and more generally

$$[n]_q = \begin{cases} \frac{1-q^n}{1-q}, & \text{if } q \neq 1, \\ n, & \text{if } q = 1. \end{cases}$$

This is introduced in section 1.2 and used extensively there, but much less after that except in sections 9.5 and 9.6. Several of the problems in Chapter 1 involve

$$[n]_{q^2} = 1 + q^2 + q^4 + \cdots + q^{2n-2} = \frac{1-q^{2n}}{1-q^2}.$$

Some older papers use $[n]$ instead of $[n]_q$, and I may have used this a few times, but I tried to catch them all.

$n!_q$ denotes the q-factorial. It is defined in terms of the q-numbers $[n]_q$ as

$$n!_q = [1]_q [2]_q \cdots [n]_q,$$

where $0!_q$ is defined to be 1. This is also introduced in section 1.2, and used often there and in section 1.3. It is used less often after that, but it does occur in the sections on the q-derivative and in sections 9.1 and 9.5. Some people prefer the notation $[n]!$.

$(q;q)_n$ is a variation on the q-factorial, and is one of the most frequently used symbols in the book. It is defined by

$$(q;q)_n = (1-q)(1-q^2)(1-q^3) \cdots (1-q^n),$$

where again $(q;q)_0$ is an empty product and hence is defined to be 1. It is related to the q-factorial by $(q;q)_n = n!_q (1-q)^n$.

$(x;q)_n$ is a generalization of $(q;q)_n$. It is sometimes called a q-shifted factorial, and it is also one of the most commonly used symbols in the book. It is defined by

$$(x;q)_n = (1-x)(1-xq)(1-xq^2)\cdots(1-xq^{n-1}),$$

where again $(x;q)_0 = 1$, and it is pronounced "x base q sub n". Sometimes q may be replaced by a power of q, so that, for example,

$$(x;q^2)_n = (1-x)(1-xq^2)(1-xq^4)\cdots(1-xq^{2n-2})$$

and

$$(q;q^2)_n = (1-q)(1-q^3)(1-q^5)\cdots(1-q^{2n-1}).$$

For the case when n isn't a nonnegative integer, see the next definition.

$(x;q)_\infty$ is an infinite version of the preceding product. Formally, the definition is

$$(x;q)_\infty = \lim_{n\to\infty}(x;q)_n.$$

This limit exists for any x as long as $|q|<1$. These products are all over the page starting with Chapter 3. They can also be used to define $(x;q)_n$ for any n as

$$(x;q)_n = \frac{(x;q)_\infty}{(xq^n;q)_\infty}.$$

This more general definition reduces to the previous one if n is a nonnegative integer. It is used in Chapters 6 and 13, and also in section 7.1.

$\binom{n}{k}_q$ is the q-binomial coefficient. These appear throughout the book, but especially in the first two chapters. They can be defined either in terms of the q-factorials by

$$\binom{n}{k}_q = \begin{cases} \frac{n!_q}{k!_q\,(n-k)!_q}, & n,k \text{ integers, } 0 \leq k \leq n, \\ 0, & \text{otherwise,} \end{cases}$$

or in terms of the q-shifted factorials by

$$\binom{n}{k}_q = \begin{cases} \frac{(q;q)_n}{(q;q)_k\,(q;q)_{n-k}}, & n,k \text{ integers, } 0 \leq k \leq n, \\ 0, & \text{otherwise.} \end{cases}$$

I have chosen a different notation than the most standard one: the q-binomial coefficient is more often denoted by $\begin{bmatrix}n\\k\end{bmatrix}$ or $\begin{bmatrix}n\\k\end{bmatrix}_q$. It does occasionally happen that one wants to change q to something else, which is usually q^2; thus

$$\binom{n}{k}_{q^2} = \frac{(q^2;q^2)_n}{(q^2;q^2)_k(q^2;q^2)_{n-k}}.$$

This is read as "n choose k sub q" or "n choose k base q".

$\binom{n}{k_1,k_2,\ldots,k_m}_q$ is the q-multinomial coefficient, a generalization of the q-binomial coefficient. It occurs in a few sections in Chapters 1, 2, and 9. As with

the q-binomial coefficient, it can be defined in either of two equivalent ways. If $k_1, \ldots, k_m$ are nonnegative integers that add up to n, then
$$\binom{n}{k_1, \ldots, k_m}_q = \frac{n!_q}{k_1!_q \ldots k_m!_q} = \frac{(q;q)_n}{(q;q)_{k_1} \ldots (q;q)_{k_m}}.$$
The q-multinomial coefficient equals zero if $k_1, \ldots, k_m$ are not all nonnegative integers, or if they do not add up to n.

$\mathbf{D}_q f(x)$ denotes the q-derivative of the function $f(x)$ (with respect to x). It occurs in sections 2.4 and 3.8, and in a few sections at the end of Chapter 9. It is defined by
$$\mathbf{D}_q f(x) = \frac{f(x) - f(qx)}{x(1-q)}$$
if $q \neq 1$. It becomes the ordinary derivative in the limit as $q \to 1$.

$|P|$ is introduced early in Chapter 3 and used a few times in Chapter 10 to denote the number that the partition P partitions, so $|14+9+6+6+5+3+2+2+1| = 48$.

$(a_1, a_2, \ldots, a_k; q)_n$ is often used to abbreviate $(a_1; q)_n (a_2; q)_n \cdots (a_k; q)_n$ in the last three chapters. Usually this happens with $n = \infty$, but in Chapter 13 it is sometimes used with a finite n.

Any other notation in the book—for example, the q-hypergeometric functions in the latter part of Chapter 5, or the q-trigonometric functions at the end of Chapter 9—is highly localized.

APPENDIX B

Infinite Products

For a given sequence $\{u_n\}$, we denote the product of the first n terms as

(B1) $$p_n = u_1 u_2 \cdots u_n = \prod_{k=1}^{n} u_k.$$

The most obvious definition of an infinite product would be the limit of (B1) as $n \to \infty$, if it exists, but this would make any expression of this form convergent if one or more of the u_k were zero. We could restrict ourselves to nonzero sequences, but in q-analysis we consider infinite products containing one or more variables, and sometimes one of the variables has a value that makes the first factor (and hence the whole product) zero. Instead we adopt the following definition.

DEFINITION 1. Suppose $\{u_n\}$ is a sequence that has only finitely many nonzero terms, so that all the terms starting with u_{m+1} are nonzero for some nonnegative integer m, and set

$$P_n(m) = u_{m+1} u_{m+2} \cdots u_{m+n}$$

for each nonnegative integer n. If

$$\lim_{n \to \infty} P_n(m) = U_m$$

exists *and is not zero*, then the number

$$P = u_1 u_2 \cdots u_m U_m,$$

which is the same for any m subject to the restriction above, is defined to be the value of

(B2) $$\prod_{k=1}^{\infty} u_k.$$

In this case, we say that this infinite product converges to P.

We insist that $U_m \neq 0$ so that the only way that an infinite product can converge to zero is for one of the factors to equal zero. Note however that this prevents many seemingly natural infinite products from converging, for example

(B3) $$\prod_{k=1}^{\infty} \frac{k}{k+1}.$$

None of the factors is zero, so we can take $m = 0$ in Definition 1, but the product of the first n factors of (B3) is

$$\frac{1}{2} \frac{2}{3} \frac{3}{4} \cdots \frac{n}{n+1} = \frac{1}{n+1},$$

which tends to zero as $n \to \infty$. Therefore we say that (B3) *diverges* to zero. This may seem strange, but it parallels what happens with infinite series. If we take the natural logarithm of (B3), we get

$$\sum_{k=1}^{\infty} \log\left(\frac{k}{k+1}\right) = \sum_{k=1}^{\infty} [\log k - \log(k+1)].$$

The first n pairs of terms are

$$\log 1 - \log 2 + (\log 2 - \log 3) + (\log 3 - \log 4) + \cdots + (\log n - \log(n+1)) = -\log(n+1)$$

by telescoping, so

$$\sum_{k=1}^{\infty} \log\left(\frac{k}{k+1}\right) \quad \text{diverges to } -\infty$$

even though the terms tend to $\log 1 = 0$.

This example suggests a further parallel between infinite products and infinite series. Returning to Definition 1, suppose (B2) converges and look at

$$\frac{P_n(m)}{P_{n-1}(m)} = \frac{u_{m+1} u_{m+2} \cdots u_{m+n}}{u_{m+1} u_{m+2} \cdots u_{m+n-1}} = u_{m+n},$$

where we know the quotient is defined since $u_k \neq 0$ for $k > m$. If we let $n \to \infty$ here we get

$$\lim_{n \to \infty} u_{m+n} = \frac{U_m}{U_m} = 1,$$

where the limit on the right side certainly exists since $U_m \neq 0$. It follows that if (B2) converges, then $u_n \to 1$ as $n \to \infty$. This is the nth term test for infinite products, which parallels the one for infinite series since $\log 1 = 0$.

Knowing this, it becomes natural to write (B2) in the form

(B4) $$\prod_{k=1}^{\infty} u_k = \prod_{k=1}^{\infty} (1 + a_k),$$

where $a_k \to 0$ as $k \to \infty$. This is particularly appropriate in q-analysis, since every infinite product we want to consider has the form

$$(x;q)_\infty = \lim_{n \to \infty} (x;q)_n = \lim_{n \to \infty} (1-x)(1-xq)(1-xq^2) \cdots (1-xq^{n-1})$$

$$= \lim_{n \to \infty} \prod_{k=1}^{n} (1 - xq^{k-1}) = \prod_{k=1}^{\infty} (1 - xq^{k-1}) = \prod_{k=0}^{\infty} (1 - xq^k)$$

for some choice of x and q, or is a product or quotient of products of this form. From the nth term test we see that unless $x = 0$, it is necessary to have $|q| < 1$ for convergence, so that $-xq^k \to 0$ as $k \to \infty$ for every x. We will prefer the right side of (B4) to the left side from now on.

Suppose $a_k \geq 0$ for all $k \geq 1$ in (B4), so that

$$(1+a_1)(1+a_2) \cdots (1+a_n) \geq 1 + a_1 + \cdots + a_n,$$

The tangent line to $y = e^x$ at $(0,1)$ is $y = 1+x$. Since e^x is concave up everywhere, it lies above its tangent line, so we have $e^x \geq 1+x$ for all real x, with equality only if $x = 0$. Combining this with the previous inequality we have

$$1 + a_1 + \cdots + a_n \leq (1+a_1) \cdots (1+a_n) \leq e^{a_1 + \cdots + a_n}.$$

Letting $n \to \infty$ we get
$$1 + \sum_{n=1}^{\infty} a_n \leq \prod_{n=1}^{\infty} (1 + a_n) \leq \exp\left(\sum_{n=1}^{\infty} a_n\right),$$
which implies that if either of
$$\sum_{n=1}^{\infty} a_n \quad \text{and} \quad \prod_{n=1}^{\infty} (1 + a_n)$$
converges, then so does the other. The nonnegativity condition only needs to hold for all sufficiently large n, so we have proved the following theorem.

THEOREM 98. *If $a_n \geq 0$ for all sufficiently large n, then*
$$\prod_{n=1}^{\infty} (1 + a_n) \quad \text{converges if and only if} \quad \sum_{n=1}^{\infty} a_n \quad \text{converges.}$$

While this theorem is not immediately applicable to the convergence of $(x; q)_\infty$, it will be once we develop the notion of absolute convergence of an infinite product. Recall that the absolute value of the complex number $a + bi$ (where a and b are real) is $\sqrt{a^2 + b^2}$. This is just the length of the line segment that connects $a + bi$ to the origin in the complex plane.

DEFINITION 2. Let $\{a_k\}$ be any sequence of real or complex numbers such that only finitely many a_k equal -1. We say that
$$\prod_{k=1}^{\infty} (1 + a_k) \quad \text{converges absolutely if and only if} \quad \prod_{k=1}^{\infty} (1 + |a_k|) \quad \text{converges.}$$

Combining this with the previous theorem, we see that
$$\prod_{k=1}^{\infty} (1 + a_k) \quad \text{converges absolutely if and only if} \quad \sum_{k=1}^{\infty} a_k \quad \text{converges absolutely.}$$
Now we just need one more theorem, again parallel to a theorem we know for infinite series.

THEOREM 99. *Let $\{a_k\}$ be any sequence of real or complex numbers such that only finitely many a_k equal -1. If*
$$\prod_{k=1}^{\infty} (1 + a_k)$$
converges absolutely, then it converges.

To prove this, it will be convenient to have the following lemma.

LEMMA 19. *If z is any complex number, then $1 + |z| \geq |1 + z|$, with equality only if z is real and positive.*

If a and b are real, then clearly $\sqrt{a^2 + b^2} \geq a$, with equality only if $b = 0$ and $a \geq 0$. Multiplying this by 2 and adding some terms to both sides we get
$$1 + 2\sqrt{a^2 + b^2} + a^2 + b^2 \geq 1 + 2a + a^2 + b^2,$$
which is
$$\left(1 + \sqrt{a^2 + b^2}\right)^2 \geq (a+1)^2 + b^2,$$

or
$$(1+|z|)^2 \geq |1+z|^2,$$
where $z = a + bi$. Taking the positive square root of both sides we get the lemma.

Suppose that
$$\prod_{k=1}^{\infty}(1+a_k)$$
converges absolutely. Set
$$P_n(m) = (1+a_{m+1})\cdots(1+a_{m+n}),$$
as in Definition 1, where none of the a_{m+k} equals -1, and further set
$$Q_n(m) = (1+|a_{m+1}|)\cdots(1+|a_{m+n}|).$$
Note that
$$P_n(m) - P_{n-1}(m) = (1+a_{m+1})\cdots(1+a_{m+n-1})\, a_{m+n}$$
and
$$Q_n(m) - Q_{n-1}(m) = (1+|a_{m+1}|)\cdots(1+|a_{m+n-1}|)\,|a_{m+n}|.$$
Using the lemma, we have
$$|P_n(m) - P_{n-1}(m)| \leq (1+|a_{m+1}|)\cdots(1+|a_{m+n-1}|)\,|a_{m+n}| = Q_n(m) - Q_{n-1}(m).$$
Now
$$\sum_{k=1}^{n}(Q_k(m) - Q_{k-1}(m)) = Q_n(m) - Q_0(m) = Q_n(m) - 1$$
by telescoping (as an empty product, $Q_0(m) = 1$), and $Q_n(m)$ has a limit as $n \to \infty$ by the assumption that
$$\prod_{k=1}^{\infty}(1+a_k)$$
converges absolutely. It follows that
$$\sum_{k=1}^{\infty}(Q_k(m) - Q_{k-1}(m))$$
converges, so
$$\sum_{k=1}^{\infty}|P_k(m) - P_{k-1}(m)|$$
converges, so
$$\sum_{k=1}^{\infty}(P_k(m) - P_{k-1}(m))$$
converges. This means that $P_n(m)$ has a limit as $n \to \infty$, which means that
$$\prod_{k=1}^{\infty}(1+a_k)$$
has a limit, but in order to be able to say that it converges, we have to be sure that this limit is not zero.

We know that
$$\sum_{k=1}^{\infty}|a_{m+k}|$$

converges, that $a_{m+k} \to 0$ as $k \to \infty$, and that no a_{m+k} equals -1. It follows from the limit comparison test that

$$\sum_{k=1}^{\infty} \left| \frac{a_{m+k}}{1+a_{m+k}} \right|$$

converges, and hence that

$$\sum_{k=1}^{\infty} \left| -\frac{a_{m+k}}{1+a_{m+k}} \right|$$

does. By what we proved above, this means that

$$\prod_{k=1}^{n} \left(1 - \frac{a_{m+k}}{1+a_{m+k}}\right)$$

has a limit as $n \to \infty$. But

$$\prod_{k=1}^{n} \left(1 - \frac{a_{m+k}}{1+a_{m+k}}\right) = \prod_{k=1}^{n} \left(\frac{1}{1+a_{m+k}}\right) = \frac{1}{P_n(m)}$$

therefore has a limit as $n \to \infty$, which means that $P_n(m)$ must have a *nonzero* limit as $n \to \infty$. Therefore

$$\prod_{k=1}^{\infty} (1+a_k)$$

converges, which is what we wanted to prove.

It is now easy to prove that

$$(x;q)_\infty = \prod_{n=0}^{\infty} (1 - xq^n)$$

converges for any x if $|q| < 1$. We have

$$\sum_{n=0}^{\infty} |-xq^n| = \sum_{n=0}^{\infty} |x||q|^n = \frac{|x|}{1-|q|},$$

so

$$\sum_{n=0}^{\infty} -xq^n$$

converges absolutely, so

$$(x;q)_\infty = \prod_{n=0}^{\infty} (1-xq^n)$$

converges absolutely, so it converges.

Exercises

1. (i) Show that

$$\prod_{k=1}^{\infty} \left(1 + \frac{(-1)^{k-1}}{k}\right) = 1.$$

(ii) Show that this product does not converge absolutely.

(iii) Explain why (ii) implies the divergence of the harmonic series $\sum_{k=1}^{\infty} \frac{1}{k}$.

2. We showed that (B3) diverges to zero. Explain why this implies the divergence of the harmonic series.

3. Show that
$$\prod_{k=2}^{\infty} \frac{k^3 - 1}{k^3 + 1} = \frac{2}{3}.$$
   Hint: Show that
$$\prod_{k=2}^{n} \frac{k^3 - 1}{k^3 + 1} = \frac{2}{3}\left(1 + \frac{1}{n(n+1)}\right)$$
   for any integer $n \geq 2$ (and even for $n = 1$ since an empty product equals 1).

4. Show that
$$\prod_{k=1}^{\infty} \cos\left(\frac{x}{2^k}\right) = \frac{\sin x}{x}$$
   for any $x \neq 0$ (and even in the limit as $x \to 0$). **Hint:** Show that
$$\prod_{k=1}^{n} \cos\left(\frac{x}{2^k}\right) = \frac{\sin x}{2^n \sin\left(\frac{x}{2^n}\right)} = \frac{\sin x}{x} \cdot \frac{\frac{x}{2^n}}{\sin\left(\frac{x}{2^n}\right)}$$
   for any nonnegative integer n.

5. How would an infinite product behave if the general term tended to 0? How would it behave if the general term tended to a number between -1 and 1? How would it behave if the general term tended to a number greater than 1? Less than -1? Equal to -1?

6. If we formally let $n \to \infty$ in the identity of problem 5 in section 3.1, we seem to get
$$\left(x^{-1} + 1 + x\right)\left(x^{-3} + 1 + x^3\right)\left(x^{-9} + 1 + x^9\right)\left(x^{-27} + 1 + x^{27}\right)\cdots = \sum_{j=-\infty}^{\infty} x^j.$$
   Show that the series on the right does not converge for any x. This is why we worked with a finite product instead in that problem.

7. Show that the product in problem 6 actually does converge for $x = i$, because every factor equals 1. What if $x = -i$?

8. Explain why the product in problem 6 couldn't possibly converge unless x was on the unit circle in the complex plane (as i and $-i$ are).

9. If x is on the unit circle in the complex plane, then $x = e^{i\theta}$ for some real angle θ. Show that for such an x the product in problem 6 is
$$(1 + 2\cos\theta)(1 + 2\cos 3\theta)(1 + 2\cos 9\theta)(1 + 2\cos 27\theta)\cdots = \prod_{k=0}^{\infty}\left(1 + 2\cos 3^k\theta\right).$$

10. Show that the product in problem 9 converges to $0 = \frac{k\pi}{6}$ for any integer k. (For some values of k it converges to zero.) Can you think of any other values of θ for which it converges?

Some good references for infinite products are Chapter VI of Bromwich's book [63], Chapter VII of Knopp's book [158], and sections 1.4–1.44 of Titchmarsh's book [237]. Our definition of an infinite product is essentially that of Knopp, and our proof that an absolutely convergent product converges is essentially that of Titchmarsh. Problem 3 was problem B-1 on the 1977 Putnam exam, and appeared long before that in [158].

APPENDIX C

Tannery's Theorem

In this brief appendix we prove a theorem from real analysis that is often useful in q-analysis. First I want you to construct a lovely example that you may have seen in calculus or real analysis.

EXERCISE 1. For $x > 0$, draw the graph of $y = \frac{1}{x}$, and divide the area under the curve from $x = 1$ to $x = n$ (for a generic positive integer n) into $n - 1$ strips of width 1.

(i) By approximating each strip by a rectangle drawn from the left endpoint, show that
$$1 + \frac{1}{2} + \cdots + \frac{1}{n-1} > \log n \quad \text{for } n \geq 2,$$
where we use $\log x$ for the natural logarithm of x, as elsewhere in this book.

(ii) By approximating each strip by a rectangle drawn from the right endpoint instead, show that
$$\log n > \frac{1}{2} + \cdots + \frac{1}{n-1} + \frac{1}{n} \quad \text{for } n \geq 2.$$

(iii) From (i), it follows that
$$S_n = 1 + \frac{1}{2} + \cdots + \frac{1}{n-1} - \log n$$
is a positive sequence for $n \geq 2$, and it is also increasing because it is the sum of the first $n - 1$ "upper corners" by which the rectangles differ from the area under the curve. From (ii) it similarly follows that
$$T_n = \log n - \left(\frac{1}{2} + \cdots + \frac{1}{n-1} + \frac{1}{n} \right)$$
is a positive increasing sequence for $n \geq 2$, since it is the sum of the first $n - 1$ "lower corners". Explain why $\lim_{n \to \infty} (S_n + T_n) = 1$.

(iv) For several reasons, the result in (iii) among them, it is interesting to look at
$$U_n = 1 - T_n = 1 + \frac{1}{2} + \cdots + \frac{1}{n} - \log n,$$
where this makes sense for $n \geq 1$. Note that $U_n - S_n = \frac{1}{n}$ for $n \geq 2$, or for $n \geq 1$ if we make the natural definitions $S_1 = 0 = T_1$. Explain why U_n is a positive decreasing sequence.

(v) Explain why $U_n > S_m$ for any positive integers m and n. (If $m = n$ this is obvious. Consider S_n if $m < n$ and U_m if $m > n$.)

(vi) U_n is a decreasing sequence bounded below, and S_n is an increasing sequence bounded above, so both sequences must converge. (This is intuitively clear and is a standard result from real analysis called the Monotone Convergence Theorem.) Explain why they must have the same limit.

(vii) This common limit is called *Euler's constant* and denoted by γ, so we have proved that $1 + \frac{1}{2} + \cdots + \frac{1}{n} - \log n$ decreases to γ as $n \to \infty$, and that $1 + \frac{1}{2} + \cdots + \frac{1}{n-1} - \log n$ increases to γ as $n \to \infty$. Explain why this can be rephrased as the double inequality

$$\log n + \gamma < 1 + \frac{1}{2} + \cdots + \frac{1}{n} < \log(n+1) + \gamma \quad \text{for } n \geq 1.$$

This gives incredibly accurate estimates of the partial sums of the harmonic series for large n, as long as one has a good value of γ.

(viii) Explain how we know that $\gamma > \frac{1}{2}$. (**Hint:** $S_n + T_n \to 1$.) In fact $\gamma = .577215664901\ldots$. It is widely believed to be transcendental (not a root of any polynomial equation with integer coefficients), but this has not been proved. It is slightly less than $\frac{1}{\sqrt{3}}$. Of course, it could not be equal to $\frac{1}{\sqrt{3}}$, because then it would be a root of $3x^2 = 1$, and we would know it was not transcendental.

(ix) We can use (vii) to prove that

(C1) $$\sum_{k=1}^{\infty} \frac{(-1)^{k-1}}{k} = \log 2.$$

Consider the nth even partial sum

$$V_n = 1 - \frac{1}{2} + \frac{1}{3} - \frac{1}{4} + \cdots + \frac{1}{2n-1} - \frac{1}{2n}.$$

Explain why we can rewrite

$$V_n = 1 + \frac{1}{2} + \frac{1}{3} + \frac{1}{4} + \cdots + \frac{1}{2n-1} + \frac{1}{2n} - 2\left(\frac{1}{2} + \frac{1}{4} + \cdots + \frac{1}{2n}\right)$$

$$= 1 + \frac{1}{2} + \frac{1}{3} + \frac{1}{4} + \cdots + \frac{1}{2n-1} + \frac{1}{2n} - \left(1 + \frac{1}{2} + \cdots + \frac{1}{n}\right).$$

(x) Resisting the temptation to cancel (a point we will return to), explain why we can further rewrite

$$V_n = 1 + \frac{1}{2} + \frac{1}{3} + \frac{1}{4} + \cdots + \frac{1}{2n-1} + \frac{1}{2n} - \log(2n) - \left(1 + \frac{1}{2} + \cdots + \frac{1}{n} - \log n\right) + \log 2.$$

(xi) Explain why (x) implies that $V_n \to \log 2$ as $n \to \infty$. This shows that the even partial sums of (C1) converge to $\log 2$. Explain why this also holds for the odd partial sums.

EXERCISE 2. The previous exercise is all that we need for this appendix, but one can go further. For two positive integers r and s, suppose we rearrange the sum in (C1) so that the first r positive terms $1 + \frac{1}{3} + \cdots + \frac{1}{2r-1}$ are listed first, then the first s negative terms $-\frac{1}{2} - \frac{1}{4} - \cdots - \frac{1}{2s}$, then the next r positive terms, then the next s negative terms, and so on. Show that the rearranged series converges to $\log 2 + \frac{1}{2} \log \frac{r}{s}$.

In q-analysis we are often in the following situation. Suppose we have a finite sum identity of the form

$$\text{(C2)} \qquad \sum_{k=0}^{p} f_k(n) = F(n),$$

where $p = p(n)$ is some simple function of n that tends steadily to infinity as n does. Often $p(n)$ will just be n, but we want to be able to handle other cases like $p(n) = n - 1$ or $p(n) = \lfloor \frac{n}{2} \rfloor$ or $p(n) = \binom{n}{2}$. Suppose we know that $f_k(n) \to a_k$ as $n \to \infty$. We hope that in that case

$$\text{(C3)} \qquad F(n) \to \sum_{k=0}^{\infty} a_k \quad \text{as } n \to \infty,$$

but in general this is false. If we had made an obvious cancellation in part (ix) above, we would have had

$$V_n = \frac{1}{n+1} + \frac{1}{n+2} + \cdots + \frac{1}{2n} = \sum_{k=1}^{n} \frac{1}{n+k}.$$

If we let $n \to \infty$ here, every term goes to zero, but we know from part (xi) above that the sum does not tend to zero, but rather to $\log 2$. Even without this knowledge, we can see that

$$V_n > \frac{1}{n+n} + \frac{1}{n+n} + \cdots + \frac{1}{n+n} = \frac{n}{2n} = \frac{1}{2},$$

so the sum could not be less than $\frac{1}{2}$. We can further observe that V_n is a Riemann sum for $\int_n^{2n} \frac{dx}{x}$, so practically the entire subject of Riemann sums furnishes counterexamples to (C3).

However, with one mild extra assumption, (C3) does follow from (C2).

THEOREM 100 (Tannery's theorem). *If $|f_k(n)| < C_k$ for each k, where C_k is independent of n, and if $\sum_{k=0}^{\infty} C_k$ converges, then (C3) follows from (C2).*

This result has a peculiar status in mathematics: it is an undergraduate real analysis theorem that rarely appears in undergraduate real analysis books. Readers who have studied real analysis may be reminded of the Weierstrass M-test. For those who have not, what follows is a pretty typical real analysis argument. If we denote the sum on the right side of (C3) by S, then we have to show that $F(n) \to S$ as $n \to \infty$; or that $|F(n) - S|$ becomes arbitrarily small as $n \to \infty$. Cauchy has taught us to rephrase this as follows: given an arbitrary positive number ϵ (which one thinks of as very small), $|F(n) - S| < \epsilon$ if n is sufficiently large. This quantifies "arbitrarily small". To quantify "sufficiently large", we say that there is a number N (which one thinks of as very large) such that if $n > N$, then $|F(n) - S| < \epsilon$. So this is what we have to argue.

Since $\sum_{k=0}^{\infty} C_k$ converges, its "tail" must be arbitrarily small in the same sense: there must be a number M such that

$$C_{m+1} + C_{m+2} + \cdots < \frac{\epsilon}{3}$$

whenever $m > M$, for the same ϵ as above. Since $p = p(n) \to \infty$ as n does, we will have $p > M$ for a sufficiently large n, say $n > N_1$. Therefore, by the triangle inequality,

$$|f_{m+1}(n) + \cdots + f_p(n)| \leq |f_{m+1}(n)| + \cdots + |f_p(n)| < C_{m+1} + \cdots + C_p < \frac{\epsilon}{3},$$

which means that

$$|F(n) - (f_0(n) + f_1(n) + \cdots + f_m(n))| < \frac{\epsilon}{3}.$$

Moreover $a_k \leq C_k$ for each k, so

$$a_{m+1} + a_{m+2} + \cdots < \frac{\epsilon}{3}$$

as well. Now

$$\begin{aligned}|F(n) - S| &= |f_0(n) + f_1(n) + \cdots + f_p(n) - (a_0 + a_1 + \ldots)| \\ &= \left| \begin{aligned} &f_0(n) + \cdots + f_m(n) - (a_0 + \cdots + a_m) \\ &+ (f_{m+1}(n) + \cdots + f_p(n)) - (a_{m+1} + a_{m+2} + \ldots) \end{aligned} \right| \\ &\leq |f_0(n) + \cdots + f_m(n) - (a_0 + \cdots + a_m)| + |f_{m+1}(n) + \cdots + f_p(n)| \\ &\quad + |a_{m+1} + a_{m+2} + \ldots| \end{aligned}$$

by the triangle inequality again. Using our bounds on the last two groups of terms, we therefore have

$$|F(n) - S| < |f_0(n) + \cdots + f_m(n) - (a_0 + \cdots + a_m)| + \frac{\epsilon}{3} + \frac{\epsilon}{3}$$

for $m > M$ and $n > N_1$. Since $f_k(n) \to a_k$ as $n \to \infty$, the first term becomes very small if n is very large. In other words, if $n > N_2$ for some number N_2, then

$$|f_0(n) + \cdots + f_m(n) - (a_0 + \cdots + a_m)| < \frac{\epsilon}{3}.$$

Now let N be the larger of N_1 and N_2, and we have

$$|F(n) - S| < \frac{\epsilon}{3} + \frac{\epsilon}{3} + \frac{\epsilon}{3} = \epsilon$$

whenever $n > N$. This proves Tannery's theorem.

For an application of Tannery's theorem in q-analysis we recall the partial fractions expansion

(C4) $$\frac{(ax;q)_n \left(\frac{q}{ax};q\right)_n}{(x;q)_{n+1} \left(\frac{q}{x};q\right)_n} = \sum_{k=-n}^{n} \frac{(a;q)_{n-k} \left(\frac{q}{a};q\right)_{n+k}}{(q;q)_{n-k} (q;q)_{n+k}} \frac{a^k}{1 - xq^k}$$

from sections 2.8 and 7.1. It is most convenient to use Tannery's theorem when $0 < q < a < 1$, so that

$$\frac{(a;q)_{n-k} \left(\frac{q}{a};q\right)_{n+k}}{(q;q)_{n-k} (q;q)_{n+k}} < \frac{1}{(q;q)_\infty^2},$$

but in the more general case where $|q| < |a| < 1$ we can use the hideous expression

$$\frac{(-|a|;|q|)_\infty \left(-\left|\frac{q}{a}\right|;|q|\right)_\infty}{(|q|;|q|)_\infty^2}$$

as an upper bound. For sufficiently large nonnegative k and $|q| < 1$ we have
$$\left|\frac{a^k}{1-xq^k}\right| \le \frac{|a|^k}{1-|x||q|^k},$$
and
$$\sum_{k=0}^{\infty} \frac{|a|^k}{1-|x||q|^k}$$
converges by the ratio test when $|a| < 1$. For the terms with k negative we can rewrite
$$\sum_{k=-\infty}^{-1} \frac{a^k}{1-xq^k} = \sum_{k=1}^{\infty} \frac{a^{-k}}{1-xq^{-k}} \left(\frac{q}{a}\right)^k \left(\frac{a}{q}\right)^k = \sum_{k=1}^{\infty} \frac{\left(\frac{q}{a}\right)^k}{q^k - x}.$$
Here we have
$$\left|\frac{\left(\frac{q}{a}\right)^k}{q^k - x}\right| \le \frac{\left|\frac{q}{a}\right|^k}{|x|-|q|^k}$$
if $x \ne 0$, k is sufficiently large, and $|q| < 1$, and
$$\sum_{k=1}^{\infty} \frac{\left|\frac{q}{a}\right|^k}{|x|-|q|^k}$$
converges by the ratio test if $\left|\frac{q}{a}\right| < 1$. It follows from Tannery's theorem that the limit as $n \to \infty$ of (C4) is
$$\frac{(ax;q)_\infty \left(\frac{q}{ax};q\right)_\infty}{(x;q)_\infty \left(\frac{q}{x};q\right)_\infty} = \frac{(a;q)_\infty \left(\frac{q}{a};q\right)_\infty}{(q;q)_\infty^2} \sum_{k=-\infty}^{\infty} \frac{a^k}{1-xq^k}$$
for $|q| < |a| < 1$, provided that x is neither zero nor an integer power of q. If we further restrict x to $|q| < |x| < 1$, then we have proved that
$$\frac{(ax;q)_\infty \left(\frac{q}{ax};q\right)_\infty (q;q)_\infty^2}{(x;q)_\infty \left(\frac{q}{x};q\right)_\infty (a;q)_\infty \left(\frac{q}{a};q\right)_\infty} = \sum_{k=-\infty}^{\infty} \frac{a^k}{1-xq^k} = \sum_{k=-\infty}^{\infty} \frac{x^k}{1-aq^k},$$
where the last step is by the symmetry in a and x. This is the key identity in Chapter 7, Cauchy's special case of Ramanujan's $_1\psi_1$ sum.

If we have some complex analysis, we could give a less ugly argument by using the convenient range $0 < q < a < 1$ and then appealing to analytic continuation for other values of q and a for which the sum remains convergent. This example is more delicate than most. Very often, as for example in the finite forms of the Rogers–Ramanujan identities in Chapter 10, there is a quadratic exponent of q that makes the series converge rapidly for $|q| < 1$.

Tannery's theorem appears on pp. 292–293 of [**233**] in 1904. I have not found it in the first edition of 1886. The second edition also extends to a second volume [**234**], which was completed by Tannery's great student Jacques Hadamard in 1910, the year of Tannery's death. It is so much improved from the first edition that one suspects Hadamard's influence throughout the revision, but he is unaccredited in the first volume. The name "Tannery's theorem" was bestowed by T. J. I'a. Bromwich in section 49 of [**63**] in 1908. We have more or less followed Bromwich's proof.

Euler's constant γ makes its debut in [**92**], published in 1740 but written at least 5 years earlier. Like so many of Euler's great papers, it is in volume 14 of his

Opera Omnia, 1st series. (His papers on q-analysis are mostly in volume 2.) The best reference for Euler's constant is Julian Havil's beautiful book [**136**].

Bibliography

[1] Niels Henrik Abel, *Untersuchungen über die Reihe* $1+\frac{m}{1}x+\frac{m(m-1)}{1.2}+\frac{m(m-1)(m-2)}{1.2.3}+\ldots$, Journal für die reine und angewandte Mathematik **1** (1826), 311–339; *Recherches sur la série* $1+\frac{m}{1}x+\frac{m(m-1)}{1.2}+\frac{m(m-1)(m-2)}{1.2.3}+\ldots$, Œuvres Complètes, 2nd edition, vol. 1, Grøndahl & Søn, Christiania, 1881, 219–250. MR1577619

[2] R. P. Agarwal, *On the partial sums of series of hypergeometric type*, Proc. Cambridge Philos. Soc. **49** (1953), 441–445. MR56754

[3] Donald J. Albers, Gerald L. Alexanderson, and William Dunham, eds., *The G. H. Hardy Reader*, Cambridge University Press and Mathematical Association of America, 2015.

[4] Krishnaswami Alladi, *Some new observations on the Göllnitz-Gordon and Rogers-Ramanujan identities*, Trans. Amer. Math. Soc. **347** (1995), no. 3, 897–914, DOI 10.2307/2154877. MR1284910

[5] Krishnaswami Alladi, *Analysis of a generalized Lebesgue identity in Ramanujan's Lost Notebook*, Ramanujan J. **29** (2012), no. 1-3, 339–358, DOI 10.1007/s11139-012-9380-z. MR2994106

[6] George E. Andrews, *A Simple Proof of Jacobi's triple product identity*, Proceedings of the American Mathematical Society **16** (1965), 333–334; *The Selected Works of George E Andrews*, Imperial College Press, London, 2013, 56–57. MR171725

[7] G. E. Andrews, *On basic hypergeometric series, mock theta functions, and partitions. I*, Quart. J. Math. Oxford Ser. (2) **17** (1966), 64–80, DOI 10.1093/qmath/17.1.64. MR193282

[8] George E. Andrews, *On generalizations of Euler's partition theorem*, Michigan Math. J. **13** (1966), 491–498. MR202617

[9] George E. Andrews, *An Analytic Proof of the Rogers-Ramanujan-Gordon Identities*, American Journal of Mathematics **88** (1966), 844–846; *The Selected Works of George E Andrews*, Imperial College Press, London, 2013, 406–408. DOI 10.2307/2373082, MR202616

[10] George E. Andrews, *On Ramanujan's summation of* $_1\psi_1(a;b;z)$, Proc. Amer. Math. Soc. **22** (1969), 552–553, DOI 10.2307/2037098. MR241703

[11] George E. Andrews, *A Polynomial Identity Which Implies the Rogers-Ramanujan Identities*, Scripta Mathematica **28** (1970), 297–305.

[12] George E. Andrews, *Partition identities*, Advances in Math. **9** (1972), 10–51, DOI 10.1016/0001-8708(72)90028-X. MR306105

[13] George E. Andrews, *On the q-analog of Kummer's theorem and applications*, Duke Mathematical Journal **40** (1973), 525–528; *The Selected Works of George E Andrews*, Imperial College Press, London, 2013, 58–61. MR320375

[14] George E. Andrews, *Applications of Basic Hypergeometric Functions*, SIAM Review **16**(4) (1974), 441–484; *The Selected Works of George E Andrews (With commentary)*, edited by Andrew V. Sills, Imperial College Press, London, 2013, 879–922, DOI 10.1137/1016081, MR352557.

[15] George E. Andrews, *The theory of partitions*, Addison-Wesley Publishing Co., Reading, Mass.-London-Amsterdam, 1976. Encyclopedia of Mathematics and its Applications, Vol. 2. MR0557013

[16] George E. Andrews, *Partitions: yesterday and today*, New Zealand Mathematical Society, Wellington, 1979. With a foreword by J. C. Turner. MR557539

[17] George E. Andrews, *Multiple series Rogers–Ramanujan type identities*, Pacific Journal of Mathematics **114**(2) (1984), 267–283; *The Selected Works of George E Andrews*, Imperial College Press, London, 2013, 113–129. MR757501

[18] George E. Andrews, *Generalized Frobenius partitions*, Mem. Amer. Math. Soc. **49** (1984), no. 301, iv+44, DOI 10.1090/memo/0301. MR743546

[19] George E. Andrews, *q-series: their development and application in analysis, number theory, combinatorics, physics, and computer algebra*, CBMS Regional Conference Series in Mathematics, vol. 66, Published for the Conference Board of the Mathematical Sciences, Washington, DC; by the American Mathematical Society, Providence, RI, 1986. MR858826

[20] George E. Andrews, *J. J. Sylvester, Johns Hopkins and partitions*, A century of mathematics in America, Part I, Hist. Math., vol. 1, Amer. Math. Soc., Providence, RI, 1988, pp. 21–40. MR1003160

[21] George E. Andrews, *Fibonacci Numbers and the Rogers-Ramanujan Identities*, The Fibonacci Quarterly **42** (2004), 3–19; *The Selected Works of George E. Andrews*, Imperial college Press, London, 2013, 710–726. MR2060558

[22] George E. Andrews, *Bailey chains*, in *The Selected Works of George E Andrews*, Imperial College Press, London, 2013, 107–108.

[23] George E. Andrews and Richard Askey, *A simple proof of Ramanujan's summation of the $_1\psi_1$*, Aequationes Math. **18** (1978), no. 3, 333–337, DOI 10.1007/BF03031684. MR522519

[24] George E. Andrews, Richard Askey, and Ranjan Roy, *Special functions*, Encyclopedia of Mathematics and its Applications, vol. 71, Cambridge University Press, Cambridge, 1999. MR1688958

[25] George E. Andrews and Bruce C. Berndt, *Ramanujan's Lost Notebook, Part I*, Springer-Verlag, New York, 2005.

[26] George E. Andrews and Bruce C. Berndt, *Ramanujan's lost notebook. Part II*, Springer, New York, 2009. MR2474043

[27] George E. Andrews and Bruce C. Berndt, *Ramanujan's lost notebook. Part III*, Springer, New York, 2012. MR2952081

[28] George E. Andrews and Bruce C. Berndt, *Ramanujan's lost notebook. Part III*, Springer, New York, 2012. MR2952081

[29] George E. Andrews and Bruce C. Berndt, *Ramanujan's lost notebook. Part V*, Springer, 2018. MR3838409

[30] George E. Andrews and Kimmo Eriksson, *Integer partitions*, Cambridge University Press, Cambridge, 2004. MR2122332

[31] George E. Andrews, Arnold Knopfmacher, and Peter Paule, *An infinite family of Engel expansions of Rogers-Ramanujan type*, Adv. in Appl. Math. **25** (2000), no. 1, 2–11, DOI 10.1006/aama.2000.0686. MR1773190

[32] George E. Andrews and S. Ole Warnaar, *The product of partial theta functions*, Adv. in Appl. Math. **39** (2007), no. 1, 116–120, DOI 10.1016/j.aam.2005.12.003. MR2319567

[33] Richard Askey, *Ramanujan's extensions of the gamma and beta functions*, Amer. Math. Monthly **87** (1980), no. 5, 346–359, DOI 10.2307/2321202. MR567718

[34] Richard Askey, *An elementary evaluation of a beta type integral*, Indian J. Pure Appl. Math. **14** (1983), no. 7, 892–895. MR714840

[35] Richard Askey, *The very well poised $_6\psi_6$. II*, Proc. Amer. Math. Soc. **90** (1984), no. 4, 575–579, DOI 10.2307/2045033. MR733409

[36] Richard Askey, *Ramanujan's $_1\psi_1$ and formal Laurent series*, Indian J. Math. **29** (1987), no. 2, 101–105. MR919888

[37] Richard Askey and Mourad E. H. Ismail, *The very well poised $_6\psi_6$*, Proc. Amer. Math. Soc. **77** (1979), no. 2, 218–222, DOI 10.2307/2042642. MR542088

[38] Richard Askey and James Wilson, *Some basic hypergeometric orthogonal polynomials that generalize Jacobi polynomials*, Mem. Amer. Math. Soc. **54** (1985), no. 319, iv+55, DOI 10.1090/memo/0319. MR783216

[39] W. N. Bailey, *Generalized Hypergeometric Series*, Cambridge Tracts in Mathematics and Mathematical Physics No. 32, Cambridge University Press, London, 1935; reprinted by Hafner, New York, 1964.

[40] W. N. Bailey, *Series of Hypergeometric Type which are Infinite in Both Directions*, Quarterly Journal of Mathematics (Oxford) **7** (1936), 105–115.

[41] W. N. Bailey, *An Algebraic Identity*, J. London Math. Soc. **11** (1936), no. 2, 156–160, DOI 10.1112/jlms/s1-11.2.156. MR1574766

[42] W. N. Bailey, *Some Identities Connected with Representations of Numbers*, J. London Math. Soc. **11** (1936), no. 4, 286–289, DOI 10.1112/jlms/s1-11.4.286. MR1574926

[43] W. N. Bailey, *A note on certain q-identities*, Quart. J. Math. Oxford Ser. **12** (1941), 173–175, DOI 10.1093/qmath/os-12.1.173. MR5964

[44] W. N. Bailey, *A note on two of Ramanujan's formulae*, Quart. J. Math. Oxford Ser. (2) **3** (1952), 29–31, DOI 10.1093/qmath/3.1.29. MR46380

[45] W. N. Bailey, *A further note on two of Ramanujan's formulae*, Quart. J. Math. Oxford Ser. (2) **3** (1952), 158–160, DOI 10.1093/qmath/3.1.158. MR49226

[46] Alexander Berkovich, *On the Difference of Partial Theta Functions*, unpublished manuscript.

[47] Bruce C. Berndt, *Ramanujan's Notebooks, Part I*, Springer-Verlag, New York, 1985.

[48] Bruce C. Berndt, *Ramanujan's notebooks. Part II*, Springer-Verlag, New York, 1989. MR970033

[49] Bruce C. Berndt, *Ramanujan's notebooks. Part III*, Springer-Verlag, New York, 1991. MR1117903

[50] Bruce C. Berndt, *Ramanujan's notebooks. Part IV*, Springer-Verlag, New York, 1994. MR1261634

[51] Bruce C. Berndt, *Ramanujan's notebooks. Part V*, Springer-Verlag, New York, 1998. MR1486573

[52] Bruce C. Berndt, *Number theory in the spirit of Ramanujan*, Student Mathematical Library, vol. 34, American Mathematical Society, Providence, RI, 2006. MR2246314

[53] Bruce C. Berndt, Song Heng Chan, Boon Pin Yeap, and Ae Ja Yee, *A reciprocity theorem for certain q-series found in Ramanujan's lost notebook*, Ramanujan J. **13** (2007), no. 1-3, 27–37, DOI 10.1007/s11139-006-0241-5. MR2281155

[54] M. V. Bouniakowsky, *Recherches sur différentes lois nouvelles relatives a la somme des diviseurs des nombres*, Mémoires de l'Académie impériale des sciences de Saint-Pétersbourg, 6th series, Sciences mathématiques et physiques **4** (1850), 259–295.

[55] Douglas Bowman, *q-difference operators, orthogonal polynomials, and symmetric expansions*, Mem. Amer. Math. Soc. **159** (2002), no. 757, x+56, DOI 10.1090/memo/0757. MR1921582

[56] Douglas Bowman, James Mc Laughlin, and Andrew V. Sills, *Some more identities of the Rogers-Ramanujan type*, Ramanujan J. **18** (2009), no. 3, 307–325, DOI 10.1007/s11139-007-9109-6. MR2495550

[57] David M. Bressoud, *A combinatorial proof of Schur's 1926 partition theorem*, Proc. Amer. Math. Soc. **79** (1980), no. 2, 338–340, DOI 10.2307/2043263. MR565367

[58] David M. Bressoud, *Analytic and combinatorial generalizations of the Rogers-Ramanujan identities*, Mem. Amer. Math. Soc. **24** (1980), no. 227, 54, DOI 10.1090/memo/0227. MR556608

[59] D. M. Bressoud, *Some identities for terminating q-series*, Math. Proc. Cambridge Philos. Soc. **89** (1981), no. 2, 211–223, DOI 10.1017/S0305004100058114. MR600238

[60] D. M. Bressoud, *An easy proof of the Rogers-Ramanujan identities*, J. Number Theory **16** (1983), no. 2, 235–241, DOI 10.1016/0022-314X(83)90043-4. MR698167

[61] David M. Bressoud, *Proofs and confirmations*, MAA Spectrum, Mathematical Association of America, Washington, DC; Cambridge University Press, Cambridge, 1999. The story of the alternating sign matrix conjecture. MR1718370

[62] David M. Bressoud and Doron Zeilberger, *A short Rogers-Ramanujan bijection*, Discrete Math. **38** (1982), no. 2-3, 313–315, DOI 10.1016/0012-365X(82)90298-9. MR676546

[63] T. J. I'A. Bromwich, *An Introduction to the Theory of Infinite Series*, 2nd edition, London, 1926; reprinted by AMS Chelsea, 1991.

[64] L. Carlitz, *Sequences and inversions*, Duke Math. J. **37** (1970), 193–198. MR252237

[65] L. Carlitz, *A combinatorial property of q-Eulerian numbers*, Amer. Math. Monthly **82** (1975), 51–54, DOI 10.2307/2319133. MR366683

[66] L. Carlitz and J. Riordan, *Congruences for Eulerian numbers*, Duke Math. J. **20** (1953), 339–343. MR56008

[67] Augustin-Louis Cauchy, *Mémoire sur les fonctions dont plusieurs valeurs sont liées entre elles par une équation linéaire*, Comptes Rendus de l'Académie des Sciences (Paris) **17** (1843), 523–531; Œuvres Complètes d'Augustin Cauchy, 1st series, vol. 8, Gauthier-Villars, Paris, 1893, 42–50.

[68] Augustin-Louis Cauchy, *Second Mémoire sur les fonctions dont plusieurs valeurs sont liées entre elles par une équation linéaire*, Comptes Rendus de l'Académie des Sciences (Paris) **17**

(1843), 567–572; Œuvres Complètes d'Augustin Cauchy, 1st series, vol. 8, Gauthier-Villars, Paris, 1893, 50–55.

[69] Augustin-Louis Cauchy, *Mémoire sur l'application du calcul des résidus au développement des produits composés d'un nombre infini de facteurs*, Comptes Rendus de l'Académie des Sciences (Paris) **17** (1843), 572–581; Œuvres Complètes d'Augustin Cauchy, 1st series, vol. 8, Gauthier-Villars, Paris, 1893, 55–64.

[70] Arthur Cayley, *On a theorem of M. Lejeune-Dirichlet's*, Cambridge and Dublin Mathematical Journal **9** (1854), 163–165; The Collected Mathematical Papers of Arthur Cayley, vol. 2, Cambridge University Press, 1889, 47–48.

[71] Arthur Cayley, *Researches on the partition of numbers*, Philosophical Transactions of the Royal Society of London **145** (1855), 127–140; The Collected Mathematical Papers of Arthur Cayley, vol. 2, Cambridge University Press, Cambridge, 1890, 235–249.

[72] Arthur Cayley, *On a q-formula leading to an expression for E_1*, Messenger of Mathematics **6** (1877), 63–66, The Collected Mathematical Papers of Arthur Cayley, vol. 10, Cambridge University Press, 1896, 25–27.

[73] A. Cayley, *Note on a Partition-Series*, Amer. J. Math. **6** (1883/84), no. 1-4, 63–64, DOI 10.2307/2369210. MR1505342

[74] Robin Chapman, *A new proof of some identities of Bressoud*, Int. J. Math. Math. Sci. **32** (2002), no. 10, 627–633, DOI 10.1155/S0161171202110155. MR1953600

[75] William Y. C. Chen, Wenchang Chu, and Nancy S. S. Gu, *Finite form of the quintuple product identity*, J. Combin. Theory Ser. A **113** (2006), no. 1, 185–187, DOI 10.1016/j.jcta.2005.04.002. MR2192776

[76] William Y. C. Chen and Gian-Carlo Rota, *q-analogs of the inclusion-exclusion principle and permutations with restricted position*, Discrete Math. **104** (1992), no. 1, 7–22, DOI 10.1016/0012-365X(92)90622-M. MR1171787

[77] William Y. C. Chen and Deheng Xu, *Labeled partitions and the q-derangement numbers*, SIAM J. Discrete Math. **22** (2008), no. 3, 1099–1104, DOI 10.1137/06066326X. MR2424839

[78] J. Cigler, *Operatormethoden für q-Identitäten* (German, with English summary), Monatsh. Math. **88** (1979), no. 2, 87–105, DOI 10.1007/BF01319097. MR551934

[79] Th. Clausen, *Beitrag zur Theorie der Reihen* (German), J. Reine Angew. Math. **3** (1828), 92–95, DOI 10.1515/crll.1828.3.92. MR1577683

[80] Shaun Cooper, *The quintuple product identity*, Int. J. Number Theory **2** (2006), no. 1, 115–161, DOI 10.1142/S1793042106000401. MR2217798

[81] Shaun Cooper, *Ramanujan's theta functions*, Springer, Cham, 2017. MR3675178

[82] Shaun Cooper and Heung Yeung Lam, *Sums of two, four, six and eight squares and triangular numbers: an elementary approach*, Indian J. Math. **44** (2002), no. 1, 21–40. MR1982053

[83] Sylvie Corteel and Jeremy Lovejoy, *Frobenius partitions and the combinatorics of Ramanujan's $_1\psi_1$ summation*, J. Combin. Theory Ser. A **97** (2002), no. 1, 177–183, DOI 10.1006/jcta.2001.3205. MR1879133

[84] Gabriel Cramer, *Introduction à l'Analyse des Lignes Courbes algébriques*, Geneva, 1750.

[85] J. A. Daum, *The basic analogue of Kummer's theorem*, Bull. Amer. Math. Soc. **48** (1942), 711–713, DOI 10.1090/S0002-9904-1942-07764-0. MR7079

[86] Leonard Eugene Dickson, *History of the Theory of Numbers*, vol. 2, The Carnegie Institute, Washington, D.C., 1919; reprinted by Dover, New York, 2005.

[87] Gustav Lejeune Dirichlet, *Recherches sur diverses applications de l'analyse infinitésimale à la théorie des nombres*, Journal für die reine und angewandte Mathematik **19** (1839), 324–369, **21** (1840), 1–12, 134–155; G. Lejeune Dirichlet's Werke, vol. 1, Georg Reimer, Berlin, 1889, 411–496.

[88] J. M. Dobbie, *A simple proof of some partition formulae of Ramanujan's*, Quart. J. Math. Oxford Ser. (2) **6** (1955), 193–196, DOI 10.1093/qmath/6.1.193. MR72896

[89] F. Gotthold Eisenstein, *Transformations remarquables de quelques séries*, Journal für die reine und angewandte Mathematik **27** (1844), 193–197; Mathematische Werke, vol. 1, Chelsea Publishing Company, New York, 1969, 35–39. DOI 10.1515/crll.1844.27.193, MR1578393.

[90] Ferdinand G. Eisenstein, *Neuer Beweis und Verallgemeinerung des Binomischen Lehrsatzes*, Journal für die reine und angewandte Mathematik **28** (1844), 44–48; Mathematische Werke, 2nd ed., vol. 1, Chelsea Publishing Company, New York, 1989, 117–121. DOI 10.1515/crll.1844.28.44, MR1578415.

[91] Ferdinand G. Eisenstein, *Neue Theoreme der höheren Arithmetik*, Journal für die reine und angewandte Mathematik **35** (1847), 117–136; *Mathematische Werke*, 2nd ed., vol. 1, Chelsea Publishing Company, New York, 1989, 483–502. DOI 10.1515/crll.1847.35.117, MR1578592.

[92] Leonhard Euler, *De progressionibus harmonicis observationes*, Commentarii academiae scientiarum Petropolitanae, **7** (1734/5), 150–161; *Opera Omnia*, Series Prima, vol. 14, B. G. Teubner, Berlin, 1925, 87–100.

[93] Leonhard Euler, *Observationes analyticae variae de combinationibus*, Commentarii academiae scientiarum Petropolitanae, **13** (1751), 64–93; *Opera Omnia*, Series Prima, vol. 2, B. G. Teubner, Berlin, 1915, 163–193.

[94] Leonhard Euler, *Introductio in Analysin Infinitorum*, vol. 1, Marcum-Michaelum Bousquet, Lausanne,1748; *Opera Omnia*, Series Prima, vol. 8, B. G. Teubner, Berlin, 1922; English translation by John D. Blanton, Springer-Verlag, New York, 1988.

[95] Leonhard Euler, *Découverte d'une loi tout extraordinare par rapport à la somme de leurs diviseurs*, Bibliothèque impartiale **3** (1751), 10–31; *Opera Omnia*, Series Prima, vol. 2, B. G. Teubner, Berlin, 1915, 241–253.

[96] Leonhard Euler, *Consideratio quarumdam serierum, quae singularibus proprietatibus sunt praeditae*, Novi commentarii academiae scientiarum Petropolitanae **3** (1753), 86–108; *Opera Omnia*, Series Prima, vol. 14, B. G. Teubner, Berlin, 1925, 516–541.

[97] Leonhard Euler, *De partitione numerorum*, Novi commentarii academiae scientiarum Petropolitanae **3** (1753), 125–169; *Opera Omnia*, Series Prima, vol. 2, B. G. Teubner, Berlin, 1915, 254–294.

[98] Leonhard Euler, *Institutiones Calculi Differentialis*, Academiae Imperialis Scientiarum Petropolitanae, 1755; *Opera Omnia*, Series Prima, vol. 10, B. G. Teubner, Leipzig, 1913.

[99] Leonhard Euler, *Observatio de summis divisorum*, Novi commentarii academiae scientiarum Petropolitanae **5** (1760), 59–74; *Opera Omnia*, Series Prima, vol. 2, B. G. Teubner, Berlin, 1915, 373–389.

[100] Leonhard Euler, *Demonstratio theorematis circa ordinem in summis divisorum observatum*, Novi commentarii academiae scientiarum Petropolitanae **5** (1760), 75–83; *Opera Omnia*, Series Prima, vol. 2, B. G. Teubner, Berlin, 1915, 390–398.

[101] Leonhard Euler, *Evolutio producti infiniti* $(1-x)(1-xx)(1-x^3)(1-x^4)(1-x^5)(1-x^6)$ *etc. in serierum simplicem* Acta academiae scientiarum Petropolitanae 1780 **1** (1783), 47–55; *Opera Omnia*, Series Prima, vol. 3, B. G. Teubner, Berlin, 1917, 472–479.

[102] Nathan J. Fine, *Some new results on partitions*, Proc. Nat. Acad. Sci. U.S.A. **34** (1948), 616–618, DOI 10.1073/pnas.34.12.616. MR27798

[103] Nathan J. Fine, *Basic hypergeometric series and applications*, Mathematical Surveys and Monographs, vol. 27, American Mathematical Society, Providence, RI, 1988. With a foreword by George E. Andrews. MR956465

[104] Dominique Foata, *On the Netto inversion number of a sequence*, Proc. Amer. Math. Soc. **19** (1968), 236–240, DOI 10.2307/2036179. MR223256

[105] Dominique Foata, *Rearrangements of Words*, Combinatorics on Words, Encyclopedia of mathematics and its applications **17**, Cambridge University Press, 1983, 184–212; Cambridge Mathematical Library, 1997.

[106] Dominique Foata, *Une démonstration combinatoire de l'identité de Pfaff-Saalschütz* (French, with English summary), C. R. Acad. Sci. Paris Sér. I Math. **297** (1983), no. 4, 221–224. MR727174

[107] Dominique Foata and Marcel-Paul Schützenberger, *Major index and inversion number of permutations*, Math. Nachr. **83** (1978), 143–159, DOI 10.1002/mana.19780830111. MR506852

[108] Fabian Franklin, *Sur le développement du produit infini* $(1-x)(1-x^2)(1-x^3)\ldots$, Comptes rendus de l'Académie des sciences Paris **82** (1881), 448–450.

[109] Robert Fricke, *Die Elliptischen Funktionen und ihre Anwendungen*, vol. 1, B. G. Teubner, Leipzig, 1916.

[110] Georg Frobenius, *Über die Charaktere der symmetrischen Gruppe*, Sitzungsberichte der Preussischen Akademie der Wissenschaften 1900, 516–534.

[111] Jean Guillaume Garnier, *Analyse Algébrique, faisant suite à la première section de l'algèbre*, 2nd edition, Paris, 1814.

[112] Kristina Garrett, Mourad E. H. Ismail, and Dennis Stanton, *Variants of the Rogers-Ramanujan identities*, Adv. in Appl. Math. **23** (1999), no. 3, 274–299, DOI 10.1006/aama.1999.0658. MR1722235

[113] A. M. Garsia and S. C. Milne, *A Rogers-Ramanujan bijection*, J. Combin. Theory Ser. A **31** (1981), no. 3, 289–339, DOI 10.1016/0097-3165(81)90062-5. MR635372

[114] George Gasper and Mizan Rahman, *Basic Hypergeometric Series*, 2nd edition, Encyclopedia of Mathematics and Its Applications **35**, Cambridge University Press, 2004.

[115] Carl Friedrich Gauss, *Summatio quarundam serierum singularium*, Commentationes societatis regiae scientiarum Gottingensis recentiores **1** (1811); *Werke*, vol. 2, 10–45, Königlichen Gesellschaft der Wissenschaften zu Göttingen, 1863.

[116] Carl Friedrich Gauss, *Theorematis fundamentalis in doctrina de residuis quadraticis*, Commentationes societatis regiae scientiarum Gottingensis recentiores **4** (1818); *Werke*, vol. 2, 47–64, Königlichen Gesellschaft der Wissenschaften zu Göttingen, 1863.

[117] Carl Friedrich Gauss, *Zur Theorie der neuen Transscendenten*, Werke, vol. 3, Königliche Gesellschaft der Wissenschaften zu Göttingen, 1876, 433–480.

[118] Carl Friedrich Gauss, *Zur Theorie der Transscendenten Functionen I*, Werke, vol. 10, Königliche Gesellschaft der Wissenschaften zu Göttingen, 1917, 287–307.

[119] Joseph Diez Gergonne, *Analise élémentaire. Développement de la théorie donnée par M. Laplace, pour l'élimination au premier degré* (French), Ann. Math. Pures Appl. [Ann. Gergonne] **4** (1813/14), 148–155. MR1555735

[120] Ira Martin Gessel, *Generating Functions and Enumeration of Sequences*, ProQuest LLC, Ann Arbor, MI, 1977. Thesis (Ph.D.)–Massachusetts Institute of Technology. MR2940769

[121] Ira M. Gessel, *Counting permutations by descents, greater index, and cycle structure*, unpublished manuscript.

[122] J. W. L. Glaisher, *Arithmetical Proof of Clausen's Identity*, Messenger of Mathematics **5** (1875), 83.

[123] J. W. L. Glaisher, *A theorem in partitions*, Messenger of Mathematics **12** (1883), 158–170.

[124] Heinz Göllnitz, *Einfache partitionen*, Diplomarbeit, Göttingen, 1960.

[125] Heinz Göllnitz, *Partitionen mit Differenzenbedingungen* (German), Dissertation zur Erlangung des Doktorgrades der Mathematisch-Naturwissen schaftlichen Fakultät der Georg-August-Universität zu Göttingen, Dissertation, Göttingen, 1963. MR0183699

[126] H. Göllnitz, *Partitionen mit Differenzenbedingungen* (German), J. Reine Angew. Math. **225** (1967), 154–190, DOI 10.1515/crll.1967.225.154. MR211973

[127] Basil Gordon, *Some continued fractions of the Rogers-Ramanujan type*, Duke Math. J. **32** (1965), 741–748. MR184001

[128] Basil Gordon, *A combinatorial generalization of the Rogers-Ramanujan identities*, Amer. J. Math. **83** (1961), 393–399, DOI 10.2307/2372962. MR123484

[129] Johann Philipp Gruson, *Neuer analytischen Lehrsatz*, Abhandlungen der Königlichen Preußischen Akademie der Wissenschaften zu Berlin 1814–1815, Mathematische Klasse, Realschul-Bachhadlung, Berlin, 1818, 36–41.

[130] G. H. Hardy, *A course of pure mathematics*, Centenary edition, Cambridge University Press, Cambridge, 2008. Reprint of the tenth (1952) edition with a foreword by T. W. Körner. MR2400109

[131] G. H. Hardy, *Srinivasa Ramanujan*, Proceedings of the London Mathematical Society, 2nd series **19** (1921), xl–lviii; *Collected Papers of Srinivasa Ramanujan*, Cambridge University Press, 1927, xi–xxxvi, reprinted by AMS Chelsea, 2000; *Collected Papers of G. H. Hardy*, volume VII, Clarendon Press, Oxford, 1979, 702–720.

[132] G. H. Hardy, *The Indian Mathematician Ramanujan*, Amer. Math. Monthly **44** (1937), no. 3, 137–155, DOI 10.2307/2301659. MR1523880

[133] G. H. Hardy, *Ramanujan. Twelve lectures on subjects suggested by his life and work*, Cambridge University Press, Cambridge, England; Macmillan Company, New York, 1940; reprinted by AMS Chelsea, 1999. MR0004860

[134] G. H. Hardy, *A Mathematician's Apology*, Cambridge University Press, 1940; reprinted with a Foreword by C. P. Snow, 1967.

[135] G. H. Hardy and E. M. Wright, *An introduction to the theory of numbers*, 6th ed., Oxford University Press, Oxford, 2008. Revised by D. R. Heath-Brown and J. H. Silverman; With a foreword by Andrew Wiles. MR2445243

[136] Julian Havil, *Gamma*, Princeton University Press, Princeton, NJ, 2003. With a foreword by Freeman Dyson. MR1968276
[137] E. Heine, *Untersuchungen über die Reihe* (German), J. Reine Angew. Math. **34** (1847), 285–328, DOI 10.1515/crll.1847.34.285. MR1578577
[138] M. D. Hirschhorn, *Polynomial identities which imply identities of Euler and Jacobi*, Acta Arith. **32** (1977), no. 1, 73–78, DOI 10.4064/aa-32-1-73-78. MR441851
[139] Michael D. Hirschhorn, *Ramanujan's "most beautiful identity"*, Amer. Math. Monthly **118** (2011), no. 9, 839–845, DOI 10.4169/amer.math.monthly.118.09.839. MR2854006
[140] Michael D. Hirschhorn, *The power of q*, Developments in Mathematics, vol. 49, Springer, Cham, 2017. With a foreword by George E. Andrews. MR3699428
[141] Olga Holtz, Volker Mehrmann, and Hans Schneider, *Potter, Wielandt, and Drazin on the matrix equation $AB = \omega BA$: new answers to old questions*, Amer. Math. Monthly **111** (2004), no. 8, 655–667, DOI 10.2307/4145039. MR2091542
[142] Mourad E. H. Ismail, *A simple proof of Ramanujan's $_1\psi_1$ sum*, Proc. Amer. Math. Soc. **63** (1977), no. 1, 185–186, DOI 10.2307/2041093. MR508183
[143] Mourad E. H. Ismail, *Classical and quantum orthogonal polynomials in one variable*, Encyclopedia of Mathematics and its Applications, vol. 98, Cambridge University Press, Cambridge, 2005. With two chapters by Walter Van Assche; With a foreword by Richard A. Askey. MR2191786
[144] F. H. Jackson, *Transformations of q-series*, Messenger of Mathematics **39** (1910), 145–153.
[145] F. H. Jackson, *Summation of a q-hypergeometric series*, Messenger of Mathematics **50** (1921), 101–112.
[146] M. Jackson, *On Lerch's transcendant and the basic bilateral hypergeometric series $_2\Psi_2$*, J. London Math. Soc. **25** (1950), 189–196, DOI 10.1112/jlms/s1-25.3.189. MR36882
[147] Carl G. Jacobi, letter to A. M. Legendre of 12 April 1828, *Gesammelte Werke*, vol. 1, Reimer, Berlin, 1881, 409–416; reprinted by Chelsea, New York, 1969.
[148] Carl G. Jacobi, *Fundamenta nova theoriae functionum ellipticarum*, Regiomonti, 1829; *Gesammelte Werke*, vol. 1, Reimer, Berlin, 1881, 49–239; reprinted by Chelsea, New York, 1969.
[149] Carl G. Jacobi, *Beweis des Satzes, dass jede nicht fünfeckige Zahl ebenso oft in eine gerade als ungerade Anzahl verschiedener Zahlen zerlegt werden kann*, Journal für die reine und angewandte Mathematik **32** (1846), 164–175; *Gesammelte Werke*, vol. 6, Reimer, Berlin, 1891, 303–317; reprinted by Chelsea, New York, 1969; DOI 10.1515/crll.1846.32.164, MR1578520.
[150] Carl G. Jacobi, *Über einige der Binomialreihe analoge Reihen*, Journal für die reine und angewandte Mathematik **32** (1846), 197–204; *Gesammelte Werke*, vol. 6, Reimer, Berlin, 1891, 163–173; reprinted by Chelsea, New York, 1969; DOI 10.1515/crll.1846.32.197, MR1578524.
[151] Carl G. Jacobi, *Über unendliche Reihen, deren Exponenten zugleich in zwei verschiedenen quadratischen Formen enthalten sind*, Journal für die reine und angewandte Mathematik **37** (1849), 61–94, 221–254; *Gesammelte Werke*, vol. 2, Reimer, Berlin, 1882, 217–288; reprinted by Chelsea, New York, 1969.
[152] Warren P. Johnson, *q-extensions of identities of Abel-Rothe type*, Discrete Math. **159** (1996), no. 1-3, 161–177, DOI 10.1016/0012-365X(95)00108-9. MR1415291
[153] Warren P. Johnson, *How Cauchy missed Ramanujan's $_1\psi_1$ summation*, Amer. Math. Monthly **111** (2004), no. 9, 791–800, DOI 10.2307/4145190. MR2104050
[154] Victor Kac and Pokman Cheung, *Quantum calculus*, Universitext, Springer-Verlag, New York, 2002. MR1865777
[155] Clark Kimberling, *The origin of Ferrers graphs*, The Mathematical Gazette **83** (1999), 194–198.
[156] Gabriel Klambauer, *Summation of series*, Amer. Math. Monthly **87** (1980), no. 2, 128–130, DOI 10.2307/2321992. MR559150
[157] Konrad Knopp, *Theory of Functions, Part I*, Dover Publications, Inc., New York, 1945.
[158] Konrad Knopp, *Theory and Application of Infinite Series*, Dover Publications, Inc., New York, 1990.
[159] Donald E. Knuth, *Two Notes on Notation*, The American Mathematical Monthly **99** (1992), 403–422; *Selected Papers on Discrete Mathematics*, CSLI Lecture Notes **106**, Stanford, CA, 2003, 15–44; DOI 10.2307/2325085, MR1163629.

[160] Johann Heinrich Lambert, *Anlage zur Architectonic, oder Theorie des ersten und des einfachen in der philosophischen und mathematischen Erkenntnis*, vol. 2, Johann Friedrich Hartenoch, Riga, 1771; *Philosophische Schriften*, vol. 4, Georg Olm, Hildesheim, 1965.

[161] Pierre Simon Laplace, *Recherches sur le calcul intégral et sur le système du monde*, Œuvres complètes de Laplace, vol. 8, Gauthier-Villars, Paris, 1772, pp. 369–477.

[162] V. A. Lebesgue, *Sommation de quelques séries*, Journal de Mathématiques pures et appliquées **5** (1840), 42–71.

[163] Adrien Marie Legendre, *Traité des fonctions elliptiques et des intégrales eulériennes*, vol. 3, Huzard-Courcier, Paris, 1828.

[164] Adrien Marie Legendre, *Theorie des nombres*, 3rd edition, Paris, 1830.

[165] Jesper Lützen, *Joseph Liouville 1809–1882: master of pure and applied mathematics*, Studies in the History of Mathematics and Physical Sciences, vol. 15, Springer-Verlag, New York, 1990. MR1066463

[166] Percy A. MacMahon, *The Indices of Permutations and the Derivation therefrom of Functions of a Single Variable Associated with the Permutations of any Assemblage of Objects*, American Journal of Mathematics **35** (1913), 281–322; *Percy Alexander MacMahon: Collected Papers*, vol. 1, George E. Andrews, ed., MIT Press, Cambridge, MA, 1978, 508–549; DOI 10.2307/2370312, MR1506186.

[167] Percy A. MacMahon, *The Superior and Inferior Indices of Permutations*, Transactions of the Cambridge Philosophical Society **29** (1914), 55–60; *Percy Alexander MacMahon: Collected Papers*, George E. Andrews, ed., vol. 1, MIT Press, Cambridge, MA, 1978, 550–555.

[168] Percy A. MacMahon, *Combinatory Analysis*, vol. 1 & 2, Cambridge University Press, Cambridge, 1915, 1916; Chelsea Publishing Company, New York, 1960, MR0141605.

[169] Percy A. MacMahon, *Two applications of general theorems in combinatory analysis: (1) to the theory of inversions of permutations; (2) to the ascertainment of the numbers of terms in the development of a determinant which has amongst its elements an arbitrary number of zeros*, Proceedings of the London Mathematical Society, 2nd series **15** (1916), 314–321; *Percy Alexander MacMahon: Collected Papers*, George E. Andrews, ed., vol. 1, MIT Press, Cambridge, MA, 1978, 556–563; DOI 10.1112/plms/s2-15.1.314, MR1576566.

[170] Meghana Madhyastha, Solution of AMM 11908, The American Mathematical Monthly **125** (2018), 279.

[171] Pierre Rémond de Montmort, *Essay d'analyse sur les jeux de hazard* (French), 2nd ed., J. Quillau, Paris, 1713, Chelsea Publishing Co., New York, 1980. MR605303

[172] Sir Thomas Muir, *Determination of the sign of a single term of a determinant*, Proceedings of the Royal Society of Edinburgh **22** (1899), 441–477.

[173] Sir Thomas Muir, *The theory of determinants in the historical order of development* vol. I, Macmillan and Co., Limited, London, 1923.

[174] Eugen Netto, *Lehrbuch der Combinatorik*, 2nd edition, B. G. Teubner, Leipzig, 1927; Chelsea Publishing Company, New York, 1958.

[175] François Nicole, *Méthode pour sommer une infinité de Suites nouvelles, dont on ne peut trouver les Sommes par les Méthodes connuës*, Mémoires de l'Academie Royale des Sciences Paris, 1727, 257–268.

[176] Hideyuki Ohtsuka, Problem 12078, The American Mathematical Monthly **125** (2018), 944.

[177] Igor Pak, *On Fine's partition theorems, Dyson, Andrews, and missed opportunities*, Math. Intelligencer **25** (2003), no. 1, 10–16, DOI 10.1007/BF02985633. MR1962923

[178] Igor Pak, *Partition bijections, a survey*, Ramanujan J. **12** (2006), no. 1, 5–75, DOI 10.1007/s11139-006-9576-1. MR2267263

[179] Karen Hunger Parshall, *James Joseph Sylvester: Life and work in letters*, The Clarendon Press, Oxford University Press, New York, 1998. MR1674190

[180] Karen Hunger Parshall, *James Joseph Sylvester: Jewish mathematician in a Victorian world*, Johns Hopkins University Press, Baltimore, MD, 2006. MR2216541

[181] Peter Paule, *On identities of the Rogers-Ramanujan type*, J. Math. Anal. Appl. **107** (1985), no. 1, 255–284, DOI 10.1016/0022-247X(85)90368-3. MR786027

[182] Herbert Pieper, *Carl Gustav Jacob Jacobi*, Mathematics in Berlin, Birkhäuser, Berlin, 1998, pp. 41–48, DOI 10.1007/978-3-322-81035-9. MR1648675

[183] Herbert Pieper, ed., *Korrespondenz zwischen Legendre und Jacobi, Correspondance mathématique entre Legendre et Jacobi*, Teubner-Archiv, Leipzig, 1998.

[184] G. Polya, *Induction and analogy in mathematics. Mathematics and plausible reasoning, vol. I*, Princeton University Press, Princeton, N. J., 1954. MR0066321

[185] G. Pólya, *On the number of certain lattice polygons*, J. Combinatorial Theory **6** (1969), 102–105. MR236031

[186] G. Pólya, *Gaussian binomial coefficients and the enumeration of inversions*, Proc. Second Chapel Hill Conf. on Combinatorial Mathematics and its Applications (Univ. North Carolina, Chapel Hill, N.C., 1970), Univ. North Carolina, Chapel Hill, N.C., 1970, pp. 381–384. MR0269521

[187] G. Pólya and G. L. Alexanderson, *Gaussian binomial coefficients*, Elem. Math. **26** (1971), 102–109. MR299490

[188] G. Pólya and G. Szegő, *Problems and theorems in analysis. Vol. I: Series, integral calculus, theory of functions*, Springer-Verlag, New York-Berlin, 1972. Translated from the German by D. Aeppli; Die Grundlehren der mathematischen Wissenschaften, Band 193. MR0344042

[189] H. S. A. Potter, *On the latent roots of quasi-commutative matrices*, Amer. Math. Monthly **57** (1950), 321–322, DOI 10.2307/2306202. MR34745

[190] Hans Rademacher, *Topics in analytic number theory*, Springer-Verlag, New York-Heidelberg, 1973. Edited by E. Grosswald, J. Lehner and M. Newman. MR0364103

[191] V. Ramamani and K. Venkatachaliengar, *On a partition theorem of Sylvester*, Michigan Math. J. **19** (1972), 137–140. MR304323

[192] S. Ramanujan, *Some definite integrals [Messenger Math. **44** (1915), 10–18]*, Collected papers of Srinivasa Ramanujan, AMS Chelsea Publ., Providence, RI, 2000, pp. 53–58. MR2280854

[193] S. Ramanujan, *Proof of certain identities in combinatory analysis [Proc. Cambridge Philos. Soc. **19** (1919), 214–216]*, Collected papers of Srinivasa Ramanujan, AMS Chelsea Publ., Providence, RI, 2000, pp. 214–215. MR2280869

[194] S. Ramanujan, *Some properties of p(n), the number of partitions of n [Proc. Cambridge Philos. Soc. **19** (1919), 207–210]*, Collected papers of Srinivasa Ramanujan, AMS Chelsea Publ., Providence, RI, 2000, pp. 210–213. MR2280868

[195] John Riordan, *An introduction to combinatorial analysis*, Wiley Publications in Mathematical Statistics, John Wiley & Sons, Inc., New York; Chapman & Hall, Ltd., London, 1958. MR0096594

[196] Olinde Rodrigues, *Note sur les Inversions, ou dérangements produits dans les permutations*, Journal de Mathématiques Pures et Appliquées **4** (1839), 236–240.

[197] L. J. Rogers, *Note on the Transformation of an Heinean Series*, Messenger of Mathematics **23** (1893), 28–31.

[198] L. J. Rogers, *On a Three-fold Symmetry in the Elements of Heine's Series*, Proc. Lond. Math. Soc. **24** (1892/93), 171–179, DOI 10.1112/plms/s1-24.1.171. MR1577123

[199] L. J. Rogers, *On the Expansion of some Infinite Products*, Proc. Lond. Math. Soc. **24** (1892/93), 337–352, DOI 10.1112/plms/s1-24.1.337. MR1577136

[200] L. J. Rogers, *Second Memoir on the Expansion of certain Infinite Products*, Proc. Lond. Math. Soc. **25** (1893/94), 318–343, DOI 10.1112/plms/s1-25.1.318. MR1576348

[201] L. J. Rogers, *Third Memoir on the Expansion of certain Infinite Products*, Proc. Lond. Math. Soc. **26** (1894/95), 15–32, DOI 10.1112/plms/s1-26.1.15. MR1575886

[202] L. J. Rogers, *On two theorems of Combinatory Analysis and some allied identities*, Proceedings of the London Mathematical Society, 2nd series **16** (1917), 315–336.

[203] L. J. Rogers, *Proof of certain identities in combinatory analysis*, Proceedings of the Cambridge Philosophical Society **19** (1919), 211–214; *Collected Papers of Srinivasa Ramanujan*, G. H. Hardy, P. V. Seshu Aiyar and B. M. Wilson, eds., Cambridge University Press, 1927, 345–346; reprinted with commentary by Bruce C. Berndt by AMS Chelsea Publishing, Providence, 2000.

[204] Gian-Carlo Rota, *The Phenomenology of Mathematical Truth*, in *Indiscrete Thoughts*, Birkhäuser, Boston, 1997, 108–120.

[205] Gian-Carlo Rota, *The Phenomenology of Mathematical Beauty*, in *Indiscrete Thoughts*, Birkhäuser, Boston, 1997, 121–133.

[206] Heinrich August Rothe, *Formulae de serierum reversione*, Leipzig, 1793.

[207] Heinrich August Rothe, *Ueber Permutationen, in Beziehung auf die Stellen ihrer Elemente. Anwendungen der daraus abgeleiten Sätze auf das Eliminationsproblem*, Sammlung combinatorisch-analytischer Abhandlungen, herausg. v. C. F. Hindenburg **2** (1800), 263–305.

[208] Heinrich August Rothe, *Systematisches Lehrbuch der Arithmetik*, Leipzig, 1811.

[209] Michael Rowell, *A new exploration of the Lebesgue identity*, Int. J. Number Theory **6** (2010), no. 4, 785–798, DOI 10.1142/S1793042110003204. MR2661280

[210] Ranjan Roy, *Sources in the development of mathematics: Infinite series and products from the fifteenth to the twenty-first century*, Cambridge University Press, Cambridge, 2011. MR2807493

[211] H. F. Scherk, *Bemerkungen über die Lambertsche Reihe*

$$\frac{x}{1-x} + \frac{x^2}{1-x^2} + \frac{x^3}{1-x^3} + \frac{x^4}{1-x^4} +$$

etc (German), J. Reine Angew. Math. **9** (1832), 162–168, DOI 10.1515/crll.1832.9.162. MR1577901

[212] Anne Schilling and S. Ole Warnaar, *Conjugate Bailey pairs: from configuration sums and fractional-level string functions to Bailey's lemma*, Recent developments in infinite-dimensional Lie algebras and conformal field theory (Charlottesville, VA, 2000), Contemp. Math., vol. 297, Amer. Math. Soc., Providence, RI, 2002, pp. 227–255, DOI 10.1090/conm/297/05100. MR1919820

[213] Michael Schlosser, *Abel-Rothe type generalizations of Jacobi's triple product identity*, Theory and applications of special functions, Dev. Math., vol. 13, Springer, New York, 2005, pp. 383–400, DOI 10.1007/0-387-24233-3_17. MR2132472

[214] Issai Schur, *Ein Beitrag zur additiven Zahlentheorie und zur Theorie der Kettenbrüche*, Sitzungsberichte der Preussischen Akademie der Wissenschaften 1917, Physikalisch-Mathematische Klasse, 302–321; *Gesammelte Abhandlungen*, vol. 2, Springer-Verlag, Berlin, 1973, 117–136.

[215] Issai Schur, *Zur additiven Zahlentheorie*, Sitzungsberichte der Preussischen Akademie der Wissenschaften 1926, Physikalisch-Mathematische Klasse, 488–495; *Gesammelte Abhandlungen*, vol. 3, Springer-Verlag, Berlin, 1973, 43–50.

[216] Marcel Paul Schützenberger, *Une interprétation de certaines solutions de l'équation fonctionnelle: $F(x+y) = F(x)F(y)$* (French), C. R. Acad. Sci. Paris **236** (1953), 352–353. MR53402

[217] H. A. Schwarz, *Formeln und Lehrsätze zum Gebrauche der Elliptischen Funktionen nach Vorlesungen und Aufzeichnungen des Herrn Prof. K. Weierstrass*, Springer-Verlag, Berlin, 1893.

[218] Ferd Schweins, *Analysis*, Mohr und Winter, Heidelberg, 1820.

[219] Atle Selberg, *Über einige arithmetische Identitäten*, Avhandlinger utgitt av Det Norske Videnskaps-Akademi i Oslo I. Mat.-Naturv. Klasse **8** (1936), 1–23; *Collected Papers I*, Springer-Verlag, Berlin, 1989, 1–23.

[220] Andrew V. Sills, *An invitation to the Rogers-Ramanujan identities*, CRC Press, Boca Raton, FL, 2018. With a foreword by George E. Andrews. MR3752624

[221] Lucy Joan Slater, *Generalized hypergeometric functions*, Cambridge University Press, Cambridge, 1966. MR0201688

[222] Henry J. S. Smith, *Report on the Theory of Numbers*, Part VI, *The Collected Mathematical Papers of Henry John Stephen Smith*, vol. 1, 1894, 289–364; reprinted by Chelsea Publishing Company, New York, 1979.

[223] Richard P. Stanley, *Binomial posets, Möbius inversion, and permutation enumeration*, J. Combinatorial Theory Ser. A **20** (1976), no. 3, 336–356, DOI 10.1016/0097-3165(76)90028-5. MR409206

[224] Richard P. Stanley, *Enumerative combinatorics. Vol. 1*, Cambridge Studies in Advanced Mathematics, vol. 49, Cambridge University Press, Cambridge, 1997. With a foreword by Gian-Carlo Rota; Corrected reprint of the 1986 original. MR1442260

[225] Richard P. Stanley, *Enumerative combinatorics. Vol. 2*, Cambridge Studies in Advanced Mathematics, vol. 62, Cambridge University Press, Cambridge, 1999. With a foreword by Gian-Carlo Rota and appendix 1 by Sergey Fomin. MR1676282

[226] M. Stern, *Aufgaben* (German), J. Reine Angew. Math. **18** (1838), 100, DOI 10.1515/crll.1838.18.100. MR1578180

[227] James Joseph Sylvester, *On Mr Cayley's impromptu demonstration of the rule for determining at sight the degree of any symmetrical function of the roots of an equation expressed*

in terms of the coefficients, Philosophical Magazine **5** (1853), 199–202; *The Collected Mathematical Papers of James Joseph Sylvester*, vol. 1, Cambridge University Press, Cambridge, 1904, 595–598; reprinted by Chelsea, New York, 1974.

[228] James Joseph Sylvester, *Outlines of seven lectures on the Partitions of Numbers*, Proceedings of the London Mathematical Society **28** (1897), 33–96; *The Collected Mathematical Papers of James Joseph Sylvester*, vol. 2, Cambridge University Press, Cambridge, 1908, 119–175; reprinted by Chelsea, New York, 1974.

[229] James Joseph Sylvester, *Note on the paper of Mr Durfee's*, Johns Hopkins University Circulars **2** (1883), 23–24, 42–43; *The Collected Mathematical Papers of James Joseph Sylvester*, vol. 3, Cambridge University Press, Cambridge, 1909, 661–663; reprinted by Chelsea, New York, 1974.

[230] James Joseph Sylvester, *On Dr. F. Franklin's proof of Euler's theorem concerning the form of the infinite product* $(1-x)(1-x^2)(1-x^3)\ldots$, Johns Hopkins University Circulars **2** (1883), 42–44; *The Collected Mathematical Papers of James Joseph Sylvester*, vol. 3, Cambridge University Press, 1909, 664–666; reprinted by Chelsea, New York, 1974.

[231] James Joseph Sylvester, *Note on the graphical method in partitions*, Johns Hopkins University Circulars **2** (1883), 70–71; *The Collected Mathematical Papers of James Joseph Sylvester*, vol. 3, Cambridge University Press, Cambridge, 1909, 683–684; reprinted by Chelsea, New York, 1974.

[232] James Joseph Sylvester, *A constructive theory of partitions in three acts, an interact, and an exodion*, American Journal of Mathematics **5** (1882), 251–330, **6** (1884), 334–336; *The Collected Mathematical Papers of James Joseph Sylvester*, vol. 4, Cambridge University Press, Cambridge, 1912, 1–83; reprinted by Chelsea, New York, 1974.

[233] Jules Tannery, *Introduction à la théorie des fonctions d'une variable*, 2nd edition, volume 1, Librarie Scientifique A. Hermann, Paris, 1904.

[234] Jules Tannery, *Introduction à la théorie des fonctions d'une variable*, 2nd edition, volume 2, Librarie Scientifique A. Hermann, Paris, 1910.

[235] Olry Terquem, *Solution d'un Problème de combinaison*, Journal de Mathématiques Pures et Appliquées **3** (1838), 559–560.

[236] Thotsaporn Thanatipanonda, *Inversions and Major Index for Permutations*, Math. Mag. **77** (2004), no. 2, 136–140. MR1573738

[237] E. C. Titchmarsh, *The theory of functions*, 2nd ed., Oxford University Press, Oxford, 1939. MR3728294

[238] Isaac Todhunter, *A History of the Mathematical Theory of Probability*, Macmillan and Company, Cambridge, 1865; Chelsea Publishing Company, New York, 1949.

[239] Keisuke Uchimura, *An identity for the divisor generating function arising from sorting theory*, J. Combin. Theory Ser. A **31** (1981), no. 2, 131–135, DOI 10.1016/0097-3165(81)90009-1. MR629588

[240] Sam Vandervelde, *Balanced partitions*, The Ramanujan Journal **23** (2010), 297–306; *Combinatory Analysis*, Krishnaswami Alladi, Peter Paule, James Sellers, and Ae Ja Yee, eds., Springer, New York, 2013, 295—-304; DOI 10.1007/s11139-009-9206-9, MR2739218.

[241] K. Venkatachaliengar, *Development of elliptic functions according to Ramanujan*, Monographs in Number Theory, vol. 6, World Scientific Publishing Co. Pte. Ltd., Hackensack, NJ, 2012. Edited, revised, and with a preface by Shaun Cooper. MR3113525

[242] Michelle L. Wachs, *On q-derangement numbers*, Proc. Amer. Math. Soc. **106** (1989), no. 1, 273–278, DOI 10.2307/2047402. MR937015

[243] S. Ole Warnaar, *Partial theta functions. I. Beyond the lost notebook*, Proc. London Math. Soc. (3) **87** (2003), no. 2, 363–395, DOI 10.1112/S002461150201403X. MR1990932

[244] G. N. Watson, *A New Proof of the Rogers-Ramanujan Identities*, J. London Math. Soc. **4** (1929), no. 1, 4–9, DOI 10.1112/jlms/s1-4.1.4. MR1574904

[245] G. N. Watson, *Theorems Stated by Ramanujan (VII): Theorems on Continued Fractions*, J. London Math. Soc. **4** (1929), no. 1, 39–48, DOI 10.1112/jlms/s1-4.1.39. MR1574903

[246] G. N. Watson, *Ramanujans Vermutung über Zerfällungszahlen* (German), J. Reine Angew. Math. **179** (1938), 97–128, DOI 10.1515/crll.1938.179.97. MR1581588

[247] Karl Weierstrass, *Über die Entwicklung der Modular-Functionen*, Mathematische Werke, vol. 1, Mayer & Müller, Berlin, 1894, 1–49.

[248] Ae Ja Yee, *Combinatorial proofs of Ramanujan's $_1\psi_1$ summation and the q-Gauss summation*, J. Combin. Theory Ser. A **105** (2004), no. 1, 63–77, DOI 10.1016/j.jcta.2003.10.002. MR2030140

[249] Doron Zeilberger, *A q-Foata proof of the q-Saalschütz identity*, European J. Combin. **8** (1987), no. 4, 461–463, DOI 10.1016/S0195-6698(87)80054-9. MR930183

[250] Doron Zeilberger, *Kathy O'Hara's constructive proof of the unimodality of the Gaussian polynomials*, Amer. Math. Monthly **96** (1989), no. 7, 590–602, DOI 10.2307/2325177. MR1008789

[251] Doron Zeilberger, *A one-line high school algebra proof of the unimodality of the Gaussian polynomials $[{n \atop k}]$ for $k < 20$*, q-series and partitions (Minneapolis, MN, 1988), IMA Vol. Math. Appl., vol. 18, Springer, New York, 1989, pp. 67–72, DOI 10.1007/978-1-4684-0637-5_6. MR1019843

[252] Doron Zeilberger, *James Joseph Sylvester: the GREATEST Mathematician of ALL TIMES*, Opinions of Doron Zeilberger **75**, October 27, 2006 (online).

[253] Jiang Zeng, *Pfaff-Saalschütz revisited*, J. Combin. Theory Ser. A **51** (1989), no. 1, 141–143, DOI 10.1016/0097-3165(89)90088-5. MR993660

Index of Names

Abel, Niels Henrik, 37, 274, 288, 370
Adams, John Couch, 148
Agarwal, R. P., 56
Alladi, Krishnaswami, 189, 261, 262, 269, 403, 406, 407, 414, 416
Andrews, George E., 364
Andrews, George E., xv, 49, 54, 92, 109, 139, 148, 160, 163, 185, 189, 223, 233, 235, 244, 252, 255, 262, 268, 349, 365, 370, 375, 394, 408, 416, 434, 439, 462, 471, 476, 481, 483
Askey, Richard A., xiv, 47, 254, 255, 269, 383, 394, 449, 454, 457, 459, 461, 465, 467, 468, 482

Bailey, W. N., 175, 178, 189, 217, 231, 233, 237, 248, 250, 437, 449, 451, 452, 471, 481
Bartlett, Nick, 156
Benkart, Georgia, 44
Berkovich, Alexander, 441, 442
Berndt, Bruce C., 189, 244, 268
Bernoulli, Nicolas, 349
Blanton, John, 93
Bowman, Douglas, 416
Bressoud, David, 416
Bressoud, David, 189, 357, 365, 375, 385–387, 402
Bromwich, T. J. I'a., 493, 499
Bunyakowsky, Viktor, 199

Carlitz, Leonard, 20, 38, 334, 349
Cauchy, Augustin-Louis, xiv, 74, 92, 122, 132, 146, 148, 179, 196, 221, 231, 243, 247, 258, 259, 263, 265, 269, 271, 276, 284, 288, 289, 301, 497, 499
Cayley, Arthur, 38, 116, 123, 148, 163, 274, 275, 281, 287, 288
Chapman, Robin, xiv, 357, 375
Chen, William Y. C., 88, 308, 324, 349
Chu, Wenchang, 88
Church, Charles A., 38
Cigler, Johann, 49, 54, 92
Claus, Santa, 27

Clausen, Thomas, 173, 189
Cooper, Shaun, 244, 268, 481
Corteel, Sylvie, 268
Cramer, Gabriel, 37
Crelle, August Leopold, 1
Crelle, August Leopold, 37

Daum, John Andrew, 233
de Moivre, Abraham, 349
Deutsch, Emeric, 6, 109, 120
Dirichlet, Gustav Peter Lejeune, 274, 275, 280, 288
Dobbie, J. M., 437, 481
Dougall, John, 478, 482
Durfee, W. P., 157

Eisenstein, F. Gotthold, 37, 132, 172, 189, 482
Eriksson, Kimmo, 148
Euler, Leonhard, xi, 82, 87, 92, 93, 99, 121, 126, 129, 130, 132, 136, 142, 148, 149, 154, 165, 167, 182, 189, 191, 221, 224, 264, 267, 287, 310, 343, 349, 392, 394, 405, 422, 496, 500

Ferrers, Norman, 106, 148
Fine, Nathan J., 167, 182, 189, 237, 244
Foata, Dominique, xiv, 33, 316, 349
Franklin, Fabian, xiii, 117, 148, 159, 165, 189, 351
Fricke, Robert, 244
Frobenius, Georg, 201, 351

Göllnitz, Heinz, 403, 409, 416
Garnier, Jean Guillaume, 37
Garrett, Kristina, 365
Garsia, Adriano, 385, 416
Gasper, George, 227, 245
Gauss, Carl Friedrich, 1, 5, 16, 21, 37, 50, 61, 74, 91, 113, 124, 153, 189, 191, 196, 199, 204, 206–213, 218–221, 223, 225, 243, 244, 258, 282, 284, 287, 288, 394, 395, 403, 423, 445, 446
Gergonne, Joseph Diez, 37
Gessel, Ira, 324, 347, 349

INDEX OF NAMES

Glaisher, James Whitbread Lee, 100, 148, 189
Gordon, Basil, 403, 411, 416, 434
Gruson, Johann Philipp, 45, 47, 91
Gu, Nancy S., 88

Hadamard, Jacques, 499
Hall, Newman, 233, 236
Hammond, Christopher N. B., 12
Hardy, G. H., xiii, 136, 188, 247, 268, 290, 292, 377, 421, 436
Havil, Julian, 500
Hayes, Brian, 38
Heine, Eduard, 124, 132, 135, 227, 243
Henderson, Greg, 43
Hirschhorn, Michael, 78, 92, 131, 207, 244, 292, 303
Holtz, Olga, 42

Ismail, Mourad, 247, 250, 268, 365, 461, 482

Jackson, F. H., 241, 243, 250, 478, 479, 482
Jackson, Margaret, 250, 268, 481
Jacobi, Carl Jacob, 37, 71, 74, 92, 97, 106, 130, 132, 134, 148, 156, 158, 189, 191, 197, 198, 200, 211, 221, 225, 228, 241, 243, 244, 259, 261, 263, 269, 274, 281, 288, 405, 444, 482
Jayawant, Pallavi, 38
Johnson, Robert Shepard, 209
Jordan, Camille, 288

Kant, Immanuel, xiii
Knopfmacher, Arnold, 364
Knopp, Konrad, 493
Knuth, Donald E., 38
Kronecker, Leopold, 288

Lagrange, Joseph Louis, 286
Lalov, Emil, 174
Lambert, Johann Heinrich, 167, 189
Laplace, Pierre Simon, 37
Lebesgue, Henri, 221
Lebesgue, Victor-Amédée, 221, 225, 244
Legendre, Adrien-Marie, 155, 165, 243, 284, 288
Leibniz, Gottfried, 148
Liouville, Joseph, 2
Littlewood, John E., 292, 377
Lovejoy, Jeremy, 268

MacMahon, Percy Alexander, xiv, 29, 37, 74, 92, 189, 258, 292, 305, 308, 316, 319, 324, 349, 355, 377, 383, 416
Milne, Stephen C., 416
Milne, Stephen C., 385
Moss, Eric, 156
Muir, Sir Thomas, 37

Netto, Eugen, 37

O'Hara, Kathy, 38

Pólya, George, 18, 189
Pak, Igor, 167, 189, 244, 375, 436
Parshall, Karen Hunger, 188
Paule, Peter, xi, 92, 364
Perron, Oskar, 377
Pfaff, Johann Friedrich, 228, 240
Pieper, Herbert, 243
Potter, H. S. A., 41, 45, 91, 147

Rademacher, Hans, 189
Rahman, Mizan, 227
Ramamani, V., 186
Ramanujan, Srinivasa, xiii, xiv, 136, 148, 189, 195, 228, 230, 244, 245, 247, 268, 289, 290, 292, 377, 381, 416, 417, 421, 436, 443, 451, 457, 479, 481, 499
Riordan, John, 349
Rodrigues, Olinde, xii, 4, 310
Rogers, Leonard James, xiv, 237, 292, 377, 388, 392, 394, 402, 416–418, 421, 436
Rota, Gian-Carlo, xiii, 44, 349
Rothe, Heinrich August, 4, 47, 91, 92
Rowell, Michael, 69
Roy, Ranjan, xv, 244
Rudolph the Red-Nosed Reindeer, 27

Saalschütz, Louis, 240
Sagan, Bruce, 44
Schützenberger, Marcel-Paul, 41, 91, 147, 316
Schaller, Sarah, 225
Scherk, Heinrich Ferdinand, 189
Schilling, Anne, 441
Schlosser, Michael, xiv, 247, 256, 258, 269
Schur, Issai, xiv
Schur, Issai, 351, 355, 357, 361, 363, 365, 370, 372, 375, 377, 382, 383, 436
Schwarz, Hermann Amadeus, 244
Schweins, Ferdinand, 48, 91
Selberg, Atle, xiv, 223, 416–418, 420, 421, 423, 424, 427–429, 432, 433, 436
Sills, Andrew, 416
Slater, Lucy Joan, 481
Smith, Henry John Stephen, 482
Smoot, Nicolas Allen, 110, 148
Snow, C. P., 188, 436
Stanley, Richard P., 347, 349
Stanton, Dennis, 365
Stein, Paul R., 44
Stern, Moritz Abraham, 1, 38
Sylvester, James Joseph, xiii, 117, 123, 148, 157, 159–161, 180, 189, 223, 420

Tannery, Jules, 497, 499
Terquem, Olry, xii, 2, 6, 37
Titchmarsh, E. C., 493
Todhunter, Isaac, 349

Uchimura, Keisuke, 171, 174, 189

Venkatachaliengar, K., 186, 268, 481

Wachs, Michelle, 324, 349
Warnaar, S. Ole, 439, 441
Warters, Elena, 113
Watson, G. N., 244, 478, 481, 482
Weierstrass, Karl, 207, 244, 481, 497
Wilson, James, 465

Xu, Deheng, 308, 324, 349

Yang, Mingjia, 110, 148
Yee, Ae Ja, 268

Zeilberger, Doron, 33, 38, 188, 385, 416
Zeng, Jiang, 33, 36

Index of Topics

q-Eulerian numbers, 334–338, 349
q-Fibonacci numbers, 66, 127
　r^{th} order, 361, 363
q-Pascal recurrences, 17–18, 26–27, 37, 41, 45–46
　for q-multinomial coefficients, 30, 321–323
　iterated, 21–22
q-Pfaff–Saalschütz identity, xiv, 36, 240–243, 256, 472, 478
q-Taylor theorem, 59, 142, 144
q-binomial coefficients, xi, 15–29, 311–315
　Fundamental Property of, xii, 16–19, 26, 29, 37, 41
q-binomial series, 133, 227
q-derangement numbers, 324–330
q-derivative, xii, xiv, 57–61, 141–148, 331–333, 340–341
q-exponential functions, 141–148, 329, 339
q-factorials, 8–14, 305–310
q-multinomial coefficients, 29–36, 316–323
　Fundamental Property of, 29
　set partition property, 30
q-trigonometric functions, xiv, 339–348

Abel's limit theorem, 370, 372, 375
Agarwal's q-binomial theorem, 56, 92, 472
AKP identity, 364, 375
Andrews's q-binomial theorem, 54, 92
Andrews's q-Pfaff transformation, 233
Andrews's lemma, 235
Askey–Wilson integral, 465

Bachet's theorem, 288
Bailey chain, 476, 482
Bailey pair, 471, 472, 475, 476
Bailey's formula, 437–439, 442, 444, 449
Bailey's lemma, 474–476
Bailey's transform, 471
Bailey's very well poised $_6\phi_5$ sum, 452, 478
Bailey's very well poised $_6\psi_6$ sum, xiv, 449–465, 475
Bailey–Daum summation formula, 233, 236, 237, 244, 274

calculus of finite differences, 57
Carlitz coefficients, 20
Carlitz's theorem, 334–338
Cauchy sequence, 429
Cauchy's mistaken identity, 263–266
Cauchy/Crelle series, xiii, 132–140, 145, 148, 221, 227, 259, 260, 264, 268
Cayley's theorem, 116, 161
Chen–Chu–Gu identity, xiii, 90–92, 214
crossing diagrams, 12
cyclic derivative, 44, 92

derangements, 323–330
　reduced, 324, 327
divisor sums, 167–175
Double Lebesgue identity, 233, 236
double Lebesgue identity, 223, 226, 244
Durfee square, 157–167, 352, 354, 355

Euler numbers, 343
Euler's "odd equals distinct" theorem, 99, 102, 217, 223, 275, 276, 423, 432, 470
　Sylvester's refinement of, 182
Euler's constant, 496, 500
Euler's divisor sum theorem, 170
Euler's lemma, 115, 148, 150, 153
Euler's pentagonal number theorem, xiii, 149–157, 160, 165–167, 169, 194, 205, 290, 293, 351, 420
Eulerian numbers, 311
Eulerian polynomials, 310, 349

Ferrers diagram, 106, 116, 158, 165, 180, 187, 201, 286, 309, 320, 352, 353, 386–389
Fibonacci numbers, 292
Foata alphabet, 34
Franklin triangle, 159–167, 352, 354, 355
Franklin's rule, 165
Frobenius symbol, 202, 205, 244
　modified, 203

Göllnitz–Gordon continued fraction, 412
Göllnitz–Gordon identities, xiii, xiv, 406–416

Gauss's q-binomial theorems, xii, 61–66, 124, 185, 188, 342
Gessel's theorem, 324, 330, 349
GIS identity, 365, 375
Glaisher's bijection, 100, 181, 187, 366, 370
Gordon's theorem, 436

Heine's q-Euler transformation, 229, 239
Heine's q-Gauss summation formula, 228, 240
Heine's q-Pfaff transformation, 228, 232
Heine's intermediate transformation, 229, 231, 232, 372

identity theorem for analytic functions, 461, 470, 482
inclusion-exclusion principle, 329, 349
inversions, 1–37, 40, 313, 316–319, 346–348
  between-set, 25–30, 348

Jackson's q-Dougall identity, 478, 482
Jacobi's q-binomial theorem, xii, 71–73, 85, 92, 134
Jacobi's q-Gauss summation formula, 228
Jacobi's q-Gauss summation formula, 234, 251, 252
Jacobi's cube identity, 205, 290, 301
Jacobi's Durfee square identity, 158–159, 162, 197–199, 259–260, 389
Jacobi's eight square theorem, xiv, 445, 447–448, 462, 482
Jacobi's four square theorem, xiv, 283–288, 465
Jacobi's triple product identity, xiii–xiv, 131, 158, 191–213, 217, 218, 247, 250, 253, 265, 266, 268, 275, 276, 282, 303, 351, 354, 379, 390–392, 404, 410, 417, 440, 462
Jacobi's two square theorem, xiv, 278–282, 288, 464
Jordan–Kronecker function, 288

Lagrange's four square theorem, 286, 288
Lebesgue's identity, 221–227, 233, 237, 244, 407
Legendre's four square theorem, 285, 286, 288
little Göllnitz identities, 407
Lucas numbers, 292

MacMahon's q-binomial theorem, xiii, 74–79, 90, 92, 201, 207, 243, 258
MacMahon's theorem, xiv, 316, 319, 321, 349
major index, 305–309, 311–323, 338

Nicole's identity, 115, 148

Pólya's property, 18, 19, 38, 116

partitions, xiii, 94–132, 136, 148, 152, 157–167, 180–189, 319–321, 351–356, 365–375, 383–388, 399–403, 412–415
  broken, 326
  conjugate, 107, 202
  deranged, 327
  excesses, 117
  fixed, 326
  labeled, 309, 325
  Schur, 366, 370, 374
  self-conjugate, 108, 113
  standard labeled, 308, 325
permutations, 1–26, 29, 30, 33, 305–349
  conjugate of, 4
  down-up, 344
  falls, 305–338
  self-conjugate, 7, 13, 32
  up-down, 344
Potter–Schützenberger q-binomial theorem, xii, 41, 47, 49, 57, 91, 147
problème des rencontres, 323, 349
Putnam exam, 148, 173, 493
Pythagorean triples, 279–280, 288

quintuple product identity, xiii–xv, 214–221, 244, 266, 269, 294, 461, 462

Ramanujan's q-Gauss summation formula, 228, 268
Ramanujan's $_1\psi_1$ summation formula, xiii, xiv, 247–269, 271, 457, 461, 482, 499
Ramanujan's "lost notebook", 244, 269, 436, 481
Ramanujan's "most beautiful" identity, xiv, 292–298, 303, 443, 481
Ramanujan's congruences, 289–291, 301, 303
Ramanujan's notebooks, 244, 268, 443
Ramanujan's reciprocity theorem, 266, 269
Ramanujan's transformation, 231, 244
reciprocal polynomials, 13, 19, 334
Rodrigues's theorem, 9, 29
Rogers–Fine identity, 237, 244
Rogers–Ramanujan continued fraction, 381
Rogers–Ramanujan identities, xiv, 292, 351–363, 375, 377–394, 396, 416, 421–423, 432, 434–436, 478, 479, 482, 499
Rogers–Selberg function, 417–436
  fundamental recurrence, 419, 420, 424, 427, 434
Rothe diagram, 5, 7
Rothe's q-binomial theorem, xii, 47–57, 59, 60, 72, 74, 78, 91, 129, 130, 134, 136, 224, 261, 406
Rowell's identity, 69, 92, 224

Schur shape, 352–355
Stanley–Gessel theorem, 347–349

Stern's problem, 1, 12, 31, 37, 120, 309
Sylvester's fishhook bijection, xii, 180–189, 370

Tannery's theorem, xiii, 201, 207, 214, 224, 258, 271, 361, 479, 497–499
theta product, 194
totally noncommutative binomial theorem, 40
triangular numbers, 109, 286, 446–447

unimodal, 13, 19

Wachs's theorem, 325, 349
Warnaar's formula, 440
Watson's transformation, xiv, 471, 477–478, 482

Z-identity, 33, 241